Fungal Infections and Immune Responses

INFECTIOUS AGENTS AND PATHOGENESIS

Series Editors: Mauro Bendinelli, *University of Pisa*
Herman Friedman, *University of South Florida*

COXSACKIEVIRUSES
A General Update
Edited by Mauro Bendinelli and Herman Friedman

FUNGAL INFECTIONS AND IMMUNE RESPONSES
Edited by Juneann W. Murphy, Herman Friedman, and Mauro Bendinelli

MYCOBACTERIUM TUBERCULOSIS
Interactions with the Immune System
Edited by Mauro Bendinelli and Herman Friedman

NEUROPATHOGENIC VIRUSES AND IMMUNITY
Edited by Steven Specter, Mauro Bendinelli, and Herman Friedman

PSEUDOMONAS AERUGINOSA AS AN OPPORTUNISTIC PATHOGEN
Edited by Mario Campa, Mauro Bendinelli, and Herman Friedman

VIRUS-INDUCED IMMUNOSUPPRESSION
Edited by Steven Specter, Mauro Bendinelli, and Herman Friedman

Fungal Infections and Immune Responses

Edited by
Juneann W. Murphy
University of Oklahoma Health Sciences Center
Oklahoma City, Oklahoma
Herman Friedman
University of South Florida
Tampa, Florida
and
Mauro Bendinelli
University of Pisa
Pisa, Italy

Springer Science+Business Media, LLC

Library of Congress Cataloging-in-Publication Data

Fungal infections and immune responses / edited by Juneann W. Murphy, Herman Friedman, and Mauro Bendinelli.
p. cm. -- (Infectious agents and pathogenesis)
Includes bibliographical references and index.
ISBN 978-1-4899-2402-5 ISBN 978-1-4899-2400-1 (eBook)
DOI 10.1007/978-1-4899-2400-1
1. Mycoses--Immunological aspects. I. Murphy, Juneann W. II. Friedman, Herman, 1931- . III. Bendinelli, Mauro. IV. Series.
[DNLM: 1. Host-Parasite Relations. 2. Infection--Immunology. 3. Mycoses. QW 700 F981]
QR201.M98F85 1993
616.9'68079--dc20
DNLM/DLC
for Library of Congress 92-49043
CIP

ISBN 978-1-4899-2402-5

Originally published by Plenum Press, New York in 1993
Softcover reprint of the hardcover 1st edition 1993

Contributors

BLAINE L. BEAMAN • Department of Medical Microbiology and Immunology, University of California School of Medicine, Davis, California 95616

HERMAN CHMEL • Division of Infectious and Tropical Diseases, Department of Internal Medicine, College of Medicine, University of South Florida, Tampa, Florida 33612. *Present address*: Department of Medicine, St. Francis Medical Center, Trenton, New Jersey 08629.

CHESTER R. COOPER, JR. • Laboratories for Mycology, Wadsworth Center for Laboratories and Research, New York State Department of Health, Albany, New York 12201-0509

REBECCA A. COX • Department of Research Immunology, San Antonio State Chest Hospital, San Antonio, Texas 78223

JULIE Y. DJEU • Department of Medical Microbiology and Immunology, College of Medicine, University of South Florida, Tampa, Florida 33612

JUDITH E. DOMER • Department of Microbiology and Immunology, Tulane University School of Medicine, New Orleans, Louisiana 70112

JOHN E. EDWARDS, JR. • Department of Medicine, Division of Infectious Diseases, Harbor-UCLA Medical Center, Torrance, California 90509, and UCLA School of Medicine, Los Angeles, California 90024

MOHAMAD M. EL-ZAATARI • Department of Pathology, University of Texas Medical Branch, Galveston, Texas, 77555

SCOTT G. FILLER • Department of Medicine, Division of Infectious Diseases, Harbor-UCLA Medical Center, Torrance, California 90509, and UCLA School of Medicine, Los Angeles, California 90024

JORDAN N. FINK • Research Service, Zablocki Veterans Affairs Medical Center and Allergy–Immunology Division, Medical College of Wisconsin, Milwaukee, Wisconsin 53295

GARY E. HOLLICK • Department of Pathology and Laboratory Medicine, Microbiology Laboratory, Rochester General Hospital, Rochester, New York 14621

D. H. HOWARD • Department of Microbiology and Immunology, UCLA School of Medicine, Los Angeles, California 90024-1747

THOMAS R. KOZEL • Department of Microbiology and the Cellular and Molecular Biology Program, University of Nevada School of Medicine, Reno, Nevada 89557-0046

VISWANATH P. KURUP • Research Service, Zablocki Veterans Affairs Medical Center and Allergy–Immunology Division, Medical College of Wisconsin, Milwaukee, Wisconsin 53295

ROBERT I. LEHRER • Department of Medicine, UCLA School of Medicine, Los Angeles, California 90024

MICHAEL R. McGINNIS • Department of Pathology, University of Texas Medical Branch, Galveston, Texas 77555

HAROLD G. MUCHMORE • Departments of Medicine and Microbiology and Immunology, Infectious Disease Section, College of Medicine, University of Oklahoma Health Sciences Center, Oklahoma City, Oklahoma 73190, and Department of Veterans Affairs Medical Center, Oklahoma City, Oklahoma 73104

JUNEANN W. MURPHY • Department of Microbiology and Immunology, University of Oklahoma Health Sciences Center, Oklahoma City, Oklahoma 73190

R. SCOTT PORE • Department of Microbiology and Immunology, West Virginia University School of Medicine, Morgantown, West Virginia 26506-9177

ERROL REISS • Molecular Mycology Section, Mycotic Diseases Branch, Division of Bacterial and Mycotic Diseases, National Center for Infectious Diseases, Centers for Disease Control and Prevention, Atlanta, Georgia 30333

ANGELA RESTREPO-MORENO • Mycology Section, Corporación para Investigaciones Biologicas (CIB), Hospital Pablo Tobon Uribe, Medellin, Colombia

JUDITH C. RHODES • Department of Pathology and Laboratory Medicine, University of Cincinnati College of Medicine, Cincinnati, Ohio 45267-0529

IRA F. SALKIN • Laboratories for Mycology, Wadsworth Center for Laboratories and Research, New York State Department of Health, Albany, New York 12201-0509

E. NAN SCOTT • Departments of Medicine and Microbiology and Immunology, Infectious Disease Section, College of Medicine, University of Oklahoma Health Sciences Center, Oklahoma City, Oklahoma, 73190, and Department of Veterans Affairs Medical Center, Oklahoma City, Oklahoma 73104

PETER G. SOHNLE • Section of Infectious Diseases, Department of Medicine, the Medical College of Wisconsin, Milwaukee, Wisconsin 53226, and the Medical and Research Services, Milwaukee VA Medical Center, Milwaukee, Wisconsin 53295

W. G. SORENSON • National Institute for Occupational Safety and Health, Morgantown, West Virginia 26505, and Department of Microbiology and Immunology, West Virginia University, Morgantown, West Virginia 26506

ALAN M. SUGAR • Evans Memorial Department of Clinical Research, The University Hospital, and Department of Medicine, The Thorndike Memorial Laboratory, Boston City Hospital, Boston, Massachusetts 02118

BETTY WU-HSIEH • Department of Microbiology and Immunology, UCLA School of Medicine, Los Angeles, California 90024-1747

Preface to the Series

The mechanisms of disease production by infectious agents are presently the focus of an unprecedented flowering of studies. The field has undoubtedly received impetus from the considerable advances recently made in the understanding of the structure, biochemistry, and biology of viruses, bacteria, fungi, and other parasites. Another contributing factor is our improved knowledge of immune responses and other adaptive or constitutive mechanisms by which hosts react to infection. Furthermore, recombinant DNA technology, monoclonal antibodies, and other newer methodologies have provided the technical tools for examining questions previously considered too complex to be successfully tackled. The most important incentive of all is probably the regenerated idea that infection might be the initiating event in many clinical entities presently classified as idiopathic or of uncertain origin.

Infectious pathogenesis research holds great promise. As more information is uncovered, it is becoming increasingly apparent that our present knowledge of the pathogenic potential of infectious agents is often limited to the most noticeable effects, which sometimes represent only the tip of the iceberg. For example, it is now well appreciated that pathological processes caused by infectious agents may emerge clinically after an incubation of decades and may result from genetic, immunologic, and other indirect routes more than from the infecting agent itself. Thus, there is a general expectation that continued investigation will lead to the isolation of new agents of infection, the identification of hitherto unsuspected etiologic correlations, and eventually, more effective approaches to prevention and therapy.

Studies on the mechanisms of disease caused by infectious agents demand a breadth of understanding across many specialized areas, as well as much cooperation between clinicians and experimentalists. The series *Infectious Agents and Pathogenesis* is intended not only to document the state of the art in this fascinating and challenging field but also to help lay bridges among diverse areas and people.

M. Bendinelli
H. Friedman

Preface

Biomedical scientists widely acknowledge that individuals' immune responsiveness is important in resistance to infections by microorganisms, including fungi. Because of the devastating acquired immunodeficiency syndrome (AIDS) epidemic, caused by the human immunodeficiency retrovirus, it is now accepted that suppressed immune responses, especially cellular immunity, are important contributors to increased individual susceptibility to opportunistic infections—including infections caused by fungi which were at one time thought to be very lowly or nonpathogenic. Within the last few years, there has been an almost explosive increase in interest and studies concerning the nature and mechanisms of the immune response to fungal infections. Many immunologists who are not well versed in mycology have begun to study the nature and mechanisms of antifungal immunity using a wide variety of newer as well as more conventional immunologic technologies, both *in vivo* and *in vitro*. Up to the 1980s, however, there was little interest among basic immunologists concerning fungal immunity. This situation has changed dramatically in the past half decade, mainly because of AIDS.

Two of the major difficulties, according to basic immunologists, in using fungal antigens in immunologic studies have been their "impurity" and the lack of knowledge concerning specific fungal immunogens. Thus, it was reasoned, it would be difficult to investigate in detail the broader mechanisms whereby the immune system, including highly defined components such as lymphokines and immunocytes, interacts with fungi and the antigens from these organisms, especially since these antigens are essentially "impure." The last few years, however, have seen a number of significant advances concerning our understanding of the nature and mechanisms of the interaction of the immune response system with fungi and their antigens. The frequency of severe fungal infections in immunocompromised patients, especially those with AIDS, has permitted a new evaluation and appreciation of the interaction of the immune system with fungi and their antigens. The purpose of this book, therefore, is to present new as well as established information concerning important fungi and the host immune response to these organisms. "Classic" mycologists are quite familiar with these fungi and their pathogenic characteristics, whereas basic immunologists have been using fungi mainly as "tools" to help understand the role of the immune system in the control of these organisms. The editors as well as the contributors to this volume believe that a detailed presentation not only of the various classes of fungi but also of the nature and mechanism whereby these organisms either elicit an immune response or are controlled by such a response will be of importance to the biomedical community in general and especially to mycologists and immunologists.

The first purpose of this book and the other books in this series is to present recent advances in our understanding of the biology and pathogenesis of infections caused by various microorganisms and their relationship to the immune response system. Leading investigators active in this area of mycology are authors of one or more chapters. The second purpose is to stimulate a renewed interest concerning these important microorganisms, especially among immunologists and microbiologists in general, as well as mycologists, particularly in understanding the nature of the immune mechanisms that influence their pathogenicities. In order to control these organisms, it is essential to gain a full appreciation of their clinical importance. Thus, it is also necessary for the evaluation of public health procedures and the development of appropriate control mechanisms for fungal infection that we understand the interaction of these organisms with a host's immune system. The exciting developments taking place in understanding the complexity of immunologic involvement in host–parasite relations, including those of bacteria and viruses, make it likely that vaccines will be useful in building protection against various fungi, especially opportunistic ones that are important in causing disease in immunosuppressed individuals, such as those with AIDS. It is anticipated that a presentation of these topics covering various fungi in relation to the immune response mechanisms will be valuable not only to active investigators in the field, but also to students and scientists in general.

Juneann W. Murphy
Herman Friedman
Mauro Bendinelli

Contents

Fungal Infections and Immune Responses

1

Introduction

JUNEANN W. MURPHY

This volume provides current information on mycotic diseases as well as on infections with certain bacteria classified in the order Actinomycetales. The latter have been included because frequently the diseases caused by these "higher" forms of eubacteria are similar to and must be differentiated from mycotic infections. The major portion of the text, however, is devoted to fungal diseases. Recently, there has been a renewed interest in mycotic diseases because of our increased ability to diagnose these diseases and, more importantly, because of an upsurge in the incidence of life-threatening mycoses. It is expected that the information covered in this volume will be useful to a broad range of individuals, including medical and graduate students studying infectious diseases, practicing physicians, and research scientists.

Mycotic diseases can be divided into three major categories. Infections with fungal agents comprise the first and largest division. Allergies to fungal agents constitute a second category, and mycotoxicoses, which result from ingestion of toxic metabolites of fungi, make up the third. Diseases in all three categories have been recognized for centuries; however, during the last decade, mycotic infections have been posing significantly greater problems to man than previously. Although the incidence of the less serious fungal infections, such as the superficial and cutaneous mycoses, has decreased over the past 20 or more years, the frequency of the more serious mycotic diseases, such as the systemic infections, has increased. For instance, ringworm infections are less frequently encountered today than they were in the past. This is because of the general improvement in personal hygiene, reduction in num-

---

JUNEANN W. MURPHY • Department of Microbiology and Immunology, University of Oklahoma Health Sciences Center, Oklahoma City, Oklahoma 73190.

Fungal Infections and Immune Responses, edited by Juneann W. Murphy *et al.* Plenum Press, New York, 1993.

bers of individuals living in crowded conditions, and the development of more effective treatment procedures. In spite of these advances, there remains a substantial number of individuals who have persistent chronic dermatophyte infections that affect their lives adversely. In contrast to dermatophytoses, the incidence of systemic mycotic infections has increased during the same time period. This increase can be attributed only in part to the fact that we are better able to recognize mycotic diseases today. A major contributor to the increase is the growing number of susceptible individuals in the population. With the immense advances in medicine, individuals with defective resistance mechanisms, who would have died relatively young in the past, are now living longer and form a subset of individuals with increased susceptibility to fungal infections. The expanding use of immunosuppressive drugs in transplant patients and individuals with autoimmune diseases has resulted in another subset of individuals that is highly susceptible to fungal infections. Other groups at risk of acquiring serious mycotic infections are cancer patients, individuals on long-term antibacterial antibiotics, and individuals with other infectious diseases. The best illustration of the impact of immunodeficiency on the incidence of mycotic infections is seen in the population with acquired immunodeficiency syndrome (AIDS). The dramatic rise in numbers of AIDS patients has been accompanied by a sharp increase in the incidence of mycotic infections. Candidiasis and cryptococcosis account for most of the fungal infections in AIDS patients; however, several other mycotic infections in these patients have been reported as well. The swelling number of people infected with the human immunodeficiency virus (HIV) will continue to provide a large pool of individuals who are highly susceptible to fungal infections. This means that mycoses are going to be significant public health problems for some time to come. Therefore, gaining a better understanding of the host–fungi interactions and developing effective procedures and drugs for control of the fungal infections are going to be essential activities of the new generation of medical scientists and physicians.

For a full appreciation of the subjects covered in this book, it is necessary to have some general knowledge of the fungi. Fungi are eukaryotic organisms based on their nuclear characteristics and their mitochondria; however, they are sufficiently different from other eukaryotic organisms so that they are grouped in a separate kingdom, Mycota. Fungi have rigid cell walls which contain chitin, glucans, mannans, and peptidoglycans, and their plasma membranes have ergosterol as the major sterol. Fungi synthesize lysine; however, they do this through a pathway that requires L-α amino adipic acid, making them different from plants and other organisms that synthesize lysine. Of course, the medically important organisms in the kingdom Mycota have the characteristics just described, but they are diverse in other characteristics and are scattered, therefore, throughout the various subdivisions of the kingdom.

Fungi that cause disease in man are typically saprobes and reproduce in their natural habitat as yeasts or molds. In fact, most fungi are not able to survive in the environment provided by human tissue and therefore cannot cause disease in humans unless they produce toxic products that are subsequently ingested by humans. Mycotoxicoses are the diseases caused by fungi of this type. In sharp contrast, however, are the anthropophilic dermatophytes, which are parasites on human skin, and infections occur by the transmission of the organism directly from human to human. Between these two extremes in host–fungi interactions, there are numerous examples of saprophytic fungi which have adapted to survive in the human host and thus are capable of causing infections in humans. The niches in nature of the organisms in this latter group of disease-causing fungi range from being ubiquitous to being found in defined geographic regions. For instance, *Cryptococcus neoformans* is a ubiquitous yeastlike organism, whereas *Coccidioides immitis* and *Histoplasma capsulatum* are mold forms in their natural habitats. The latter two organisms are also excellent examples of organisms found in defined geographic locations. *C. immitis* is prevalent in the southwestern United States, whereas, *H. capsulatum* is found predominantly in the Mississippi and Ohio river valleys. A few fungi, such as *Candida* and *Pityrosporum*, are endogenous in humans. Both of these organisms are present in the yeast form as normal flora in humans, and they cause disease only when there is some disruption or abnormality in the host's physiology, flora, or immune defenses.

Certain fungi are capable of infecting apparently normal, healthy humans and thus are considered primary pathogens. The two best examples are *C. immitis* and *H. capsulatum*. These two organisms have adapted to survive effectively in the human host by converting from the form in which they normally exist in nature (mycelial or saprophytic form) to a parasitic or tissue form (spherules/endospores for *C. immitis* and yeast form for *H. capsulatum*). The infectious particles are produced during mold form growth in nature and once those particles gain entrance into the host tissue, they change to forms that can more effectively survive and multiply in the new environment. Fungi that are capable of changing form in this manner are referred to as dimorphic fungi. In addition to *C. immitis* and *H. capsulatum*, *Blastomyces dermatitidis*, *Paracoccidioides brasiliensis*, and *Sporothrix schenckii* are dimorphic fungi. In fact, *C. albicans* is also considered to be dimorphic; however, it reproduces in the yeast form in its natural habitat of the oral and intestinal tracts. If appropriate conditions occur in the host, then *C. albicans* converts to the mycelial form and establishes an infection. Other fungi, which are not so well adapted to the human host, infect only immunologically compromised or debilitated individuals and are considered opportunistic organisms. *Cryptococcus neoformans* is an excellent example of a ubiquitous, saprophytic organism that causes disease primarily in individuals with reduced immunological function.

Humans acquire mycotic diseases by implantation of fragments or conidia of a fungus through trauma or damage to the skin or mucous membranes, by inhalation of viable fungal cells, or by ingestion of toxic products of the fungus. Typically, the organisms that cause superficial, cutaneous, and subcutaneous mycoses enter through breaks in the skin or mucous membranes, remain in the tissue initially infected, and do not penetrate into deeper tissues. The organisms that enter the body through the pulmonary route, however, often disseminate to extrapulmonary sites. The latter organisms comprise the group of systemic mycotic agents.

The fungi which are capable of causing infection generally do not possess virulence factors that allow them to rapidly invade tissues. Some do produce enzymes, however, that may enhance their ability to establish disease in the host. Dermatophytes, for example, produce keratinases which permit them to penetrate the outer most layer of the skin and possibly to spread in the skin. *C. albicans* and *C. immitis* produce proteases that may contribute to virulence.

Typically, mycotic infections have an insidious onset and often develop into chronic disease states. These characteristics of fungal disease are most likely due to a combination of the multiple, rather weak virulence factors of the mycotic agent and the numerous host resistance mechanisms engaged in response to the invading fungus. Such host–parasite interactions tend to result in slow, progressive deterioration of the host if the organism has the advantage or in gradual improvement in the host's condition if the host's resistance mechanisms maintain control of the organism.

The morphology, size, and chemical characteristics of fungi and the higher eubacteria, i.e., *Actinomyces* and *Nocardia*, often affect the manner in which the host's natural and immune defenses interact with the etiological agent. Usually the actinomyces and fungi are larger than and structurally different from most bacteria. The size, shape, and chemical composition of the outer layers of the organisms influence the manner in which the organisms are cleared from infected tissues. If the organisms do not present appropriate surface ligands to which the phagocytic cells can bind, then it is unlikely that the organism will be adversely affected by an effector cell. In cases where the effector cells are capable of binding to the organism, it is possible that the particular receptor–ligand interaction causing binding does not signal the effector cell to phagocytize the organism or to exocytose toxic components. In this type of interaction, the organism would not be damaged by the effector cell. There are also examples in which the fungus is recognized by and taken into the phagocytic cells but once in the host cell, the fungus avoids being killed and may, in fact, use the host cells as a means of getting to a new location in the body. *H. capsulatum* is one mycotic agent that does this. Frequently, however, fungal organisms are too large to be phagocytized by the effector cells even though the effector cells can bind to the fungal cell. Under these conditions, if the effector cell is not able to kill by extracellular mechanisms, the organism will go unharmed. The complex and diverse characteristics of the actinomycetes and fungi not only challenge host resistance mechanisms but also make these organisms challenging to study.

The chemical and physiological characteristics of the organisms reviewed here also dictate the type of antimicrobial agents that will be effective. Since fungal and host cells are both eukaryotic cells, there are only a few physiological differences to exploit in developing drugs to eliminate the fungus without also damaging the host. One key difference, used to an advantage in developing antifungal therapeutic agents, is the type of sterol in the plasma membrane. Fungal plasma membranes contain ergosterol, whereas, host cell plasma membranes have cholesterol as the predominant sterol. Thus, drugs such as amphotericin B, which preferentially bind to ergosterol rather than to cholesterol in plasma membranes, disrupt membrane integrity in fungal cells more effectively than in host cells. Other antimycotic agents interfere with ergosterol synthesis and thereby alter fungal cell membrane integrity and effect selective transport of substances across the membrane without seriously affecting host cells. The azoles such as ketoconazole, miconazole, and fluconazole are drugs that inhibit ergosterol synthesis. The number of antifungal drugs is small, yet the needs for such therapeutic agents are rapidly expanding. Development of new effective antifungal agents is a challenging and difficult task, but one that is worthy of pursuit.

The chapters to follow convey specific, current information about each of the mycotic and actinomycotic diseases and the complex organisms which cause those diseases. From the comprehensive reviews that the authors have prepared, it is clear that there are still numerous unknowns regarding the diseases, the etiologic agents, and the host responses to the etiologic agents.

2

Superficial Fungal Infections

PETER G. SOHNLE

1. DESCRIPTION OF THE DISEASES

There are a number of fungi capable of infecting various superficial structures, including hair, nails, stratum corneum of the skin, the cornea, and the lining of the external ear canal. These infections are to be distinguished from deeper forms of cutaneous mycoses, including blastomycosis, sporotrichosis, mycetomas, and the like. Superficial candidiasis and dermatophytosis are described in other chapters, so this discussion will cover tinea versicolor, Malassezia folliculitis, tinea nigra, white and black piedra, fungal keratitis of the cornea, and fungal otitis externa. The portal for all these diseases is usually the surface that is infected. Deep and/or disseminated infections with Malassezia and Trichosporon also occur in certain predisposed patients, and with this kind of infection the portal may not be as obvious. The various pathogens and associated conditions are listed in Table I.

1.1. Tinea Versicolor

This condition is a chronic superficial fungal infection of the skin, generally affecting the trunk or proximal parts of the extremities. It is caused by a yeastlike organism known as *Malassezia furfur* (*Pityrosporum orbiculare*). The lesions of tinea versicolor are macules, which may coalesce into large, irregular areas that are characterized by scaling and changes in pigmenta-

PETER G. SOHNLE • Section of Infectious Diseases, Department of Medicine, the Medical College of Wisconsin, Milwaukee, Wisconsin 53226, and the Medical and Research Services, Milwaukee VA Medical Center, Milwaukee, Wisconsin 53295.

Fungal Infections and Immune Responses, edited by Juneann W. Murphy *et al.* Plenum Press, New York, 1993.

TABLE I
The Pathogens of Superficial Fungal Infections and the Conditions Associated with Them

Pathogens	Associated conditions
Malessezia furfur (*Pityrosporum orbiculare*)	Tinea (pityriasis) versicolor, Malassezia folliculitis, Malassezia sepsis; *possibly*: Seborrheic dermatitis, atopic dermatitis, psoriasis, confluent and reticulate papillomatosis
Phaeoannellomyces wernickii (*Exophiala wernickii*), and occasionally, *Stenella araguata*	Tinea nigra
Trichosporon beigelii, and occasionally, *Blastoschizomyces capitatus* (*Trichosporon capitatum*)	White piedra, disseminated trichosporonosis
Piedraia hortae	Black piedra
Fusarium solani, *Aspergillus fumagatus*, *Candida albicans*	Fungal keratitis
Aspergillus niger, *Candida albicans*	Fungal otitis externa

tion. Although this infection can persist for years unless treated appropriately, the major symptoms associated with it are cosmetic. Pruritus, however, is produced occasionally by the lesions and can be bothersome for some patients. Diagnosis of tinea versicolor can be made by a potassium hydroxide (10%) preparation of skin scrapings to reveal the typical yeast and hyphae of the causative fungus.

Electron microscopic studies of tinea versicolor have demonstrated that the organisms actually invade the cornified cells of the stratum corneum, producing a true intracellular infection.[1] Tinea versicolor may be found in unusual patterns in certain individuals. For example, the infection may occasionally be located in sites such as the scalp, genitalia, or flexural areas (a condition termed inverse tinea versicolor). Infection on the face has been reported in immunosuppressed patients,[2] and it may present as intertrigo in other patients without obvious underlying conditions.[3] Although tinea versicolor usually affects adults, it has been reported in infants.[4] This condition, however, is much more common after puberty,[5] presumably because of the increased activity of the sebaceous glands at this time.

1.2. Malassezia Folliculitis

This entity resembles to some extent acne vulgaris, the macronodular lesions of disseminated candidiasis in immunosuppressed patients, and the candidal papular folliculitis of heroin addicts.[6] It may also be confused with graft-versus-host disease (GVHD) in bone marrow transplant recipients.[7] Like tinea versicolor, Malassezia folliculitis is caused by *M. furfur* and tends to infect the same areas—the upper shoulders, back, and chest. The lesions, however, begin as an inflammation of the hair follicles, rather than of the macules typical of tinea versicolor, and then progress to papules and even frank pustules. These lesions are generally quite pruritic. The content of the infected hair follicles includes keratinaceous material, amorphous cellular debris, inflammatory cells, and the yeast forms (but not the hyphal forms) of the infecting organisms.[8,9]

Malassezia folliculitis may develop in immunosuppressed patients,[7,8,10] but this condition is also quite common in patients with presumably normal immunological function.[11] Experimental infections with Malassezia have been produced under occlusive dressings, and from these studies it appears that certain patients (i.e., those with seborrheic dermatitis) develop folliculitis more frequently than do others.[12] Therefore, there may be other host factors, in addition to obvious immunosuppression, that are related to the pathogenesis of this condition.

1.3. Deep Malassezia Infections

Malassezia can cause serious systemic infections, in addition to the superficial ones described above. The most common type of systemic infection is sepsis in neonates receiving intravenous hyperalimentation; the hyperalimentation fluids appear to supply the necessary lipids required for the growth of the organism, which presumably gains access to the vascular system through contamination of the intravascular device.[13–15] This type of infection has also been reported in two adults who were receiving total parenteral nutrition supplemented with lipids,[16] and in one patient who developed Malassezia septicemia in the absence of either a central venous catheter or a history of intravenous lipid supplementation.[17]

In infants receiving total parenteral hyperalimentation supplemented with lipids, the infection may manifest itself as pulmonary vasculitis, with the organisms being associated with deposits of lipids from the intravenous infusions.[15] The pattern of clinical findings in such cases is generally one of fever, leucocytosis, thrombocytopenia, and on occasion, bilateral pulmonary infiltrates.[6] As discussed above, unless appropriate lipid-containing media are used for culture, Malassezia will generally not be isolated from the blood or from the infected catheter. Malassezia yeasts, however, have been shown to grow in standard media supplemented only with the lipids used in parenteral hyperalimentation,[18] so it is possible that blood or catheter cultures will reveal the organism if enough of this lipid material is present in the specimens removed for culture.[19]

1.4. Tinea Nigra

Tinea nigra is a superficial fungal infection located most often on the palms; it is caused by *Phaeoannellomyces werneckii* (*Exophiala werneckii*) and occasionally by other dematiaceous fungi. The lesions are generally dark colored, non scaling macules. They are usually asymptomatic but can be confused with melanomas and perhaps result in unnecessary surgery.[20] The causative fungus can readily be demonstrated by potassium hydroxide preparations of skin scrapings. Tinea nigra is seen most often in tropical or semitropical areas of Central and South America, Africa, and Asia, although some cases do occur in North America.

1.5. White Piedra

White piedra is an asymptomatic fungal infection of the hair shafts; it is caused by *Trichosporon beigelii*, which produces light-colored, soft nodules that are attached to the hairs and may cause the hair shafts to break. Otherwise, the condition appears to be asymptomatic, although the causative fungus can produce serious infections in immunocompromised patients (as discussed below). White piedra is most often seen in warm climates; however, it may be acquired more commonly in the United States than previously suspected,[21] and may also be imported from other countries.[22] This condition usually affects the hairs on the scalp or face, axilla, or in the groin. It may be found as a genital infection,[21,23,24] and the organism may also be carried asymptomatically.[25] Cutaneous trichosporonosis, presenting generally as genitocrural intertrigo, has been reported from India.[26] Additionally, Trichosporon may also produce some cases of onychomycosis.[27]

1.6. Deep Trichosporon Infections

Trichosporon can produce more severe infections in certain types of patients. These infections may be either disseminated ones in immunocompromised patients or deep infections of a particular site in other situations.[28] *T. beigelii* is the usual pathogen, although *Blastoschizomyces capitatus* (formerly called *Trichosporon capitatum*) can cause very similar infections. As discussed below, Trichosporon and Cryptococcus contain common antigens, so the Cryptococcal latex agglutination test may be useful for diagnosing serious Trichosporon infections.

Disseminated trichosporonosis occurs most frequently in patients with underlying neoplastic diseases, especially those treated with cytotoxic chemotherapy or corticosteroids. The patients are frequently granulocytopenic, and because some strains of the organism are relatively resistant to amphotericin B, symptoms of disseminated Trichosporon infection may appear while the patients are receiving this drug empirically for suspected fungal infection.[28] The portal for this organism appears to be either the lungs or the alimentary tract.[29] Cutaneous entry has also been postulated.[30] The most common signs and symptoms of disseminated Trichosporon infection are fever, pulmonary infiltrates, skin lesions (multiple papular lesions progressing to either ulcers or bullae), renal involvement (with hematuria, proteinuria, and renal failure), and isolation of the organism from the blood.[28–32] The patients have often had preceding polymicrobial infections,[31] and the majority have died in spite of antifungal therapy.[28,29,31] The diagnosis is usually made by isolation of the organism from sterile body fluids or tissues as a germ-tube negative yeast that grows well on standard fungal media.[28]

Trichosporon can also produce deep infections of localized sites. A facial granuloma caused by this organism in one patient was attributed to the use of local corticosteroids under occlusion during the early stages of the infection.[33] Trichosporon has also produced infections of prosthetic heart valves,[34] fungemia in the setting of intravenous drug abuse,[35] infections of intravascular catheters,[36] fungal peritonitis complicating continuous ambulatory peritoneal dialysis,[37,38] and a postsurgical wound infection in a diabetic patient.[39]

1.7. Black Piedra

Black piedra is similar to white piedra in that it is a nodular, generally asymptomatic fungal infection of the hair shafts; however, the nodules are darker in color, harder, and more firmly attached to the hairs. The causative organism is *Piedraia hortae* and the condition is most commonly seen in tropical climates. It usually affects the scalp hair.

1.8. Fungal Keratitis

Fungal keratitis usually occurs because of some kind of local predisposition, as discussed below. The fungi most commonly involved are *Fusarium solani*, *Aspergillus fumigatus*, *Candida albicans*, and several other genera of filamentous fungi. This condition generally produces irritation of the eye and the production of a fungal ulcer in the cornea. In some cases, the keratitis may progress to a generalized fungal endopthalmitis.[40]

1.9. Fungal Otitis Externa

Fungal infection of the external auditory canal can be caused by several species of *Aspergillus* (most often *A. niger*), but *Candida albicans* is also capable of infecting this site. The major symptoms are itching and a feeling of fullness in the ear. Grayish membranes in the external auditory canal are typical of *A. niger* infection.[41]

2. ETIOLOGY

2.1. *Malassezia furfur*

The organism causing tinea versicolor and Malassezia folliculitis is known as either *Malassezia furfur* or *Pityrosporum orbiculare*.[42] There are two lipophilic species that have been described for the genus Pityrosporum, *P. orbiculare* and *P. ovale*. There is, however, good evidence to believe that these really represent different morphological forms of the same species, as demonstrated by nutritional studies and genomic analysis.[43,44] Therefore, because the name *M. furfur* has priority, it should probably be retained[44] and will be used here, although some authors prefer use of the name *P. orbiculare*.[42]

On artificial media, *M. furfur* grows primarily as small, oval, budding yeast, although filaments may also be seen. Filamentous forms are much more evident when scrapings of infected skin are examined in potassium hydroxide preparations. This organism cannot synthesize medium to long chain fatty acids, so lipids must be added to the culture medium for growth to occur. Therefore, *M. furfur* is usually not identified in hospital microbiology laboratories unless special media are used, such as Sabouraud's dextrose agar overlaid with sterile olive oil. When cultured on this medium, the organism will grow as soft, light colored colonies within about 3 days of incubation at room temperature. The genus Malassezia also contains a nonlipophilic species, *M. pachydermatis*, which can grow on standard fungal media. It produces external ear infections in dogs,[45,46] and has caused systemic infections in human infants.[47]

2.2. *Phaeoannellomyces werneckii*

This organism was initially called *Cladosporium werneckii*, then *Exophiala werneckii*, and now is known as *P. werneckii*. It can be isolated on standard fungal media containing chloramphenicol and cycloheximide and grows initially as two-celled yeast, producing brown or black colonies. After a longer period of culture, pigmented mycelial growth begins to occur and will eventually become predominant. Several genera, including Aureobasidium and Wangiella, as well as Phaeoannellomyces, grow in this manner, and they are generally distinguishable on the basis of their characteristic conidia-bearing cells after the conversion to the mycelial phase. Some isolates do not readily convert to this phase, however, so identification may be difficult. For this reason, an immunologic procedure has been developed for the rapid identification of *P. werneckii*.[48] The organism has been isolated not only from the lesions of tinea nigra, but also from other kinds of superficial lesions in humans, and from a variety of environmental sites.[49] Another similar organism, *Stenella araguata*, also causes some cases of tinea nigra. *P. werneckii* and *S. araguata* are both members of a group of organisms known as the dematiaceous fungi, that live in soil, sewage, decaying vegetation, and similar sites in warm, humid environments.

2.3. *Piedraia hortae*

P. hortae is the etiologic agent of black piedra. This organism can be isolated on standard fungal media where it grows (slowly) as dark brown to black mycelial colonies. On microscopic examination the cultured organisms appear as irregular septate hyphae with chlamydospores. The sexual phase occurs on infected hairs, but is difficult to culture on artificial media.

2.4. *Trichosporon beigelii*

T. beigelii, the agent of white piedra, is a fungus that is commonly found in the soil and other environmental sites and has been found to colonize a small percentage of throat and rectal cultures of hospitalized patients.[50] It has also been implicated as a cause of hypersensitivity pneumonitis (summer-type hypersensitivity pneumonitis) in Japan.[51,52] The organism can be isolated on standard fungal media (although it may be inhibited by cycloheximide) and grows as light-colored colonies that develop radiating furrows as they age. Microscopic examination reveals yeast and septate hyphae that readily break up into arthroconidia, approximately 3 by 7 microns in size.

2.5. *Fusarium solani*

A variety of fungi can produce fungal keratitis in humans, but *F. solani* appears to be the most common cause of this condition.[40,53] This species also seems to be the most pathogenic at this site when compared experimentally with other organisms.[54,55] *F. solani* is a filamentous fungus that grows well on standard fungal media, such as Sabouraud's dextrose agar, and develops bluish to brown pigmented colonies with curved macroconidia. This organism has caused disseminated infections in cancer patients and is relatively resistant to most antifungal agents. As discussed earlier, various other fungal organisms, including several species of Aspergillus and Candida, may also produce keratitis under certain situations; these organisms are discussed in other chapters.

3. EPIDEMIOLOGY

M. furfur can be cultured from scrapings in infected skin in patients with tinea versicolor,[56,57] the uninvolved skin of patients with this condition,[57] and the skin from normal persons.[58,59] The organism therefore appears to be part of the normal human skin flora. Presumably, actual infection of the stratum corneum to produce tinea versicolor is encouraged by some predisposing factor in the host. Since more filamentous forms of Malassezia are found in the lesions of tinea versicolor than on normal skin, the change to pathogenicity may involve conversion to the hyphal form.[60] As discussed above, however, the yeasts of this organism appear to predominate in Malassezia folliculitis and in sepsis associated with intravenous hyperalimentation. Therefore, other factors may be involved in promoting invasion, perhaps in addition to those that cause germination of the yeast into the filamentous form.

Tinea versicolor has been reported to be more frequent in steroid treated patients, those with Cushing's disease, those undergoing immunosuppression for renal transplantation, and in persons who are severely malnourished.[61–63] Additionally, experimental tinea versicolor was reported in one study to be more easily produced in patients with Cushing's disease or in those treated with steroids.[64] Also, as discussed above, Malassezia folliculitis may have an increased incidence in immunosuppressed patients. Therefore, systemic host defenses would also appear to be involved in preventing cutaneous Malassezia infections. Genetic susceptibility has also been postulated for tinea versicolor in that infections in relatives occur at a higher than expected rate.[57,64,65] Warm climates definitely seem to be associated with an increased incidence of this condition.[66] Since some conjugal cases of tinea versicolor do occur,[57] there may be some degree of infectivity of the responsible organism; however, inasmuch as Malassezia is clearly part of the normal human skin flora, the other factors discussed earlier would seem to be more important.

With respect to disseminated Malassezia infections, the major predisposing causes appear to be the presence of a central venous catheter, infusion of lipid-containing hyperalimentation solutions, and perhaps a young age. As discussed earlier, however, these infections may occur in the absence of such conditions. The source of the organism in these infections is presumably the patient's own skin. Infants are probably less likely to have this organism than are older persons, but in one study, Malassezia was found on the skin of 8 of 27 infants in a neonatal unit.[14] The probable portal of the organism in this type of infection is contamination of the central venous catheter.

Malassezia has also been suggested to play a role in certain cutaneous conditions where it presumably is not the primary cause. These include seborrheic dermatitis, atopic dermatitis, psoriasis (particularly of the scalp), and confluent and reticulate papillomatosis.[67] This kind of association has been made generally on the basis of a beneficial effect of oral antifungal therapy with ketoconazole, and the ability to demonstrate Malassezia organisms in the cutaneous lesions. More information will be needed to determine if this organism is really associated with these or other diseases, but if so, then antifungal therapy may represent an additional form of treatment for certain diseases which may otherwise be difficult to control.

Tinea nigra, and both black and white piedra, are caused by fungi that are frequently found in the environment. These conditions all seem to be more frequent in warm climates, perhaps because the organisms may be more common there. Additionally, Trichosporon, the cause of white piedra, may be found on the skin or in rectal cultures from hospitalized patients without evidence of piedra or disseminated disease. Therefore, it may be part of the normal skin flora in a relatively small number of patients.

Deep Trichosporon infections are virtually always associated with well-defined predisposing factors. Most commonly, the patients are severely immunosuppressed.[28–32] On the other hand, deep or disseminated infections with this organism have also been related to other predisposing factors in nonimmunocompromised patients.[33–37] As with Malassezia infections, the patient's skin is the most likely source of the infection in many cases; however, a pulmonary or gastrointestinal source has been implicated in some of the disseminated cases in immunosuppressed patients.[29]

For fungal keratitis and otitis externa, the ultimate source of the organism is probably the environment, and the mode of infection would seem to be by direct implantation in most cases. Infections of the cornea are associated with a number of factors, including extended-wear contact lenses,[68,69] foreign bodies in the eye,[70] topical therapy with corticosteroids and antibiotics,[71] and dry, windy conditions in subtropical climates.[71]

4. HOST DEFENSES

For the most part, host defenses and immune responses against the organisms under discussion in this chapter have not been as well studied as have those of other fungi. There is some information, however, on humoral and cell-mediated immunity against Malassezia, and some on humoral immunity as a diagnostic aid in deep Trichosporon infections. In general, little specific information is available on the immunology of the organisms causing tinea nigra, black piedra, fungal keratitis, or fungal otitis externa, except for the use of specific antibodies to aid in the identification of isolates of *P. werneckii* and similar organisms.[48,72]

4.1. Innate Resistance

Since most of the conditions discussed above involve superficial sites, the barrier function of the cutaneous structures at these sites probably represents a major form of host defense. Trauma, however, seems to play a relatively minor role as a predisposing factor in these infections. Hot, humid conditions and various forms of immunosuppression appear to be relatively more important. One might suspect that the organisms involved, particularly Malassezia, are present in large enough numbers on the skin that inoculation into the deeper layers by minor trauma is a regular occurrence. Since this kind of inoculation seems to be followed rarely by infection with cutaneous organisms, one might presume that such standard mechanisms of defense as phagocytic cell function and the immunological system are readily able to eliminate the organisms when they are inoculated more deeply into the skin. Thus, significant infections with these fungal organisms seem to occur only when there is a prolonged abnormality of some important defense mechanism, be it either local or systemic. One example is Malassezia folliculitis in bone marrow transplant recipients in which the cutaneous lesions generally resolve when the patient's leucopenia begins to improve.[7]

As discussed above, deep infections with either Malassezia or Trichosporon require both a mode of entry into the deep tissues and then some factor allowing for continued infection of the deep site. For Malassezia, infection of an intravascular device appears to be the usual mode of entry, whereas the means by which Trichosporon enters the deep tissues in immunosuppressed patients is not clear. Having attained access to the vascular system, Malassezia appears to be capable of maintaining itself there when lipid-containing hyperalimentation fluids are infused—even though the patient is not undergoing immunosuppressive therapy. On the other hand, disseminated Trichosporon infections generally occur in patients who are granulocytopenic and/or treated with pharmacologic doses of corticosteroids. In both cases, infection of a foreign body may also be a way for these organisms to persist at deep sites in the presence of relatively normal phagocytic cell function.

With the exception of fungal keratitis and possibly of fungal otitis externa, the superficial fungal infections under discussion in this chapter generally produce relatively little inflammation. Tinea versicolor does produce mild acanthosis, hyperkeratosis, and a modest cellular infiltrate, but the inflammatory reaction elicited by Malassezia is generally much less than that seen with Candida or the dermatophytes. Malassezia, however, does have the capacity to increase immunity nonspecifically against *Salmonella typhimurium* infections in mice when injected intraperitoneally,[73] perhaps because these organisms can activate complement directly via the alternative pathway.[74] Therefore, in the light of this potential for eliciting host responses, it is somewhat surprising that the lesions of tinea versicolor should be associated with so little inflammation. As discussed below, however, the reason may be related to the relative antigenic content of the organism itself or perhaps to the type of immune response it elicits.

4.2. Immune Responses

4.2.1. Humoral Immunity

Several studies have demonstrated that serum antibody to *M. furfur* is present in a large percentage of normal persons who do not show evidence of infection with this organism.[75–78] The antibody to this organism in normal subjects is predominantly of the IgG class, although there are also some from the IgM and IgA classes.[77] Titers of this antibody have been shown to be lower in children under five years of age, perhaps because these children are less likely to be colonized with Malassezia and therefore not yet sensitized.[79]

In elderly individuals, the titers have also been found to be reduced,[77,79] and this decrease can perhaps be related to decreased numbers of the organism on the skin of older individuals.[79] Antibody titers have also been compared in homosexual men who were either seropositive or seronegative for HIV but not found to be different between the two groups.[80]

Humoral responses to Malassezia have been studied in patients with tinea versicolor and found to be either slightly elevated as compared to controls in one study[76] or to be approximately equivalent in another.[81] In Malassezia folliculitis, IgG antibodies against this organism, as measured using an immunofluorescence technique, were found to be markedly higher in the patients in normal subjects, although immediate hypersensitivity to the antigens of this organism seemed to be lacking when prick tests were done with a *M. furfur* extract.[79]

Antibody to Trichosporon antigens has been shown by an immunoblot technique to be present not only in patients with invasive infections, but also in uninfected controls.[82] Additionally, both IgG and IgA antibodies have been demonstrated by an indirect immunofluorescence technique in the vaginal washings from three women with asymptomatic carriage of *T. beigelii*.[83] In that study, washings from women having vaginitis caused by other organisms generally had low titers against Trichosporon. Trichosporon also has been shown to share antigens with *Cryptococcus neoformans*, and the latex agglutination test for Cryptococcal antigen may be positive in patients with disseminated Trichosporon infections.[84]

4.2.2. *Cell-Mediated Immunity*

As demonstrated by lymphocyte transformation tests and lymphokine assays, cell-mediated immunity to Malassezia antigens is present in a majority of uninfected human subjects, whereas tinea versicolor patients may have deficient lymphokine responses to these antigens.[85] Also, a modest, but statistically significant, reduction found in the magnitude of tinea versicolor patients' lymphocyte transformation responses to Malassezia antigens has been shown to be caused by a decrease in the number of cells initially responding to the antigen.[86] On the other hand, proportions of B and T lymphocytes and subsets of T lymphocytes have been found to be normal in tinea versicolor patients,[87] as have lymphocyte transformation responses against nonspecific mitogens and unrelated antigens.[85,87]

As a possible explanation for the relatively modest degree of inflammation present in the lesions of tinea versicolor, the relative antigenicity of this organism was compared to that of *C. albicans*, which produces highly inflammatory lesions, and was found to be markedly less by a number of assays.[88] Therefore, two possible reasons for the minimal inflammation in tinea versicolor are a reduction in the responsiveness of lymphokine-producing cells of these patients to Malassezia antigens and a relative lack of antigenic material in the organism itself.

4.2.3. *Regulation of Immune Responses*

Relatively little is known about the regulation of immunological responses to the organisms causing the superficial fungal infections under discussion here. In one study, the cells infiltrating into the infected skin of patients with either tinea versicolor or Malassezia folliculitis were characterized as helper/inducer or suppressor/cytotoxic cells by staining with monoclonal antibodies. In both types of infection, the majority of cells was demonstrated to be of the helper/inducer group.[81,87]

5. IMMUNODIAGNOSIS AND PROGNOSIS

5.1. Serological Aspects

For the most part, the diagnosis of the infections discussed in this chapter is best made from the clinical appearance of the lesions and appropriate stains and cultures of the infected tissues. In the case of deep *T. beigelii* infections, however, immunodiagnostic methods may be of some use. A monoclonal antibody has been employed for the identification of this organism in paraffin-embedded tissue sections of infected material, primarily so that the Trichosporon yeast can be distinguished from Candida.[89] Additionally, because *T. beigelii* shares antigenic determinants with the capsular polysaccharide of *C. neoformans*, the latex agglutination test can be used in the diagnosis of disseminated Trichosporon infections in immunocompromised patients.[84] Humoral responses against other antigens of this organism do occur in patients with disseminated infections, but because similar antibodies are common in normal subjects this type of assay will probably be difficult to use as a diagnostic test.[82]

5.2. Cellular Aspects

Cell-mediated immunity has been examined to some extent for Malassezia, as described above; however, it is doubtful that any of the assays employed could ever be used as a diagnostic test for infections caused by this organism because of the high degree of reactivity present in the normal human population. This type of immune response has not been evaluated in the other types of superficial fungal infections under consideration here, but one would suspect that more direct diagnostic methods would be of more value in these infections as well. Similarly, the significance of negative delayed skin test responses in the face of active infection has not been examined with respect to prognosis for this group of diseases.

6. PREVENTION AND THERAPY

6.1. Immunization or Vaccines

At the present time, there are no useful vaccines for any of the superficial fungal infections discussed above. Indeed, for Malassezia and Trichosporon, the two organisms with some potential for producing serious, disseminated infections, the normal population appears to become sensitized to the antigens of these organisms naturally. Infections are apparently related to abnormalities in the host defenses of the patient rather than to an initial lack of exposure to the organism.

6.2. Therapy

6.2.1. Chemotherapy

Tinea versicolor can be treated topically with lotions or creams containing selenium, sodium thiosulfate, clotrimazole, miconazole, or ketoconazole. Sulfur-salicylic acid shampoo can also be used. The treatment should be continued for two to three weeks and is generally successful. Relapses are frequent, however, presumably because the patient's predisposition to the infection have not been reversed, and also because the organisms may persist within hair follicles.[6] Oral ketoconazole is a very effective treatment for tinea versicolor and may sometimes be effective in a single dose of 400 mg.[90] It should probably be reserved for patients with extensive involvement of the skin, however, because of the potential for hepatic toxicity with this agent, particularly if prolonged courses are needed. Relapses may occur after oral ketoconazole therapy, so that continued prophylaxis with this drug may be necessary. Itraconazole has also been shown to be effective in treating tinea versicolor.[91]

Malassezia folliculitis can be treated using topical clotrimazole or miconazole cream for two to three weeks and may also respond to 50% propylene glycol in water applied twice a day for three weeks.[11] Treatment with a preparation containing 15% propylene glycol has also been reported to be effective in seborrheic dermatitis of the scalp, and this treatment was associated with a reduction in the number of *M. furfur* that could be cultured from this site.[92] If topical therapy fails, Malassezia folliculitis can be treated with 200 mg of oral ketoconazole given orally for two to three weeks.[11,93]

The treatment of Malassezia sepsis consists primarily of removing the central venous catheter and temporarily discontinuing infusions of lipids; these interventions appear to be successful in terminating the infection in most cases,[13,14] although it may be prudent to administer antifungal therapy as well. Amphotericin B has been used generally, although Malassezia may be more sensitive to ketoconazole.

Tinea nigra can be treated effectively with either keratinolytic agents or topical antifungal preparations. Whitfield's ointment, topical 10% thiabendazole, tincture of iodine, and miconazole, clotrimizole, and ketoconazole creams have all been used successfully. The treatment should be continued for two to three weeks. Scraping of the affected skin, followed by treatment with ciclopiroxolamine or an imidazole applied topically, has also been recommended.[94]

In both black and white piedra, shaving of the infected hairs appears to be an important part of therapy. Topical application of imidazoles has also been used with some success, but the effectiveness of this therapy is difficult to evaluate because some spontaneous remissions do occur in patients with white piedra—although relapses are also quite frequent.[21]

The treatment of deep Trichosporon infections may be difficult because of the severe underlying diseases present in most of the patients. The mortality rate in this type of infection has been reported to be 64% and 74%, respectively, in two studies.[28,31] Some, although not all, strains of *T. beigelii* are sensitive to amphotericin B,[95] and therapy should be initiated with this agent in a dose of 0.5 mg/kg intravenously once a day; if the patient does not respond, the dose should be raised to 1.0 mg/kg.[28] If the patient still fails to respond, other options may include adding flucytosine or possibly using an experimental antifungal agent. The results of one study indicated that in neutropenic patients with disseminated Trichosporon infections, resolution of the neutropenia was required for the patient's survival.[31]

Fungal keratitis may be associated with the use of contact lenses, particularly extended-wear, soft contact lenses.[68,69] Heat or chemical disinfection systems may not kill all the fungal spores, so it is desirable to attempt to avoid infection by not wearing contact lenses in moldy environments, using sterile solutions, washing the hands before manipulating the lenses, and replacing lenses that are contaminated with fungi instead of attempting to decontaminate them.[69] Once a fungal keratitis does develop, it may be difficult to treat. *F. solani* is generally resistant to antifungals other than natamycin. The latter can be used topically in a solution of 50 mg/ml, administered hourly. Additionally, systemic antifungal therapy with amphotericin B or ketoconazole may be used if the organisms are susceptible. The cornea, however, may become damaged by the infection, necessitating debridement and/or corneal grafting.[71]

With fungal infections of the external ear, it is generally required that the ear canal be thoroughly cleaned to remove the membranes characteristic of this process, before treatment with an antifungal agent can be successful.

Clotrimizole drops applied twice a day, combined with cleaning of the ear canal by a physician once or twice a week, may be successful in treating this kind of infection.[41]

6.2.2. *Immunotherapy*

Although many of the superficial fungal infections discussed above seem to be related to some kind of predisposing factor, which in some cases may be immunologic in nature, immunotherapy does not play a role in the treatment of these infections at the present time.

REFERENCES

1. Montes, L. F., 1970, Systemic abnormalities and the intracellular site of infections of the stratum corneum, *J. Am. Med. Assoc.* **213:**1469–1472.
2. Hughes, B. R., 1988, Tinea versicolor in immunosuppressed patients, *J. Am. Acad. Dermatol.* **19:**357–358.
3. Katoh, T., Kagawa, S., and Ishimoto, M., 1988, *Malassezia* intertrigo, a new clinical entity, *Mycoses* **31:**558–562.
4. Nanda, A., Kaur, S., Bhakoo, O. N., Kaur, I., and Vaishnavi, C., 1988, Pityriasis (tinea) versicolor in infancy, *Pediat. Dermatol.* **5:**260–262.
5. Faergemann, J., and Fredriksson, T., 1980, Age incidence of *Pityrosporum orbiculare* on human skin, *Acta Derm. Venereol.* **60:**531–533.
6. Klotz, S. A., 1989, *Malassezia furfur*, *Infect. Dis. Clin. North America* **3:**53–64.
7. Bufill, J. A., Lum, L. A., Caya, J. G., Chitambar, C. R., Ritch, P. S., Anderson, T., and Ash, R. C., 1988, Pityrosporum folliculitis after bone marrow transplantation—Clinical observations in five patients, *Ann. Int. Med.* **108:**560–563.
8. Klotz, S. A., Drutz, D. J., Huppert, M., and Johnson, J. E., 1982, Pityrosporum folliculitis: Its potential for confusion with skin lesions of systemic candidiasis, *Arch. Int. Med.* **142:**2126–2129.
9. Potter, B. S., Burgoon, C. F., Jr., and Johnson, W. C., 1973, Pityrosporum folliculitis—Report of seven cases and review of the *Pityrosporum* organism relative to cutaneous disease, *Arch. Dermatol.* **107:**388–391.
10. Yohn, J. J., Lucas, J., and Camisa, C., 1985, *Malassezia* folliculitis in immunocompromised patients, *Cutis* **35:**536–538.
11. Back, O., Faergemann, J., and Hornqvist, R., 1985, Pityrosporum folliculitis: A common disease of the young and middle-aged, *J. Am. Acad. Dermatol.* **12:**56–61.
12. Goodfield, M. J. D., Saihan, E. M., and Crowley, J., 1987, Experimental folliculitis with *Pityrosporum orbiculare*: The influence of host response, *Acta Derm. Venereol.* **67:**445–447.
13. Long, J. G., and Keyserling, H. L., 1985, Catheter-related infection in infants due to an unusual lipophilic yeast—*Malassezia furfur*, *Pediatrics* **76:**896–900.
14. Surmont, I., Gavilanes, A., Vandepitte, J., Devlieger, and Eggermont, E., 1989, *Malassezia furfur* fungaemia in infants receiving intravenous lipid emulsions. A rarity or just underestimated? *Eur. J. Pediatr.* **148:**435–438.
15. Redline, R. W., and Dahms, B. B., 1982, Malassezia pulmonary vasculitis in an infant on long-term intralipid therapy, *New Engl. J. Med.* **305:**1395–1398.
16. Garcia, C. R., Johnson, B. L., Corvi, G., Walker, L. J., and George, W. L., 1987, Intravenous catheter-associated *Malassezia furfur* fungemia, *Am. J. Med.* **83:**790–792.

17. Wurtz, R. M., and Knospe, W. N., 1988, *Malassezia furfur* fungemia in a patient without the usual risk factors, *Ann. Int. Med.* **109:**432–433.
18. Powell, D. A., Durrell, D. E., and Marcon, M. J., 1986, Growth of *Malassezia furfur* in parenteral fat emulsions, *J. Infect. Dis.* **153:**640–641.
19. Marcon, M. J., Powell, D. A., and Durrell, D. E., 1986, Methods for optimal recovery of *Malassezia furfur* from blood culture, *J. Clin. Microbiol.* **24:**696–700.
20. Palmer, S. R., Bass, J. W., Mandojana, R., and Wittler, R. R., 1989, Tinea nigra palmaris and plantaris: A black fungus producing black spots on the palms and soles, *Pediatr. Infect. Dis. J.* **8:**48–50.
21. Kalter, D. C., Tschen, J. A., Cernoch, P. L., McBride, M. E., Sperber, J., Bruce, S., and Wolfe, J. E., Jr., 1986, Genital white piedra: Epidemiology, microbiology, and therapy, *J. Am. Acad. Dermatol.* **14:**982–993.
22. Benson, P. M., Lapins, N. A., and Odom, R. B., 1983, White piedra, *Arch. Dermatol.* **119:** 602–604.
23. Steinman, H. K., and Pappenfort, R. B., 1984, White piedra—a case report and review of the literature, *Clin. Exp. Dermatol.* **9:**591–598.
24. Fishman, H. C., 1987, White piedra, *Int. J. Dermatol.* **26:**538.
25. Stenderup, A., Schonheyder, H., Ebbesen, P., and Melbye, M., 1986, White piedra and *Trichosporon beigelii* carriage in homosexual men, *J. Med. Vetern. Mycol.* **24:**401–406.
26. Kamalan, A., Senthamilselvi, G., Ajithadas, K., and Thambiah, A. S., 1988, Cutaneous trichosporosis, *Mycopathol.* **101:**167–175.
27. Fusaro, R. M., and Miller, N. G., 1984, Onychomycosis caused by *Trichosporon beigelii*, *J. Am. Acad. Dermatol.* **11:**747–749.
28. Walsh, T. J., 1989, Trichosporonosis, *Infect. Dis. Clin. North America* **3:**43–52.
29. Walsh, T. J., Newman, K. R., Moody, M., Wharton, R. C., and Wade, J. C., 1986, Trichosporonosis in patients with neoplastic disease, *Medicine* **65:**268–279.
30. Haupt, H. M., Merz, W. G., Beschorner, W. E., Vaughn, W. P., and Saral, R., 1983, Colonization and infection with *Trichosporon* species in the immunosuppressed host, *J. Infect. Dis.* **147:**199–203.
31. Hoy, J., Hsu, K. C., Rolston, K., Hopfer, R. L., Luna, M., and Bodey, G. P., 1986, *Trichosporon beigelii* infection: A review, *Rev. Infect. Dis.* **8:**959–967.
32. LeBlond, V., Saint-Jean, O., Datry, A., Lesco, G., Frances, C., Bellefiqh, S., Gentilini, M., and Binet, J. L., 1986, Systemic infections with *Trichosporon beigelii* (Cutaneum)—Report of three new cases, *Cancer* **58:**2399–2405.
33. Otsuka, F., Seki, Y., Takizawa, K., Ohara, K., and Ishibashi, Y., 1986, Facial granuloma associated with *Trichosporon cutaneum* infection, *Arch. Dermatol.* **122:**1176–1179.
34. Reinhart, H., Urbanski, D. M., Harrington, S. D., and Sobel, J. D., 1988, Prosthetic valve endocarditis caused by *Trichosporon beigelii*, *Am. J. Med.* **84:**355–358.
35. Moreno, S., Buzon, L., and Sanchez-Sousa, A., 1987, *Trichosporon capitatum* fungemia and intravenous drug abuse, *Rev. Infect. Dis.* **9:**1202–1203.
36. Finkelstein, R., Singer, P., and Lefler, E., 1989, Catheter-related fungemia caused by *Trichosporon beigelii* in non-neutropenic patients, *Am. J. Med.* **86:**133.
37. Carr, M. E., O'Brien, A. A. J., Moore, D., and Keogh, J. A. B., 1987, *Trichosporon cutaneum* C.A.P.D. peritonitis, *Postgrad. Med. J.* **63:**1008.
38. Pritchard, R. C., and Muir, D. B., 1985, *Trichosporon beigelii*: Survey of isolates from clinical material, *Pathology* **17:**20–23.
39. Greenberg, R. G., and Berger, T. G., 1989, Postoperative *Trichosporon beigelii* soft tissue infection, *J. Dermatol. Surg. Oncol.* **15:**432–434.
40. Pflugfelder, S. C., Flynn, H. W., Jr., Zwickey, T. A., Forster, R. K., Tsiligianni, A., Culbertson, W. W., and Mandelbaum, S., 1988, Exogenous fungal endopthalmitis, *Opthalmology* **95:** 19–30.
41. Caruso, V. G., and Meyerhoff, W. L., 1980, Trauma and infections of the external ear, in:

Otolaryngology, 2nd Ed. (M. M. Paparella and D. A. Shumrick, eds.), W. B. Saunders Company, Philadelphia, pp. 1345–1350.

42. Fredriksson, T., and Faergemann, J., 1984, Semantics—Tinea versus pityriasis versicolor and *Pityrosporum orbiculare* versus *Malassezia furfur*. Which is proper? *Int. J. Dermatol.* **23:** 110–111.
43. Caprilli, F., Mercantini, R., Nazzoro Porro, M., Passi, S., and Tonolo, A., 1973, Studies of the genus *Pityrosporum* in submerged culture, *Mycopathol. Mycol. Appl.* **51:**171–189.
44. Gueho, E., and Meyer, S. A., 1989, A reevaluation of the genus *Malassezia* by means of genome comparison, *Antonie van Leewenhoek* **55:**245–251.
45. Gabal, M. A., Chastain, C. B., and Hogle, R. M., 1979, *Pityrosporum pachydermatis* ('canis') as a major cause of otitis externa in dogs, *Mykosen* **22:**192–199.
46. Gabal, M. A., 1988, Preliminary studies on the mechanism of infection and characterization of *Malassezia pachydermatis* in association with canine otitis externa, *Mycopathol.* **104:**93–98.
47. Mickelsen, P. A., Viano-Paulson, M. C., Stevens, D. A., and Diaz, P. S., 1988, Clinical and microbiological features of infection with *Malassezia pachydermatis, J. Infect. Dis.* **157:**1163–1168.
48. Standard, P. G., Kaufman, L., and Padhye, A. A., 1988, An immunological procedure for the rapid identification of *Phaeoannellomyces werneckii, Diag. Clin. Immunol.* **5:**367–370.
49. Matsumoto, T., Padhye, A. A., and Ajello, L., 1987, Medical significance of the so-called black yeasts, *Eur. J. Epidemiol.* **3:**87–95.
50. Rose, H. D., and Kurup, V. P., 1977, Colonization of hospitalized patients with yeast-like organisms, *Sabouraudia* **15:**251–256.
51. Yoshida, K., Ando, M., Sakata, T., and Araki, S., 1988, Environmental mycological studies on the causative agent of summer-type hypersensitivity pneumonitis, *J. Allergy Clin. Immunol.* **81:**475–483.
52. Soda, K., Ando, M., Shimazu, K., Sakata, T., Yoshida, K., and Araki, S., 1986, Different classes of antibody activities to *Trichosporon cutaneum* antigen in summer-type hypersensitivity pneumonitis by enzyme-linked immunosorbent assay, *Am. Rev. Respir. Dis.* **133:**83–87.
53. Zapater, R. C., and Arrechea, A., 1975, Mycotic keratitis by Fusarium: A review and report of two cases, *Opthalmologica* **170:**1–12.
54. Ishibashi, Y., Kaufman, H. E., Matsumoto, T., and Kagawa, S., 1986, The pathogenicities of *Cylindrocarpon tonkinense* and *Fusarium solani* in the rabbit cornea, *Mycopathol.* **94:**145–152.
55. Ishibashi, Y., Kaufman, H. E., and Kagawa, S., 1986, Comparison of the pathogenicities of *Fusarium solani* and *Candida albicans* in the rabbit cornea, *J. Med. Vet. Mycol.* **24:**369–376.
56. Gordon, M. A., 1951, Lipophilic yeast-like organisms associated with tinea versicolor, *J. Invest. Dermatol.* **17:**267–272.
57. Roberts, S. O. B., 1969, Pityriasis versicolor: A clinical and mycological investigation, *Br. J. Dermatol.* **81:**315–326.
58. Roberts, S. O. B., 1969, Pityrosporum orbiculare: Incidence and distribution on clinically normal skin, *Br. J. Dermatol.* **81:**264–269.
59. Leeming, J. P., and Notman, F. H., 1987, Improved methods for isolation and enumeration of *Malassezia furfur* from human skin, *J. Clin. Microbiol.* **25:**2017–2019.
60. McGinley, K. J., Lantis, L. R., and Marples, R. R., 1970, Microbiology of tinea versicolor, *Arch. Dermatol.* **102:**168–171.
61. Boardman, C. R., and Malkinson, F. D., 1962, Tinea versicolor in steroid-treated patients, *Arch. Dermatol.* **85:**84–92.
62. Findling, J. W., Tyrrell, J. N., Aron, D. C., Fitzgerald, P. A., Young, C. N., and Sohnle, P. G., 1981, Fungal infections in Cushing's syndrome, *Ann. Int. Med.* **95:**392.
63. Koranda, F. C., Dehmel, E. M., Kahn, G., and Penn, I., 1974, Cutaneous complications in immunosuppressed renal homograft recipients, *J. Am. Med. Assoc.* **229:**419–424.
64. Burke, R. C., 1961, Tinea versicolor: Susceptibility factors and experimental infection in human beings, *J. Invest. Dermatol.* **36:**389–401.

65. Hafez, M., and El-Shamy, S., 1985, Genetic susceptibility in pityriasis versicolor, *Dermatologica* **171**:86–88.
66. Marples, M. J., 1950, The incidence of certain skin diseases in Western Samoa: A preliminary study, *Trans. R. Soc. Med. Hyg.* **44**:319–332.
67. Faergemann, J., and Maibach, H. I., 1984, The *Pityrosporon* yeasts—Their role as pathogens, *Int. J. Dermatol.* **23**:463–465.
68. Wilson, L. A., and Ahearn, D. G., 1986, Association of fungi with extended-wear soft contact lenses, *Am. J. Opthalmol.* **101**:434–436.
69. Wilhelmus, K. R., Robinson, N. M., Font, R. A., Hamil, M. B., and Jones, D. B., 1988, Fungal keratitis in contact lens wearers, *Am. J. Opthalmol.* **106**:708–714.
70. Hemo, I., Peer, J., and Polacheck, I., 1989, *Fusarium oxysporum* keratitis, *Opthalmologica* **198**: 3–7.
71. Newell, F. W., 1986, *Opthalmology, Principles and Concepts*, 6th Ed., C. V. Mosby Co., St. Louis, pp. 231–239.
72. Espinel-Ingroff, A., Shadomy, S., Kerkering, T. M., and Shadomy, H. J., 1984, Exoantigen test for differentiation of *Exophiala jeanselmei* and *Wangiella dermatitidis* isolates from other dematiaceous fungi, *J. Clin. Microbiol.* **20**:23–27.
73. Takahashi, M., Ushijima, T., and Ozaki, Y., 1984, Biological activity of *Pityrosporum*. I. Enhancement of resistance in mice stimulated by *Pityrosporum* against *Salmonella typhimurium*, *Immunol.* **51**:697–702.
74. Sohnle, P. G., and Collins-Lech, C., 1983, Activation of complement by *Pityrosporum orbiculare*, *J. Invest. Dermatol.* **80**:93–97.
75. Alexander, S., 1967, Loss of hair and dandruff, *Br. J. Dermatol.* **79**:549–552.
76. DaMert, G. J., Kirkpatrick, C. H., and Sohnle, P. G., 1980, Comparison of antibody responses in chronic mucocutaneous candidiasis and tinea versicolor, *Int. Arch. Allergy Appl. Immunol.* **63**:97–104.
77. Sohnle, P. G., Collins-Lech, C., and Huhta, K. E., 1983, Class-specific antibodies in young and aged humans against organisms producing superficial fungal infections, *Br. J. Dermatol.* **108**:69–76.
78. Faergemann, J., 1983, Antibodies to *Pityrosporum orbiculare* in patients with tinea versicolor and controls of various ages, *J. Invest. Dermatol.* **80**:133–135.
79. Bergbrant, I. M., and Faergemann, J., 1988, Variations of *Pityrosporum orbiculare* in middle-aged and elderly individuals, *Acta Derm. Venereol.* **68**:537–540.
80. Hakansson, C., Faergemann, J., and Lowhagen, G. B., 1988, Studies on the lipophilic yeast *Pityrosporum ovale* in HIV-seropositive and HIV-seronegative homosexual men, *Acta Derm. Venereol.* **68**:422–426.
81. Faergemann, J., Johansson, S., Back, O., and Scheynius, A., 1986, An immunologic and cultural study of *Pityrosporum* folliculitis, *J. Am. Acad. Dermatol.* **14**:429–433.
82. Matthews, R. C., Burnie, J. P., Fox, A., Woods, M., and Tabaqchali, S., 1986, Immunoblot analysis of the serological response in invasive *Trichosporon beigelii* and *Blastoschizomyces capitus* infections, *J. Clin. Microbiol.* **23**:395–397.
83. Quindos, G., Schneider, J., Alvarez, M., Ponton, J., and Cisterna, R., 1989, Antibodies against *Trichosporon beigelii* in vaginal washings from asymptomatic women, *J. Med. Microbiol.* **28**: 223–225.
84. McManus, E. J., Bozdech, M. J., and Jones, J. M., 1985, Role of the latex agglutination test for Cryptococcal antigen in diagnosing disseminated infections with *Trichosporon beigelii*, *J. Infect. Dis.* **151**:1167–1169.
85. Sohnle, P. G., and Collins-Lech, C., 1978, Cell-mediated immunity to *Pityrosporum orbiculare* in tinea versicolor, *J. Clin. Invest.* **62**:45–53.
86. Sohnle, P. G., and Collins-Lech, C., 1982, Analysis of the lymphocyte transformation response to *Pityrosporum orbiculare* in patients with tinea versicolor, *Clin. Exp. Immunol.* **49**: 559–564.

87. Scheynius, A., Faergemann, J., Forsum, U., and Sjoberg, O., 1984, Phenotypic characterization *in situ* of inflammatory cells in pityriasis (tinea) versicolor, *Acta Derm. Venereol.* **64:** 473–479.
88. Sohnle, P. G., and Collins-Lech, C., 1980, Relative antigenicity of *P. orbiculare* and *C. albicans*, *J. Invest. Dermatol.* **75:**279–283.
89. Takeuchi, T., Kobayashi, M., Moriki, T., and Miyoshi, I., 1988, Application of a monoclonal antibody for the detection of *Trichosporon beigelii* in paraffin-embedded tissue sections, *J. Pathol.* **156:**23–27.
90. Borelli, D., 1980, Treatment of pityriasis versicolor with ketoconazole, *Rev. Infect. Dis.* **2:** 592–595.
91. Delescluse, J., Cauwenbergh, G., and Degreef, H., 1986, Itraconazole, a new orally active antifungal, in the treatment of pityriasis versicolor, *Br. J. Dermatol.* **114:**701–703.
92. Faergemann, J., 1988, Propylene glycol in the treatment of seborrheic dermatitis of the scalp: A double blind study, *Cutis* **42:**69–71.
93. Ford, G. P., Ive, F. A., and Midgley, G., 1982, Pityrosporum folliculitis and ketoconazole, *Br. J. Dermatol.* **107:**691–695.
94. Sayegh-Carreno, R., Abramovits-Ackerman, W., and Planas Giron, G., 1989, Therapy of tinea nigra plantaris, *Int. J. Dermatol.* **28:**46–48.
95. McBride, M. E., Kalter, D. C., and Wolf, J. E., Jr., 1988, Antifungal susceptibility testing of *Trichosporon beigelii* to imidazole compounds, *Can. J. Microbiol.* **34:**850–854.

3

Dermatophytosis

PETER G. SOHNLE

1. DESCRIPTION OF THE DISEASES

Dermatophytosis is the infection of keratinized structures, including the nails, hair shafts, and stratum corneum of the skin, by organisms of three genera of fungi termed the dermatophytes. These organisms are particularly well adapted to this location because they can use keratin as a source of nutrients; therefore, they differ from most of the other pathogenic fungi in that they are not opportunists. The portal for the dermatophytes is the superficial surface that becomes infected. These superficial sites can also be infected by a variety of other fungi, including *Candida albicans*, *Malassezia furfur* (*Pityrosporum orbiculare*), *Phaeoannellomyces wernickii*, *Trichosporon beigelii*, *Fusarium solani*, and *Aspergillus fumigatus* and *niger*; however, these organisms and the infections they produce will be discussed in other chapters.

The different types of dermatophytosis can probably be best classified according to body site, using the word "tinea," followed by a term for the particular body site. Table I lists the major types of dermatophytosis and the most frequent organisms associated with them. The word "tinea" itself comes from the Roman word for the clothes moth because of the similarity of their effect on wool garments to the shape of the fungal skin infections.[1] The degree of inflammation produced in the lesions appears to depend primarily on the particular organism and perhaps also to some extent on the immunological competence of the patient. In any event, the more chronic forms of dermatophytosis are generally associated with less inflammation and tend to

PETER G. SOHNLE • Section of Infectious Diseases, Department of Medicine, the Medical College of Wisconsin, Milwaukee, Wisconsin 53226, and the Medical and Research Services, Milwaukee VA Medical Center, Milwaukee, Wisconsin 53295.

Fungal Infections and Immune Responses, edited by Juneann W. Murphy *et al.* Plenum Press, New York, 1993.

TABLE I
Types of Dermatophytosis and the Responsible Pathogens

Type of infection	Pathogens
Tinea pedis	*T. rubrum, T. mentagrophytes, E. floccosum*
Tinea cruris	*E. floccosum, T. rubrum*
Tinea corporis	*T. rubrum, T. mentagrophytes, T. concentricum, T. verrucosum, M. canis, M. gypseum*
Tinea barbae	*T. rubrum, T. verrucosum*
Tinea unguium (onychomycosis)	*T. rubrum, T. mentagrophytes*
Tinea capitis	*T. tonsurans, T. schoenleini* (favus), *M. audouini, M. canis*

be produced most often by organisms usually associated with humans (anthropophilic) rather than those associated with animals (zoophilic) or the soil (geophilic).

Of the different types of dermatophytosis, tinea pedis (athletes foot) is probably the most common. This condition is a chronic toe web infection that can be scaly, vesicular, or ulcerative in form and can sometimes produce hyperkeratosis of the sole of the foot. Secondary bacterial infections frequently complicate tinea pedis. Tinea cruris is a dry, expanding dermatophyte infection in the flexural areas of the groin; it occurs much more frequently in males than in females. Tinea corporis is a term for dermatophytosis of the major surface areas of the body. This type of infection can sometimes be extensive, particularly in patients with some kind of preexisting immunological deficiency, and it frequently takes the classical annular, or "ringworm," shape. Infections of the beard area in men, a condition known as tinea barbae, is often caused by zoophilic organisms such as *Trichophyton verrucosum*. These infections can be highly inflammatory because of the organism involved or because of secondary bacterial infections. Tinea unguium is a form of onychomycosis, or fungal infection of the nails, and is most frequently caused by *T. rubrum*. Of all the varieties of dermatophytosis, nail infections, particularly of the toenails, are the most difficult to treat. Infections of the hair and skin on the scalp, or tinea capitis, is more common in children than in adults. The degree of inflammation produced may vary considerably, but in some cases a raised, tender mass of inflamed tissue, called a kerion, may occur and generally indicates that a spontaneous cure will result. In other cases (black dot ringworm) there may be minimal inflammation, but hair loss may occur with the scattered stumps of broken hairs being left behind in the infected area. Favus (or tinea favosum) is a now rare, distinctive type of tinea capitis that lasts for many years and is characterized by the development of extensive, cup-shaped crusts called scutula. Of interest is the fact that the description in 1837 of the fungal origin of favus marked the first time that a microorganism had been incriminated as causing a human disease.[1]

One interesting characteristic of dermatophytosis is that some patients develop secondary cutaneous eruptions known as dermatophytids or id reactions. These eruptions generally occur at a site distant from the primary dermatophyte infection; they are generally vesicular but can also be papular, eczematous, or have the appearance of erythema multiforme. Dermatophytids are most commonly associated with inflammatory dermatophyte infections. It is a characteristic of dermatophytids that the organism cannot usually be cultured from them, although it is easily recoverable from the original site of dermatophytosis.[2] There are some indications that an immunologic mechanism may be involved in the production of dermatophytids, in that most patients can be shown to have delayed hypersensitivity to dermatophyte antigens[3] and some have also been demonstrated to have vigorous lymphocyte transformation responses to these antigens.[4] In the past, dermatophytids have been treated effectively with corticosteroids.[3]

Even in severely immunocompromised patients, dermatophytes rarely invade the deep tissues or produce systemic infections. On infrequent occasions, however, granulomas, draining sinuses, or actual abscesses can be produced by these organisms.[5] As discussed below, competition for iron by serum proteins, such as transferrin, may preclude the dermatophytic fungi from invading deep tissues in most patients, even if they are immunosuppressed.

2. ETIOLOGY

The three genera of dermatophytes, *Trichophyton*, *Epidermophyton*, and *Microsporum*, together contain over 40 species. These organisms are relatively similar but can be differentiated from one another by colony morphology, microscopic appearance, and some biochemical tests. Classification of the dermatophytes has been complicated by the discovery of the perfect or sexual reproductive stage of several species that have been reclassified into two new genera called *Arthroderma* (containing the perfect *Trichophyton* species) and *Nannizzia* (containing the perfect *Microsporum* species). The two new genera are included in the subdivision Ascomycotina, which also includes the etiologic agents of blastomycosis and histoplasmosis, whereas the original three genera are included in the subdivision Deuteromycotina.

The distribution of the various dermatophyte species varies considerably throughout the world. For example, tinea imbricata is mainly a tropical disease that is found in certain areas of the West Pacific and in the Amazon basin in Brazil. Additionally, the organisms found in a particular area of the world are not always constant, as evidenced by the increase in cases of tinea capitis caused by *T. tonsurans* in the United States and the decrease of *M. audouinii* as a cause of this condition in Europe. The species causing dermatophytosis in Africa and Asia may also be different from those producing similar infections in other parts of the world.

The many species of dermatophytes can be classified as anthropophilic (found mostly in humans), zoophilic (found mostly in animals), or geophilic (found mostly in the soil). This classification is important because the groupings help to describe the reservoirs of these organisms (and hence how exposure to them may occur) and also because members of the different groups may have certain clinical characteristics in common. In particular, the anthropophilic species tend to produce less inflammatory and more chronic infections than do members of the other two groups. This characteristic suggests that the former are better adapted to the human host than are the zoophilic or geophilic species. On the other hand, even the anthropophilic species of dermatophytes probably do not represent part of the normal skin flora of humans.[6] This group of organisms is therefore different from the agents of the other two major types of superficial fungal infections, cutaneous candidiasis and tinea versicolor, in that *Candida albicans* and *Malassezia furfur* are normal inhabitants of humans. Like *C. albicans*, the various species of dermatophytes produce an extracellular keratinase that can digest keratin of the stratum corneum, hair, and nails, so that the organisms can use these substances as a source of nutrients.

3. EPIDEMIOLOGY

The classification of dermatophytes as anthropophilic, zoophilic, or geophilic is an important consideration in understanding the epidemiology of these infections. Those dermatophytes found in the soil are generally unusual causes of dermatophytosis, although *M. gypseum* may produce some cases. On the other hand, the zoophilic organisms are more important as human pathogens. *T. mentagrophytes*, *T. verrucosum*, and *M. canis* are the most common zoophilic dermatophytes producing infections in humans. In fact, *T. verrucosum* is an important enough cause of infection in cattle, with resultant transmission to humans, that an extensive effort has been made to produce vaccines against it for veterinary use.[7] Also, *M. canis*, a common cause of dermatophytosis in dogs and cats, is widely distributed throughout the world and causes many human cases.

Of the three groups, the anthropophilic dermatophytes produce the largest number of infections in humans. These infections are sometimes spread from person to person through contact with infected skin scales shed from an infected individual. This type of transmission often occurs in common facilities such as shower rooms, swimming pools, or in military barracks. The anthropophilic organisms causing tinea capitis in children are probably also spread from individual to individual in many cases, although the exact route of transmission may not be evident.

Exposure to a dermatophyte may not always result in an infection. There is a variety of factors, either local or systemic in nature, that affect the outcome of a particular exposure. Additionally, once an infection does occur, it may either heal spontaneously or persist for years as a chronic infection. The outcome of a particular infection may depend on immunological or nonimmunological characteristics of the host, but it may also be determined by the particular organism involved.

Local cutaneous factors appear to be very important in determining the outcome of an exposure to the dermatophytic fungi. In general, experimental infections with these organisms are rather difficult to produce in humans unless measures are taken to make the skin in the inoculation area more susceptible to the infection.[8,9] A number of factors seems to make the local skin conditions more receptive to dermatophytosis. Warmth and moisture appear to enhance susceptibility to either naturally occurring or experimental dermatophytosis. For example, the warm and moist conditions in the swampy areas of Vietnam were felt to be responsible in part for the high incidence of dermatophytosis in combat troops there.[10] Also, occlusion appears to be very important in enhancing susceptibility to experimental dermatophyte infections in both humans and animals.[8–13] Occlusion has been postulated to increase hydration of the underlying skin and to increase the emission of carbon dioxide which could favor dermatophyte growth.[14] Abrasion of the inoculated skin may enhance susceptibility to dermatophyte infection,[15] although bleeding may decrease the incidence of experimental infections,[16] perhaps because of competition for iron by serum transferrin (as discussed below).

There are other characteristics of the skin itself that may affect susceptibility to dermatophytosis. A good example is the high frequency of tinea capitis in children as compared to that in adults—a phenomenon that has been attributed to the higher quantities of fungistatic lipids present in adult sebum.[17,18] Disorders of keratinization and underlying peripheral vascular disease have both been found to be associated with chronic dermatophytosis in the absence of other complicating factors.[19]

A variety of systemic conditions has been found to be related to the persistence of chronic dermatophyte infections. Most cases of chronic dermatophytosis in relatively normal patients are caused by *T. rubrum*; however, if this kind of infection is caused by other dermatophytes, the patients usually have some underlying disease, such as collagen vascular disease, systemic corticosteroid therapy or Cushing's disease, hematologic malignancy, chronic mucocutaneous candidiasis, diabetes mellitus, or atopy.[19–22] Advanced age has also been related to an increased incidence of dermatophytosis,[23,24] and these infections are often asymptomatic.[25] Obviously, many of these individual factors may adversely affect immunologic function, as will be discussed later.

There is some evidence that chronic dermatophytosis, or at least susceptibility to this condition, may occur as an inherited trait. The best example is tinea imbricata, where predisposition to the infection has been demonstrated to be transmitted as an autosomal recessive trait.[26] Additionally, tinea capitis appears to be more frequent in black adults than in Caucasian adults.[27] Because some of the predisposing factors listed above may have a tendency toward familial transmission, the predisposition to chronic dermatophytosis that results from them may be inherited as well.

4. HOST DEFENSES

4.1. Innate Resistance

4.1.1. Nonimmunological Cutaneous Defenses

The physical structure of the skin represents a mechanism of defense against pathogens such as the dermatophytes. Because of exposure to ultraviolet light, drying, and competition from the normal bacterial flora, the skin itself presents a hostile environment to potentially infectious agents. The stratum corneum is made up of keratin, which is a protein that most microorganisms cannot use as a source of nutrients. Dermatophytes, however, produce keratinases that allow them to grow in the stratum corneum. Additionally, this site of infection may protect them from direct contact with at least some of the effector cells of the immune system.

The process of keratinization, by which the stratum corneum is continually renewed, may also present a form of defense against organisms infecting this site. The basal epidermal cells, or keratinocytes, are responsible for continued growth of the epidermis; these cells lie in a single layer along the dermal–epidermal junction and undergo continued cell divisions that move the resulting daughter cells outward, toward the surface. As these cells grow outward, they lose their nuclei and become flattened to form the keratinized cells. This process results in continuous shedding of the stratum corneum, with the possible result of removal of organisms residing there. If a microorganism cannot grow down into the deeper layers of the skin, it will eventually be removed by this mechanism. Additionally, inflammation of immunologic or nonimmunologic types increases the epidermal activity and therefore the rate at which the stratum corneum is shed.[28] Berk et al.[29] found that in the periphery of annular dermatophytosis (i.e., ringworm), the epidermal cells divide at a much faster rate than they do in normal skin. Immunological factors may also enhance this process, resulting in increased rates of transit of epidermal cells from the basal layer to the stratum corneum.[30] The epidermis appears to be capable of enhancing its turnover rate in response to a variety of stimuli, and a number of authors have suggested that this process may be important in the defense against dermatophytosis.[16,29–31] Similarly, the growth of hairs may also be important in susceptibility to dermatophytosis in that hairs in the late stages of the hair growth cycle (i.e., late anagen) appear to be the only ones that can be infected.[32]

Tinea capitis is more common in children than in adults, and as discussed above, this difference may relate to the presence of fungistatic lipids in the sebum of adults. Rothman et al.[17] have reported that lipids extracted from adult hair contain saturated fatty acids, particularly those with odd numbered carbon chains from C7 to C11, which can significantly inhibit the growth of *M. audouini*, a common cause of hair and scalp infections. Whereas the sebum of adults may not actually be significantly more fungistatic than that from children, it was found in another study that adults produce quantitatively more of this material than do children.[33]

The dermatophytic fungi appear to be relatively incapable of infecting deep tissues, even in severely immunosuppressed patients. Therefore, infections with these organisms are generally confined to the dead, keratinized layers. This phenomenon has been related to the growth inhibiting effect of a serum factor on dermatophyte growth,[34] a factor that has subsequently been identified as unsaturated transferrin.[35] This substance appears to be capable of competing successfully with the dermatophytes for the iron they need in order to grow.

4.1.2. The Inflammatory Response

Dermatologists have long been aware of the inverse relationship between the degree of inflammation produced by a particular dermatophyte and the chronicity of that infection. The anthropophilic organisms *T. rubrum* and *M. audouini* generally produce little inflammation in their cutaneous lesions and frequently cause infections that persist for long periods. In contrast, many of the zoophilic or geophilic dermatophytes produce highly inflammatory infections that are usually self-limited. These observations suggest that the local inflammatory processes are indeed involved in the defense against this group of pathogens.

In spite of the fact that some dermatophyte infections are highly inflammatory, the site of infection is a very superficial one in the stratum corneum. The organisms can generally be demonstrated in the upper cutaneous layers using tape stripping. The dermatophytes also readily infect hairs or hair follicles. Infections of hairs can be classified as either ectothrix (arthrospores produced on the outside of the hairs), or endothrix (arthrospores produced inside the hairs). A variety of gross changes can be observed in the infected skin, including scaling, vesicles, pustules, annular dermatitis, and severe inflammatory reactions (kerions). Microscopically, the lesions are characterized initially by accumulation of neutrophils in the infected skin, followed later by epidermal proliferation and a mononuclear cell infiltrate in the dermis. Experimental data showing that neutrophils can kill *Trichophyton sp.* in vitro through oxidative processes suggests that these cells may play an active role in the defense against acute dermatophytosis.[36]

The mechanism by which neutrophils are attracted into skin infected by dermatophytes has received some attention. *T. rubrum* and *T. mentagrophytes* have both been shown to be capable of activating complement by the alternative pathway to produce chemotactic activity for neutrophils,[37,38] and *T. mentagrophytes* has been found to produce a low-molecular weight chemotactic factor analogous to similar ones made by growing bacteria.[39] *T. rubrum* extracts have been shown to both produce and destroy chemotactic activity from serum,[40] perhaps explaining why this organism usually results in minimally inflammatory lesions. In addition, patients with dermatophytosis may have antibodies cross reactive with epithelial antigens,[41,42] and it has been postulated that some of the inflammation in the infected skin may be caused by this antibody. Finally, keratinocytes can generate the chemotactic cytokine, interleukin I, when properly stimulated,[43–45] and it is possible that this substance could be responsible for some of the neutrophil infiltration in the lesions of dermatophytosis.

4.2. Immune Responses

4.2.1. Immunity to Reinfection

Most of the early animal work with experimental dermatophytosis involved guinea pigs and generally demonstrated that some kind of immunity, either local or systemic, resulted from a preceding infection. Bloch[46] demonstrated generalized immunity in guinea pigs after experimental dermatophyte infections and found that this immunity was manifested by an accelerated course of a second infection at a different site. On the other hand, Greenbaum,[47] De Lamater and Benham,[48] and Bonk et al.[49] all found that the accelerated response to a second infection was greatest at the site of the first infection, suggesting some form of local immunity. Memory cells in the healed skin at the site of a first infection have been demonstrated after experimental *T. mentagrophytes* infections in guinea pigs.[50] In rats and mice, the success of experimental dermatophyte inoculations appears to be dependent upon the stage of hair and skin growth cycles, so that immunity to reinfection may not be as easy to demonstrate as it is with guinea pigs.[32] In cattle, a previous infection appears to afford partial immunity to a second infection at the same site as well as at different sites.[51] Because dermatophytosis in cattle is an important economic problem in some countries, and because these infections represent potential sources for human dermatophytosis, attempts have been made to produce vaccines against *T. verrucosum* for veterinary use. One of these, a live vaccine (LTF-130) based on a less virulent strain of the organism, appears to show considerable promise for controlling infections with wild type strains of this organism.[7]

The findings with respect to immunity to a second experimental dermatophytic infection in humans generally parallel those in animals; that is, most workers have found some indication of immunity, even though it may be only local in nature.[16,47,48,52] Although the findings reported in various studies on experimental dermatophyte infections in humans have differed somewhat, in general it appears as if this immunity is not complete and is probably less than that found in animals.[3,8]

4.2.2. *Antigens of the Dermatophytes*

The cell wall of the dermatophytes is made up primarily of chitin and glucan in addition to the glycopeptides that represent the major antigens of these organisms. The dermatophytes, like other fungi, have a very complicated antigenic makeup. The term "trichophytin" has been used to refer to relatively crude antigenic extracts of *Trichophyton sp*; such preparations contain a variety of antigens. Antigenic substances of the dermatophytes may be glycopeptides, peptides, or carbohydrates, and the individual types of antigens may elicit different types of responses.[53,54] Additionally, Grappel and Blank[55] have found using experimental infections in guinea pigs that the keratinase of *T. mentagrophytes* may be an important antigen, not only because the animals produced a strong delayed hypersensitivity response against it, but also because antibodies were produced that could inhibit the proteolytic activity of this enzyme.

Extracts of various dermatophyte species have usually been found to contain a mixture of antigens that are either species specific or broadly cross reactive with those of other dermatophytes.[56,57] Additionally, these antigens may be shared with other fungi.[58] It is possible that this phenomenon may relate to the susceptibility of atopic patients to chronic dermatophytosis because these patients may develop immediate hypersensitivity to airborne molds that cross react with the dermatophyte antigens, and this immediate hypersensitivity may later interfere with the development of delayed responses to these antigens.[59,60] Cross reactivity between penicillin and dermatophyte antigens has also been postulated as a possible cause of penicillin hypersensitivity in patients who have never received this drug before.[61] Similar cross reactivity between the antigens of *T. rubrum* and those of epidermal cells could also be a factor in the development of inflammatory responses in dermatophytosis.[41,42]

4.2.3. *Humoral Immunity*

Even though dermatophyte infections are very superficial in location, they do appear to sensitize the host. In fact, low levels of antibody against *T.*

rubrum have been found in uninfected humans using an enzyme-linked immunosorbent assay (ELISA), although it is possible that such antibodies represent cross reactivity with other microorganisms,[62] Rabbits inoculated with *T. verrucosum* have been shown to develop precipitins against extracts of the organisms, although the titers of these antibodies do not seem to correlate with susceptibility to reinfection.[63] Complement fixing antibodies to dermatophyte antigens have been found in the serum of experimentally infected guinea pigs and naturally infected horses.[64]

Humoral immunity has been demonstrated in human dermatophytosis using a variety of techniques. Although precipitins are produced relatively infrequently,[65,66] more sensitive methods may show antibody against the infecting organism in larger numbers of patients with dermatophytosis.[67–69] Antibody titers, however, are usually higher in patients than in control subjects, although some patients may have low titers.[69] Therefore, the humoral responses in these patients appear to be variable and perhaps related to the type of infection involved.

Immediate hypersensitivity to dermatophyte antigens has been well documented in dermatophytosis and may be involved in the pathogenesis of this condition, as suggested from the evidence that this type of response may interfere with the development of a protective delayed hypersensitivity response. Both immediate hypersensitivity to dermatophyte antigens and a higher than expected incidence of atopy have been found in patients with dermatophytosis in a number of studies.[19,70,71] Whether chronic dermatophytosis stimulates IgE production, or this condition tends to develop in persons who are already atopic, has been the subject of some speculation.[59] A frequent pattern seen in these patients is the presence of immediate, but not delayed hypersensitivity to trichophytin.[72–75] In one patient with Cushing's disease and widespread, chronic dermatophytosis, however, treatment of the underlying disease by adrenalectomy resulted in restoration of a positive delayed trichophytin skin test and appeared to improve the patient's cutaneous infections, even though the patient continued to manifest strong immediate hypersensitivity to Trichophyton antigens.[76] Additionally, many patients with chronic dermatophytosis are neither atopic nor manifest immediate hypersensitivity to trichophytin, indicating that other factors must also be involved in producing susceptibility to this type of infection.

4.2.4. *Cell-Mediated Immunity*

Delayed hypersensitivity responses to intradermal injections of trichophytin are fairly common in the normal population. For example, Wood and Cruickshank[77] found that 30% of normal, uninfected control subjects responded with a positive delayed hypersensitivity response to this antigen, and 23% had immediate responses. Because trichophytin can be obtained commercially, it can be used as one of a battery of delayed hypersensitivity skin tests to assess immunological competence.[78,79] However, since only 30% or so of normal subjects will have a positive delayed response to this antigen, it is not particularly useful for this purpose. Although these responses are probably caused by prior episodes of self-limited dermatophytosis, it is also possible that cross reactivity with another organism is responsible.

Several studies have demonstrated that experimental dermatophyte infections in animals do result in cell-mediated immunity to the antigens of the infecting organism.[80–82] In guinea pigs with experimental *T. mentagrophytes* infections, maximal erythema in the infected skin occurred at the time the animals developed cell-mediated immunity to trichophytin, as manifested by delayed hypersensitivity skin test responses and positive *in vitro* lymphocyte transformation responses.[80] Similar findings have been reported in cattle experimentally infected with *T. verrucosum*.[83] Additionally, in the bovine model of dermatophytosis, delayed hypersensitivity reactions in the infected skin appeared to promote clearance of the infection by increasing the rate of desquamation of the stratum corneum.[30]

In humans, the type and duration of dermatophytosis appear to be important factors in determining whether or not delayed hypersensitivity to trichophytin will be present. Positive delayed hypersensitivity responses to trichophytin have been found to be more frequent in patients with acute dermatophytosis caused by *T. mentagrophytes* than in patients with chronic infections caused by *T. rubrum*.[71,72,84] In children with tinea capitis, positive delayed skin tests to trichophytin were found in one study to be present in 15 of 16 of those with highly inflammatory infections (kerions) but in none of 36 with minimally inflammatory (black dot) infections.[85] Patients with chronic dermatophytosis are generally fairly healthy and have been found to have only a mild depression in their ability to develop delayed hypersensitivity skin test responses to antigens unrelated to trichophytin.[86] The importance of cell-mediated immunity to trichophytin has been demonstrated in studies of experimental infections in humans in that patients with positive delayed trichophytin skin tests were found to be resistant to inoculation with *T. mentagrophytes*, whereas those with negative skin tests were susceptible.[87]

In vitro lymphocyte transformation assays have been used to demonstrate cell mediated immunity to dermatophyte antigens in both animals and humans. Guinea pigs with experimental *T. mentagrophytes* infections have been shown to develop this type of response approximately 9 to 11 days after inoculation with the fungus.[80,82] In human patients with dermatophytosis, positive *in vitro* lymphocyte transformation to trichophytin have also been found in a number of studies.[4,65,72,88,89] Generally the *in vitro* response correlates with the presence of a positive delayed skin test to trichophytin, although a few patients with chronic *T. rubrum* infections have been found to have negative skin tests along with positive *in vitro* lymphocyte transformation responses.[88] This finding suggests some kind of local suppression of the skin test response, as discussed below. Other tests that have been used to assess cell-mediated immunity to trichophytin in humans with dermatophytosis are the leukocyte adherence inhibition test[90,91] and the leukocyte migration inhibition test.[92,93] The results of these assays also appear to correlate quite well with the delayed trichophytin skin test response.

4.2.5. *Regulation of Immune Responses*

Suppression of immunological responses has been demonstrated in acute *T. mentagrophytes* infections produced experimentally in animals. In guinea pigs with this type of infection, Green and Balish[94] found suppression of lymph node lymphocyte blastogenic responses to nonspecific B- and T-cell mitogens in addition to antigenic extracts of the infecting organisms. This suppression appeared to be due to a serum factor inasmuch as the responses were higher if the medium contained fetal calf serum instead of autologous guinea pig serum. In Balb/c mice with experimental dermatophytosis, Calderon and Hay[95] found suppression of *in vitro* blastogenic responses to mitogens, and that this suppression was mediated by Thy 1.2 positive suppressor T lymphocytes. These authors also demonstrated that lymph node cells from sublethally irradiated mice with chronic dermatophytosis could confer adoptive immunity to irradiated recipients, and that this type of transfer could be blocked by serum from chronically infected donors or by dermatophyte antigen. Using an ELISA system, they were able to demonstrate circulating antigen in the serum of the infected animals, with greater amounts present in those with chronic infections.

Studies in humans also suggest that serum factors may be involved in the suppression of cell-mediated immunity to fungal antigens in patients with chronic dermatophytosis. Walters *et al.*[91] found that the *in vitro* cell-mediated immune responses to dermatophyte antigens, as measured by the leukocyte adherence inhibition test, could be blocked in an antigen-specific manner by serum from chronically infected, but not acutely infected, patients. Phosphorylcholine-like antigens of *T. rubrum* and *T. concentricum* have been found in the serum of patients chronically infected with these organisms using an immunoradiometric assay.[96] Sherwin *et al.*[97] have reported a patient with widespread dermatophytosis who had a serum factor that suppressed his *in vitro* blastogenic responses to mitogens and that disappeared after the infection was cleared with griseofulvin. The authors suggested that this factor could be an immune-complex-containing fungal antigen.

Antagonism between immediate and delayed hypersensitivity is another possible mechanism of immunosuppression in dermatophytosis. The common pattern of positive immediate and negative delayed skin test responses to trichophytin in patients with chronic dermatophytosis, along with the frequent finding of atopy in these patients, suggests some sort of relationship between these two kinds of immune responses. Because atopic children often show immediate hypersensitivity to trichophytin before they develop dermatophytosis,[60] these immediate responses could be involved in the pathogenesis of chronic dermatophyte infections. Jones *et al.*[87] have demonstrated that subjects with immediate hypersensitivity to trichophytin were more easily infected experimentally with *T. mentagrophytes*.

Local interference between immediate and delayed reactions is one possible explanation for the frequent lack of delayed hypersensitivity in patients with chronic dermatophytosis. Marcussen[98] found that an immediate skin test response to trichophytin produced by passive transfer could block delayed reactions in patients who had previously had only the delayed response. Jones *et al.*[31] used suppression of the immediate reactions with injections of antihistamines into the site to uncover delayed responses in patients who had had only immediate responses beforehand. Hay[74] was unable to use similar methods to uncover delayed hypersensitivity to trichophytin in patients with chronic dermatophytosis who had failed to respond to griseofulvin; however, the immediate reactions appeared to be quite strong in this group of patients and were not completely suppressed by the methods used.

It is also possible that immediate hypersensitivity may interfere with delayed hypersensitivity on a systemic basis in patients with chronic dermatophytosis. One study has found an inverse correlation between total serum IgE levels and T-lymphocyte numbers in this type of patient,[99] although most others have found normal T-cell numbers and relatively normal cell-mediated immunity to nonspecific mitogens and antigens other than trichophytin.[65,72,73,86] On the other hand, there is evidence that elevated IgE levels can be correlated with susceptibility to infection in patients with atopic dermatitis[100] and the syndrome of dermatitis, increased IgE levels, and infections.[101] Additionally, there is a type of suppressor cell that appears to be activated by histamine,[102,103] so treatment with antihistamines might seem to be a reasonable approach in patients where antagonism between the two types of responses is possible.

5. IMMUNODIAGNOSIS AND PROGNOSIS

As discussed earlier, prior sensitization to dermatophyte antigens, as measured using either serological techniques or by delayed hypersensitivity skin testing, is relatively common in uninfected adults. Also, significant numbers of patients with dermatophytosis will have negative trichophytin skin tests and titers of antibody against dermatophyte antigens that are not significantly higher than those in uninfected controls. Therefore, serological methods of skin testing are of little use in diagnosing dermatophytosis, particularly when the diagnosis can usually be made from the appearance of the lesions and cultures from the infected surface. On the other hand, delayed hypersensitivity skin testing to trichophytin may have some prognostic significance since a negative skin test in the face of active dermatophytosis might suggest difficulty in handling the infection.

6. PREVENTION AND THERAPY

6.1. Immunization or Vaccines

Guinea pigs immunized to trichophytin in Freund's complete adjuvant have been found to demonstrate immediate and delayed skin test responses to this antigen similar to those seen in animals immunized by an experimental infection.[81] By immunizing guinea pigs with an extract of *T. mentagrophytes*, Keeney and Huppert[104] were able to partially protect the animals from experimental dermatophytosis. In another study, Hussin and Smith[105] found that a live vaccine of *T. mentagrophytes* was more effective than extracts of the organism for immunizing guinea pigs against experimental dermatophytosis. Also, rabbits immunized with a *T. rubrum* vaccine were found to be resistant to experimental infections for many months thereafter.[106] As discussed earlier, an apparently useful live vaccine (LTF-130) against *T. verrucosum*, the major cause of cattle ringworm, has been developed and used successfully in reducing the infections in cattle herds.[7,107]

In spite of the success in vaccinating animals against dermatophytosis, the use of vaccines against this type of infection is much more difficult in humans. Generally, dermatophytosis has not been considered a serious enough disease to warrant large scale immunization programs in children. Therefore, the types of human patients that have been vaccinated in the past include those with intractable, chronic dermatophytosis and those with significant dermatophytid reactions. As will be discussed later, this type of therapy has not been particularly effective and has been supplanted in recent years by the use of antifungal drugs.

6.2. Therapy

6.2.1. *Chemotherapy*

The treatment of dermatophytosis has improved markedly in recent years with the development of newer antifungal agents for topical application or oral administration. The initial approach to most cases of dermatophytosis is to try topical therapy with keratinolytics like Whitfield's ointment (a compound of salicylic and benzoic acids) and/or creams or powders containing specific antifungal agents such as chlorphenesin, undecylenate, tolnaftate, azole drugs (including miconazole, clotrimazole, or ketoconazole), and cyclopiroxolamine. Generally, topical therapy with azole-containing topical preparations is applied once daily for 2 to 4 weeks. Certain kinds of dermatophytosis, however, including widespread infections, and those of hair and nails, will often respond poorly to topical therapy and will require prolonged courses of an oral antifungal agent.

Griseofulvin is the oral agent used most often in treating dermatophytosis. It is generally given in a dose of 500 to 1000 mg of the micronized form per day (for adults) for at least 4 weeks. Scalp infections may require 6 to 12 weeks of griseofulvin therapy, and nail infections are even more difficult to treat—often requiring 6 to 12 months of therapy, with a significant failure rate for infections of the toenails. Adverse effects of griseofulvin include headache, nausea, hypersensitivity reactions (including photosensitivity), and occasionally, proteinuria and leucopenia.

Ketoconazole can be used to treat dermatophytosis, although the risk of potentially fatal hepatitis makes this drug a less attractive choice for this condition. Two new agents in the azole class, itraconazole and fluconazole, as well as the allylamine drug, terbinafine, appear to be effective in dermatophytosis and have relatively fewer adverse reactions than ketoconazole does.[108,109]

6.2.2. *Immunotherapy*

Vaccination with trichophytin has been tried in the past as a form of therapy for patients with chronic dermatophytosis and for those with dermatophytid eruptions. The situation with these patients, however, is quite different from that in animals that have never been infected with a dermatophyte or sensitized to the antigens of these organisms. Hence, in the human patients who are infected at the time they are treated, an inadequate immune response is most likely caused by the dysfunction of some element of the immune system rather than by a lack of primary exposure to the antigens involved; this type of immune dysfunction would generally not be expected to respond to additional amounts of the antigen. Even so, some workers have reported beneficial effects of trichophytin injections in human dermatophytosis,[110,111] although others have not.[112–114] In the studies reporting success, it is difficult to tell if the infections were indeed chronic or if they might have cleared spontaneously. Additionally, trichophytin injections in patients with dermatophytosis have been associated with significant toxicity, particularly local reactions at the injection site. Because we now have much better antifungal therapy, trichophytin injections would seem to have little role in treating patients already infected with these organisms.

Attempts have also been made to treat dermatophytid reactions using trichophytin immunotherapy. A beneficial effect of this type of therapy was reported in 10 of 15 of such patients by Sulzberger and Wise,[115] whereas Traub and Tolmach[114] reported little effect. Because patients with dermatophytids usually have strong delayed hypersensitivity to dermatophyte antigens,[3,77] attempts to desensitize them by injections of trichophytin makes some sense immunologically. Immunotherapy with trichophytin has been shown to reduce the intensity of delayed skin tests with this material while not affecting immediate skin test responses.[70,115] As with therapy of the infection itself, however, dermatophytid eruptions are probably treated best with appropriate antifungal therapy at the present time.

Other immunomodulatory methods have some potential in the treatment of patients with chronic and intractable dermatophytosis. Suppression of immediate reactions to dermatophyte antigens could benefit this type of patient, provided the immediate reactions were indeed suppressing a protective delayed hypersensitivity response. Repeated injections of trichophytin, however, appear to suppress only delayed hypersensitivity, the opposite of the desired effect. In one study,[74] treatment with oral cimetidine partially suppressed immediate hypersensitivity to trichophytin in patients with chronic dermatophytosis but did not restore the delayed hypersensitivity responses. Because the infection itself may directly suppress cell-mediated immunity to the antigens of the infecting organism, it is possible that successful antifungal therapy by itself could restore the protective immunological responses and perhaps prevent future relapses. Such a result has been seen in a few cases of chronic mucocutaneous candidiasis[116] and in some patients treated successfully for chronic dermatophytosis.[59] Finally, a variety of immunomodulatory agents, including interleukins, interferons, colony-stimulating factors, monoclonal antibodies, and synthetic immunomodulators, have been used to treat other conditions associated with depressed immunological function,[117] and it is possible that, in the future, this type of therapy may prove useful as adjunctive therapy for patients with dermatophytosis not responsive to the standard antifungal agents.

REFERENCES

1. Ajello, L., 1974, Natural history of the dermatophytes and related fungi, *Mycopathol. Mycol. Appl.* **53**:93–110.
2. Williams, C. M., 1926, Dermatophytids complicating dermatophytosis of the glabrous skin, *Arch. Dermatol.* **13**:661–669.
3. Grappel, S. F., Bishop, C. T., and Blank, F., 1974, Immunology of dermatophytes and dermatophytosis, *Bacteriol. Rev.* **38**:222–250.
4. Svejgaard, E., Thomsen, M., Morling, N., and Christiansen, A. H., 1976, Lymphocyte transformation *in vitro* in dermatophytosis, *Acta Path. Microbiol. Scand.* Sec. C., **84**:511–519.
5. Faergemann, J., Gisslen, H., Dahlberg, E., Westin, J., and Roupe, G., 1989, Trichophyton rubrum abscesses in immunocompromised patients, *Acta Derm. Venereol.* **69**:244–247.
6. Cornbleet, T., 1926, Cultures from the skin of apparently normal feet, *Arch. Dermatol.* **13**: 670–671.
7. Segal, E., 1989, Vaccines for the management of dermatophyte and superficial yeast infections, *Curr. Top. Med. Mycol.* **3**:36–49.
8. Knight, A. G., 1972, A review of experimental human fungus infections, *J. Invest. Dermatol.* **59**:354–358.

9. Baer, R. L., Rosenthal, S. A., and Furnari, D., 1955, Survival of dermatophytes applied on the feet. *J. Invest. Dermatol.* **24:**619–622.
10. Blank, H., Taplin, D., and Zaias, N., 1969, Cutaneous *Trichophyton mentagrophytes* infections in Vietnam, *Arch. Dermatol.* **99:**135–144.
11. Strauss, J. S., and Kligman, A. M., 1957, An experimental study of tinea pedis and onychomycosis of the foot, *Arch. Dermatol.* **76:**70–79.
12. Tagami, H., Watanabe, S., and Ofugi, S., 1973, Trichophytin contact sensitivity in guinea pigs with experimental dermatophytosis induced by a new inoculation method, *J. Invest. Dermatol.* **61:**237–241.
13. Greenberg, J. H., King, R. D., Krebs, S., and Field, R., 1976, A quantitative dermatophyte infection model in the guinea pig: A parallel to the quantitated human infection model, *J. Invest. Dermatol.* **67:**704–708.
14. King, R. D., Cunico, R. L., Maibach, H. I., Greenberg, J. H., West, M. L., and Jeppsen, J. C., 1978, The effect of occlusion on carbon dioxide emission from human skin, *Acta. Derm. Venereol.* **58:**135–138.
15. White, C., 1929, Autoinoculation dermatophytosis from toe cultures of human volunteers, *Arch. Dermatol.* **20:**315–319.
16. Kligman, A. M., 1952, The pathogenesis of tinea capitis due to *Microsporum audouini* and *Microsporum canis*, *J. Invest. Dermatol.* **18:**231–246.
17. Rothman, S., Smiljanic, A., Shapiro, A. L., and Weitkamp, A. W., 1947, The spontaneous cure of tinea capitis in puberty, *J. Invest. Dermatol.* **8:**81–98.
18. Laur, W. E., 1951, Spontaneous cure of tinea capitis due to *Microsporum audouini*, *Arch. Dermatol. Syph.* **64:**364–366.
19. Hay, R. J., 1982, Chronic dermatophyte infections. I. Clinical and mycological features, *Br. J. Dermatol.* **106:**1–7.
20. Nelson, L. M., and McNiece, K. J., 1959, Recurrent Cushing's syndrome with *Trichophyton rubrum* infection, *Arch. Dermatol.* **80:**700–704.
21. Canizares, O., Shatin, H., and Kellert, A. J., 1959, Cushing's syndrome and dermatomycosis, *Arch. Dermatol.* **80:**705–712.
22. Findling, J. W., Tyrrell, J. N., Aron, D. C., Fitzgerald, P. A., Young, C. W., and Sohnle, P. G., 1981, Fungal infections in Cushing's syndrome, *Ann. Intern. Med.* **95:**392.
23. Gilchrest, B. A., 1979, Some gerontologic considerations in the practice of dermatology, *Arch. Dermatol.* **155:**1343–1346.
24. Peachey, R. D. G., and English, M. P., 1974, An outbreak of *Trichophyton rubrum* infection in a geriatric hospital, *Br. J. Dermatol.* **4:**389–398.
25. Tindall, J. P., and Smith, J. G., 1963, Skin lesions of the aged and their association with internal changes, *J. Am. Med. Assoc.* **186:**1039–1042.
26. Serjeantson, S., and Lawrence, G., 1977, Autosomal recessive inheritance of susceptibility to tinea imbricata, *Lancet* **1:**13–15.
27. Blank, F., Mann, S. J., and Reale, R. A., 1974, Distribution of dermatophytosis according to age, ethnic group, and sex, *Sabouraudia* **12:**352–361.
28. Sohnle, P. G., and Kirkpatrick, C. H., 1978, Epidermal proliferation in the defense against experimental cutaneous candidiasis, *J. Invest. Dermatol.* **70:**130–133.
29. Berk, S. H., Penneys, N. S., and Weinstein, G. D., 1976, Epidermal activity in annular dermatophytosis, *Arch. Dermatol.* **112:**485–488.
30. Lepper, A. W. D., and Anger, H. S., 1976, Experimental bovine *Trichophyton verrucosum* infection. Comparison of the rate of epidermal cell proliferation and keratinization in non-infected and reinoculated cattle, *Res. Vet. Sci.* **20:**117–121.
31. Jones, H. E., Reinhardt, J. H., and Rinaldi, M. G., 1974, Immunologic susceptibility to chronic dermatophytosis, *Arch. Dermatol.* **110:**213–220.
32. Kligman, A. M., 1956, Pathophysiology of ringworm infections in animals with skin cycles, *J. Invest. Dermatol.* **27:**171–185.

33. Kligman, A. M., and Ginsberg, D., 1950, Immunity of the adult scalp to infection with *Microsporum audouini*, *J. Invest. Dermatol.* **14**:345–356.
34. Blank, H., Sagami, S., Boyd, C., and Roth, F. J. Jr., 1959, The pathogenesis of superficial fungal infections in cultured human skin, *Arch. Dermatol.* **79**:524–535.
35. King, R. D., Khan, H. A., Foye, J. C., Greenberg, J. H., and Jones, H. E., 1975, Transferrin, iron, and dermatophytes. I. Serum dermatophyte inhibitory component definitively identified as unsaturated transferrin, *J. Lab. Clin. Med.* **86**:204–212.
36. Calderon, R. A., and Hay, R. J., 1987, Fungicidal activity of human neutrophils and monocytes on dermatophytic fungi, *Trichophyton quinckeanum* and *Trichophyton rubrum*, *Immunology* **61**:289–995.
37. Davies, R. R., and Zaini, F., 1984, *Trichophyton rubrum* and the chemotaxis of polymorphonuclear leucocytes, *Sabouraudia* **22**:65–71.
38. Swan, J. W., Dahl, M. V., Coppo, P. A., and Hammerschmidt, D. E., 1983, Complement activation by *Trichophyton rubrum*, *J. Invest. Dermatol.* **80**:156–158.
39. Tagami, H., Natsume, N., Aoshima, T., Inoue, F., Suehisa, S., and Yamada, M., 1982, Analysis of transepidermal leucocyte chemotaxis in experimental dermatophytosis in guinea pigs, *Arch. Dermatol. Res.* **273**:205–217.
40. Davies, R. R., and Zaini, F., 1984, Enzymatic activities of *Trichophyton rubrum* and the chemotaxis of polymorphonuclear leucocytes, *Sabouraudia* **22**:235–241.
41. Hopfer, R. L., Grappel, S. F., and Blank, F., 1975, Antibodies with affinity for epithelial tissue in chronic dermatophytosis, *Dermatologica* **151**:135–143.
42. Peck, S. M., Osserman, K. E., and Rule, A. H., 1972, Intercellular antibodies: presence in a *Trichophyton rubrum* infection, *J. Invest. Dermatol.* **58**:133–138.
43. Sauder, D. N., Carter, C. S., Katz, S. I., and Oppenheim, J. J., 1982, Epidermal cell production of thymocyte activating factor (ETAF), *J. Invest. Dermatol.* **79**:34–39.
44. Luger, T. A., and Oppenheim, J. J., 1983, Characteristics of interleukin 1 and epidermal cell derived thymocyte activating factor, *Adv. Inflammation Res.* **5**:1–25.
45. Kupper, T. S., Ballard, D. W., Chua, A. O., McGuire, J. S., Flood, P. M., Horowitz, M. C., Langdor, R., Lightfoot, L., and Gubler, U., 1987, Human keratinocytes contain mRNA indistinguishable from monocyte interleukin 1 α and β mRNA. Keratinocyte epidermal cell derived thymocyte activating factor is identical to interleukin 1, *J. Exp. Med.* **164**:2095–2100.
46. Bloch, B., 1908, Zur Lehre von den Dermatomykosen, *Arch. Dermatol. Syph.* **93**:157–220.
47. Greenbaum, S. S., 1924, Immunity in ringworm infections. I. Active acquired immunity with a note on complement fixation tests in superficial ringworm infections, *Arch. Dermatol.* **10**:279–288.
48. De Lamater, E. D., and Benham, R. W., 1938, Experimental studies with the dermatophytes. II. Immunity and hypersensitivity produced in laboratory animals, *J. Invest. Dermatol.* **1**:469–488.
49. Bonk, A. F., Friedman, L., and Derbes, V. J., 1962, Experimental dermatophytosis, *J. Invest. Dermatol.* **39**:281–286.
50. Poulain, D., Tronchin, G., Vernes, A., Delabre, M., and Biguet, J., 1980, Experimental study of resistance to infection by *Trichophyton mentagrophytes*: Demonstration of memory skin cells, *J. Invest. Dermatol.* **74**:205–209.
51. Lepper, A. W. D., 1972, Experimental bovine *Trichophyton verrucosum* infection. Preliminary clinical, immunological, and histological observations in primarily infected and reinoculated cattle, *Rev. Vet. Sci.* **13**:105–115.
52. Jones, H. E., Reinhardt, J. H., and Rinaldi, M. G., 1974, Acquired immunity to the dermatophytes, *Arch. Dermatol.* **109**:840–847.
53. Moser, S. A., and Pollack, J. D., 1978, Isolation of glycopeptides with skin test activity from dermatophytes, *Infect. Immun.* **19**:1031–1046.
54. Barker, S. A., Cruickshank, C. N. D., Morris, J. H., and Wood, S. R., 1962, The isolation of

trichophytin glycopeptide anJ its structure in relationship to the immediate and delayed hypersensitivity reactions, *Immunology* **5**:627–632.
55. Grappel, S. F., and Blank, F., 1972, Role of keratinases in dermatophytosis. I. Immune responses of guinea pigs infected with *Trichophyton mentagrophytes* and guinea pigs immunized with keratinases, *Dermatologica* **145**:245–255.
56. Grappel, S. F., Blank, F., and Bishop, C. T., 1968, Immunological studies on dermatophytes. III. Further analyses of the reactivities of neutral polysaccharides with rabbit antisera to *Microsporum quinckeanum, Trichophyton schoenleini, Trichophyton rubrum, Trichophyton interdigitale,* and *Trichophyton granulosum, J. Bacteriol.* **96**:70–75.
57. Grappel, S. F., Buscavage, C. A., Blank, F., and Bishop, C. T., 1970, Comparative serological reactivities of twenty-seven polysaccharides from nine species of dermatophytes, *Sabouraudia* **8**:116–125.
58. Reyes, E. C., and Friedman, L., 1966, Concerning the specificity of dermatophyte-reacting antibody in human and experimental animal sera, *J. Invest. Dermatol.* **47**:27–34.
59. Jones, H. E., 1980, The atopic-dermatophytosis syndrome, *Acta Derm. Venereol. Suppl.* **92**: 81–85.
60. Jones, H. E., Rinaldi, M. G., Chai, H., and Kahn, G., 1973, Apparent cross-reactivity of airborne molds and the dermatophyte fungi, *J. Allergy Clin. Immunol.* **53**:346–351.
61. Peck, S. M., Siegel, S., and Bergamini, R., 1947, Immunologic relationships of the antibiotics and trichophytin. Clinical observations and animal experiments. *J. Invest. Dermatol.* **9**:165–185.
62. Sohnle, P. G., Collins-Lech, D., and Huhta, K. E., 1983, Class-specific antibodies in young and aged humans against organisms producing superficial fungal infections, *Br. J. Dermatol.* **108**:69–76.
63. Cox, W. A., and Moore, J., 1968, Experimental *Trichophyton verrucosum* infections in laboratory animals, *J. Comp. Pathol.* **78**:35–41.
64. Brocq-Rousseu, P., Urbain, A., and Barotte, J., 1927, Etude des teignes du cheval et de l'immunite dans les teignes experimentales, *Ann. Inst. Pasteur Paris* **41**:513–551.
65. Hay, R. J., and Shennan, G., 1982, Chronic dermatophyte infections. II. Antibody and cell-mediated immune responses, *Br. J. Dermatol.* **106**:191–198.
66. Grappel, S. F., Blank, F., and Bishop, C. T., 1971, Circulating antibodies in human favus, *Dermatologica* **143**:271–276.
67. Grappel, S. F., Blank, F., and Bishop, C. T., 1972, Circulating antibodies in dermatophytosis, *Dermatologica* **144**:1–11.
68. Papini, M., and Simonetti, S., 1985, Humoral immunofluorescent antibodies in subjects with dermatophytosis, *Mykosen* **28**:419–429.
69. Kaaman, T., von Stedingk, L. V., von Stedingk, M., and Wasserman, J., 1981, ELISA-determined serological reactivity against purified trichophytin in dermatophytosis, *Acta Derm. Venereol.* **61**:313–317.
70. Jillson, O. F., and Huppert, M., 1949, The immediate wheal and 24–48 hour tuberculin type edematous reactions to trichophytin, *J. Invest. Dermatol.* **12**:179–185.
71. Jones, H. E., Reinhardt, J. H., and Rinaldi, M. G., 1973, A clinical, mycological, and immunological survey for dermatophytosis, *Arch. Dermatol.* **108**:61–65.
72. Hanifin, J. M., Ray, L. F., and Lobitz, W. C., 1974, Immunological reactivity in dermatophytosis, *Br. J. Dermatol.* **90**:1–8.
73. Hunziker, N., and Brun, R., 1980, Lack of delayed reaction in presence of cell-mediated immunity in trichophytin hypersensitivity, *Arch. Dermatol.* **116**:1266–1268.
74. Hay, R. J., 1979, Failure of treatment in chronic dermatophyte infections, *Postgrad. Med. J.* **55**:608–610.
75. Kaaman, T., 1978, The clinical significance of cutaneous reactions to trichophytin in dermatophytosis, *Acta Derm. Venereol.* **58**:139–143.
76. Espiritu, B. R., Szpindor-Watson, A., Zeitz, H. J., and Thomas, L. L., 1988, IgE-mediated

sensitivity to *Trichophyton rubrum* in a patient with chronic dermatophytosis and Cushing's syndrome, *J. Allergy Clin. Immunol.* **81**:847–851.
77. Wood, S. R., and Cruickshank, C. N. D., 1962, The relationship between trichophytin sensitivity and fungal infection, *Br. J. Dermatol.* **74**:329–335.
78. Grossman, J., Baum, J., Gluckman, J., Fusner, J., Condemi, J. J., 1975, The effect of aging and acute illness on delayed hypersensitivity, *J. Allergy Clin. Immunol.* **55**:268–275.
79. Palmer, D. L., and Reed, W. P., 1974, Delayed sensitivity skin testing. I. Response rates in a hospitalized population, *J. Infect. Dis.* **130**:132–137.
80. Krebs, S., Greenberg, J., and Jesrani, K., 1977, Temporal correlation of lymphocyte blastogenesis, skin tests responses, and erythema during dermatophyte infections, *Clin. Exp. Immunol.* **27**:526–530.
81. Cruickshank, C. N. D., Trotter, M. D., and Wood, S. R., 1960, Studies on trichophytin sensitivity, *J. Invest. Dermatol.* **35**:219–223.
82. Green, F., and Balish, E., 1980, *Trichophyton mentagrophytes* dermatophytosis in germfree guinea pigs, *J. Invest. Dermatol.* **75**:476–480.
83. Lepper, A. W. D., 1974, Experimental bovine *Trichophyton verrucosum* infection. The cellular responses in primary lesions of the skin resulting from surface or intradermal inoculation, *Res. Vet. Sci.* **16**:287–298.
84. Kaaman, T., 1981, Cell-mediated reactivity in dermatophytosis: Differences in skin responses to purified trichophytin in tinea pedis and tinea cruris, *Acta Derm. Venereol.* **61**:119–123.
85. Rasmussen, J. E., and Ahmed, A. R., 1978, Trichophytin reactions in children with tinea capitis, *Arch. Dermatol.* **114**:371–372.
86. Sorensen, G., and Jones, H. E., 1976, Immediate and delayed hypersensitivity in chronic dermatophytosis, *Arch. Dermatol.* **112**:40–42.
87. Jones, H. E., Reinhardt, J. H., and Rinaldi, M. G., 1974, Model dermatophytosis in naturally-infected subjects, *Arch. Dermatol.* **110**:369–374.
88. Kaaman, T., Petrini, B., and Wasserman, S., 1979, *In vivo* and *in vitro* immune responses to trichophytin in dermatophytosis, *Acta Derm. Venereol.* **59**:229–233.
89. Stahl, D., and Svejgaard, E., 1982, Lymphocyte transformation *in vitro* in acute dermatophytosis: A follow-up study, *Acta Derm. Venereol.* **62**:289–293.
90. Walters, B. A. J., Beardmore, G. L., and Halliday, W. J., 1976, Specific cell-mediated immunity in the laboratory diagnosis of dermatophyte infections, *Br. J. Dermatol.* **94**:55–61.
91. Walters, B. A. J., Chick, J. E. D., and Halliday, W. J., 1974, Cell-mediated immunity and serum blocking factors in patients with chronic dermatophyte infections, *Int. Arch. Allergy Appl. Immunol.* **46**:849–857.
92. Hay, R. J., Reid, S., Talwat, E., and MacNamara, K., 1983, Immune responses of patients with tinea imbricata, *Br. J. Dermatol.* **108**:581–586.
93. Helander, I., 1978, Leukocyte migration agarose test in dermatophytosis, *Mykosen* **21**:167–175.
94. Green, F., and Balish, E., 1979, Suppression of *in vitro* lymphocyte transformation during an experimental dermatophyte infection, *Infect. Immun.* **26**:554–562.
95. Calderon, R. A., and Hay, R. J., 1984, Cell-mediated immunity in experimental murine dermatophytosis. I. Temporal aspects of T-suppressor activity caused by *Trichophyton quinckeanum*, *Immunology* **53**:457–464.
96. Calderon, R. A., Hay, R. J., and Shennan, G. I., 1987, Circulating antigens and antibodies in human and mouse dermatophytosis: Use of monoclonal antibody to phosphorylcholine-like epitopes, *J. Gen. Microbiol.* **133**:2699–2705.
97. Sherwin, W. K., Ross, T. H., Rosenthal, C. M., and Petrozzi, J. W., 1979, An immunosuppressive serum factor in widespread cutaneous dermatophytosis, *Arch. Dermatol.* **115**:600–604.
98. Marcussen, P. V., 1937, Relationship of the urticarial to the inflammatory reaction to trichophytin, *Arch. Dermatol.* **36**:494–514.
99. Balogh, E., Fiorizs, E., Debreczeni, M., and Szabolcsy, M., 1980, Serum IgE level and T-cell count in chronic dermatophytosis, *Mykosen* **24**:84–89.

100. McGready, S. J., and Buckley, R. H., 1975, Depression of cell mediated immunity in atopic eczema, *J. Allergy Clin. Immunol.* **56**:393–406.
101. Dahl, M. V., Greene, W. H., and Quie, P. G., 1976, Infection, dermatitis, increased IgE, and impaired neutrophil chemotaxis, *Arch. Dermatol.* **112**:1387–1390.
102. Rocklin, R. E., 1976, Modulation of cellular-immune responses *in vivo* and *in vitro* by histamine receptor-bearing lymphocytes, *J. Clin. Invest.* **57**:1051–1058.
103. Rocklin, R. E., 1977, Histamine induced suppressor factor (HSF): effect on migration inhibition factor (MIF) production and proliferation, *J. Immunol.* **118**:1734–1738.
104. Keeney, E. L., and Huppert, M., 1959, Immunization against superficial fungus infection, *J. Invest. Dermatol.* **32**:73–76.
105. Hussin, Z., and Smith, J., 1983, Vaccination procedures and the infectivity of dermatophyte lesions, *Mycopathologica* **81**:71–75.
106. Wharton, M. L., Reiss, F., and Wharton, D. R. A., 1950, Active immunization against *Trichophyton purpureum*, *J. Invest. Dermatol.* **14**:291–303.
107. Sarkisov, A. K. H., 1979, Specific prophylaxis of trichophytosis in cattle, *Veterinarii* **1**:40–41.
108. Hay, R. J., 1988, New oral treatments for dermatophytosis, *Ann. N. Y. Acad. Sci.* **544**:580–585.
109. Savin, R., 1988, Successful treatment of chronic tinea pedis (moccasin type) with terbinafine (Lamisil), *Clin. Exp. Dermatol.* **14**:116–119.
110. Tager, A., Lass, N., Avigard, J., and Breemer, A. M., 1973, Immunotherapy of superficial dermatomycoses, *Dermatologica* **147**:123–129.
111. Strickler, A., 1915, The vaccine treatment of ringworm of the scalp, *J. Amer. Med. Assoc.* **65**: 224–227.
112. Lewis, G. M., and Hopper, M. E., 1937, Ringworm of the scalp. IV. a) Comparative reactions to cutaneous tests with trichophytin in children with and without ringworm of the scalp; b) evaluation of therapy with stock vaccines in types of infection resistant to treatment, *Arch. Dermatol.* **36**:821–832.
113. Miller, H. E., Stewart, R. A., and Kimura, F., 1941, Undenatured trichophytin. Preparation and clinical application, *Arch. Dermatol.* **44**:804–815.
114. Traub, E. F., and Tolmach, J. A., 1935, Dermatophytosis. Its treatment with trichophytin, *Arch. Dermatol.* **32**:413–421.
115. Sulzberger, M. B., and Wise, F., 1932, Ringworm and trichophytin: newer developments, including practical and theoretical considerations, *J. Amer. Med. Assoc.* **99**:1759–1764.
116. Kirkpatrick, C. H., and Smith, T. K., 1974, Chronic mucocutaneous candidiasis: Immunologic and antibiotic therapy, *Ann. Int. Med.* **80**:310–320.
117. Galbraith, G. M. P., 1988, Therapeutic immunomodulation, *Dermatol. Clin.* **6**:561–568.

4

Introduction to Candida

Systemic Candidiasis

JUDITH E. DOMER and ROBERT I. LEHRER

1. INTRODUCTION TO CANDIDA

1.1. Description of the Disease

Candidiasis is the prototypic opportunistic fungal disease. *Candida* spp. reside most commonly in the gastrointestinal tract[1,2] and vagina, where they are normally held in check by as-yet incompletely defined local factors that include competition with the resident bacterial flora and innate effectors of nonspecific mucosal resistance.[3,4] Infrequently, but under a wide variety of clinical circumstances, these normally inconspicuous commensals can multiply on or invade through the mucosal surface to initiate local or systemic disease. Various immunologic and nonimmunologic conditions can predispose to such opportunistic infections.[5] Among the former are certain malignancies[6–8] and the use of cytotoxic or immunosuppressive therapy.[9,10] Nonimmunologic predisposing factors include diabetes,[11] trauma,[12] pregnancy,[13] antibiotic therapy,[7] and hyperalimentation.[14] Not infrequently, such predisposing factors occur in combination.

Significant risk factors for nosocomial candidiasis in patients with cancer include previous surgery, neutropenia, central catheterization, chemotherapy, specific antibiotic treatment, and peripheral cultures positive for

JUDITH E. DOMER • Department of Microbiology and Immunology, Tulane University School of Medicine, New Orleans, Louisiana 70112. ROBERT I. LEHRER • Department of Medicine, UCLA School of Medicine, Los Angeles, California 90024.

Fungal Infections and Immune Responses, edited by Juneann W. Murphy *et al.* Plenum Press, New York, 1993.

Candida spp.[15] Additional risk factors may include the presence of a bladder catheter, treatment with two or more antibiotics, azotemia, transfer from another hospital, diarrhea, and candiduria.[16] The potential roles of dietary, mechanical or iatrogenic factors have been reviewed by Odds,[17] and the clinical and microbiological aspects of nosocomial candidiasis have been reviewed by Pfaller.[18]

Although the syndromes of candidiasis can be broadly categorized into cutaneous, mucocutaneous, and systemic forms, more specialized classifications based on the predominant site(s) of infection are useful clinically. Systemic infections can include prosthetic valve disease, endophthalmitis, hepatosplenic candidiasis, peritonitis, endocarditis, meningitis, arthritis, and osteomyelitis, whereas mucocutaneous infections can include esophagitis, vulvovaginitis, cystitis, oral thrush, and denture stomatitis.[17,19]

In recent years, mucocutaneous candidiasis has garnered much attention because an especially noxious form that affects the mouth, tongue, and esophagus is very common among patients infected by human immunodeficiency virus (HIV).[20–22] In general, it appears that granulocytopenia or defective phagocyte function promotes systemic disease, whereas reduced T-cell/macrophage function promotes mucocutaneous disease. As the immunology of cutaneous and mucocutaneous disease is covered in Chapter 5, this chapter will center on systemic candidiasis. Although results obtained from animal models are featured prominently in this chapter, where possible, we will highlight results obtained with human materials. An excellent review of the animal models used to study various forms of candidiasis was published by Guentzel *et al.*[23]

1.2. Etiology

There are almost 200 recognized species in the genus *Candida*,[24] including those organisms formerly classified as *Torulopsis* spp., but only a few species are part of the normal flora and/or regularly cause human disease. The two most common pathogenic *Candida* spp. are *C. albicans* and *C. tropicalis*. Although precise figures are difficult to obtain, *C. albicans* appears to be responsible for approximately 60–75% of all candidal disease, *C. tropicalis* for 15–25%, *C. glabrata* for 10–15%, *C. parapsilosis* for 5–15% and *C. krusei* for about 2%.[6,25] *C. tropicalis* appears to be less pathogenic than *C. albicans* under most circumstances, but disease caused by *C. tropicalis* has become more prevalent, especially in individuals with hematologic malignancies and those undergoing bone marrow transplantation.[26,27] Based on their histopathologic observations of the GI tracts of individuals with *C. tropicalis* and *C. albicans* infections, Walsh and Merz[28] proposed that *C. tropicalis* was more enteroinvasive than *C. albicans* in susceptible hosts. Howlett[29] reported that *Candida* spp. differ in their ability to penetrate keratinized rat tongue mucosa (*C. albicans* > *C. tropicalis* and *C. krusei* > *C. parapsilosis* and *C. guilliermondii*), and a more recent study suggests that clinical isolates of *C. albicans* differ in their pathogenicity for dorsal rat tongue mucosa.[30] The ability of *C. albicans* to form hyphae is believed to be important for colonization and invasion of the oral and vaginal mucosa.[31,32] Species with less pathogenic potential, e.g., *C. parapsilosis*, *C. krusei* and *C. guilliermondii*, are often associated with a more restricted range of clinical conditions. For example, *C. parapsilosis* is found in paronychia, endocarditis, endophthalmitis and otitis media; *C. guilliermondii* causes endocarditis, cutaneous candidiasis and onychomycosis; and *C. krusei* occasionally causes endocarditis.[18,19,33] Since the three organisms mentioned last are more commonly found among the flora of skin than in the gastrointestinal tract,[19] this ecological niche might position them to initiate the clinical syndromes mentioned previously. *C. lusitaniae*, an organism which the diagnostic laboratory can easily confuse with *C. tropicalis*, and *C. parapsilosis* are also being identified more frequently as a cause of disease in immunosuppressed individuals.[34] Infections due to *C. lusitaniae* and *C. guilliermondii*, *Candida* spp. of low intrinsic virulence, can be difficult management problems since either organism may be resistant to polyene antibiotics.[35,36]

1.3. Epidemiology

Although various species of *Candida* can be isolated from lower animals and from numerous environmental sources, summarized in Odds,[17] most human candidal disease was considered to be of endogenous origin until a few years ago. It has been known for some time that there are two primary serotypes of *C. albicans*, A and B.[37] Since all *C. albicans* strains have identical sugar assimilation and fermentation patterns when tested with the usual limited battery of substrates, and since disease was considered to be endogenous, little effort was expended in attempting to determine intraspecific strain differences until recently. During the last decade, at least eight typing methods that are potentially applicable to the epidemiology of *C. albicans* infections have been developed. The methods include biotyping,[38,39] resistotyping,[40] sensitivity to killer yeasts,[41] extracellular enzyme production,[42] immunoblotting,[43] morphotyping,[44] DNA restriction fragment analysis,[45,46] and electrophoretic karyotyping.[47,48]

Such methods have been used to investigate outbreaks,[49,50] to determine the dominance of a strain type in a clinical syndrome,[51,52] or simply to survey the biotypes predominant in an institutional or geographic environment.[53–55] In two of these studies,[53,55] isolates of *C. albicans* from different anatomic sites in the same patient appeared to be the same, although Soll *et al.*[56] isolated multiple strains and species of *Candida* from a single patient from different anatomic sites at different times during the course of disease. The studies to date have been limited, but some candidal infections may prove to be acquired from exogenous (possibly nosocomial) sources rather than from the patient's own endogenous yeast-cell flora.[54]

2. SYSTEMIC CANDIDIASIS

2.1. Innate Immunity

2.1.1. Nutritional Control Mechanisms

Although *C. albicans* requires biotin for growth,[57,58] its other nutritional needs can be satisfied by simple, fully-defined media that contain any of a wide range of carbon sources; ammonium, sulfate and phosphate salts; and trace elements. Notwithstanding this metabolic versatility, some of the innate mechanisms responsible for inhibiting *Candida* spp. proliferation *in vivo* may operate on the level of substrate deprivation or sequestration. The best established of these mechanisms involves the sequestration of iron by two host-derived iron-binding proteins, transferrin and lactoferrin. Indeed, the observations that normal human serum inhibited the growth of *C. albicans in vitro* and that this effect was reversed by addition of iron were directly responsible for the discovery of transferrin,[59] the major iron-binding protein of serum. The specific granules of neutrophils contain large amounts of an iron-binding protein, apolactoferrin, that is structurally homologous to transferrin.[60–62] Not surprisingly, apolactoferrin has also been shown to be candidastatic by virtue of its ability to bind iron.[63] Recent evidence[64–66] suggests that *in vitro* and *in vivo* growth restriction of *C. albicans* may also be mediated by a calcium and/or zinc-binding protein that has been found in the cytoplasm of neutrophils and certain epithelial cells by a number of workers. These studies have been reviewed by Steinbakk *et al.*[66] Ongoing work by Wang and Lehrer (unpublished data) suggests that competition for biotin between *C. albicans* and various bacteria requiring biotin, including oral streptococci, may play a role in limiting the population density of *C. albicans* in the mouth. An increased local population of *C. albicans* occurs in the oral cavity of many diabetics and may be attributable to the higher concentrations of glucose in oral secretions from hyperglycemic individuals.[67] Other host factors that may regulate the growth of oral *C. albicans* have been reviewed by Epstein *et al.*[68]

The factors responsible for maintaining relatively low concentrations of *C. albicans* within the gastrointestinal tract are understood only in part, and include competition with other members of the intestinal flora and host immune factors. The areas of colonization are highly selective and may be related to the surface characteristics of the affected epithelium and its keratin.[69] In rats reared conventionally, the administration of broad-spectrum antibiotics[70] and corticosteroids[71] both predispose to long lasting intestinal carriage of *C. albicans*. Cantorna and Balish[72] examined the ability of *C. albicans* to colonize the gastrointestinal tract of germfree mice. Doubly immunodeficient beige and athymic (bg/bg and nu/nu) animals developed extensive yeast and hyphal infections of the palate, tongue, esophagus, and stomach, whereas singly immunodeficient beige or athymic mice and their normal nu/+, bg/+ littermates developed only moderate mucosal infections. The doubly immunodeficient mice later also developed progressive systemic infections caused by the *C. albicans*. The possibility that the small intestine might contain and secrete antifungal components that might affect *Candida* spp. population density has not been examined systematically. It warrants attention, however, since murine Paneth cells, which line the base of the small intestine's numerous crypts, contain large amounts of mRNA for cryptdin,[73] a member of a family of peptides (defensins) many of whose members are strongly candidacidal.[74]

2.1.2. Intravascular Access and Clearance

Clinical disseminated (systemic) candidiasis results when *C. albicans* enters the vascular system in numbers that exceed the ability of host defense mechanisms to clear them and prevent their subsequent growth in the tissues. *C. albicans* probably most often gains entry to the vascular system from its reservoir in the gastrointestinal tract, but the organisms may also be introduced into the circulation from the skin via indwelling intravenous lines or by injection of contaminated materials. In a historic experiment that is not likely to be repeated soon, Krause *et al.*[75] administered a large number of viable *C. albicans* blastospores intragastrically to a normal human subject. Some of the fungi appeared in the circulating blood within 30 minutes, evidently having traversed the epithelial barriers of the normal human gastrointestinal tract to do so. Sass *et al.*[76] have provided evidence that the transgastrointestinal migration of particles, such as latex spheres, fungal spores, and pollen grains, is mediated by M cells in Peyer's patches and that subsequent intravascular access is provided via the abdominal lymphatics. Since even an intact gastrointestinal tract evidently permits some intraluminal *C. albicans* yeast cells to gain access to the blood, intestines damaged by disease, drugs, or trauma are likely to be even more pervious. Since bloodstream seeding by intestinal *C. albicans* undoubtedly is initiated by fungal organisms adherent to the epithelial surface, factors (e.g., broad spectrum antibiotics) that enhance *Candida* spp. growth by decreasing microbial competition for nutrients or anchorage sites probably increase the relative likelihood of hematogenous seeding.

Clearance of intravenously injected organisms from rabbit blood shows biphasic kinetics. Approximately 99% of the organisms leave the blood within 10 minutes, followed by a longer period of low-grade fungemia. Depending on the route of injection, different organs predominate in clearance. After peripheral venous injections, lungs take up 45% of the total inoculum; after mesenteric vein injection, liver takes up almost 60%.[77] Mice injected with live, but not formalin-killed, *C. albicans* evidenced accelerated reticuloendothelial clearance of carbon particles.[78]

2.1.3. *Polymorphonuclear Leukocytes*

2.1.3a. Protective Role of Polymorphonuclear Leukocytes in Vivo. The polymorphonuclear leukocytes (neutrophils, PMNL) are extremely important in defense against systemic infection by *C. albicans*, as evidenced by the fact that a) neutropenic patients are highly susceptible to the development of candidiasis,[79,80] b) the disappearance of organisms from tissue parallels the appearance of PMNL,[81] and c) patients with leukocyte defects, such as those seen in hereditary or acquired myeloperoxidase (MPO) deficiency or chronic granulomatous disease (CGD), may experience fungal infections caused by *Candida* spp. and other organisms. (MPO is one component of the oxidative pathways involved in killing microorganisms within phagocytic cells. It is discussed in more detail in Section 2.1.3c.) The importance of the leukocyte, whose numbers may have been reduced by bone marrow suppressive disease or therapy, has been shown in animal models where granulocyte transfusions resulted in partial reversal of the effects of leukopenia.[82,83] For example, although normal dogs tolerate 10^7 *C. albicans* organisms administered intravenously, widespread disseminated candidiasis results when only 10^6 organisms are given to leukopenic animals. Granulocyte transfusions to the immunosuppressed dogs, however, afforded substantial protection against the fungus.[82] Studies with diffusion chambers placed intraperitoneally in mice have also provided evidence for the importance of phagocytes in clearing sites of *C. albicans in vivo*.[84]

Hereditary MPO deficiency, usually discovered after the routine use of certain automated blood count analyzers, appears to be a common but infrequently diagnosed condition. In a population of 60,000 patients tested at Stamford Hospital in Stamford, Connecticut over 40 months, 28 MPO-deficient persons were identified. Half of these had completely MPO-deficient PMNL and the others had partial deficiencies. Most MPO-deficient patients had no clinical sequelae, but several had experienced serious systemic infections caused by *Candida* spp. or other organisms. The most serious of these occurred in a 30-year-old woman who also had insulin-dependent diabetes mellitus and developed fatal pneumonia caused by *C. albicans* and *C. parapsilosis*. Overall, 4 of the first 29 reported individuals with complete (homozygous) hereditary MPO deficiency had systemic candidal infections.[85–89] Three of these persons had coexisting diabetes mellitus, which may also have contributed to their susceptibility to systemic candidiasis. If the frequency of homozygous hereditary MPO deficiency is as high as suggested by the Stamford Hospital study (14/60,000 = 0.00023) then, assuming an autosomal recessive pattern of inheritance, it can be calculated that the heterozygous carrier state is likely to exist in approximately 1.5% of the population. Although PMNL from subjects that were heterozygous for hereditary MPO deficiency contained decreased levels of MPO, their ability to kill ingested *C. albicans* was intact.[85] Consequently, it seems unlikely that partial (heterozygous) MPO deficiency will prove to be a significant risk factor for systemic candidiasis in humans.

The association of acquired MPO deficiency with leukemic and preleukemic states is now well recognized.[90–92] In one such patient studied at UCLA, 99.2% of blood PMNL were completely MPO deficient, and the cells showed greatly impaired killing of ingested *C. albicans*, *C. tropicalis* and *S. aureus in vitro*.[93] The patient succumbed to fungal infections caused by *A. fumigatus* and *C. albicans*. Although PMNL from patients with hereditary complete MPO deficiency can phagocytize the organisms normally, they are greatly impaired in their ability to kill *C. albicans* and certain other *Candida* species, including *C. tropicalis*, *C. stellatoidea*, and *C. krusei*.[85,87,94] MPO-deficient PMNL also killed many, but not all, bacteria at a reduced rate.[86,95] Although MPO appears to be essential for effective killing of *C. albicans* by human PMNL and mononuclear (MN) cells,[85,96,97] other *Candida* species such as *C. parapsilosis*, *C. pseudotropicalis*, and *C. krusei* were killed effectively by MPO-deficient PMNL.[94] Certain of these (*C. parapsilosis* and *C. pseudotropicalis*) are also killed normally by PMNL from subjects with CGD.[94] It can be concluded from these observations that *C. albicans* is relatively resistant to human PMNL and that its eradication requires both MPO and hydrogen peroxide (H_2O_2). Other *Candida* species are killed effectively by PMNL mechanisms that require neither MPO nor H_2O_2. These nonoxidative mechanisms appear to be relatively ineffective in killing *C. albicans*, although some of them might contribute to candidastasis.

The effects of a number of additional underlying illnesses and their treatments on the candidacidal function of PMNL have been examined as well. Patients with neoplasms are predisposed to systemic candidal infections especially while receiving immunosuppressive therapy. PMNL from patients with acute leukemias are generally able to ingest *C. albicans* normally but frequently manifest decreased candidacidal activity.[98–100] Similar results have been described in patients with large burns.[101] PMNL from patients with chronic granulocytic leukemia have been reported to kill *C. albicans*[100] but not *C. guilliermondii*[102] normally. Subjects with myeloid metaplasia, polycythemia vera, untreated metastatic solid tumors, Hodgkin's disease, and other lymphomas had PMNL with normal candidacidal activity for *C. albicans*. Patients receiving chemotherapy or radiation therapy for these conditions often had PMNL with diminished candidacidal activity.[99] Another study using *C. guilliermondii* to study PMNL from patients with various leukemias and lymphomas noted the frequent occurrence (12 of 24 patients) of candidacidal defects, often corrected *in vitro* by addition of levamisole.[103] The PMNL of patients with AIDS-related complex (ARC) and AIDS associated with deep candidiasis were reported to have a normal ability to kill *C. albicans*.[104]

Although neonates are generally considered to be relatively immuno-compromised hosts and were reported to have PMNL with decreased candidacidal activity,[105] this finding has not been confirmed.[106,107] Malnourished children, known to harbor large concentrations of *Candida* spp. in their upper intestinal tracts, were found to have PMNL with significantly decreased candidacidal activity.[108] The PMNL of selenium-deficient rats were reported to display impaired *in vitro* fungicidal activity for *C. albicans*.[109] Hyperosmolar conditions decreased ingestion and killing of *C. albicans* by human PMNL *in vitro*.[110] Normal phagocytosis but decreased cidal activity against *C. albicans* has been reported in PMNL obtained from children with thalassaemia.[111] Amphotericin B-resistant *C. tropicalis* developed in a patient whose PMNL displayed diminished candidacidal activity against both *C. tropicalis* and *C. albicans*, especially when tested in autologous serum.[112]

In addition to their fungicidal activities, there is some evidence that PMNL may protect some tissues from destruction by the fungus as well. Edwards *et al.*,[113] using an *in vitro* assay involving the development of a monolayer of endothelial cells,[114] noted that the endothelial cells were protected from the harmful effects of candidal hyphae if PMNL were present. The mechanism(s) has not been delineated, but PMNL were attracted to the sites of hyphal attachment and spread over the hyphal elements, both within and beneath the endothelial cells.

In the healthy host, there is no shortage of PMNL to fend off *Candida* spp. PMNL are the most abundant phagocytes of circulating blood. In man, they are normally present at a concentration of approximately $2.5–5.0 \times 10^6$/ml of blood, but this can increase several-fold in response to infection. Circulating mature PMNL are short-lived cells, $T_{1/2} = 6–7$ hours, with a minimal capacity for new protein synthesis. MN are only about 10% as numerous as PMNL in the peripheral blood, but after entering tissues, they undergo further differentiation which enables them to survive for days or weeks as macrophages.

2.1.3b. Interactions between Candida spp. and PMNL in Vitro. Although animal leukocytes are often used as surrogates for their human counterparts, the validity of this practice is more an article of faith than of science. Fortunately, the availability of human PMNL and MN has allowed their interactions with *Candida* spp. to be studied directly *in vitro*. Human blood is fungicidal for *C. albicans*, killing both blastoconidia and organisms bearing pseudohyphae up to 200 mm in length.[115] The candidacidal activity of blood can be attributed directly to its leukocytes, since neither plasma nor serum alone kills *C. albicans*.[116,117]

The relationship between *C. albicans* and the phagocytic process itself and associated events, such as chemotaxis, are incompletely understood, but some data have been gathered on chemotaxis and the opsonic requirements for phagocytosis of *C. albicans*. Viable or heat-killed *C. albicans* were chemotactic for human PMNL,[118] as was mannan, a major component of the candidal cell wall.[119,120] Intact cells and purified mannan activated the alternative pathway of complement,[118,121] suggesting that *Candida*-induced chemotaxis could result from complement activation. Cutler[122] described a soluble factor from seven of eight isolates of *C. albicans* that was chemotactic for guinea pig PMNL in the absence of serum, a condition that precluded complement activation.

The role of complement in uptake and subsequent killing of *C. albicans* is somewhat controversial. Most studies have shown that complement enhances phagocytosis and suggested that both classical and alternative systems are involved.[123–129] While there is agreement that serum complement factors facilitate phagocytosis[85,130–132] and that intracellular killing of the organism is not influenced by the presence or absence of specific antibody,[123,127,132] there is disagreement on specific aspects of the role of C3. Kernbaum[126] reported that phagocytosis and killing were normal in the absence of C3, Yamamura and Valdimarsson[124] reported that C3 was not necessary for ingestion of the organism but was necessary for killing, and Morrison and Cutler[127] reported that C3 was necessary for ingestion to occur. In one study,[133] the heat- stable opsonic factor of normal human serum was believed to be naturally occurring, low-level anti-*Candida* IgG. High antibody concentrations, typically present in serum from individuals with chronic mucocutaneous candidiasis or some patients with disseminated candidiasis, inhibited candidacidal activity by normal PMNL without significantly impairing phagocytic uptake.[134–136]

Although most research emphasis has centered on the opsonic properties of complement and/or antibody, other factors may also influence ingestion of *Candida* spp. Ginsburg *et al.*[131] reported that polycationic substances, such as nuclear histones, that accumulate in lesions after cell death and tissue destruction might function as efficient opsonins. Fleischmann *et al.*[137] reported that defensins, which are also polycationic peptides, were nonspecific opsonins that promoted the uptake of *C. albicans* and various bacteria by rabbit alveolar macrophages. The ability of defensins to opsonize *C. albicans* for uptake by PMNL has not been reported.

Several *in vivo* studies have provided evidence that complement participates in resistance to systemic candidiasis. Mice with various genetic defects involving complement were more susceptible to systemic challenge than those without such defects,[123,138,139] as were normal guinea pigs depleted of the alternative pathway of complement activation by treatment with cobra venom factor.[140] Since two of the mouse strains were deficient in C5, these data lend support to the *in vitro* work of Morelli and Rosenberg[123] on the importance of C5 in the phagocytic process.

Unlike most bacteria, only a portion of *C. albicans* yeast cells ingested by human PMNL are killed intracellularly during the first few hours. This was first shown with a quantitative vital staining (methylene blue) assay, which revealed that PMNL from normal human donors killed 29% ± 7.4% of ingested yeast phase *C. albicans* after 60 min.[116] These data have been confirmed by assay systems based on Giemsa staining,[141] isotope incorporation or chromium release,[142] and colony counting,[130,143] although the latter generally provide higher estimates of killing (approximately 50% after 60 minutes) by PMNL and MN. Serum, although required for efficient phagocytosis, did not facilitate killing of ingested organisms in one study[116] but did so in another.[144] The presence of PMNL substantially enhanced the ability of purified human MN to kill *C. albicans*.[96] The ability of human PMNL to kill ingested *C. albicans* was abolished when the incubations were performed under anaerobic conditions or in the presence of MPO inhibitors (cyanide or azide), although ingestion of the yeasts was not inhibited by these treatments.

It is clear that *Candida* spp. differ in their susceptibility to the fungicidal mechanisms of the PMNL. Yeast forms of *C. albicans*, *C. tropicalis*, and *C. viswanathii*, three of the more pathogenic species of *Candida*, were killed more slowly and at higher murine peritoneal (thioglycollate-induced) PMNL:target ratios than less pathogenic species such as *C. guilliermondii*, *C. krusei*, and *C. parapsilosis*.[145] Nevertheless, all *Candida* spp. were killed. It is also clear that both yeast and filamentous (hyphal and pseudohyphal) forms of *C. albicans* can be damaged by PMNL, although the extent of damage and the mechanics of the killing process may differ, in particular, with those filamentous forms too large to be ingested by the PMNL. Scherwitz and Martin[146] reported 90% phagocytosis of yeasts versus 10% phagocytosis of germinated forms; Schuit[147] reported 35% killing of yeasts versus 10% killing of pseudohyphae; and Cockayne and Odds[148] reported that yeasts, germinated forms and hyphae, were all ingested to a similar extent but that the filamentous forms were killed to a greater extent than the other two forms. Additionally, Schuit[147] noted that both ingestion and killing were more rapid when MN were used instead of PMNL. The inconsistencies between the data from Schuit[147] and Cockayne and Odds[148] emphasize the difficulties in comparing data when different assay conditions are used. Brummer *et al.*[149] compared the ability of murine peritoneal and circulating PMNL to kill *C. albicans* and reported that PMNL obtained after intraperitoneal placement of proteose peptone and caseinate did not differ significantly from peripheral blood PMNL in their ability to kill *C. albicans*, whereas cells recovered after thioglycollate instillations were defective.

Various antibiotics may affect the ability of phagocytes to ingest and kill *C. albicans*, an important observation when one considers that antibiotic treatment is one of the two most important predisposing factors to the development of systemic candidiasis, the other factor being neutropenia. Sulfonamides interfered with killing of ingested *C. albicans* by PMNL by inhibiting myeloperoxidase(MPO)-mediated reactions.[143] Penicillin, tetracycline, chloramphenicol, and gentamicin did not interfere with PMN candidacidal activity in one study,[150] although five aminoglycosides (including gentamicin) inhibited this function in another.[151] Ketoconazole and Sc-39304 were reported to enhance the ability of human PMNL to kill ingested *C. albicans*,[152] although the methods used may have assessed only those events transpiring in unsealed phagocytic vacuoles. Cilofungin (LY 12109) was reported to impair PMNL candidacidal activity by interfering with opsonization and subsequent ingestion.[153] High levels of glucose and β-hydroxy butyrate were reported to inhibit the candidacidal activity of PMNL from diabetic, but not normal, subjects.[154] Amphotericin B has been reported to inhibit phagocytosis of *C. albicans* by PMNL.[155]

There have been relatively few ultrastructural observations of phagocytosis and killing of *C. albicans* by PMNL. Belcher *et al.*,[156] and Richardson and Smith,[157] working with human and murine PMNL respectively, observed the postphagocytic fusion of PMNL lysosomes with tight and loose phagosomes, as well as the presence of *C. albicans* cells with intact cell walls but disintegrating plasma membranes and intracellular structures. Shortly after the observations of Richardson and Smith[157] appeared, Cech and Lehrer[158] reported that *C. albicans* ingested by human PMNL were found in two types of phagolysosomes: "unsealed vacuoles," which maintained communication with the extracellular compartment and "sealed vacuoles," which did not. Approximately 40% of all PMNL vacuoles were unsealed, and only 15% of the *C. albicans* blastospores within them had been killed within 1 hour. In contrast, approximately 72% of *C. albicans* inside the sealed vacuoles had been killed. The sequestration of *C. albicans* in sealed and unsealed PMNL vacuoles was confirmed by Levitz *et al.*,[159] who also extended these observations to MN. Many *C. albicans* blastospores within unsealed vacuoles remained viable after their ingestion by PMNL. Perhaps leakage of potentially fungicidal molecules (e.g., MPO, defensins, H_2O_2, etc.) from unsealed vacuoles, or ingress of inhibitory ions and proteins from serum, or the combined effects of both processes contributed to the observed fungicidal activity.

2.1.3c. Oxidative Candidacidal Mechanisms of PMNL. The fungicidal mechanisms of human PMNL are of two general types: oxidative and nonoxidative. The former, which appear to be most important with respect to *C. albicans*,[85,116,141] depend on the neutrophil's ability to synthesize a group of potent oxidants, including H_2O_2, hydroxyl radicals, hypochlorous acids and chloramines. In normal PMNL, H_2O_2 is derived from the dismutation of superoxide anions, which are produced by an NADPH oxidase activated in PMNL after phagocytosis or stimulation by a variety of soluble mediators. The production of hypochlorous acid and chloramines by PMNL results from the ability of its MPO, an enzyme also found in MN, to catalyze the oxidation of chloride ion by H_2O_2.[160] MPO occurs in the primary (azurophil) granules of normal human PMNL and is released into their phagocytic vacuoles and extracellularly when PMNL are exposed to microbes, including *C. albicans*. Although MPO is present in bone marrow promonocytes and most blood MN, it is lost when these cells mature into macrophages. A distinctly different peroxidase that exists in human eosinophils can generate hypobromous acid from bromide ions and H_2O_2.[161] Human eosinophils can also kill ingested *C. albicans* by oxidative mechanisms.[162] Much evidence indicates that phagocytic normal human granulocytes produce hypochlorous acid[163,164] and that hypochlorite, chloramines, and cell-free MPO-H_2O_2-halide systems kill *Candida* spp. and other fungi *in vitro*.[165–169] Blastoconidia from opaque phenotypes of *C. albicans* strain WO-1 were more susceptible to these oxidants than were white phenotypes.[170]

The mechanisms involved in intracellular killing of *C. albicans* by PMNL have been studied actively for many years. That oxidative mechanisms, specifically those mediated by MPO and H_2O_2, are very important, was first reported over twenty years ago.[85,116] The mechanisms of oxidant production and their significance for neutrophil-mediated antifungal activity have been clarified by studies of PMNL from patients with CGD and hereditary MPO deficiency. The PMNL of CGD patients are profoundly deficient in their ability to generate superoxide anion (O_2^-) and H_2O_2 via the NADPH oxidase pathway as a consequence of various distinct molecular lesions. Most cases of CGD show X-linked inheritance and deficient PMNL function that arises from abnormalities affecting the large subunit of cytochrome b_{558}, a critical component of the activated NADPH oxidase complex. Other forms of CGD are autosomal and result from abnormalities of cytosolic activators of NADPH oxidase or from abnormalities affecting the small subunit of cytochrome b_{558}.[171,172] In their review of 245 patients with CGD, Cohen *et al.*[173] reported that severe fungal infections, especially *Aspergillus* spp. pneumonia, occurred in 50 cases.

Since the infected host must deal with both yeast and hyphal forms in candidal lesions, a few investigators[146–148] have compared the interaction of PMN with the yeast and hyphal forms of *C. albicans*, but many more have examined exclusively the yeast form[85,116,123,124,130,141,147,149,156–158,174–177] or hyphal/pseudohyphal forms.[97,113,153,159,167,168,178–184] Unfortunately, comparisons of data generated from *in vitro* systems in different laboratories are complicated by the differences in their effector:target ratios, incubation times, methods for determination of ingestion and/or killing, animal and anatomic sources of effector cells, and incubation conditions—especially with respect to the presence and concentration of serum in the assay. Many such studies involving either PMNL or mononuclear phagocytes and the yeast form of *C. albicans* have been tabulated in another recent review.[5]

Diamond and his colleagues have provided detailed observations concerning the contribution of oxidative mechanisms to the ability of human PMNL and MN to kill *C. albicans* hyphae or pseudohyphae,[97,113,153,159,167,168,178–184] Using radiolabeled metabolic precursors, chemiluminescence, viable and nonviable *Candida*, and electron microscopy, Diamond *et al.*[97] were the first to demonstrate nonphagocytic damage to pseudohyphal forms of *C. albicans* by PMNL in the absence of serum. Initially, the PMNL spread over the hyphal surface to maximize membrane contact between the two forms. Fresh or heat-inactivated serum enhanced the process, but, in contrast to yeast forms, opsonization was not a requirement for the initiation of microbicidal processes by PMNL. The enhancement was related to anti-*Candida* antibody that was present in the sera. In subsequent reports, it was shown that damage to hyphae was mediated by oxidative means, in particular by the MPO system.[167,168,178] Involvement of MPO was concluded from the following observations: 1) pseudohyphal damage was inhibited by sodium cyanide, sodium azide, catalase, superoxide dismutase and 1,4 diazobicyclo (2,2,2) octane, all agents known to act on PMNL oxidative mechanisms, 2) PMNL from 3 of 4 patients with CGD, a disease in which PMNL are produced that have defective microbicidal activity, and from 2 patients with hereditary MPO deficiency, were unable to damage hyphae, 3) pseudohyphal forms were severely damaged in a PMNL-free *in vitro* system containing MPO, H_2O_2 and iodide or chloride, and the H_2O_2 in the system could be replaced by an H_2O_2-generating system, and 4) methionine, tryptophan and alanine, compounds capable of inhibiting the effects of hypochlorous acid and chloramines, inhibited killing of pseudohyphae.

Induction of the oxidative burst and production of reactive oxygen microbicidal intermediates is a complex process involving events initiated at the PMNL membrane with subsequent involvement of secondary messengers and intracellular responses necessary to the effective function of the cell.[172,185] Delineating these aspects of the PMNL response to unopsonized and opsonized hyphae has been the principal focus of recent research on PMNL–*C. albicans* interactions by Diamond and his co-workers.[97,153,167,178–184] A number of these studies are summarized in Table I. Attachment of PMNL to opsonized or unopsonized hyphae, and spreading over them, appears to be independent of opsonins since both events occur at the same rate in the presence or absence of opsonins. Some of the events following attachment and spreading, however, occur at different rates and one of them, PMNL membrane depolarization, occurs only when opsonized hyphae are attacked.[180] Moreover, PMNL exposed to unopsonized hyphae manifest a delayed rise of cytosolic free Ca^{++},[180] and alterations in phospholipase C activation,[153] actin polymerization,[182] and initiation of superoxide anion release.[180] PMNL degranulation results in low levels of β-glucuronidase release and in lactoferrin release. Much of the released lactoferrin is found bound to hyphal surfaces and does not appear in the supernatant. Whether *in vivo* hyphae of *C. albicans* correspond more closely to the opsonized or unopsonized forms used in these *in vitro* experiments remains to be determined.

2.1.3d. Nonoxidative Anticandidal Mechanisms of PMNL. The ability of human leukocytes to kill some *Candida* spp. by nonoxidative mechanisms most likely reflects the presence of proteins and peptides with intrinsic antifungal activity. The major antimicrobial components of human PMNL are present in a subset of the cell's cytoplasmic granules known as the "primary" or "azurophil" granules. They include defensins, lysozyme, cathepsin G (a serine protease) and several proteins that are homologous to it.[186] Of these, only defensins and several of the serine protease congeners are active against *Candida* spp.

TABLE I
Selected Biochemical and Morphological Observations of Opsonized Zymosan and Opsonized or Unopsonized Hyphae of *C. albicans* during Interaction with Human PMNL

	Opsonized zymosan	Opsonized hyphae	Unopsonized hyphae
Attachment/ spreading[179]	Same kinetics	Same kinetics	Same kinetics
Intracellular Ca^{++}[179]	Peaked within 1 min; greatest concentration in periplasm	Peaked within 1 min; uniform distribution throughout hyphae	Peaked at 4 min; uniform distribution throughout hyphae
Membrane depolarization[180]	Ca^{++} increase first, then membrane depolarization, then respiratory burst	Ca^{++} increase first, then membrane depolarization, then respiratory burst	Ca^{++} increase delayed, no membrane depolarization, respiratory burst delayed
Role of GTP-binding proteins, sensitivity to pertussin-toxin[181]	Partial abrogation of normal Ca^{++}, membrane and respiratory burst responses	Partial abrogation of normal Ca^{++}, membrane and respiratory burst responses	Ca^{++} increase completely eliminated, respiratory burst intact
Inositol triphosphate generation[153]	Transient rise in cellular inositol phosphates	Transient rise in cellular inositol phosphates	Transient rise in cellular inositol phosphates
Actin polymerization[182]	Rapid, initially diffuse, then collarlike, then cylinderlike	Rapid, initially diffuse, then collarlike, then cylinderlike	Delayed, delayed appearance of collarlike, cylinderlike
Degranulation[182]	Low levels of β-glucuronidase detected	Low levels of β-glucuronidase detected	Low levels of β-glucuronidase detected
Specific granule marker[182]		Lactoferrin release into supernatant minimal, but it was found bound to surface	Lactoferrin release into supernatant minimal, but it was found bound to surface
Effects of chelation on respiratory burst[183]	Eliminated respiratory burst completely	Partially inhibited (60%) respiratory burst if both inracellular and extracellular Ca^{++} chelated, superoxide response inhibited	Partially inhibited (23%) respiratory burst; if both intracellular and extracellular Ca^{++} chelated, superoxide response unchanged

The human PMNL contains four defensins, which are designated HNP-1,2,3, and 4. Homologous peptides, called "lysosomal cationic proteins" were initially discovered approximately 25 years ago by Zeya and Spitznagel in the PMNL of rabbits and guinea pigs.[187] The human defensins are small (Mr < 4,000) peptides that contain 29 (HNP-2), 30 (HNP 1 and 3) or 33 (HNP-4) residues.[188] The known defensins and their primary amino acid sequences are presented in Table II. All defensins contain an invariant cysteine motif composed of six cysteine residues. These form three intramolecular disulfide bonds that fold the peptides into a complexly shaped cyclic structure, as reviewed elsewhere.[188,189] Although HNP-1, 2 and, 3 have identical primary sequences with the exception of their amino-terminal residues, only HNP-1 and HNP-2 kill *C. albicans* effectively *in vitro*.[190] HNP-1 and HNP-2 are remarkably abundant, and together they constitute approximately 4–5% of the total protein in human PMNL and 25–40% of the total protein in their azurophil granules.[191] HNP-4, which is substantially less abundant that HNP-1 and HNP-2, also exerts candidacidal activity *in vitro*.[192]

Rabbit[193] and rat[194,195] PMNL contain six and four defensins, respectively. Several of these defensins include *C. albicans* among their broad antimicrobial spectrum.[74,194,195] Although defensins have not yet been demonstrated in mouse PMNL, mRNA that codes for a typical defensin (cryptdin) is abundantly transcribed by cells that reside in the murine small intestine.[73] Based on *in situ* hybridization studies, the principal cells containing defensin message are epithelial (Paneth) cells located at the base of the intestinal crypts. Interstitial cells, presumably macrophages, in the small intestine's lamina propria also hybridized with the cryptdin probe.

The binding of defensins to *C. albicans* and the correlation of binding with candidacidal activity have been reported.[176] Several human, rabbit, and rat defensins kill *C. albicans* directly,[176,177] while some others were relatively ineffective alone but caused synergistic interactions when combined with submicromolar concentrations of the active peptides.[177] HNP-1 was rendered ineffective against *C. albicans* in an anaerobic environment and its candidacidal effects were also blocked by chemicals that blocked mitochondrial respiration by the yeast cell targets. These observations suggest that oxidative metabolism by *C. albicans* sensitizes it to human defensins. Although the reasons for the links between mitochondrial respiration and defensin-susceptibility remain to be established, recent studies showing that human and rabbit defensins produced voltage-sensitive membrane channels in artificial membrane bilayers may provide some clues.[196] If the *Candida* cell's own transmembrane potential provides the chemical energy used by defensins to create membrane damaging channels, the protective effects of mitochondrial inhibitors such as m-chlorophenylhydrazone (CCCP) or azide might result from a depolarization of the target cell membrane which prevents defensin channel formation.

TABLE II
Primary Amino Acid Sequences of Defensins Extracted from the Leukocytes of Various Species, Illustrating the Consensus of Seven Invariant Amino Acid Residues (Six Cysteines and a Glycine), as Shown by the Standard Single-Letter Code; Generally Conserved Residues are Indicated by Asterisks

Human	HNP-1	A C Y C R I P	A C I A G E R R Y G T C I Y Q G R L W A F C C
	HNP-2	C Y C R I P	A C I A G E R R Y G T C I Y Q G R L W A F C C
	HNP-3	D C Y C R I P	A C I A G E R R Y G T C I Y Q G R L W A F C C
	HNP-4	V C S C R L V	F C R R T E L R V G N C L I G G V S F T Y C C T R V
Guinea pig	GPNP	R R C I C T T R	T C R F P Y R R L G T C I F Q N R V Y T F C C
Rabbit	NP-1	V V C A C R R A	L C L P R E R R A G F C R I R G R I H P L C C R R
	NP-2	V V C A C R R A	L C L P L E R R A G F C R I R G R I H P L C C R R
	NP-3A	G I C A C R R R	F C P N S E R F S G Y C R V N G A R Y V R C C S R R
	NP-3B	G R C V C R K Q L	L C S Y R E R R I G D C K I R G V R F P F C C P R
	NP-4	V S C T C R R F	S C G F G E R A S G S C T V N G V R H T L C C R R
	NP-5	V F C T C R G F	L C G S G E R A S G S C T I N G V R H T L C C R R
Rat	RatNP-1	V T C Y C R R T	R C G F R E R L S G A C G Y R G R I Y R L C C R
	RatNP-2	V T C Y C R S T	R C G F R E R L S G A C G Y R G R I Y R L C C R R
	RatNP-3	C S C R T S	S C R F G E R L S G A C R L N G R I Y R L C C
	RatNP-4	A C Y C R I G	A C V S G E R L T G A C G L N G R I Y R L C C R
Consensus		C C *	C * * G C * C C

Abstracted from Lehrer, *et al.*[188]

In addition to its defensins, the azurophil granules of human PMNL contain several other polypeptides with antibacterial and antifungal activity.[186] The best studied of these are several isozymes of cathepsin G, a 25–28,000-dalton, highly cationic protein with chymotrypsin-like protease activity,[197,198] and a more recently described congener of this protein that has been named "azurocidin."[199–201] Although both azurocidin and cathepsin G are homologous to a large group of serine proteases, the fungicidal activity of CCCP was retained after inactivation of its protease activity,[174,197] and azurocidin itself is not enzymatically active. Recently, Shafer *et al.*[202] described the presence of several antimicrobial domains in cathepsin G and suggested that they contribute to the antimicrobial effects of the holoprotein. Extension of these studies to azurocidin and to *C. albicans* would be of considerable interest.

Other granulocyte constituents that may also contribute to host defense against fungi include lactoferrin and lysozyme (muramidase). Iron-free (apo)lactoferrin can, by tightly complexing iron necessary for optimal fungal growth, inhibit the growth of *C. albicans in vitro*.[63] Apolactoferrin, prepared from human colostrum, has been reported to kill *C. albicans in vitro*.[203] *Coccidioides immitis*[204] and *Cryptococcus neoformans*[205] are reportedly susceptible to killing by lysozyme, a hydrolase found in granulocytes and macrophages of most species, including man.[206] Killing of *C. albicans* by lysozyme is optimal in solutions of low ionic strength,[207] and may be facilitated by glucose.[208]

Finally, nonoxidative killing of *C. albicans* in skin may differ somewhat from that of other tissues. The importance of PMNL (and complement) in host defense against experimental cutaneous candidiasis was established by Ray and Wuepper,[209,210] but McNamara *et al.*[64] obtained evidence for the role of a small cytoplasmic protein in antifungal activity. They studied candidal lesions in the skin of mice. A clear zone existed between many of the more superficial pseudohyphae and the PMNL infiltrate in the deep dermis, suggesting that PMNL produced a soluble factor that inhibited the hyphae. Subsequently, they determined that lysates of PMNL contained a 30 kDa protein that was fungistatic for *C. albicans*.[64,65] The factor differs in several respects from previously described antimicrobial components of PMNL, e.g., it is found in the cytoplasm, not granules, and it appears extracellularly after the death or disruption of the PMNL.[64] Moreover, it appears to be fungistatic and not fungicidal,[65] although this has been contested.[66] It has been postulated that it may play a role in controlling the growth of *Candida* spp. in lesions prior to the time that intact PMNL can exert candidacidal effects.

2.1.3e. Alternation of PMNL Function by C. albicans. Despite the fact that normal PMNL are relatively effective candidacidal agents, the fungus may have considerable potential to subvert PMNL activity. Louria and Brayton[211] demonstrated that phagocytized *C. albicans* yeasts that were not killed by PMNL subsequently germinated, penetrated the cell membrane and actually grew out of the leukocyte. Diamond *et al.*[97,212] demonstrated that *C. albicans* hyphae released small peptides, 2500–3000 kDa, which inhibit the adherence process for not only *C. albicans*, but also *Rhizopus* spp., *Aspergillus* spp., and some bacteria. Smail *et al.*[213,214] reported that a crude hyphal inhibitory product (CHIP) produced by *C. albicans* inhibited superoxide anion production and release of azurophil and specific granule components by FMLP-stimulated PMNL. It is not clear whether or not CHIP and the small peptide(s) described earlier are the same molecules. Hilger and Danley[215] and Danley *et al.*[216] noted that PMNL released decreasing amounts of H_2O_2 when incubated with viable yeasts, whereas increasing amounts were released in response to dead organisms. Although inconclusive, the data suggest the possibility that *C. albicans* may somehow modulate PMNL activity. Finally, the intriguing observations of complement receptors on the surface of *C. albicans*[217–221] have led to various speculations regarding their role in pathogenesis. Receptor expression was reportedly increased when the organisms were grown at glucose levels that prevail in hyperglycemic subjects.[222,223] A relationship between complement receptors and virulence has been established, but their role in pathogenicity has not been determined definitively.

Finally, PMNL appear to promote phenotypic switching in *C. albicans*,[170] an event which may allow the fungus to escape destruction by the PMNL. Phenotypic switching, a phenomenon first described by Slutsky *et al.*,[224] involves inheritable spontaneous changes in cellular and colonial morphology which occur at a relatively high frequency. The relationship between switching and pathogenicity is problematic at this stage, but, since changes in coagglutination and adherence to epithelial cells occur during switching, it has been suggested that switching may play a role in disease manifestations.[225]

2.1.4. Eosinophils

Human eosinophils ingest and kill *C. albicans in vitro* at a rate similar to neutrophils and their phagocytic activity also depends on immunoglobulin and complement components.[162,226,227] Eosinophils from subjects with CGD have an impaired ability to kill *C. albicans.*[162]

2.1.5. Monocytes and Macrophages

2.1.5a. Interactions between Human Mononuclear Cells and C. albicans. In contrast to the strong evidence that identifies PMNL as key cells in innate defenses against systemic *C. albicans* infections, the role of monocyte (MN) and macrophage (MØ) in resistance is not as well defined. Data on interactions of human MØ with *Candida* spp. are relatively limited. The candidacidal activity of MN in patients with various disorders has been studied by relatively few investigators, which is unfortunate given the ease whereby such studies can be performed.[96,228] MN from patients with lepromatous leprosy[229] and AIDS[230] were reported to be markedly deficient in their ability to kill *C. pseudotropicalis*. These interesting findings warrant verification and further exploration. Normal human MN were reported to kill 63.4 ± 10.2% of ingested *C. albicans* in 2.5h when they were tested in the presence of normal PMNL.[96] Under these conditions, normal human MN also killed 46.5 ± 3.7% of ingested *C. parapsilosis* and 65.7 ± 2.9% of ingested *C. pseudotropicalis* in 3h. When normal MN were tested under PMNL-free conditions, they were less effective against *C. albicans* but still killed them about as effectively as purified PMNL. These data indicate that human blood MN have the capacity to contribute significantly to host defense against *C. albicans*.

Human alveolar MØ utilize energy provided by both their oxidative and glycolytic metabolism to ingest *C. albicans*.[231] Although they were able to kill ingested *Listeria monocytogenes*, they failed to kill yeast cells of *C. albicans*.[231,232] Human alveolar MØ killed 18% of *C. pseudotropicalis* in 90 min, with equivalent performances given by cells derived from smokers and nonsmokers.[233] Alveolar MØ from human bone marrow transplant recipients[234] and from subjects with pulmonary alveolar proteinosis[235] were reported to have a decreased ability to kill ingested *C. pseudotropicalis*. MØ from human milk could ingest *C. albicans* in the presence of milk, but their candidacidal activity was not determined.[236] Studies of tissues from patients with systemic candidiasis have shown damaged-appearing *Candida* within fixed MØ.[237] Although such findings are consistent with the intrinsic fungicidal competence of MØ, they could also result from the uptake of *Candida* cells that had been killed by other host factors. MØ that were derived by cultivating human blood MN *in vitro* retained their ability to take up and digest heat-killed *C. albicans*[238] but lost their ability to kill viable *C. albicans* yeast cells.[239,240]

Prostaglandin E_1 and theophylline, substances that modulate intracellular cAMP levels, inhibited the candidacidal activity (*C. albicans*) of human PMNL[241] and MN.[96] *In vitro*, phenylbutazone was a potent inhibitor of human PMNL[94] and monocyte candidacidal[96] and digestive activity.[242] As hydrocortisone lacked such activity in granulocytes,[94,110] the ability of such corticosteroids to inhibit chemotactic entry of leukocytes into *Candida*-infected tissues may underlie its *in vivo* impairment of anti-*Candida* defenses.[243,244] Corticosteroids were reported to inhibit candidacidal activity by human MN *in vitro* in some experiments[245] but not in others.[96] *Escherichia coli* endotoxin reportedly reduced ingestion and degradation of ^{125}I-labeled *C. albicans* by human MN.[246]

The candidacidal mechanisms of human MN are partially understood, based largely on studies with cells that were obtained from patients with hereditary MPO-deficiency or CGD.[96] The ability of MPO-deficient MN to kill *C. albicans* was greatly impaired, but such cells killed both *C. parapsilosis* and *C. pseudotropicalis* more effectively than did normal MN. In sharp contrast, MN that were unable to generate O_2-(superoxide anion) and H_2O_2

(obtained from patients with CGD) failed to kill any of these three *Candida* species. Human MN were also shown to contain components that killed *C. parapsilosis in vitro*,[96] but these have not been identified precisely and their participation in MN-mediated anticandidal mechanisms remains to be demonstrated. In summary, the available data indicate that a) normal human MN can kill *C. albicans* at least as effectively as PMNL, b) their candidacidal mechanisms are linked to the production of oxidants, c) the candidacidal mechanisms of human MN are evidently mediated by both MPO-dependent and MPO-independent pathways, and d) human MN contain potentially candidacidal components. Based on the studies with MPO-deficient MN, MPO-dependent oxidative mechanisms appear to be most important with respect to *C. albicans*, while MPO-independent oxidative mechanisms suffice to enable MN to kill *C. parapsilosis* and *C. pseudotropicalis*. It should be noted that the nonoxidative candidacidal effectors discussed previously with respect to human PMNL, including defensins[247] and cathepsin G[248] are not present, or are present only in small amounts in blood MN.[248]

2.1.5b. In Vivo Studies of Mononuclear Cell–Candida spp. Interactions in Murine Models. Several investigators have used mice treated with macrophage suppressants such as carrageenan[249] or silica[250,251] to abrogate reticuloendothelial activity, or mice with genetically defective macrophage activity,[252] to determine the role of MØ in resistance to systemic candidiasis. Unfortunately, most suppressive treatments induce multiple abnormalities and responses in many genetically altered animals are not limited to a single defect, so that interpretation becomes problematic. For example, carrageenan may increase resistance against systemic candidiasis[249] by virtue of the fact that it stimulates granulocytosis. Athymic nude mice, animals that reportedly have increased numbers of activated MØ,[253] are more resistant to an initial challenge with *C. albicans*,[254] and, further, Cutler and Poor,[84] using a unique chamber implanted subcutaneously into mice,[255] determined that MØ from nude mice were candidacidal, whereas those from normal mice were only candidastatic. The enhanced resistance was only temporary, and death ultimately occurred from infection. Lee and Balish,[250] however, comparing nude mice and their normal littermates treated with silica found no differences between nude and normal mice, i.e., the nude mice were neither more resistant nor more susceptible. Finally, Bistoni *et al.*[256] using splenocytes from mice immunized with an avirulent *C. albicans*, transferred protection to recipients challenged with a virulent *C. albicans* with an adherent spleen cell population.

Other experimental approaches have also been adopted in an attempt to define the contributions of MN and MØ in resistance to candidal infections. In many investigations, cyclophosphamide (CY)[82,257–261] or the related compound, nitrogen mustard[262,263] were used as the suppressive agents. In all cases cited, CY or nitrogen mustard promoted enhanced mortality or tissue destruction when animals were challenged with *C. albicans*. In one of the studies,[258] the investigators were able to demonstrate striking similarities between *in vivo* susceptibility and depressed *in vitro* measures of immunologic function, including assays for PMNL function, natural killer (NK) cell function, and alloreactivity. Moreover, although CY-treated mice were highly susceptible to *C. albicans* challenge a few days following CY administration, 2–3 weeks following CY administration the animals were actually more resistant than normal to challenge with the fungus. This resistance could be explained on the basis of the rebound effect, i.e., as CY is cleared from the system and cell division proceeds, PMNL are produced at a greater-than-normal rate so that their levels exceed normal values for that interval. Since CY and nitrogen mustard are alkylating agents which nonspecifically damage dividing cells, data based on their use must be interpreted with care. PMNL would certainly be the most severely affected, but MN and other lymphoid cells are affected as well.

A number of other investigations have focused on the ability of macrophage activators such as LPS and *Bordetella pertussis*,[264] *Listeria monocytogenes*,[265] Bacille Calmette Guerin (BCG),[251,266,267] *Corynebacterium parvum*,[266,268] or muramyl dipeptide[269–271] when administered *in vivo* to increase resistance against an intravenous challenge with a potentially lethal dose of *C. albicans*. On balance, in most studies the stimulator induced resistance to *C. albicans* that was demonstrable immediately after the candidal challenge, but once a chronic infection was initiated, the disease progressed and many additional deaths occurred. In several cases,[265,270,271] disease progression could be halted if the stimulator was administered repeatedly throughout the course of the experiment.

Some recent observations suggest that immature macrophage precursors may protect mice during experimental *C. albicans* infections. Approximately two weeks after receiving CY, mice developed a highly candidacidal population of spleen cells which were nonadherent to nylon wool and appeared to be macrophage precursors.[258,272,273] Moreover, a similar population was found in the liver of CY-treated animals[274] and in bone marrow cultures established with the L-929 cell line.[275] Using these data and observations concerning cell-surface antigens and the effects of colony-stimulating factor, these investigators concluded that a highly candidacidal but immature precursor macrophage population occurred in the spleen, liver and bone marrow. When they were allowed to differentiate into mature MØ, the cells lost their candidacidal properties.[275,276] The investigators speculated that if colony-stimulating factors were generated *in vivo* during infection, the cells could be maintained in a perpetual, immature (and candidacidal) state that could contribute significantly to anti-*Candida* activity.

Some data concerning *in vivo* candidacidal activity of MØ is available. By 15 min after injection into the tail vein, *C. albicans* could be found within murine hepatic MØ, and after 48h the ingested yeasts displayed altered staining characteristics, and granulomas, possibly glucan-induced, persisted for weeks.[277] Normal rat livers cleared *C. albicans* from perfusates, but their candidacidal activity was negligible. Livers from animals that received prior vaccination with *Corynebacterium parvum* killed approximately 40% of the cleared *C. albicans*. This fungicidal activity was reversed by silica, phenylbutazone, and iodoacetate.[268] Phenylbutazone, but not iodoacetate, was reported to inhibit killing of *C. albicans* by human blood MN.[96] In contrast, iodoacetate, but not phenylbutazone, blocked killing of *C. albicans* by rabbit lung MØ.[278] Such discrepancies emphasize the heterogeneity of mononuclear phagocyte fungicidal mechanisms according to host species and/or tissue location.

2.1.5c. Pulmonary Resistance Mediated by Mononuclear Cells in Rabbit and Murine Models. Because *C. albicans* normally resides in the mouth, aspiration of this organism into the lungs probably occurs with some frequency. Nevertheless, pneumonias caused by *C. albicans* are extremely rare, suggesting that the lungs and respiratory tract are well defended against this organism. Blastospores that are injected into peripheral veins of rabbits, which are highly resistant to infection by *C. albicans*, are cleared principally in the lung, while those injected into mesenteric veins are chiefly removed by the liver.[77] Such hepatic clearance is probably responsible for the initiation of hepatic candidiasis in neutropenic subjects with cancer or leukemia.[279–281]

Experimental models have been developed in rabbits and mice to investigate the potential for the alveolar MØ to deal with *C. albicans*. Rabbits that receive up to 10^8 blastospores intratracheally develop a bronchopneumonia that clears within 5 days without leaving residual damage.[282] Although the mechanisms responsible for pulmonary resistance are not certain, rabbit PMNL contain large concentrations of defensins,[174] several of which are potently candidacidal.[74] Rabbit alveolar MØ can also exert candidastatic and candidacidal activity[283] and possess lysosomal components that are inhibitory to *C. albicans*.[284] Peritoneal and alveolar MØ recovered from rabbits that had been stimulated *in vivo* with complete Freund adjuvant (CFA) showed enhanced candidacidal activity relative to their control, unstimulated counterparts.[278] Patterson-Delafield *et al.*[285] identified two peptides, initially designated as MCP-1 and MCP-2 (macrophage cationic peptide), in rabbit alveolar MØ. The peptides were potently candidacidal *in vitro* against *C. albicans* as well as against a variety of gram-positive and gram-negative bacteria. MCP-1 and MCP-2 were later sequenced[286] and shown to be structurally identical to two of the six defensins (NP-1 and NP-2) that are expressed in rabbit PMNL.[193] Neither MCP-1 nor MCP-2 was present in rabbit blood MN or peritoneal MØ, and the peptides were present in greatly reduced amounts in the pulmonary MØ of neonatal rabbits,[287] which have a reduced ability to kill ingested *C. albicans*.[288] Cell-free fluid recovered from rabbit lungs by bronchoalveolar lavage contains a 10 kDa agglutinin for *C. albicans*[289] whose significance for resistance to pulmonary candidiasis is uncertain.

Other investigators have examined candidacidal activity in rabbit MØ as well. Arai *et al.*[290] compared alveolar MØ obtained from normal and *C. albicans*-immunized rabbits. They reported that immune MØ in immune serum displayed the most effective ingestive powers, and that immune and control MØ displayed equivalent candidacidal activity. Another study[278] compared resident MØ with MØ elicited by prior *in vivo* injection of CFA. Resident (unelicited) alveolar MØ destroyed approximately 28% of ingested *C. albicans* after 4h, whereas resident peritoneal MØ killed 15% after 4h. Peritoneal MØ from animals pretreated with CFA manifested enhanced candidacidal activity (28% in 4h), and CFA-elicited alveolar MØ killed 32% of ingested *C. albicans* after 4h. In a limited study[291] of antecedent BCG stimulation, BCG-elicited rabbit peritoneal and alveolar MØ showed candidacidal activity for *C. albicans* that was similar to that of their unstimulated counterparts. In the study in which intracellular vital staining with methylene blue was employed, 42–45% of ingested organisms were killed in 1h. Using an assay based on [^{3}H] leucine incorporation, there was 71–93% inhibition of macromolecular synthesis by *C. albicans* that had been ingested by alveolar MØ. The latter findings were consistent with fungistatic and/or fungicidal activity.[283]

In mice, Evans[292] reported that *C. albicans* blastospores and hyphae injected intravenously were killed in lungs more effectively than any other organ. Hyphae were eliminated from lung sections by 24h without detectable inflammatory response. Swiss-Webster mice uniformly survived the intratracheal administration of 3×10^6 blastospores, a dosage that killed >85% of the mice if injected intravenously.[293] After a 6h delay, the instilled organisms were cleared rapidly (>90% in 24h), and this rapid pulmonary clearance was inhibited in steroid-treated mice. Intratracheal immunization with *C. albicans* two weeks prior to intratracheal challenge was reported to enhance pulmonary clearance,[294,295] which was thought to be mediated largely by PMNL recruited into the lungs during the infection.

A potentially important contribution to understanding pulmonary resistance to *Candida* infections was made by Nugent and Fick,[296] who reported that cell-free bronchoalveolar lavage fluid from Swiss-Webster mice contained a protein that killed *C. albicans*, *C. tropicalis*, and *C. glabrata*. The protein had an apparent molecular weight of 29 kDa and was not active against *C. parapsilosis* and *C. krusei*. The precise identity of the protein, its cell of origin, and its relationship to similarly sized candidacidal effectors, such as cathepsin G and azurocidin, that have been described in human PMNL remain to be determined.

Several factors contribute to the overall ability of pulmonary mononuclear cells to deal with *C. albicans*. For example, antifungal activity of alveolar MØ may be impaired by antecedent viral infections. Normal mouse alveolar MØ killed 54.5% of *C. krusei* in the presence of nonimmune serum and complement factors, but those from mice with parainfluenza-1 virus infections showed significantly diminished candidacidal ability, despite intact powers of ingestion.[297] Moreover, alveolar MØ obtained from neonatal rhesus monkeys were markedly impaired in their ability to kill *C. albicans* compared to MØ obtained from adult monkeys.[298] Impaired candidacidal activity has also been reported in alveolar MØ obtained from neonatal rabbits[288] and may be a reflection of the low levels of defensins (MCP-1 and MCP-2) and defensin mRNA in these cells.[287]

Brummer and Stevens[299] demonstrated that although resident murine pulmonary MØ lacked the ability to kill ingested *C. albicans* blastospores, MØ that had been cultured overnight with lymph node cells and concanavalin A (con A) or con A-splenocyte supernatants killed 24–43% of the organisms. Recombinant interferon-gamma (IFN-γ) failed to activate candidacidal activity in these MN, although it markedly enhanced the candidacidal performance of peritoneal MØ obtained from the same animals.

2.1.5d. Interaction of Murine Peritoneal Mononuclear Cells with C. albicans. Murine peritoneal MØ have been used in the majority of studies dealing with the activities of animal MØ against *C. albicans*. In an early study with resident mouse peritoneal MØ that had been grown in tissue culture medium for 24h, Stanley and Hurley[300] found that the cells rapidly ingested six species of *Candida* in the presence of 10–20% homologous serum and that within 2h, intracellular replication by germ tubes (*C. albicans*) or budding (other *Candida* spp.) had occurred. By 24h, MØ cultured with *C. albicans* or *C. tropicalis* had been destroyed, while those incubated with *C. parapsilosis*, *C. pseudotropicalis*, *C. krusei*, and *C. guilliermondii* remained viable for an additional 24–48h. The authors concluded that MØ were unlikely to contribute substantially toward host defenses against disseminated candidal infections. Ozato and Uesake[301] used [3H] leucine or [3H] uridine uptake with radioautography to study the ability of caseinate-elicited mouse peritoneal MØ to inhibit *C. albicans*. By 2h, ingested organisms displayed no isotope uptake, despite significant incorporation by extracellular fungi. Intracellular growth of *C. albicans* commenced by 3h and resulted in rapid destruction of the MØ.

In a murine model[84] wherein diffusion chambers with different pore sizes (0.45 and 3 microns) were implanted intraperitoneally, killing of *C. albicans* required phagocytic cells. Thioglycollate-induced peritoneal MØ from BALB/c mice restricted growth of *C. albicans* but did not kill them. MØ from nude mice reduced the number of *C. albicans* by 80–90% in 24h when added at high (> 40:1) macrophage:yeast cell ratios.

An *in vitro* colony-counting assay was used to compare resident peritoneal MØ with cells obtained from LPS- or BCG-treated mice.[302] Resident MØ killed 10% of *C. albicans* in 3h, compared with 27% for LPS-elicited cells and 23% for BCG-elicited cells. In contrast, *C. parapsilosis* was killed much more efficiently, as follows: resident MØ, 72%; LPS-elicited cells, 84%; and BCG-elicited cells, 81%. Muramyl dipeptide, a synthetic adjuvant, increased fungicidal activity of overnight-cultured mouse peritoneal MØ against *C.*

albicans in vivo[269] and *C. parapsilosis in vitro*.[303] Maiti *et al.*[304] also showed that peritoneal MØ obtained from BCG-vaccinated Swiss mice more effectively resisted the intracellular germination of ingested *C. albicans* than did MØ obtained from control mice and that exposure of the MØ to crude lymphokine mixtures enhanced their candidastatic properties. The ability of most ingested *C. albicans* blastospores to survive after their ingestion by resident peritoneal mouse MØ cannot be ascribed to the failure of phagosome-lysosome fusion.[305]

Brummer *et al.*[306] demonstrated that murine MØ pulsed overnight with recombinant murine IFN-γ, but not with LPS, exhibited significantly enhanced fungicidal activity against *C. albicans*, relative to control MØ that had been incubated overnight with medium (44% versus 0% killing). Enhanced candidacidal activity was also observed when the MØ were treated with supernatants from con A-treated spleen cells or with con A itself. This effect was abrogated by including antibody to IFN-γ in the incubation mixtures. Kolotila *et al.*[307] demonstrated that *in vivo* (intraperitoneal) administration of con A to mice enhanced the candidacidal activity of their peritoneal MØ and described sex and strain-specific factors that influenced macrophage performance.

2.1.5e. Summary. In summary, most of the *in vitro* data suggest that activated MØ are more candidacidal than resident cells.[299,302,304,307–310] Unheated serum, with or without *Candida*-specific antibody, promoted phagocytosis,[77,127,268,290,309] but did not appear to increase killing capacity.[127,290] Despite the fact that activated cells appear to be more candidacidal than resident cells, the level of killing observed is modest, even by activated cells, and suggests that many *Candida* survive their initial encounter with these cells. The implications of these seemingly "low" levels of leukocyte-mediated killing on clearing foci of *C. albicans* from infected tissues is considered in Section 2.1.7. Although much remains to be learned about the mechanisms by which MØ kill *Candida* spp., they are likely to include oxygen-dependent[96,240,302,311] and oxygen-independent pathways[96,278,284,285,311–313] whose primacy may vary according to the cell's site and species of origin[261] and their state of activation. Genetic or sex differences such as those observed among mouse strains may also contribute.[272,307] Consequently, the often divergent data summarized in these studies may have multiple causes in addition to those mentioned previously, which include use of different assay conditions, effector: target ratios, incubation periods, viability tests, etc.

2.1.6. NK Cells

Evidence describing potential interactions of NK cells with other host cells and *C. albicans* is reviewed elsewhere in this volume and will not be discussed here.

2.1.7. *Overview of Cellular Defenses*

Elimination of infections caused by *C. albicans* will occur over time if the organisms are killed more rapidly than they reproduce. The experimental data reviewed above demonstrates that phagocytic cells, specifically PMNL, MN, and MØ, have the ability to kill a fraction of the *C. albicans* that they encounter. This fraction may vary from 10 to 50% or more, depending on the type of phagocyte and the experimental conditions employed. A feature (some would say defect) of the *in vitro* assays used to derive these data is that they measure only the initial results of the interaction between the fungus and phagocyte(s). *In vivo*, fungi that survive the initial round of this struggle undoubtedly continue to encounter additional phagocytes until the issue is resolved. Although a net reduction in the number of fungi by 15% or 20% per phagocytic encounter may not seem a highly promising foundation for host defense, the theoretical calculations illustrated in Fig. 1 suggest otherwise. In composing this figure, it was assumed that the average battle cycle between an individual *C. albicans* and a phagocyte lasted for 6h, and that four such interactions occurred each day for each surviving organism. The curves indicate overall *C. albicans* survival, and are predicated on net reductions of 5%, 10%, or 15% of the organisms/cycle. Note that as little as 15% net killing/cycle would effectively clear the infection within a week. Although Figure 4.1 is theoretical, it suggests how relatively small degrees of enhanced candidacidal activity could reap large "dividends" in overall host resistance and why a system that combines candidastatic and modestly effective candidacidal components normally has such high efficacy.

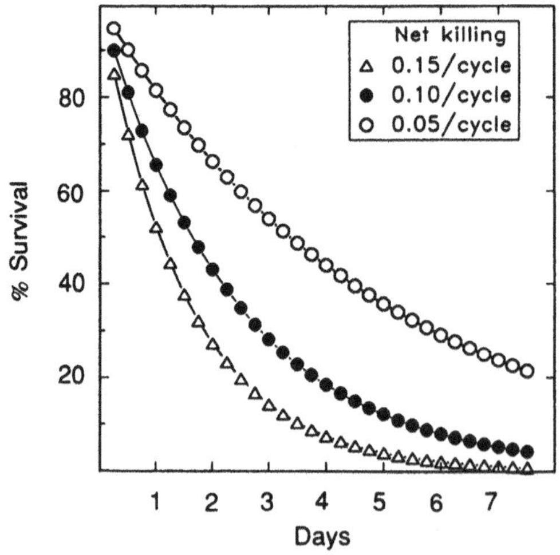

FIGURE 1. This figure is a hypothetical curve for the clearance kinetics of a hypothetical population of *C. albicans* subjected to 4 daily cycles of combat with host phagocytes. To construct these curves, it was assumed that each cycle of *Candida*-phagocyte interaction lasted for 6 h (4 cycles/day) and resulted in the net killing of 5% (○), 10% (●), or 15% (△) of the surviving organisms/cycle. The curves show that "nibbling away" at organisms in this manner can eliminate them and also that relatively small increases in net killing efficiency would greatly accelerate clearance.

2.2. Acquired Immune Responses

2.2.1. Protective Immunity

Despite the fact that a number of investigators have been attempting for many years to dissect specific immune responses to *Candida* spp. to determine their relationship to protection, protective immunity in candidiasis remains an enigma. Confusion over this issue relates to several problems. First, many researchers have investigated innate immunity and attempted to extrapolate the data to acquired immunity. Secondly, several investigations have been carried out wherein mice have been immunized with nonviable blastoconidia[294,314–316] or subcellular preparations[314,315,317,318] of *C. albicans*, which likely have no relevance to the development of protective immunity in the intact individual. In general, greater protective responses have been achieved when animals were immunized with viable preparations of virulent[319,320] or avirulent[256] organisms. Thirdly, it has been recognized only recently that protective immunity against mucocutaneous candidiasis appears to differ fundamentally from that against systemic disease. Fourthly, since defense against primary fungal pathogens such as *Histoplasma capsulatum* and *Coccidioides immitis*, organisms that produce systemic disease, was determined early on to be primarily cell-mediated, there was a built-in bias that immunity in candidiasis should also be cell mediated. The accumulated evidence, however, as summarized below, favors the interpretation that protective immunity in candidiasis is a complex response involving both cellular and humoral mechanisms in conjunction with innate immunity, especially the PMNL. Cellular immunity appears to be more critical in defense against mucocutaneous disease, as summarized in Chapter 5, than against systemic disease.

Evidence supporting a role for antibody in protection against systemic candidiasis has been provided at several levels, namely, immunization designed to elicit antibody prior to challenge,[315,321] reports of transfer of protective responses to naive recipients with serum from immunized animals,[321–324] treatments intended to abrogate the ability of animals to produce antibody prior to immunization and challenge,[325,326] and the correlation of clinical conditions in humans with an antibody response to a specific antigen detected by the Western blotting technique.[327] None of the data have provided definitive proof of the assertion that antibody is protective, however.

For example, in the studies by Mourad and Friedman,[322] antibody was administered repeatedly, and when antibody administration was terminated, mice began dying at a rate similar to that of the controls. In the studies by Pearsall *et al.*,[324] assessment of protection involved the measurement of the thickness of the thigh, comparing treated and control animals; no cultural assays were performed. Since Moser and Domer[325] have shown that two lesions of equal volume can contain very different numbers of organisms, it is not clear whether or not lesion size is an accurate reflection of protection.

Further, Scheld *et al.*[321] were not actually investigating systemic candidiasis, they were working with a model of experimental endocarditis. In such a model, where attachment to the target tissue would be critical to the establishment of disease, one can readily envision the prevention of disease in the presence of antibody to surface antigens. The CY model employed by Moser and Domer,[325] wherein the CY was administered in such a way as to abrogate antibody production but leave cellular immunity intact, provides circumstantial data supporting the role of antibody, as does the study of Matthews *et al.*[327]

Perhaps the model that held the most promise initially in experimental efforts to determine the role of antibody in protection was that involving anti-μ treatment of mice.[326] Here again, however, the data did not provide a definitive answer to the question. Female mice treated with anti-μ were partially protected after immunization but male mice were not. Female mice are innately more resistant to *C. albicans* than males,[328] and the data support the hypothesis that the innate defenses, antibody and cellular immunity are all necessary to complete protection against systemic candidiasis. Several studies involving immunosuppressed animals performed in a single laboratory are summarized in Table III to illustrate the fact that both T cells and antibody appear to be important in successful defense against *C. albicans*, but that the specific role of each has not yet been identified.

Although the studies described above would suggest a role for antibody in protection, there are examples in the literature of studies that illustrate just the opposite.[329–331] Moreover, attempts to illustrate the involvement of antibody in killing *in vitro* have yielded conflicting data. Morelli and Rosenberg[123] and Morrison and Cutler[127] found no evidence for enhanced intracellular killing in the presence of *C. albicans*-specific antibody, whereas Maiti *et al.*[304] and Pereira and Hosking[132] found that immune serum enhanced killing.

Included in Table III are data from a study involving mice thymectomized and irradiated at an early age, then reconstituted with bone marrow.[332] Such mice were neither innately more susceptible nor more resistant to *C. albicans*, but neither were they protected against systemic challenge following immunization. It is clear from this study and others[330,333–336] that the T lymphocyte is involved in acquired immunity, but its precise role remains to be elucidated. In the two most recent publications,[335,336] mice immunized with an avirulent mutant and then challenged with a virulent mutant were given anti-L3T4 or anti-Lyt-2.2 after immunization and immediately prior to reinfection for assessment of a protective response. As assessed by quantitative culture of kidneys, the antibody-treated mice were significantly more susceptible to the virulent *C. albicans* than were mice treated with medium alone. Taking all the data into consideration, however, the effect of the anti-T cell reagents or T-cell products seemed to be one involving the kinetics of the response not the final outcome, in that even the animals treated with anti-T cell reagents eventually cleared *C. albicans* from their tissues.

TABLE III
Correlation of Protective Immune Response to *C. albicans* with Demonstrable Antibody to Cytoplasmic Substances or DH

	Sex	14–28 days post-IV CIE +/TOTAL	14–28 days post-IV ELISA	DH	Protection	Reference
Normal						320
Imm	M	36/39	2.73 (0.17)	Yes	Yes	
Uunimm	M		1.70 (0.00)	No	No	
Thymectomized (TX)						332
TX Imm	M	2/30		No	No	
sham Imm	M	13/30		Yes	Yes	
normal Imm	M	41/90		Yes	Yes	
Cyclophosphamide (CY)-treated						325
CY Imm	M	3/36		Yes	No	
normal Imm	M	35/42		Yes	Yes	
Anti-mu(μ)-treated						326
μ Imm	M	0/5	0.30 (0.30)	Yes	No	
μ Imm	F	0/6	0.17 (0.17)	Yes	Partial	
μ Unimm	M	0/6	0	No	No	
μ Unimm	F	0/10	0.17 (0.17)	No	No	
Mock-Imm	M	7/7	2.88 (0.09)	Yes	Yes	
Mock-Imm	F	7/7	2.68 (0.20)	Yes	Yes	
Mock-Unimm	M	0/10	0.30 (0.30)	No	No	
Mock-Unimm	F	0/11	0	No	No	

In conjunction with studies attempting to detail the role of lymphoid cells in resistance to infection or reinfection, some investigators have considered lymphokine production,[335,337] lymphokine administration,[338] or transfer of immunity with crude mixtures such as dialyzable leukocyte extract.[339] Neutralization of endogenous IFN-γ slowed the rate of clearance of *C. albicans* in immunized animals,[336] but the administration of exogenous IFNγ to unimmunized mice challenged with *C. albicans* was somewhat suppressive, resulting in increased growth of the organism in selected tissues.[338] Administration of IFNγ to immunized mice had no effect, however. The administration of human dialyzable leukocyte extract to mice, effective in only a few instances, appeared to result from nonspecific effects of the extracts.[339] Studies involving the administration of various cytokines to animals in attempts to modify their responses to *C. albicans* are in their infancy, and data generated by such studies are difficult to interpret. It is clear that multiple cytokines are produced during immune responses and several have been shown to have synergistic effects *in vitro*. Only after considerably more is known of *in vivo* cytokine production, can the effects of exogenously applied cytokines be evaluated effectively.

Since the delayed hypersensitivity (DH) response is a measure of one aspect of cellular immunity, several investigators have based their conclusions on the role of T lymphocytes in protection on correlations between DH levels and protection. Kagaya *et al.*,[340] Sinha *et al.*,[341] Cenci *et al.*,[335] and Ashman[342] have all suggested that since DH levels parallel resistance, cell-mediated immunity is clearly involved in protection. Hurtrel *et al.*,[343] however, noted an absence of correlation between DH and protective immunity in the murine system. Unfortunately, the reagents that are used to abrogate T-cell activity do not appear to be highly selective for subsets of T cells, and until it has been clearly established that the T cells involved in DH are the same as those involved in protection, conclusions regarding the relationship between DH and protection are premature.

As indicated previously, a number of investigators have attempted to determine the role of lymphoid cells in protection by assessing response to a primary exposure to *C. albicans* in immunologically modified animals.[72,134,254,333,341,344–351] In many instances, the defects observed, both in genetically deficient animals and in artificially induced immunologically modified animals, are multiple and interpretation of the data is complicated. For example, Cutler[254] and Rogers *et al.*[346] both reported enhanced resistance to primary challenge in congenitally athymic mice, thus suggesting that T cells were not critical to defense against *C. albicans*. On closer inspection, however, the resistance observed was evaluated after challenge with either low or high levels of *C. albicans*, and the data could be explained on the basis of increased numbers of activated macrophages observed in athymic mice, not necessarily T lymphocytes. Interestingly, Mahanty *et al.*,[349] working with the severe combined immunodeficient mouse model, noted little difference between normal and immunodeficient mice and concluded that neither T nor B cells were necessary for innate resistance to a primary challenge with *C. albicans*. Recently, Cantorna and Balish[72] found that only congenitally deficient mice that were defective in both T-cell activity and phagocytic activity were susceptible to the development of mucosal candidiasis that spread systemically.

Finally, several groups of researchers[138,347,350,352–357] have taken a genetic approach to investigations of innate murine susceptibility to *C. albicans*. Not surprisingly, resistant and susceptible mouse strains have been identified and genes affecting the response have been mapped both within[347,356] and outside[354] the major histocompatibility complex. Despite the fact that there are clear differences in innate susceptibility, protective responses following immunization could be demonstrated in all 6 mouse strains investigated by Hector *et al.*[138] An interesting new development in studies of this type is the observation that one strain of genetically susceptible mouse, the CBA/H strain, develops not only weaker pathologic and immune responses to *C. albicans*, but their cells can also be stimulated *in vitro* to develop autoreactive cells, i.e., cells that, when then administered *in vivo*, are cytotoxic to apparently unaltered syngeneic host cells.[357] They suggest that "molecular mimicry" exists between *C. albicans* and host tissue antigens, which may contribute to pathogenesis. Since in-depth genetic studies of resistance to *Candida* spp. are a relatively recent innovation, it is not yet clear how they apply specifically to innate versus acquired defenses.

2.2.2. *Immunoregulation*

It is clear from clinical data[358] that some patients with candidiasis have depressed immune function, especially cellular immunity, demonstrable *in vivo* as skin testing for DH, or *in vitro* by lymphocyte stimulation assays. The most profound immunosuppressive effects are usually *Candida* specific, but some patients show generalized or partial anergy when tested with noncandidal antigens. The relationship between suppressed activity and disease is not clear, in that cause and effect relationships have not been defined. Most patients in whom suppressed responses have been described have had mucocutaneous, not systemic, disease, either in the form of chronic mucocutaneous candidiasis[358] or recurrent vaginitis.[359] The fact that the clinical observations are skewed toward mucocutaneous disease may be misleading with respect to the candidal immunoregulatory phenomena in general, however, in that patients with systemic candidiasis, in contrast to those with mucocutaneous disease, are often not evaluated immunologically.

The working hypothesis for researchers investigating immunoregulatory phenomena is that one or more components of *C. albicans* seeps into the circulation and initiates a series of events that culminate in reduced immunologic activity. In fact, serum factors, presumably responsible for some or all of the immunosuppression observed, have been found in many patients. The nature of the putative inhibitors has not been defined clearly in each case, but there is evidence[358] that a polysaccharide, in some instances mannan specifically, may have been the cause of the immunosuppression in a subset of the patients. Contrary to the polysaccharide theory, however, in the one instance where patients with disseminated disease were investigated,[135] the immunosuppressive agent in the sera of seven patients appeared to be immunoglobulin. The inhibitor in the sera of those patients was demonstrated in a PMNL candidacidal assay. When patient sera were included in the assay, candidacidal activity was reduced from a mean of approximately 18% to $< 5\%$ in most cases. Moreover absorption of patient sera with heat-killed *C. albicans* abrogated suppressive effects.

Experimental, as opposed to clinical, data in support of a role for *C. albicans* or components thereof in immunomodulation, have been generated in both *in vitro* lymphoproliferative[360–369] or antibody generating assays,[365,370,371] and in *in vivo* systems.[372–379] The *in vivo* studies were designed to investigate the immunomodulatory effects of various cell wall extracts on both *Candida*-specific[376,379,380] and noncandidal immune responses.[372–375,377,378]

Two types of *in vitro* assays have been used to demonstrate immunoregulatory phenomena associated with *C. albicans* or extracts thereof. First, human lymphocytes have been employed in lymphocyte stimulation (LS) assays,[366,381] and secondly, murine lymphoid cells have been used in LS or antibody plaque-forming cell (PFC) assays.[364,365] The assays with the human lymphocytes have been difficult to interpret in some instances because many humans have cellular immunity specific for *C. albicans*, and frequently investigators did not determine the underlying sensitivity patterns of their patients. For example, Piccolella *et al.*[381] reported on the "mitogenic" effect of an extract designated MPPS (Mangion purified polysaccharide) for human lymphocytes, which was in all likelihood *Candida*-specific stimulation. With that caveat in mind, however, using the human lymphocyte-MPPS system, the generation of suppressor cells was demonstrated,[363] as well as the production of a nonspecific inhibitor, which blocked interleukin-1 and depressed interleukin-2 production in 5–6-day cultures.[382]

In vivo and *in vitro* murine assays used for investigating immunoregulatory phenomena associated with *C. albicans* or its components are summarized in Table IV. The *in vitro* studies, done initially with formalin-killed cells as modulators, and later with a dithiothreitol cell wall extract, have resulted in the identification of both suppressor B lymphocytes[364,365] and antigen-nonspecific suppressor T lymphocytes.[370,371] The suppressor T cells were capable of inhibiting both primary and secondary *in vitro* antibody responses to the T-dependent antigen sheep erythrocytes (SRBC), and the lymphocytes were determined to be L3T4+, Ly-1+ and Lyt-2−. The antibody response to T-independent antigens, trinitrophenyl-LPS (type I) or trinitrophenyl-Ficoll (type II), were unaffected by the suppressor cell.

The *in vivo* studies have been targeted to investigations of immunoregulatory phenomena associated with both *Candida*-specific and non-*Candida*-specific responses. The initial studies were with whole fungal cells predominantly involving evaluations of nonspecific activity against T-cell dependent antigens or mitogens. Conflicting data were observed, in that Vardinon and Segal[373] noted suppression of anti-SRBC antibody responses, whereas Cutler and Lloyd[375] noted enhancement. Only suppression was observed with respect to T cell mitogens.[364] When soluble extracts were employed, enhancement of antibody responses to noncandidal antigens was noted in all cases where the extracts were glucomannoprotein complexes,[375,378] but when cell wall extracts devoid of the glucan component, i.e., mannan or mannoprotein,[383] were tested, antibody responses were either enhanced or suppressed.[377] The specific type of modulation observed was dependent on the size and charge of the component molecules within the mannan extract. Podzorski *et al.*[368] and Podzorski *et al.*[369] have obtained similar data *in vitro* with human lymphocytes and mannan extracts, i.e., extracts containing saccharides with > 7 saccharide units were stimulatory to lymphocytes from *C. albicans*-sensitized individuals, whereas oligosaccharides with $\leqslant 6$ mannose residues were suppressive.

TABLE IV
Murine Model Systems used for the Demonstration of the Immunodulatory Capacity of *C. albicans* or Components thereof

Modulator	Extraction	Assay	Antigen/mitogen	Effect
In vivo models				
Live cells[373]	N/A	PFC	SRBC	Suppression
Live or killed[375] cells, cell walls,	N/A	PFC	SRBC, CGG-HE	Enhancement in all cases
wall polysaccharide	Phenol			
Formalin-killed[364] cells	N/A	LS	PHA	Suppression
			Con A	Suppression
			LPS	No effect
Cell-wall glycoprotein[376]	Ethylenediamine	DH	Cell-wall glycoprotein	Suppression
Mannan[377]	Peat method	PFC	SSS-III and SRBC	Enhancement and suppression to both antigens
Cell-wall glycoprotein, and peptidoglucomannan[378]	Ethylenediamine	PFC	SSS-III and SRBC	Enhancement in all cases
	NaOH			
Mannan[379]	Peat method	DH	Mannan	Suppression
In vitro models				
Formalin-killed cells[365]	N/A	LS	PHA	Suppression
			LPS	No effect
		PFC	SRBC	Suppression
Formalin-killed cells[370]	N/A	PFC	SRBC	Suppression
			TNP-LPS	No effect
			TNP-Ficoll	No effect
		MLR		No effect
Cell-wall extract[371]	Dithiothreitiol	PFC	SRBC	Suppression

Abbreviations: N/A, not applicable; PFC, antibody plaque-forming cell assay; SRBC, sheep erythrocytes; CGG-HE, chicken gamma globulin-horse erythrocytes; LS, lymphocyte stimulation; PHA, phytohemagglutinin; Con A, concanavalin A; LPS, lipopolysaccharide; DH, delayed hypersensitivity; SSS-III, pneumococcal polysaccharide type III; TNP-LPS, trinitrophenyllipopolysaccharide; TNP-Ficoll, trinitrophenyl-Ficoll; MLR, mixed lymphocyte reaction.

When soluble extracts, specifically glucomannoprotein[376] or mannoprotein,[379] were investigated for their potential to modulate *C. albicans*-specific cellular immunity, mannan-specific responses were suppressed but immune responses detectable with antigens that did not contain mannan were unaffected. The cell responsible for mannan-specific suppression is Thy 1.2+, Ly-1−, and Ly2.1+.[380] Since mannan has been shown to bind to human T lymphocytes,[384] binding of the polysaccharide to murine T cells may initiate the suppressor cell phenomenon. Mannan may actually exert its effect at several levels of the immune response, however. For example, Fischer *et al.*[385] have shown defective handling of mannan by MN, and Wright *et al.*[386] demonstrated inhibitory effects of mannan on PMNL function.

In summary, there is good evidence that subcellular components of *C. albicans* circulate during bouts of serious candidiasis and that suppression of *Candida*-specific and, occasionally, noncandidal immune responses occurs. The nature of the subcellular component may vary, but mannan is a good candidate for at least some of the immunosuppressive activity noted. Mannan can be immunosuppressive in experimental models, depending on the size and charge of the molecules involved, but it can also enhance immune responses, as can all cell wall extracts known to contain glucan as well as mannan tested to date. The mechanisms responsible for suppression have been identified as B and T lymphocytes, depending on the modulator and assay conditions but may also involve mononuclear cells and PMNL. Absolute correlation between the phenomena observed during clinical disease and those observed in model systems cannot be made at this time; additional data from both humans and animal models are needed.

2.3. Immunodiagnosis and Prognosis

The value of serology in the diagnosis of systemic candidiasis remains an enigma, despite the accumulation of 30 years of data on antibody responses and antigen detection. It was recognized early that both colonized and infected individuals regularly produced antibody, i.e., agglutinins, to surface immunogens of *C. albicans*.[387] Since it appeared that agglutinin production would be of little value in separating colonized individuals from those with systemic disease, Taschdjian *et al.*[388] proposed that cytoplasmic antigens would be released from phagocytic cells following ingestion and killing or after senescent extracellular cells died in deep tissues. Considerable effort has been expended, therefore, on attempting to diagnose invasive disease by detecting anticytoplasmic antibodies. Unfortunately, it has proven difficult to correlate cytoplasm-specific antibody levels with disease state, and the search for the unequivocal serological test for the diagnosis of systemic candidiasis continues.

There have been a number of reviews in recent years devoted to, or including, serodiagnosis of candidiasis,[5,17,389–391] some of which contain comprehensive coverage of the historical aspects of the development and application of various immunologic assays. More recently, emphasis has been placed on qualitative rather than quantitative antibody assays and on antigen detection; these aspects of candidal serology will be emphasized below.

In order to develop a definitive test for the serodiagnosis of candidiasis, the investigator must keep several aspects of the disease firmly in mind:

1. Although there are many divisions into which one can place the clinical manifestations of candidiasis, basically there are two broad categories, i.e., cutaneous/mucocutaneous and systemic, and any serodiagnostic technique developed must distinguish between systemic and more superficial forms of the disease.
2. In both types of disease the immune system is exposed to several different forms of the fungus, namely, blastospores (yeasts), pseudohyphae, and hyphae. There is good evidence to suggest that antigens unique to blastospores or hyphae occur at both the cytoplasmic[392–394] and cell wall levels.[395–398] and that the expression of surface antigens is variable under various conditions of growth[399–401] as well as during morphogenesis.[400] Moreover, it was shown many years ago that at least two serotypes of *C. albicans* occur based on the surface antigen, mannan,[37,402] and since *C. albicans* serotype A is not the only *Candida* spp. capable of causing disease, there are concerns that serodiagnostic tests developed with *C. albicans* serotype A may not be applicable to all *Candida* spp. Therefore, not only is the selection of antigen for the serodiagnostic test critical, but the time during the course of disease at which sera are obtained from patients is also critical.
3. The patients from whom sera are being obtained for serodiagnostic tests are frequently immunosuppressed and may have low or nonexistent levels of antibody.
4. The results of many studies in which investigators were attempting to determine the validity of cytoplasmic rather than cell wall extracts as antigens are called into question by the observation that cytoplasmic extracts, whether commercially obtained or prepared in-house, often contain mannan.[403,404]
5. The interpretation of data from many studies is complicated because each investigator tends to use his or her own definition of clinical conditions. For example, if candidemia, a condition that may or may not lead to systemic disease, is included in the category of systemic candidiasis, the conclusions drawn from the data may be quite different than if candidemia is considered separately from truly invasive disease.

2.3.1. *Antibody Assays*

A number of different techniques have been employed to detect anticytoplasmic antibodies in the sera of patients suspected of having, or with proven, systemic candidiasis, including double immunodiffusion (ID) and latex agglutination (LA),[405] counterimmunoelectrophoresis (CIE),[406] crossed immunoelectrophoresis (XIE),[407] and enzyme immunoassay (ELISA).[408,409] Odds[17] catalogued the known studies up to 1988. It is clear that most healthy individuals do not have antibodies to cytoplasmic antigens, even when a particularly sensitive assay such as ELISA was used. To the contrary, many individuals with other diseases, fungal and nonfungal, have antibodies detectable by each of these assays, the highest number of false negatives occurring when the LA assay was used. The assay showing the least false negatives was XIE; it is also the most difficult of the assays listed to perform and adapt to the routine diagnostic laboratory.

The one assay that has probably been most helpful as a diagnostic aid has been CIE.[410–416] Therefore, several laboratories have developed modifications of this technique in attempts to improve its usefulness further. First, Syverson *et al.*[417] reported that the positive predictive value could be increased from 31% to 71% when using XIE. This is a technique involving electrophoresis in two directions. The antigenic mixture is separated in one direction first, then electrophoresed through two additional gels in sequence; the first contains con A to remove mannan and the second contains the patient's serum. Additional studies with this technique supported the claim that if precipitins were detected in the sera of patients after its passage through con A, those patients had high likelihoods of having systemic candidiasis.[418]

A second modification of CIE, designated co-CIE, which has proven useful in some hands[419,420] but not others,[421] was developed in France. This assay involves the demonstration of lines of identity between precipitin reactions resulting from positive control antisera and selected antigenic extracts and those resulting from patient sera tested against the same extracts. The precipitin reaction which appeared most promising is that between a "somatic" antigen, i.e, a soluble antigen prepared as a water extract of whole disrupted cells, and an antiserum generated against germ tubes of *C. albicans*. Unfortunately, in the studies designed to evaluate this technique, a distinction was not made between candidal septicemia and true systemic candidiasis and the value of this technique as a diagnostic and prognostic aid remains to be confirmed.

Because of the difficulty in interpretation of the serologic assays available commercially as kits or for which component reagents are available, one of the major emphases in the last ten years has been on attempting to identify one or more immunodominant antigens to which all or most individuals with systemic candidiasis, but not with superficial colonization or candidemia, would produce antibody. To demonstrate such antigens, most investigators have used immunoblotting, i.e, the Western blotting technique, wherein candidal antigens are electrophoresed in agarose gel and then incubated in patient sera. Reactions between patient sera and selected antigens in the electrophoretic profile are detected by radioimmunoassay or enzyme-linked immunoassay. Jones[422] was the first to report on a candidate antigen of this type in yeast extracts, and several additional groups subsequently reported similar findings in either yeast[423–425] or mycelial[425,426] extracts. There seems to be general agreement among these five investigative groups that most patients with systemic candidiasis produce antibody to an antigen in the 43–54 kDa range, and Porsius *et al.*[427] recently reported that intense reactions to a 47 kDa and a 68 kDa antigen were more often associated with sera from 15 patients with disseminated candidiasis than with sera from 12 individuals who were only colonized. It is not clear whether or not each group of investigators has identified the same antigen in the 43–54 kDa range since no comparative studies have been reported, although the data obtained by Strockbine *et al.*[426] would seem to indicate that their 48 kDa antigen is different from the 54 kDa antigen of Greenfield and Jones.[428] Strockbine *et al.*[426] isolated their antigen from mycelial cells, whereas Greenfield and Jones[428] isolated theirs from yeast cells.

Greenfield and Jones[428] purified their major cytoplasmic antigen, the source of which was yeast cells, using column chromatography with an ion exchange resin followed by con A-sepharose. The purified protein had a molecular weight of 54,300, had 432 amino acid residues, was electrophoretically pure, but had some low level contaminating proteins demonstrable by XIE. When tested for enzymic activity, it was inactive against 19 substrates for common cellular enzymes. Strockbine *et al.*[426] have been attempting to characterize their 48 kDa antigen as well. Initially, it was thought to be actin because preparations of the 48 kDa antigen obtained contained actin. When purified actin was compared to the 48 kDa antigen, however, it was clear that the two proteins were not identical.[429] Subsequently, they demonstrated that the protein was enolase.[430] In support of the enolase data, Franklyn *et al.*[431] prepared cDNA from *C. albicans* yeasts which encoded a portion of the 48 kDa antigen and determined from the deduced amino acid sequence that the 48 kDa protein was an enolase. The 47 kDa antigen of Matthews *et al.*[432] was isolated from patient sera by affinity chromatography using rabbit antisera prepared against a cell homogenate, and by using affinity-purified antibody and immunoelectron microscopy, Matthews *et al.*[433] located the 47 kDa antigen in selected areas within the cell wall and in the peripheral aspects of the cytoplasm.

Despite the studies suggesting a correlation between production of antibody to specific antigens and disease classification, there are reports from other laboratories that are not in agreement with that hypothesis. Manning-Zweerink *et al.*,[434] Reen,[435] and Weller *et al.*,[436] for example, were unable to correlate qualitative antibody responses with clinical condition. In fact, Weller *et al.*[436] observed that both controls and patients made antibody to the 47 kDa antigen. Their primary objective was to evaluate sera from individuals with superficial candidiasis, and they concluded that such individuals responded to antigens of 29, 32, 38, and 65 kDa. Moreover, Greenfield *et al.*,[437] following a series of patients undergoing induction chemotherapy for acute leukemia, found few instances of antibody to their 54 kDa antigen during episodes of invasive candidiasis. It is clear that the identification of immunodominant antigens and their diagnostic and prognostic value for systemic candidiasis remains controversial, although it does hold some promise for the future, perhaps as antigen to which antibodies can be generated for detection of antigenemia.[438]

It has been suggested that antibodies to germ tube antigens may be more diagnostic of active disease than those specific for yeasts only, and an indirect immunofluorescence assay was developed for that specific purpose.[439] Antibodies specific for the yeast phase must be removed from specimens under consideration before testing for the germ tube-specific antibodies. Quindos *et al.*[440,441] used this assay in a retrospective study of various groups of patients and concluded that the test was able to discriminate between systemic candidiasis and other invasive mycoses. Its usefulness as a diagnostic aid for detection of systemic candidiasis remains to be determined in prospective studies including colonized patients, as well as those with superficial disease.

Finally, a novel assay has been described for the detection of antibodies to different antigens in a single assay, the usefulness of which is unknown at this stage, but it has potential for future development. McHugh *et al.*[442] coated polystyrene beads of varying diameters with three separate antigenic preparations of *C. albicans*, namely, a whole cell extract, a cytoplasmic extract and a cell wall polysaccharide, and tested patient sera in an assay that allowed for the separation of beads by flow cytometry. There are all sorts of possibilities for alterations in the protocol presented that could ultimately lead to a very specific assay for diagnosis of systemic candidiasis based on antibody responses to specific antigens.

It appears to be the consensus of those most heavily involved in serodiagnostic assay performance and development that antibody detected in any one of these assays, when evaluated in the context of the patient's clinical condition and other laboratory parameters, such as culture, can be helpful diagnostically and prognostically, but when evaluating only a single specimen, antibody data generated by these techniques are nearly impossible to interpret.

2.3.2. Antigen Detection

Early diagnosis of systemic candidiasis is necessary if effective therapy is to be instituted. It is clear that detection of antibody has not provided appropriate assays for early diagnosis of this clinical condition, and it was reasoned that detection of antigen might provide more rapid and definitive diagnostic data. Emphasis for antigen detection has been with two types of preparations, namely, antimannan antibody and antibody directed against a heat-labile antigen(s) that does not appear to be mannan. The first antigen targeted for detection was mannan, and hemagglutination inhibition,[443]

CIE,[444] radioimmunoassay,[445] ELISA-inhibition,[409,446] and ELISA[447] assays were developed for trials. In the original study,[443] only 4 of 14 individuals with proven systemic candidiasis had demonstrable mannan antigenemia. Kerkering *et al.*[444] had somewhat better results, in that eight out of 13 patients who eventually developed systemic candidiasis were positive. In yet another study, only five of 11 patients with systemic candidiasis had circulating mannan detectable by RIA.[445]

Since mannan can be present in sera in immune complexes, several investigators have suggested dissociative methods[447] or heat extraction[448] for increasing the likelihood of mannan detection. Although investigators do use these dissociative steps,[449,450] it is not clear that the sensitivity of the test is greatly increased. Others[450,451] have suggested that serial assays for mannan are important, perhaps because mannan has a relatively short half-life in serum[422] and is not released continuously from sites of infection.[450] There is a single mannan-detection kit available commercially in the United States, the LA-*Candida* Antigen Detection System, Immuno-Mycologics, Inc., Norman, Oklahoma. That kit appears to be relatively insensitive in that none of the patients tested by Phillips *et al.*,[452] which included 33 patients with candidemia, none of those tested by Fung *et al.*,[453] which included 9 patients with invasive candidiasis, and none tested by Bisbe *et al.*,[416] which included 36 heroin addicts diagnosed as having systemic candidiasis, were positive. Many of those same patients tested for a putatively different antigen (see below) were positive. Several groups[449,454] have detected mannan by ELISA with considerably more success in terms of higher titers, but increasing the sensitivity decreased the specificity with regard to invasive candidiasis and its distinction from colonization. For example, Fujita *et al.*[454] detected mannan in nine of 10 patients with systemic candidiasis, but it was also detected in two of four patients who had superficial candidiasis. Fujita *et al.*[454] correlated their antigenemia studies with observations of antibody; antigenemia preceded rising antibody titers by 6 to 23 days.

A second antigen detection system, originally developed by Gentry *et al.*[455] and now marketed as the CAND-TEC™ by Ramco Laboratories, Inc., Houston, Texas, has been evaluated rather extensively by several different laboratories.[449,452,453,456–458] In one study,[452] only patients with candidemia were evaluated, along with patients at risk and healthy subjects, but in all others,[449,453,456–458] patients with proven invasive candidiasis were included. Piens *et al.*,[456] Ness *et al.*,[457] and Escuro *et al.*[458] determined the sensitivity to be 38%, 55%, and 76%, the specificity to be 90%, 29%, and 93%, the positive predictive value to be 50%, 17%, and 50%, and the negative predictive value to be 85%, 71%, and 100%, respectively.

The data available suggest that when using the CAND-TEC system, a titer of $\geq$ 1:4 is equivocal and $\geq$ 1:8 is evidence of disseminated disease.[453,455] If that criterion was applied, sensitivity in the data reported by Piens *et al.*[456] would fall to 0%, since none of their patients had titers $>$ 1:4. In a comparative study of the CAND-TEC and ELISA assays, only four of 14 patients with disseminated candidiasis were positive and 4/22 patients were positive by ELISA using 1:8 as the cutoff point. Others have suggested that rheumatoid factor[449] and high creatinine levels[457] interfere with the test, but Price and Gentry[459] found no interference with the test in patients undergoing dialysis, and they[460] determined in their survey that patients with rheumatoid factor and no antigen had titers of $\leq$ 1:2. The latter investigators, as well as others,[450] suggested also that the testing of sequential specimens would increase the likelihood of detecting antigen, but Escuro *et al.*,[458] using weekly sampling, were not able to improve their detection rate. Pretreatment of serum with protease and heat improved results here as well.[461] One clinical condition where the CAND-TEC system would appear to be very valuable is in the diagnosis of pulmonary candidiasis,[462] a condition that, in general, is very difficult to diagnose. Fifteen of 16 patients with clinical and laboratory evidence of *C. albicans* pneumonia had positive CAND-TEC titers when bronchoalveolar lavage fluids were tested.

At least two additional antigen detection systems have been described but neither appears to have been tested in other laboratories, and their potential as candidate tests for general use is unknown.[463,464] Moreover, Western blotting has been used recently to detect candiduria in patients with multiple positive blood cultures.[465] The results of this one study suggest that detection of antigenuria may be useful diagnostically and prognostically, but more data are clearly needed to confirm the hypothesis.

2.3.3. *Summary*

It is obvious from the above discussion that the development of serodiagnostic tests for the diagnosis and prognosis of systemic candidiasis is a high priority in a number of laboratories. It is also obvious that no one test has yet been devised that is appropriately sensitive and specific for application to the routine clinical laboratory. For those laboratories equipped to do it, CIE or XIE appear the most reliable antibody assays, whereas an ELISA assay for mannan appears to be the most sensitive for antigen detection. As stated by Odds,[17] however, "[Serodiagnosis] is one of the fastest developing aspects of candidal research. Those who devise novel serological approaches to diagnosis usually present their work in a positive fashion: those who apply them in clinical practice are usually more or less disappointed."

2.3.4. *Cell-Mediated Immunity (CMI)*

2.3.4a. CMI in Humans. A high percentage of normal individuals, ranging from 54 to 94%, depending on the survey,[466–470] have DH demonstrable by skin testing with *C. albicans* antigens. In one of the more extensive studies, Shannon *et al.*[467] noted 80%, 89%, and 83% positive rates for children 7–12 months of age, 1–5 years of age, and 5–15 years of age, respectively, and 94%

in adults. Because of the high response rate in the general population, skin testing is of little value for the diagnosis or prognosis of systemic candidiasis, although skin test response and response rates in *in vitro* cell-mediated assays have been used to categorize patients with chronic mucocutaneous candidiasis.[471]

Since there is such a high rate of responders in the population-at-large, a *C. albicans* extract is usually included in the battery of common recall antigens administered to patients in whom immunodeficiency is suspected. The antigen most frequently employed is a poorly defined *C. albicans* extract obtained from Hollister-Stier (Spokane, WA). Caution should be exercised in its use if applied repeatedly to the same patient, however, in that Hogan *et al.*[470] have noted increased responsiveness after sequential skin test applications. It would be ideal to have available a reliable *in vitro* assay that could be employed to determine cellular immunity to *C. albicans*, and to that end, several groups of investigators have compared skin test results in normal patients with *in vitro* correlations of DH such as LS[469,472] and migration inhibition assay.[468,472] Close correlations were found between skin test reactivity and LS assays, but contradictory data were obtained using the migration inhibition assay.

The Hollister-Stier antigen has been used for LS *in vitro*,[469,473] and other investigators have used other commercial antigens[472] or extracts prepared in the laboratory.[474] Most of these extracts are poorly defined complex mixtures. The development of new antigenic preparations for *in vivo* or *in vitro* testing has moved quite slowly, in fact, although Esch and Buckley[475] reported several years ago on a new type of extract they tested in humans. It was an ammonium sulfate fractionation of an aqueous extract of lyophilized *C. albicans* yeast. Fifty percent of a cohort of the normal population responded with reactions of $\geq$ 5 mm at 48h when tested with 1 μg of the preparation.

2.3.3b. CMI in Experimental Models. Several different preparations, including heat-killed cells,[476,477] cell wall extracts,[379,478–482] and cytoplasmic components[481,483] have been used in animal models in attempts to develop skin test antigens or correlate DH with other immune responses in animals. Evaluation of these antigens as potential candidates for human use has been complicated by the fact that, in a number of instances, the antigens were tested in animals immunized with nonviable extracts rather than infected with viable organisms.[478–480,483] Further, skin test results with particulate antigens, as when heat-killed cells are employed, are difficult to interpret because of the possibility of B lymphocytes producing antibody *in situ*,[484] and/or foreign body reactions resulting in complement activation, the latter of which could result in enhanced antigen retention and slowed antigen degradation. Moreover, in one human study,[485] the injection of whole cells elicited a strong eosinophil response that persisted through 24h.

Cell wall and membrane components, as opposed to cytoplasmic preparations, appear to be the best antigens for *in vivo* testing described to date in infected animals.[379,479,481,482] Several types of cell wall preparations have been used, ranging from glycoproteins extracted with ethylenediamine (EDA)[481,482] and cold dilute alkali,[479] to mannans extracted by the Peat *et al.* method.[379,383] Domer and Moser[481] compared many of these extracts, cell wall and cytoplasmic, the latter of which included both ribosomes and soluble cytoplasmic substances, in a murine model. A membrane extract and several cell wall preparations elicited the best responses. Many responses, however, were complicated by large 4h reactions. In subsequent studies, however, it was possible to demonstrate that if the cell wall glycoprotein extracted with EDA was subjected to extensive dialysis with 1 M NaCl to ensure removal of all ethylenediamine,[379] it no longer elicited the large early responses. Moreover, testing of infected mice with mannan extracted by the Peat *et al.* method[383] did not result in large early responses either, although such responses had been demonstrated in guinea pigs.[480] The lack of early response in infected mice may reflect the fact that mice respond poorly to mannan with respect to antibody production.[486] The response to the EDA-extracted glycoprotein was destroyed by treatment with proteolytic enzymes and partially ablated by periodate oxidation of the preparation.[481] Moreover, the *in vivo* responses to that glycoprotein, as well as to a membrane preparation extracted with hot phosphate-buffered saline, could be transferred with T lymphocytes.[481,482]

Many of the preparations used for *in vivo* testing have been tested for their applicability in *in vitro* tests, especially for LS assays. Moser *et al.*,[482] for example, compared EDA-extracted cell wall glycoprotein, a soluble membrane extract, and soluble cytoplasmic substances, for their ability to stimulate lymphocytes from infection mice. Responses to the cell wall and membrane extracts correlated well with DH responses *in vivo*, but surprisingly, the cytoplasmic preparation, a preparation which was poorly reactive *in vivo*, induced strong proliferative responses *in vitro*. These studies confirmed an earlier study with human cells in which cytoplasmic and cell wall antigens were compared, in that the EDA-extracted material elicited the best proliferative responses.[474] Although the EDA-extracted glycoprotein induced lymphocytes to proliferate *in vitro*,[482,487] as did a phosphorylated mannoprotein complex from the cell wall of *C. albicans*[488] and an ethanol-precipitated and deproteinized extract,[489] mannan did so poorly, if at all, with human[488,490,491] or murine lymphocytes.[487] Tollemar *et al.*[491] noted a much better response in peripheral blood lymphocytes to cytoplasmic antigens, as did Gettner and Mackenzie.[490] The conflict between the *in vivo* and *in vitro* data with mannan, may relate to the extraction procedure. Most investigators have used the method of Peat *et al.*[383] Podzorski *et al.*[369] tested a mannan extracted with cetryltrimethylammonium bromide and found it to be a potent stimulator of *in vitro* lymphoproliferation for human lymphocytes. Small oligosaccharides derived from that mannan were not stimulatory, however. Perhaps smaller molecular weight components in the Peat *et al.*[383] extract act to depress the response *in vitro* but not *in vivo*.

2.3.4c. Summary. It is clear from many studies that a high percentage of humans develop cellular immunity demonstrable *in vivo* by skin testing and *in vitro* by testing for proliferation of lymphocytes when testing is done with a number of different antigenic mixtures. It is also quite clear that detection of cellular immune responses is hampered by a lack of standardization in methodology and in the antigenic preparations employed. The preparations available commercially are poorly defined complex mixtures, and those developed in the laboratory and tested under experimental conditions have not yet been tested in prospective studies with normal subjects and human patients with various forms of candidiasis. It is difficult, in fact, based on the experimental work, to make generalizations as to which of the many preparations tested should be targeted for development for *in vivo* or *in vitro* testing, although it would appear that mannan should not be considered. Other cell wall antigens appear more promising, but since many normal individuals have circulating antibody to cell wall antigens, large, immediate hypersensitivity reactions might be expected with cell wall antigens.

2.4. Therapy of Systemic Candidiasis

The treatment of systemic candidiasis generally consists of measures directed against the underlying condition(s) (e.g., neutropenia or leukemia) and administration of an antifungal agent. Because difficulties are often encountered in establishing the diagnosis and because the infection often occurs in complex clinical settings, clinicians must be guided by the "art" of medicine, as well as its "literature," in deciding when to initiate antifungal treatment, what to use, and how long to continue the therapy.

Amphotericin B, a polyene antibiotic produced by *Streptomyces nodosus*, is the current mainstay of therapy. Amphotericin B interacts with fungal plasma membrane sterols and renders the membrane leaky.[492] Resistant *Candida* spp. organisms are uncommon but have been described. Because amphotericin B is insoluble, the intravenous form of this drug is prepared as a complex with deoxycholate. The intravenous administration of amphotericin B-deoxycholate is associated with considerable host toxicity and regularly causes fever, chills, impaired renal function, and anemia. Bennett[493] has provided excellent guidelines for its clinical use. Considerable interest attends the development of alternative, less toxic liposomal formulations of amphotericin B, some of which are in current clinical trials.

When used in combination with amphotericin B, 5-fluorocytosine may have utility in treating candidal meningitis.[494] This agent is converted by deamination and phosphorylation to 5-fluorodeoxyuridylic acid, a competitive inhibitor of thymidylate synthetase,[495] and is also incorporated extensively into fungal RNA.[496] Primary resistance of *C. albicans* to the agent is not uncommon.[496]

Imidazoles (e.g. ketoconazole) and the closely related triazoles (e.g., fluconazole) inhibit fungal ergosterol production by inhibiting cytochrome P_{450}-dependent 14-α-demethylase.[497] The drugs are effective after oral administration and are generally much less toxic than amphotericin B-deoxycholate complex. Fluconazole attains excellent penetration into the cerebrospinal fluid.[498] Although it has given promising results in the treatment of oropharyngeal candidiasis in patients with cancer[499] and AIDS,[500] its efficacy in systemic candidiasis is unproven. Perhaps, as with the earlier imidazole drug ketoconazole,[501–504] its primary utility will be in chemoprophylaxis.

For those seeking a broader description of the clinical syndromes of systemic candidiasis and their treatments, the excellent recent reviews by Edwards[504] and Meunier[505] are highly recommended.

ACKNOWLEDGMENTS. This work was supported in part by Public Health Service grants AI-12806 and AI-22839 from the National Institute of Allergy and Infectious Diseases. We thank Ersebet Huffman and Sally Grogono for expert assistance in manuscript preparation.

This chapter was completed in January 1991.

REFERENCES

1. Cohen, R., Roth, F. J., Delgado, E., Ahearn, D. G., and Kalser, M. H., 1969, Fungal flora of the normal human small and large intestine, *N. Engl. J. Med.* **280**:638–641.
2. Draser, B. S., and Hill, M. J., 1974, *Human Intestinal Flora*, Academic Press, New York.
3. Kennedy, M. J., 1989, Regulation of *Candida albicans* populations in the gastrointestinal tract: mechanisms and significance in GI and systemic candidiasis, *Curr. Top. Med. Mycol.* **3**: 315–402.
4. Budtz-Jorgensen, E., 1990, Etiology, pathogenesis, therapy, and prophylaxis of oral yeast infections, *Acta Odontol. Scand.* **48**:61–69.
5. Domer, J. E., and Carrow, E. W., 1989, Candidiasis, in: *Immunology of the Fungal Diseases* (R. A. Cox, ed.), CRC Press, Inc., Boca Raton, Florida, pp. 57–91.
6. Young, R. C., Bennett, J. E., Geelhoed, G. W., and Levine, A. S., 1974, Fungemia with compromised host resistance. A study of 70 cases, *Ann. Intern. Med.* **80**:605–612.
7. Parker, J. C., Jr., McCloskey, J. J., and Knauer, K. A., 1976, Pathobiologic features of human candidiasis. A common deep mycosis of the brain, heart and kidney in the altered host, *Am. J. Clin. Pathol.* **65**:991–1000.
8. Horn, R., Wong, B., Kiehn, T. E., and Armstrong, D., 1985, Fungemia in a cancer hospital: changing frequency, earlier onset and results of therapy, *Rev. Infect. Dis.* **7**:646–655.
9. Pirsch, J. D., and Maki, D. G., 1986, Infectious complications in adults with bone marrow transplantation and T-cell depletion of donor marrow, *Ann. Intern. Med.* **104**:619–631.
10. Tollemar, J., Ringden, O., Bostrom, L., Nilsson, B., and Sundberg, B., 1989, Variables predicting deep fungal infections in bone marrow transplant recipients, *Bone Marrow Transplantation* **4**:635–641.
11. Tapper-Jones, L., Aldred, M. J., Walker, D. M., and Hayes, T. M., 1981, Candidal infections and populations of *Candida albicans* in the mouths of diabetics, *J. Clin. Pathol.* **34**:706–711.
12. Prasad, J. K., Feller, I., and Thomson, P. D., 1987, A ten-year review of *Candida* sepsis and mortality in burn patients, *Surgery* **101**:213–216.
13. Ryley, J. F., 1986, Pathogenicity of *Candida albicans* with particular reference to the vagina, *J. Med. Vet. Mycol.* **24**:5–22.

14. Montgomerie, J. Z., and Edwards, J. E., Jr., 1978, Association of infection due to *Candida albicans* with intravenous hyperalimentation, *J. Infect. Dis.* **137**:197–201.
15. Karabinis, A., Hill, C., Leclercq, B., Tancrede, C., Baume, D., and Andremont, A., 1988, Risk factors for candidemia in cancer patients: a case-control study, *J. Clin. Microbiol.* **26**: 429–432.
16. Bross, J., Talbot, G. H., Maislin, G., Hurwitz, S., and Strom, B. L., 1989, Risk factors for nosocomial candidemia: a case-control study in adults without leukemia, *Am. J. Med.* **87**: 614–620.
17. Odds, F. C., 1988, *Candida and Candidosis*, Bailliere Tindall, London, pp. 93–114.
18. Pfaller, M. A., 1988, Nosocomial *Candida* infections, *Curr. Opin. Infect. Dis.* **1**:764–771.
19. Rippon, J. W., 1988, *Medical Mycology, The Pathogenic Fungi and the Pathogenic Actinomycetes*, W. B. Saunders Company, Philadelphia, PA, pp. 532–581.
20. Spencer, P. M., and Jackson, G. G., 1989, Fungal and mycobacterial infections in patients infected with the human immunodeficiency virus, *J. Antimicrob. Chemother.* **23**(suppl. A): 107–125.
21. Klein, R. S., Harris, C. A., Small, C. B., Moll, B., Lesser, M., and Friedland, G. H., 1984, Oral candidiasis in high-risk patients as the initial manifestation of the acquired immunodeficiency syndrome, *N. Engl. J. Med.* **311**:354–358.
22. Tavitian, A., Raufman, J.-P., and Rosenthal, L. E., 1986, Oral candidiasis as a marker for esophageal candidiasis in the acquired immunodeficiency syndrome, *Ann. Intern. Med.* **104**: 54–55.
23. Guentzel, M. N., Cole, G. T., and Pope, L. M., 1985, Animal models for candidiasis, *Curr. Top. Med. Mycol.* **1**:57–116.
24. Kreger-van Rij, N. J. W., 1984, *The Yeasts. A Taxonomic Study*, Elsevier Science Publishers B. V., Amsterdam, pp. 585–844.
25. Kiehn, T. E., Edwards, F. F., and Armstrong, D., 1980, The prevalence of yeasts in clinical specimens from cancer patients, *Am. J. Clin. Path.* **73**:518–521.
26. Wingard, J. R., Merz, W. G., and Saral, R., 1979, *Candida tropicalis*: a major pathogen in immunocompromised patients, *Ann. Intern. Med.* **91**:539–543.
27. Meunier-Carpentier, F., Kiehn, T. E., and Armstrong, D., 1981, Fungemia in the immunocompromised host. Changing patterns, antigenemia, high mortality, *Am. J. Med.* **71**:363–370.
28. Walsh, T. J., and Merz, W. G., 1986, Pathologic features in the human alimentary tract associated with invasiveness of *Candida tropicalis*, *Am. J. Clin. Pathol.* **85**:498–502.
29. Howlett, J. A., 1976, The infection of rat tongue mucosa *in vitro* with five species of *Candida*, *J. Med. Microbiol.* **9**:309–316.
30. Allen, C. M., and Beck, F. M., 1987, Differences in mucosal reaction related to *Candida albicans* isolates, *J. Oral Pathol.* **16**:89–93.
31. Martin, M. V., Craig, G. T., and Lamb, D. J., 1984, An investigation of the role of true hypha production in the pathogenesis of experimental oral candidosis, *Sabouraudia* **22**:471–476.
32. Sobel, J. D., Muller, G., and Buckley, H. R., 1984, Critical role of germ tube formation in the pathogenesis of candidal vaginitis, *Infect. Immun.* **44**:576–580.
33. Walsh, T. J., and Pizzo, P. A., 1988, Nosocomial fungal infections: a classification for hospital-acquired fungal infections and mycoses arising from endogenous flora or reactivation, *Ann. Rev. Microbiol.* **42**:517–545.
34. Hadfield, T. L., Smith, M. B., Winn, R. E., Rinaldi, M. G., and Guerra, C., 1987, Mycoses caused by *Candida lusitaniae*, *Rev. Infect. Dis.* **9**:1006–1012.
35. Dick, J. D., Rosenguard, B. R., Merz, W. G., Stuart, R. K., Hutchins, G. M., and Saral, R., 1985, Fatal disseminated candidiasis due to amphotericin B-resistant *Candida guilliermondii*, *Ann. Intern. Med.* **102**:67–68.
36. Pappagianis, D., Collins, M. S., Hector, R., and Remington, J., 1979, Development of resistance to amphotericin B in *Candida lusitaniae* infecting a human, *Antimicrob. Agents Chemother.* **16**:123–126.

37. Hasenclever, H.F ., and Mitchell, W. O., 1961, Antigenic studies of *Candida*. I. Observations of two antigenic groups in *Candida albicans*, *J. Bacteriol.* **82:**570–573.
38. Odds, F. C., and Abbott, A. B., 1980, A simple system for the presumptive identification of *Candida albicans* and differentiation of strains within the species, *Sabouraudia* **18:**301–317.
39. Odds, F. C., and Abbott, A. B., 1983, Modification and extension of tests for differentiation of *Candida* species and strains, *Sabouraudia* **21:**79–81.
40. McCreight, M. C., and Warnock, D. W., 1982, Enhanced differentiation of isolates of *Candida albicans* using a modified resistogram method, *Mykosen* **25:**589–598.
41. Polonelli, L., Archibusacci, C., Sestito, M., and Morace, G., 1983, Killer system: a simple method for differentiating *Candida albicans* strains, *J. Clin. Microbiol.* **17:**774–780.
42. Roman, M. C., and Linares Sicilia, M. J., 1983, Preliminary investigation of *Candida albicans* biovars, *J. Clin. Microbiol.* **18:**430–431.
43. Lee, W., Burnie, J., and Matthews, R., 1986, Fingerprinting *Candida albicans*, *J. Immunol. Methods* **93:**177–182.
44. Phongpaichit, S., Mackenzie, D. W. R., and Fraser, C., 1987, Strain differentiation of *Candida albicans* by morphotyping, *Epidemiol. Infect.* **99:**421–428.
45. Magee, B. B., D'Souza, T. M., and Magee, P. T., 1987, Strain and species identification by restriction fragment length polymorphisms in the ribosomal DNA repeat of *Candida* species, *J. Bacteriol.* **169:**1639–1643.
46. Scherer, S., and Stevens, D. A., 1987, Application of DNA typing methods to epidemiology and taxonomy of *Candida* species, *J. Clin. Microbiol.* **25:**675–679.
47. Magee, B. B., and Magee, P. T., 1987, Electrophoretic karyotypes and chromosome numbers in *Candida* species, *J. Gen. Microbiol.* **133:**425–430.
48. Merz, W. G., Connelly, C., and Hieter, P., 1988, Variation of electrophoretic karyotypes among clinical isolates of *Candida albicans*, *J. Clin. Microbiol.* **26:**842–845.
49. Vaudry, W. L., Tierney, A. J., and Wenman, W. M., 1988, Investigation of a cluster of systemic *Candida albicans* infections in a neonatal intensive care unit, *J. Infect. Dis.* **158:**1375–1379.
50. Burnie, J. P., Odds, F. C., Lee, W., Webster, C., and Williams, J. D., 1985, Outbreak of systemic *Candida albicans* in intensive care unit caused by cross infection, *Brit. Med. J. (Clin. Res.)* **290:**746–748.
51. Odds, F. C., Palacio-Hernanz, A., Cuadra, J., and Sanchez, J., 1987, Disseminated *Candida* infection syndrome in heroin addicts—dominance of a single *Candida albicans* biotype, *J. Med. Microbiol.* **32:**275–277.
52. Neely, A. N., Odds, F. C., Basatia, B. K., and Holder, I. A., 1988, Characterization of *Candida* isolates from pediatric burn patients, *J. Clin. Microbiol.* **26:**1645–1649.
53. Fox, B. C., Mobley, H. L. T., and Wade, J. C., 1989, The use of a DNA probe for epidemiological studies of candidiasis in immunocompromised hosts, *J. Infect. Dis.* **159:** 488–494.
54. Pfaller, M., 1989, Infection control: opportunistic fungal infections—the increasing importance of *Candida* species, *Infect. Control Hosp. Epidemiol.* **10:**270–273.
55. Stevens, D. A., Odds, F. C., and Scherer, S., 1990, Application of DNA typing methods to *Candida albicans* epidemiology and correlations with phenotype, *Rev. Infect. Dis.* **12:** 258–266.
56. Soll, D. R., Staebell, M., Langtimm, C., Pfaller, M., Hicks, J., and Gopala Rao, T. V., 1988, Multiple *Candida* strains in the course of a single systemic infection, *J. Clin. Microbiol.* **26:** 1448–1459.
57. Burkholder, P. R., 1943, Vitamin deficiencies in yeasts, *Am. J. Botany* **30:**206–211.
58. Drouhet, E., and Vieu, M., 1957, Facteurs vitaminiques de croissance des *Candida*, *Ann. Inst. Pasteur* **92:**825–831.
59. Caroline, L., Taschdjian, C. L., Kozinn, P. J., and Schade, A. L., 1964, Reversal of serum fungistasis by addition of iron, *J. Invest. Dermatol.* **42:**415–419.

60. Masson, P. L., Heremans, J. F., and Schonne, E., 1969, Lactoferrin, an iron-binding protein in neutrophilic leukocytes, *J. Exp. Med.* **130:**643–658.
61. Bretz, U., and Baggiolini, M., 1974, Biochemical and morphological characterization of azurophil and specific granules of human neutrophilic polymorphonuclear leukocytes, *J. Cell Biol.* **63:**251–269.
62. Rado, T. A., Wei, X., and Benz, E. J., Jr., 1987, Isolation of lactoferrin cDNA from a human myeloid library and expression of mRNA during normal and leukemic myelopoiesis, *Blood* **70:**989–993.
63. Kirkpatrick, C. H., Green, I., Rich, R. R., and Schade, A. L., 1971, Inhibition of growth of *Candida albicans* by iron-unsaturated lactoferrin: relation to host-defense mechanisms in chronic mucocutaneous candidiasis, *J. Infect. Dis.* **124:**539–544.
64. McNamara, M. P., Wiessner, J. H., Collins-Lech, C., and Hahn, B. L., 1988, Neutrophil death as a defence mechanism against *Candida albicans* infections, *Lancet* **2:**1163–1165.
65. Sohnle, P. G., and Collins-Lech, C., 1990, Comparison of candidacidal and candidastatic activities of human neutrophils, *Infect. Immun.* **58:**2696–2698.
66. Steinbakk, M., Naess-Andresen, C.-F., Lingaas, E., Dale, I., Brandtzaeg, P., and Fagerhol, M. K., 1990, Antimicrobial actions of calcium binding leucocyte L1 protein, calprotectin, *Lancet* **336:**763–765.
67. Knight, L., and Fletcher, J., 1971, Growth of *Candida albicans* in saliva; stimulation by glucose associated with antibiotics, corticosteroids, and diabetes mellitus, *J. Infect. Dis.* **123:** 371–377.
68. Epstein, J. B., Truelove, E. L., and Izutzu, K. T., 1984, Oral candidiasis: pathogenesis and host defense, *Rev. Infect. Dis.* **6:**96–106.
69. Fisker, A. V., Schiott, C. R., and Philipsen, H. P., 1982, Short-term oral candidosis in rats, with special reference to the site of infection, *Acta Pathol. Microbiol. Immunol. Scand. B* **90:** 49–57.
70. Russell, C., Jones, J. H., and Gibbs, A. C. C., 1976, The carriage of *Candida albicans* in the mouths of rats treated with tetracycline briefly or for a prolonged period, *Mycopathologia* **58:**125–129.
71. DeMaria, A., Buckley, H., and von Lichtenberg, F., 1976, Gastrointestinal candidiasis in rats treated with antibiotics, cortisone and azathioprine, *Infect. Immun.* **13:**1761–1770.
72. Cantorna, M. T., and Balish, E., 1990, Mucosal and systemic candidiasis in congenitally immunodeficient mice, *Infect. Immun.* **58:**1093–1100.
73. Ouellette, A. J., Greco, R. M., James, M., Frederick, D., Naftilan, J., and Fallon, J. T., 1989, Developmental regulation of cryptdin, a corticostatin/defensin precursor mRNA in mouse small intestinal crypt epithelium, *J. Cell Biol.* **108:**1687–1695.
74. Selsted, M. E., Szklarek, D., Ganz, T., and Lehrer, R. I., 1985, Activity of rabbit leukocyte peptides against *Candida albicans*, *Infect. Immun.* **49:**202–206.
75. Krause, W., Matheis, H., and Wulf, K., 1969, Fungaemia and funguria after oral administration of *Candida albicans*, *Lancet* **1:**598–599.
76. Sass, W., Dreyer, H. P., Bockeler, W., Hamelmann, H., and Seifert, J., 1987, Prinzipien der partikelresorption im magen-darm-trakt, *Z. Gastroenterol.* **25:**306–315.
77. Baine, W. B., Koenig, M. G., and Goodman, J. S., 1974, Clearance of *Candida albicans* from the bloodstream of rabbits, *Infect. Immun.* **10:**1420–1425.
78. Bird, D. C., and Sheagren, J. N., 1970, Evaluation of reticuloendothelial system phagocytic activity during systemic *Candida albicans* infection in mice, *Proc. Soc. Exp. Biol. Med.* **133:**34–37.
79. Cho, Y. S., and Choi, H. Y., 1979, Opportunistic fungal infection among cancer patients: a ten year autopsy study, *Am. J. Clin. Pathol.* **72:**617–620.
80. Bach, M. C., Sahyoun, A., Adler, J. L., Schlesinger, R. M., Breman, J., Madras, P., P'eng, F. K., and Monaco, A. P., 1973, High incidence of fungus infections in renal transplantation patients treated with antilymphocyte and conventional immunosuppression, *Transplant. Proc.* **5:**549–553.

81. Louria, D. B., Brayton, R. G., and Finkel, G., 1963, Studies on the pathogenesis of experimental *Candida albicans* infections in mice, *Sabouraudia* **2**:271–283.
82. Ruthe, R. C., Andersen, B. R., Cunningham, B. L., and Epstein, R. B., 1978, Efficacy of granulocyte transfusions in control of systemic candidiasis in leukopenic host, *Blood* **52**: 493–498.
83. Hurtrel, B., Lagrange, P. H., and Michel, J.-C., 1980, Systemic candidiasis in mice. II. Main role of polymorphonuclear leukocytes in resistance to infection, *Ann. D'Immunol.* **131**:105–118.
84. Cutler, J. E., and Poor, A. H., 1981, Effect of mouse phagocytes on *Candida albicans* in in vivo chambers, *Infect. Immun.* **31**:1110–1116.
85. Lehrer, R. I., and Cline, M. J., 1969, Leukocyte myeloperoxidase deficiency and disseminated candidiasis: the role of myeloperoxidase in resistance to *Candida* infection, *J. Clin. Invest.* **48**:1478–1488.
86. Cech, P., Papathanassiou, A., Boreux, G., Roth, P., and Miescher, P. A., 1979, Hereditary myeloperoxidase deficiency, *Blood* **53**:403–411.
87. Cech, P., Stalder, H., Widmann, J.-J., Rohner, A., and Miescher, P. A., 1979, Leukocyte myeloperoxidase deficiency and diabetes mellitus associated with *Candida albicans* liver abscess, *Am. J. Med.* **66**:149–153.
88. Moosman, K., and Bojanovsky, A., 1975, Rezidivierende candidosis bei myeloperoxydase mangel, *Monatschr. Kinderheilkd.* **123**:408–409.
89. Parry, M. F., Root, R. K., Metcalf, J. A., Delaney, K. K., Kaplow, L. S., and Richar, W. J., 1981, Myeloperoxidase deficiency. Prevalence and clinical significance, *Ann. Intern. Med.* **95**: 293–301.
90. Catovsky, D., Galton, D. A., Robinson, J., 1972, Myeloperoxidase-deficient neutrophils in acute myeloid leukaemia, *Scand. J. Haematol.* **9**:142–148.
91. Bendix-Hansen, K., and Nielsen, H. K., 1983, Myeloperoxidase-deficient polymorphonuclear leucocytes (II). Longitudinal study in acute myeloid leukaemia, untreated, in remission and in relapse, *Scand. J. Haematol.* **31**:5–8.
92. Bendix-Hansen, K., Kerndrup, G., and Pedersen, B., 1986, Myeloperoxidase-deficient polymorphonuclear leucocytes (VI): Relation to cytogenetic abnormalities in primary myelodysplastic syndromes, *Scand. J. Haematol.* **36**:3–7.
93. Lehrer, R. I., Goldberg, L. S., Apple, M. A., and Rosenthal, N. P., 1972, Refractory megaloblastic anemia with myeloperoxidase-deficient neutrophils, *Ann. Intern. Med.* **76**: 447–453.
94. Lehrer, R. I., 1972, Functional aspects of a second mechanism of candidacidal activity by human neutrophils, *J. Clin. Invest.* **51**:2566–2572.
95. Lehrer, R. I., Hanifin, J., and Cline, M. J., 1969, Defective bactericidal activity in myeloperoxidase-deficient human neutrophils, *Nature (Lond.)* **223**:78–79.
96. Lehrer, R. I., 1975, The fungicidal mechanisms of human monocytes. I. Evidence for myeloperoxidase-linked and myeloperoxidase-independent candidacidal mechanisms, *J. Clin. Invest.* **55**:338–346.
97. Diamond, R. D., Krzesicki, R., and Jao, W., 1978, Damage to pseudohyphal forms of *Candida albicans* by neutrophils in the absence of serum *in vitro*, *J. Clin. Invest.* **61**:349–359.
98. Rosner, F., Valmont, I., Kozinn, P. J., and Caroline, L., 1970, Leukocyte function in patients with leukemia, *Cancer* **25**:835–842.
99. Lehrer, R. I., and Cline, M. J., 1971, Leukocyte candidacidal activity and resistance to systemic candidiasis in patients with cancer, *Cancer* **27**:1211–1217.
100. Goldman, J. M., and Th'ng, K. H., 1973, Phagocytic function of leukocytes from patients with acute myeloid and chronic granulocytic leukaemia, *Br. J. Haematol.* **25**:299–308.
101. Bjerknes, R., Vindenes, H., and Laerum, O. D., 1990, Altered neutrophil functions in patients with large burns, *Blood Cells* **16**:127–141.
102. El-Maalem, H., and Fletcher, J., 1976, Defective neutrophil function in chronic granulocytic leukaemia, *Br. J. Haematol.* **66**:95–103.

103. Child, J. A., Martin, S., Cawley, J. C., and Ghoniem, A. T. M., 1978, Defective microbicidal function of neutrophils in haematological malignancies and lymphomas: correction by levamisole in vitro, *Biomedicine* **29**:159–161.
104. Uberti-Foppa, C., Franzetti, F., Hegri, C., Novati, R., Cinque, P., Privitera, G., Parente, F., Crocchiolo, P., and Lazzarin, A., 1987, Role of PMN's candidacidal activity in the pathogenesis of AIDS-related systemic candidiasis, *Bull. Ist. Sieroter. Milan* **66**:471–478.
105. Xanthou, M., Valassi-Adam, E., Kintsonidou, E., and Matsaniotis, N., 1975, Phagocytosis and killing ability of *Candida albicans* by blood leucocytes of healthy term and preterm babies, *Arch. Dis. Child* **50**:72–75.
106. Quie, P. G., and Chilgren, R. A., 1971, Acute disseminated and chronic mucocutaneous candidiasis, *Semin. Hematol.* **8**:227–242.
107. Oseas, R., and Lehrer, R. I., 1978, A micromethod for measuring neutrophil candidacidal activity in neonates, *Pediatr. Res.* **12**:828–829.
108. Tuck, R., Burke, V., Gracey, M., Malajczuk, A., and Sunoto, 1979, Defective *Candida* killing in childhood malnutrition, *Arch. Dis. Child.* **54**:445–447.
109. Boyne, R., Arthur, J. R., and Wilson, A. B., 1986, An in vivo and in vitro study of selenium deficiency and infection in rats, *J. Comp. Pathol.* **96**:379–386.
110. Kernbaum, S., and Vilde, J.-L., 1974, Etude du pouvoir candidacide des polynucleaires neutrophiles humains, *Pathol. Biol. (Paris)* **22**:61–66.
111. El Falaky, I. H., and Shoukry, H. M., 1977, Phagocytosis and intracellular killing of *Candida albicans* by polymorphonuclear leucocytes in thalassaemia, *J. Egyptian Public Health Assoc.* **52**:317–329.
112. Drutz, D. J., and Lehrer, R. I., 1978, Development of amphotericin B-resistant *Candida tropicalis* in a patient with defective leukocyte function, *Am. J. Med. Sci.* **276**:77–92.
113. Edwards, J. E., Jr., Rotrosen, D., Fontaine, J. W., Haudenschild, C. C., and Diamond, R. D., 1987, Neutrophil-mediated protection of cultured human vascular endothelial cells from damage by growing *Candida albicans* hyphae, *Blood* **69**:1450–1457.
114. Retrosen, D., Edwards, J. E., Jr., Gibson, T. R., Moore, J. C., Cohen, A. H., and Green, I., 1985, Adherence of *Candida* to cultured vascular endothelial cells: mechanisms of attachment and endothelial cell penetration, *J. Infect. Dis.* **152**:1264–1274.
115. Davies, R. R., and Denning, T. J. V., 1972, *Candida albicans* and the fungicidal activity of the blood, *Sabouraudia* **10**:301–312.
116. Lehrer, R. I., and Cline, M. J., 1969, Interaction of *Candida albicans* with human leukocytes and serum, *J. Bacteriol.* **98**:996–1004.
117. Edwards, J. E., Jr., Lehrer, R. I., Stiehm, E. R., Fischer, T. J., and Young, L. S., 1978, Severe candidal infections. Clinical perspective, immune defense mechanisms, and current concepts of therapy, *Ann. Intern. Med.* **89**:91–106.
118. Ray, T. L., and Wuepper, K. D., 1976, Activation of the alternative (properdin) pathway of complement by *Candida albicans* and related species, *J. Invest. Derm.* **67**:700–703.
119. Denning, T. J. V., and Davies, R. R., 1973, *Candida albicans* and the chemotaxis of polymorphonuclear neutrophils, *Sabouraudia* **11**:210–221.
120. Weeks, B. A., Escobar, M. R., Hamilton, P. B., and Fueston, V. M., 1976, Chemotaxis of polymorphonuclear neutrophilic leukocytes by mannan-enriched preparations of *Candida albicans*, *Adv. Exp. Med. Biol.* **73**:161–169.
121. Ray, T. L., Hanson, A., Ray, L. F., and Wuepper, K. D., 1979, Purification of a mannan from *Candida albicans* which activates serum complement, *J. Invest. Dermatol.* **73**:269–274.
122. Cutler, J. E., 1977, Chemotactic factor produced by *Candida albicans*, *Infect. Immun.* **18**:568–573.
123. Morelli, R., and Rosenberg, L. T., 1971, The role of complement in the phagocytosis of *Candida albicans* by mouse peripheral blood leukocytes, *J. Immunol.* **107**:476–480.
124. Yamamura, M., and Valdimarsson, H., 1977, Participation of C3 in intracellular killing of *Candida albicans*, *Scand. J. Immunol.* **6**:591–594.

125. Ferrante, A., and Thong, Y. H., 1979, Requirement of heat-labile opsonins for maximal phagocytosis of *Candida albicans*, *Sabouraudia* **17**:293–297.
126. Kernbaum, S., 1975, Pouvoirs phagocytaire et fongicide envers *Candida albicans* des polynucleaires neutrophiles humains en presence de serum depourvu de C3 et C4, *Ann. Microbiol. (Inst. Pasteur).* **126A**:75–81.
127. Morrison, R. P., and Cutler, J. E., 1981, *In vitro* studies of the interaction of murine phagocytic cells with *Candida albicans*, *J. Reticuloendothel. Soc.* **29**:23–24.
128. Sinski, J. T., Reed, G. L., Kelley, L. M., and Le Vebvre, R., 1973, Macrophage migration technique using coccidioidin, *Infect. Immun.* **7**:226–230.
129. Solomkin, J. S., Mills, E. L., Giebink, G. S., Nelson, R. D., Simmons, R. L., and Quie, P. G., 1978, Phagocytosis of *Candida albicans* by human leukocytes: opsonic requirements, *J. Infect. Dis.* **137**:30–37.
130. Leijh, P. C., van den Barselaar, M. T., and van Furth, R., 1977, Kinetics of phagocytosis and intracellular killing of *Candida albicans* by human granulocytes and monocytes, *Infect. Immun.* **17**:313–318.
131. Ginsburg, I., Sela, M. N., Morag, A., Ravid, Z., Duchan, Z., Ferne, M., Rabinowitz-Bergner, S., Thomas, P. P., Davies, P., Niccols, J., Humes, J., and Bonney, R., 1981, Role of leukocyte factors and cationic polyelectrolytes in phagocytosis of group A streptococci and *Candida albicans* by neutrophils, macrophages, fibroblasts and epithelial cells: modulation by anionic polyelectrolytes in relation to pathogenesis of chronic inflammation, *Inflammation* **5**:289–312.
132. Pereira, H. A., and Hosking, C. S., 1984, The role of complement and antibody in opsonization and intracellular killing of *Candida albicans*, *Clin. Exp. Immunol.* **57**:307–314.
133. Sandhu, D. K., Sandhu, R. S., Damodaran, V. N., and Randhawa, H. S., 1970, Effect of cortisone on bronchopulmonary aspergillosis in mice exposed to spores of various *Aspergillus* species, *Sabouraudia* **8**:32–38.
134. Hurley, D. L., Balow, J. E., and Fauci, A. S., 1975, Experimental disseminated candidiasis. II. Administration of glucocorticosteroids, susceptibility to infection and immunity, *J. Infect. Dis.* **132**:393–398.
135. Laforce, F. M., Mills, D. M., Iverson, K., Cousins, R., and Everett, E. D., 1975, Inhibition of leukocyte candidacidal activity by serum from patients with disseminated candidiasis, *J. Lab. Clin. Med.* **86**:657–666.
136. Walker, S. M., and Urbaniak, S. J., 1980, A serum-dependent defect of neutrophil function in chronic mucocutaneous candidiasis, *J. Clin. Pathol.* **33**:370–372.
137. Fleischmann, J., Selsted, M. E., and Lehrer, R. I., 1985, Opsonic activity of MCP-1 and MCP-2, cationic peptides from rabbit alveolar macrophages, *Diagn. Microbiol. Infect. Dis.* **3**: 233–242.
138. Hector, R. F., Domer, J. E., and Carrow, E. W., 1982, Immune responses to *Candida albicans* in genetically-distinct mice, *Infect. Immun.* **38**:1020–1028.
139. Lyon, F. L., Hector, R. F., and Domer, J. E., 1986, Innate and acquired immune responses against *Candida albicans* in congenic B10.D2 mice with deficiency of the C5 complement component, *J. Med. Vet. Mycol.* **24**:359–367.
140. Gelfand, J. A., Hurley, D. L., Fauci, A. S., and Frank, M. M., 1978, Role of complement in host defense against experimental disseminated candidiasis, *J. Infect. Dis.* **138**:9–16.
141. Lehrer, R. I., 1970, Measurement of candidacidal activity of specific leukocyte types in mixed cell populations. I. Normal, myeloperoxidase-deficient, and chronic granulomatous disease neutrophils, *Infect. Immun.* **2**:42–47.
142. Husseini, R. H., Hoadley, M. E., Hutchinson, J. J. P., Penn, C. W., and Smith, H., 1985, Intracellular killing of *Candida albicans* by human polymorphonuclear leucocytes: comparison of three methods of assessment, *J. Immunol. Methods* **81**:215–221.
143. Lehrer, R. I., 1971, Inhibition by sulfonamides of the candidacidal activity of human neutrophils, *J. Clin. Invest.* **50**:2498–2505.

144. Schmid, L., and Brune, K., 1974, Assessment of phagocytic and antimicrobial activity of human granulocytes, *Infect. Immun.* **10:**1120–1126.
145. Vecchiarelli, A., Bistoni, F., Cenci, E., Perito, S., and Cassone, A., 1985, In-vitro killing of *Candida* species by murine immunoeffectors and its relationship to the experimental pathogenicity, *Sabouraudia* **23:**377–387.
146. Scherwitz, C., and Martin, R., 1979, The phagocytosis of *Candida albicans* blastospores and germ tubes by polymorphonuclear leukocytes, *Dermatologia* **159:**12–23.
147. Schuit, K. E., 1979, Phagocytosis and intracellular killing of pathogenic yeast by human monocytes and neutrophils, *Infect. Immun.* **24:**932–938.
148. Cockayne, A., and Odds, F. C., 1984, Interactions of *Candida albicans* yeast cells, germ tubes, and hyphae with human polymorphonuclear leucocytes in vitro, *J. Gen. Microbiol.* **130:** 465–471.
149. Brummer, E., McEwen, J. G., and Stevens, D. A., 1986, Fungicidal activity of murine inflammatory polymorphonuclear neutrophils: comparison with murine peripheral blood PMN, *Clin. Exp. Immunol.* **66:**681–690.
150. Kernbaum, S., 1975, Pouvoir candidacide des polynucleaires neutrophiles humains et chimiotherapie antibacterienne, *Pathol. Biol.* **22:**789–794.
151. Ferrari, F. A., Pagani, A., Marconi, M., Stefanoni, R., and Siccardi, A. G., 1980, Inhibition of candidacidal activity of human neutrophil leukocytes by aminoglucoside antibiotics, *Antimicrob. Agents Chemother.* **17:**87–88.
152. Roilides E., Walsh, T. J., Rubin, M., Venzon, D., and Pizzo, P. A., 1990, Effects of antifungal agents on the function of human neutrophils in vitro, *Antimicrob. Agents Chemother.* **34:** 196–201.
153. Meshulam, T., Diamond, R. D., Lyman, C. A., Wysong, D. R., and Melnick, D. A., 1988, Temporal association of calcium mobilization, inositol triphosphate generation, and superoxide anion release by human neutrophils activated by serum opsonized and nonopsonized particulate stimuli, *Biochem. Biophys. Res. Commun.* **150:**532–539.
154. Wilson, R. M., and Reeves, W. G., 1986, Neutrophil phagocytosis and killing in insulin-dependent diabetes, *Clin. Exp. Immunol.* **63:**478–484.
155. Chan, C. K., and Balish, E., 1978, Inhibition of granulocyte phagocytosis of *Candida albicans* by amphotericin B, *Can. J. Microbiol.* **24:**363–364.
156. Belcher, R. W., Carney, J. F., and Monahan, F. G., 1973, An electron microscopic study of phagocytosis of *Candida albicans* by polymorphonuclear leukocytes, *Lab. Invest.* **29:** 620–627.
157. Richardson, M., and Smith, H., 1983, Ultrastructural features of phagocytosis and intracellular killing of *Candida albicans* by mouse polymorphonuclear phagocyte monolayers, *Mycopathologia* **83:**97–102.
158. Cech, P., and Lehrer, R. I., 1984, Heterogeneity of human neutrophil phagolysosomes: functional consequences for candidacidal activity, *Blood* **64:**147–151.
159. Levitz, S. M., DiBenedetto, D. J., and Diamond, R. D., 1987, A rapid fluorescent assay to distinguish attached from phagocytized yeast particles, *J. Immunol. Methods* **101:**37–42.
160. Klebanoff, S. J., 1980, Oxygen metabolism and the toxic properties of phagocytes, *Ann. Intern. Med.* **93:**480–489.
161. Weiss, S. J., Test, S. T., Eckmann, C. M., Roos, D., and Regiani, S., 1986, Brominating oxidants generated by human eosinophils, *Science* **234:**200–203.
162. Lehrer, R. I., 1971, Measurement of candidacidal activity of specific leukocyte types in mixed cell populations: II. Normal and chronic granulomatous disease eosinophils, *Infect. Immun.* **3:**800–802.
163. Foote, C. S., Goyne, T. E., and Lehrer, R. I., 1983, Assessment of chlorination by human neutrophils, *Nature* **301:**715–716.
164. Weiss, S. J., Klein, R., Slivka, A., and Wei, M., 1982, Chlorination of taurine by human neutrophils: evidence for hypochlorous acid generation, *J. Clin. Invest.* **70:**598–607.

165. Lehrer, R. I., 1969, Antifungal effects of peroxidase systems, *J. Bacteriol.* **99**:361–365.
166. Lehrer, R. I., and Jan, R. G., 1970, Interaction of *Aspergillus fumigatus* spores with human leukocytes and serum, *Infect. Immun.* **1**:345–350.
167. Diamond, R. D., Clark, R. A., and Haudenschild, C. C., 1980, Damage to *Candida albicans* hyphae and pseudohyphae by the myeloperoxidase system and oxidative products of neutrophil metabolism *in vitro*, *J. Clin. Invest.* **66**:908–917.
168. Wagner, D. K., Collins-Lech, C., and Sohnle, P. G., 1986, Inhibition of neutrophil killing of *Candida albicans* pseudohyphae by substances which quench hypochlorous acid and chloramines, *Infect. Immun.* **51**:731–735.
169. Beilke, M. A., Collins-Lech, C., and Sohnle, P. G., 1989, Candidacidal activity of the neutrophil myeloperoxidase system can be protected from excess hydrogen peroxide by the presence of ammonium ion, *Blood* **73**:1045–1049.
170. Kolotila, M. P., and Diamond, R. D., 1990, Effects of neutrophils and *in vitro* oxidants on survival and phenotypic switching of *Candida albicans* W0-1, *Infect. Immun.* **58**:1174–1179.
171. Babior, B. M., 1988, The respiratory burst oxidase, in, Phagocytic defects II (Curnutte, J. J., ed). *Hematol. Oncol Clin. North Am.* **2**:201–212.
172. Clark, R. A., 1990, Perspective: The human neutrophil respiratory burst oxidase, *J. Infect. Dis.* **161**:1140–1147.
173. Cohen, M. S., Isturiz, R. E., Malech, H. L., Root, R. K., Wilfert, C. M., Gutman, L., and Buckley, R. H., 1981, Fungal infection in chronic granulomatous disease—the importance of the phagocyte in defense against fungi, *Am. J. Med.* **71**:59–66.
174. Lehrer, R. I., Ladra, K. M., and Hake, R. B., 1975, Nonoxidative fungicidal mechanisms of mammalian granulocytes: demonstration of components with candidacidal activity in human, rabbit, and guinea pig leukocytes, *Infect. Immun.* **11**:1226–1234.
175. Levitz, S. M., and Diamond, R. D., 1984, Killing of *Aspergillus fumigatus* spores and *Candida albicans* yeast phase by the iron-hydrogen peroxide-iodide cytotoxic system: comparison with the myeloperoxidase-hydrogen peroxide-halide system, *Infect. Immun.* **43**:1100–1102.
176. Lehrer, R. I., Szklarek, D., Ganz, T., and Selsted, M. E., 1985, Correlation of binding of rabbit granulocyte peptides to *Candida albicans* with candidacidal activity, *Infect. Immun.* **49**: 207–211.
177. Lehrer, R. I., Szklarek, D., Ganz, T., and Selsted, M. E., 1986, Synergistic activity of rabbit granulocyte peptides against *Candida albicans*, *Infect. Immun.* **52**:902–904.
178. Diamond, R. D., and Krzesicki, R., 1978, Mechanism of attachment of neutrophils to *Candida albicans* pseudohyphae in the absence of serum, and of subsequent damage to pseudohyphae by microbicidal processes of neutrophils *in vitro*, *J. Clin. Invest.* **61**:360–369.
179. Levitz, S. M., Lyman, C. A., Murata, T., Sullivan, J. A., Mandell, G. L., and Diamond, R. D., 1987, Cytosolic calcium changes in individual neutrophils stimulated by opsonized and unopsonized *Candida albicans* hyphae, *Infect. Immun.* **55**:2783–2788.
180. Lyman, C. A., Simons, E. R., Melnick, D. A., and Diamond, R. D., 1987, Unopsonized *Candida albicans* hyphae stimulate a neutrophil respiratory burst and a cytosolic calcium flux without membrane depolarization, *J. Infect. Dis.* **156**:770–776.
181. Lyman, C. A., Simons, E. R., Melnick, D. A., and Diamond, R.D ., 1988, Induction of signal transduction in human neutrophils by *Candida albicans* hyphae: the role of pertussis toxin-sensitive guanosine triphosphate-binding proteins, *J. Infect. Dis.* **158**:1056–1064.
182. Kolotila, M. P., and Diamond, R. D., 1988, Stimulation of neutrophil actin polymerization and degranulation by opsonized and unopsonized *Candida albicans* hyphae and zymosan, *Infect. Immun.* **56**:2016–2022.
183. Wysong, D. R., Lyman, C. A., and Diamond, R. D., 1989, Independence of neutrophil respiratory burst oxidant generation from the early cytosolic calcium response after stimulation with unopsonized *Candida albicans* hyphae, *Infect. Immun.* **57**:1499–1505.
184. Diamond, R. D., and Noble, L., 1990, Patterns of guanine nucleotide exchange reflecting

disparate neutrophil activation pathways by opsonized and unopsonized *Candida albicans* hyphae, *J. Infect. Dis.* **162**:262–265.
185. Sawyer, D. W., Donowitz, G. R., and Mandell, G. L., 1989, Polymorphonuclear neutrophils: an effective antimicrobial force, *Rev. Infect. Dis.* **11**:S1532–S1544.
186. Lehrer, R. I., and Ganz, T., 1990, Antimicrobial polypeptides of human neutrophils, *Blood* **76**:2169–2181.
187. Zeya, H. I., and Spitznagel, J. K., 1966, Cationic proteins of polymorphonuclear leukocyte lysosomes. I. Resolution of antibacterial and enzymatic activities, *J. Bacteriol.* **91**:750–754.
188. Lehrer, R. I., Ganz, T., and Selsted, M. E., 1990, Defensins: natural peptide antibiotics from neutrophils, *ASM News* **56**:315–318.
189. Ganz, T., Selsted, M. E., and Lehrer, R. I., 1990, Defensins, *Eur. J. Haematol.* **44**:1–8.
190. Lehrer, R. I., Ganz, T., Szklarek, D., and Selsted, M. E., 1988, Modulation of the *in vitro* candidacidal activity of human neutrophil defensins by target cell metabolism and divalent cations, *J. Clin. Invest.* **81**:1829–1835.
191. Rice, W. G., Ganz, T., Kinkade, J. M., Jr., Selsted, M. E., Lehrer, R. I., and Parmley, R. T., 1987, Defensin-rich dense granules of human neutrophils, *Blood* **70**:757–765.
192. Wilde, C. G., Griffith, J. E., Marra, M. N., Snable, J. L., and Scott, R. W., 1989, Purification and characterization of human neutrophil peptide 4, a novel member of the defensin family, *J. Biol. Chem.* **264**:11200–11203.
193. Selsted, M. E., Brown, D. M., DeLange, R. J., Harwig, S. S., and Lehrer, R. I., 1985, Primary structures of six antimicrobial peptides of rabbit peritoneal neutrophils, *J. Biol. Chem.* **260**: 4579–4584.
194. Eisenhauer, P. B., Harwig, S. S. L., Szklarek, D., Ganz, T., Selsted, M. E., and Lehrer, R. I., 1989, Purification and antimicrobial properties of three defensins from rat neutrophils, *Infect. Immun.* **57**:2021–2027.
195. Eisenhauer, P., Harwig, S. S. S. L., Szklarek, D., Ganz, T., and Lehrer, R. I., 1990, Polymorphic expression of defensins in neutrophils from outbred rats, *Infect. Immun.* **58**: 3899–3902.
196. Kagan, B. L., Selsted, M. E., Ganz, T., and Lehrer, R. I., 1990, Antimicrobial defensin peptides form voltage-dependent ion-permeable channels in planar lipid bilayer membranes, *Proc. Natl. Acad. Sci. USA* **87**:210–214.
197. Odeberg, H., and Olsson, I., 1975, Antibacterial activity of cationic proteins from human granulocytes, *J. Clin. Invest.* **56**:1118–1124.
198. Odeberg, H., and Olsson, I., 1976, Microbicidal mechanisms of human granulocytes: synergistic effects of granulocyte elastase and myeloperoxidase or chymotrypsin-like cationic protein, *Infect. Immun.* **14**:1276–1283.
199. Gabay, J. E., Scott, R. W., Campanelli, D., Griffith, J., Wilde, C., 1989, Antibiotic proteins of human polymorphonuclear leukocytes, *Proc. Natl. Acad. Sci. USA* **86**:5610–5614.
200. Campanelli, D., Detmers, P. A., Nathan, C. F., Gabay, J. E., 1990, Azurocidin and a homologous serine protease from neutrophils. Differential antimicrobial and proteolytic properties, *J. Clin. Invest.* **85**;904–915.
201. Wilde, C. G., Snable, J. L., Griffith, J. E., Scott, R. W., 1990, Characterization of two azurophil granule proteases with active-site homology to neutrophil elastase, *J. Biol. Chem.* **265**:2038–2041.
202. Shafer, W. M., Pohl, J., Onunka, V. C., Bangalore, N., and Travis, J., 1990, Human lysosomal cathepsin G and granzyme B share a functionally conserved broad spectrum antibacterial peptide, *J. Biol. Chem.* **266**:112–116.
203. Arnold, R. R., Brewer, M., and Gauthier, J. J., 1980, Bactericidal activity of human lactoferrin: sensitivity of a variety of microorganisms, *Infect. Immun.* **28**:893–898.
204. Collins, M. S., and Pappagianis, D., 1973, Effects of lysozyme and chitinase on the spherules of *Coccidioides immitis in vitro, Infect. Immun.* **7**:817–822.

205. Gadebusch, H. H., and Johnson, A. G., 1966, Natural host resistance to infection with *Cryptococcus neoformans*. IV. The effect of some cationic proteins on the experimental disease, *J. Infect. Dis.* **116:**551–565.
206. Rausch, P. G., and Moore, T. G., 1975, Granule enzymes of polymorphonuclear neutrophils: a phylogenetic comparison, *Blood* **46:**913–919.
207. Marquis, G., Montplaisir, S., Garzon, S., Strykowski, H., and Auger, P., 1982, Fungitoxicity of muraminidase: ultrastructural damage to *Candida albicans*, *Lab. Invest.* **46:**627–636.
208. Kamaya, T., 1970, Lytic action of lysozyme on *Candida albicans*, *Mycopathol. Mycol. Appl.* **42:** 197–207.
209. Ray, T. L., and Wuepper, K. D., 1976, Experimental cutaneous candidiasis in rodents, *J. Invest. Derm.* **66:**29–33.
210. Ray, T. L., and Wuepper, K. D., 1978, Experimental cutaneous candidiasis in rodents II. Role of the stratum corneum barrier and serum complement as a mediator of a protective inflammatory response, *Arch. Dermatol.* **114:**539–543.
211. Louria, D. B., and Brayton, R. G., 1964, Behavior of *Candida* cells within leukocytes, *Proc. Soc. Exp. Biol. Med.* **115:**93–98.
212. Diamond, R. D., Oppenheim, F., Nakagawa, Y., Krzesicki, R., and Haudenschild, C. C., 1980, Properties of a product of *Candida albicans* hyphae and pseudohyphae that inhibits contact between the fungi and human neutrophils *in vitro*, *J. Immunol.* **125:**2797–2804.
213. Smail, E. H., Melnick, D. A., Ruggeri, R., and Diamond, R. D., 1988, A novel natural inhibitor from *Candida albicans* hyphae causing dissociation of the neutrophil respiratory burst response to chemotactic peptides from other post-activating events, *J. Immunol.* **140:** 3893–3899.
214. Smail, E. H., Kolotila, M. P., Ruggeri, R., and Diamond, R. D., 1989, Natural inhibitor from *Candida albicans* blocks release of azurophil and specific granule contents by chemotactic peptide-stimulated human neutrophils, *Infect. Immun.* **57:**689–692.
215. Hilger, A. E., and Danley, D. L., 1980, Alteration of polymorphonuclear leukocyte activity by viable *Candida albicans*, *Infect. Immun.* **27:**714–720.
216. Danley, D. L., Hilger, A. E., Winkel, C. A., 1983, Generation of hydrogen peroxide by *Candida albicans* and influence on murine polymorphonuclear leukocyte activity, *Infect. Immun.* **40:**97–102.
217. Heidenreich, F., and Dierich, M. P., 1985, *Candida albicans* and *Candida stellatoidea* in contrast to other *Candida* species bind iC3b and C3d but not C3b, *Infect. Immun.* **50:**598–600.
218. Edwards, J. E., Jr., Gaither, T. A., O'Shea, J. J., Rotrosen, D., Lawley, T. J., Wright, S. A., Frank, M. M., and Green, I., 1986, Expression of specific binding sites on *Candida* with functional and antigenic characteristics of human complement receptors, *J. Immunol.* **137:** 3577–3583.
219. Calderone, R. A., Linehan, L., Wadsworth, E., and Sandberg, A. L., 1988, Identification of C3d receptors on *Candida albicans*, *Infect. Immun.* **56:**252–258.
220. Gilmore, B. J., Retsinas, E. M., Lorenz, J. S., and Hostetter, M. K., 1988, An iC3b receptor on *Candida albicans*: structure, function and correlates for pathogenicity, *J. Infect. Dis.* **157:** 38–46.
221. Eigentler, A. Schulz, T. F., Larcher, C., Breitwieser, E. M., Myones, B. L., Petzer, A. L., and Dierich, M. P., 1989, C3bi-binding protein on *Candida albicans*: temperature-dependent expression and relationship to human complement receptor type 3, *Infect. Immun.* **57:** 616–622.
222. Hostetter, M. K., 1990, Handicaps to host defense. Effects of hyperglycemia on C3 and *Candida albicans*, *Diabetes* **39:**271–275.
223. Hostetter, M. K., Lorenz, J. S., Preus, L., Kendrick, K. E., 1990, The iC3b receptor on *Candida albicans*: subcellular localization and modulation of receptor expression by glucose, *J. Infect. Dis.* **161:**761–768.

224. Slutsky, B., Buffo, J., and Soll, D. R., 1985, High-frequency switching of colony morphology in *Candida albicans*, *Science* **230**:666–669.
225. Soll, D. R., Slutsky, B., Mackenzie, S., Langtimm, C., and Staebell, M., 1987, Two newly discovered switching systems in *Candida albicans* and their possible role in oral candidiasis, in: *Oral mucosal diseases: biology, etiology, and therapy*, (I. C. Mackenzie, C. A. Squier, and E. Dabelsteen, eds.), Laegeforeningen Forlag, Copenhagen, pp. 52–59.
226. Ishikawa, T., Dalton, A. C., and Arbesman, C. E., 1972, Phagocytosis of *Candida albicans* by eosinophilic leukocytes, *J. Allergy Clin. Immunol.* **49**:311–315.
227. Ishikawa, T., Yu, M. C., and Arbesman, C. E., 1972, Electron microscopic demonstration of phagocytosis of *Candida albicans* by human eosinophilic leukocytes, *J. Allergy Clin. Immunol.* **50**:183–187.
228. Lehrer, R. I., 1981, Ingestion and destruction of *Candida albicans*, in: *Methods for Studying Mononuclear Phagocytes*, (D. A. Adams, H. Koren, and P. Edelson, eds.), Academic, New York, pp. 693–708.
229. Estevez, M. E., Sen, L., Vadez, R., and Balina, L. M., 1979, Defective blood mononuclear phagocyte function in patients with leprosy, *Int. J. Leprosy* **47**:575–579.
230. Estevez, M. E., Ballart, I. J., Diez, R. A., Planes, N., Scaglione, C., and Sen, L., 1986, Early defect of phagocytic cell function in subjects at risk for acquired immunodeficiency syndrome, *Scand. J. Immunol.* **24**:215–221.
231. Cohen, A. B., and Cline, M. J., 1971, The human alveolar macrophage: isolation, cultivation *in vitro*, and studies of morphologic and functional characteristics, *J. Clin. Invest.* **50**:1390–1398.
232. Kemmerich, B., Rossing, T. H., and Pennington, J. E., 1987, Comparative oxidative microbicidal activity of human blood monocytes and alveolar macrophages and activation by recombinant gamma interferon, *Am. Rev. Respir. Dis.* **136**:266–270.
233. Territo, M. C., and Golde, D. W., 1979, The function of human alveolar macrophages, *J. Reticuloendothel. Soc.* **25**:111–120.
234. Winston, D. J., Territo, M. C., Ho, W. G., Miller, M. J., Gale, R. P., and Golde, D. W., 1982, Alveolar macrophage dysfunction in human bone marrow transplant recipients, *Am. J. Med.* **73**:859–866.
235. Golde, D. W., Territo, M., Finley, T. N., and Cline, M. J., 1976, Defective lung macrophages in pulmonary alveolar proteinasis, *Ann. Intern. Med.* **85**:304–309.
236. Robinson, J. E., Harvey, B. A. M., and Soothill, J. F., 1978, Phagocytosis and killing of bacteria and yeast by human milk cells after opsonisation in aqueous phase of milk, *Br. Med. J.* **1**:1443–1445.
237. Taschdjian, C. L., Toni, E. F., Hsu, K. C., Seelig, M. S., Cuesta, M. B., and Kozinn, P. J., 1971, Immunofluorescence studies of *Candida* in human reticuloendothelial phagocytes: implications for immunogenesis and pathogenesis of systemic candidiasis, *Am. J. Clin. Pathol.* **56**:50–58.
238. Odegaard, A., Viken, K. E., and Lamvik, J., 1974, Structural and functional properties of blood monocytes cultured *in vitro*, *Acta Pathol. Microbiol. Scand.* Sect. G. **82**:223–234.
239. Lehrer, R. I., 1970, The fungicidal activity of human monocytes: a myeloperoxidase-linked mechanism, *Clin. Res.* **18**:408–413.
240. Sasada, M., Kubo, A., Nishimura, T., Kakita, T., Moriguchi, T., Yamamoto, K., and Uchino, H., 1987, Candidacidal activity of monocyte-derived human macrophages: relationship between *Candida* killing and oxygen radical generation by human macrophages, *J. Leukoc. Biol.* **41**:289–294.
241. Bourne, H. R., Lehrer, R. I., Cline, M. J., and Melmon, K. L., 1971, Cyclic 3′,5′-adenosine monophosphate in the human leukocyte: synthesis, degradation, and effects on neutrophil candidacidal activity, *J. Clin. Invest.* **50**:920–929.
242. Odegaard, A., and Lamvik, J., 1976, The effect of phenylbutazone and chloramphenicol on

phagocytosis of radiolabelled *Candida albicans* by human monocytes cultured *in vitro*, *Acta Pathol. Microbiol. Scand*, Sect. C **84**:37–44.
243. Louria, D. B., Fallon, N., and Browne, H. G., 1960, The influence of cortisone on experimental fungus infections in mice, *J. Clin. Invest.* **39**:1435–1449.
244. Diamond, R. D., 1981, Mechanisms of host resistance to *Candida albicans*. In *Microbiology, 1981*, American Society for Microbiology, Washington, D. C., pp. 200–204.
245. Rinehart, J. J., Sagone, A. L., Balcerzak, S. P., Ackerman, G. A., and LoBuglio, A. F., 1975, Effects of corticosteroid therapy on human monocyte function, *N. Engl. J. Med.* **292**:236–241.
246. Hammerstrom, J., and Unsgaard, G., 1979, *In vitro* influence of endotoxin on human mononuclear phagocyte structure and function, *Acta Pathol. Microbiol. Scand.* **Sect. C 87**: 381–389.
247. Ganz, T., Selsted, M. E., Szklarek, D., Harwig, S. S. L., Daher, K., Bainton, D. F., and Lehrer, R. I., 1985, Defensins: natural peptide antibiotics of human neutrophils, *J. Clin. Invest.* **76**: 1427–1435.
248. Campbell, E. J., Silverman, E. K., and Campbell, M. A., 1989, Elastase and cathepsin G of human monocytes. Quantification of cellular content, release in response to stimuli and heterogeneity in elastase-mediated proteolytic activity, *J. Immunol.* **143**:2961–2968.
249. Hurtrel, B., and Lagrange, P. H., 1981, Comparative effects of carrageenan on systemic candidiasis and listeriosis in mice, *Clin. Exp. Immunol.* **44**:355–358.
250. Lee, K. W., and Balish, E., 1983, Systemic candidosis in silica-treated athymic and euthymic mice, *Infect. Immun.* **41**:902–907.
251. Sinha, B. K., Monga, D. P., and Prasad, S., 1987, Studies of the role of macrophages in experimental candidosis in mice, *Mykosen* **30**:105–112.
252. Baghian, A., and Lee, K. W., 1988, Role of activated macrophages in resistance to systemic candidosis, *J. Leukoc. Biol.* **44**:166–171.
253. Cheers, C., and Waller, R., 1975, Activated macrophages in congenitally athymic "nude" mice and in lethally irradiated mice, *J. Immunol.* **115**:844–847.
254. Cutler, J. E., 1976, Acute systemic candidiasis in normal and congenitally thymic-deficient (nude) mice, *J. Reticuloendothel. Soc.* **19**:121–124.
255. Poor, A. H., and Cutler, J.E ., 1981, Analysis of an *in vivo* model to study the interaction of host factors with *Candida albicans*, *Infect. Immun.* **31**:1104–1109.
256. Bistoni, F., Vecchiarelli, A., Cenci, E., Puccetti, P., Marconi, P., and Cassone, A., 1986, Evidence for macrophage-mediated protection against lethal *Candida albicans* infection, *Infect. Immun.* **51**:668–674.
257. Chow, H. S., Sarpel, S. C., and Epstein, R. B., 1980, Pathophysiology of *Candida albicans* meningitis in normal, neutropenic and granulocyte transfused dogs, *Blood* **55**:546–551.
258. Bistoni, F., Baccarini, M., Blasi, E., Marconi, P., Puccetti, P., and Garaci, E., 1983, Correlation between *in vivo* and *in vitro* studies of modulation of resistance to experimental *Candida albicans* infection by cyclophosphamide in mice, *Infect. Immun.* **40**:46–55.
259. Fromtling, R. A., Abruzzo, G. K., and Giltinan, D. M., 1987, *Candida tropicalis* infection in normal, diabetic, and neutropenic mice, *J. Clin. Microbiol.* **25**:1416–1420.
260. van't Wout, J. W., Linde, I., Leijh, P. C. J., and van Furth, R., 1988, Contribution of granulocytes and monocytes to resistance against experimental disseminated *Candida albicans* infection, *Eur. J. Clin. Microbiol. Infect. Dis.* **7**:736–741.
261. Sawyer, R. T., and Harmsen, A. G., 1989, The relative contribution of resident pulmonary alveolar macrophage and inflammatory polymorphonuclear neutrophils in host resistance to pulmonary infection by *Candida albicans*, *Mycopathologia* **108**:95–105.
262. Oblack, D. L., Schwarz, J., and Holder, I. A., 1979, Enhancement of tissue damage by *Candida albicans* in cortisone or nitrogen mustard-treated mice, *Can. J. Microbiol.* **25**:1015–1020.
263. Nugent, K. M., and Onofrio, J. M., 1984, Pulmonary clearance of *Candida albicans* in neutropenic mice, *J. Infect. Dis.* **149**:1020.

264. Hasenclever, H. F., and Corley, E. J., 1968, Enhancement of acquired resistance in murine candidiasis by *Bordetella pertussis* vaccine, *Sabouraudia* **6**:289–295.
265. Marra, S., and Balish, E., 1974, Immunity to *Candida albicans* induced by *Listeria monocytogenes*, *Infect. Immun.* **10**:72–82.
266. Sher, N. A., Chaparas, S. D., Greenberg, L. E., and Bernard, S., 1975, Effects of BCG, *Corynebacterium parvum*, and methanol-extraction residue in the reduction of mortality from *Staphylococcus aureus* and *Candida albicans* infection in immunosuppressed mice, *Infect. Immun.* **12**:1325–1330.
267. Rogers, T. J., and Balish, E., 1977, The role of activated macrophages in resistance to experimental renal candidiasis, *J. Reticuloendothel. Soc.* **22**:309–318.
268. Sawyer, R. T., Moon, R. J., and Beneke, E. S., 1981, Trapping and killing of *Candida albicans* by *Corynebacterium parvum*-activated livers, *Infect. Immun.* **32**:945–950.
269. Cummings, N. P., Pabst, M. J., and Johnston, R. B., Jr., 1980, Activation of macrophages for enhanced release of superoxide anion and greater killing of *Candida albicans* by injection of muramyl dipeptide, *J. Exp. Med.* **152**:1659–1669.
270. Fraser-Smith, E. B., Waters, R. V., and Matthews, T. R., 1982, Correlation between *in vivo* anti-*Pseudomonas* and anti-*Candida* activities and clearance of carbon by the reticuloendothelial system for various muramyl dipeptide analogs, using normal and immunosuppressed mice, *Infect. Immun.* **35**:105–110.
271. Fraser-Smith, E. B., Eppstein, D. A., Larsen, M. A., and Matthews, T. R., 1983, Protective effect of a muramyl dipeptide analog encapsulated in or mixed with liposomes against *Candida albicans* infection, *Infect. Immun.* **39**:172–178.
272. Baccarini, M., Blasi, E., Puccetti, P., and Bistoni, F., 1983, Phagocytic killing of *Candida albicans* by different murine effector cells, *Sabouraudia* **21**:271–286.
273. Baccarini, M., Bistoni, F., and Lohmann-Matthes, M. L., 1986, Organ-associated macrophage precursor activity: isolation of candidacidal and tumoricidal effectors from the spleens of cyclophosphamide-treated mice, *J. Immunol.* **136**:837–843.
274. Decker, T., Baccarini, M., and Lohmann-Matthes, M. L., 1986, Liver-associated macrophage precursors as natural cytotoxic effectors against *Candida albicans* and Yac-1 cells, *Eur. J. Immunol.* **16**:693–699.
275. Baccarini, M., Bistoni, F., and Lohmann-Matthes, M.-L., 1985, *In vitro* natural cell-mediated cytotoxicity against *Candida albicans*: macrophage precursors as effector cells, *J. Immunol.* **134**:2658–2665.
276. Decker, T., Lohmann-Matthes, M. L., and Baccarini, M., 1986, Heterogeneous activity of immature and mature cells of the murine monocyte-macrophage lineage derived from different anatomical districts against yeast-phase *Candida albicans*, *Infect. Immun.* **54**: 477–486.
277. Meister, H., Heymer, B., Schafer, H., and Haferkamp, O., 1977, Role of *Candida albicans* in granulomatous tissue reactions. II. *In vivo* degradation of *C. albicans* in hepatic macrophages of mice, *J. Infect. Dis.* **135**:235–242.
278. Lehrer, R. I., Ferrari, L. G., Patterson-Delafield, J., and Sorrell, T., 1980, Fungicidal activity of rabbit alveolar and peritoneal macrophages against *Candida albicans*, *Infect. Immun.* **28**: 1001–1008.
279. Thaler, M., Pastakia, B., Shawker, T. H., O'Leary, T., Pizzo, P. A., 1988, Hepatic candidiasis in cancer patients: the evolving picture of the syndrome, *Ann. Intern. Med.* **108**:88–100.
280. Carstensen, H., Widding, E., Storm, K., Ostergaard, E., Herlin, T., 1990, Hepatosplenic candidiasis in children with cancer. Three cases in leukemia children and a literature review, *Pediatr. Hematol. Oncol.* **7**:3–12.
281. Pizzo, P. A., and Walsh, T. J., 1990, Fungal infections in the pediatric cancer patient. *Semin. Oncol.* **17**:6–9.
282. Damodaran, V. N., and Chakravarty, S. C., 1973, Mechanism of production of *Candida* lesions in rabbits, *J. Med. Microbiol.* **6**:287–292.

283. Peterson, E. M., and Calderone, R. A., 1977, Growth inhibition of *Candida albicans* by rabbit alveolar macrophages, *Infect. Immun.* **15**;910–915.
284. Peterson, E. M., and Calderone, R. A., 1978, Inhibition of specific amino acid uptake in *Candida albicans* by lysosomal extracts from rabbit alveolar macrophages, *Infect. Immun.* **21:** 506–513.
285. Patterson-Delafield, J., Szklarek, D., Martinez, R. J., and Lehrer, R. I., 1981, Microbicidal cationic proteins of rabbit alveolar macrophages: amino acid composition and functional attributes, *Infect. Immun.* **31:**723–731.
286. Selsted, M. E., Brown, D. M., DeLange, R. J., and Lehrer, R. I., 1983, Primary structures of MCP-1 and MCP-2, natural peptide antibiotics of rabbit lung macrophages, *J. Biol. Chem.* **258:**14485–14489.
287. Ganz, T., Sherman, M. P., Selsted, M. E., and Lehrer, R. I., 1985, Newborn rabbit alveolar macrophages are deficient in two microbicidal cationic peptides, MCP-1 and MCP-2, *Am. Rev. Respir. Dis.* **132:**901–904.
288. D'Ambola, J. B., Sherman, M. P., Tashkin, D. P., Gong, H., Jr., 1988, Human and rabbit newborn lung macrophages have reduced anti-*Candida* activity, *Pediatr. Res.* **24:**285–290.
289. LaForce, F. M., Sharrar, R. G., and Arai, G., 1979, Characterization of yeast agglutinins in lavage fluid from lungs of rabbits, *J. Infect. Dis.* **140:**96–104.
290. Arai, T., Mikami, Y., and Yokoyama, K., 1977, Phagocytosis of *Candida albicans* by rabbit alveolar macrophages and guinea pig neutrophils, *Sabouraudia* **15:**171–177.
291. Gontijo, P. P., Jr., and Wayne, L. G., 1978, Phagocytic and antimicrobial activities of rabbit peritoneal and alveolar macrophages activated with *Mycobacterium bovis* BCG on *Candida albicans*, *Rev. Latinoam. Microbiol.* **20:**41–44.
292. Evans, Z. A., 1980, Tissue responses to the blastospores and hyphae of *Candida albicans* in the mouse, *J. Med. Microbiol.* **14:**307–319.
293. Nugent, K. M., and Onofrio, J. M., 1983, Pulmonary tissue resistance to *Candida albicans* in normal and in immunosuppressed mice, *Am. Rev. Respir. Dis.* **128:**909–914.
294. Lal, S., Mitsuyama, M., Amako, K., and Nomoto, K., 1986, Enhanced clearance of *Candida albicans* from lung after intratracheal immunization, *J. Clin. Lab. Immunol.* **21:**23–30.
295. Lal, S., Mitsuyama, M., Miyata, M., Ogata, N., Amako, K., and Nomoto, K., 1986, Pulmonary defense mechanism in mice. A comparative role of alveolar macrophages and polymorphonuclear cells against infection with *Candida albicans*, *J. Clin. Lab. Immunol.* **19:** 127–133.
296. Nugent, K. M., and Fick, R. B., Jr., 1987,Candidacidal factors in murine bronchoalveolar lavage fluid, *Infect. Immun.* **55:**541–546.
297. Warr, G. A., and Jakab, G. J., 1979, Alterations in lung macrophage antimicrobial activity associated with viral pneumonia, *Infect. Immun.* **26:**492–497.
298. Kurland, G., Cheung, A. T., Miller, M. E., Ayin, S. A., Cho, M. M., and Ford, E. W., 1988, The ontogeny of pulmonary defenses: alveolar macrophage function in neonatal and juvenile rhesus monkeys, *Pediatr. Res.* **23:**293–297.
299. Brummer, E., and Stevens, D. A., 1987, Activation of pulmonary macrophages for fungicidal activity by gamma-interferon or lymphokines, *Clin. Exp. Immunol.* **70:**520–528.
300. Stanley, V. C., and Hurley, R., 1969, The growth of *Candida* species in cultures of mouse peritoneal macrophages, *J. Pathol.* **97:**357–366.
301. Ozato, K., and Uesaka, I., 1974, The role of macrophages in *Candida albicans* infection *in vitro*, *Jap. J. Microbiol.* **18:**29–35.
302. Sasada, M., and Johnston, R. B., Jr., 1980, Macrophage microbicidal activity: correlation between phagocytosis-associated oxidative metabolism and the killing of *Candida* by macrophages, *J. Exp. Med.* **152:**85–98.
303. Nozawa, R. T., Sekiguchi, R., and Yokota, T., 1980, Stimulation by conditioned medium of L-929 fibroblasts, *E. coli* lipopolysaccharide, muramyl dipeptide of candidacidal activity of mouse macrophages, *Cell. Immunol.* **53:**116–124.

304. Maiti, P. K., Kumar, R., and Mohapatra, L. N., 1980, Candidacidal activity of mouse macrophages *in vitro*, *Infect. Immun.* **29**:477–482.
305. Mor, N., and Goren, M. B., 1987, Discrepancy in assessment of phagosome-lysosome fusion with two lysosomal markers in murine macrophages infected with *Candida albicans*, *Infect. Immun.* **55**:1663–1667.
306. Brummer, E., Morrison, C. J., and Stevens, D. A., 1985, Recombinant and natural gamma-interferon activation of macrophages *in vitro*: different dose requirements for induction of killing activity against phagocytizable and nonphagocytizable fungi, *Infect. Immun.* **49**: 724–730.
307. Kolotila, M. P., Smith, C. W., and Rogers, A. L., 1987, Candidacidal activity of macrophages from three mouse strains as demonstrated by a new method: neutral red staining, *J. Med. Vet. Mycol.* **25**:283–290.
308. Lehrer, R. I., Ferrari, L. G., and Patterson-Delafield, J., 1977, The fungicidal activity of rabbit macrophages, *Clin. Res.* **25**:490A.
309. Kagaya, K., and Fukazawa, Y., 1981, Murine defense mechanisms against *Candida albicans* infection. II. Opsonization, phagocytosis and intracellular killing of *C. albicans*, *Microbiol. Immunol.* **25**:807–818.
310. Brummer, E., and Stevens, D. A., 1989, Candidacidal mechanisms of peritoneal macrophages activated with lymphokines or gamma-interferon, *J. Med. Microbiol.* **28**:173–181.
311. Diamond, R. D., and Haudenschild, C. C., 1981, Monocyte-mediated serum independent damage to hyphal and pseudohyphal forms of *Candida albicans in vitro*, *J. Clin. Invest.* **67**: 173–182.
312. Patterson-Delafield, J., Martinez, R. J., and Lehrer, R. I., 1980, Microbicidal cationic proteins in rabbit alveolar macrophages: a potential host defense mechanism, *Infect. Immun.* **30**:180–192.
313. Lehrer, R. I., Szklarek, D., Selsted, M. E., and Fleischmann, J., 1981, Increased content of microbicidal cationic peptides in rabbit alveolar macrophages elicited by complete Freund adjuvant, *Infect. Immun.* **33**:775–778.
314. Mourad, S., and Friedman, L., 1961, Active immunization of mice against *Candida albicans*, *Proc. Soc. Exp. Biol. Med.* **106**:570–572.
315. Oblack, D. L., and Holder, I. A., 1979, Active immunization of mice against muscle damage mediated by *Candida albicans*, *J. Med. Microbiol.* **12**:503–506.
316. Banerjee, U., Mohapatra, L. N., and Kumar, R., 1985, Effect of immunization with formalin killed cells in complete Freund's adjuvant in experimental candidosis, *Indian J. Med. Res.* **81**:454–458.
317. Levy, R., Segal, E., and Eylan, E., 1981, Protective immunity against murine candidiasis elicited by *Candida albicans* ribosomal fractions, *Infect. Immun.* **31**:874–878.
318. Kuttin, E. S., Muller, J., Douchet, C., and Vogt, A., 1983, Immunological and pathological observations with *Candida albicans*-infected animals, *Sabouraudia* **21**:185–194.
319. Hasenclever, H. F., and Mitchell, W. O., 1963, Acquired immunity to candidiasis in mice, *J. Bacteriol.* **86**:401–406.
320. Giger, D. K., Domer, J. E., and McQuitty, J. T., Jr., 1978, Experimental murine candidiasis: pathologic and immune responses to cutaneous inoculation with *Candida albicans*, *Infect. Immun.* **19**:499–509.
321. Scheld, W. M., Calderone, R. A., Brodeur, J. P., and Sande, M. A., 1983, Influence of preformed antibody on the pathogenesis of experimental *Candida albicans* endocarditis, *Infect. Immun.* **40**:950–955.
322. Mourad, S., and Friedman, L., 1968, Passive immunization of mice against *Candida albicans*, *Sabouraudia* **6**:103–105.
323. Al-Doory, Y., 1970, An immune factor in baboon anti-*Candida* serum, *Sabouraudia* **8**: 41–47.
324. Pearsall, N. N., Adams, B. L., and Bunni, R., 1978, Immunologic responses to *Candida*

albicans. III. Effect of passive transfer of lymphoid cells or serum on murine candidiasis, *J. Immunol.* **120**:1176–1180.

325. Moser, S. A., and Domer, J. E., 1980, Effects of cyclophosphamide on experimental murine candidiasis, *Infect. Immun.* **27**:376–386.
326. Kuruganti, U., Henderson, L. A., Garner, R. E., Asofsky, R., Baker, P. J., and Domer, J. E., 1988, Nonspecific and *Candida*-specific immune responses in mice suppressed by chronic administration of anti-μ, *J. Leukoc. Biol.* **44**:422–433.
327. Matthews, R., Burnie, J., Smith, D., Clark, I., Midgley, J., Conolly, M., and Gazzard, B., 1988, *Candida* and AIDS: evidence for protective antibody, *Lancet* **2**:263–266.
328. Domer, J. E., and Hector, R. H., 1987, Enhanced immune responses in mice treated with penicillin/tetracycline or trimethoprim-sulfamethoxazole when colonized intragastrically with *Candida albicans*, *Antimicrob. Agents Chemother.* **31**:691–697.
329. Hurd, R. C., and Drake, C. H., 1953, *Candida albicans* infections in actively and passively immunized animals, *Mycopathol. Mycol. Appl.* **6**:290–297.
330. Miyake, T., Takeya, K., Nomoto, K., and Muraoka, S., 1977, Cellular elements in the resistance to *Candida* infection in mice. I. Contribution of T lymphocytes and phagocytes at various stages of infection, *Microbiol. Immunol.* **21**:703–725.
331. Banerjee, U., Mohapatra, L. N., and Kumar, R., 1984, Role of antibody in defence against murine candidosis, *Indian J. Med. Res.* **79**:760–765.
332. Giger, D. K., Domer, J. E., Moser, S. A., and McQuitty, J. T., Jr., 1978, Experimental murine candidiasis: pathological and immune responses in T-depleted mice, *Infect. Immun.* **21**: 729–737.
333. Grubek, H., Szymanska, D., Trenkner, E,. Weyman-Rzucidlo, D., and Okon, M., 1970, Effect of anti-lymphocyte serum on experimental infection with *Candida albicans* in guinea pigs, *Acta Microbiol. Pol. Ser.* A **2**:21–30.
334. Sohnle, P. G., Frank, M. M., and Kirkpatrick, C. H., 1976, Mechanisms involved in elimination of organisms from experimental cutaneous *Candida albicans* infections in guinea pigs, *J. Immunol.* **117**:523–530.
335. Cenci, E., Romani, L., Vecchiarelli, A., Puccetti, P., and Bistoni, F., 1989, Role of L3T4+ lymphocytes in protective immunity to systemic *Candida albicans* infection in mice, *Infect. Immun.* **59**:3581–3587.
336. Cenci, E., Romani, L., Vecchiarelli, A., Puccetti, P., and Bistoni, F., 1990, T cell subsets and IFN-gamma production in resistance to systemic candidosis in immunized mice, *J. Immunol.* **144**:4333–4339.
337. Vecchiarelli, A., Cenci, E., Puliti, M., Blasi, E., Puccetti, P., Cassone, A., and Bistoni, F., 1989, Protective immunity induced by low-virulence *Candida albicans*: cytokine production in the development of the anti-infectious state, *Cell. Immunol.* **124**:334–344.
338. Garner, R. E., Kuruganti, U., Czarniecki, C. W., Chiu, H. H., and Domer, J. E., 1989, *In vivo* immune responses to *Candida albicans* modified by treatment with recombinant murine gamma interferon, *Infect. Immun.* **57**:1800–1808.
339. Sen, P., Smith, J. K., Buse, M., Hsieh, H. C., Lavenhar, M. A., Lintz, D., and Louria, D. B., 1982, Modification of an experimental mouse *Candida* infection by human dialyzable leukocyte extract, *Sabouraudia* **20**:85–93.
340. Kagaya, K., Shinoda, R., and Fukazawa, Y., 1981, Murine defense mechanisms against *Candida albicans* infection. I. Collaboration of cell-mediated and humoral immunities in protection against systemic *C. albicans* infection, *Microbiol. Immunol.* **25**:647–654.
341. Sinha, B. K., Prasad, S., and Monga, D. P., 1987, Studies of experimental candidiasis in T-cell-deficient mice, *Zbt. Bacteriol. Hyg.* **A 256**:203–209.
342. Ashman, R. B., 1990, Murine candidiasis: cell-mediated immune responses correlate directly with susceptibility and resistance to infection, *Immunol. Cell Biol.* **68**:15–20.
343. Hurtrel, B., Lagrange, P. H., and Michel, J.-C., 1981, Absence of correlation between delayed-type hypersensitivity and protection in experimental systemic candidiasis in immunized mice, *Infect. Immun.* **31**:95–101.

344. Salvin, S. B., Peterson, R. D. A., and Good, R. A., 1965, The role of the thymus in resistance to infection and endotoxin toxicity, *J. Lab. Clin. Med.* **65**:1004–1022.
345. Gordee, R. S., and Simpson, P. J., 1967, Relationships of x-irradiation to the enhancement of *Candida albicans* infections, *J. Bacteriol.* **94**:6–12.
346. Rogers, T. J., Balish, E., and Manning, D. D., 1976, The role of thymus-dependent cell-mediated immunity in resistance to experimental disseminated candidiasis, *J. Reticuloendothel. Soc.* **20**:291–298.
347. Ashman, R. B., 1987, Mouse candidiasis. II. Host responses are T-cell dependent and regulated by genes in the major histocompatibility complex, *Immunogenetics* **25**: 200–203.
348. Sinha, B. K., Prasad, S., and Monga, D. P., 1987, Studies on the role of B-cells in the resistance of mice to experimental candidiasis, *Zbl. Bakt. Hyg.* **A 266**:316–322.
349. Mahanty, S., Greenfield, R. A., Joyce, W. A., and Kincade, P. W., 1988, Inoculation candidiasis in a murine model of severe combined immunodeficiency syndrome, *Infect. Immun.* **56**:3162–3166.
350. Marquis, G., Montplaisir, S., Pelletier, M., Auger, P., and Lapp, W. S., 1988, Genetics of resistance to infection with *Candida albicans* in mice, *Br. J. Exp. Path.* **69**:651–660.
351. Vecchiarelli, A., Cenci, E., Marconi, P., Rossi, R., Riccardi, C., and Bistoni, F., 1989, Immunosuppressive effect of cyclosporin A on resistance to systemic infection with *Candida albicans*, *J. Med. Microbiol.* **30**;183–192.
352. Bistoni, F., Marconi, P., Frati, L., Bonmassar, E., and Garcia, E., 1982, Increase of mouse resistance to *Candida albicans* infection by thymosin a1, *Infect. Immun.* **36**:609–614.
353. Marquis, G., Montplaisir, S., Pelletier, M., Mousseau, S., and Auger, P., 1986, Strain-dependent differences in susceptibility of mice to experimental candidosis, *J. Infect. Dis.* **154**:906–909.
354. Ashman, R. B., and Papadimitriou, J. M., 1987, Murine candidiasis. Pathogenesis and host responses in genetically distinct inbred mice, *Immunol. Cell Biol.* **65**:163–171.
355. Ashman, R. B., 1990, Murine candidiasis: susceptibility is associated with the induction of T cell-mediated, strain-specific autoreactivity, *Immunol. Cell Biol.* **68**:179–185.
356. Ashman, R. B., 1987, Murine candidiasis. III. Host inflammatory responses are regulated in part by class I MHC genes, *J. Immunogenet.* **14**:317–321.
357. Ashman, R. B., and Papadimitriou, J. M., 1989, Genetic regulation of pathogenesis and host responses in fungal infection, in *Immunology of Fungal Diseases* (E. Kurstak, ed.), Marcel Dekker, Inc., New York, pp. 347–371.
358. Fischer, A., Ballet, J. J., and Griscelli, C., 1978, Specific inhibition of in vitro *Candida*-induced lymphocyte proliferation by polysaccharidic antigens present in serum of patients with chronic mucocutaneous candidiasis, *J. Clin. Invest.* **62**:1005–1013.
359. Witkin, S. S., Yu, R., and Ledger, W. J., 1983, Inhibition of *Candida albicans*-induced lymphocyte proliferation by lymphocytes and sera from women with recurrent vaginitis, *Am. J. Obstet. Gynecol.* **147**:809–811.
360. Borkowsky, W., and Valentine, F. T., 1979, The proliferative response of human lymphocytes to antigen is suppressed preferentially by lymphocytes precultured with the same antigen, *J. Immunol.* **122**:1867–1873.
361. Rogers, T. J., and Balish, E., 1978, Effect of systemic candidiasis on blastogenesis of lymphocytes from germfree and conventional rats, *Infect. Immun.* **20**:142–150.
362. Skerl, K. G., Scheld, W. M., Alliegro, G. M., and Calderone, R. A., 1980, Lymphocyte blastogenesis during experimental endocarditis caused by *Candida albicans*, *J. Reticuloendothel. Soc.* **28**:495–506.
363. Piccolella, E., Lombardi, G., and Morelli, R., 1981, Generation of suppressor cells in the response of human lymphocytes to a polysaccharide from *Candida albicans*, *J. Immunol.* **126**: 2151–2159.
364. Rivas, V., and Rogers, T. J., 1983, Studies on the cellular nature of *Candida albicans*-induced suppression, *J. Immunol.* **130**:376–379.

365. Cuff, C. F., Rogers, C. M., Lamb, B. J., and Rogers, T. J., 1986, Induction of suppressor cells in vitro by *Candida albicans*, *Cell. Immunol.* **100**:47–56.
366. Nelson, R. D., Herron, M. J., McCormack, R. T., and Gehrz, R. C., 1984, Two mechanisms of inhibition of human lymphocyte proliferation by soluble yeast mannan polysaccharide, *Infect. Immun.* **43**:1041–1046.
367. Witkin, S. S., 1986, Inhibition of *Candida*-induced lymphocyte proliferation by antibody to *Candida albicans*, *Obstet. Gynecol.* **68**:696–699.
368. Podzorski, R. P., Herron, M. J., Fast, D. J., and Nelson, R. D., 1989, Pathogenesis of candidiasis. Immunosuppression by cell wall mannan catabolites, *Arch. Surg.* **124**:1290–1294.
369. Podzorski, R. P., Gray, G. R., and Nelson, R. D., 1990, Different effects of native *Candida albicans* mannan and mannan-derived oligosaccharides on antigen-stimulated lymphoproliferation in vitro, *J. Immunol.* **144**;707–716.
370. Cuff, C. F., Packer, B. J., and Rogers, T. J., 1989, A further characterization of *Candida albicans*-induced suppressor B-cell activity, *Immunology* **68**:80–86.
371. Cuff, C. F., Taub, D. D., and Rogers, T. J., 1989, The induction of T-suppressor cells with a soluble extract of *Candida albicans*, *Cell. Immunol.* **122**:71–82.
372. Segal, E., Schwartz, J., Altboun, Z., Vardinon, N., and Eylan, E., 1977, Suppressive action of *Candida albicans* on the immune response in guinea pigs, *Microbios* **19**:79–87.
373. Vardinon, N., and Segal, E., 1979, Suppressive action of *Candida albicans* on the immune response in mice, *Exp. Cell Biol.* **47**:275–280.
374. Segal, E., Sandovsky-Losica, H., and Vardinon, N., 1980, Suppressive action of cytoplasmic and metabolic extracts of *Candida albicans* on the immune response in guinea pigs, *Mycopathologia* **72**:121–128.
375. Cutler, J. E., and Lloyd, R. K., 1982, Enhanced antibody responses induced by *Candida albicans* in mice, *Infect. Immun.* **38**:1102–1108.
376. Carrow, E. W., and Domer, J. E., 1985, Immunoregulation in experimental murine candidiasis: specific suppression induced by *Candida albicans* cell wall glycoprotein, *Infect. Immun.* **49**:172–181.
377. Domer, J. E., Stashak, P. W., Elkins, K., Prescott, B., Caldes, G., and Baker, P. J., 1986, Separation of immunomodulatory effects of mannan from *Candida albicans* into stimulatory and suppressive components, *Cellular Immunol.* **101**:403–414.
378. Domer, J. E., Elkins, K. L., Ennist, D. L., Stashak, P. W., Garner, R. E., and Baker, P. J., 1987, Immunoenhancement of non-*Candida* antibody responses by *Candida albicans* cell wall glycoprotein, *Infect. Immun.* **55**:2619–2624.
379. Domer, J., Garner, R., and Befidi-Mengue, R., 1989, Mannan as an antigen in cell-mediated immune (CMI) assays and as a modulator of mannan-specific CMI, *Infect. Immun.* **57**: 693–700.
380. Garner, R. E., Childress, A. M., Human, L. G., and Domer, J. E., 1990, Characterization of *Candida albicans* mannan-induced, mannan-specific delayed hypersensitivity suppressor cells, *Infect. Immun.* **58**:2613–2620.
381. Piccolella, E., Lombardi, G., and Morelli, R., 1980, Human lymphocyte-activating properties of a purified polysaccharide from *Candida albicans*: B and T cell cooperation in the mitogenic response, *J. Immunol.* **125**:2082–2088.
382. Lombardi, G., Vismara, D., Piccolella, E., Colizzi, V., and Asherson, G. L., 1985, A nonspecific inhibitor produced by *Candida albicans* activated T cells impairs cell proliferation by inhibiting interleukin-1 production, *Clin. Exp. Immunol.* **60**:303–310.
383. Peat, S., Whelan, W. J., and Edwards, T. E., 1961, Polysaccharide of baker's yeast. Part IV. Mannan, *J. Chem. Soc.* (London) **1**:29–34.
384. Durandy, A., Fischer, A., Charron, D., and Griscelli, G., 1986, Specific binding of antigen onto human T lymphocytes, *J. Clin. Invest.* **77**:1557–1564.
385. Fischer, A., Pichat, L., Audinot, M., and Griscelli, C., 1982, Defective handling of mannan

by monocytes in patients with chronic mucocutaneous candidiasis resulting in a specific cellular unresponsiveness, *Clin. Exp. Immunol.* **47**:653–660.
386. Wright, C. D., Herron, M. J., Gray, G. R., Holmes, B., and Nelson, R. D., 1981, Influence of yeast mannan on human neutrophil functions: inhibition of release of myeloperoxidase related to carbohydrate-binding property of the enzyme, *Infect. Immun.* **32**:731–738.
387. Winner, H. I., and Hurley, R., 1964, *Candida Albicans*, J. & A. Churchill Ltd, London.
388. Taschdjian, D. L., Kozinn, P. J., and Caroline, L., 1964, Immune studies in candidiasis. III. Precipitating antibodies in systemic candidiasis, *Sabouraudia* **3**:312–320.
389. de Repentigny, L., and Reiss, E., 1984, Current trends in immunodiagnosis of candidosis and aspergillosis, *Rev. Infect. Dis.* **6**:301–312.
390. Bennett, J. E., 1987, Rapid diagnosis of candidiasis and aspergillosis, *Rev. Infect. Dis.* **9**: 398–402.
391. de Repentigny, L., 1989, Serological techniques for diagnosis of fungal infection, *Eur. J. Clin. Microbiol. Infect. Dis.* **81**:362–375.
392. Evans, E. G. V., Richardson, M. D., Odds, F. C., and Holland, K. T., 1973, Relevance of antigenicity of *Candida albicans* growth phases to diagnosis of systemic candidiasis, *Br. Med. J.* **4**:86–87.
393. Syverson, R. E., Buckley, H. R., and Campbell, C. C., 1975, Cytoplasmic antigens unique to the mycelial or yeast phase of *Candida albicans*, *Infect. Immun.* **12**:1184–1188.
394. Manning, M., and Mitchell, T. G., 1980, Morphogenesis of *Candida albicans* and cytoplasmic proteins associated with differences in morphology, strain, or temperature, *J. Bacteriol.* **144**: 258–273.
395. Ho, Y. M., Ng, M. H., and Huang, C. T., 1979, Antibodies to germinating and yeast cells of *Candida albicans* in human and rabbit sera, *J. Clin. Pathol.* **32**:399–405.
396. Smail, E. H., and Jones, J. M., 1984, Demonstration and solubilization of antigens expressed primarily on the surfaces of *Candida albicans* germ tubes, *Infect. Immun.* **45**:74–81.
397. Sundstrom, P. M., and Kenny, G. E., 1984, Characterization of antigens specific to the surface of germ tubes of *Candida albicans* by immunofluorescence, *Infect. Immun.* **43**: 850–855.
398. Ponton, J., and Jones, J. M., 1986, Analysis of cell wall extracts of *Candida albicans* by sodium dodecyl sulfate-polyacrylamide gel electrophoresis and western blot techniques, *Infect. Immun.* **53**:565–572.
399. Brawner, D. J., and Cutler, J. E., 1984, Variability in expression of a cell surface determinant as evidenced by an agglutinating monoclonal antibody, *Infect. Immun.* **43**:966–972.
400. Brawner, D. J., and Cutler, J. E., 1986, Variability in expression of cell surface antigens of *Candida albicans* during morphogenesis, *Infect. Immun.* **51**;337–343.
401. Chaffin, W. L., Skudlarek, J., and Morrow, K. J., 1988, Variable expression of a surface determinant during proliferation of *Candida albicans*, *Infect. Immun.* **56**:302–309.
402. Summers, D. F., Grollman, A. P., and Hasenclever, H. F., 1964, Polysaccharide antigens of *Candida* cell wall, *J. Immunol.* **92**:491–499.
403. Syverson, R. E., and Buckley, H. R., 1977, Cell wall antigens in soluble cytoplasmic extracts of *Candida albicans* as demonstrated by crossed-immunoelectrophoresis with concanavalin A, *J. Immunol. Methods* **18**:149–156.
404. Munoz, M., Estes, G., Kilpatrick, M., DiSalvo, A., and Virella, G., 1980, Purification of cytoplasmic antigens from the mycelial phase of *Candida albicans*: possible advantages of its use in *Candida* serology, *Mycopathologia* **72**:47–53.
405. Taschdjian, C. L., Dobbin, G. B., Kozinn, P. J., and Caroline, L., 1964, Experimental and preliminary clinical studies on antibody formation in systemic candidiasis, *Sabouraudia* **3**: 129–139.
406. Remington, J. S., Gaines, J. D., and Gilmer, M. A., 1972, Demonstration of *Candida* precipitins in human sera by counter-immunoelectrophoresis, *Lancet* **1**:413.
407. Axelsen, N. H., Buckley, H. R., Drouhet, E., Budtz-Jorgensen, E., Hattel, T., and Andersen,

P. L., 1975, Crossed immunoelectrophoretic analysis of precipitins to *Candida albicans* in deep *Candida* infection, *Scand. J. Immunol.* **4**:217–230.
408. Lehmann, P. F., and Reiss, E., 1980, Comparison by ELISA of serum anti-*Candida albicans* mannan IgG levels of a normal population and in diseased patients, *Mycopathologia* **70**: 89–93.
409. Meckstroth, K. L., Reiss, E., Keller, J. W., and Kaufman, L., 1981, Detection of antibodies and antigenemia in leukemic patients with candidiasis by enzyme-linked immunosorbent assay, *J. Infect. Dis.* **144**:24–32.
410. Odds, F. C., Evans, E. G. V., and Holland, K. T., 1975, Detection of *Candida* precipitins. A comparison of double diffusion and counter immunoelectrophoresis, *J. Immunol. Methods* **7**: 211–218.
411. Harding, S. A., Sanford, G. R., and Merz, W. G., 1976, Three serologic tests for candidiasis. Diagnostic value in distinguishing deep or disseminated infection from superficial infection or colonization, *Am. J. Clin. Pathol.* **65**:1001–1009.
412. Dee, T. H., and Rytel, M. W., 1977, Detection of *Candida* serum precipitins by counterimmunoelectrophoresis: an adjunct in determining significant candidiasis, *J. Clin. Microbiol.* **5**:453–457.
413. Merz, W. G., Evans, G. L., Shadomy, S., Anderson, S., Kaufman, L., Kozinn, P. J., Mackenzie, D. W., Protzman, W. P., and Remington, J. S., 1977, Laboratory evaluation of serological tests for systemic candidiasis: a cooperative study, *J. Clin. Microbiol.* **5**:596–603.
414. Kozinn, P. J., Taschdjian, C. L., Goldberg, P. K., Protzmann, W. P., Mackenzie, D. W. R., Remington, J. S., Anderson, S., and Seelig, M. S., 1978, Efficacy of serological tests in the diagnosis of systemic candidiasis, *Am. J. Clin. Pathol.* **70**:893–898.
415. Dee, T. H., Johnson, G. M., and Berger, C. S., 1981, Sensitivity specificity, and predictive value of anti-*Candida* serum precipitin and agglutinin quantification: comparison of counterimmunoelectrophoresis and latex agglutination, *J. Clin. Microbiol.* **13**:750–753.
416. Bisbe, J., Miro, J. M., Torres, J. M., Latorre, X., Alia, C., Amaral, M., Estivill, D., Mallalas, J., Trilla, A., and Soriano, E., 1989, Diagnostic value of serum antibody and antigen detection in heroin addicts with systemic candidiasis, *Rev. Infect. Dis.* **11**:310–315.
417. Syverson, R. E., Buckley, H. R., and Gibian, B. A., 1978, Increasing the predictive value positive of the precipitin test for the diagnosis of deep-seated candidiasis, *Am. J. Clin. Pathol.* **70**:826–831.
418. Glew, R. H., Buckley, H. R., Rosen, H. M., Moellering, R. C., and Fischer, J. E., 1978, Serologic tests in the diagnosis of systemic candidiasis. Enhanced diagnostic accuracy with crossed immunoelectrophoresis, *Am. J. Med.* **64**:586–591.
419. Poulain, D., and Pinon, J. M., 1986, Diagnosis of systemic candidiasis: development of co-counterimmunoelectrophoresis, *Eur. J. Clin. Microbiol.* **5**:420–426.
420. Poulain, D., and Pinon, J. M., 1986, Application of co-counterimmunoelectrophoresis, *Eur. J. Clin Microbiol.* **5**:427–434.
421. Bougnoux, M.-E., Hill, C., Moissenet, D., de Chavin, M. F., Bonnay, M., Vicens-Sprauel, I., Pietri, F., McNeil, M., Kaufman, L., Dupouy-Camet, J,. Bohuon, C., and Andremont, A., 1990, Comparison of antibody, antigen, and metabolite assays for hospitalized patients with disseminated or peripheral candidiasis, *J. Clin. Microbiol.* **28**:905–909.
422. Jones, J. M., 1980, Quantitation of antibody against cell wall mannan and a major cytoplasmic antigen of *Candida* in rabbits, mice and humans, *Infect. Immun.* **30**:78–89.
423. Matthews, R. C., Burnie, J. P., and Tabaqchali, S., 1984, Immunoblot analysis of the serologic response in systemic candidosis, *Lancet* **2**:1415–1418.
424. Au-Young, J. K., Troy, F. A., and Goldstein, E., 1985, Serologic analysis of antigen-specific reactivity in patients with systemic candidiasis, *Diag. Microbiol. Infect. Dis.* **3**:419–432.
425. Gatermann, S., Heesemann, J., and Laufs, R., 1986, Identification of *Candida albicans* antigens recognized by sera of patients with candidiasis, *Mykosen* **29**:343–354.
426. Strockbine, N. A., Largen, M. T., Zweibel, S. M., and Buckley, H. R., 1984, Identification

and molecular weight characterization of antigens from *Candida albicans* that are recognized by human sera, *Infect. Immun.* **43**:715-721.

427. Porsius, J. C., van Vleit, H. J., van Zeijl, J. H., Goessens, W. H., and Michel, M. F., 1990, Detection of an antibody response in immunocompetent patients with systemic candidiasis or *Candida albicans* colonization, *Eur. J. Clin. Microbiol. Infect. Dis.* **9**:352–355.
428. Greenfield, R. A., and Jones, J. M., 1981, Purification and characterization of a major cytoplasmic antigen of *Candida albicans*, *Infect. Immun.* **34**:469–477.
429. Fiss, E., and Buckley, H. R., 1987, Purification of actin from *Candida albicans* and comparison with the *Candida* 48,000-Mr Protein, *Infect. Immun.* **55**:2324–2326.
430. Mason, A. B., Brandt, M. E., and Buckley, H. R., 1989, Enolase activity associated with *Candida albicans* cytoplasmic antigen, *Yeast* **5**:231–239.
431. Franklyn, K. M., Warmington, J. R., Ott, A. K., and Ashman, R. B., 1990, An immunodominant antigen of *Candida albicans* shows homology to the enzyme enolase, *Immunol. Cell Biol.* **68**:173–178.
432. Matthews, R. C., Burnie, J. P., and Tabaqchali, S., 1987, Isolation of immunodominant antigens from sera of patients with systemic candidiasis and characterization of serological response to *Candida albicans*, *J. Clin. Microbiol.* **25**:230–237.
433. Matthews, R., Wells, C., and Burnie, J. P., 1988, Characterization and cellular localization of the immunodominant 47-Kda antigen of *Candida albicans*, *J. Med. Microbiol.* **27**:227–232.
434. Manning-Zweerink, M., Maloney, C. S., Mitchell, T. G., and Weston, H., 1986, Immunoblot analyses of *Candida albicans*-associated antigens and antibodies in human sera, *J. Clin. Microbiol.* **23**:46–52.
435. Reen, D. J., 1988, Assessment of immunoblotting in the investigation of specific immune responses during infection, *Biochem. Soc. Trans.* **16**:132–134.
436. Weller, B. I., Simmons, P. D., Ivanyl, L., 1990, Identification of immunodominant antigens of *Candida albicans* in patients with superficial candidosis, *Clin. Immunol. Immunopathol.* **54**: 347–353.
437. Greenfield, R. A., Bussey, M. J., Stephens, J. L., and Jones, J. M., 1983, Serial enzyme-linked immunosorbent assays for antibody to *Candida* antigens during induction chemotherapy for acute leukemia, *J. Infect. Dis.* **148**:275–283.
438. Strockbine, N. A., Largen, M. T., and Buckley, H. R., 1984, Production and characterization of three monoclonal antibodies to *Candida albicans* proteins, *Infect. Immun.* **43**:1012–1018.
439. Ho, Y. M., Ng, M. H., Teoh-Chan, C. H., Yue, P. C. K., and Huang, C. T., 1976, Indirect immunofluorescence assay for antibody to germ tube of *Candida albicans*—a new diagnostic test, *J. Clin. Pathol.* **29**:1007–1010.
440. Quindos, G., Ponton, J., and Cisterna, R., 1987, Detection of antibodies to *Candida albicans* germ tube in the diagnosis of systemic candidiasis, *Eur. J. Clin. Microbiol.* **6**:142–146.
441. Quindos, G., Ponton, J., Cisterna, R., and Mackenzie, D. W. R., 1990, Value of detection of antibodies to *Candida albicans* germ tube in the diagnosis of systemic candidosis, *Eur. J. Clin. Microbiol. Infect. Dis.* **9**:178–183.
442. McHugh, T. M., Wang, Y. J., Chong, H. O., Blackwood, L. L., and Stites, D. P., 1989, Development of a microsphere-based fluorescent immunoassay and its comparison to an enzyme immunoassay for the detection of antibodies to three antigen preparations from *Candida albicans*, *J. Immunol. Methods* **116**:213–219.
443. Weiner, M. H., and Yount, W. J., 1976, Mannan antigenemia in the diagnosis of invasive *Candida* infections, *J. Clin. Invest.* **58**:1045–1053.
444. Kerkering, T. M., Espinel-Ingroff, A., and Shadomy, S., 1979, Detection of *Candida* antigenemia by counterimmunoelectrophoresis in patients with invasive candidiasis, *J. Infect. Dis.* **140**:659–664.
445. Weiner, M. H., and Coats-Stephen, M., 1979, Immunodiagnosis of systemic candidiasis: mannan antigenemia detected by radioimmunoassay in experimental and human infections, *J. Infect. Dis.* **140**:989–993.

446. Segal, E., Berg, R. A., Pizzo, P. A., and Bennett, J. E., 1979, Detection of *Candida* antigen in sera of patients with candidiasis by an enzyme-linked immunosorbent assay-inhibition technique, *J. Clin. Microbiol.* **10:**116–118.
447. Reiss, E., Stockman, L., Kuykendall, R. J., and Smith, S. J., 1981, Dissociation of mannan-serum complexes and detection of *Candida albicans* mannan by enzyme immunoassay variations, *Clin. Chem.* **28:**306–310.
448. Lew, M. A., Siber, G. R., Donahue, D. M., and Maiorca, F., 1982, Enhanced detection with an enzyme-linked immunosorbent assay of *Candida* mannan in antibody-containing serum after heat extraction, *J. Infect. Dis.* **145:**45–56.
449. Lemieux, C., St-Germain, G., Vincelette, J., Kaufman, L., and deRepentigny, L., 1990, Collaborative evaluation of antigen detection by a commercial latex agglutination test and enzyme immunoassay in the diagnosis of invasive candidiasis, *J. Clin. Microbiol.* **28:** 249–253.
450. Kahn, F. W., and Jones, J. M., 1986, Latex agglutination test for detection of *Candida* antigens in sera of patients with invasive candidiasis, *J. Infect. Dis.* **153:**579–585.
451. deRepentigny, L., Marr, L. D., Keller, J. W., Carter, A. W., Kuykendall, R. J., Kaufman, L., and Reiss, E., 1985, Comparison of enzyme immunoassay and gas–liquid chromatography for the rapid diagnosis of invasive candidiasis in cancer patients, *J. Clin. Microbiol.* **21:** 972–979.
452. Phillips, P., Dowd, A., Jewesson, P., Radigan, G., Tweeddale, M. G., Clarke, A., Geere, I., and Kelly, M., 1990, Nonvalue of antigen detection immunoassays for diagnosis of candidemia, *J. Clin. Microbiol.* **28:**2320–2326.
453. Fung, J. C., Donta, S. T., and Tilton, R. C., 1986, *Candida* detection system (CAND-TEC) to differentiate between colonization and disease, *J. Clin. Microbiol.* **24:**542–547.
454. Fujita, S., Matsubara, F., and Matsuda, T., 1986, Enzyme-linked immunosorbent assay measurement of fluctuations in antibody titer and antigenemia in cancer patients with and without candidiasis, *J. Clin. Microbiol.* **23:**568–575.
455. Gentry, L. O., Wildinson, I. D., Lea, S. S., and Price, M. F., 1983, Latex agglutination test for detection of *Candida* antigen in patients with disseminated disease, *Eur. J. Clin. Microbiol.* **2:** 122–128.
456. Piens, M.-A., Guyotat, D., Archimbaud, E., Plotton, C., Maupas, J., Mojon, M., and Fiere, D., 1988, Evaluation of a *Candida* antigen detection test (Cand-TecTM) in the diagnosis of deep candidiasis in neutropenic patients, *Eur. J. Cancer Clin. Oncol.* **24:**1655–1659.
457. Ness, M. J., Vaughan, W. P., and Woods, G. L., 1989, *Candida* antigen latex test for detection of invasive candidiasis in immunocompromised patients, *J. Infect. Dis.* **159:**495–502.
458. Escuro, R. S., Jacobs, M., Gerson, S. L., Machicao, A. R., and Lazarus, H. M., 1989, Prospective evaluation of a *Candida* antigen detection test for invasive candidiasis in immunocompromised adult patients with cancer, *Am. J. Med.* **87:**621–627.
459. Price, M. F., and Gentry, L. O., 1990, *Candida* antigen latex test, *J. Infect. Dis.* **161:**807.
460. Price, M. F., and Gentry, L. O., 1986, Incidence and significance of *Candida* antigen in low-risk and high-risk patient populations, *Eur. J. Clin. Microbiol.* **5:**416–419.
461. Bailey, J. W., Sada, E., Brass, C., and Bennett, J. E., 1985, Diagnosis of systemic candidiasis by latex agglutination for serum antigen, *J. Clin. Microbiol.* **21:**749–752.
462. Ness, M. J., Rennard, S. I., Vaughn, W. P., Ghafouri, M. A., and Linder, J. A., 1988, Detection of *Candida* antigen in bronchoalveolar lavage fluid, *Acta Cytol.* **32:**347–352.
463. Araj, G. F., Hopfer, R. L., Chesnut, S., Fainstein, V., and Bodey, G. P., 1982, Diagnostic value of the enzyme-linked immunosorbent assay for detection of *Candida albicans* cytoplasmic antigen in sera of cancer patients, *J. Clin. Microbiol.* **16:**46–52.
464. Burnie, J., 1985, A reverse passive latex agglutination test for the diagnosis of systemic candidosis, *J. Immunol. Methods* **82:**267–280.
465. Ferraira, R. P., Yu, B., Niki, Y., and Armstrong, D., 1990, detection of *Candida* antigenuria in disseminated candidiasis by immunoblotting, *J. Clin. Microbiol.* **28:**1075–1078.

466. Longbottom, J. J., Murray, I. G., and Pepys, J., 1963, Diagnosis of fungal disease, in: *Clinical Aspects of Immunology*, (P. L. Gell and R. R. A. Coombs, eds.), Blackwell Scientific Publications, Oxford, p. 71.
467. Shannon, D. C., Gohnson, G., Rosen, F. S., and Austen, K. F., 1966, Cellular reactivity to *Candida albicans* antigen, *N. Engl. J. Med.* **275**:690–693.
468. Hassett, A. M., Woods, R. J., Temperley, I. J., and Mullins, G. M., 1977, Cell mediated immunity to recall antigens in vivo and in vitro, *Irish J. Med. Sci.* **146**:167–174.
469. Ferguson, A. C., Kershnar, H. E., Collin, W. K., and Stiehm, E. R., 1977, Correlation of cutaneous hypersensitivity with lymphocyte response to *Candida albicans*, *Am. J. Clin. Pathol.* **68**:499–540.
470. Hogan, T. F., Borden, E. C., Freeberg, B. L., Vieau, J., Hsieh, F.-Y., and Crowley, J., 1980, Enhancement in recall antigen responses by frequent, repetitive skin testing with *Candida*, mumps, and streptokinase-streptodornase in 38 normal adults, *Cancer Immunol. Immunother.* **10**:27–31.
471. Valdimarsson, H., Higgs, J. M., Wells, R. S., Yamamura, M., Hobbs, J. R., and Holt, P. J. L., 1973, Immune abnormalities associated with chronic mucocutaneous candidiasis, *Cellular Immunol.* **6**:348–361.
472. Bice, D. E., Lopez, M., Rothschild, H., and Salvaggio, J., 1974, Comparison of *Candida*-delayed hypersensitivity skin test size with lymphocyte transformation, migration inhibitory factor production and antibody titer, *Int. Arch. Allergy* **47**:54–62.
473. Alford, R. H., 1973, Transformation of lymphocytes of normal and hospitalized adults by *Candida albicans* extract, *Proc. Soc. Exp. Biol. Med.* **144**:826–829.
474. Foroozanfar, N., Yamamura, M., and Hobbs, J. R., 1974, Standardization of lymphocyte transformation to *Candida* immunogen, *Clin. Exp. Immunol.* **16**:301–310.
475. Esch, R. E., and Buckley, C. E., 1988, A novel *Candida albicans* skin test antigen: efficacy and safety in man, *J. Biol. Standardization* **16**:33–43.
476. Hurtrel, B., and Lagrange, P. H., 1978, Delayed type hypersensitivity reactions to *Candida albicans* in mice, *Ann. D'Immunol.* **C129**:653–668.
477. Ponton, J., Quindos, G., Regulez, P., and Cisterna, R., 1985, Cellular and humoral immune responses to *Candida-albicans* in subcutaneously infected mice, *Mycopathologia* **92**:11–18.
478. Suzuki, M., and Hayashi, Y., 1975, Skin reaction and macrophage migration inhibition tests for polysaccharides from *Aspergillus fumigatus* and *Candida albicans*, *Jap. J. Microbiol.* **19**: 355–362.
479. Reiss, E., Stone, S. H., and Hasenclever, H. R., 1974, Serologic and cellular immune activity of peptidoglucomannan fractions of *Candida albicans* cell walls, *Infect. Immun.* **9**:881–890.
480. Ellsworth, J. H., Reiss, E., Bradley, R. L., Chmel, H., and Armstrong, D., 1977, Comparative serological and cutaneous reactivity of candidal cytoplasmic proteins and mannans separated by affinity for concanavalin A., *J. Clin. Microbiol.* **5**:91–99.
481. Domer, J. E., and Moser, S. A., 1978, Experimental murine candidiasis: cell-mediated immunity after cutaneous challenge, *Infect. Immun.* **20**:88–98.
482. Moser, S. A., Domer, J. E., and Mather, F. J., 1980, Experimental murine candidiasis: cell mediated immunity after cutaneous challenge, *Infect. Immun.* **27**:140–149.
483. Levy, R., Segal, E., Eylan, E., and Barr-Nea, L., 1983, Cell-mediated immunity following experimental vaccinations with *Candida albicans* ribosomes, *Mycopathologia* **83**:161–168.
484. Crowle, A. J., 1975, Delayed hypersensitivity in the mouse, *Adv. Immunol.* **20**:197–264.
485. Dikeacou, R., and Drouhet, E., 1981, Cellular inflammatory response to fungal antigen studied with the skin window technique—correlation of the exudate with the evolution of the mycotic infection, *Agents Actions* **11**:631–633.
486. Winterrowd, G. E., and Cutler, J. E., 1983, *Candida albicans*-induced agglutinin and immunoglobulin E responses in mice, *Infect. Immun.* **41**:33–38.
487. Domer, J. E., 1989, *Candida* cell wall mannan: a polysaccharide with diverse immunologic properties, *CRC Crit. Rev. Microbiol.* **17**:33–51.

488. Ausiello, C. M., Spagnoli, G. C., Boccanera, M., Casalinuovo, I., Malavasi, F., Casciani, C. U., and Cassone, A., 1986, Proliferation of human peripheral blood mononuclear cells induced by *Candida albicans* and its cell wall fractions, *J. Med. Microbiol.* **22**:195–202.
489. Piccolella, E., Lombardi, G., and Morelli, R., 1980, Human lymphocyte-activating properties of a purified polysaccharide from *Candida albicans*: B and T cell cooperation in the mitogenic response, *J. Immunol.* **125**:2082–2088.
490. Gettner, S. M. M., and Mackenzie, D. W. R., 1981, Responses of human peripheral lymphocytes to soluble and insoluble antigens of *Candida albicans*, *J. Med. Microbiol.* **14**: 333–340.
491. Tollemar, J., Ringden, O., and Holmberg, K., 1989, *Candida albicans*: mannan and protein activation of cells from various human lymphoid organs, *Scand. J. Immunol.* **30**:473–480.
492. Kerridge, D., and Whelan, W. L., 1984, The polyene macrolide antibiotics and 5-fluorocytosine Molecular actions and interactions, in: *Mode of Action of Antifungal Agents*, (A. P. J. Trinci, and J. F. Riley, eds.), London, British Mycological Society, pp. 343–375.
493. Bennett, J. E., 1990, Antifungal agents, in: *Principles and Practice of Infectious Diseases*, third edition, (G. L. Mandell, R. G. Douglas, and J. E. Bennett, eds.), Churchill Livingstone, New York, pp. 361–370.
494. Smego, R. A., Jr., Perfect, J. R., Durack, D. T., 1984, Combined therapy with amphotericin B and 5-fluorocytosine for *Candida* meningitis, *Rev. Infect. Dis.* **6**:791–801.
495. Diasio, R. B., Bennett, J. E., and Myers, C. E., 1978, Mode of action of 5-fluorocytosine, *Biochem. Pharmacol.* **27**:703–707.
496. Polak, A., and Scholer, H. J., 1975, Mode of action of 5-fluorocytosine and mechanisms of resistance, *Chemother.* **21**:113–130.
497. Vanden Bossche, H., Bellens, D., Cools, W., Gorrens, J., Marichal, P., Verhoeven, H., Willemsens, G., De Coster, R., Beerens, D., Haelterman, C., Coene, M.-C., Lauwers, W., and Le Jeune, L., 1986, Cytochrome P-450: target for itraconazole, *Drug Dev. Res.* **8**:287–298.
498. Dismukes, W. E., 1988, Azole antifungal drugs: old and new, *Ann. Intern. Med.* **109**: 177–179.
499. Meunier, F., Gerain, J., Snoeck, R., Libotte, F., Lambert, C., Cueppers, A. M., 1987, Fluconazole therapy of oropharyngeal candidiasis in cancer patients, in: *Recent Trends in the Discovery, Development and Evaluation of Antifungal Agents*, (R. A. Fromtling, ed.), J. R. Prouss, Barcelona, pp. 169–174.
500. Dupont, B., and Drouhet, E., 1988, Fluconazole in the management of oropharyngeal candidosis in a predominantly HIV antibody-positive group of patients, *J. Med. Vet. Mycol.* **26**:67–71.
501. Meunier-Carpentier, F., Cruciani, M., Klastersky, J., 1983, Oral prophylaxis with miconazole or ketoconazole of invasive fungal disease in neutropenic cancer patients, *Eur. J. Cancer Clin. Oncol.* **19**:43–48.
502. Jones, P. G., Kauffmann, C. A., McAuliffe, L. S., Liepman, M. K., and Bergman, A. G., 1984, Efficacy of ketoconazole versus nystatin in prevention of fungal infections in neutropenic patients, *Arch. Intern. Med.* **144**:549–551.
503. Meunier, F., Leleux, A., Snoeck, R., Gerain, J., Lambert, C,. Cueppers, A. M., 1989, Chemoprophylaxis of fungal infections, in: *Diagnosis and Therapy of Systemic Fungal Infections*, (K. Holmberg, and R. Meyer, eds.), Raven Press, New York, pp. 167–177.
504. Edwards, J. E., Jr., 1990, *Candida* species, in: *Principles and Practice of Infectious Diseases*, (G. L. Mandell, R. G. Douglas, and J. E. Bennett, eds.), Churchill-Livingstone, New York, pp. 1943–1958.
505. Meunier, F., 1989, Candidiasis, *Eur. J. Clin. Microbiol. Infect. Dis.* **8**:438–447.

5

Chronic Mucocutaneous Candidiasis

SCOTT G. FILLER and JOHN E. EDWARDS, JR.

1. DESCRIPTION OF THE DISEASE

Chronic mucocutaneous candidiasis (CMC) is a relatively rare, heterogenous disorder characterized by chronic and recurrent infections of the skin, mucous membranes, and nails with *Candida* spp. Many afflicted patients have associated endocrinopathy or autoimmune phenomena. The primary immune defect usually involves either the thymus-dependent lymphocytes or the monocyte/macrophage system.

1.1. Candidal Infections

The disease generally begins either in infancy or before age 20. Onset after age 30, however, does occur, frequently in association with thymoma.[1] Oral thrush is often the first manifestation of CMC,[2] followed by skin and nail involvement.[3] In infants, the disease may begin as diaper rash, with candidal involvement of the nails appearing later in childhood.[4] The severity ranges from isolated involvement of the nails or intertriginous areas to disfiguring, exophytic *Candida* granulomas which may be located on the trunk, limbs, or scalp.[5] Other sites of infection include the larynx[6] and esophagus.[2] Complications of persistent mucocutaneous disease such as erosion through the

SCOTT G. FILLER and JOHN E. EDWARDS, JR. • Department of Medicine, Division of Infectious Diseases, Harbor-UCLA Medical Center, Torrance, California 90509, and UCLA School of Medicine, Los Angeles, California 90024.

Fungal Infections and Immune Responses, edited by Juneann W. Murphy *et al.* Plenum Press, New York, 1993.

esophagus with development of an esophageal-pleural fistula,[2] and oral[7,8] and esophageal[9] cancer at the site of chronic inflammation can occur. Disseminated *Candida* infections, however, are very rare.[10]

Histopathology of the infected areas reveals hyperkeratosis and parakeratosis of the stratum corneum with a dense, chronic inflammatory infiltrate of plasma cells and lymphocytes in the dermis. Occasional multinucleated giant cells are present. Both the yeast and mycelial phases of the organism may be seen, and the organism is limited to the stratum corneum; dermal invasion is very uncommon.[11]

1.2. Noncandidal Infections

In addition to chronic mucocutaneous infections with *Candida*, approximately 20% of patients have chronic dermatophyte infections with organisms such as *Trichophyton* and *Microsporum* spp. The symptoms caused by the dermatophyte infection may be more severe than those due to *Candida* and the cutaneous lesions caused by the two different types of fungi may appear identical. Fungal culture should be performed to distinguish dermatophyte from candidal lesions, since therapy differs significantly for the various fungi.[9,12–16]

Bacterial infections remain a major cause of morbidity and mortality in CMC. Approximately 50–80% of patients have significant noncandidal infections.[2,11,17] The most common problems are recurrent bacterial infections of soft tissues and the upper and lower respiratory tract. *Staphylococcus aureus*, *Streptococcus pneumoniae*, and *Haemophilus influenzae* are the most frequently isolated pathogens. There is also an increased incidence of severe viral infections including disseminated herpesvirus, and rubeola and varicella pneumonia. Apparently because of their defects in cell-mediated immunity (see below), patients with CMC have developed opportunistic infections such as disseminated cryptococcosis, histoplasmosis and *Mycobacterium avium*, as well as pneumonias due to *Pneumocystis*, *Nocardia*, *Aspergillus*, and *Actinomyces*. Although reminiscent of those seen in AIDS patients, these infections have been reported either before the era of widespread HIV disease[17] or in HIV negative individuals.[2]

1.3. Endocrinopathies

Approximately 50% of patients with CMC have an endocrinopathy.[2,12,17] Conversely, it is estimated that 73–78% of patients with type 1 autoimmune polyendocrinopathy have CMC.[18] Endocrine disorders associated with CMC include hypoparathyroidism, Addison's disease, hypothyroidism, hypopituitarism,[19] pernicious anemia, gonadal failure, and diabetes mellitus. Some patients have more than one endocrine defect. Hypoparathyroidism has been described as the most common endocrine defect.[12,20] In at least one report, however, patients with congenital absence of the thymus and para-thyroid glands (third and fourth branchial pouch or DiGeorge syndrome) were included with patients with CMC.[12] In a more recent study where such patients with more generalized immune defects were excluded from analysis, hypoparathyroidism occurred in conjunction with CMC less frequently.[2]

The endocrine defects usually occur in juvenile-onset CMC and have an autosomal recessive pattern of genetic transmission (see below).[12,20] They may develop either before, during, or up to 20 years after the onset of CMC.[2] Because of this late expression, it is recommended that patients with CMC have annual endocrinologic evaluations.[21] Siblings of patients with the endocrinopathy–candidiasis syndrome may have endocrine defects without CMC.[20] Correction of the endocrine abnormalities does not alter the severity of the candidal infections.

The cause of the endocrinopathy associated with CMC is unclear. Many of these patients have autoantibodies to adrenal, thyroid, or gastric tissues.[11] Thus, some investigators hypothesize that the endocrinopathy is due to circulating antibodies directed against various endocrine organs,[20] but others believe that the autoantibodies are only markers of immune-mediated organ damage.[18,22]

1.4. Autoimmune Diseases

Chronic mucocutaneous candidiasis is associated with other diseases of presumed autoimmune origin. These include Coombs positive hemolytic anemia, immune thrombocytopenic purpura, neutropenia, alopecia, and juvenile rheumatoid arthritis.[2] Some people with CMC have autoantibodies to melanocytes and many of these individuals have vitiligo. In contrast, people with either vitiligo alone or vitiligo associated with other diseases do not have antimelanocyte antibodies.[23]

Other diseases found in patients with CMC are less clearly autoimmune in etiology. For instance, patients with the endocrinopathy–candidiasis syndrome have a high incidence of keratoconjunctivitis.[12,24] The cause of this ocular disease is not known, but it is caused by neither hypoparathyroidism nor direct *Candida* infection of the eye.[25] Similarly, many of these patients have dysplastic teeth with enamel hypoplasia unrelated to hypoparathyroidism.[11,24,26,27] Finally, there have been reports of myopathy[28] and aminoaciduria[12] occurring in individuals with CMC.

1.5. Thymoma

Chronic mucocutaneous candidiasis associated with thymoma occurs in a unique subset of patients.[29] In contrast to the majority of cases of CMC where the disease begins in infancy or childhood, CMC with thymoma occurs typically late in life. For example, in one review of 27 patients who had thymoma and CMC,[1] the age of onset of CMC ranged from 34–71 years. The onset of mucocutaneous infections with *Candida* preceded the discovery of thymoma by an average of three years. Some thymomas, however, were diagnosed either simultaneously with or after the onset of candidal infections. There was no sex predominance. Of the thymomas that were examined histologically, 9 were benign and 11 were malignant.

As expected with thymoma, 30% of patients had myasthenia gravis and 11% had myositis. Additionally, 30% had hematologic abnormalities, including aplastic anemia, neutropenia, pancytopenia, red cell aplasia, pernicious anemia, and hypogammaglobulinemia. Fifteen percent had autoimmune cutaneous abnormalities; alopecia areata was the most common, but exfoliative erythrodermatitis and vitiligo were also seen. Serologic evidence of autoimmune disease was found in 41%. Antinuclear antibodies were detected most frequently, occurring in 26% of patients. Other individuals, however, had a positive rheumatoid factor or autoantibodies directed against thyroid tissue, thyroglobulin, gammaglobulin, or smooth muscle. Interestingly, no endocrine defects were seen in these patients.[1]

It is unknown whether the thymoma causes defects in the immune system which lead to the development of CMC or the thymic tumor is a pathologic compensation for a failure in the cell-mediated immune response mechanism.

1.6. Genetics

Approximately 25% of patients have other family members with the disease.[2,12] Both autosomal dominant[27,30] and autosomal recessive[31] modes of inheritance have been described. Chronic mucocutaneous candidiasis with endocrinopathy tends to be autosomal recessive,[31,32] whereas autosomal dominant disease is rarely associated with endocrine abnormalities or autoantibodies.[2,27,30] Some sporadic cases of CMC with endocrinopathy do occur, however.[12] The heterozygous parents of patients with autosomal recessive disease are not affected, and tests of their lymphocyte functions are usually normal.[33] Although CMC is associated with an increased frequency of HLA-B7, the HLA type is not associated with any specific syndrome.[21]

2. ETIOLOGY

The majority of patients are infected with *C. albicans*; infections with other species, such as *C. parapsilosis*, *C. tropicalis*, and *C. guilliermondii*, are occasionally seen.[3,14,16]

3. HOST DEFENSE

Chronic mucocutaneous candidiasis represents the common end result of a variety of different diseases. Some patients with persistent mucocutaneous candidal infections have severe, generalized defects in cell-mediated immunity.[12] Such diseases include AIDS,[34] Swiss-type agammaglobulinemia, and thymic dysplasia[35] or aplasia (as seen in the DiGeorge[36,37] and Nezelof-Allibone syndromes).[38] Hyperimmunoglobulin E (Job's syndrome) is also associated with chronic superficial *Candida* infections.[39] In order to meet the criteria for CMC, however, patients must have a narrower defect in cell-mediated immunity associated predominantly with chronic candidiasis (see below).[2,11]

3.1. Cell-Mediated Immunity

Total numbers of T cells are usually normal.[40] In a recent study,[2] however, 5 of 20 patients with CMC in childhood had T4/T8 ratios of ≤ 1.0. These patients had negative HIV antibody tests.

3.1.1. *Skin Testing*

Three patterns of delayed cutaneous hypersensitivity have been found in patients with CMC. Some individuals are completely anergic with no delayed cutaneous response to *Candida* extracts or unrelated antigens, such as streptokinase–streptodornase, mumps antigen, purified protein derivative, and histoplasmin. These individuals cannot be sensitized with 1-chloro-2,4 dinitrobenzene. A second subset exhibit no delayed response to *Candida* but have a positive skin test to other antigens. Finally, a third group of patients (18–40% of those tested) has a normal delayed cutaneous response to all antigens tested, including *Candida*.[2,11,41] This last finding provides evidence that the delayed cutaneous hypersensitivity response to *Candida* may not be a good marker for the immune defect(s) causing CMC. The weakness of this association is emphasized by the fact that 16–35% of normal individuals without CMC have negative *Candida* skin tests.[42,43]

3.1.2. *In Vitro Tests of Cell-Mediated Immunity*

Most patients with CMC and negative delayed cutaneous hypersensitivity to *Candida* also exhibit decreased lymphocyte transformation and/or diminished lymphocyte production of macrophage inhibitory factor (MIF) in response to *Candida* antigens *in vitro*. Production of MIF usually parallels the degree of lymphocyte transformation. A few patients have a discordance between their cutaneous and lymphocyte responses. For example, some individuals have a positive skin test and a negative lymphocyte transformation response to *Candida*, while others have the opposite findings.[11,17,41,42,44] In addition to being unresponsive to *Candida* antigens, a minority of individuals has diminished lymphocyte transformation to other stimuli such as phytohemagglutinin.[17] Lymphocyte cytotoxicity has been reported to be either normal[41] or decreased in CMC.[42]

The above patterns of abnormalities of delayed cutaneous hypersensitivity response and *in vitro* lymphocyte function are unrelated to the clinical manifestations of the disease. For example, one cannot differentiate the subset of patients with CMC and thymoma from other patients with CMC on the basis of *Candida* skin test reactivity or the degree of lymphocyte unresponsiveness to *Candida* antigens *in vitro*.[1] Additionally, even patients who might be expected to have the same underlying immune defect, such as those with the autosomal recessive form of CMC associated with endocrinopathy, have different patterns of *in vitro* and *in vivo* immune responses.[33] Therefore, it seems likely that these relatively insensitive methods of immunologic evaluation do not accurately define the defects in cell-mediated immunity responsible for the clinical manifestations of CMC.

Studies that compare the immunologic responses of patients with CMC to individuals without the disease have emphasized the lack of correlation between the immunologic defect(s) and the clinical manifestations of the disease. For instance, when the lymphocyte transformation responses to *Candida* and herpesvirus antigens were examined, patients with CMC had the same frequency of negative tests (50%) as did normal controls. In this study, although 92% of patients with CMC had decreased lymphocyte production of MIF in response to *Candida*, 62% of normals also had a low MIF response to this antigen.[42] Similarly, when the kindred of two siblings with autosomal recessive CMC without polyendocrinopathy were evaluated, both the patients and three asymptomatic relatives had diminished *Candida*-induced lymphocyte transformation, so this defect alone could not have been responsible for the patients' symptoms.[43] Finally, a third study found that 10 of 26 patients with CMC had completely normal immune evaluations including a positive *Candida* skin test, normal immunoglobulin profile, and positive lymphocyte transformation and MIF release in response to *Candida*.[41] Thus, in many individuals, the immunologic abnormality causing the disease cannot be determined by such simple tests. Further investigations into the cause of CMC have identified a variety of abnormalities that emphasize the heterogeneity of this syndrome. When interpreting these data, however, it must be remembered that the prevalence of many of these abnormalities in asymptomatic individuals is unknown.

The cause of decreased lymphocyte responsiveness to *Candida* and other antigens has been examined in some detail. It has been found that the abnormal tests of immune response resolve after treatment with antifungal agents in some,[45–48] but not all patients,[49] suggesting that the organism itself may inhibit immune function. Similarly, a minority of patients appear to have a circulating factor that inhibits the response of normal lymphocytes to *Candida*.[1,41,50] In many of these individuals, this factor becomes undetectable when antifungal therapy induces remission.[51–53] Further studies have revealed that this factor is not a protein and it may be a *Candida* mannan.[54] Mannan from *C. albicans* has been shown to cause activation of T-suppressor cells from normal individuals *in vitro*.[53,55] In some patients with CMC, mannan catabolism by monocytes is decreased. Based on this knowledge, it has been suggested that abnormally high levels of mannan may accumulate and stimulate specific suppressor T-cell activity.[56]

A possible test of this hypothesis occurred in a patient with CMC who underwent plasmapheresis. This patient had aplastic anemia, thymoma, and a circulating inhibitor of lymphocyte function. Following the procedure, the patient had transient resolution of his cutaneous anergy and the circulating inhibitor was temporarily undetectable. His candidal skin lesions became inflamed and painful after plasmapheresis but did not resolve, suggesting a partial, but incomplete return of immune function.[57] The concentration of circulation mannan was not reported in this study, so it is not clear if the salutary effect of plasmapheresis on immune function was caused by the removal of this factor.

3.2. Phagocytic Cells

Although it is believed that the majority of patients have normal neutrophil[41,58] and monocyte function,[24] some individuals with CMC and monocyte and/or neutrophil dysfunction have been described. The monocyte disorders have generally been restricted to abnormal responses to *C. albicans*, with normal responses to other stimuli. Reported abnormalities have included decreased chemiluminescence,[59] chemotaxis,[60] and candidacidal activity.[29,58,61] In a few patients, the decreased ability of monocytes to kill *Candida* has been caused by a circulating inhibitor.[29,58]

There are only a few reports of decreased neutrophil candidacidal activity in CMC. Three patients have been described with a circulating factor that inhibited *Candida* killing by both their neutrophils and leukocytes from normal individuals.[62] One of these patients had a thymoma and his serum also inhibited monocyte candidacidal activity.[29]

These abnormalities of monocyte or neutrophil function are probably exceptions; in the majority of individuals with CMC, these cells appear to be functionally intact and only T-cell abnormalities can be identified.[21] Furthermore, some patients have been found to have supra-normal monocyte function,[58] suggesting a compensatory response to chronic candidal infection.

3.3. Humoral Immunity

Elevated serum levels of IgG and IgA, as well as an increased IgA concentration in parotid fluid, are frequently seen in CMC.[11,13] A few patients, however, have been found to have either low IgG, IgA, or IgM levels.[2,16,17,32] Anti-*candida* antibody titers are usually elevated and may remain so even after therapy.[63,64] The few patients who lack detectable anti-*Candida* antibodies in either their serum[2,17] or saliva[42] are clinically indistinguishable from other patients with CMC, so it is unclear how variations in antibody responses relate to the different manifestations of the disease.

Even though gross tests of antibody function are often normal or elevated, there is some evidence that the antibody response to new antigens may be inhibited in CMC. For example, in one report, four children with CMC had positive Schick tests despite vaccination with diphtheria, indicating deficient antibody response to this antigen. They appeared to have normal IgM synthesis to the vaccine but diminished IgG response.[65]

One *in vitro* study also suggests that B-cell function may be abnormal in CMC. Although 18 patients with CMC had normal B-cell responses to pokeweed mitogen, the B-cells of all 14 patients with active disease were unable to produce antibody to *Candida* mannan *in vitro*. In contrast, all 10 patients whose B cells were tested during clinical remission showed a positive antibody response to this antigen. This lack of B-cell response to *Candida* during active disease was thought to be due to the suppressive action of T cells. Co-cultivation of T lymphocytes from these patients with normal B lymphocytes also resulted in decreased antibody response to mannan. Both CD-8(+) and CD-8(−) cells inhibited the normal B cells from producing antibodies, and this effect could be abrogated by irradiating the T cells from the CMC donor prior to co-cultivation. Suppression of B-cell response to *Candida* mannan could be induced in normal T cells by preincubating the T lymphocytes in high concentrations of mannan prior to mixing with B cells.[53] Similarly, inhibition of *in vivo* antibody response to nonfungal antigens has been demonstrated in mice infected with *C. albicans*.[66] Nevertheless, the clinical relevance of this phenomenon is unclear since, as stated previously, most patients with CMC have normal to high levels of circulating anti-*Candida* antibodies.

The majority of patients has normal complement activity,[21] although one family has been described as having CMC and low total hemolytic complement levels.[67] Sohnle *et al.*[68] have proposed that complement plays an important role in preventing disseminated *Candidiasis* in CMC. They examined skin biopsy specimens from the lesions of patients with CMC. Using immunofluorescence they found dermal deposits of C3 and properdin in 3 of 6 patients. *Candida* antigens were detected in the dermis of one patient, but neither immunoglobulins nor C4 were seen in any lesion. They suggested that soluble antigens from the superficial *Candida* infection diffuse into the dermis and directly activate C3, leading to the intense inflammatory infiltrate in this region. This inflammatory infiltrate does not cure the infection but prevents deeper invasion by the fungus.

It is not clear whether it is complement or the action of cytokines that cause recruitment of mononuclear cells to the area of the candidal lesions. The fact that disseminated candidiasis occurs rarely in CMC despite a massive mucocutaneous fungal burden, however, provides insight into the host defense against superficial and invasive candidiasis. First, the portion of the immune system that is responsible for the defense against disseminated candidiasis must be normal or upregulated in CMC. The cell types active in this defense are probably neutrophils and/or monocytes and macrophages. Second, the presence of extensive mucocutaneous candidal infections and the lack of disseminated candidiasis seen in both CMC and AIDS suggests that T cells play a critical role in mucocutaneous defense against this organism but are less important in preventing invasive disease. Nevertheless, since it appears that multiple types of defects in cell-mediated immunity can result in the syndrome of CMC, there may be several different mucocutaneous defense mechanisms against *Candida*, all of which are essential for preventing superficial infection. Alternatively, all the immune abnormalities that occur in CMC may affect one common pathway, thereby predisposing the individual to CMC. Further immunologic investigations into the cause of CMC, with emphasis on determining abnormalities of immune response on the molecular level, are needed to provide answers to these questions.

4. DIAGNOSIS AND PROGNOSIS

4.1. Diagnosis

The following criteria are used to diagnose CMC: 1) persistent mucous membrane and/or cutaneous *Candida* infections for at least two years, and 2) lack of evidence of generalized immunosuppression such as the DiGeorge syndrome or AIDS.[2,17]

4.2. Outcome

There have been few long-term studies to examine the outcome in patients with CMC. Generally, mucocutaneous infections remain a life-long problem, necessitating either frequent courses of antifungal agents or chronic suppressive therapy.[69] Bacterial infections, especially of the lung, are a major cause of morbidity and mortality in this group of patients.[9] For example, in a recent study, 10 of 43 patients with CMC in childhood had chronic pulmonary disorders such as bronchiectasis and obstructive lung disease. There were 6 deaths during the study; all occurred before the age of 22 and all were due to noncandidal infections.[2] These findings emphasize that the immune defects in CMC are not limited solely to an abnormal response to *Candida*.

5. TREATMENT

5.1. Immunotherapy

Until the development of ketoconazole, amphotericin B was the only effective agent for the long-term treatment of CMC (see section 5.2). Because of the need for hospitalization and the toxicities associated with prolonged amphotericin B therapy, considerable efforts were made to correct the underlying immune defects in CMC.

Since many patients had defects in thymus-dependent lymphocyte function, initial attempts at immunotherapy focused on thymus transplants and thymic hormone replacement. Treatment of CMC with implants of fetal thymic tissue have been occasionally successful.[19,70,71] There is a danger, however, of graft-versus-host reaction in patients with severely decreased cell-mediated immunity, and often the graft is rejected in patients with relatively intact immune systems.[71,72] In an effort to effect more specific thymic replacement, two patients with CMC were treated with weekly injections of thymosin fraction V. Following therapy, they improved both clinically and immunologically.[73] In another study, however, culturing lymphocytes from patients with CMC in media supplemented with thymosin did not improve the diminished lymphocyte response to *Candida* antigens.[74] Additionally, serum levels of thymopoetin have been shown to be normal in most patients with CMC, even those with thymoma.[40] For these reasons, immunotherapy of CMC based on augmentation of thymic function is infrequently used.

Other investigators have transplanted other cell lines to enhance cell-mediated immunity in CMC. At least one patient has had immunologic and clinical improvement after receiving a leukocyte transfusion from his HLA identical brother.[75] Another patient had remission of his disease after undergoing bone marrow transplantation for aplastic anemia.[76] While these observations reinforce the conclusion that CMC is caused by defects in leukocyte function, these procedures are impractical for most patients.

In addition to reconstitution of the immune system by transplantation, immunostimulatory agents have been evaluated in CMC. Both levamisole and cimetidine improve the immune response to *Candida* antigens as measured by delayed cutaneous hypersensitivity and *in vitro* lymphocyte proliferation response to *Candida* antigens. The clinical response to these agents, however, has been disappointing.[10,44,77,78] There is also one report of transient clinical and immunologic improvement in a patient with CMC who was vaccinated with heat killed *C. albicans*,[50] but this finding has not been replicated.

Some patients with CMC have been found to have iron deficiency due to decreased iron absorption. Replacement of iron stores has been associated with clinical improvement.[72] Similarly, Montes *et al.*[79] reported that seven of 12 patients with CMC had hypovitaminosis A even though their β-carotene levels were normal. This vitamin deficiency was considered to cause abnormal keratinization. No mention was made, however, of the patients' clinical status after the hypovitaminosis was corrected and follow-up studies have not been published.

The most successful form of immunotherapy for CMC is the use of transfer factor. This is a cell-free, dialyzable extract of leukocytes that can transfer delayed cutaneous hypersensitivity reactions to anergic individuals.[5]

How transfer factor works is unclear. Transfer factor contains a variety of proteins as determined by analysis with gel chromatography. Some of these fractions can transfer specific delayed hypersensitivity responses in humans while others act as nonspecific enhancers of lymphocyte function (reviewed by Kirkpatrick *et al.*).[80] Optimally, transfer factor should be obtained from donors who are *Candida* skin-test positive in order to be effective in treating CMC.[81]

Long-term remissions have been induced with either transfer factor alone,[82] or in conjunction with implantation of fetal thymic tissue.[83,84] At least one patient, however, has required bimonthly doses of transfer factor to sustain clinical remission.[85] The best results have been obtained when transfer factor is combined with specific treatment with antifungal agents.[21,46] In one retrospective study, 13 of 18 patients with successful correction of their immune defects with transfer factor in conjunction with antifungal antibiotics had long-term clinical remissions lasting up to 16 years.[4]

Although some forms of immunotherapy for CMC appear promising, they remain experimental. Since antifungal agents alone are often unable to induce long-term remission in CMC, and since drug resistance may develop when chronic suppressive therapy is used (see section 5.2), therapies to correct the immunologic abnormalities in CMC would be extremely useful, both in this disease and in AIDS. Thus, investigations aimed at defining and correcting the specific immune defects in CMC need to be continued.

5.2. Antifungal Agents

Mild mucosal involvement can be treated with topical nystatin or clotrimazole early in the disease.[10,72] Topical amphotericin B has been used, but it is only marginally effective.[11] The keratoconjunctivitis that is associated with CMC should be treated with topical steroids and surgical intervention should be avoided.[25]

Ketoconazole has been the mainstay of therapy for CMC. Cutaneous and mucosal lesions generally clear within one month, while nail involvement may take as long as one year to resolve with this drug.[47,69,86,87] One group of investigators has reported that patients with concurrent dermatophyte infections do not respond as well to ketoconazole,[14] but another group has been unable to replicate this finding.[69] Because relapse frequently occurs following cessation of treatment, maintenance therapy is usually required.[48,87–89] Patients with achlorhydria often require high doses of ketoconazole to achieve adequate serum levels. As might be expected, the bacterial infections that are frequently seen are usually unaffected by antifungal therapy.[87] Toxicities associated with ketoconazole use have included suppression of testosterone synthesis[87] and hepatitis.[69,90] Additionally, resistant *C. albicans* have been isolated from patients who have relapsed while on continuous ketoconazole maintenance therapy.[69] ketoconazole-resistant organisms isolated from some of these individuals have also been resistant, both *in vivo* and *in vitro*, to all azoles tested, including itraconazole and fluconazole.[91]

Other oral antifungal agents that have been used include 5-flurocytosine (5-FC), fluconazole, and itraconazole. Rapid development of resistance by the organism has been a problem with 5-FC, and it is generally ineffective for prolonged therapy when used as a single agent.[92] Fluconazole and itraconazole have been tried in a limited number of patients who have either failed ketoconazole therapy[16,93,94] or have poor ketoconazole absorption due to achlorhydria.[95] In one study, fluconazole was effective in treating 8 patients whose *Candida* isolates were still sensitive to ketoconazole.[93] Although there are two instances of treatment success with itraconazole in patients who did not respond to ketoconazole, the *in vitro* sensitivities of the organisms were not stated in these reports.[16,94] In contrast, a patient with CMC whose *Candida* isolate was resistant to ketoconazole failed to improve on itraconazole.[91] These data, combined with the information that cross-resistance among azoles may be common in ketoconazole-resistant organisms,[91] suggests that the newer azoles may be of limited efficacy in patients with CMC who fail ketoconazole therapy.

Intravenous amphotericin B remains the treatment of choice in patients whose infections are unresponsive to other antifungal agents. This drug may cause complete resolution and prolonged remission of disfiguring granulomas, but oral thrush often recurs earlier.[11] Some patients may respond to a total dose of as little as 25–40 mg.[96]

6. SUMMARY

Chronic mucocutaneous candidiasis is a heterogeneous disease that represents the common manifestation of a variety of defects in cell-mediated immunity. Although abnormal T-cell function appears to be common in this disease, 25% of patients have normal tests of immune function. The standard tests of immunologic competence are insensitive in this disease; the development of more sophisticated immunologic evaluations may lead to new insights into the cause and perhaps treatment. Since antifungal therapy is only palliative in CMC, and because of the high incidence of morbidity and mortality from noncandidal infections, it is clear that future research efforts should be continued toward the goal of restoring immunocompetence in these patients.

REFERENCES

1. Kirkpatrick, C. H., and Windhorst, D. B., 1979, Mucocutaneous candidiasis and thymoma, *Am. J. Med.* **66**:939–945.

2. Herrod, H. G., 1990, Chronic mucocutaneous candidiasis in childhood and complications of non-*Candida* infection: a report of the pediatric immunodeficiency study group, *J. Pediatr.* **116:**377–382.
3. Crislip, M. A., and Edwards, J. E., Jr., 1989, Candidiasis, *Infect. Dis. Clin. N. Am.* **3:**103–133.
4. Kirkpatrick, C. H., 1988, Chronic mucocutaneous candidiasis. Antibiotic and immunologic therapy, *Ann. N. Y. Acad. Sci.* **544:**471–480.
5. Jorizzo, J. L., 1982, Chronic mucocutaneous candidiasis: an update, *Arch. Dermatol.* **118:** 963–965.
6. Dudley, J. P., Kobayashi, R., Rosenblatt, H. M., Byrne, W. J., Ament, M. E., and Stiehm, E. R., 1980, *Candida* laryngitis in chronic mucocutaneous candidiasis: its association with *Candida* esophagitis, *Ann. Otol.* **80:**574–575.
7. Richman, R. A., Rosenthal, I. M., Solomon, L. M., and Karachorlu, K. V., 1975, Candidiasis and multiple endocrinopathy with oral squamous cell carcinoma complications, *Arch. Dermatol.* **111:**625–627.
8. McGurk, M., and Holmes, H., 1988, Chronic mucocutaneous candidiasis and oral neoplasia, *J. Laryngol. Otol.* **102:**643–645.
9. Kugelman, T. P., Cripps, D. J., and Harrell, E. R., 1963, *Candida* granuloma with epidermophytosis, *Arch. Dermatol.* **88:**150–157.
10. Edwards, J. E., Jr., Lehrer, R. I., Stiehm, E. R., Fischer, T. J., Young, L. S., 1978, Severe candidal infections: clinical perspective, immune defense mechanisms, and current concepts of therapy, *Ann. Intern. Med.* **89:**91–106.
11. Kirkpatrick, C. H., Rich, R. R., and Bennett, J. E., 1971, Chronic mucocutaneous candidiasis: model building in cellular immunity, *Ann. Intern. Med.* **74:**955–978.
12. Hermans, P. E., Ulrich, J. A., and Markowitz, H., 1969, Chronic mucocutaneous candidiasis as a surface expression of deep-seated abnormalities, *Am. J. Med.* **47:**503–519.
13. Chappler, R. R., Maibach, H. I., Conant, M. A., and Aly, R., 1978, Mucocutaneous candidiasis or mucocutaneous microbiosis? *JAMA* **239:**428–429.
14. Hay, R. J., Wells, R. S., Clayton, Y. M., and Wingfield, H. J., 1980, Treatment of chronic mucocutaneous candidosis with ketoconazole: a study of 12 cases, *Rev. Infect. Dis.* **2:** 600–605.
15. Shama, S. K., and Kirkpatrick, C. H., 1980, Dermatophytosis in patients with chronic mucocutaneous candidiasis, *J. Am. Acad. Dermatol.* **2:**285–294.
16. Burke, W. A., 1989, Use of itraconazole in a patient with chronic mucocutaneous candidiasis, *J. Amer. Acad. Derm.* **21:**1309–1310.
17. Chipps, B. E., Saulsbury, F. T., Hsu, S. H., Hughes, W. T., and Winkelstein, J. A., 1979, Non-candidal infections in children with chronic mucocutaneous candidiasis, *John Hopkins Med. J.* **144:**175–179.
18. Neufeld, M., Maclaren, N., and Blizzard, R., 1980, Autoimmune polyglandular syndromes, *Pediatr. Ann.* **9:**154–162.
19. Arvanitakis, C., and Knouss, R. F., 1973, Selective hypopituitarism, impaired cell-mediated immunity and chronic mucocutaneous candidiasis, *JAMA* **225:**1492–1495.
20. Blizzard, R. M., and Gibbs, J. H., 1968, Candidiasis: studies pertaining to its association with endocrinopathies and pernicious anemia, *Pediatrics* **42:**231–237.
21. Kirkpatrick, C. H., 1984, Host defense against fungal infections, *Am. J. Med.* **77**(suppl 4D): 1–12.
22. Kirkpatrick, C. H., 1989, Chronic mucocutaneous candidiasis, *Eur. J. Clin. Microbiol. Infect. Dis.* **8:**448–456.
23. Howanitz, N., Nordlund, J. L., Lerner, A. B., and Bystryn, J., 1981, Antibodies to melanocytes: occurrence in patients with chronic mucocutaneous candidiasis, *Arch. Dermatol.* **117:**705–708.
24. Gass, J. D. M., 1962, The syndrome of keratoconjunctivitis, superficial moniliasis, idiopathic hypoparathyroidism and Addison's disease, *Am. J. Ophthal.* **54:**660–674.

25. Wagman, R. D., Kazdan, J. J., Kooh, S. W., and Fraser, D., 1987, Keratitis associated with multiple endocrine deficiency, autoimmune disease, and candidiasis syndrome, *Am. J. Ophthalmol.* **103:**569–575.
26. Myllärniemi, S., and Perheentupa, J., 1978, Oral findings in the autoimmune polyendocrinopathy-candidosis syndrome (APECS) and other forms of hypoparathyroidism, *Oral Surg. Oral Med. Oral Path.* **45:**721–729.
27. Sams, W. M., Jr., Jorizzo, J. I., Snyderman, R., Jegasothy, B. V., Ward, F. E., Weiner, M., Wilson, J. G., Yount, W. J., and Dillard, S. B., 1979, Chronic mucocutaneous candidiasis. Immunologic studies of three generations of a single family, *Am. J. Med.* **67:**948–959.
28. Evans, R. A., Carter, J. H., Shenston, B., Smith, A., Hills, E., Walls, R., and Corbett, A., 1989, Candidiasis-endocrinopathy syndrome with progressive myopathy, *Quart. J. Med.* **70:**139–144.
29. Rycroft, R. J. G., Valdimarsson, H., Bannister, L. H., and Wells, R. S., 1976, Chronic mucocutaneous candidiasis of late onset, thymoma and myopathy. A report of four cases, *Clin. Exp. Dermatol.* **1:**59–74.
30. Okamoto, G. A., Hall, J. G., Ochs, H., Jackson, C., Rodaway, K., and Chandler, J., 1977, New syndrome of chronic mucocutaneous candidiasis, *Birth Defects: Original Article Series* **13**(3B)**:**117–125.
31. Ahonen, P., 1985, Autoimmune polyendocrinopathy-candidosis-ectodermal dystrophy (APECED): autosomal recessive inheritance, *Clin. Genetics* **27:**535–542.
32. Arulanantham, K., Dwyer, J. M., and Genel, M., 1979, Evidence for defective immunoregulation in the syndrome of familial candidiasis endocrinopathy, *N. Engl. J. Med.* **300:** 164–168.
33. Higgs, J. M., and Wells, R. S., 1972, Chronic muco-cutaneous candidiasis: associated abnormalities of iron metabolism, *Br. J. Dermatol.* **86**(suppl. 8):88–102.
34. Matthews, R., Smith, D., Midgley, J., Burne, J., Clark, I., Conolly, M., and Gazzard, B., 1988, *Candida* and AIDS: evidence for protective antibody, *Lancet* **2:**263–266.
35. Miller, M. E., and Schieken, R. M., 1967, Thymic dysplasia. A separable entity from "Swiss agammaglobulinemia," *Amer. J. Med. Sci.* **253:**741–750.
36. DiGeorge, A. M., 1968, Congenital absence of the thymus and its immunological consequences: concurrence with congenital hypoparathyroidism, *Birth Defects: Original Article Series* **4:**116–121.
37. Kretschmer, R., Say, B., Brown, D., and Rosen, F., 1968, Congenital aplasia of the thymus gland (DiGeorge's Syndrome), *N. Engl. J. Med.* **279:**1295–1301.
38. Nezelof, C., 1968, Thymic dysplasia with normal immunoglobulins and immunologic deficiency: pure alymphocytosis, *Birth Defects: Original Article Series* **4:**104–112.
39. Van Scoy, R. E., Hill, H. R., Ritts, R. E., Jr., and Quie, P. G., 1975, Familial neutrophil chemotaxis defect, recurrent bacterial infections, mucocutaneous candidiasis, and hyperimmunoglobulin E, *Ann. Intern. Med.* **82:**766–771.
40. Kirkpatrick, C. H., Greenberg, L. E., Chapman, S. W., Goldstein, G., Lewis, M., and Twomey, J. J., 1978, Plasma thymic hormone activity in patients with chronic mucocutaneous candidiasis, *Clin. Exp. Immunol.* **34:**311–317.
41. Valdimarsson, H., Higgs, J. M., Wells, R. S, Yamamura, M., Hobbs, J. R., and Holt, P. J. L., 1973, Immune abnormalities associated with chronic mucocutaneous candidiasis, *Cell. Immunol.* **6:**348–361.
42. Lehner, T., Wilton, J. M. A., and Ivanyi, L., 1972, Immunodeficiencies in chronic mucocutaneous candidosis, *Immunology* **22:**775–787.
43. Rothschild, H., Wilson, M., Lopez, M., Salvaggio, J., and Bice, D., 1976, An immunological investigation of a family with chronic mucocutaneous candidiasis, *Int. Arch. Allergy Appl. Immun.* **52:**291–296.
44. Ferguson, A. C., Kershnar, H. E., Collin, W. K., and Stiehm, E. R., 1977, Correlation of cutaneous hypersensitivity with lymphocyte response to *Candida albicans*, *Am. J. Clin. Pediatr.* **68:**499–504.

45. Imperato, P. J., Buckley, C. E., III, and Callaway, J. L., 1968, Candida granuloma: a clinical and immunological study, *Arch. Dermatol.* **97**:139–146.
46. Kirkpatrick, C. H., and Smith, T. K., 1974, Chronic mucocutaneous candidiasis: immunologic and antibiotic therapy, *Ann. Intern. Med.* **80**:310–320.
47. Drouhet, E., and Dupont, B., 1980, Chronic mucocutaneous candidosis and other superficial and systemic mycoses successfully treated with ketoconazole, *Rev. Infect. Dis.* **2**:606–619.
48. Montagnani, A., Negosanti, M., and Patrizi, A., 1986, Ketoconazole treatment of chronic mucocutaneous candidiasis, *Drugs Exp. Clin. Res.* **12**:409–412.
49. Mobacken, H., Lindholm, L., and Moberg, S., 1987, Immunologic studies in chronic mucocutaneous candidiasis before and after ketoconazole treatment, *Acta Derm. Venereol.* **67**: 257–260.
50. Burford-Mason, A. P., Matthews, R. C., and Williams, J. R. B., 1987, Transient abrogation of immunosuppression in a patient with chronic mucocutaneous candidiasis following vaccination with *Candida albicans*, *J. Infect.* **14**:147–157.
51. Aronson, I. K., Rieger, C. H. L., Soltani, K., Tkalcevic, V., Chang, W. C., Lorincz, A. L., and Matz, G., 1979, Late onset chronic mucocutaneous candidiasis with lymphoma and specific inhibitory factor, *Cancer* **43**:101–108.
52. Paterson, P. Y., Semo, R., Blumenschein, G., and Swelstarl, J., 1971, Mucocutaneous candidiasis, anergy and a plasma inhibitor of cellular immunity, *Clin. Exp. Immunol.* **9**:595–602.
53. Durandy, A., Fischer, A., Le Deist, F., Drouhet, E., and Griscelli, C., 1987, Mannan-specific and mannan-induced T-cell suppressive activity in patients with chronic mucocutaneous candidiasis, *J. Clin. Immunol.* **7**:400–409.
54. Fischer, A., Ballet, J., and Griscelli, C., 1978, Specific inhibitor of in vitro *Candida*-induced lymphocyte proliferation by polysaccharide antigens present in the serum of patients with chronic mucocutaneous candidiasis, *J. Clin. Invest.* **62**:1005–1013.
55. Lombardi, G., Di Massimo, A. M., Del Gallo, F., Vismar, D., Piccolella, E., Pugliese, O., and Colizzi, V., 1986, Mechanism of action of an antigen nonspecific inhibitory factor produced by human T cells stimulated by MPPS and PPD, *Cell. Immunol.* **98**:434–443.
56. Fischer, A., Pichat, L., Audinot, M., and Griscelli, C., 1982, Defective handling of mannan by monocytes in patients with chronic mucocutaneous candidiasis resulting in specific cellular unresponsiveness, *Clin. Exp. Immunol.* **47**:653–660.
57. Twomey, J. T., Waddel, C. C., Krantz, S., O'Rielly, R., L'Esperence, P., and Good, R. A., 1975, Chronic mucocutaneous candidiasis with macrophage dysfunction, a plasma inhibitor, and co-existent aplastic anemia, *J. Lab. Clin. Med.* **85**:968–977.
58. Van der Meer, J. W. M., Leijh, P. C. J., Van den Barselaar, M., and Van Furth, R., 1978, Functions of phagocytic cells in chronic mucocutaneous candidiasis, *Br. Med. J.* **1**:147–148.
59. Bortolussi, R., Faulkner, G., Lee, S. H. S., and Ozere, R., 1981, Phagocytosis of *Candida albicans* in chronic mucocutaneous candidiasis, *Pediatr. Res.* **15**:1287–1292.
60. Snyderman, R., Altman, L. C., Frankel, A., and Blaese, R. M., 1973, Defective mononuclear leukocyte chemotaxis: a previously unrecognized immune dysfunction, *Ann. Intern. Med.* **78**:509–513.
61. Yamazaki, M., Yasui, K., Kawaii, H., Miyagawa, Y., Komiyama, A., and Akstanl, T., 1984, A monocyte disorder in siblings with chronic candidiasis, *Am. J. Dis. Child.* **138**:192–196.
62. Chilgren, R. A., Oh, M. K., Hong, R., and Quie, P. G., 1969, Inhibition of PMN leukocyte killing of *Candida albicans*, *Fed. Proc.* **28**:498.
63. Axelsen, N. H., Kirkpatrick, C. H., and Buckley, R. H., 1974, Precipitins to *Candida albicans* in chronic mucocutaneous candidiasis studied by crossed immunoelectrophoresis with intermediate gel, *Clin. Exp. Immunol.* **17**:385–394.
64. Takeya, K., Nomoto, K., Matsumoto, T., Miyake, T., and Himeno, K., 1976, Chronic mucocutaneous candidiasis accompanied by enhanced antibody production, *Clin. Exp. Immunol.* **25**:497–500.
65. Cahill, L. T., Ainbender, E., and Glade, P. R., 1974, Chronic mucocutaneous candidiasis: T

cell deficiency associated with B cell dysfunction in man, *Cell. Immunol.* **14**:215–225.

66. Domer, J. E., Stashak, P. W., Elkins, K., Prescott, B., Caldes, G., and Baker, P. J., 1986, Separation of immunomodulatory effects of mannan from *Candida albicans* into stimulatory and suppressive components, *Cell. Immunol.* **101**:403–414.
67. Drew, J. H., 1973, Chronic mucocutaneous candidiasis with abnormal function of serum complement, *Med. J. Aust.* **1973**:77–79.
68. Sohnle, P. G., Frank, M. J., and Kirkpatrick, C. H., 1976, Deposition of complement components in the cutaneous lesions of chronic mucocutaneous candidiasis, *Clin. Immunol. Immunopath.* **5**:340–350.
69. Horsburgh, C. R., Jr., and Kirkpatrick, C. H., 1983, Long-term therapy of chronic mucocutaneous candidiasis with ketoconazole: experience with twenty-one patients, *Amer. J. Med.* **74**(Suppl 4D):23–29.
70. Levy, R. L., Bach, M. L., Huang, S., Bach, F. H., Hong, R., Ammann, A. J., Bortin, M., and Kay, H. E. M., 1971, Thymic transplantation in a case of chronic mucocutaneous candidiasis, *Lancet* **2**:898–900.
71. Hong, R., 1986, Reconstitution of T-cell deficiency by thymic hormone or thymus transplantation therapy, *Clin. Immunol. Immunopathol.* **40**:136–141.
72. Higgs, J. M., and Wells, R. S., 1973, Chronic mucocutaneous candidiasis: new approaches to treatment, *Br. J. Dermatol.* **89**:179–190.
73. Wara, D. W., and Ammann, A. J., 1978, Thymosin treatment of children with primary immunodeficiency disease, *Transplant. Proc.* **10**:203–209.
74. Akhter, J., Merker, R. I., Rieger, C. H. L., and Rothberg, R. M., 1980, Effect of thymosin on lymphocytes from patients with chronic mucocutaneous candidiasis and endocrinopathies, *J. Allergy Clin. Immunol.* **65**:34–40.
75. Valdimarsson, H., Holt, P. J. L., Moss, P. D., and Hobbs, J. R., 1972, Treatment of chronic mucocutaneous candidiasis with leukocytes from HL-A compatible sibling, *Lancet* **1**:469–472.
76. Deeg, H. J., Lum, L. G., Sanders, J., Levy, G. J., Sullivan, K. M., Beatty, P., Thomas, E. D., and Storb, R., 1986, Severe aplastic anemia associated with chronic mucocutaneous candidiasis. Immunologic and hematologic reconstitution after allogeneic bone marrow transplantation, *Transplantation* **41**:583–586.
77. Hay, R. J., 1981, Management of chronic mucocutaneous candidiasis, *Clin. Exp. Dermatol.* **6**: 515–519.
78. Jorizzo, J. L., Sams, W. M., Jegasothy, B. V., and Olansky, A. J., 1980, Cimetidine as an immunomodulator: chronic mucocutaneous candidiasis as a model, *Ann. Intern. Med.* **92**: 192–195.
79. Montes, L. F., Krumdieck, C., and Cornwell, P. E., 1973, Hypovitaminosis A in patients with chronic mucocutaneous candidiasis, *J. Infect. Dis.* **2**:227–230.
80. Kirkpatrick, C. H., Greenberg, L. E., and Petersen, E. A., 1983, Transfer Factor, *Lymphokines* **8**:1–39.
81. Littmann, B. H., Rocklin, R. E., Parkman, R., and David, J. R., 1978, Transfer factor treatment of chronic mucocutaneous candidiasis: requirement for donor reactivity to *Candida* antigen, *Clin. Immunol. Immunopath.* **9**:97–110
82. Valdimarsson, H., Wood, C. B. S., Hobbs, J. R., and Holt, P. J. L., 1972, Immunologic features in a case of chronic mucocutaneous candidiasis and its treatment with transfer factor, *Clin. Exp. Immunol.* **11**:151–163.
83. Ballow, M., and Hyman, L. R., 1977, Combination immunotherapy in chronic mucocutaneous candidiasis: synergism between transfer factor and fetal thymus tissue, *Clin. Immunol. Immunopath.* **8**:504–512.
84. Kirkpatrick, C. H., Ottenson, E. A., Smith, T. K., Wells, S. A., and Burdick, J. F., 1976, Reconstitution of defective cellular immunity with foetal thymus and dialysable transfer

factor: long-term studies in a patient with chronic mucocutaneous candidiasis, *Clin. Exp. Immunol.* **23**:414–428.

85. MacKie, R. M., 1979, Mucocutaneous candidiasis responsive to transfer factor therapy, *J. Royal Soc. Med.* **72**:926–927.
86. Graybill, J. R., Herndon, J. H., Jr., Kniker, W. T., and Levine, H. B., 1980, Ketoconazole treatment of chronic mucocutaneous candidiasis, *Arch. Dermatol.* **116**:1137–1141.
87. Mobacken, H., and Moberg, S., 1986, Ketoconazole treatment of 13 patients with chronic mucocutaneous candidiasis, *Dermatologica* **173**:229–236.
88. Hay, R. J., and Clayton, Y. M., 1982, The treatment of patients with chronic mucocutaneous candidiasis and *Candida* onychomycosis with ketoconazole, *Clin. Exp. Immunol.* **7**:155–162.
89. Rosenblatt, H. M., and Stiehm, E. R., 1983, Therapy of chronic mucocutaneous candidiasis, *Am. J. Med.* **74**(Suppl 1B):20–22.
90. Janssen, P. A. J., and Symoens, J. E., 1983, Hepatic reactions during ketoconazole treatment, *Am. J. Med.* **74**(Suppl 1B):80–85.
91. Smith, K. J., Warnock, D. W., Kennedy, C. T. C., Johnson, E. M., Hopwood, V., Van Cutsem, J., and Vanden Bossche, H., 1986, Azole resistance in *Candida albicans*, *J. Med. Vet. Mycology* **24**:133–144.
92. Bennett, J. E., 1977, Flucytosine, *Ann. Intern. Med.* **86**:319–322.
93. Hay, R. J., and Clayton, Y. M., 1988, Fluconazole in the management of patients with chronic mucocutaneous candidosis, *Br. J. Dermatol.* **119**:683–685.
94. Phillips, P., Fetchick, R., Weisman, I., Foshee, S., and Graybill, J. R., 1987, Tolerance to and efficacy of itraconazole in treatment of systemic mycoses: preliminary results, *Rev. Infect. Dis.* **9**(suppl 1):87–93.
95. Drew, R. H., Perfect, J. R., and Gallis, H. A., 1988, Use of fluconazole in a patient with documented malabsorption of ketoconazole, *Clin. Pharm.* **7**:622–623.
96. Medoff, G., Dismukes, W. E., Meade, R. H., 1972, A new therapeutic approach to *Candida* infections, *Arch. Intern. Med.* **130**:241–245.

6

Sporotrichosis

E. NAN SCOTT and HAROLD G. MUCHMORE

1. DESCRIPTION OF THE DISEASE

Sporotrichosis is a chronic disease caused by a soil fungus, *Sporothrix schenckii*. Typically, the disease produces suppurating nodules along the lymphatics of the skin and subcutaneous tissues with new lesions forming proximally as the older distal lesions heal slowly. Rarely, the process extends directly or via the blood stream beyond the skin to deeper structures, especially bone and synovial tissues. Lesions occur in other organs even less frequently.

1.1. Portal of Entry

S. schenckii gains entrance into the host by traumatic implantation into the skin or by inhalation into the lungs. The initial cutaneous lesion begins many weeks (1–10 or longer) after the penetrating skin injury and develops into a suppurating granuloma consisting of histiocytes and giant cells with neutrophils accumulating in the center, the whole surrounded by lymphocytes and plasma cells.[1] The pulmonary entrance route rarely produces clinical disease but probably serves frequently as a sufficient antigenic stimulus to establish immunity.

From the initial lesion, the fungus spreads along local lymphatic channels forming the chain of indolent nodular and ulcerating granulomata, which typify lymphocutaneous sporotrichosis. This form of the disease is the most common clinical presentation. Other tissues can become involved by

---

E. NAN SCOTT and HAROLD G. MUCHMORE • Departments of Medicine and Microbiology and Immunology, Infectious Disease Section, College of Medicine, University of Oklahoma Health Sciences Center, Oklahoma City, Oklahoma 73190, and Department of Veterans Affairs Medical Center, Oklahoma City, Oklahoma 73104.

Fungal Infections and Immune Responses, edited by Juneann W. Murphy *et al.* Plenum Press, New York, 1993.

direct extension and, less frequently, by hematogenous dissemination. Bones, joints, and other synovial tissues, including tendon sheaths and bursae, are the structures most frequently involved after the lymphocutaneous form. Disseminated sporotrichosis rarely produces lung lesions when the primary infection is at some other body site. Primary pulmonary sporotrichosis probably results from inhalation of conidia, in a manner similar to the infection sequence in primary pulmonary histoplasmosis or coccidioidomycosis.[1–5] Central nervous system sporotrichosis is distinctly rare and only 15 patients have been reported[6] in whom *S. schenckii* was recovered and clearly identified. This form of the disease results from hematogenous dissemination, and, because it is so difficult to establish a definitive diagnosis, it seems reasonable to believe that additional unrecognized cases occur.

1.2. Clinical Manifestations

1.2.1. Cutaneous Sporotrichosis

The primary lesion develops in the skin at the site of inoculation, typically the hand or fingers, but can be located on any (exposed) part of the body, including the face. The initial small nodule enlarges slowly, reddens, becomes pustular, and ulcerates, releasing purulent material from which the organism is readily cultured. In the lymphocutaneous form of the disease, a chain of nodules develop along lymphatic channels of the skin, with the older, distal lesions ulcerating and draining, and the younger, more proximal lesions forming subcutaneous nodules that attach to the skin as they age and begin to ulcerate. The lesions are usually not painful, but extensive disease may cause functional impairment. Some patients exhibit no lymphangitic spread, and the disease presents as an indolent ulcerating plaque which persists for years if untreated. This is called fixed cutaneous or plaque sporotrichosis. The frequency of this form of the disease varies from 10% in some series from the United States[7] to more than 50% in patients reported from other geographic areas.[8] We speculate that immunity to the fungus previously established by earlier contact(s) plays a role in restricting the lesion to the area of a new traumatic reinoculation.

The thermal dimorphism of this fungus is also a factor in the restriction of lesions to skin and subcutaneous lymphatics, and many *S. schenckii* isolates from lymphocutaneous lesions are unable to grow well at temperatures above 35°C.[9,10] Spontaneous healing of the cutaneous forms of sporotrichosis has been reported,[11,12] but more often, without treatment, the lesions persist for years, progressing slowly with draining and scarring.

1.2.2. Pulmonary

Pulmonary sporotrichosis probably results from inhalation of the fungus into the lung, and most patients with this form of the disease have underlying structural abnormalities of the lungs, such as chronic obstructive pulmonary disease and a history of smoking.[13] The usual clinical presentation of pulmonary sporotrichosis is a low-grade pneumonitis[14,15] that resembles that seen in other granulomatous diseases, such as tuberculosis or histoplasmosis. The pulmonary form of the disease tends to follow an indolent, progressive course, but cavitation is not frequent. Lung biopsies may reveal only acute and chronic granulomatous inflammation, and the yeast cells are often difficult to visualize in tissue sections.[13,16] The organism grows readily, however, and cultures of bronchial washings and biopsy tissue will usually recover *S. schenckii*.[13] Pulmonary disease caused by *S. schenckii* remains uncommon, with fewer than 100 verified culturally cases in world literature.[17] This form of the disease is most frequently confined solely to the lung, with no evidence of sporotrichosis elsewhere in the body.[18]

1.2.3. Extracutaneous Sporotrichosis

Lesions of deeper tissues may occur in almost any organ but there is a distinct predilection for localization in the joints especially of the extremities and in the long bones adjacent to these joints. The resulting arthritis is often confused with rheumatoid arthritis until bone destruction and draining fistulae encourage culture efforts to establish the etiology of the osteomyelitis. Lymphocutaneous lesions and lymphangitic spread is not prominent in these patients, although multiple joints may become involved in succession.

The inflammatory response in sporotrichosis of bones and joints is most often low grade and chronic, but the patient usually has pain on motion and the involved areas may be warm and red. Functional impairment in joint sporotrichosis may be very severe. Bone and joint destruction may require drainage and other procedures, including joint fusion, but these may fail because of continuing osteomyelitis.

Sporotrichotic lesions occur infrequently in many other organs, such as the eye, prostate, oral mucosa, or larynx, and the clinical manifestations in these patients is dependent on the organ involved. Involvement of the central nervous system and meninges is decidedly rare.[6] Because lesions in these organs all develop without contiguous skin lesions and without direct trauma to the infected site, each is presumed to result from hematogenous dissemination. Recovery of the fungus from extracutaneous lesions may be difficult, particularly in central nervous system disease, possibly related to low numbers of organisms present in the involved tissue. Elevated serum antibody titers are usual in these patients, and this finding should encourage repeated culture attempts.[6]

2. ETIOLOGY

S. schenckii is a dimorphic fungus capable of growing in yeast form at 37°C in culture and in the infected host or as a mold at 25°C. At room temperature, rapidly growing, off-white mold colonies develop, and these become pigmented with age, turning yellow, brown, or black. Oval hyaline conidia (2–3 × 3–6 μm) are borne on conidiophores, often forming a cluster resembling a "daisy" or "palm tree." This flowerlike cluster of conidia is characteristic of *S. schenckii* and is not produced by any other pathogenic fungus. In some isolates, a second type of conidia is seen, consisting of thick-walled, often triangular, black conidia arising laterally from the hyphae. Although usually occurring in low numbers and difficult to find in infected tissues of patients, when seen, *S. schenckii* cells appear as round, oval, or cigar-shaped yeast cells that vary in size from 1–3 × 3–10 μm. The teleomorph (sexual) state of this organism remains unknown. It has been suggested that *S. schenckii* is a member of the Ceratocystis fungal complex,[19] but recently Shadomy[20] has demonstrated that cross-matings of *S. schenckii* form perithecia and ascospores similar to the Chaetomiaceae.

3. EPIDEMIOLOGY

S. schenckii is widely distributed in nature growing on plant debris in soil and on the bark of trees, shrubs, and garden plants.[1] It has also been recovered from air, water, and a variety of other substrates. The organism is widely, but not uniformly, distributed in the environment, and the conditions that determine its natural occurrence or nonoccurrence have not been delineated precisely. It has been suggested that the presence of other molds may promote or inhibit the growth of *S. schenckii*.[21] The fungus and the disease occurs throughout much of the world, primarily in the temperate and tropical zones, but the abundance of the organism and the reported incidence of the disease shows great geographic variation, reflecting, in part, the location of interested medical investigators.

From its various environmental locations, *S. schenckii* is readily available for traumatic inoculation into and beneath the skin of man and other mammals. The penetrating trauma that introduces the conidia of the fungus is accomplished most often by splinters, thorns, or woody fragments of plants, but any activity involving contact with plant or plant products (e.g., sphagnum moss) and minor skin trauma may be adequate to initiate the infectious process.[22–24] These activities include occupational exposure, such as basketweaving, adobe brick making, horticulture and farming, and outdoor activities, such as hiking, camping, sports, and berry picking. The infecting trauma is often minor, and it is apparent from skin test studies that the majority of *S. schenckii* inoculations only promote immunity and do not lead to clinical disease.

3.1. Skin Test Surveys

Sporotrichin, ethanol-precipitated supernate of old-yeast-form cultures, has been used in epidemiological skin test surveys. Delayed cutaneous hypersensitivity is usually present in lymphocutaneous sporotrichosis. Skin test surveys in normal humans and those hospitalized with no evidence of fungal disease estimate the prevalence of subclinical exposure in nonendemic areas at 10% or less.[25,26] Occupational exposure of plant nursery workers and grass handlers is higher (32–100%), and the development of positive skin tests correlates with the length of time spent at these occupations.[27,28] In some of the endemic areas of Mexico, skin testing demonstrates a 26% prevalence of *S. schenckii* infection.[28] This indicates that this procedure is a valuable epidemiologic tool for assessing the prevalence of the disease or exposure to the organism.[29] A single sero-epidemiologic study, utilizing an enzyme immunoassay against a crude soluble *S. schenckii* antigen, showed that 18% of 300 healthy residents of Oklahoma had measurable antibody,[6,30] and distribution of positive sera by age group cohorts demonstrated that the acquisition of antibody to *S. schenckii* began as early as age 5 years. The incidence of clinical disease is very low compared to the frequency of cellular and humoral immunity demonstrated by these studies, which suggests that there is high innate resistance to this organism.

4. HOST DEFENSES

4.1. Innate Resistance

4.1.1. *Introduction*

The high immunogenicity and other properties of the cell surface components of *S. schenckii* probably combine to halt the progress and spread of infection by this fungus before any noticeable symptoms appear in the immune-competent host. In endemic areas, immunity can be demonstrated in those individuals known to have repeated contacts with *S. schenckii*.[27] Therefore, in areas where the disease is endemic, there must be other factors in addition to exposure that contribute to the development of clinical sporotrichosis. Lymphocutaneous sporotrichosis is the most frequent form of the disease, and poverty and malnutrition appear to be compromising factors in many of these patients.[31] In over one half of extracutaneous sporotrichosis cases there is ready evidence of host compromise, alcohol abuse being the most common,[1,32] followed by pulmonary tuberculosis, diabetes, sarcoidosis, steroid use, chronic obstructive pulmonary disease (COPD), and other chronic medical conditions.[33,34] In contrast with other mycoses, sporotrichosis has been a singularly infrequent, complicating fungus disease in patients with acquired immunodeficiency syndrome (AIDS).[35–38]

4.1.2. *Monocyte/Macrophage*

The phagocytosis of yeast cells from two strains of *S. schenckii* was studied using thioglycollate-elicited mouse peritoneal macrophages.[39] Opsonization of *S. schenckii* yeast cells treated with concanavalin A increased phagocytosis as compared to untreated cells. Reactivity with this lectin probably involved the peptidorhamnomannan and galactomannan components of the cell wall of *S. schenckii*. Prior treatment with neuraminidase (which hydrolysed N-glycoylneuraminic acid) increased ingestion of unopsonized yeasts. Sialic acid residues are located in the peptidopolysaccharide and on the polar glycolipid layers of the cell wall.[40] The relationship between surface components containing N-glycolyl-neuraminic acid and infection is unclear. It should be noted, however, that N-glycolyl groups are antigenic in man[41] and may be the epitopes reacting with serum sickness antibodies.[42]

Mannose, rhamnose, and galactose, which are major constituents of *S. schenckii* surface antigens, reduced phagocytosis indexes in mice.[39] Both a purified rhamnomannan and a partially purified galactomannan inhibited phagocytosis by about 60%. N-acetylglucosamine inhibited binding of yeasts to macrophages at 4°C as well as phagocytosis. This latter inhibition may be caused by binding of N-acetylglucosamine to a receptor common to other sugars, similar to the (mannose/glucosamine)-reacting receptor identified in alveolar macrophages.[43]

All these described interactions may be of little significance in advanced sporotrichotic infection when antibodies are already present but may be quite relevant at the very beginning of a new infection when the unchallenged proliferation of yeast cells determine the initial course of the mycosis.

4.1.3. *Polymorphonuclear Leukocytes*

Human peripheral polymorphonuclear leukocytes (PMNL) are capable of efficient *in vitro* phagocytosis and killing of *S. schenckii* yeast cells.[44] At an effector:target ratio of 2:1, 50% of the yeast cells were phagocytized in the presence of 10% normal human serum, but the effect of *S. schenckii*-immune serum on phagocytosis was not studied. In the presence of leukocytes, 80% of the yeast cells were killed within one hour. The possibility of lysis of *S. schenckii* by T lymphocytes or natural killer cells was not excluded because the peripheral blood leukocyte population was not fractionated. Components of the myeloperoxidase-hydrogen peroxide-halide microbicidal mechanism have been tested singly and in combination against *S. schenckii* yeast cells. Human PMNL effectively phagocytize and kill *S. schenckii* yeast cells when myeloperoxidase, peroxide, and potassium iodide are present; however, chloride ions could not be substituted for iodide, in contrast to results with *C. albicans*.[45] This provides a potential explanation for the specific efficacy of potassium iodide therapy in cutaneous sporotrichosis. The role of nonimmune serum in phagocytosis is presumably opsonization of yeast cells through direct activation of the alternative complement pathway.[46]

Ramos-Zepeda and Gonzalez-Mendoza[47] found that PMNL from normal gerbils readily phagocytized and killed yeast cells, whereas those from infected animals had impaired function because of diminished myeloperoxidase activity.

4.1.4. Complement Activation

Fungal cell wall components are able to interact with the complement system and this results in leukocyte chemotaxis and increased phagocytosis caused by macrophage recognition of bound C3 or its degradation products bound to the yeast cells. Activation of complement by fungal cells proceeds by either the classical or alternative pathways.

Initial studies have been performed that address the interactions of complement and *S. schenckii*, one with yeast cells and one with the mycelial form. Washed yeast cells grown in brain–heart infusion were used at a concentration of 10^9 cells/ml corresponding, for comparison, to similar wet weight of zymosan. These cells were able to consume all hemolytic complement activity in the presence of Mg-EGTA.[46] Incubation of yeasts with nonimmune, nonchelated serum resulted in deposition of C3 but not IgG on the yeast cell surface. Therefore, like most fungi,[48] *S. schenckii* activates complement through the alternative pathway. Opsonization by C3 may be an important factor in the host defense mechanisms against *S. schenckii*.

Torinuki and Tagami[49] used a boiled homogenized mycelial mat colony of *S. schenckii* for complement activation of serum. In the presence of *S. schenckii*, C3 was converted from β_1C to β_1A, and there was generation of C5a, which was unaffected by addition of Mg-EGTA. The formation of C5a may explain the chemotactic activity that occurs in nonimmune serum.

4.2. Immune Responses

4.2.1. Humoral

The major surface antigen of *S. schenckii* is a peptidorhamnomannan (percent composition: L-rhamnose:mannose:protein = 33.5:50:18.2). Cell walls of *S. schenckii* have a higher protein and lipid content than those of other fungal pathogens. The protein component contains high concentrations of hydroxy- and acidic-amino acids involved in glycopeptide bonds. All the glucans of the cell wall are β-linked containing 1,3, 1,4, and 1,6 bonds. The rhamnomannan consists of a linear 1,3-mannan backbone with single 1,3-rhamnose antennae in yeast forms and dirhamnosyl side chains in conidia-producing mycelial forms.

Both immediate- and delayed-type hypersensitivity reactions are elicited by the peptidopolysaccharide complexes.[50] Determinants of delayed-type hypersensitivity are probably located in the peptide portion of the molecule. A peptidopolysaccharide from *S. schenckii* eliciting both immediate and delayed-type reactions was treated with periodate and with 0.01% papain.

Periodate treatment did not inhibit the delayed-type reactions, but papain treatment inhibited both peritoneal cell migration and delayed cutaneous hypersensitivity tests.[51] Specific peptides eliciting delayed-type hypersensitivity have not yet been characterized.

Digestion of the peptidorhamnomannan with papain caused a 50% reduction in the nitrogen content.[51] The entire polysaccharide component was labile to periodate oxidation, indicating major 1,6- and 1,2 glycosidic bonds. The serologic and passive cutaneous anaphylactic activity is removed by periodate oxidation but not by proteinase treatment. From this data, it appears that the carbohydrate component of the peptidorhamnomannan is essential for the triggering of degranulation of mast cells. This carbohydrate is either the epitope recognized by IgE or serves as a carrier for protein determinants unaffected by papain digestion.

The surface location of peptidorhamnomannan and its ability to evoke specific precipitin and IgE responses provide good evidence that it is the major antigen responsible for specific humoral immunity to *S. schenckii*. On the basis of a much smaller amount of data, it is reasonable to propose that the peptido-portion of the rhamnomannan confers a dual ability to evoke T-cell-mediated reactions.

Experimental sporotrichosis induced in congenitally athymic (nude) mice (nu/nu) showed a great susceptibility of these animals to intravenous inoculation of *S. schenckii*. Elimination of the fungus began on day 8, but delayed-type hypersensitivity to a soluble *S. schenckii* antigen was not detected in phenotypically normal litter mates (nu/+) before day 17 after inoculation, indicating that host resistance does not depend solely on cellular immunity.[52]

Several observations in humans and experimental animals suggest that humoral immunity is important as a defense mechanism in early infection. Mixed granulomata of human sporotrichosis contain plasma cells.[53] The dermal infiltrate of guinea pigs inoculated with *S. schenckii* had decreased T-cell and increased B-cell numbers.[54] A self-limited lymphatic disease was obtained in Syrian hamsters by cutaneous footpad inoculation. Primary infection increased the resistance to subsequent reinfection, and sera from infected animals had persistent agglutinin titers against the yeast phase.[55] Resistance to disseminated infection, as induced by increasing the inoculum, was enhanced by subcutaneous inoculation of crude ribosomal preparations or trypsinized cell walls from *S. schenckii* yeast cells.

Lymphocytes from infected guinea pigs are capable of binding several *S. schenckii* yeast cells, forming rosettes *in vitro* and demonstrating specific immunocytoadherence. The lineage of these lymphocytes, however, has not been established.[57] The antibody response to *S. schenckii* on the cellular level was estimated by immunocytoadherence of yeast cells to lymphocytes from infected guinea pigs.[57] The nonimmune guinea pig is generally quite resistant to *S. schenckii*, but infection was induced by a large dose (7×10^8) of yeast cells injected intratesticularly. The animals were stressed by maintaining them at 10°C. The resulting orchitis suppurated, drained, and healed spontaneously by 4 weeks after the initiating infection. Peripheral blood lymphocytes were purified on nylon wool columns and the nonadherent cell population was used in the experiments. This fraction should be enriched in T cells, but the ratio of T:B lymphocytes was not reported. Immunocytoadherence, judged by the formation of *S. schenckii* yeast rosettes, correlated positively with agglutinins, peaking at 2 weeks after infection, and declined sharply or disappeared after four weeks. No precipitins or complement-fixing antibodies could be detected later in the course of the infection. The rosetted lymphocytes were specific for *S. schenckii* and no rosettes were formed with yeast cells of *C. albicans*, *H. capsulatum*, or *B. dermatitidis* when mixed with lymphocytes from these *S. schenckii* infected animals.

4.2.2. Cell-Mediated Immunity

Cell-mediated immunity (CMI) in patients with sporotrichosis was evaluated by lymphocyte transformation assays.[58] Most patients had positive CMI tests *in vitro* in response to yeast cells of *S. schenckii*. A mycelial antigen also induced lymphocyte transformation in a patient with sporotrichotic arthritis.[59] Six patients with systemic sporotrichosis, however, had deficient CMI as determined by decreased lymphocyte transformation in response to phytohemagglutinin A, as compared to normal individuals. In contrast, patients with cutaneous sporotrichosis exhibited normal lymphocyte blastogenesis to phytohemagglutinin A.

In a survey of 143 healthy individuals, 10% had blastogenic responses to *S. schenckii*, while only 3% had positive delayed-typed hypersensitivity to sporotrichin.[59] Healthy lymphocyte donors showing positive blastogenic indexes for *S. schenckii* antigen, however, also showed high blastogenic responses to *Candida albicans*. Of those 14 whose lymphocytes were transformed to blasts by *S. schenckii* antigens, 11 were women. There was a high correlation coefficient (0.89) between the blastogenic index to *C. albicans* antigen and to *S. schenckii*. The authors[59] proposed that women, who have a greater antigenic stimulus than men do to *C. albicans* because of its frequent carriage on the vaginal mucosa, are cross-protected against *S. schenckii*, and this state of immunity is evidenced by the higher incidence of blastogenic responses to *S. schenckii* in normal women. By this view, the sex-related difference in the incidence of sporotrichosis is a product of heightened innate immunity in women and perhaps increased opportunity for infection through occupational exposure to thorny or jagged plant fragments in men.

In cats infected with a nematode *Brugia malayi*, which produces lymphatic obstruction, inoculation of *S. schenckii* resulted in more severe lesions and rapid spread to other sites in the body, as compared to cats infected only with *S. schenckii*.[60] Macdonald *et al.*[61] observed that subcutaneous and skin lesions that developed in cats inoculated with *S. schenckii* regressed within 31–

88 days to complete resolution. To assess whether or not spontaneous healing of lesions also corresponded to complete elimination of the fungus, animals were inoculated with corticosteroids to suppress immune responses. Lesions reappeared at the same site or at sites close to that of the initial infection 4–9 weeks after treatment with the suppressor. Results suggest that viable fungal cells can be sequestered in tissues for up to 6 months and that healed lesions may be reactivated after immunosuppression.

Hachisuka and Sasai[62] inoculated *S. schenckii* intracutaneously in cyclophosphamide-treated mice. Cyclophosphamide damages the humoral immune system severely but has little effect on CMI. The lesions that developed in cyclophosphamide-treated mice inoculated with *S. schenckii* enlarged until the 8th day. After 12 days, the size of the lesions were similar to those of untreated mice, suggesting that a nonhumoral immune mechanism became effective in checking growth of the lesions in both treated and control mice. Results presented in this section appear to indicate that both humoral and cellular immune responses are effective inhibitors of the increasing size of individual lesions and to the spread of sporotrichosis.

5. IMMUNODIAGNOSIS AND PROGNOSIS

5.1. Serological Aspects

Tests for antibody are diagnostically important in the extracutaneous forms of *S. schenckii* infections (particularly of the central nervous system) when clinical features are lacking and tissues, exudates, or fluids are more difficult to obtain for direct examination and culture. Since rhamnomannans, the chief *S. schenckii* antigens, are unique to this pathogenic fungus and do not occur in any other human fungal species, they are the obvious antigens of choice for the diagnosis of sporotrichosis. These antibody tests are based on the antibody response to the peptidorhamnomannan of the outer cell wall of the organism.[63] Strains of *S. schenckii* isolated from human and animal disease and from soil are antigenically similar by cross-agglutination tests.[64] To date, only soluble, crude antigens from culture filtrates have been used, while purified antigens have yet to be similarly tested.[56,65,66] The yeast cell and latex particle agglutination tests are the most reliable and sensitive for all forms of the disease.[67] The latex slide agglutination test was positive in 94% of 80 cases of sporotrichosis of various types.[67] The immunodiffusion (ID) test[68] is positive in approximately 95% of patients with extracutaneous forms of sporotrichosis. The value of ID as a screen in pulmonary, articular, osseous, and disseminated disease is well established.[1,5] Furthermore, the latex agglutination and immunodiffusion tests are very specific, in that they do not cross-react with sera from other mycotic, bacterial, and parasitic infections. In contrast to the early return to negativity of serological tests (latex or ID) in the patient with treated and cured cutaneous sporotrichosis, ID reactions may remain positive in patients with the extracutaneous forms for several years after clinical and cultural cure.[66] These tests are available both commercially and from reference laboratories. An enzyme-linked immunosorbent assay has been developed that is more sensitive than any of these previous tests and may prove to be valuable, particularly in cerebrospinal fluid, where the latex agglutination test may be negative.[6,30] Western blotting may provide differentiation between cutaneous and extracutaneous disease, but further data are needed to confirm these results.[69]

5.2. Cellular Aspects

Demonstration of *S. schenckii* is required for definitive diagnosis of sporotrichosis, and this is best accomplished by cultural recovery and identification of the fungus. Infection caused by *S. schenckii* may be recognized by direct examination and culture of infected material. The fungus can be identified in biopsied tissue by culture and direct immunofluorescent or immunoperoxidase staining techniques,[70] but these tests are not widely available.

6. PREVENTION AND THERAPY

6.1. Prevention

Sporotrichosis can be prevented when environmental sources of *S. schenckii* are identified and avoided or eliminated. This has been accomplished in specific situations such as those involving sphagnum moss[24] and mine timbers.[21] For the most part, however, the precautions needed to avoid infection in agricultural workers are difficult to achieve because of the cost of the protective clothing needed for the hands, feet, arms, and legs.

6.2. Therapy

6.2.1. Chemotherapy

The common lymphocutaneous form of sporotrichosis is almost invariably cured by prolonged oral administration of a saturated solution of potassium iodide. Initially a dose of 10 drops diluted in liquid, such as fruit juice, is given three times daily and increased dropwise to the maximum tolerated by each individual patient (usually fewer than 60 drops). This latter dose is continued for a month or longer after the lesions appear to be fully healed. The mechanism of action of potassium iodide is unknown, but the material has no effect *in vitro* against the fungus. Perhaps the iodide facilitates phagocytic killing by promoting the respiratory burst associated with superoxide formation in the phagolysosome, since it has been shown that *in vitro* phagocytic killing is increased in the presence of potassium iodide as the halide in the myeloperoxidase system.[44]

Extracutaneous sporotrichosis does not respond to potassium iodide treatment but usually improves with the intravenous administration of amphotericin B. A total dose of 2.0 grams is usual, but this dose or more may not achieve a cure, and relapses are frequent, particularly of bone and joint sporotrichosis. Both flucytosine and rifampin usually show *in vitro* synergism with amphotericin B and either of these drugs may be administered simultaneously, but additional clinical benefit may not be apparent.[17] Several imidazoles have activity against *S. schenckii*, but results with ketoconazole have been disappointing and the use of miconazole is limited by toxicity. Two other imidazoles, fluconazole and itraconazole (investigational), show activity against *S. schenckii in vitro*. Oral itraconazole has produced cures of human lymphocutaneous sporotrichosis in clinical trials,[71] but the ultimate role of the imidazoles in the management of this mycosis is not yet established.

Local heat to the lymphocutaneous lesions has been advocated as an adjunct to drug therapy.[10,72] Many *S. schenckii* isolates recovered from the superficial forms of sporotrichosis grow poorly at temperatures above 35°C.[9] Raising the temperature within the lesion by hot soaks or heat pads not only inhibits growth of the fungus but also increases blood flow, bringing more of the therapeutic agent and more phagocytic cells to the involved area. The phagocytic killing process is also more efficient at higher temperatures.[72]

6.2.2. *Immunotherapy*

Linear insoluble glucans (mixed D-glucopyranose units) of *S. schenckii*, given at 10 mg/kg per day,[73,74] caused regression of sarcoma 180 implanted in mice. This antitumor activity is similar to that observed with other fungal polysaccharides.[75] *S. schenckii* rhamnomannans were also tested for their ability to reduce growth of sarcoma 180 in SW55 mice.[76] A monorhamnosyl–rhamnomannan at 200 mg/kg in 10 doses inhibited the tumor growth in 62% of the animals. All effects depended on the dosage and therapy schedule used. Rhamnomannans increased the phagocytic indexes of mouse peritoneal macrophages for normal and antibody-sensitized sheep erythrocytes, and this effect was more pronounced than that produced by *S. cerevisiae* mannan.[77] The ability of the alkali-insoluble wall fractions (glucans and rhamnomannans) of *S. schenckii* to cause regression of tumors and increased phagocytosis in mice indicates that these antigens have some immunomodulating activity.

Use of transfer factor in two patients with extracutaneous sporotrichosis showed improvement in parameters of CMI and control of their disease.[58] The transfer factor was prepared from the plasma of donors who had recovered from cutaneous sporotrichosis or normal individuals with high lymphocyte blastogenesis indexes for sporothrix antigen.

REFERENCES

1. Rippon, J. W. (ed.), 1988, Sporotrichosis, in: *Medical Mycology. The Pathogenic Fungi and the Pathogenic Actinomycetes*, 3rd ed., W. B. Saunders, Company, Philadelphia, pp. 325–352.
2. Arango, C., Davila, E., and Greenma, R., 1981, Multifocal sporotrichosis presenting with lobar pulmonary consolidation, *South Med. J.* **74:**755–756.
3. Kinas, H. Y., and Smulewicz, J. J., 1976, Primary pulmonary sporotrichosis, *Respiration* **33:** 468–474.
4. Rohatgi, P. K., 1980, Pulmonary sporotrichosis, *South Med. J.* **73:**1611–1617.
5. Zvetina, J. R., Rippon, J. W., and Daum, V., 1978, Chronic pulmonary sporotrichosis, *Mycopathologia* **64:**53–57.
6. Scott, E. N., Kaufman, L., Brown, A. C., and Muchmore, H. G., 1987, Serologic studies in the diagnosis and management of meningitis due to *Sporothrix schenckii*, *N. Engl. J. Med.* **317:** 935–940.
7. Dellatorre, D. L., Lattanand, A., Buckley, H. R., Urbach, F., 1982, Fixed cutaneous sporotrichosis of the face, Successful treatment of a case. Review of the literature, *J. Am. Acad. Dermatol.* **6:**97–100.
8. Honbo, S., Yamano, T., Masaki, J., and Urabe, H., 1985, Analytical studies on peculiar cases of sporotrichosis, the lesions of which contained numerous fungal elements, *Proc. IXth. Congr. Int. Soc. Human Anim. Mycol.* R11–12.
9. Kwon-Chung, K. J., 1979, Comparison of isolates of *Sporothrix schenckii* obtained from fixed cutaneous lesions with isolates from other types of lesions, *J. Inf. Dis.* **139:**424–431.
10. MacKinnon, J. E., and Conti-Diaz, I. A., 1962, The effect of temperature on sporotrichosis, *Sabouraudia* **2:**56–59.
11. Iwatsu, T., Nishmura, K., and Miyaji, M., 1985, Spontaneous disappearance of cutaneous sporotrichosis, *Int. J. Dermatol.* **24:**524–525.
12. Pueringer, R. J., Iber, D., Deike, M. A., and Davies, S. F., 1986, Spontaneous remission of extensive pulmonary sporotrichosis, *Ann. Intern. Med.* **104:**366–367.
13. Pluss, J. L., and Opal, S. M., 1986, Pulmonary sporotrichosis: Review of treatment and outcome, *Medicine* **65:**143–152.
14. Michelson, E., 1977, Primary pulmonary sporotrichosis, *Ann. Thorac. Surg.* **24:**83–86.
15. Ridgeway, N. A., Whitcomb, R. C., Erickson, E. E., and Law, S. W., 1962, Primary pulmonary sporotrichosis, *Am. J. Med.* **32:**153–160.
16. Smith, L. M., 1945, Sporotrichosis. Report of four clinically atypical cases, *South Med. J.* **38:** 504–509.
17. Winn, R. E., 1988, Sporotrichosis, in: *Systemic Fungal Infections: Diagnosis and Treatment I. Infectious Disease Clinics of North America*, Volume 2, (D. J. Drutz, ed.), W. B. Saunders, Co., Philadelphia, pp. 899–911.
18. Mohr, J. A., Patterson, C. D., Eaton, B. G., Rhoades, E. R., and Nichols, N. B., 1972, Primary pulmonary sporotrichosis, *Am. Rev. Respir. Dis.* **106:**260–264.
19. Mariat, F., 1977, Taxonomic problems related to the fungal complex *Sporothrix schenckii/ Ceratocystis spp.*, in: *Recent Advances in Medical and Veterinary Mycology* (K. Iwata, ed.), University of Tokyo Press, Tokyo, pp. 265–270.
20. Shadomy, H. J., and Davis, B. A., 1989, The teleomorphic state of *Sporothrix schenckii*, *Proc. Xth. Congr. Int. Soc. Human Anim. Mycol.*, Abstr #P-55, p. 64.
21. Simson, F. W., 1947, Sporotrichosis infection in mines of the Witwatersrand. A symposium, *Proc. Transv. Mine Med. Officers Assoc.* pp. 1–67.

22. CDC, 1988, Multistate outbreak of sporotrichosis in seedling handlers, *M.M.W.R.* **37:** 652–653.
23. D'Alessio, D. J., Leavens, L. J., Strumpf, G. B., and Smith, C. D., 1965, An outbreak of sporotrichosis in Vermont associated with sphagnum moss as the source of infection, *N. Engl. J. Med.* **272:**1054–1058.
24. Powell, K. E., Taylor, A., Phillips, B. J., Blakey, D. L., Campbell, G. D., Kaufman, L., and Kaplan, W., 1978, Cutaneous sporotrichosis in forestry workers, *J.A.M.A.* **240:**232–235.
25. Ingrish, F. M., and Schneidau, J. D., 1967, Cutaneous hypersensitivity to sporotrichin in Maricopa County, Arizona, *J. Invest. Dermatol.* **49:**146–149.
26. Schneidau, J. D., Lamar, L. M., and Hairston, M. A., 1964, Cutaneous hypersensitivity to sporotrichin in Louisiana, *J.A.M.A.* **188:**371–373.
27. Gonzalez-Ochoa, A., 1965, Contribuciones recientes al conociemiento de la esporotrichosis, *Gac. Med. Mex.* **95:**463–474.
28. Gonzalez-Ochoa, A., Ricoy, E., 1970, Valoracion comparatira de los antigenos polisacarido y cellular de *Sporothrix schenckii, Rev. Invest. Salud. Publica.* **30:**303–315.
29. Rocha-Posada, H., 1968, Prueba cutanea con esporotricina, *Mycopathologia* **36:**42–54.
30. Scott, E. N., Muchmore, H. G., and Parkinson, A. J., 1982, Enzyme and radio immunoassays in human sporotrichosis, *Proc. VIIIth. Congr. Int. Soc. Human Anim. Mycol.*, pp. 212–215.
31. Mariat, F., 1968, The epidemiology of sporotrichosis, in: *Systemic Mycoses* (G. E. W. Wolstenholme, ed.), A. Churchill, London, pp. 144–159.
32. Kedes, L. H., Siemienski, J., and Braude, A. I., 1964, The syndrome of the alcoholic rose gardener: sporotrichosis of radial tendon sheath. Report of a case cured with amphotericin B, *Ann. Intern. Med.* **61:**1139–1141.
33. England, D. M., and Hochholzer, L., 1985, Primary pulmonary sporotrichosis: Report of eight cases with clinicopathologic review, *Am. J. Surg. Pathol.* **9:**193–203.
34. Jay, S. J., Platt, M. R., and Reynolds, R. C., 1977, Primary pulmonary sporotrichosis, *Am. Rev. Respir. Dis.* **115:**1051–1056.
35. Bibler, M. R., Luber, H. J., Glueck, H. I., and Estes, S. A., 1986, Disseminated sporotrichosis in a patient with HIV infection after treatment for acquired factor VIII inhibitor, *J.A.M.A.* **256:**3125–3126.
36. Fitzpatrick, J. E., 1988, Acquired immunodeficiency syndrome presenting as disseminated cutaneous sporotrichosis, *Int. J. Dermatol.* **27:**406–407.
37. Kurosawa, A., 1988, *Sporothrix schenckii* endophthalmitis in a patient with human immunodeficiency virus infection, *Arch. Ophthal.* **106:**376–380.
38. Lipstein-Kresch, E., Isenberg, H. D., Singer, C., Cooke, O., and Greenwald, R. A., 1985, Disseminated *Sporothrix schenckii* infection with arthritis in a patient with acquired immunodeficiency syndrome, *J. Rheumatol.* **12:**805–808.
39. Oda, L. M., Kubelka, C. F., Alviano, C. S., and Travassos, L. R., 1983, Ingestion of yeast forms of *Sporothrix schenckii* by mouse peritoneal macrophages, *Infect. Immun.* **39:**497–504.
40. Alviano, C. S., Pereira, M. E. A., Souza, W., Oda, L. M., and Travassos, L. R., 1982, Sialic acids are surface components of *Sporothrix schenckii* yeast forms, *FEMS Microbiol. Lett.* **15:** 223–227.
41. Merrick, J. M., Zadarlik, K., and Milgrom, F., 1978, Characterization of the Hanganutziu-Deicher (serum sickness) antigen as gangliosides contain N-glycolylneuraminic acid, *Intern. Arch. Allergy Appl. Immunol.* **57:**477–480.
42. Smith, D. F., and Ginsburg, V., 1980, Antibodies against sialyloligosaccharides coupled to protein, *J. Biol. Chem.* **255:**55–59.
43. Warr, G. A., 1980, A macrophage receptor for (mannose/glucosamine) glycoproteins of potential importance in phagocytic activity, *Biochem. Biophys. Res. Commun.* **93:**737–745.
44. Cunningham, K. M., Bulmer, G. S., Rhoades, E. R., 1979, Phagocytosis and intracellular fate of *Sporothrix schenckii, J. Infect. Dis.* **140:**815–817.
45. Lehrer, R. I., 1969, Antifungal effects of peroxidase systems, *J. Bacteriol.* **99:**361–365.

46. Scott, E. N., Muchmore, H. G., and Fine, D. P., 1986, Activation of the alternative complement pathway by *Sporothrix schenckii*, *Infect. Immun.* **51**:6–9.
47. Ramos-Zepeda, R., and Gonzalez-Mendoza, A., 1986, Metabolic activity of phagocytes in experimental sporotrichosis, *Mycopathologia* **93**:109–112.
48. Fine, D. P. (ed.), 1981, *Complement and Infectious Diseases*, CRC press, Inc., Boca Raton.
49. Torinuki, W., and Tagami, H., 1985, Complement activation by *Sporothrix schenckii*, *Arch. Dermatol. Res.* **277**:332–333.
50. Noguchi, T., 1972, Immunochemical studies on *Sporotrichum schenckii*, I. Extraction and chemical composition of the immunologically active substances, *Acta Sch. Med. Univ. Gifu* **20**: 335–343.
51. Shimonaka, H., Noguchi, T., Kawai, K., Hasegawa, I., Nasawa, I., and Ito, Y., 1975, Immunochemical studies on the human pathogen *Sporothrix schenckii*: effects of chemical and enzymatic modification of the antigenic compounds upon immediate and delayed reactions, *Infect. Immun.* **11**:1187–1194.
52. Shiraishi, A., Nakagaki, K., and Arai, T., 1979, Experimental sporotrichosis in congenitally athymic (nude) mice, *RES J. Reticuloendothel. Soc.* **26**:333–336.
53. Lurie, H. I., 1971, Sporotrichosis, in: *The Pathologic Anatomy of Mycoses: Human Infection with Fungi, Actinomyces and Algae* (E. Uehlinger, ed.), Springer–Verlag, New York, pp. 614–675.
54. Hachisuka, H., and Sasai, Y., 1980, Subpopulations of lymphocytes in the infiltrate of experimental sporotrichosis, *Mycopathologia* **76**:167–169.
55. Charoenvit, Y., and Taylor, R. L., 1979, Experimental sporotrichosis in Syrian hamsters, *Infect. Immun.* **23**:366–372.
56. Roberts, G. D., and Larsh, H. W., 1971, The serologic diagnosis of extracutaneous sporotrichosis, *Am. J. Clin. Pathol.* **56**:597–600.
57. Roberts, G. D., and Larsh, H. W., 1971, A study of the immunocytoadherence test in the serology of experimental sporotrichosis, *J. Infect. Dis.* **124**:264–269.
58. Plouffe, J. F., Jr., Silva, J., Jr., Fekety, R., Reinhalter, E., and Brown, R., 1979, Cell-mediated immune responses in sporotrichosis, *J. Infect. Dis.* **139**:152–157.
59. Steele, R. W., Cannady, P. B., Jr., Moore, W. L., Jr., and Gentry, L. O., 1976, Skin test and blastogenic responses to *Sporotrichum schenckii*, *J. Clin. Invest.* **57**:156–160.
60. Barbee, W. C., Ewert, A., and Folse, D., 1977, The combined effect of a cutaneo-lymphatic fungus, *Sporothrix schenckii*, and a lymphatic-dwelling nematode, *Brugia malayi*, *Trop. Geogr. Med.* **29**:65–73.
61. Macdonald, E., Ewert, A., and Rietmeyer, J. C., 1980, Reappearance of *Sporothrix schenckii* lesions after administration of Solu-Medrol® to infected cats, *Sabouraudia* **18**:295–300.
62. Hachisuka, H., and Sasai, Y., 1981, Development of experimental sporotrichosis in normal and modified animals, *Mycopathologia* **76**:79–82.
63. Travassos, L. R., and Lloyd, K. O., 1980, *Sporothrix schenckii* and related species of Ceratocystis, *Microbiol. Rev.* **44**:683–721.
64. Andrieu, S., Giquest, J., and Massamba, S., 1971, Etude immunologique comparee de *Sporothrix schenckii* et des souches saprophytes voisines, *Sabouraudia* **9**:206–209.
65. Karlin, J. V., and Nielsen, H. S., Jr., 1970, Serologic aspects of sporotrichosis, *J. Infect. Dis.* **121**: 316–327.
66. McMillen, S., 1974, The serology of sporotrichosis, in: *The Diagnosis and Treatment of Fungal Infections*, (H. M. Robinson, Jr., ed.), Charles C. Thomas, Springfield, pp. 401–409.
67. Kaufman, L., and Reiss, E., 1986, Serodiagnosis of Fungal Diseases, in: *Manual of Clinical Laboratory Immunology*, 3rd Ed., (N. R. Rose, H. Friedman, and J. L. Fahey, eds.), *American Society for Microbiology*, Washington, D. C., pp. 446–466.
68. de Alborno, M. B., Villaneuva, E., and de Torres, E. D., 1984, Application of immunoprecipitation techniques to the diagnosis of cutaneous and extracutaneous forms of sporotrichosis, *Mycopathologia* **85**:177–183.

69. Scott, E. N., and Muchmore, H. G., 1989, Immunoblot analysis of antibody responses to *Sporothrix schenckii*, *J. Clin. Microbiol.* **27**:300–304.
70. Russell, B., Beckett, J. H., and Jacobs, P. H., 1979, Immunoperoxidase localization of *Sporothrix schenckii* and *Cryptococcus neoformans*, *Arch. Dermatol.* **115**:433–435.
71. Restrepo, A., Robledo, J., Gomez, I., Tabares, A. M., and Gutierrez, M. T., 1986, Itraconazole therapy in lymphangitic and cutaneous sporotrichosis, *Arch. Dermatol.* **122**:413–417.
72. Hiruma, M., Katoh, T., Yamamoto, I., and Kagawa, S., 1987, Local hyperthermia in the treatment of sporotrichosis, *Mykosen* **30**:315–321.
73. Tarnowski, G. S., 1975, Approaches to the immunotherapy of experimental tumours, in: *Host Defense Against Cancer and its Potentiation*, (D. Mizuna, ed.), University Park Press, Baltimore, pp. 389–396.
74. Previato, J. O., Gorin, P. A. J., Haskins, R. H., and Travassos, L. R., 1979, Soluble and insoluble glucans from different cell types of the human pathogen *Sporothrix schenckii*, *Exp. Mycol.* **3**:92–105.
75. Whistler, R. L., Bushway, A. A., Singh, P. P., Nakahara, W., and Tokuzen, R., 1976, Non-cytotoxic, antitumor polysaccharides, *Adv. Carbohydr. Chem.* **32**:235–275.
76. Kubelka, C. F., 1980, Inibiacao de Sarcoma 180 e aumento de endocitose de macrogos peritoneais de camundongos por polissacaridios de fungos, Thesis, M. Sc., Federal University of Rio de Janeiro, Brasil.
77. Travassos, L. R., 1989, Antigenic structure of *Sporothrix schenckii*, in: *Immunology of Fungal Diseases*, Volume 47, (E. Kurstak, ed.), Marcel Dekker, Inc., New York, pp. 193–242.

7

Blastomycosis

ALAN M. SUGAR

1. DESCRIPTION OF THE DISEASE

Blastomycosis is the disease caused by the dimorphic fungus *Blastomyces dermatitidis*. Recognized by T. C. Gilchrist in 1894, in a biopsy specimen from a cutaneous lesion found on a patient, the organism was subsequently identified as a fungus when seen in a second patient the following year.[1,2] Similar to coccidioidomycosis and histoplasmosis, blastomycosis is contracted following inhalation of conidia into the lungs.[3] Thus, the initial host interaction with the organism occurs within the lungs. The events occurring over the several days following deposition of conidia within the alveoli, however, are unclear and whether conidia disseminate to extrapulmonary organs or whether transition to the yeast form precedes dissemination is unknown. Similarly, a successful host response depends on either blocking the conidia to yeast transition or killing yeast following their formation. Which of these occurs is also unknown. Mechanisms of dissemination, whether via blood or through the lymphatics, are other unresolved issues. While these problems are of fundamental importance in understanding the pathogenesis of any of the mycoses caused by the dimorphic fungi, including blastomycosis, much more is known about the clinical manifestations, immune mechanisms of

ALAN M. SUGAR • Evans Memorial Department of Clinical Research, The University Hospital, and Department of Medicine, The Thorndike Memorial Laboratory, Boston City Hospital, Boston, Massachusetts 02118.

Fungal Infections and Immune Responses, edited by Juneann W. Murphy *et al.* Plenum Press, New York, 1993.

established infection, and treatment of the disease. This chapter will review blastomycosis from these perspectives.

1.1. Clinical Manifestations

Most people who inhale the organism probably do not manifest recognizable symptoms.[4–6] However, because a skin test reagent such as coccidioidin or histoplasmin has never been reliable and is now unavailable for use, it is difficult, if not impossible, to identify subclinically infected individuals. The availability of lymphocyte transformation tests, such as those performed in a research setting, may address this question and allow one to estimate the number of asymptomatically infected people in specific endemic areas.[7] In fact, recent evidence favors the notion that subclinical infection may occur, but the actual number of subclinically infected individuals is not known.[5,8] Unfortunately, the lymphocyte transformation test has not been rigorously tested in large numbers of people with histoplasmosis and other mycoses, so that the specificity of this test is not known with a high degree of certainty.

When symptoms in pulmonary blastomycosis occur, they most often consist of fever, chills, headache, cough, chest pain, and sputum production. Hemoptysis is rare. In some patients, acute pulmonary blastomycosis is a self-limited disease requiring neither extensive clinical testing to make the diagnosis nor specific antifungal therapy.[9] There is some disagreement, however, whether clinically evident pulmonary blastomycosis can be a self-limited disease such as is commonly seen with coccidioidomycosis and histoplasmosis, or whether progressive infection, including dissemination is the usual outcome.[9] This is an important concept, since patients with self-limited disease should not require antifungal therapy, whereas, if spontaneous resolution of pulmonary blastomycosis is unusual, antifungal drugs would seem to be indicated. Unfortunately, it is not yet possible to identify those patients with self-limited disease and those with progressive infection. Many practitioners confronted with a patient with obvious blastomycosis pneumonia will institute therapy, usually with an orally administered azole, such as ketoconazole, itraconazole, or fluconazole. Chronic pulmonary blastomycosis, often with cavity formation, is one sequela of *B. dermatitidis* pneumonia and requires antifungal treatment.

Extrapulmonary dissemination occurs in 40–80% of patients with clinically recognizable blastomycosis. While any organ can be involved in disseminated disease, skin, bones, and the male urogenital tract, especially the prostate are most often involved. The reasons for the usual organ distribution of involvement in disseminated blastomycosis are unknown but are of basic interest since the other dimorphic pathogens often disseminate to other areas of the body (e.g., meninges in coccidioidomycosis or liver, spleen, or bone marrow in histoplasmosis). Cutaneous blastomycosis may occur as an isolated finding but almost always reflects disseminated disease. The sole exception to this is the cutaneous lesion arising from direct inoculation of the fungus, such as would occur following penetration of the skin with a sharp object in a laboratory. In this instance, however, the disease remains localized to the skin and spontaneously resolves over weeks to months. Primary inoculation blastomycosis, in contrast to systemic blastomycosis, is associated with regional lymphadenopathy.

1.2. Diagnosis

As there are no pathognomonic features of pulmonary or disseminated blastomycosis, diagnosis is based on demonstration of the typical yeast forms in tissue or growth of the organism from patient material submitted to the microbiology laboratory. Serologic assays are not available, although recent advances in identification of *B. dermatitidis* specific antigens may result in some progress in this area.[10] There is no role for skin testing in diagnosis of blastomycosis.

Blastomycosis does not lead to any pathognomonic changes in routine laboratory tests. White blood counts may be elevated as a nonspecific indicator of infection, but no consistent abnormalities occur in other tests. Chest X rays may demonstrate infiltrates in involved areas of the lungs; infiltrates may be unilateral or bilateral and there is no distinguishing feature that would indicate that the chest X ray has been taken from a patient with blastomycosis. Cavitation may occur in chronic pulmonary blastomycosis. Abnormalities of X rays, nuclear medicine scans, and computed tomographic scans will reflect the presence of infection in the affected organ, most commonly bone, and rarely liver or spleen. Definitive diagnosis will, however, require biopsy for culture and histopathologic examination of involved sites.

Biopsies typically reveal combined suppurative and granulomatous changes.[11] Hematoxylin and eosin stained tissue sections show evidence for infiltration of neutrophils in areas occupied by yeast phase organisms. Experimental studies have documented the ability of conidia and yeasts to elicit neutrophil recruitment into lungs of mice inoculated by the respiratory route with these fungal elements.[12,13] The mechanisms responsible for neutrophil accumulation in these areas have not been completely elucidated. *B. dermatitidis* yeasts have been shown to produce a neutrophil chemoattractant,[14,15] but the clinical relevance of this observation has yet to be established. In fact, patients with blastomycosis produce a serum inhibitor of neutrophil chemotaxis, the identity and source of which are unclear.[15a] As lesions mature, giant cells are formed and typical granulomas may be seen.[16,17] The mechanisms responsible for giant cell formation and the orchestration of the granulomatous reaction have also not been elucidated, although it is of interest that congenitally athymic mice infected with *B. dermatitidis* yeasts can also develop granulomas, suggesting that mature T lymphocytes are not required for this chronic inflammatory response.[17]

2. ETIOLOGY

Blastomyces dermatitidis is the only species in the genus. While strictly not technically a correct name, *Blastomyces*, by virtue of its widespread usage, will undoubtedly remain the medical name for the organism responsible for causing blastomycosis.[18]

B. dermatitidis grows in the environment in the mycelial phase and produces one type of conidium (Table I). Conidia are approximately 2–6 μm and may be spherical or pear shaped. Typically, one conidiophore arises at right angles from the main hyphal strand and bears a single terminal conidium. Conidia are released from the conidiophore by rhexolysis, a process that is enhanced following exposure of the fungus to water.[19] Conidia, however, can be removed from conidiophores by mechanical means in the laboratory.[20]

When exposed to temperatures in excess of 35° C, conidia convert to the pathognomonic yeastlike form. The primary trigger for phase transition is temperature.[21] A recent report indicates, however, that manipulation of nutritional factors can force phase transition at a lower temperature than typically supports yeastlike growth.[22] The mechanisms responsible for this novel modulation of growth phase by nutrients in a synthetic medium require further investigation. *Blastomyces* appears in tissue as round to oval yeast forms, 8–15 μm in size and with buds that originate from the mother cell with broad based attachment sites. These yeasts stain well with silver and periodic acid Schiff stains.

3. EPIDEMIOLOGY

It is most interesting that the precise microfocus in the environment for *B. dermatitidis* has not been repeatedly demonstrated as it has for *Coccidioides immitis* and *Histoplasma capsulatum*.[5] Studies have shown that both conidia and yeasts are fragile in the soil, though clearly soil inhabitants, and are destroyed by other microorganisms that live in the same ecological niche.[23,24] Thus,

TABLE I
LD_{50} of H_2O_2 for Two Strains of *B. dermatitidis*

	$[H_2O_2]$, mM	
ATCC strain	Conidia	Yeast
26199	3	>50
26197	3	7

From references[33,44]

when *B. dermatitidis* is inoculated into the soil or once it is recovered from a particular location, it is usually impossible to reculture the fungus from the same specimens only a week or two later.[5,25] Even the most recent outbreak associated with a beaver dam along the Eagle River was characterized by the inability of investigators to repeatedly isolate the fungus over a period of more than several months.[5]

One other explanation for the difficulties in isolating the fungus from the soil that has received little, if any, attention by previous investigators concerns the methods commonly used to recover the fungus from soil samples. Virtually all studies have employed mouse inoculation as the major modality in recovering *B. dermatitidis* isolates from environmental samples.[5] Therefore, in order for the fungus to grow, mycelia must convert to yeast within the mouse and then the yeast must grow when murine tissue is cultured. If there is any reason for phase transition to be delayed or blocked, the fungus will not grow, since mycelia do not proliferate at the body temperature of the laboratory mouse. Thus, determination of the presence of *B. dermatitidis* in the environment using the published methodology depends on phase transition and not on the ability of conidia or mycelia to continue to produce hyphae. Since phase transition and germination can be dissociated and phase transition seems to be more sensitive to certain noxious phenomena,[26] it is possible that mycelia capable of continued germination and hyphae production may not grow as the yeast form of the fungus at these higher temperatures. The end result is the failure to detect yeast phase *B. dermatitidis* in specimens containing the mycelial form of the fungus. Fortunately, this hypothesis is amenable to experimental study.

On a larger scale, an endemic area for the occurrence of blastomycosis cases in humans and in animals clearly exists.[5] In the United States, the endemic region encompasses the Ohio and Mississippi river valleys, overlapping to a great extent the endemic region for histoplasmosis. Until more specific antigenic reagents are available, the precise extent of exposure to the fungus cannot be calculated.

Blastomycosis also exists in Africa, but African blastomycosis seems to be a less common disease than its counterpart in the Western hemisphere. Whether the organisms from North America and Africa constitute a single species or at least two distinct organisms is currently an area of controversy.[27] Conversion of the mycelial to the yeast phase of the African isolate seems to be less readily achieved in the laboratory than with North American isolates of *B. dermatitidis*.[28] Reasons for this laboratory phenomenon are not clear.

Sporadic cases of blastomycosis are more common than those occurring as part of an outbreak, but 8 well-defined outbreaks have been reported.[5] Much of the information about incubation period and clinical course has derived from study of the patients infected during these outbreaks. Since skin tests reagents have not been reliable, there is no satisfactory way to determine the total number of *B. dermatitidis* infected people.

4. HOST DEFENSES

4.1. Innate Resistance

Given the relative rarity of symptomatic blastomycosis, either exposure to *B. dermatitidis* is not common or host innate defenses are remarkably well suited for eliminating the organism from the lungs. As already discussed, because of the difficulties researchers have encountered in establishing the natural habitat of *B. dermatitidis*, the true extent and degree of exposure of individuals to the fungus is unknown. Until the true distribution of *B. dermatitidis* in Nature and, thus, the extent of exposure to the fungus is established, study of immune system factors responsible for resistance to infection may shed light on the ability of the host to prevent the development of blastomycosis.

In evaluating the role of innate resistance to the development of blastomycosis, one must consider the pathogenesis of naturally acquired disease and the nature of the bulk of published research on the topic. As reviewed above, blastomycosis develops after the inhalation of conidia[3] (and perhaps small mycelial fragments). In contrast, the bulk of published literature, much of it reviewed in this chapter, reflects a preoccupation with host interactions with the yeast form of the fungus. Since the initial contact of host with *B. dermatitidis* is not with the yeast form, studies with *B. dermatitidis* yeast may not be appropriate for determining factors responsible for a successful host response to naturally acquired infection. In fact, one could argue that following a successful host–fungus interaction, there are no yeasts present, the conidia having been cleared prior to phase transition. This is a testable hypothesis and one that merits attention by workers in the area of pathogenesis of dimorphic fungal infections.

With the recent description of murine models of blastomycosis based on inhalation of conidia[12,29–31] and *in vitro* studies with the infectious fungal elements,[32–35] the foundation has been laid for the continued evaluation of this pivotal aspect in the pathogenesis of blastomycosis.

One correlate of intrinsic host resistance to the development of clinically evident blastomycosis is age. While most humans with the disease are adults, this may be more a function of exposure than of age-related differences in ability to control *B. dermatitidis*. In the mouse pulmonary blastomycosis model, however, age clearly appears to be a determinant of susceptibility to infection with the yeast phase, with mice less than 5 weeks old developing fatal infections with lower inocula of yeast phase cells than those older mice.[36] The nature of the host factors responsible for these observations remain unstudied. The role of age in susceptibility to conidia remains unevaluated.

One study evaluating the susceptibility of inbred mouse strains to infection with yeast phase *B. dermatitidis* has been reported.[37] A hierarchy of susceptibility to lethal infection was demonstrated, with C3H/HeJ mice the most, and DBA/1J mice the least susceptible to pulmonary infection of the 8 mouse strains tested. No differences in *in vitro* and *in vivo* tests of delayed hypersensitivity were observed and the mice most susceptible demonstrated higher antibody titers to the fungus than the other more resistant animals. This latter observation may just reflect an increase humoral response to a greater fungal load, since other mechanisms of host defense had failed to control the growth of the organism. The reason underlying innate host susceptibility to blastomycosis pneumonia have yet to be discerned but were speculated to be secondary to differences in macrophage activity.[37]

Neutrophils and natural killer (NK) cells are presumably the cells responsible for innate resistance to infection. There are no reports, however, of experiments designed to assess the interaction of NK cells with either *B. dermatitidis* conidia or yeasts. While these cells may participate at some point in host response to inhalation of the fungus, the results of such encounters are yet to be determined, and any role for these cells in host response to inhalation of *B. dermatitidis* is speculative.

Several studies have evaluated the ability of murine and human neutrophils to ingest and kill *B. dermatitidis* with differing results. The work by Brummer, Stevens, and colleagues represents the most comprehensive evaluation of murine neutrophils and their interactions with *B. dermatitidis* yeast forms.[38–42] Neutrophils were initially obtained from the peritoneum following injection of thioglycollate.[39] These elicited neutrophils were found to be unable to kill yeast from *B. dermatitidis*[38,41,42] and this correlated with a decrease in the ability of elicited neutrophils to mount a brisk oxidative burst.[42] Interferon treatment of these elicited neutrophils resulted in the acquisition of modest fungicidal activity (21%), compared to 71% killing of *C. albicans* blastospores. When circulating peripheral blood neutrophils (nonelicited cells) were evaluated for their ability to kill *B. dermatitidis* yeasts, they would kill this fungus (38%), but this was significantly less than the fungicidal activity towards *Candida albicans* (84%).[38] Apparently, thioglycollate, the neutrophil-eliciting agent used in the earlier studies, has an adverse effect on neutrophil respiratory burst and on fungicidal activity, since resident, circulating neutrophils had better activity of these two functions. Interestingly, in all these studies, the yeast form of *B. dermatitidis* was found to be noningestible.

The interactions between human neutrophils and *B. dermatitidis* yeasts were investigated by Sixbey *et al.*[43] In contrast to the murine studies, they found that the yeasts used (obtained from their collection of clinical isolates) were readily ingested and killed to a modest extent (29%) by human neutrophils in suspension cultures.

In the most comprehensive study to date, Drutz and Frey evaluated the effects of human neutrophils and monocytes on mycelia, conidia, and yeasts of several different strains of *B. dermatitidis*.[32] This approach has great merit, since the infection is initiated by inhalation of conidia and perhaps mycelial fragments and, only after several days, are yeast forms present in tissue. Ingestion of conidia was rapid (maximal in 2 hours) and virtually complete. In contrast, within one hour, approximately 20% of yeast forms were said to be ingested. Using a combination of methods to determine fungal cell viability, including colony counting, radioactive chromium release, and methylene blue dye exclusion, they determined that normal, unstimulated human phagocytes could extensively damage mycelial fragments and conidia but that the yeast phase was more resistant to these cells. Human neutrophils ingested 90% of conidia and killed 50% of the initial inoculum in 3 hours. In contrast, as measured by a methylene blue assay, approximately 20% of yeasts were killed by neutrophils under conditions identical to those used to assess neutrophil fungicidal activity against conidia, and this figure did not increase over the next hour. Similar data for the 3-hour co-incubation time point were not provided. Fresh human serum was required for the demonstration of maximal phagocytosis and fungicidal activity, presumably reflecting the need for active complement components, since heat inactivation of the serum reduced its enhancing effect on phagocytosis and killing. Using a variety of methods, these workers concluded that oxidative mechanisms were important mediators of killing of both conidia and yeasts, although conidia appeared to be more sensitive to oxidants than yeasts.

In two separate *in vitro* studies of the susceptibility of a mouse virulent (ATCC 26199) and an avirulent (ATCC 26197) strain of *B. dermatitidis*, the LD_{50}s to H_2O_2 were calculated (Table I). From these data, it appears that the susceptibility of conidia of these two strains to H_2O_2 is similar, but the yeasts demonstrate markedly disparate susceptibility. At least with one strain, ATCC 26199, conidia appeared to be more sensitive to the lethal effects of H_2O_2 than were yeasts, a result similar to that obtained by Drutz and Frey.[32] In both *in vitro* studies, however, the combined H_2O_2-peroxidase-halide system was rapidly lethal for conidia and yeasts of both strains of *B. dermatitidis* tested.[33,44]

In a study of the interactions of humans neutrophils with various fungi, Schaffner *et al.*, using a limiting dilution assay to measure fungicidal activity, were unable to demonstrate killing of two different strains of *B. dermatitidis* yeasts.[45] However, hyphal fragments of the same strains were killed (about 50% of wells were sterilized when inoculated with 3–12 colony-forming units/well). As one explanation for these findings, these investigators demonstrated that *B. dermatitidis* yeasts elicit less of a respiratory burst, as measured by chemiluminescence, than do other stimuli. In contrast to the other dimorphic fungi evaluated in that study, *B. dermatitidis* yeasts were found to produce H_2O_2, a finding compatible with data obtained in another study, demonstrating the production of O_2^- by yeasts of the same strain used by Schaffner *et al.*[34]

While not usually considered a mediator of innate resistance to infection, macrophages appear to have a major role in the host response to infection with *B. dermatitidis*. In one study, for example, Frey *et al.* demonstrated that athymic nude mice infected with conidia or yeast died more rapidly than euthymic mice.[12] This study also provided evidence that the pace of infection in mice infected with conidia or yeasts was different, with conidia producing a more chronic, widely disseminated disease than yeast forms. However, differences in the fungicidal activity of resident macrophages obtained from mice demonstrating differing susceptibilities to infection with *B. dermatitidis* have not been reported. More discussion of the role of macrophages in host defense against *B. dermatitidis* will be provided in section 4.3.

4.2. Humoral Immunity

The study of the humoral response to infection with *B. dermatitidis* has been hampered by the lack of suitable antigens with which to measure an antibody response. Since specificity of the assay in experimental models of infection are not that critical for appropriate interpretation in animals known to be infected only with *B. dermatitidis*, some work has been done in experimental models of blastomycosis. Morozumi *et al.* studied the cellular and humoral immune responses after a subcutaneous nonlethal infection with *B. dermatitidis*.[46] Mice injected subcutaneously with a nonlethal dose of *B. dermatitidis* yeast forms become resistant to subsequent, otherwise lethal, pulmonary challenge with the fungus. After one week following subcutaneous injection, antibody to *B. dermatitidis* was detectable by enzyme immunoassay and titers continued to rise until 1 month after infection, when a plateau was reached. After 8–12 weeks, titers were said to decline, but the data on the kinetics of disappearance of specific antibody were not presented. The development of a specific humoral response to this self-limited subcutaneous infection was accompanied by the development of delayed hypersensitivity as measured by induration to a crude *B. dermatitidis* lysate antigen, *in vitro* lymphocyte proliferation of spleen cells and (several weeks later) of lymph node cells, and the ability of peritoneal macrophages to inhibit growth of yeast phase *B. dermatitidis*, within one week following the subcutaneous infection.

A follow-up study by these same workers expanded on the relative roles of cell-mediated versus humoral immunity in protection against pulmonary blastomycosis.[47] While transfer of lymphocytes harvested from *B. dermatitidis*-immune mice resulted in a dose-dependent protective effect, similar transfer of immune serum (EIA titer = 1:80) was ineffective, suggesting that host protection is the result of cell-mediated immune mechanisms. Thus, while antibody is formed in response to infection with *B. dermatitidis*, its role in protection of the host appears minimal.

4.3. Cell-Mediated Immunity

A large literature has been produced over the past decade concerning the cellular immune response to infection with yeasts. The two largest bodies of information concern experimental murine blastomycosis and naturally acquired human blastomycosis.

4.3.1. Murine Blastomycosis

In 1976, Harvey *et al.* reported on the development of a murine model of blastomycosis pneumonia[48] that was reproducible and suitable for further immunological studies.[36–38,40–42,49–54] In the aggregate, these studies have demonstrated that there is a genetic basis for the resistance of mice to infection with yeast phase *B. dermatitidis*[37] but the determinants of which remain undefined; older mice are more resistant to infection than younger mice;[55] and murine peritoneal macrophages can inhibit the growth of virulent strains of *B. dermatitidis* and this antifungal activity can be enhanced by nonspecific macrophage stimuli such as con A or thioglycollate or stimulated macrophage supernatants.[49,50] Actual macrophage fungicidal activity was demonstrated using murine peritoneal macrophages and a short 4-hour incubation period,[53] in contrast to the 72-hour incubations used in the previously reviewed studies. This fungicidal effect was also enhanced by activation of macrophages with con A or BCG. In that study, antibody or complement failed to modulate macrophage activity against the organism. Indeed, subsequent work has shown that cell-mediated immune mechanisms are the primary host defense against the yeast phase of the fungus.[46] Moreover, *in vitro* attempts at augmenting macrophage fungicidal activity with gamma interferon or crude lymphokines have been successful, furthering the notion that macrophages are important in controlling the growth of *B. dermatitidis* yeasts and that this activity can be modulated by endogenously produced cytokines.

The mechanisms responsible for the actual antifungal effects on *B. dermatitidis* yeasts cells have not yet been completely defined. As noted above, *B. dermatitidis* conidia and yeasts have been noted to be susceptible to oxidants that are produced by phagocytic cells[33,44,56] and oxidative mechanisms were noted to correlate with activated neutrophil fungicidal activity in an *in vitro* assay.[42] The ability of *B. dermatitidis* conidia and yeasts to elicit a respiratory burst in murine bronchoalveolar macrophages[34,35] and neutrophils,[57] however, is suboptimal compared to a variety of soluble and particulate stimuli. Thus, the *in vivo* importance of oxidative versus nonoxidative mechanisms of host defense is still unresolved.

A murine model based on intravenous infection with *B. dermatitidis* yeast forms was used by Cozad and colleagues to investigate the role of delayed hypersensitivity in response to infection with the fungus.[58–62] Using subcutaneous infection with a low dose of *B. dermatitidis* yeast, these investigators were able to demonstrate protection from subsequent lethal intravenous inoculation with larger numbers of yeast forms, which correlated with the development of delayed hypersensitivity as measured by footpad induration following injection with killed *B. dermatitidis* yeast.[62] Similar to the results of Brummer *et al.*, these workers were also able to passively transfer delayed hypersensitivity to naive mice with spleen cells obtained from mice previously subcutaneously injected with viable *B. dermatitidis* yeast cells. A followup study, demonstrated that enhanced survival following lethal inoculation with *B. dermatitidis* yeasts paralleled the development of cell-mediated immune responses.[59] Transfer of immune serum had no effect on footpad reactivity to *B. dermatitidis* antigens. While no delayed hypersensitivity to mycobacterial antigens could be demonstrated in that study, McDaniel and Cozad subsequently reported that injection of mice with thimerosal killed *B. dermatitidis* yeast cells could protect the mice from the development of lymphoma following challenge with EL 4 lymphoma cells.[62] These results point to a nonspecific enhancement of host defense to tumors and raise the possibility of immunoadjuvant effects of *B. dermatitidis* derived antigens. Interestingly, in that study, protection of mice with *B. dermatitidis* injections was noted, even when the adjuvant, *Corynebacterium parvum* (now called, *Propionobacterium acnes*) was not effective.

Deepe *et al.* studied a murine model of disseminated blastomycosis and demonstrated that a series of immunological changes ensue following intravenous injection of yeast phase organisms.[16] The histopathology of infection in this model was characterized by progressive infection of lung, brain, superficial fascia, liver, and spleen. Combined neutrophilic and granulomatous inflammation was present. Overall, lymph nodes and spleen were relatively free of infection. Generalized suppression of cell-mediated immunity was found starting the fifth week following infection, as measured by skin testing and spleen cell proliferation *in vitro* to mitogens and antigens. Thus, it appears that infection with *B. dermatitidis*, at least in this murine model, is capable of regulating immune responses, but the mediators of these effects remain to be identified.

The role of previous exposure to *B. dermatitidis* and, thus, of acquired cell-mediated immunity to conidia has not been studied in detail but may offer insights into the importance of "activated" bronchoalveolar macrophages in growth inhibition or killing of conidia. It is clear, however, that conidia do not stimulate a significant respiratory burst in resident murine bronchoalveolar macrophages.[34,35] These murine macrophages, however, are capable of inhibiting the conidia to yeast phase transition, thus effectively eliminating pathogenicity of the fungus once in the host lung.[26] While high concentrations of H_2O_2 can duplicate this effect, the mediator of the macrophage activity has yet to be defined.

4.3.2. Naturally Acquired Human Blastomycosis

In vitro studies of patients exposed to *B. dermatitidis*, with or without clinically evident disease have been performed in an attempt to detect cell-mediated immune responses to the fungus. Lymphocyte transformation to an alkali-soluble/water-soluble *B. dermatitidis* antigen (B-ASWS)[63] has been used in most studies.[5,8,64] The results from these studies clearly indicate that peripheral blood circulating lymphocytes obtained from patients with histories of blastomycosis respond to this. Reactivity was detected up to 16 years following the infection, indicating that memory T cells produced in response to this infection may be long lived.[8] The contribution of reexposure to the organism to maintenance of lymphocyte proliferation and overall host immunity to *B. dermatitidis*, however, has been unexamined and remains unknown.

The subpopulations of lymphocytes present in blood and alveolar lavage in patients with blastomycosis was studied by Jacobs *et al.*[65] Both the absolute numbers of lymphocytes and helper T lymphocytes obtained from patients with active pulmonary blastomycosis prior to therapy were decreased compared to those obtained from controls. The absolute number of suppressor cells was decreased only in patients with resolving extrapulmonary blastomycosis. In contrast, in the bronchoalveolar lavage samples, absolute T cells were increased compared to those from controls, T-helper cells were decreased, and T-suppressor cells were decreased, resulting in a marked suppression of the T_H:T_S ratio in patients compared to controls. The functional concomitants to these observations remain to be defined.

In a study of a large outbreak of blastomycosis associated with an exposure on the Eagle River (Wisconsin) at a beaver dam, Klein *et al.* measured several parameters of cell-mediated immune function in patients suspected of clinical and subclinical infection with the fungus.[5,5a] Blastomycin skin testing was insensitive as an indicator of exposure to *B. dermatitidis*, since only 41% of patients overall reacted to this antigen. Better results were seen with *in vitro* lymphocyte transformation to B-ASWS antigen. Sixty percent of definite cases and 84% of probable cases reacted in this test. The authors emphasize that the lymphocyte transformation test was the only evidence of blastomycosis in 8 of the probable cases (21%). The specificity of reactivity of lymphocytes to B-ASWS in this *in vitro* assay, however, has not been rigorously documented in large numbers of control patients, especially in those with histoplasmosis. Therefore, the utility of this assay as a routine indicator of blastomycosis in the typical clinical situation remains controversial. Another variable, not addressed in the original report, was the effect of histoplasmin skin testing on lymphocyte transformation and other *in vitro* assays performed on patients with blastomycosis or histoplasmosis or other medical conditions.

In a series of studies, Bradsher and colleagues reported on the phagocytosis and killing of *B. dermatitidis* yeast cells by human-monocyte-derived macrophages and bronchoalveolar macrophages.[64,66] Patients recovering from blastomycosis following therapy demonstrated increased phagocytosis and inhibition of yeast-phase growth compared to normal controls, presumably reflecting enhancement of cell-mediated immune mechanisms following infection with this fungus.[64] Similar increases in phagocytosis and ability to inhibit fungal growth were seen with bronchoalveolar macrophages recovered from patients recovering from blastomycosis.[66] The fungal isolate used in these studies was obtained from a clinical specimen and, given the ability to be internalized into the phagocytes, may be smaller than the isolates used by other investigators, since other groups have noted that *B. dermatitidis* yeast cells are rarely internalized within phagocytes.[13,34,50,51,67] These studies demonstrate that cell-mediated immune mechanisms are apparently engaged in the host response to *B. dermatitidis* infection and support the data, reviewed above, derived from the investigations with murine blastomycosis.

4.4. Regulation of the Immune Response

As reviewed above, it is apparent that sublethal infection with *B. dermatitidis* in mice is associated with the development of cell-mediated immune responses, which may be generalized to protection against tumors.[60] In contrast, the observations of Deepe *et al.*[16] reflect the development of a state of anergy at one point during the course of acute blastomycosis in mice. A multiplicity of effects of different *B. dermatitidis* antigens interacting at different immunologically active sites undoubtedly plays a role in mediating these disparate effects.

The first indication that murine macrophages could be "activated" to demonstrate enhanced antifungal activity against *B. dermatitidis* appeared in 1980 in a report showing that macrophages incubated with various nonspecific agents could inhibit the growth of *B. dermatitidis* yeast cells more than could resident macrophages.[50] Subsequent reports suggested that mitogens and other antigens could enhance the fungicidal activity of resident murine peritoneal macrophages.[53] Similarly, murine neutrophils elicited by an intraperitoneal injection of killed *B. dermatitidis* yeasts could be activated for enhanced killing of *B. dermatitidis* by supernatants obtained from antigen stimulated spleen cells.[68] These data suggested that macrophage and/or lymphocyte derived products, that is, cytokines, could modulate resident phagocyte fungicidal activity. Indeed, when purified interferon gamma was employed in *in vitro* experiments, resident murine peritoneal macrophages, bronchoalveolar macrophages, and thioglycollate elicited neutrophils were stimulated to greater fungicidal activity than their corresponding controls.[42,51,52,69] Some of these effects seemed to be specific for *B. dermatitidis* yeast cells, as candidacidal activity of murine bronchoalveolar macrophages could not be enhanced by incubation with interferon gamma under the study conditions.[52] While these results suggest *in vitro* activation of phagocyte function, the *in vivo* relevance and clinical importance of these observations remains to be determined. Preliminary results in mice, however, suggest that *in vivo* administration of interferon can activate bronchoalveolar macrophages[40] and neutrophils for enhanced killing of *B. dermatitidis* yeasts. Perhaps, more importantly, the interactions of the various cytokines elaborated during the course of natural or experimental infection need to be addressed.

In vitro studies of the ability of interferon gamma to augment resident macrophage fungicidal activity comprises the bulk of published information on immunoregulation in blastomycosis.[41,52,69] Similar studies with other cytokines (e.g., tumor necrosis factor [TNF], IL-2) have yet to be reported.

5. IMMUNODIAGNOSIS AND PROGNOSIS

5.1. Serological Aspects

As previously discussed, the availability of suitable antigens for use in immunodiagnostic assays has been a serious limitation to workers in this area. However, many different *in vitro* tests designed to measure an antibody response to various *B. dermatitidis* derived antigens have been reported[7,45,70–79,79a] Some, such as the immunodiffusion and complement fixation tests have not been very sensitive and cross reactions with sera obtained from patients with histoplasmosis have been common.[71,79] If appropriate reference antisera are used, specificity can be enhanced by identifying lines of identity in the immunodiffusion test. The relatively low sensitivity of this test, however, precludes its usefulness as a routine test in the usual clinical setting.

More recent attempts at increasing sensitivity have turned to the use of enzyme immunoassay (EIA) and radioimmune assay (RIA) based techniques.[10,70,72–78] Unfortunately, in most of these reports, relatively small numbers of patients have been evaluated and, therefore, the general utility of these tests cannot be assessed. Typically, when serodiagnostic tests are moved from a research to the clinical practice setting, sensitivity and specificity fall. Since the antigens used in most of these assays are not pure and not specific to *B. dermatitidis*, expectations for the successful use of the reported tests are not high. The A antigen, identified over 10 years ago,[80] has not proved to be as specific as one might have suspected, based on early reports of its utility.[72,80] An enzyme capture EIA, using the A antigen, however, has been reported to enhance specificity of the method for detecting cases of blastomycosis and excluding patients with histoplasmosis or coccidioidomycosis.[74] Because the number of patients with other mycoses was small in that study, more work will need to be done to confirm these preliminary observations.

Taking a different approach, Klein and Jones have reported on the isolation and purification of a 120 kilodalton surface protein (named WI-1) found on the surface of *B. dermatitidis* yeast forms.[10] This antigen was not found on the surfaces of either *H. capsulatum* or *C. albicans*. In evaluating this antigen in patients with other mycoses (n = 73), these investigators found that 2 patients (3%) with histoplasmosis demonstrated relatively low titers in an antigen inhibition RIA. Thus, even this antigen, which could not be demonstrated on *H. capsulatum* yeasts, generated apparent positive antibody titers in the sensitive assays employed in these investigations. Further work using the purified 120 kD protein will determine whether or not this molecule will become a clinically useful reagent for immunodiagnosis of blastomycosis.

5.2. Cellular Aspects

In vitro lymphocyte transformation testing has been the most commonly employed method for making the diagnosis of exposure to *B. dermatitidis*. In most of the published experience, peripheral blood mononuclear cells from blastomycosis and controls have been incubated with B-ASWS, an antigen isolated by Cox *et al.*[63] This test has been noted to be the most sensitive indicator of immunological contact of a person with *B. dermatitidis*. In one outbreak, however, only 60% of confirmed cases demonstrated a positive test. Thirty-two of 39 people (84%) said to have "probable blastomycosis" also reacted with a positive lymphocyte transformation test. The definition of "probable blastomycosis," however, required that at least one of the following markers be positive: chest X ray, blastomycin skin test, serological studies (EIA, immunodiffusion, or complement fixation tests), and *in vitro* lymphocyte transformation. As is widely appreciated, blastomycin skin test reactivity has limited specificity, as does an abnormal chest X ray. Moreover, the EIA may be positive in patients with histoplasmosis and other mycoses.[72,77] Thus, the "probable" cases in this relatively large study might represent a heterologous group of individuals with a variety of underlying diseases. Until a large study demonstrating the range of cross reactivity of *in vitro* lymphocyte transformation tests (and other serodiagnostic assays) is performed, conclusions derived from the published experience must be considered tentative. Routine clinical use for these tests is not yet at hand.

6. PREVENTION AND THERAPY

6.1. Prevention

There is currently no method to prevent exposure to the organism, other than avoiding outdoor activities. Because of the extremely slow progress in identifying *B. dermatitidis* specific antigens, there is also no vaccine or candidate vaccines in study that one could predict to be useful in the 1990s. Klein and Jones, however, have identified several new *B. dermatitidis* derived antigens that could be useful as immunodiagnostic reagents as well as immunizing substances.[10]

6.2. Therapy

Several options are available for treating patients with blastomycosis. For years, amphotericin B was the standard of therapy, especially for disseminated disease, primarily because no other drug was available. Intravenous amphotericin B still finds use in critically ill patients who require parenteral therapy and in those who have contraindications for the use of azole derivatives. Specific practical recommendations on the administration of amphotericin B have been presented elsewhere.[81]

Several studies have indicated the usefulness of ketoconazole in the treatment of blastomycosis.[82–85] The largest was performed by members of the National Institutes of Health sponsored Mycoses Study group and suggests that 400 mg of ketoconazole is effective in the therapy of nonmeningeal blastomycosis in immunocompetent patients.[85] Eight hundred milligrams was also effective, but there was an increased incidence of side effects, especially gastrointestinal intolerance and endocrine system abnormalities. Other smaller studies have confirmed the utility of ketoconazole in patients with blastomycosis.[82,84] Failures with ketoconazole have also been reported either in an immunosuppressed patient,[86] in patients with severe initial disease,[87] or following apparently successful therapy in an immunocompetent patient.[83] This latter experience suggests that, in some cases, prolonged therapy (in excess of 6 months to 1 year) may be required in some patients to effect cure. Moreover, failure to respond to ketoconazole should prompt evaluation of the patient for poor absorption of the drug by drawing blood for serum ketoconazole concentrations. Pending results of ongoing clinical trials with other azole derivatives, amphotericin B remains the only proven effective alternative to ketoconazole for progressive blastomycosis.

Immunosuppressed patients with blastomycosis may present with disseminated disease, but the illness in these patients is not as aggressive as is coccidioidomycosis or histoplasmosis in similarly immunologically abnormal patients. The reasons for this are not clear. Additionally, response to ketoconazole may be inferior to that seen in patients with normal immune system function. This may be explained, in part, by more erratic absorption of ketoconazole in seriously ill patients. The general consensus is that ketoconazole should be reserved for treating patients without immediately life-threatening blastomycosis. Moreover, the immunoadjuvant effects of amphotericin B may provide additional beneficial effects in critically ill patients that are not apparent with ketoconazole therapy.

Additional azole derivatives have been synthesized and are being evaluated for use in patients with different mycoses, including blastomycosis. Itraconazole and fluconazole are two of these compounds furthest along in the investigative process. In fact, itraconazole was approved in late 1992 for treatment of patients with blastomycosis. More information concerning their utility in treating patients with blastomycosis should be available by the middle of the 1990s.

Stilbamidine is no longer used, even in mild disease, because more effective alternatives exist in the azole derivatives.

There are no clinically useful modes of immunotherapy as of the time of this writing in early 1990. As discussed above, however, interferon gamma has been shown in animal models to augment the ability of macrophages to dispatch *B. dermatitidis* yeasts. How these observations will translate into clinically useful management strategies is not apparent and pilot studies have yet to be done in infected patients.

6.3. Immunization or Vaccines

Until the identification of apparently specific *B. dermatitidis* antigens by Klein and Jones,[10] there have been no suitable antigens available for use as immunizing reagents or vaccines. Mice can be successfully "immunized," however, with low doses of virulent yeast forms or with dead *B. dermatitidis* yeasts.[46] These "immunized" mice are resistant to inoculation with otherwise lethal doses of yeast phase *B. dermatitidis*. Given the negative results using a crude antigen preparation in the Coccidioidomycosis Vaccine Trial,[88] it is unlikely that a similar undertaking will be attempted to prevent blastomycosis. More practically, blastomycosis is not as common as coccidioidomycosis and the multiplicity of problems that must be overcome in order to conduct a vaccine trial once a suitable antigen is available might be sufficient to abort any attempt to perform such a study.

7. SUMMARY

While the endemic region for blastomycosis overlaps that of histoplasmosis, the scope of the clinical problem of subclinical and spontaneously resolving blastomycosis remains unknown. Once infecting a mammalian host, usually humans or dogs, a combined suppurative–granulomatous reaction ensues. Studies evaluating mechanisms of host defenses against *B. dermatitidis* conidia are in their infancy, but it is well established that the primary host response to yeast phase *B. dermatitidis* involves the cell-mediated immune system. Systematic investigation into the mechanisms of phase transition and its modulation by host phagocytic cells may have more general applicability to other mycoses caused by dimorphic fungi. Recent advances in the identification of *B. dermatitidis* specific antigens will enhance basic research and may be helpful in devising more sensitive and specific diagnostic assays for clinical use. The development and introduction into clinical practice of new azole antifungals will lead to safer therapeutic options in treating patients with blastomycosis than have been available in years past.

REFERENCES

1. Gilchrist, T. C., and Stokes, W. R., 1898, A case of pseudo-lupus vulgaris caused by a *Blastomyces*, *J. Exper. Med.* **3**:53.
2. Gilchrist, T. C., 1896, The presence of an oidium in the tissues of a case of pseudo-lupus vulgaris, *Johns Hopkins Hops. Bull.* **7**:129.
3. Garrison, R. G., and Boyd, K. S., 1978, Role of the conidium in dimorphism of *Blastomyces dermatitidis*, *Mycopathologia* **64**:29.
4. Witorsch, P., and Utz, J. P., 1968, North American blastomycosis: A study of 40 patients, *Medicine* **47**:169.
5. Klein, B. S., Vergeront, J. M., and Davis, J. P., 1986, Epidemiologic aspects of blastomycosis, the enigmatic systemic mycosis, *Sem. Respir. Infect.* **1**:29.

5a. Klein, B. S., Vergeront, J. M., Weeks, R. J., Kumar, U. N., Mathai, G., Varkey, B., Kaufman, L., Bradsher, R. W., Stoebig, J. F., Davis, J. P., and The Investigation Team, 1986, Isolation of *Blastomyces dermatitidis* in soil associated with a large outbreak of blastomycosis in Wisconsin, *N. Engl. J. Med.* **314**:529.

6. Sarosi, G. A., and Davies, S. F., 1979, Blastomycosis, *Amer. Rev. Resp. Dis.* **120**:911–938.
7. Bradsher, R. W., 1984, Development of specific immunity in patients with pulmonary or extrapulmonary blastomycosis, *Am. Rev. Respir. Dis.* **129**:430.
8. Bradsher, R. W., and Alford, R. H., 1981, *Blastomyces dermatitidis* antigen-induced lymphocyte reactivity in human blastomycosis, *Infect. Immun.* **33**:485.
9. Sarosi, G. A., Davies, S. F., and Phillips, J. R., 1986, Self-limited blastomycosis: A report of 39 cases. *Sem. Respir. Infect.* **1**:40.
10. Klein, B. S., and Jones, J. M., 1990, Isolation, purification, and radiolabeling of a novel 120-kD surface protein on *Blastomyces dermatitidis* yeasts to detect antibody in infected patients, *J. Clin. Invest.* **85**:152.
11. Baker, R. D., 1942, Tissue reactions in human blastomycosis. An analysis of tissue from twenty-three cases, *Amer. J. Pathol.* **18**:479.
12. Frey, C. L., Demarsh, P. L., and Drutz, D. J., 1989, Divergent patterns of pulmonary blastomycosis induced by conidia and yeasts in athymic and euthymic mice, *Am. Rev. Respir. Dis.* **140**:118.
13. Sugar, A. M., and Field, K. G., 1985, Characteristics of the pulmonary cellular immune response to two strains of *Blastomyces dermatitidis* in the mouse, *Am. Rev. Respir. Dis.* **132**:1319.
14. Mitchell, T. G., and Turner, S. R., 1984, Effects of culture filtrates of *Blastomyces dermatitidis* on neutrophil locomotion, *Sabouraudia* **22**:145.
15. Thurmond, L. M., and Mitchell, T. G., 1984, *Blastomyces dermatitidis* chemotactic factor: Kinetics of production and biological characterization evaluated by a modified neutrophil chemotaxis assay, *Infect. Immun.* **46**:87.

15a. Repine, J. E., Clawson, C. C., Rasp, F. L., Jr., Sarosi, G. A., and Hoidal, J. R., 1978, Defective neutrophil locomotion inhuman blastomycosis: Evidence for a serum inhibitor, *Am. Rev. Respir. Dis.* **118**:325.

16. Deepe, G. S., Jr., Taylor, C. L., and Bullock, W. E., 1985, Evolution of inflammatory response and cellular immune responses in a murine model of disseminated blastomycosis, *Infect. Immun.* **50**:183.
17. Miyaji, M., and Nishimura, K., 1983, Granuloma formation and killing functions of granuloma in congenitally athymic nude mice infected with *Blastomyces dermatitidis* and *Paracoccidioides brasiliensis*, *Mycopathologia* **82**:129.

18. Sugar, A. M., 1992, Taxonomy and Biology of *Blastomyces dermatitidis*, in: *Blastomycosis* (Y. Al-Doory and A. F. DiSalvo, eds.), Plenum, New York, pp. 9–29.
19. McDonough, E. S., Wisniewski, T. R., Penn, L. A., Chan, D. M., and McNamara, W. J., 1976, Preliminary studies on conidial liberation of *Blastomyces dermatitidis* and *Histoplasma capsulatum, Sabouraudia* **14:**199.
20. Sugar, A. M., and Field, K. G., 1985, A rapid method for the preparation of *Blastomyces dermatitidis* conidia, *J. Med. Micro.* **3:**345.
21. Salvin, S.B , 1949, Phase-determining factors in *Blastomyces dermatitidis, Mycologia* **41:**311.
22. Kane, J., 1984, Conversion of *Blastomyces dermatitidis* to the yeast form at 37C and 26C, *J. Clin. Micro.* **20:**594.
23. McDonough, E. S., 1963, Effects of natural soils on *Blastomyces dermatitidis*, Histoplasma capsulatum and Allescheria Boydii, *Amer. J. Hygiene.* **77:**66.
24. McDonough, E. S., Dubats, J. J., and Wisniewski, T. R., 1973, Soil streptomycetes and bacteria related to lysis of *Blastomyces dermatitidis, Sabouraudia* **11:**244.
25. Denton, J. F., and DiSalvo, A. F., 1964, Isolation of *Blastomyces dermatitidis* from natural sites at Augusta Georgia, *Amer. J. Trop. Med. Hyg.* **13:**716.
26. Sugar, A. M., and Picard, M., 1991, Macrophage- and oxidant-mediated inhibition of the ability of live *Blastomyces dermatitidis* conidia to transform to the pathogenic yeast phase: implications for the pathogenesis of dimorphic fungal infections, *J. Infect. Dis.* **163:**371.
27. Editorial, 1989, Blastomycosis—One disease or two? *Lancet* **i:**25.
28. Lombardi, G., Padhye, A. A., and Ajello, L., 1988, In vitro conversion of African isolates of *Blastomyces dermatitidis* to their yeast form, *Mycoses* **31:**447.
29. Moser, S. A., Koker, P. J., and Williams, J. E., 1987, Fungal-strain-dependent alterations in the time course and mortality of chronic murine pulmonary blastomycosis, *Infect. Immun.* **56:**34.
30. Sugar, A. M., and Picard, M., 1988, Experimental blastomycosis pneumonia in mice by infection with conidia, *J. Med. Vet. Mycol.* **26:**321.
31. Williams, J. E., and Moser, S. A., 1987, Chronic murine pulmonary blastomycosis induced by intratracheally inoculated *Blastomyces dermatitidis* conidia, *Am. Rev. Respir. Dis.* **135:**17.
32. Drutz, D. J., and Frey, C. L., 1985, Intracellular and extracellular defenses of human phagocytes against *Blastomyces dermatitidis* conidia and yeasts, *J. Lab. Clin. Med.* **105:**737.
33. Sugar, A. M., and Field, K. G, 1988, Susceptibility of *Blastomyces dermatitidis* conidia to products of oxidative metabolism, *Exper. Mycol.* **12:**84.
34. Sugar, A. M., and Field, K. G., 1988, Characterization of murine bronchoalveolar macrophage respiratory burst: Comparison of soluble and particulate stimuli, *J. Leukocyte. Biol.* **44:**500.
35. Sugar, A. M., and Picard, M., 1990, Attachment of *Blastomyces dermatitidis* conidia to murine bronchoalveolar macrophages: Characterization of binding and elicitation of respiratory burst. Antonie van Leeuwenhoek, *J. Microbiol.* **57:**21.
36. Brass, C., Volkmann, C. M., Klein, H. P., Halde, C. J., Archibald, R. W. R., and Stevens, D. A., 1982, Pathogen factors and host factors in murine pulmonary blastomycosis, *Mycopathologia* **78:**129.
37. Morozumi, P. A., Halpern, J. W., and Stevens, D. A., 1981, Susceptibility differences of inbred strains of mice to blastomycosis, *Infect. Immun.* **32:**160.
38. Brummer, E., McEwen, J. G., and Stevens, D. A., 1986, Fungicidal activity of murine inflammatory polymorphonuclear neutrophils: comparison with murine peripheral blood PMN, *Clin. Exp. Immunol.* **66:**681.
39. Brummer, E., and Stevens, D. A., 1983, Enhancing effect of murine polymorphonuclear neutrophils (PMN) on the multiplication of *Blastomyces dermatitidis* in vitro and in vivo, *Clin. Exp. Immunol.* **54:**587.
40. Brummer, E., and Stevens, D. A., 1984, Activation of murine polymorphonuclear neutrophils for fungicidal activity with supernatants from antigen-stimulated immune spleen cell cultures, *Infect. Immun.* **45:**447.

41. Morrison, C. J., Brummer, E., Isenberg, R. A., and Stevens, D. A., 1987, Activation of murine polymorphonuclear neutrophils for fungicidal activity by recombinant gamma interferon, *J. Leukocyte. Biol.* **41**:434.
42. Morrison, C. J., Isenberg, R. A., and Stevens, D. A., 1988, Enhanced oxidative mechanisms in immunologically activated versus elicited polymorphonuclear neutrophils: correlations with fungicidal activity, *J. Med. Microbiol.* **25**:115.
43. Sixbey, J. W., Fields, B. T., Sun, C. N., Clark, R. A., and Nolan, C. M., 1979, Interactions between human granulocytes and *Blastomyces dermatitidis*, *Infect. Immun.* **23**:41.
44. Sugar, A. M., Chahal, R. S., Brummer, E., and Stevens, D. A., 1983, Susceptibility of *Blastomyces dermatitidis* strains to products of oxidative metabolism, *Infect. Immun.* **41**:908.
45. Scalarone, G. M., 1987, Use of a urease-antibody conjugate in an enzyme immunoassay for the detection of blastomycosis, *J. Med. Vet. Mycol.* **25**:215.
46. Morozumi, P. A., Brummer, E., and Stevens, D. A., 1982, Protection against pulmonary blastomycosis: Correlation with cellular and humoral immunity in mice after subcutaneous nonlethal infection, *Infect. Immun.* **37**:670.
47. Brummer, E., Morozumi, P. A., Vo, P. T., and Stevens, D. A., 1982, Protection against pulmonary blastomycosis: Adoptive transfer with T lymphocytes, but not serum, from resistant mice, *Cell. Immunol.* **73**:349.
48. Harvey, R. P., Schmid, E. S., Carrington, C. C., and Stevens, D. A., 1978, Mouse model of pulmonary blastomycosis: Utility, simplicity, and quantitative parameters, *Am. Rev. Respir. Dis.* **117**:695.
49. Brummer, E., Morozumi, P. A., Philpott, D. E., and Stevens, D. A., 1981, Virulence of fungi: Correlation of virulence of *Blastomyces dermatitidis* in vivo with escape from macrophage inhibition of replication in vitro, *Infect. Immun.* **32**:864.
50. Brummer, E., Morozumi, P. A., and Stevens, D. A., 1980, Macrophages and fungi: In vitro effects of method of macrophage induction, activation by different stimuli, and soluble factors on Blastomyces, *J. Reticuloendothel. Soc.* **28**:507.
51. Brummer, E., Morrison, C. J., and Stevens, D. A., 1985, Recombinant and natural gamma-interferon activation of macrophages in vitro: Different dose requirements for induction of killing activity against phagocytizable and nonphagocytizable fungi, *Infect. Immun.* **49**:724.
52. Brummer, E., and Stevens, D. A., 1987, Fungicidal mechanisms of activated macrophages: Evidence for nonoxidative mechanisms for killing of *Blastomyces dermatitidis*, *Infect. Immun.* **55**:3221.
53. Brummer, E., Sugar, A. M., and Stevens, D. A., 1983, Activation of peritoneal macrophages by concanavalin A or *Mycobacterium bovis* BCG for fungicidal activity against *Blastomyces dermatitidis* and effect of specific antibody and complement, *Infect. Immun.* **39**:817.
54. Morozumi, P. A., Brummer, E., and Stevens, D. A., 1981, Strain differences in resistance to infection reversed by route of challenge: Studies in blastomycosis, *Infect. Immun.* **34**:623.
55. Brass, C., and Stevens, D. A., 1982, Maturity as a critical determinant of resistance to fungal infections: Studies in murine blastomycosis, *Infect. Immun.* **36**:387.
56. Sugar, A. M., Chahal, R. S., Brummer, E., and Stevens, D. A., 1984, The iron-hydrogen peroxide-iodide system is fungicidal: Activity against the yeast phase of *Blastomyces dermatitidis*, *J. Leukocyte. Biol.* **36**:545.
57. Schaffner, A., Davis, C. E., Schaffner, T., Markert, M., Douglas, H., and Braude, A. I., 1986, In vitro susceptibility of fungi to filling by neutrophil granulocytes discriminates between primary pathogenicity and opportunism, *J. Clin. Invest.* **78**:511.
58. Baker, R. D., 1942, Experimental blastomycosis in mice, *Amer. J. Pathol.* **18**:463.
59. Cozad, G. C., and Chang, C.-T., 1980, Cell-mediated immunoprotection in blastomycosis, *Infect. Immun.* **28**:398.
60. McDaniel, L. S., and Cozad, G. C., 1982, Immunoadjuvant effects of *Blastomyces dermatitidis* against EL 4 lymphoma in C57BL/6J mice, *JNCI* **69**:1337.

61. Scillian, J. J., Cozad, G. C., and Spencer, H. D., 1974, Passive transfer of delayed hypersensitivity to *Blastomyces dermatitidis* between mice, *Infect. Immun.* **10**:705.
62. Spencer, H. D., and Cozad, G. C., 1973, Role of delayed hypersensitivity in blastomycosis of mice, *Infect. Immun.* **7**:329.
63. Cox, R. A., and Larsh, H. W., 1974, Isolation of skin test-active preparations from yeast cells of *Blastomyces dermatitidis*, *Infect. Immun.* **10**:42.
64. Bradsher, R. W., Ulmer, W. C., Marmer, D. J., Townsend, J. W., and Jacobs, R. F., 1985, Intracellular growth and phagocytosis of *Blastomyces dermatitidis* by monocyte-derived macrophages from previously infected and normal subjects, *J. Infect. Dis.* **151**:57.
65. Jacobs, R. F., Marmer, D. J., Balk, R. A., and Bradsher, R. W., 1985, Lymphocyte subpopulations of blood and alveolar lavage in blastomycosis, *Chest.* **88**:579.
66. Bradsher, R. W., Balk, R. A., and Jacobs, R. F., 1987, Growth inhibition of *Blastomyces dermatitidis* in alveolar and peripheral macrophages from patients with blastomycosis, *Am. Rev. Respir. Dis.* **135**:412.
67. Brummer, E., and Stevens, D. A., 1982, Opposite effects of human monocytes, macrophages, and polymorphonuclear neutrophils on replication of *Blastomyces dermatitidis* in vitro, *Infect. Immun.* **36**:297.
68. Brummer, E., Sugar, A. M., and Stevens, D. A., 1984, Immunological activation of polymorphonuclear neutrophils for fungal killing: Studies with murine cells and *Blastomyces dermatitidis* in vitro, *Infect. Immun.* **36**:505.
69. Brummer, E., Hanson, L. H., Restrepo, A., and Stevens, D. A., 1988, In vivo and in vitro activation of pulmonary macrophages by IFN-gamma for enhanced killing of Paracoccidioides brasiliensis or *Blastomyces dermatitidis*, *J. Immunol.* **140**:2786.
70. Johnson, S. M., and Scalarone, G. M., 1988, Preparation and ELISA evaluation of *Blastomyces dermatitidis* yeast phase lysate antigens, *Diag. Microbiol. Infect. Dis.* **11**:81.
71. Kaufman, L., McLaughlin, D. W., Clark, M. J., and Blumer, S., 1973, Specific immunodiffusion test for blastomycosis, *Appl. Microbiol.* **26**:244.
72. Klein, B. S., Vergeront, J. M., Kaufman, L., Bradsher, R. W., Kumar, U. N., Mathai, G., Varkey, B., and Davis, J. P., 1987, Serological tests for blastomycosis: Assessments during a large point-source outbreak in Wisconsin, *J. Infect. Dis.* **155**:262.
73. Lambert, R. S., and George, R. B., 1987, Evaluation of enzyme immunoassay as a rapid screening test for histoplasmosis and blastomycosis, *Am. Rev. Respir. Dis.* **136**:316.
74. Lo, C. Y., and Notenboom, R. H., 1990, A new enzyme immunoassay specific for blastomycosis, *Am. Rev. Respir. Dis.* **141**:84.
75. Sato, H., Honigman, A. S., Scalarone, G. M., 1987, Development of a comparative Dot ELISA for the detection of antibodies in blastomycosis, *Diag. Microbiol. Infect. Dis.* **7**:37.
76. Turner, S., and Kaufman, L., 1986, Immunodiagnosis of blastomycosis, *Sem. Respir. Infect.* **1**:22.
77. Turner, S., Kaufman, L., and Jalbert, M., 1986, Diagnostic assessment of an enzyme-linked immunosorbent assay for human and canine blastomycosis, *J. Clin. Micro.* **23**:294.
78. Wheeler, J. W., Owens, R. D., and Scalarone, G. M., 1988, Enzyme immunoassay detection of antibodies in North American blastomycosis: Comparison of three *Blastomyces dermatitidis* antigens, *Mycoses* **31**:459.
79. Williams, J. E., Murphy, R., Standard, P. G., and Phair, J. P., 1981, Serologic response in blastomycosis: Diagnostic value of double immunodiffusion assay, *Am. Rev. Respir. Dis.* **123**:209.
79a. Klein, B. S., Kuritsky, J. N., Chappell, W. A., Kaufman, L., Green, J., Davies, S. F., Williams, J. E., and Sarosi, G. A., 1986, Comparison of the enzyme immunoassay, immunodiffusion, and complement fixation tests in detecting antibody in human serum to the A antigen of *Blastomyces dermatitidis*, *Amer. Rev. Resp. Dis.* **133**:144–148.
80. Green, J. H., Harrell, W. K., Johnson, J. E., and Benson, R., 1980, Isolation of an antigen

from *Blastomyces dermatitidis* that is specific for the diagnosis of blastomycosis, *Current Microbiol.* **4**:293.
81. Bennett, J. E., 1980, Treatment of cryptococcal, candidal, and coccidioidal meningitis, in: *Current Clinical Topics in Infectious Diseases vol. 2* (J. S. Remmington and J. S. Swartz, eds.), McGraw-Hill Book Co., New York, p. 54.
82. Bradsher, R. W., Rice, D. C., and Abernathy, R. S., 1985, Ketoconazole therapy for endemic blastomycosis, *Ann. Intern. Med.* **103**:872.
83. Hebert, C. A., King, J. W., and George, R. B., 1989, Late dissemination of pulmonary blastomycosis during ketoconazole therapy, *Chest* **95**:240.
84. McManus, E. J., and Jones, J. M., 1986, The use of ketoconazole in the treatment of blastomycosis, *Am. Rev. Respir. Dis.* **133**:141.
85. NIAID Mycoses Study Group, 1985, Treatment of blastomycosis and histoplasmosis with ketoconazole. Results of a prospective randomized clinical trial, *Ann. Intern. Med.* **103**:861.
86. Greene, N. B., Baughman, R. P., Kim, C. K., and Roselle, G. A., 1985, Failure of ketoconazole in an immunosuppressed patient with pulmonary blastomycosis, *Chest* **88**:640.
87. Thiele, J. S., Buechner, H. A., and Cook, E. W., Jr., 1983, Failure of ketoconazole in two patients with blastomycosis, *Am. Rev. Respir. Dis.* **128**:763.
88. Pappagianis, D., 1986, Vaccine Study Group: Evaluation of the protective efficacy of the killed *Coccidioides immitis* vaccine in man. Abstracts of the 1986 ICAAC:784.

8

Coccidioidomycosis

REBECCA A. COX

1. DESCRIPTION OF THE DISEASE

1.1. Portal of Entry

Coccidioidomycosis is a systemic mycosis caused by the dimorphic fungus *Coccidioides immitis*. The fungus is a soil saprophyte in the semiarid regions of the western hemisphere. The saprobic cells propagate in a mycelial phase that gives rise to infectious enterothallic arthroconidia. The arthroconidia are easily aerosolized and, when inhaled by a susceptible host, convert in tissue into a parasitic spherule/endospore cycle (Fig. 1). The spherule–endospore phase persists in host tissue or, if deposited in nature, converts to the mycelial–arthroconidia phase.

Although the pulmonary route is the natural route of infection, there have been infrequent reports of infection resulting from percutaneous inoculation of the fungus.[1,2] Most of these have occurred in persons working with the fungus or infected tissues and, with rare exceptions, result in a benign infection localized to the site of inoculation.

1.2. Clinical Manifestations

1.2.1. Pulmonary

The vast majority of residents in endemic areas have experienced infection with *C. immitis*.[3–5] The primary infection may be asymptomatic,

REBECCA A. COX • Department of Research Immunology, San Antonio State Chest Hospital, San Antonio, Texas 78223.

Fungal Infections and Immune Responses, edited by Juneann W. Murphy *et al.* Plenum Press, New York, 1993.

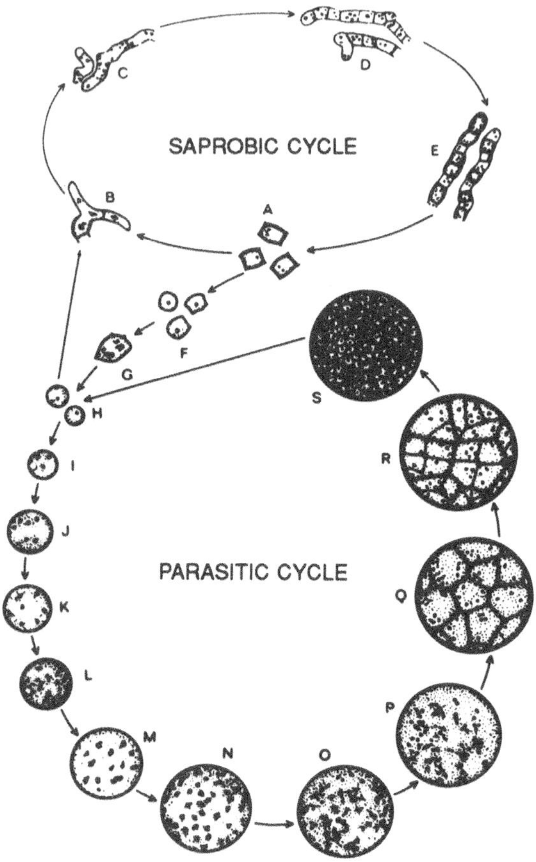

FIGURE 1. Saprobic (A–E) and parasitic (F–S) life cycles in *C. immitis*. (From Ref. 40.)

evidenced solely by the acquisition of delayed-type hypersensitivity (DTH) to the fungus or manifested as a mild bronchopneumonia.[4,5]

Approximately 5% of primary infections develop into a frank flu-like illness, accompanied by variable degrees of fever, cough, headache, pleural pain, anorexia, and malaise. The incubation period is usually 10 to 16 days, with extremes of 7 to 28 days.[4,6] Patients with acute symptomatic lung infection typically manifest pulmonary infiltrates on roentgenograms, which may be accompanied by mild or massive pleural effusion. Hilar lymphadenopathy occurs in about 20% of patients.[5] Most individuals resolve their primary infection without complication, but approximately 5% are left with pulmonary nodules (coccidiomas), abscesses, or cavities.[7,8]

Cutaneous rashes may be manifested in patients with primary coccidioidomycosis.[1,4,5,9] A generalized "toxic erythema," resembling the macular rash observed in measles, urticaria, or scarlet fever, develops in approximately 10% of primary cases and is regarded as being a nonspecific reaction in that it occurs before the acquisition of immune response to *C. immitis.* Erythema nodosum and erythema multiforme constitute the "San Joaquin fever" or "valley fever" complex, which may also include arthralgias ("desert rheumatism"). These manifestations are regarded as being "specific" in that they are temporally associated with acquired immunologic reactivity with *C. immitis.* Erythema nodosum presents as painful, red symmetrical nodules, generally located on the lower extremities. It occurs more often in women than in men, and more often in Caucasian women than in nonCaucasian women. The development of erythema nodosum generally denotes a favorable prognosis for recovery, although extrapulmonary spread has been documented in some of these patients.[1] Erythema multiforme is characterized by the development of reddish nodules, macules, and vesicles, and may accompany erythema nodosum or occur in the absence of the latter. Unlike erythema nodosum, there is no apparent relationship between the development of erythema multiforme and immune resistance to *C. immitis.*

1.2.2. Disseminated Disease

Extrapulmonary coccidioidomycosis is a relatively rare but serious complication of primary coccidioidomycosis, occurring in less than 1% of cases.[3,4,10] Disseminated disease characteristically develops within a few weeks of the primary respiratory infection. Late manifestations have been reported, but it is unclear whether these are recent foci of extrapulmonary spread or activation of previous, clinically inapparent foci. With the general exception of the gastrointestinal tract, disseminated coccidioidomycosis may involve any organ but most commonly involves the skin, bones and joints, meninges, and genitourinary system.

Predisposition to disseminated coccidioidomycosis has been documented in persons of Filipino (or other Asian), black, or Hispanic ancestry,[3,11–14] pregnant women,[13,15–19] and persons who are immunologically compromised,[20–25] including patients with the acquired immunodeficiency syndrome (AIDS).[26–36] These predisposing factors are discussed in detail in section 3.3.

2. ETIOLOGY

2.1. Etiologic Agent

Coccidioides immitis is the sole etiologic agent of coccidioidomycosis. The fungus is classified within the subdivision Deuteromycotina of the fungi Imperfecti,[37] denoting the lack of a demonstrable meiotic division. In the event that a sexual phase is discovered, the fungus will be reclassified within the Phycomycetes, Ascomycetes, or Basidiomycetes, depending on the characteristics of meiotic division.

C. immitis is a dimorphic fungus, exhibiting a mycelium–arthroconidium cycle in nature and a spherule–endospore phase in host tissue. In the saprobic phase, the fungus propagates in a mycelial form that produces asexual arthroconidia by enterothallic development.[38,39] At the initiation of this process, the hyphae undergo septation, forming chains of multinucleated compartments. Compartments that are destined to become arthroconidia show wall synthesis and condensed cytoplasm, while the adjacent cells become nonrefractile and eventually undergo autolysis, resulting in the disarticulation of the arthroconidia. The arthroconidia are multinucleated, usually containing two or three nuclei per cell, and are characteristically thick-walled, barrel-shaped cells measuring 3–6 μm in length and 2–4 μm in diameter.[40]

The mycelial phase can be grown on most laboratory media and over a wide range of temperatures. There is considerable variability in colonial morphology, but most strains produce flat, gray, membranous colonies within 3–5 days of incubation, followed by the development of white to gray aerial hyphae with a cottony appearance.[41] Pigmentation, if it occurs, is usually evident at 2 weeks and is characteristically light brown at first but becomes darker brown with age. The pigment may diffuse into the surrounding agarose medium or may be limited to the underside of the colony.

The nutritional requirements and conditions for culturing the spherule–endospore phase were delineated by Converse and co-workers.[42–45] The chemically defined medium of Converse, as modified by Levine *et al.*,[46,47] consists of basal salts with dextrose and ammonium acetate as the sole sources of organic carbon and nitrogen and Tamol N as a surfactant. Spherulation is enhanced at temperatures of 39°C to 40°C and by a 20% CO_2-atmosphere.[42–45,48,49]

In the initial phase of arthroconidia conversion, the cells swell and round up, and undergo multiple and apparently synchronous nuclear divisions.[38,39] Nuclear division ceases and is followed by a progressive segmentation of the protoplasm, eventually giving rise to uninucleated endospores. At maturity, the spherules (30–80 μm in diameter) rupture, releasing hundreds of endospores. The newly released endospores are enclosed by a hyaline membrane (Fig. 2) but, with time, detach into single cells measuring approximately 2–4 μm in diameter. Each endospore has the potential of developing into a mature, endosporulating spherule, thus perpetuating the spherule–endospore cycle *in vivo*.

2.2. Ultrastructure

The ultrastructure of *C. immitis* has been examined in detail by several investigators.[38–40,50–55] Electron micrographs of arthroconidia reveal an

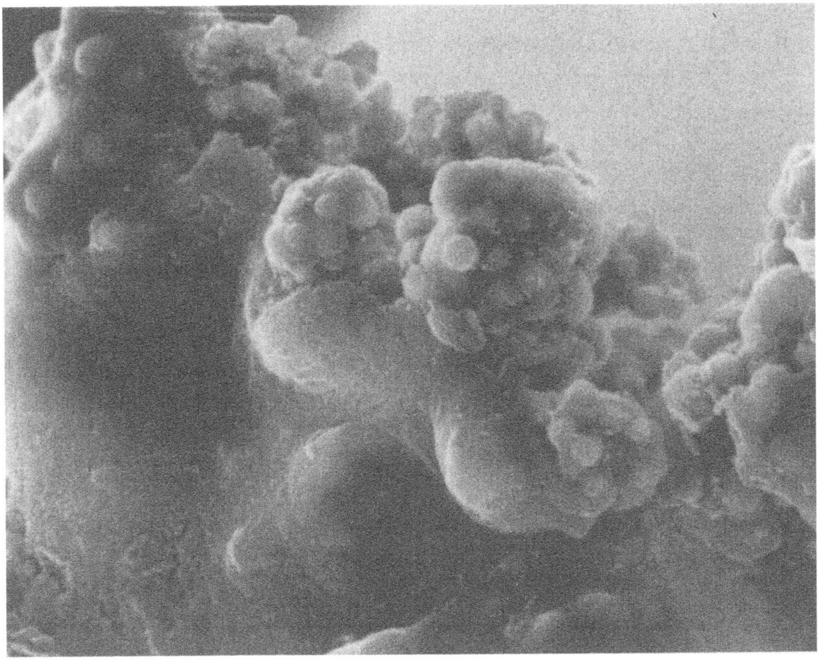

FIGURE 2. Ruptured spherules showing newly released endospores encased in a fibrillar membrane. (From Ref. 38.)

outer electron-opaque cell-wall layer that measures approximately 0.1 μm in thickness and an inner electron translucent layer that measures about 0.2 μm in thickness (Fig. 3). The outer wall layer is a remnant of the parent hyphal outer wall layer (HOWL) and, in newly disarticulated arthroconidia, is observed to extend beyond the cell as tags from the adjacent, degenerated compartments.

Electron micrographs of successive stages in the conversion of an arthroconidium into an endosporulating spherule show that the wall of the young, immature spherule is similar to that of the arthroconidium, being composed of an outer electron dense and an inner electron transparent layer (Fig. 4). During the initial stage of endosporulation, the inner wall layer of the young spherule invaginates, producing a primary segmentation, and with continued wall ingrowth and branching, eventually delineates multinucleate compartments in which the uninucleate endospores develop.[38,39,55] The cell wall of the mature spherule measures approximately 500 nm and is composed of at least two layers by electron microscopy.[50,51,55]

Limited information has accrued on the structural composition of the morphological forms of *C. immitis*. The chitin layer appears to be located within the inner cell wall, as evidenced by the digestion of this layer following

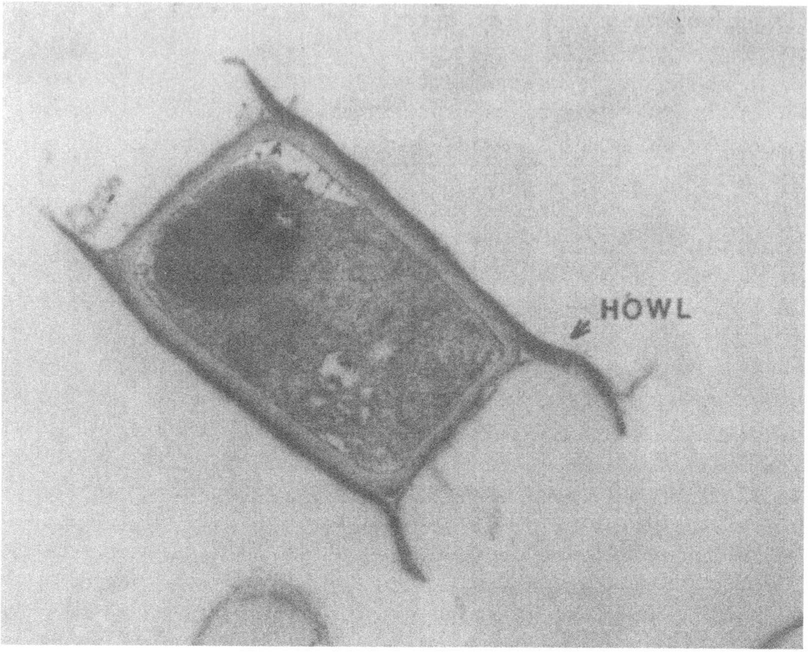

FIGURE 3. Arthroconidium showing remnants of the hyphal outer wall layer (HOWL). (From Ref. 38.)

enzymatic treatment with chitinase.[56] Treatment with pronase or mannosidase results in a buckling and detachment of the immature spherule wall, indicating the presence of a mannoprotein complex. Chemical analyses of wall fractions have established the presence of N-acetylglucosamine, amino acids, glucose, mannose, galactose, 3-0-methylmannose, and lipids, with quantitative differences being evident in walls of spherules, endospores, mycelia, and arthroconidia.[57–60] The detection of 3-0-methylmannose in wall fractions is particularly noteworthy since this methylated carbohydrate is not known to exist in the cell walls of other pathogenic fungi.

3. EPIDEMIOLOGY

3.1. Ecology

The endemic areas of coccidioidomycosis are confined to the Western Hemisphere in a region extending from 40°N, 120°W in northern California to 40°S, 65°W in Argentina.[13,61,62] The areas of highest endemicity lie within

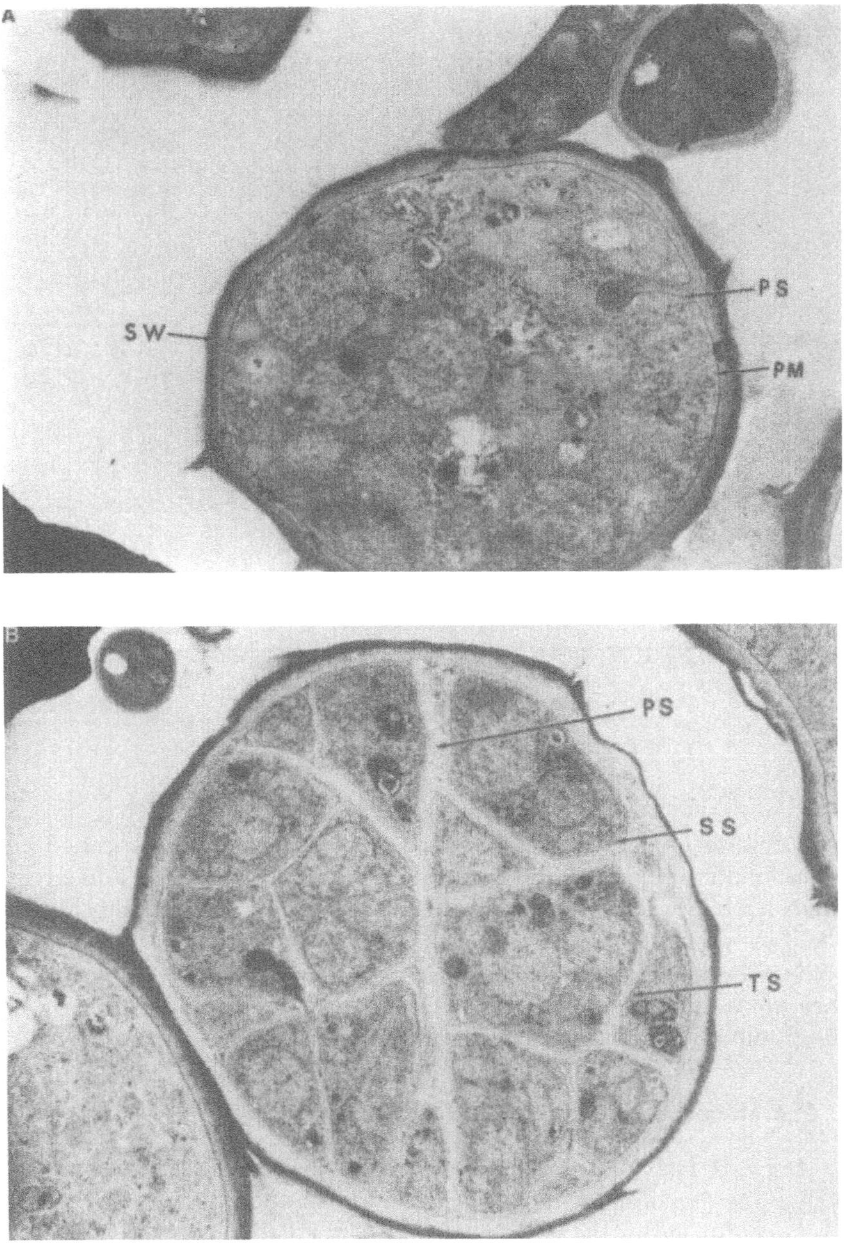

FIGURE 4. (A) Young immature spherule showing initiation of primary segmentation (PS) from inner wall layer of young spherule wall (SW). (PM, plasma membrane.) (B) Progressive segmentation of protoplasm. (PS, SS, and TS are primary, secondary and tertiary segmentation planes, respectively.) (C) Mature, endosporulating spherule. (E, endospore; N, nucleus; SW, spherule wall.) (From Ref. 55.)

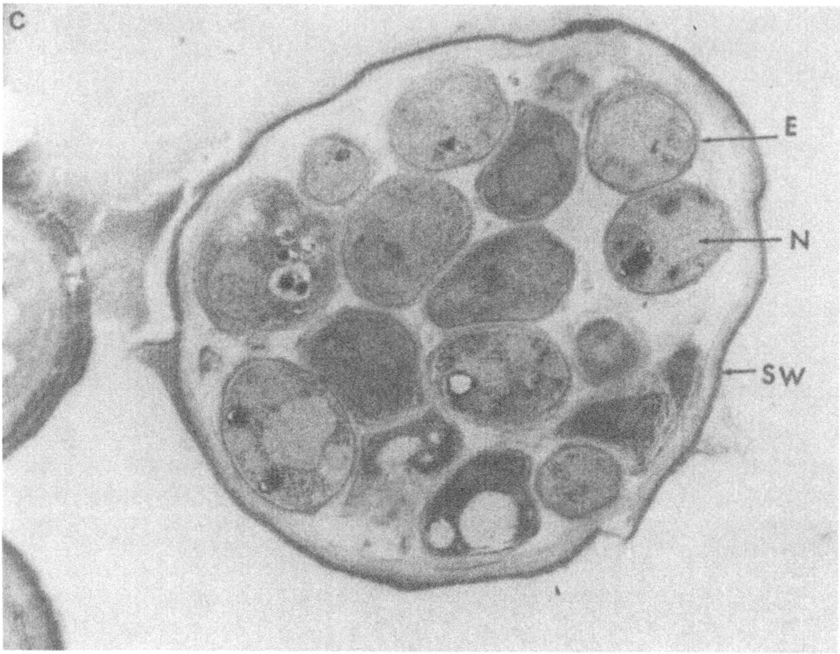

FIGURE 4. (*Cont.*)

the southwestern United States and northern Mexico and are associated with the bioclimatic regions of the Lower Sonoran Life Zone.[61,62] This area is characterized by a semiarid or arid climate and soil having a slightly alkaline pH and increased salinity.[61,63] It has been suggested that propagation of the mycelia occurs during rainy seasons and is followed by the production of arthroconidia during the hot, dry periods, thereby accounting for the increased number of new cases occurring in late summer or early fall.[64,65]

3.2. Prevalence and Incidence of Infection

Most estimates of the prevalence of coccidioidal infection have been based on the demonstration of DTH response to coccidioidin (CDN).[10,13,66] Surveys conducted in the early 1940s established that as many as 90% of persons residing in highly endemic areas were skin-test reactive to CDN.[3,66,67] More recent estimates indicate that less than one-third of residents in endemic areas have experienced primary infection.[68,69] This decreased incidence of exposure has probably resulted from the extensive land development in the endemic regions.

The incidence of coccidioidomycosis is increased in persons who, by occupation, have high risks of exposure to the fungus, for example agricultural workers, construction workers, and archaeologists.[3,6,70,71] Numerous cases of primary infection have also occurred in laboratory and research personnel[13,72,73] and, in two reports,[74,75] in hospital personnel who were exposed to the fungus while removing plastic casts from patients with bone and/or joint disease. There have been rare instances in which persons have acquired the infection as a result of handling contaminated fruits, wool, or other fomites from endemic regions.[71]

3.3. Risk Factors for Development of Disseminated Disease

3.3.1. Race

The increased susceptibility of persons of certain races to developing disseminated coccidioidomycosis was noted in early studies by Gifford and colleagues.[11] Analyses of new cases occurring between 1901 and 1936 in Kern County, California revealed that Filipinos, Blacks, and Mexican-Americans were 176, 14, and 3 times more likely to develop disseminated disease, respectively, than were Caucasians. It has been argued that the observed increase in these racial (or ethnic) groups may be ascribed to occupational exposure.[76,77] However, epidemiological studies of a recent outbreak in northern California that occurred after a windstorm carried arthroconidia from southern California provided compelling evidence to support a racial predisposition to disseminated disease.[12–14] Pappagianis,[13] in an analysis of 230 symptomatic cases resulting from the outbreak, reported that dissemination occurred in 66.6% of the Filipino patients, 53.8% of Blacks, and 17.6% of Mexican Americans as compared to 11.2% of Caucasians.

An association has been reported between disseminated coccidioidomycosis and an increased phenotype frequency of the HLA-A9 histocompatibility antigen.[78] Patients with disseminated disease have also been reported to have a high frequency of the ABO blood group B antigen.[79] These associations and their significance in terms of genetically acquired susceptibility require additional study since the observed increase in the HLA-A9 and blood group B antigens may merely reflect their increased phenotype frequencies in persons of Filipino and black descent.[80–82]

The factors that govern genetically determined susceptibility to disseminated coccidioidomycosis have not been elucidated. If susceptibility is attributable to an immunologic defect, it does not appear to reside in the inability of these persons to mount a cell-mediated immune (CMI) response to the fungus, as suggested by the finding that coccidioidomycosis patients of Filipino or Black ancestry manifest skin-test reactivity to *C. immitis* antigens, at the same frequency as do patients of Caucasian descent.[83] Likewise, conversion of coccidioidal skin-test reactivity was not found to differ among

Filipinos, Blacks, and Caucasians following vaccination with a killed spherule vaccine.[84]

3.3.2. Pregnancy

The association between pregnancy and a predisposition to developing extrapulmonary disease is well documented.[13,15–19] In areas of high endeminicity, approximately 0.1% of pregnancies are complicated by coccidioidomycosis and, of these, 20% disseminate to extrapulmonary tissues, with a resulting mortality rate of 88%. The later the gestational stage, the more likely dissemination will occur. During the first trimester, the rate of dissemination is 12.5%. This increases to 50% in the second trimester and to 91% in the third trimester. The morbidity is even higher in pregnant females of black ancestry. Although placental involvement may be present, maternal–fetal transmission is rare[17]

It has been theorized that the risk for pregnant women developing progressive, disseminated coccidioidomycosis is attributed to the immunosuppressive state that accompanies gestation.[21] More recent studies by Drutz *et al.*[85–87] offer an alternative explanation. These investigators showed that progesterone and 17 β-estradiol, at concentrations comparable to levels in sera of pregnant women, potentiate the growth of spherules *in vitro*. Although stimulation of spherule growth by sex hormones offers an attractive hypothesis to account for the unique susceptibility of gravid females to *C. immitis*, studies have not been done to establish their potentiating effects on spherules *in vivo*.

3.3.3. Immunocompromising Conditions

The risk for developing disseminated coccidioidomycosis is greatly increased in patients who are immunocompromised, in particular patients given corticosteroids or cytotoxic therapy[20–25] and those with AIDS.[26–36] Clinical onset is often insidious but, once manifested, it progresses rapidly and usually leads to death.

4. HOST DEFENSES

4.1. Innate Resistance

The high incidence of exposure to *C. immitis* in persons who reside in areas endemic for coccidioidomycosis, coupled with the low frequency of clinical disease, suggests that innate resistance is important in host defense against this fungus. Although the fungus elicits a marked cellular infiltrate of polymorphonuclear and mononuclear phagocytes, these cells appear to be ineffectual in inhibiting or restricting *C. immitis* in the nonimmune host.

4.1.1. Complement

Early investigations by McNall *et al.*[88] provided evidence that properdin, a component of the alternate complement pathway, has a role in innate resistance to *C. immitis*. Swiss-Webster mice, treated with inulin to reduce serum properdin levels, were more susceptible to the disease than nontreated mice. Halde *et al.*[89] reported that reduced serum properdin levels were associated with the development of disseminated coccidioidomycosis in humans and in progressive disease of experimentally infected mice.

One explanation for the decreased properdin levels observed in coccidioidomycosis would be the activation of the complement pathway by the fungus. Support for this is derived from the finding that a mycelial-derived polysaccharide extract directly binds properdin and complement components Cl, Cl, C3, and C4.[90] More recently, soluble autolysates from mycelia and spherules were shown to cause a decrease in total hemolytic complement (CH_{50}) and C4 levels and an activation of C3, C4, and Factor B in fresh serum from normal subjects. These results indicate that the fungus activates both the alternative and classical complement pathways in the absence of antibody.[91]

4.1.2. Natural Cellular Resistance

4.1.2a. Polymorphonuclear Leukocytes. Arthroconidia and newly released endospores elicit an intense cellular infiltrate composed predominantly of polymorphonuclear leukocytes (PMNL).[92–94] The influx of PMNL likely represents a chemotactic response to the fungus, as inferred by the finding that CDN and SPH evoke chemotaxis of PMNL *in vitro*.[95]

The effect of PMNL on arthroconidia viability has been assessed *in vitro* using leukocytes from nonimmune humans,[95–100] rhesus macaques,[101] dogs,[102] and mice.[103] The extent of phagocytosis various considerably among *C. immitis* strains,[52,100] a difference that may be attributable to quantitative differences in the hyphal outer cell-wall layer, since this wall layer has been reported to impede phagocytosis.[52]

Ingestion of the arthroconidia is followed by a respiratory burst[100,103] and, although arthroconidia are susceptible to the products of the oxidative burst[69,98] and to cationic peptides (defensins),[104] less than 20% of the phagocytized cells are killed.[52,100,101,103] Rather, PMNL appear to promote the morphologic conversion of arthroconidia into the spherule-phase cells.[69,96,105]

Maturation of arthroconidia into mature spherules is accompanied by a marked increase in cell size, rendering the cells impervious to phagocytosis by PMNL.[52,69,100] PMNL may inhibit spherules by directly attaching to and digesting the spherule wall, as suggested by electron micrograph studies of lavage cells from infected mice.[52] Direct contact between PMNL and spherules has been reported to be inhibited, however, by an extracellular fibrillar matrix on the outer spherule wall.[100] It has not been determined if spherules elicit an oxidative burst in PMNL. Exposure of spherules to 2 mM H_2O_2 is without effect upon their viability, however, indicating that the cells are resistant to the products of the oxidative burst.[69]

The release of endospores from ruptured spherules *in vivo* evokes an influx of PMNL, suggesting that endospores, like arthroconidia, are chemotactic for PMNL.[92–94] Although endospores are readily phagocytized by PMNL and induce an oxidative burst, their viability is essentially unaffected by the encounter.[100,101]

Less noted, but equally significant, is the finding that PMNL from *Blastomyces dermatitidis*-immunized mice are highly effective in reducing endospore viability.[103] The enhanced antifungal activity of these immunologically activated PMNL can likely be ascribed to activation by cytokines that are secreted by immune cells. It is noteworthy that, while immunologically activated PMNL are fungicidal to endospores, they do not exert an increased antimicrobial effect on arthroconidia.[103] The basis for the difference in the susceptibility of arthroconidia and endospores to PMNL is not known but may be related to differences in cell-wall composition or structure.

Sera from coccidioidomycosis patients or immunized animals promote phagocytosis of arthroconidia by PMNL[52,100,102] and, in one study, was reported to potentiate PMNL killing of *C. immitis*.[52] Although these effects were ascribed to antibody, they could be attributable to other components in coccidioidomycosis sera, for example PMNL-activating cytokines.[106]

4.1.2b. Monocytes/Macrophages. The outcome of the interaction between *C. immitis* and mononuclear phagocytes has been examined in studies using alveolar macrophages from rhesus macaques,[107] resident peritoneal macrophages from mice,[108] and peripheral blood monocytes from healthy persons.[98] Monocytes/macrophages effectively phagocytize arthroconidia and endospores, and the process is enhanced by immune sera.[107] The phagocytes are ineffective, however, in restricting the intracellular growth of the fungus. This lack of inhibitory activity is paralleled by a lack or reduced level of phagosome–lysosome fusion,[98,107,108] suggesting that the fungal cells may inhibit this process. In contrast to the lack of anticoccidioidal activity of naive macrophages, immunologically activated macrophages are able to restrict or kill arthroconidia and endospores, an effect that is paralleled by an increase in the level of phagosome–lysosome fusion (see section 4.2.4c.).

Spherules are reported to elicit a predominant macrophage infiltrate in infected tissues.[1,92] These cells are not able to engulf the large spherule-phase cells, and there is no evidence to suggest that macrophages exert a direct inhibitory effect on spherules.[52,96]

4.1.2c. Natural Killer Cells. Petkus and Baum[109] reported that natural killer (NK) lymphocytes may be important in host resistance to *C. immitis*. Incubation of spherules with freshly isolated human peripheral blood mononuclear cells resulted in a 70% reduction in colony forming units (CFU). Treatment of the cells with Leu-11B, a monoclonal antibody that recognizes the Fc-gamma receptor of NK cells, reduced fungicidal activity. The inhibitory effect of the effector cells was potentiated by treatment with IFN-γ, hence NK cells could also function as an effector cell in the acquired immune response, following activation by cytokines from immune T cells.

Indirect evidence to support a role of NK cells *in vivo* is derived from a study by Clemons *et al.*[110] Beige C57BL/6 mice, which have impaired NK cell function,[111] manifested increased numbers of fungal CFU in their lungs and spleens at 4 and 7 days after intravenous infection with 50 arthroconidia compared to their phenotypically normal bg/+ littermates. No differences were observed in the ability of the two mouse strains to restrict fungal growth later in the course of disease, as measured by CFU at days 14 or 21. The increased susceptibility of beige mice could be attributed to NK cells but, as pointed out by Clemons *et al.*,[110] beige mice are also defective in PMNL chemotaxis and lysosomal enzyme secretion, and the latter defects could have contributed to, or accounted for, their increased susceptibility.

4.2. Acquired Immunity

4.2.1. Antigens Used to Detect Immune Responses

Most of the early investigations of skin tests and humoral immune responses in coccidioidomycosis were performed using CDN preparations that were prepared as the broth culture filtrates of mycelia grown for several weeks in a synthetic asparagine medium.[66] CDN preparations of proven efficacy have also been produced as toluene-induced lysates or culture filtrates of 5-day old mycelia grown in 2%-glucose–1%-yeast extract broth.[112,113]

Levine *et al.*[114] produced spherulin (SPH) as an aqueous extract from spherules cultured in modified liquid Converse medium and lyzed by agitation in distilled water. In comparative analyses, SPH was reported to be more sensitive than CDN in eliciting *in vivo* and *in vitro* cell-mediated immune responses[114–119] and in detecting complement-fixing (CF) antibody.[120] Other studies have yielded contrasting results, however, with CDN having an increased sensitivity as compared to SPH, in humoral[121] and cell-mediated immune assays.[83,122,123] The specificity of SPH appears to be comparable to that of CDN in skin tests of persons with histoplasmosis and paracoccidioidomycosis,[124] but has been reported to be less than that of CDN in assays for the CF antibody.[121]

Regardless of the source, CDN and SPH are heterogeneous extracts, composed of antigens that are specific to *C. immitis* and antigens that cross react with other fungi, in particular *Histoplasma capsulatum* and *Blastomyces dermatitidis*.[125–130] Approaches to purifying the immunoreactive antigens of *C. immitis* have included subjecting CDN or SPH to conventional molecular sieve, ion-exchange, or lectin-affinity chromatography,[57,119,131,132] extracting intact cells or cell walls from mycelia or spherules using physicochemical methods,[58,133–138] and employing monospecific antibodies as ligands for purifying the antigens by immunoadsorption of CDN or SPH.[139–142] Although these studies have yielded a wealth of information on the identification and characterization of immunoreactive antigens, most immunologic investigations continue to be performed with CDN, SPH, or both.

4.2.2. *Relationship between Humoral and CMI Responses*

The profile of humoral and CMI response in human coccidioidomycosis evolved in a series of epoch studies by Smith and co-workers.[66,126,143,144] Persons with primary asymptomatic or benign infection typically manifest strong DTH response to CDN and have low to nondetectable levels of CF antibody (Fig. 5). Conversely, patients with acute or chronic progressive coccidioidomycosis have high titers of CF antibody and manifest depressed or absent skin-test reactivity to CDN. The depressed T-lymphocyte responses are, in most patients, reversed following clinical remission.

The direct relationship between disease progression and serum CF titers and the inverse correlation between CF antibody and DTH response to CDN provides indirect evidence that CMI is crucial to host defense against *C. immitis*, whereas serum CF antibody is inconsequential or perhaps detrimental. This concept has received considerable support from studies of patients and experimentally infected animal models.

4.2.3. *Humoral Immunity*

4.2.3a. Quantitative Immunoglobulin Levels. Chronic or progressive coccidioidomycosis is associated with a polyclonal B lymphocyte activation, as evidenced by elevated serum levels of IgG, IgA, and IgE.[145–150] Antibodies reactive with coccidioidal antigens are demonstrable within each of these Ig classes.[147,148–152] Serum IgG levels are increased in the majority of patients with extensive disease involving the lungs or extrapulmonary sites, and total IgG correlates directly with CF antibody titers to CDN.[146–150] IgA is elevated in sera of approximately 20% of patients with active coccidioidomycosis, being manifested most often in patients with chronic pulmonary disease.[149]

Hyperproduction of IgE has been demonstrated in approximately 23% of patients with active coccidioidomycosis patients, with the highest incidence occurring in patients with extensive multifocal disease.[149,150] Total serum IgE directly correlates with IgE antibody levels to CDN; however, IgE reactivity is also demonstrable against common allergens, such as bermuda grass and ragweed. Longitudinal studies of coccidioidomycosis patients with hyper-IgE levels revealed that, in most patients, IgE production diminished to normal or nearly normal levels after clinical regression, suggesting that hyperproduction is a consequence of the disease. This interpretation is countered, however, by the report that atopic persons are at greater risk of

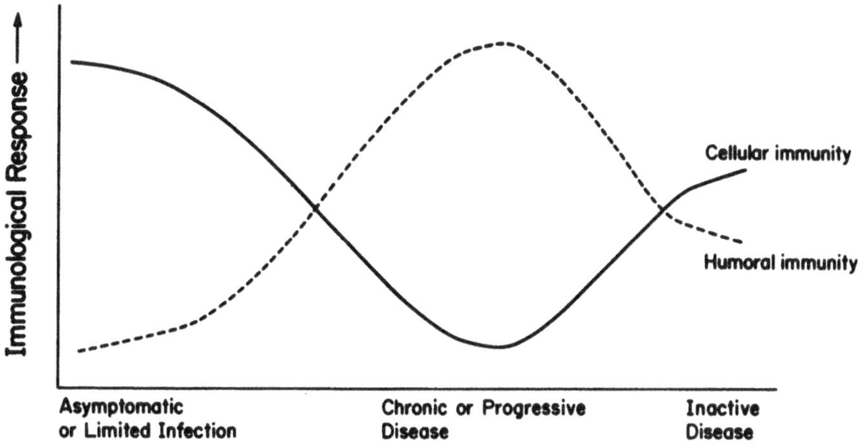

FIGURE 5. Profile of humoral and cellular immune responses in various stages of human coccidioidomycosis. (From Ref. 191.)

developing symptomatic coccidioidomycosis than are nonatopic persons.[153] Perhaps relevant to the occurrence of elevated IgE is the finding that patients with primary coccidioidomycosis and, less commonly, patients with disseminated disease may manifest a peripheral blood eosinophilia.[4,92,154–161] Tissue eosinophilia has also been noted to occur in microabscesses and granulomatous lesions of some patients with progressive pulmonary or disseminated disease.[160] No studies have been done to determine if an association exists between eosinophilia and IgE production.

4.2.3b. Immune Complexes. Circulating Clq-binding immune complexes have been detected in sera from patients with active coccidioidomycosis[162,163] and, in one study,[163] were shown to correlate with disease severity. Whereas 33% of sera from patients having disease involving a single organ system had elevated immune complex levels, 67% of sera from patients with disease involving two or more organ systems had circulating complexes. Compositional analysis of immune complexes revealed the presence of *C. immitis* antigen, anti-*Coccidioides* IgG, and Clq.[162]

It is not known if immune complexes have a role in the immunopathogenesis of coccidioidomycosis. The clinical manifestations of erythema nodosum and erythema multiforme could be mediated by immune complexes, as has been shown in other diseases.[164–167] Immune complexes have been hypothesized to have an immunosuppressive effect on host response,[162] however, no evidence exists to support this supposition (see section 4.2.5b.).

4.2.3c. Role of Antibodies in Host Defense. There is no evidence to support a role of antibody, either alone or complexed with coccidioidal antigen, in protective immunity against *C. immitis*. Rather, excessive antibody production is a salient feature of, and could contribute to, progressive coccidioidomycosis in humans. Consistent with this possibility is a report that C57BL/6 mice, which are susceptible to *C. immitis*,[168–171] produce high levels of serum antibody following infection, whereas the resistant A/J mouse strain does not manifest a detectable antibody response to the fungus.[172]

In studies to assess the passive transfer of immunity to *C. immitis*, Kong *et al.*[173] reported that spleen cells but not sera from spherule-immunized donors protected recipients against challenge. Rather, the investigators suggested that passively transferred sera may have potentiated the disease in recipient mice. Beaman *et al.*[174,175] confirmed the adoptive transfer of immunity with immune splenocytes and showed that protection was not reduced by depleting the spleen cells of B lymphocytes. The investigators also reported that sera from spherule-immunized mice did not neutralize the infectivity of arthroconidia.[175] The findings that immune sera do not transfer protection, that depletion of B cells does not ablate adoptive transfer of immunity with immune spleen cells, and that immune sera do not neutralize the infectivity of arthroconidia provide a convincing argument against a protective role of antibodies in host defense against *C. immitis*.

4.2.4. Cell-Mediated Immunity

4.2.4a. Granuloma Formation. Development of granulomas is mediated by T lymphocytes and serves to localize the invading organism.[176–179] Histologic studies have established that granulomas of patients with benign, self-limited coccidioidomycosis are well defined, consisting of a peripheral mantle of lymphocytes surrounding epithelioid cell aggregates.[1,180] By contrast, granulomas of patients with progressive coccidioidomycosis are poorly circumscribed and show lymphocytes interspersed with macrophages, and fungal elements.[1,34,180]

Analyses of T-lymphocyte subsets by immunoperoxidase staining with monoclonal antibodies specific for $CD4^+$ (helper/inducer) and $CD8^+$ (suppressor/cytotoxic) cells revealed further differences.[180] Granulomas from patients with benign pulmonary coccidioidomycosis showed $CD4^+$ cells distributed within the central area and the localization of $CD8^+$ cells at the periphery. By contrast, granulomas of patients with disseminated disease showed $CD4^+$ and $CD8^+$ cells interspersed throughout the granulomas. Differences were also evident in the numbers of $CD4^+$ and $CD8^+$ lymphocytes. Helper/inducer cells predominated in granulomas of benign coccidioidomycosis, whereas suppressor/cytotoxic lymphocytes predominated in granulomas of progressive disease.

The contrasting patterns of granuloma formation in patients with localized versus disseminated coccidioidomycosis resemble those observed in tuberculoid and lepromatous leprosy, respectively[181–184] and, as with the polar extremes of leprosy, reflect intact versus impaired CD4 lymphocyte function.[184]

4.2.4b. Delayed-Type Hypersensitivity. Early investigations by Smith *et al.*[66,144,185,186] established that 83% of persons with nonprogressive pulmonary coccidioidomycosis manifest skin-test reactivity to CDN within the first week of clinical illness, 93% are reactive by the second week, and virtually all persons are skin-test positive by the third week. Cutaneous DTH response to CDN persists in most persons who recover from primary infection and is associated with life-long immunity to exogenous reinfection.[3,4,7,10]

Depressed skin-test reactivity to CDN is a characteristic feature of severe chronic or progressive coccidioidomycosis. Whereas greater than 70% of patients with a solitary lesion involving the lung or an extrapulmonary site manifest CDN reactivity, less than one third of patients with extensive pulmonary or disseminated involvement respond to CDN skin tests.[7,8,66,187–191] Depressed or absent response denotes a poor prognosis for recovery. Smith *et al.*,[66] in a study of 100 disseminated coccidioidomycosis patients, reported that only 17% of patients who were skin-test negative to CDN 1:100 recovered from their disease, as compared to 75% of patients who were reactive to CDN. The prognosis is even worse for patients who fail to react to a 1:10 dilution of CDN.

The cutaneous anergy in most patients is specific for *C. immitis* antigens, as judged by DTH response to a panel of recall antigens such as mumps, streptokinase–streptodornase, Dermatophytin "O," and trichophyton.[66,83,187–191] Exceptions to this occur with patients who have coccidioidomycosis involving multiple organ systems. Approximately 13% of these fail to respond to coccidioidal and noncoccidioidal antigens.[190] A generalized anergy occurs at a much higher frequency in AIDS patients with coccidioidomycosis.[36]

In general, patients who manifest cutaneous DTH to CDN respond to the antigen *in vitro* in assays for lymphocyte proliferation and the production of macrophage migration inhibitory factor.[187–195] Proliferation responses to the mitogen phytohemagglutinin have been reported to be depressed in some patients but normal in others[187–195] and may vary in kinetics when compared to the mitogen responses of healthy controls.[187]

An important question that has not been resolved is whether T-lymphocyte hyporeactivity to *C. immitis* is the cause or the effect of the disease. Smith *et al.*[66] reported that one patient who was skin-test positive to CDN at the onset of clinical illness became anergic two weeks after the disease disseminated. Unfortunately, similar sequential studies of DTH responses have not been reported, so it is not known whether loss of reactivity is a common occurrence or an exception.

4.2.4c. Role of T Cells in Protection. The association between T-lymphocyte hyporeactivity or anergy and progressive coccidioidomycosis suggests that CMI is crucial to host defense in coccidioidomycosis. This concept is borne out by the finding that neonatal or adult thymectomized mice and congenitally athymic BALB/c mice are highly susceptible to challenge with *C. immitis* as compared with mice having intact T-cell function.[110,174,196] A protective role of T cells is further documented by the transfer of immunity to challenge with spleen cells from spherule-immunized mice,[173–175] and the ablation of immune transfer following depletion of T cells.[175]

One mechanism by which T lymphocytes function as effectors of immunity to *C. immitis* was elucidated in studies by Beaman and co-workers.[108,175,197,198] The investigators showed that peritoneal macrophages from normal DBA/2 mice restricted the growth of arthroconidia and endospores when cultured in the presence of immune spleen cells.[108] The immune splenocytes did not directly affect fungal viability but rather secreted a lymphokine(s) that activated macrophages to an anticoccidioidal state. Pretreatment of immune spleen cells with anti-Thy-1 or anti-Lyt-1 plus complement abrogated lymphokine production; hence, the effector cell is a Lyt-1^+2^- T cell.[197] When the immune lymphocytes were incubated with rabbit anti-IFN-γ, the killing of arthroconidia was reduced by 44%, indicating that IFN-γ was a major component of the lymphokine activating activity. In the converse experiment, recombinant IFN-γ was shown to activate alveolar and peritoneal macrophages from normal BALB/c mice to kill arthroconidia and endospores.[198] The ability of immune lymphocytes to activate macrophages has also been demonstrated with peripheral blood cells from healthy, CDN skin-test positive donors.[98] Studies have not yet been done to determine if this effector-cell function is depressed or absent in patients with active coccidioidomycosis, in which case the important question arises as to whether the impairment can be reversed by treatment with IFN-γ.

Macrophages from nonimmune hosts show a low level of phagosome–lysosome fusion following ingestion of arthroconidia or endospores.[98,107,108,197] Exposure to immune lymphocytes or supernatants from antigen-stimulated immune lymphocytes enhances the fusion of phagosomes and lysosomes and is paralleled by an increase in the anticoccidioidal activity of these phagocytic cells.[108,198]

Evidence that activated macrophages are important mediators of host defense *in vivo* is obtained from a study by Clemons *et al.*[199] These investigators evaluated the effect of silica, a reagent that impairs macrophage function, on the course of the disease in Swiss-Webster mice. Silica treatment of nonimmune mice did not alter the course of disease, as judged by the numbers of CFU at days 7, 14, or 21 postinfection when compared to the CFU of nontreated mice. By contrast, silica treatment of immunized mice significantly potentiated disease progression.

A recent report showed that that peritoneal-exudate macrophages from nonimmune mice secreted tumor necrosis factor-alpha (TNF-α) in response to viable or killed spherules.[200] More recent studies of infected BALB/c mice have established that TNF-α is produced early in the course of the disease but then diminishes to nondetectable levels with progressive (and lethal) disease (R. A. Cox, unpublished data). The production of TNF-α could be important in host defense against *C. immitis* since this pleiotrophic monokine has been shown to potentiate the antimicrobial effect of macrophages, PMNL, and NK cells.[106,201–203]

4.2.5. *Suppression of T-Cell Response*

The T-lymphocyte anergy observed in progressive coccidioidal disease and the reacquisition of T-lymphocyte responses after disease regression suggest that anergy is a consequence of active immunosuppression. Evidence to support this has been obtained in studies of cells or sera from experimentally infected animals and coccidioidomycosis patients.

4.2.5a. Cell-Mediated Suppression Ibrahim and Pappagianis[204] established that CDN-immunized guinea pigs became nonresponsive to CDN following daily intraperitoneal injections of 130 mg of the antigen. The induced anergy was reversed upon termination of antigen injections. Desensitization of immunized guinea pigs has also been induced by multiple subcutaneous injections of spherulin.[205] Ibrahim[206] showed that newborn guinea pigs injected with high doses of killed hyphae (3 mg) in complete Freund adjuvant (CFA) were rendered nonresponsive to CDN in DTH assays. The tolerant state was antigen specific, as judged by the responses of the guinea pigs to tuberculin, and was negated by the adoptive transfer of peritoneal-exudate cells from *C. immitis*-sensitized adult guinea pigs.

Levine and colleagues,[207–210] in studies of a killed spherule vaccine, established that intramuscular administration of the vaccine induced immunity against challenge, whereas administration of the vaccine via an intravenous route did not engender protection.[209] Rather, intravenous injection of the killed spherule vaccine prior to, concomitant with, or up to 35 days after intramuscular vaccination diminished the protective effects of the vaccine.[210]

Similar findings have been reported in studies of an alkali-soluble, water-soluble cell-wall extract, designated as C-ASWS. Intramuscular administration of 0.5 to 3 mg of C-ASWS in CFA induces protection in mice against challenge and is accompanied by the acquisition of DTH response to *C. immitis* antigens.[171,211] In contrast, administration of C-ASWS via an intravenous route suppresses, in a dose-dependent manner, the ability of the mice to respond to immunization with C-ASWS or CDN.[191,212] When spleen cells from mice suppressed with C-ASWS were transferred intravenously to syngeneic recipients, the recipients showed a reduced response to immunization; hence, suppression is mediated by a splenic suppressor cell or factor. Passage of the spleen cells over a nylon wool column depleted suppressor activity, suggesting that the suppressor cell is an adherent macrophage.

Evidence has also been obtained to support the activation of splenic suppressor cells *in vivo* as a consequence of the disease. BALB/c mice, which are highly susceptible to infection with *C. immitis*,[168,171] mount a DTH response to CDN early during the course of experimental disease, to a level comparable to that observed with the resistant DBA/2 mice. In contrast to

DBA/2 mice, however, BALB/c mice develop anergy later in the disease.[171] The acquired anergy is temporally associated with the activation of a splenic suppressor cell or factor, as evidenced by the finding that spleen-cell lysates (that had been filtered to remove viable fungi) suppressed the response of syngeneic recipients to immunization with CDN in CFA.[212] Passage of the spleen cells over nylon wool abrogated suppressor activity, a result that is consistent with suppression by an adherent macrophage population. In contrast to the suppression observed with spleen-cell lysates from BALB/c mice, lysates prepared from spleens of infected DBA/2 mice were without suppressor activity (R. A. Cox and W. Kennell, unpublished data). These data suggest that genetically determined susceptibility to *C. immitis* is associated with, and may be attributable to, the activation of a suppressor cell that functions to down regulate T-cell response.

Catanzaro[213] provided evidence to support the activation of suppressor cells in human coccidioidomycosis. Proliferation responses of leukocytes from patients with multifocal, disseminated disease were depressed when compared to the responses of patients with unifocal disease. The responses were augmented when the cells were assayed in the presence of indomethacin, an agent that inhibits prostaglandin synthesis. Removal of adherent cells, by passage of the cells over nylon wool, also resulted in increased proliferative responses. These results indicate that suppression of response to mitogen is mediated by an adherent, prostaglandin-producing macrophage. The important question of whether suppression occurs with coccidioidal antigens or is limited to mitogen responses was not addressed.

4.2.5b. Serum-Mediated Suppression Several investigators have reported that sera from active coccidioidomycosis patients suppress lymphocyte proliferation to coccidioidal antigens.[192,193,214] The suppressive effect was evident by demonstrating a diminished proliferation by cells from healthy, CDN-reactive donors when assays were conducted in the presence of sera from patients and, in reciprocal experiments, an augmentation of patients' responses when their cells were assayed using healthy donor serum. The suppression was specific for *C. immitis* antigens in most patients and was reversed upon clinical remission.

When patients' sera were immunoadsorbed with *Staphylococcus* protein A, suppression was demonstrable with the desorbed column eluate and not the effluent fraction.[214] These results provide evidence that suppression is mediated by IgG antibody, either alone or complexed with antigen.[215] Although immune complexes have been detected in coccidioidomycosis patients and have been shown to suppress T-cell responses in other models,[216,217] attempts to demonstrate a suppressive effect of complexes on the proliferative responses of lymphocytes were unsuccessful.[214] The possibility that suppression may be mediated by an anti-idiotypic IgG antibody,[218,219] which would have been adsorbed by protein A, was not pursued.

Suppression of T-lymphocyte response by coccidioidomycosis sera has also been demonstrated in studies of infected mice.[212] Passive transfer of sera from infected BALB/c mice suppressed the response of normal recipient mice to immunization with CDN when compared to the response of recipients given sera from noninfected donors. Unlike the antibody-mediated suppression observed in studies of human coccidioidomycosis, suppression in the murine model was shown to be attributed to circulating coccidioidal antigen.[212] This conclusion is based on the removal of suppressor activity by adsorption of the sera with hyperimmune goat anti-CDN and the concomitant reduction in the serum antigen levels.

It has not been determined whether coccidioidal antigen(s) directly suppresses T-cell function or acts indirectly by inducing a suppressor cell or factor. The kinetics of antigen-mediated suppression support the latter possibility. That is, suppression by sera containing antigen or by intravenously injected antigen was optimally demonstrable when recipients were immunized one week after treatment.[212] By contrast, suppression by spleen cells from antigen-treated mice or spleen-cell lysates from infected mice was optimally demonstrable when recipients were immunized on the day of transfer. The delayed kinetics observed with antigen-induced suppression would allow time for antigen to activate a splenic cell that, upon adoptive transfer, suppresses T-cell function, as has been shown in other suppressor-cell models.[220]

5. IMMUNODIAGNOSIS AND PROGNOSIS

5.1. Serological Aspects

Although *C. immitis* induces a spectrum of antibody responses,[151,152] detection of the tube precipitin (TP) and CF antibodies remain the gold standards for serologic tests in coccidioidomycosis. These antibody responses are of proved value in the diagnosis and management of the disease[1,4,5,221] and are detected using the classical TP and CF assays described by Smith *et al.*[126,143] or the immunodiffusion (ID) assays, designated IDTP and IDCF, respectively, developed by Huppert and co-workers.[113,222–224]

The TP antibody appears early after infection, being demonstrable in 53% of patients with primary, nondisseminated coccidioidomycosis within the first week of symptomatic illness, 91% by the second and third weeks, and 86% during the fourth week.[126] Thereafter, the response decreases rapidly, usually detectable in less than 10% of patients at 4 months after clinical onset. The early production of the TP antibody is consistent with its identification as an IgM antibody response.[147,148] The lack of a significant IgG secondary antibody response to this antigen is unexplained, but raises the possibility the TP antigen, which is a high molecular weight polysaccharide,[139,141,225–227] might be tolerogenic.

The CF antibody is predominantly of the IgG isotype,[147,148] and is produced later during coccidioidal disease, generally within 3 to 4 months of clinical onset, and persists in most patients throughout active disease. CF antibody titers provide a useful index for assessing the extent of disease involvement.[1,126,143,221] A low titer ($\leqslant$ 1:8) denotes unifocal disease involving the lungs or a solitary extrapulmonary focus, whereas titers of $\geqslant$ 1:32 signify multifocal disease. Serial monitoring of CF antibody titers is of particular value in evaluating the disease status. An increasing CF antibody titer occurs in disease progression and, conversely, a decreasing CF titer signifies disease regression or stabilization.[126,143]

Other immunoassay procedures for detecting antibodies to *C. immitis* have utilized the techniques of enzyme-linked immunosorbent and radioimmunoassays.[221,228] Although these methods have increased sensitivity, as compared to the CF or TP assays, their uses are limited by a concomitant increase in cross reactivity, owing to the use of CDN or other extracts that contain a multiplicity of antigens.

Development of an assay for detecting coccidioidal antigen in sera or other biofluids would be a valuable diagnostic and prognostic aid. This area of research has been addressed in two investigations[229,230] and, although the results appear promising, their use has not been evaluated beyond preliminary studies.

5.2. Skin Tests

Detection of skin-test reactivity to CDN has limited diagnostic value in coccidioidomycosis, except when conversions from negative to positive responses are demonstrable. This limitation is based on the high frequency of skin test reactivity in persons who reside in endemic areas and, secondly, the presence of antigens that react with histoplasmin and blastomycin.[126–128]

The CDN skin test is useful, however, in assessing the patient's prognosis for recovery (see section 4.2.4b.). A patient who manifests skin-test reactivity to CDN has a favorable prognosis, whereas anergy denotes an unfavorable outcome.[66] More recently, clinicians have utilized SPH as a skin-test reagent for coccidioidomycosis. While a negative response to both CDN and SPH predicates a poor prognosis, the prognostic significance of skin-test reactivity to SPH in the absence of a response to CDN is unclear.

6. PREVENTION AND THERAPY

6.1. Vaccines

A large body of evidence supports the feasibility of developing a vaccine for coccidioidomycosis. First, most persons who recover from benign or asymptomatic infection develop a solid state of immunity to exogenous reinfection.[1,4,144,155,185] Second, the infectious saprophytic phase of the fungus is geographically restricted, thereby delineating the areas of potential infection. Third, the target population is well defined, and includes persons who are genetically predisposed to developing disseminated coccidioidomycosis and persons who have a high probability of exposure by virtue of their occupation (see section 3.2).

6.1.1. *Viable Cells*

Rixford and Gilchrist[231] reported in 1896 that experimental cutaneous infection of a dog with the exudate of lesions from a coccidioidomycosis patient induced immunity to cutaneous challenge. The protective effects of a sublethal infection was subsequently confirmed by several groups of investigators.[232–239] The use of viable cells posed a major limitation, however, in that organisms persisted at the site of vaccination and, not infrequently, were demonstrable at distal sites, even when the vaccine was administered via a subcutaneous route.[235,237,240] In an attempt to circumvent this problem, Pappagianis *et al.*[233] vaccinated mice using a riboflavin-requiring auxotrophic mutant, induced by X irradiation of *C. immitis* strain Silveira.[241] The mutant strain engendered immunity to challenge but was shown to revert *in vivo* to a prototrophic state that was accompanied by a reacquisition of virulence. Similar results were observed using a cobalt-irradiated isolate of *C. immitis*[238] and with viable spherules that had been attenuated by *in vitro* passage.[239] Efforts to reduce the risks imposed by viable vaccines included prevaccination with killed organisms[236] and treatment of vaccinated mice with amphotericin B.[242] Both approaches ameliorated the disease but were not applicable for use in humans because of the potential adverse consequences.

6.1.2. *Nonviable Cells*

Early studies established that killed mycelial-phase cells induced a demonstrable level of immunity against challenge.[233–244] Following the development of Converse medium for propagating spherules, killed spherules were shown to be more effective in inducing immunity than saprobic-phase cells.[46,47,240,247] In a comparative study, Kong and Levine[240] reported that mice immunized with 1.5 mg of formalin-killed spherules were protected against an intranasal challenge of 10^4 arthroconidia, whereas only 50% of mice vaccinated with 1.5 mg of killed mycelia survived.[247] Immunogenicity increased with the maturation phase of the parasitic cells, with mature spherules having an increased potency as compared to endospores.[245]

The effect of dose, schedule, and route on the immunizing capacity of killed spherules were delineated in a series of investigations by Levine and coworkers.[47,207–210,245] Mice immunized with 2 to 2.7 mg of killed spherules were more resistant to challenge than were mice given lesser doses. Administration of the vaccines in three injections over a period of several days was more immunogenic than a single injection, and protection was further enhanced when the vaccine was administered at multiple sites as opposed to a single site. The route of vaccine administration proved to be a major determinant of vaccine efficacy. Administration of the vaccine by an intramuscular or subcutaneous routes induced immunity, whereas administration of the vaccine by an intravenous route was ineffective.

The efficacy of the killed spherule vaccine in the murine model prompted clinical trials to assess its toxicity and potential use in healthy, CDN skin-test negative persons.[84,246–248] Recipients of a total of 2.7 mg of the vaccine given intramuscularly in one to three injections, showed localized tenderness or induration at the site(s) of injection but otherwise tolerated the vaccine well. Doses greater than 10 mg produced marked and unacceptable levels of toxicity. Based on these results, subsequent trials were conducted by administering 1.75 mg or less of the spherule vaccine in a total of three intramuscular injections. Approximately 7% of recipients showed adverse reactions, ranging from local inflammation to systemic toxicity. Conversion of skin test reactivity was dependent on the dose of vaccine, with less than one third of persons given 4.5 mg of killed spherules showing conversion of their CDN skin-test response,[246–250] as compared to one half of more of persons given higher doses of the vaccine.[249] Serologic conversion was infrequent, occurring in less than 16% of recipients.

A double-blind, multicentered study was undertaken in 1980 to evaluate the vaccine efficacy of killed spherules.[249,250] A total of 2967 healthy, skin-test negative persons were randomized into placebo and vaccine groups. The vaccine recipients were given three intramuscular injections of 1.75 mg of formalin-killed spherules suspended in saline; the placebo group received three injections of saline alone. After a 5-year follow-up period, no significant difference was observed in the incidence or severity of coccidioidal cases that developed following vaccine or placebo injections.[250] The basis for the lack of protection in vaccine recipients is not known. It bears consideration, however, that the dosage of spherules used in humans was, on a body weight basis, less than 1% of the amount required to protect mice.[251]

6.1.3. Cell-Derived Immunogens

The toxicity of the killed spherule vaccine, at concentrations that may be required to induce protective immunity, warrants studies to purify the immunoprotective component(s) for use in human trials.

Kong *et al.*[252] subjected spherules to mechanical disruption and found that the immunoprotective component of spherules resided primarily, if not exclusively, with the cell wall fraction. Pappagianis and co-workers[251] reported the extraction of an immunoprotective spherule wall antigen by incubating the walls in phosphate-buffered saline for 5 days with constant agitation. Administration of 0.55 or 1.1 mg of the wall extract in alum adjuvant induced protection in mice against pulmonary challenge with 1000 arthroconidia, to a level comparable to that induced by the whole killed spherule vaccine. Incubation of the walls with chitinase during the extraction procedure resulted in a reduced immunogenicity of the wall-derived fraction. Whether the diminished efficacy of the extract was attributed to chitinase or, as the investigators pointed out, to a possible contamination of the chitinase preparation with β-(1-3)-glucanase was not determined.

Immunity to challenge has also been induced in mice injected subcutaneously or intramuscularly with 1 mg of C-ASWS admixed with CFA.[211] Unlike the extract studied by Pappagianis and co-workers,[251] the immunoprotective effect of C-ASWS (from mycelia or spherules) was less than that of intact spherules.[142] This reduced efficacy may be attributed to a denaturation of the immunogenic component(s) during alkali extraction or to incomplete extraction of the immunogen(s).

6.2. Therapy

6.2.1. Chemotherapy

Coccidioidomycosis is one of the most difficult fungal diseases to treat and, as yet, no uniform guidelines have been developed.[5,253,254] It is generally accepted that patients with fever, prostration, or progressive pulmonary disease lasting six weeks or longer and those with increasing CF antibody titers should be considered for treatment. Treatment may also be considered for patients who have one or more risk factors for dissemination. Patients with disseminated coccidioidomycosis should be treated aggressively and over a prolonged period of time.

The mainstay of chemotherapeutic agents for coccidioidomycosis is amphotericin B, a polyene drug that was introduced in 1959.[5,253] All isolates of *C. immitis* are uniformly susceptible to the drug and, although the drug appears to be fungicidal in *in vitro* analyses, its effect *in vivo* appears be fungistatic. Amphotericin B is administered intravenously, with the dose and duration of therapy being decided empirically, based upon clinical response or drug toxicity. The toxic side effects of the drug are numerous and include fever, chills, nausea, vomiting, nephrotoxicity, and anemia. These adverse reactions may be ameliorated by the use of salicylates, antihistamines, antiemetics, and low doses of steroids. Amphotericin B methyl ester was synthesized in an attempt to mollify the toxicity of the drug; however, the use of the methyl ester formulation was abandoned after reports of leukoencephalopathy.[255]

The synthetic azoles have offered an alternative to amphotericin B for the treatment of coccidioidomycosis in patients who have failed to respond to amphotericin B or have refused to take the latter drug because of its toxicity.[256–259] These drugs were introduced sequentially, beginning with miconazole, followed by ketoconazole and, most recently, itraconazole and fluoconazole. The results obtained with fluoconazole have been most encouraging, in particular in the treatment of patients with coccidioidal meningitis.[259]

6.2.2. *Immunotherapy*

The discovery by Lawrence in 1955[260] that DTH could be transferred using extracts of leukocytes from immune donors and reports that the active component(s) was dialyzable[261] prompted a plethora of studies on the immunotherapeutic benefits of transfer factor in diseases associated with impaired T-cell responses, including coccidioidomycosis. Administration of transfer factor was reported to restore T-cell reactivity to *C. immitis* antigens, as measured by *in vivo* and *in vitro* assays, and was accompanied by clinical improvement in their coccidioidal disease.[262–264] These promising reports led to a multicentered collaborative study by the *Coccidioides* Cooperative Treatment Group (CCTG) to evaluate the immunotherapeutic efficacy of transfer factor in coccidioidomycosis patients. The clinical responses were reported to be dramatic, with probable cures recorded in 27% of patients receiving transfer factor and improvement in an additional 25% of patients.[195,265,266] A subsequent report, however, failed to substantiate a clinical benefit of transfer factor, when given alone or adjunctive with amphotericin B.[267]

The finding that IFN-γ potentiates the anticoccidioidal properties of murine macrophages *in vitro*,[198] coupled with the known immunotherapeutic effects of IFN-γ in other infectious diseases,[268] provides a rationale for evaluating its efficacy in the treatment of coccidioidomycosis. Such studies have not yet been done, however. An investigation was undertaken to assess the immunotherapeutic potential of recombinant human interleukin 2 (IL-2) in experimental coccidioidomycosis of Swiss-Webster mice.[269] Daily intravenous injection of separate groups of mice with log-gradient concentrations of IL-2, ranging from 25 to 250,000 units, failed to ameliorate the course of disease following intratracheal challenge with 100 arthroconidia. By contrast, mice given amphotericin B without IL-2 survived the 30-day observation period. The lack of a protective effect with IL-2 is anomalous since the binding of this cytokine to its receptor on T cells is known to trigger the production of IFN-γ.[270] Additional studies are clearly warranted to assess the roles of IFN-γ, IL-2, and other cytokines that might augment host response to this pathogen.

REFERENCES

1. Drutz, D., and Catanzaro, A., 1978, Coccidioidomycosis. I. State of the art, *Am. Rev. Respir. Dis.* **117**:559–585.
2. Carroll, G. F., Haley, L. D., and Brown, J. M., 1977, Primary cutaneous coccidioidomycosis. A review of the literature and a report of a new case, *Arch. Dermatol.* **113**:933–939.
3. Smith, C. E., Beard, R. R., Whiting, E. G., and Rosenberger, H. G., 1946, Varieties of coccidioidal infection in relation to the epidemiology and control of the disease, *Am. J. Public Health* **36**:1394–1404.
4. Fiese, M. J., 1958, *Coccidioidomycosis*, Charles C. Thomas, Springfield, Illinois.

5. Drutz, D., and Catanzaro, A., 1978, Coccidioidomycosis. II. State of the art. *Am. Rev. Respir. Dis.* **117**:727–771.
6. Werner, S. B., Pappagianis, D., Heindl, I., and Mickel, A,. 1972, An epidemic of coccidioidomycosis among archeology students in northern California, *N. Engl. J. Med.* **286**:507–512.
7. Smith, C. E., Beard, R. R., and Saito, M. T., 1948, Pathogenesis of coccidioidomycosis with special reference to pulmonary cavitation, *Ann. Intern. Med.* **29**:623–655.
8. Hyde, L., 1968, Coccidioidal pulmonary cavitation, *Dis. Chest* **54**:17–21.
9. Smith, C. E., 1940, Epidemiology of acute coccidioidomycosis with erythema nodosum ("San Joaquin" or "Valley Fever"), *Am. J. Public Health* **30**:600–611.
10. Smith, C. E., Pappagianis, D., Levine, H. B., and Saito, M, 1961, Human coccidioidomycosis, *Bacteriol. Rev.* **25**:310–320.
11. Gifford, M. A., Buss, W. C., and Douds, R. J., 1937, Data on coccidioides fungus infection, Kern County, 1901–1936, Kern County Health Department Annual Report, 39–54.
12. Flynn, N. M., Hoeprich, P. D., Kawachi, M. M., Lee, K. K., Lawrence, R. M., Goldstein, E., Jordan, G. W., Kundargi, R. S., and Wong, G. A., 1979, An unusual outbreak of windborne coccidioidomycosis, *N. Engl. J. Med.* **301**:358–361.
13. Pappagianis, D., 1980, Epidemiology of coccidioidomycosis, in: Coccidioidomycosis, (D. A. Stevens, ed.), Plenum Publishing Co., New York, pp. 63–85.
14. Johnson, W. M., 1982, Racial factors in coccidioidomycosis: mortality experience in Arizona, *Ariz. Med.* **39**:18–24.
15. Smale, L. E., and Birsner, J. W., 1949, Maternal deaths from coccidioidomycosis, *JAMA* **140**: 1152–1154.
16. Vaughan, J. E., and Ramirez, H., 1951, Coccidioidomycosis as a complication of pregnancy, *Calif. Med.* **74**:121–123.
17. Harrison, H. N., 1958, Fatal maternal coccidioidomycosis: a case report and review of sixteen cases from the literature, *Am. J. Obstet. Gynecol.* **75**:813–820.
18. Harris, R. E., 1966, Coccidioidomycosis complicating pregnancy, Report of three cases and review of the literature, *Obstet. Gynecol.* **28**:401–405.
19. Harvey, R. P., 1980, Coccidioidomycosis in pregnancy, in: *Coccidioidomycosis. A Text* (D. A. Stevens, ed.), Plenum Pres, New York, pp. 241–244.
20. Rifkind, D., Marchioro, T. I., Schneck, S. A., and Hill, R. B., 1967, Systemic fungal infections complicating renal transplantation and immunosuppressive therapy, *Am. J. Med.* **43**:28–41.
21. Deresinski, S. C., and Stevens, D. A., 1974, Coccidioidomycosis in compromised hosts. Experience at Stanford University Hospital, *Medicine* **54**:377–395.
22. Schroter, G. P., Bakshandeh, K., Husberg, B. S., and Well, R., III, 1977, Coccidioidomycosis and renal transplantation, *Transplantation* **23**:485–489.
23. Rowland, V. S., Westfall, R. E., and Hinchcliffe, W. A., 1977, Fatal coccidioidomycosis, analysis of host factors, in: *Coccidioidomycosis. Current Clinical and Diagnostic Status* (L. Ajello, ed.), Symposia Specialists, Miami, Fla., pp. 91–106.
24. Rutala, P. J., and Smith, J. W., 1978, Coccidioidomycosis in potentially compromised hosts: the effect of immunosuppressive therapy in dissemination, *Am. J. Med. Sci.* **275**:283–295.
25. Cohen, I. M., Galgiani, J. N., Potter, D., and Ogden, D. A., 1982, Coccidioidomycosis in renal replacement therapy, *Arch. Intern. Med.* **142**:489–94.
26. Roberts, C. J., 1984, Coccidioidomycosis in acquired immune deficiency syndrome, *Am. J. Med.* **76**:734–736.
27. Kovacs, A., Forthal, D. N., Kovacs, J. A., and Overturf, G. D., 1984, Disseminated coccidioidomycosis in a patient with acquired immune deficiency syndrome, *West. J. Med.* **140**: 447–449.
28. Abrams, D. I., Robia, M., Blumenfeld, W., Simonson, J., Cohen, M. B., and Hadley, W. K., 1984, Disseminated coccidioidomycosis in AIDS [letter], *N. Engl. J. Med.* **310**:986.
29. Wolf, J. E., Little, J. R., Pappagianis, D., and Kobayashi, G. S., 1986, Disseminated

coccidioidomycosis in a patient with the acquired immune deficiency syndrome, *Diagn. Microbiol. Infect. Dis.* **5**:331–336.
30. Salberg, D. J., and Venkatachalam, H., 1986, Disseminated coccidioidomycosis presenting in AIDS, *VA Practitioner* **3**:89–93.
31. Prichard, J. G., Sorotzkin, R. A., and James, III, R. E., 1987, Cutaneous manifestations of disseminated coccidioidomycosis in the acquired immunodeficiency syndrome, *Cutis* **39**: 203–205.
32. Bronnimann, D. A., Adam, R. D., Galgiani, J. N., Habib, M. P., Petersen, E. A., Porter, B., and Bloom, J. W., 1987, Coccidioidomycosis in the acquired immunodeficiency syndrome, *Ann. Intern. Med.* **106**:372–379.
33. Jarvik, J. G., Hesselink, J. R., Wiley, C., Mercer, S., Robbins, B., Higginbottom, P., 1988, Coccidioidomycotic brain abscess in an HIV-infected man, *West. J. Med.* **149**:83–86.
34. Graham, A. R., Sobonya, R. E., Bronnimann, D. A., and Galgiani, J. N., 1988, Quantitative pathology of coccidioidomycosis in acquired immunodeficiency syndrome, *Human Pathol.* **19**:800–806.
35. Byrne, W. R., and Dietrich, R. A., 1989, Disseminated coccidioidomycosis with peritonitis in a patient with acquired immunodeficiency syndrome, *Arch. Intern. Med.* **149**:947–948.
36. Galgiani, J. N., and Ampel, N. M., 1990, Coccidioidomycosis in Human Immunodeficiency Virus-infected patients, *J. Infect. Dis.* **162**:1165–1169.
37. Huppert, M., and Sun, S. H., 1980, Overview of mycology, and the mycology of *Coccidioides immitis*, in: *Coccidioidomycosis. A Text* (D. A. Stevens, ed.), Plenum Press, New York, pp. 21–46.
38. Huppert, M., Sun, S. H., and Harrison, J. L., 1982, Morphogenesis throughout saprobic and parasitic cycles of *Coccidioides immitis*, *Mycopathologia* **78**:107–122.
39. Cole, G. T., and Sun, S. H., 1985, Arthroconidium–spherule–endospore transformation in *Coccidioides immitis*, in: *Fungal Dimorphism* (P. J. Szaniszlo, ed.), Plenum Press, New York, pp. 281–333.
40. Sun, S. H., and Huppert, M., 1976, A cytological study of morphogenesis in *Coccidioides immitis*, *Sabouraudia* **14**:185–198.
41. Huppert, M., Sun, S. H., and Bailey, J. W., 1967, Natural variability in *Coccidioides immitis*, in: *Coccidioidomycosis. Proceedings of the 2nd Symposium on Coccidioidomycosis* (L. Ajello, ed.), University of Arizona Press, Tucson, pp. 323–328.
42. Converse, J. L., 1955, Growth of spherules of *Coccidioides immitis* in a chemically defined liquid medium, *Proc. Soc. Exp. Biol. Med.* **90**:709–711.
43. Converse, J. L., 1956, Effect of physico-chemical environment on spherulation of *Coccidioides immitis* in a chemically defined medium, *J. Bacteriol.* **72**:784–792.
44. Converse, J. L., 1957, Effect of surface active agents on endosporulation of *Coccidioides immitis* in a chemically defined medium, *J. Bacteriol.* **74**:106–107.
45. Converse, J. L., and Besemer, A. R., 1959, Nutrition of the parasitic phase of *Coccidioides immitis* in a chemically defined liquid medium, *J. Bacteriol.* **78**:231–239.
46. Levine, H. B., Cobb, J. M., and Smith, C. E., 1960, Immunity to coccidioidomycosis induced in mice by purified spherule, arthrospore, and mycelial vaccines, *Trans. N.Y. Acad. Sci.* **22**:436–449.
47. Levine, H. B., Cobb, J. M., and Smith, C. E., 1961, Immunogenicity of spherule–endospore vaccine of *Coccidioides immitis* for mice, *J. Immunol.* **87**:218–227.
48. Lones, G. W., Peacock, C. L., and McNey, F. A., 1971, Factors affecting the reversion of *Coccidioides immitis* spherules to mycelium, *Sabouraudia* **9**:287–296.
49. Breslau, A. M., and Kubota, M., 1964, Continuous *in vitro* cultivation of spherules of *Coccidioides immitis*, *J. Bacteriol.* **87**:468–472.
50. Breslau, A. M., Henseley, T. J., and Erickson, J. O., 1961, Electron microscopy of cultured spherules of *Coccidioides immitis*, *J. Biophys. Biochem. Cytol.* **9**:627–637.
51. Donnely, W. H., and Yunis, E. J., 1974, The ultrastructure of *Coccidioides immitis*, *Arch. Pathol.* **98**:227–232.

52. Drutz, D. J., and Huppert, M, 1983, Coccidioidomycosis: factors affecting the host-parasite interaction, *J. Infect. Dis.* **147**:372–390.
53. Gabal, M. A., 1985, Pathogenesis and electron microscopic changes of spherulogenesis of *Coccidioides immitis* (valley fever), *Am. J. Vet. Res.* **46**:671–675.
54. Sun, S. H., Cole, G. T., Drutz, D. J., and Harrison, J. L., 1986, Electron-microscopic observations of the *Coccidioides immitis* parasitic cycle *in vivo*, *J. Med. Vet. Mycology* **24**:183–192.
55. Sun, S. H., Sekhon, S. S., and Huppert, M., 1979, Electron microscopic studies of saprobic and parasitic forms of *Coccidioides immitis*, *Sabouraudia* **17**:265–273.
56. Hector, R. F., and Pappagianis, D., 1982, Enzymatic degradation of the walls of spherules of *Coccidioides immitis*, *Exp. Mycol.* **6**:136–152.
57. Wheat, R. W., and Su Chung, K. S., 1977, Antigenic fractions of *Coccidioides immitis*, in: *Coccidioidomycosis: Current Clinical and Diagnostic Status* (L. Ajello), Symposia Specialists, pp. 453–460.
58. Wheat, R. W., Woodruff, W. W., III, and Haltiwanger, R. S., 1983, Occurrence of antigenic (species-specific?) partially 3-0-methylated heteromannans in cell wall and soluble cellular (nonwall) components of *Coccidioides immitis* mycelia, *Infect. Immun.* **41**:728–734.
59. Cole, G. T., Pope, L. M., Huppert, M., Sun, S. H., and Starr, P., 1983, Ultrastructure and composition of conidial wall fractions of *Coccidioides immitis*, *Exp. Mycol.* **7**:297–318.
60. Cole, G. T., Pope, L. M., Huppert, M., Sun, S. H., and Starr, P., 1985, Wall composition of different cell types of *Coccidioides immitis*, in: *Proceedings of the Fourth International Conference on Coccidioidomycosis* (H. E. Einstein and A. Catanzaro, eds.), Plenum Press, New York, pp. 281–333.
61. Maddy, K., 1957, Ecological factors possibly relating to the geographic distribution of *Coccidioides immitis*, in: *Proceedings of the Symposium on Coccidioidomycosis*, U.S. Public Health Serv. Publ. No. 575, pp. 144–157.
62. Elconin, A. F., Egeberg, R. O., and Lubarsky, R., 1957, Growth patterns of *Coccidioides immitis* in the soil of an endemic area. *Proc. Symp. Coccidioidomycosis*. U.S. Public Health Serv. Publ. 575, pp. 168–170.
63. Lacy, G. H., and Swatek, F. E., 1974, Soil ecology of *Coccidioides immitis* at Amerindian middens in California, *Appl. Microbiol.* **27**:379–388.
64. Smith, C. E., Beard, R. R., Rosenberger, H. G. and Whiting, E. G., 1946, Effect of season and dust control on coccidioidomycosis, *JAMA* **132**:833–838.
65. Hugenholtz, P., 1957, Climate and coccidioidomycosis, in: *Proceedings of the Symposium on Coccidioidomycosis*, U.S. Public Health Serv. Publ. No. 575, pp. 136–143.
66. Smith, C. E., Whiting, E. G., Baker, E. E., Rosenberger, H. G., Beard, R. R., and Saito, M. T., 1948b, The use of coccidioidin, *Am. Rev. Tuberc.* **57**:330–360.
67. Aronson, J. D., Saylor, R. M., and Parr, E. I., 1942, Relationship of coccidioidomycosis to calcified pulmonary nodules, *Arch. Pathol.* **34**:31–48.
68. Dodge, R. R., Lebowitz, M. D., Barbee, R., Burows, B., 1985, Estimates of *Coccidioides immitis* infection by skin test reactivity in an endemic community, *Am. J. Public Health* **75**: 863–865.
69. Galgiani, J. N., 1986, Inhibition of different phases of *Coccidioides immitis* by human neutrophils or hydrogen peroxide, *J. Infect. Dis.* **153**:217–222.
70. Schmelzer, L., and Tabershaw, I. R., 1968, Exposure factors in occupational coccidioidomycosis, *Am. J. Public Health* **58**:107–113.
71. Johnson, W. M., 1981, Occupational factors in coccidioidomycosis, *J. Occupational Med.* **23**: 367–374.
72. Sulkin, S. E., and Pike, R. M., 1951, Laboratory-acquired infections, *JAMA* **147**:1740–1745.
73. Johnson, J. E., Perry, J. E., Fekety, F. R. *et al.*, 1964, Laboratory-acquired coccidioidomycosis. Report of 210 cases, *Ann. Intern. Med.* **60**:941–956.
74. Eckmann, B. H., Schaefer, G. L., and Huppert, M., 1964, Bedside interhuman transmission of coccidioidomycosis via growth on fomites, *Am. Rev. Respir. Dis.* **189**:175–185.

75. Iverson, R. E., and Vistnes, L. M., 1973, Coccidioidomycosis tenosynovitis in the hand, *J. Bone Joint Surg. [Am]* **55:**413–417.
76. Huppert, H., 1979, Racism in coccidioidomycosis? *Am. Rev. Respir. Dis.* **118:**797–798.
77. Sievers, M. L., 1979, Coccidioidomycosis and race, *Am. Rev. Respir. Dis.* **119:**839.
78. Scheer, M., Opelz, G., Terasaki, P., and Hewett, W., 1973, The association of disseminated coccidioidomycosis and histocompatibility type, in: *13th Intersci. Conf. Antimicrobial Agents Chemotherapy*, American Society for Microbiology, Washington, D.C., (Abstr. 157).
79. Deresinski, S. C., Pappagianis, D., and Stevens, D. A., 1979, Association of ABO blood group and outcome of coccidioidal infection, *Sabouraudia* **17:**261–264.
80. Payne, R., Kidd, K. K., Rodvany, P., and Perkins, H. H., 1973, HL-A and other polymorphisms of the Filipino people, in: *Histocompatibility Testing* (J. Dausett and J. Colombani, eds.), Williams & Wilkins, Baltimore, pp. 203–208.
81. Albert, E. D., Mickey, M. R., and Terasaki, P. I., 1973, Genetics of the HL-A system in four populations: American-Caucasians, Japanese-Americans, American Negroes, and Mexican Americans, in: *Histocompatibility Testing* (J. Dausett and J. Colombani, eds.), Williams & Wilkins, Baltimore, pp. 233–240.
82. Mourant, A. E., Kopec, A. C., and Kazmiera, D. S., 1959, *The ABO Blood Groups*, Blackwell Scientific, Oxford.
83. Gifford, J., and Catanzaro, A., 1981, A comparison of coccidioidin and spherulin skin testing in the diagnosis of coccidioidomycosis, *Am. Rev. Respir. Dis.* **124:**440–444,
84. Williams, P. L., Sable, D. L., Sorgen, S. P., Pappagianis, D., Levine, H. B., Brodine, S. K., Brown, B. W., Grumet, F. C., and Stevens, D. A., 1984, Immunologic responsiveness and safety associated with the *Coccidioides immitis* spherule vaccine in volunteers of white, black, and Filipino ancestry, *Am. J. Epidemiol.* **119:**591–602.
85. Drutz, D. J., Huppert, M., Sun, S. H., and McGuire, W. L., 1981, Human sex hormones stimulate the growth and maturation of *Coccidioides immitis, Infect. Immun.* **32:**897–907.
86. Powell, B. L., Drutz, D. J., Huppert, M., and Sun, S. H., 1983, Relationship of progesterone- and estradiol-binding proteins in *Coccidioides immitis* to coccidioidal dissemination in pregnancy, *Infect. Immun.* **40:**478–485.
87. Powell, B. L., and Drutz, D. J., 1984, Identification of a high-affinity binder for estradiol and a low-affinity binder for testosterone in *Coccidioides immitis, Infect. Immun.* **45:**784–786.
88. McNall, E. G., Newcomer, V. D., and Sternberg, T. H., 1959, The effect of altered properdin levels in mice on the course of systemic mycotic infections, *J. Invest. Dermatol.* **32:**423–426.
89. Halde, C., McNall, E. G., Newcomer, V. D., and Sternberg, T. H., 1957–1958, Properdin levels in mice and man with coccidioidomycosis during soluble Amphotericin B administration, *Antibiot. Annu.*, pp. 598–601.
90. McNall, E. G., Sorensen, L. J., Newcomer, V. D., and Sternberg, T. H., 1960, The role of specific antibodies and properdin in coccidioidomycosis, *J. Invest. Dermatol.* **34:**213–216.
91. Galgiani, J. N., Yam, P., Petz, L. D., Williams, P., and Stevens, D. A., 1980, Complement activation by *Coccidioides immitis*: *in vitro* and clinical studies, *Infect. Immun.* **28:**944–949.
92. Forbus, W. D., and Bestebreurtje, A. M., 1946, Coccidioidomycosis: a study of 95 cases of the disseminated type with special reference to the pathogenesis of the disease, *Mil. Surg.* **99:**653–719.
93. Savage, D. C., and Madin, S. H., 1968, Cellular responses in lungs of immunized mice to intranasal infection with *Coccidioides immitis, Sabouraudia* **6:**94–102.
94. Huppert, M., Sun, S. H., Gleason-Jordan, I., and Vukovich, K. R., 1976, Lung weight parallels disease severity in experimental coccidioidomycosis, *Infect. Immun.* **14:**1356–1368.
95. Galgiani, J. N., Isenberg, R. A., and Stevens, D. A., 1978, Chemotaxigenic extracts from the mycelial and spherule phases of *Coccidioides immitis* for human polymorphonuclear leukocytes, *Infect. Immun.* **21:**862–865.
96. Galgiani, J. N., Hayden, R., and Payne, C. M., 1982, Leukocyte effects on the dimorphism of *Coccidioides immitis, J. Infect. Dis.* **146:**56–63.

97. Galgiani, J. N., Payne, C. M., and Jones, J. F., 1984, Human polymorphonuclear-leukocyte inhibition of incorporation of chitin precursors into mycelia of *Coccidioides immitis*, *J. Infect. Dis.* **149:**404–412.
98. Beaman, L., and Pappagianis, D., 1985, Fate of *Coccidioides immitis* arthroconidia in human peripheral blood monocyte cultures *in vitro*, in: *Proc. 4th Int. Conf. Coccidioidomycosis*, (H. E. Einstein and A. Catanzaro, eds.), National Foundation for Infectious Diseases, Washington, D.C., pp. 170–180.
99. Galgiani, J. N., 1985, Potential role of human polymorphonuclear leukocytes in the early host response to *Coccidioides immitis*, in: *Proc. 4th Int. Conf. Coccidioidomycosis*, (H. E. Einstein and A. Catanzaro, eds.), National Foundation for Infectious Diseases, Washington, D.C., pp. 181–191.
100. Frey, C. L., and Drutz, D. J., 1986, Influence of fungal surface components on the interaction of *Coccidioides immitis* with polymorphonuclear neutrophils, *J. Infect. Dis.* **153:** 933–943.
101. Beaman, L., and Holmberg, C. A., 1980, Interaction of nonhuman primate peripheral blood leukocytes and *Coccidioides immitis in vitro*, *Infect. Immun.* **29:**1200–1201.
102. Wegner, T. N., Reed, R. E., Trautman, R. J., and Beavers, C. D., 1972, Some evidence for the development of a phagocytic responses by polymorphonuclear leukocytes recovered from the venous blood of dogs inoculated with *Coccidioides immitis* or vaccinated with an irradiated spherule vaccine, *Am. Rev. Respir. Dis.* **105:**845–849.
103. Brummer, E., Beaman, L., and Stevens, D. A., 1985, Killing of endospores, but not arthroconidia, of *Coccidioides immitis* by immunologically activated polymorphonuclear neutrophils, in: *Proc. 4th Int. Conf. Coccidioidomycosis*, (H. E. Einstein and A. Catanzaro, eds.), National Foundation for Infectious Diseases, Washington, D.C., pp. 201–213.
104. Segal, G. P., Lehrer, R. I., and Selsted, M. E., 1985, *In vitro* effect of phagocyte cationic peptides on *Coccidioides immitis*, *J. Infect. Dis.* **151:**890–894.
105. Baker, D., and Braude, A. I., 1956, A study of stimuli leading to the production of spherules in coccidioidomycosis, *J. Lab. Clin. Med.* **47:**169–181.
106. Le, J., and Vilcek, J., 1987, Tumor necrosis factor and interleukin 1: cytokines with multiple overlapping biological activities, *Lab. Invest.* **56:**234–248.
107. Beaman, L., and Holmberg, C. A., 1980, *In vitro* response of alveolar macrophages to infection with *Coccidioides immitis*, *Infect. Immun.* **28:**594–600.
108. Beaman, L., Benjamini, E., and Pappagianis, D., 1981, Role of lymphocytes in macrophage-induced killing of *Coccidioides immitis in vitro*, *Infect. Immun.* **34:**347–353.
109. Petkus, A. F., and Baum, L. L., 1987, Natural killer cell inhibition of young spherules and endospores of *Coccidioides immitis*, *J. Immunol.* **139:**3107–3111.
110. Clemons, K. V., Leathers, C. R., and Lee, K. W., 1985, Systemic *Coccidioides immitis* infection in nude and beige mice, *Infect. Immun.* **47:**814–821.
111. Roder, J. C., 1979, The beige mutation in the mouse. I. A stem cell predetermined impairment of natural killer cell function, *J. Immunol.* **123:**2168–2173.
112. Pappagianis, D., Smith, C. E., Saito, M. T., and Kobayashi, G., 1961, Studies of antigens from young mycelia of *Coccidioides immitis*, *J. Infect. Dis.* **108:**35–44.
113. Huppert, M., Bailey, J. W., and Chitjian, P., 1967, Immunodiffusion as a substitute for complement fixation and tube precipitin tests in coccidioidomycosis, in: *Coccidioidomycosis. Current Clinical and Diagnostic Status* (L. Ajello, ed.), Symposia Specialists, Miami, Fla., pp. 221–225.
114. Levine, H. B., Cobb, J. M., and Scalarone, G. M., 1969, Spherule coccidioidin in delayed dermal sensitivity reactions of experimental animals, *Sabouraudia* **7:**20–32.
115. Levine, H. B., Gonzalez-Ochoa, A., Ten Eyck, D. R., 1973, Dermal sensitivity to *Coccidioides immitis*. A comparison of responses elicited in man by spherulin and coccidioidin, *Am. Rev. Respir. Dis.* **107:**379–386.
116. Deresinski, S. C., Levine, H. B., and Stevens, D. A., 1974, Soluble antigens of mycelia and

spherules in the *in vitro* detection of immunity to *Coccidioides immitis*, *Infect. Immun.* **10:** 700–704.

117. Stevens, D. A., Levine, H. B., and Ten Eyck, D. R., 1974, Dermal sensitivity to different doses of spherulin and coccidioidin, *Chest* **65:**530–533.
118. Stevens, D. A., Levine, H. B., Deresinski, S. C., and Blaine, L. J., 1975, Spherulin in clinical coccidioidomycosis: comparison with coccidioidin, *Chest* **68:**5–9.
119. Levine, H. B., and Scalarone, G. M., 1975, Properties of spherulin, a skin-test reagent in coccidioidomycosis, in: *3rd Int. Conf. Mycoses*, Sci. Pub. No. 304, Pan American Health Organization, Washington, D. C., pp. 101–109.
120. Scalarone, G. M., Levine, H. B., Pappagianis, D., and Chaparas, S. D., 1974, Spherulin as a complement-fixing antigen in human coccidioidomycosis, *Am. Rev. Respir. Dis.* **110:** 324–328.
121. Huppert, M., Krasnow, I., Vukovich, R., Sun, S. H., Rice, E. H., and Kutner, L. J., 1977, Comparison of coccidioidin and spherulin in complement fixation tests for coccidioidomycosis, *J. Clin. Microbiol.* **6:**33–41.
122. Cox, R. A., Brummer, E., and Lecara, G., 1977, *In vitro* lymphocyte responses of coccidioidin skin-test positive and negative persons to coccidioidin, spherulin, and a *Coccidioides* cell wall antigen, *Infect. Immun.* **15:**751–755.
123. Cox, R. A., Mead, C. G., and Pavey, E. F., 1981, Comparisons of mycelia- and spherule-derived antigens in cellular immune assays of *Coccidioides immitis*-infected guinea pigs, *Infect. Immun.* **31:**687–692.
124. Levine, H. B., Restrepto-Moreno, A., Ten Eyck, D. R., and Stevens, D. A., 1975, Spherulin and coccidioidin: cross-reactions in dermal sensitivity to histoplasmin and paracoccidioidin, *Am. J. Epidemiol.* **101:**512–516.
125. Campbell, C. C., and Binkley, G. E., 1953, Serologic diagnosis with respect to histoplasmosis, coccidioidomycosis, and blastomycosis and the problem of cross-reactions, *J. Lab. Clin. Med.* **42:**896–906.
126. Smith, C. E., Saito, M. T., and Simons, S. A., 1956, Pattern of 39,500 serologic tests in coccidioidomycosis, *J. Am. Med. Assoc.* **160:**546–552.
127. Campbell, C. C., 1960, The accuracy of serologic methods in diagnosis, *Ann. N.Y. Acad. Sci.* **89:**163–177.
128. Campbell, C. C., 1968, Use and interpretation of serologic and skin tests in the respiratory mycoses, *Dis. Chest.* **54**(Suppl. 1):49–54.
129. Huppert, M., Spratt, N. S., Vukovich, K. R., Sun, S. H., and Rice, E. H., 1978, Antigenic analysis of coccidioidin and spherulin determined by two-dimensional immunoelectrophoresis, *Infect. Immun.* **20:**542–551.
130. Cox, R. A., 1979, Cross reactivity between antigens of *Coccidioides immitis*, *Histoplasma capsulatum*, and *Blastomyces dermatitidis* in lymphocyte transformation assays, *Infect. Immun.* **25:**932–938.
131. Anderson, K. L., Wheat, R. W., and Conant, N. I., 1971, Fractionation and composition studies of skin-test active components of sensitins from *Coccidioides immitis*, *Appl. Microbiol.* **22:**294–299.
132. Kruse, D., and Cole, G. T., 1990, Isolation of tube precipitin antibody-reactive fractions of *Coccidioides immitis*, *Infect. Immun.* **58:**169–178.
133. Ward, E. R., Jr., Cox, R. A., Schmitt, J. A., Jr., Huppert, M., and Sun, S. H., 1975, Delayed-type hypersensitivity responses to a cell wall fraction of the mycelial phase of *Coccidioides immitis*, *Infect. Immun.* **12:**1093–1097.
134. Collins, M. S., Pappagianis, D., and Yee, J., 1977, Enzymatic solubilization of precipitin and complement fixing antigen from endospores, spherules, and spherule fraction of *Coccidioides immitis*, in: *Coccidioidomycosis: Current Clinical and Diagnostic Status* (L. Ajello, ed.), Symposia Specialists, pp. 429–444.
135. Wheat, R. W., Chung, K. S. S., Ornellas, E. P., and Scheer, E. R., 1978, Extraction of skin test

activity from *Coccidioides immitis* mycelial by water, perchloric acid, and aqueous phenol extraction, *Infect. Immun.* **19**:152–159.
136. Cox, R. A., Huppert, M., Starr, P., and Britt, L. A., 1984, Reactivity of alkali-soluble, water-soluble cell wall antigen of *Coccidioides immitis* with anti-*Coccidioides* immunoglobulin M precipitin antibody, *Infect. Immun.* **43**:502–507.
137. Cole, G. T., Seshan, K. R., Franco, M., Bukownik, E., Sun, S. H., and Hearn, V. M., 1988, Isolation and morphology of an immunoreactive outer wall fraction produced by spherules of *Coccidioides immitis*, *Infect. Immun.* **56**:2686–2694.
138. Cole, G. T., Kirkland, T. N., Franco, M., Zhu, S., Yuan, L., Sun, S. H., and Hearn, V. M., 1988, Immunoreactivity of a surface wall fraction produced by spherules of *Coccidioides immitis*, *Infect. Immun.* **56**:2695–2701.
139. Cox, R. A., and Britt, L. A., 1986, Isolation of a coccidioidin component that reacts with immunoglobulin M precipitin antibody, *Infect. Immun.* **53**:449–453.
140. Cox, R. A., Britt, L. A., and Michael, R. A., 1987, Isolation of *Coccidioides immitis* F antigen by immunoaffinity chromatography with monospecific antiserum, *Infect. Immun.* **55**: 227–232.
141. Zimmer, B. L., and Pappagianis, D., 1989, Immunoaffinity isolation and partial characterization of the *Coccidioides immitis* antigen detected by the tube precipitin and immunodiffusion-tube precipitin tests, *J. Clin. Microbiol.* **27**:1759–1766.
142. Cox, R. A., 1989, Antigenic structure of *Coccidioides immitis*, in: *Immunology of Fungal Diseases* (E. Kurstak, G. Marquis, P. Auger, L. de Repentigny, and S. Montplaisir, eds.), Marcel Dekker, Inc., New York, pp. 133–170.
143. Smith, C. E., Saito, M. T., Beard, R. R., Kepp, R. M., Clark, R. W., and Eddie, B. U., 1950, Serological tests in the diagnosis and prognosis of coccidioidomycosis, *Am. J. Hyg.* **52**:1–21.
144. Smith, C. E., Saito, M. T., Campbell, C. C., Hill, G. B., Saslaw, S., Salvin, S. B., Fenton, J. E., and Krupp, M. A., 1957, Comparison of complement fixation tests for coccidioidomycosis, *Public Health Rep.* **72**:888–894.
145. McKelvey, E. M., and Fahey, J. L., 1965, Immunoglobulin changes in disease: quantitation on the basis of heavy polypeptide chains, IgG ($_\gamma$G), IgA ($_\gamma$A), and IgM ($_\gamma$M), and of light polypeptide chains, type K (I) and type L (II), *J. Clin. Invest.* **44**:1778–1787.
146. Pappagianis, D., Saito, M, Hoosear, K. H. V., 1972, Antibody in cerebrospinal fluid in nonmeningitic coccidioidomycosis, *Sabouraudia* **10**:173–179.
147. Pappagianis, D., Lindsey, N. J., Smith, C. E., and Saito, M. T., 1965, Antibodies in human coccidioidomycosis: immunoelectrophoretic properties, *Proc. Soc. Exp. Biol. Med.* **118**: 118–122.
148. Sawaki, Y., Huppert, M., Bailey, J. W., and Yagi, Y., 1966, Patterns of human antibody reactions in coccidioidomycosis, *J. Bacteriol.* **91**:422–427.
149. Cox, R. A., and Arnold, D. R., 1979, Immunoglobulin E in coccidioidomycosis, *J. Immunol.* **123**:194–200.
150. Cox, R. A., Baker, B. S., and Stevens, D. A., 1982, Specificity of immunoglobulin E in coccidioidomycosis and correlation with disease involvement, *Infect. Immun.* **37**:609–616.
151. Zimmer, B. L., and Pappagianis, 1986, Comparison of immunoblot analyses of spherule-endospore phase extracellular protein and mycelial phase antigen of *Coccidioides immitis*, *Infect. Immun.* **53**:64–70.
152. Calhoun, D. L., Osir, E. O., Dugger, K. O., Galgiani, J. N., and Law, H. H., 1986, Humoral antibody responses to specific antigens of *Coccidioides immitis*, *J. Infect. Dis.* **154**:265–272.
153. Dodge, R. R., Lebowitz, M. D., Barbee, R. A., and Burrows, B., 1985, Immediate reactions to coccidioidin and spherulin: the relation of atopy to coccidioidomycosis, in *Proc. 4th Int. Conf. Coccidioidomycosis* (H. E. Einstein and A. Catanzaro, eds.), National Foundation for Infectious Diseases, Washington, D. C., pp. 214–223.
154. Dickson, E. C., 1938, Primary coccidioidomycosis—The initial acute infection which results in coccidioidal granuloma, *Am. Rev. Tuberc.* **38**:722–727.

155. Smith, C. E., 1943, Coccidioidomycosis, *Med. Clin. North. Am.* **27**:790–807.
156. Goldstein, D. M., and Louie, S., 1943, Primary pulmonary coccidioidomycosis—Report of an epidemic of seventy-five cases, *War Med.* **4**:299–317.
157. Willett, F. M., and Weiss, A., 1945, Coccidioidomycosis in southern California: Report of a new endemic area with a review of 100 cases, *Ann. Intern. Med.* **23**:349–375.
158. Alznauer, R. L., Rolle, C., Jr., and Pierce, W. F., 1955, Analysis of focalized pulmonary granulomas due to *Coccidioides immitis*, *Arch. Pathol.* **59**:641–655.
159. Huntington, R. W., Jr., 1957, Diagnostic and biologic implication of the histopathology of coccidioidomycosis, in: Proceedings of a *Symposium on Coccidioidomycosis*, U.S. Public Health Service (Pub. No. 575), Washington, D. C., pp. 36–46.
160. Echols, R. M., Palmer, D. L., and Long, G. W., 1982, Tissue eosinophilia in human coccidioidomycosis, *Rev. Infect. Dis.* **4**:656–664.
161. Schermoly, M. J., and Hinthorn, D. R., 1988, Eosinophilia in coccidioidomycosis, *Arch. Int. Med.* **148**:895–896.
162. Yoshinoya, S., Cox, R. A., and Pope, R. M., 1980, Immune complexes in coccidioidomycosis. Detection and characterization, *J. Clin. Invest.* **66**:655–663.
163. Cox, R. A., Pope, R. M., and Stevens, D. A., 1982, Immune complexes in coccidioidomycosis. Correlation with disease involvement, *Am. Rev. Respir. Dis.* **126**:439–443.
164. Waters, M. F., Turk, J. L., and Wemambu, S. N., 1971, Mechanisms of reactions in leprosy, *Int. J. Lepr.* **39**:417–424.
165. Rea, T. H., and Levan, N. E., 1977, Current concepts in the immunology of leprosy, *Arch. Dermatol.* **112**:791–800.
166. Safai, B., Good, R. A., and Day, N. K., 1977, Erythema multiforme: report of two cases and speculation on immune mechanisms involved in the pathogenesis, *Clin. Immunol. Immunopathol.* **7**:379–381.
167. Imamura, S., Yanase, K., Taniguchi, S., Ofuji, S., and Mangaoil, L,. 1980, Erythema multiforme: demonstration of immune complexes in sera and skin lesions, *Br. J. Dermatol.***102**:161–166.
168. Kirkland, T. N., and Fierer, J., 1983, Inbred mouse strains differ in resistance to lethal *Coccidioides immitis* infection, *Infect. Immun.* **40**:912–916.
169. Kirkland, T. N., and Fierer, J., 1985, Genetic control of resistance to *Coccidioides immitis*: a single gene that is expressed in spleen cells determines resistance, *J. Immunol.* **135**:548–552.
170. Kirkland, T. N., and Fierer, J., 1986, Genetics of resistance to coccidioidomycosis, in: Microbiology–1986 (L. Leive, ed.), American Society for Microbiology, Washington, D. C., pp. 169–171.
171. Cox, R. A., Kennell, W., Boncyk, L., and Murphy, J. W., 1988, Induction and expression of cell-mediated immune responses in inbred mice infected with *Coccidioides immitis*, *Infect. Immun.* **56**:13–17.
172. Segal, G. P., 1985, Genetic regulation of antibody responses to *Coccidioides immitis*: the role of antibody in mediated immunity, in: *Proc. 4th Int. Conf. Coccidioidomycosis* (H. E. Einstein and A. Catanzaro, eds.), National Foundation for Infectious Diseases, Washington, D.C., pp. 191–200.
173. Kong, Y. M., Savage, D. C., and Levine, H. B., 1966, Enhancement of immune responses in mice by a booster injection of *Coccidioides* spherules, *J. Immunol.* **95**:1048–1056.
174. Beaman, L., Pappagianis, D., and Benjamini, E., 1977, Significance of T cells in resistance to experimental murine coccidioidomycosis, *Infect. Immun.* **17**:580–585.
175. Beaman, L., Pappagianis, D., and Benjamini, E., 1979, Mechanisms of resistance to infection with *Coccidioides immitis* in mice, *Infect. Immun.* **23**:681–685.
176. Warren, K. S., Domingo, E. O., and Cowan, R. B. T., 1967, Granuloma formation around schistosome eggs as a manifestation of delayed hypersensitivity, *Am. J. Pathol.* **51**:735–741.
177. Chensue, S. W., Boros, D. L., and David, C. W., 1980, Regulation of granulomatous

inflammation in murine schistosomiasis. *In vitro* characterization of T lymphocyte subsets involved in the production and suppression of migration inhibition factor, *J. Exp. Med.* **151:** 1398–1412.
178. Wellhausen, S. R., Chensue, S. W., and Boros, D. L., Modulation of granulomatous hypersensitivity analysis by adoptive transfer of effector and suppressor T lymphocytes involved in granulomatous inflammation in murine schistosomiasis, in: *Basic and Clinical Aspects of Granulomatous Diseases* (D. L. Boros and T. Yoshida, eds.), Elsevier/North-Holland, Amsterdam, pp. 219–234.
179. Mathew, R. C., and Boros, D. L., Anti-L3T4 antibody treatment suppresses hepatic granuloma formation and abrogates antigen-induced interleukin 2 production in *Schistosoma mansoni* infection, *Infect. Immun.* **54:**820–826.
180. Modlin, R. L., Segal, G. P., Hofmann, F. M., Walley, M. S., Johnson, R. H., Taylor, C. R., and Rea, T. H., 1985, *In situ* localization of T lymphocytes in disseminated coccidioidomycosis, *J. Infect. Dis.* **151:**314–319.
181. Ridley, D. S., and Jopling, W. H., 1966, Classification of leprosy according to immunity. A five-group system, *Int. J. Lepr.* **34:**255–273.
182. Van Voorhis, W. C., Kaplan, G., Sarno, E. N., Horwitz, M. A., Stinman, R. M., Levis, W. R., Nogueira, N., Hais, L. S., Gattass, C. R., Arrick, B. A., and Cohn, Z. A., 1982, The cutaneous infiltrates of leprosy. Cellular characteristics and the predominant T cell phenotypes, *N. Engl. J. Med.* **307:**1593–1579.
183. Modlin, R. L., Gebhard, J. F., Taylor, C. R., and Rea, T. H., 1983, *In situ* characterization of T lymphocyte subsets in the reactional states of leprosy, *Clin. Exp. Immunol.* **53:**17–24.
184. Bloom, B. R., and Mehra, V., 1984, Immunological unresponsiveness in leprosy, *Immunol. Rev.* **80:**5–28.
185. Smith, C. E., 1942, Parallelism of coccidioidal and tuberculous infections, *Radiology* **38:** 643–668.
186. Smith, C. E., 1951, Diagnosis of pulmonary coccidioidal infections, *Calif. Med.* **75:**385–391.
187. Catanzaro, A., Spitler, L. E., and Moser, K. M., 1975, Cellular immune response in coccidioidomycosis, *Cell. Immunol.* **15:**360–371.
188. Cox, R. A., Vivas, J. R., Gross, A., Lecara, G., Miller, E., and Brummer, E., 1976, *In vivo* and *in vitro* cell-mediated responses in coccidioidomycosis. I. Immunologic responses of persons with primary, asymptomatic infections, *Am. Rev. Respir. Dis.* **114:**937–943.
189. Cox, R. A., and Vivas, J. R., 1977, Spectrum of *in vivo* and *in vitro* cell-mediated immune responses in coccidioidomycosis, *Cell. Immunol.* **31:**130–141.
190. Cox, R. A., 1983, Cellular immunity, in: *Fungi Pathogenic for Humans and Animals. Pathogenicity and Detection* (D. H. Howard, ed.), Marcel Dekker, New York, pp. 61–98.
191. Cox, R. A., 1989, Coccidioidomycosis, in: *Immunology of the Fungal Diseases* (R. A. Cox, ed.), CRC Press, Inc., Boca Raton, Fla., pp. 165–197.
192. Opelz, G., and Scheer, M. I., 1975, Cutaneous sensitivity and *in vitro* responsiveness of lymphocytes in patients with disseminated coccidioidomycosis, *J. Infect. Dis.* **132:**250–255.
193. Harvey, R. P., and Stevens, D. A., 1981, *In vitro* assays of cellular immunity in progressive coccidioidomycosis. Evaluation of suppression with parasitic-phase antigen, *Am. Rev. Respir. Dis.* **23:**665–669.
194. Barbee, R. A., and Hicks, M. J., 1988, Clinical usefulness of lymphocyte transformation in patients with coccidioidomycosis, *Chest* **93:**1001–1007.
195. Catanzaro, A., 1977, Development of immunologic and clinical staging for immunotherapy, in: *Coccidioidomycosis. Current Clinical and Diagnostic Status* (L. Ajello, ed.), Symposia Specialists, Miami, FL., pp. 325–334.
196. Hicks, H. R., and Northey, W. T., 1967, Studies on the response of thymectomized mice to infection with *Coccidioides immitis*, in: *Proc. 2nd Coccidioidomycosis Symp.* (L. Ajello, ed.), University of Arizona Press, Tucson, pp. 183–187.

197. Beaman, L., Benjamini, E., and Pappagianis, D., 1983, Activation of macrophages by lymphokines: enhancement of phagosome-lysosome fusion and killing of *Coccidioides immitis*, *Infect. Immun.* **39:**1201–1207.
198. Beaman, L., 1987, Fungicidal activation of murine macrophages by recombinant gamma interferon, *Infect. Immun.* **55:**2951–2955.
199. Clemons, K. V., Leathers, C. R., and Lee, K. W., 1985, Role of activated macrophages in resistance to experimental *Coccidioides immitis* infection, in: *Proc. 4th Int. Conf. Coccidioidomycosis* (H. E. Einstein and A. Catanzaro, eds.), National Foundation for Infectious Diseases, Washington, D.C., pp. 149–159.
200. Slagle, D. C., Cox, R. A., and Kuruganti, U., 1989, Induction of tumor necrosis factor alpha by spherules of *Coccidioides immitis*, *Infect. Imun.* **57:**1916–1921.
201. Klebanoff, S. J., Vadas, M. A., Harlan, J. M., Sparks, L. H., Gamble, J. R., Agosti, J. M., and Waltersdorph, A. M., 1986, Stimulation of neutrophils by tumor necrosis factor, *J. Immunol.* **136:**4220–4225.
202. Ostensen, M. E., Thiele, D. L., and Lipsky, P. E., 1987, Tumor necrosis factor-α enhances cytolytic activity of human natural killer cells, *J. Immunol.* **138:**4185–4191.
203. Livingston, D. H., Appel, S. H., Sonnenfeld, G., and Malangoni, M. A., 1989, The effect of tumor necrosis factor-α and interferon-γ on neutrophil function, *J. Surg. Res.* **46:**322–326.
204. Ibrahim, A. B., and Pappagianis, D., 1973, Experimental induction of anergy to coccidioidin by antigens of *Coccidioides immitis*, *Infect. Immun.* **7:**786–794.
205. Likholetov, S. M., 1979, Delayed hypersensitivity and transfer factor in experimental coccidioidomycosis, *Sabouraudia* **17:**251–259.
206. Ibrahim, A. B., 1974, Induction of tolerance to coccidioidin in newborn guinea pigs, *J. Immunol.* **112:**387 -391.
207. Scalarone, G. M., and Levine, H. B., 1971, Deficient resistance to *Coccidioides immitis* following intravenous vaccination. I. Distribution of spherules after intravenous and intramuscular doses, *Sabouraudia* **9:**81–89.
208. Scalarone, G. M., and Levine, H. B., 1971, Deficient resistance to *Coccidioides immitis* following intravenous vaccination. II Evidence against an immune tolerance mechanism, *Sabouraudia* **9:**90–96.
209. Scalarone, G. M., and Levine, H. B., 1969, Attributes of deficient immunity in mice receiving *Coccidioides immitis* spherule vaccine by the intravenous route, *Sabouraudia* **7:** 169–177.
210. Levine, H. B., and Kong, Y. M., 1966, Immunologic impairment in mice treated intravenously with killed *Coccidioides immitis* spherules: suppressed response to intramuscular doses, *J. Immunol.* **21:**297–305.
211. Lecara, G., Cox, R. A., and Simpson, R. B., 1983, *Coccidioides immitis* vaccine: potential of an alkali-soluble, water-soluble cell wall antigen, *Infect. Immun.* **39:**473–475.
212. Cox, R. A., and Kennell, W., 1988, Suppression of T-lymphocyte response by *Coccidioides immitis* antigen, *Infect. Immun.* **56:**1424–1429.
213. Catanzaro, A., 1981, Suppressor cells in coccidioidomycosis, *Cell. Immunol.* **64:**235–245.
214. Cox, R. A., and Pope, R. M., 1987, Serum-mediated suppression of lymphocyte transformation responses in coccidioidomycosis, *Infect. Immun.* **55:**1058–1062.
215. McDougal, J. S., Redecha, P. B., Inman, R. D., and Christian, C. L., 1979, Binding of immunoglobulin G aggregates and immune complexes in human serum to staphylococci containing protein A, *J. Clin. Invest.* **63:**627–636.
216. Sinclair, N. R., 1979, Modulation of immunity by antibody, antigen-antibody complexes and antigen, *Pharmacol. Ther.* **4:**355–432.
217. Rao, V. S., Bennett, J. A., Shen, F. W., Gershon, R. K., and Mitchell, M. S., 1980, Antigen-antibody complexes generate Lyt 1 inducers of suppressor cells, *J. Immunol.* **125:**63–67.
218. Jerne, N. K., 1984, Idiotypic networks and other preconceived ideas, *Immunol. Rev.* **79:** 5–24.

219. Phillips, S. M., Lin, J., Galal, N., Linette, G. P., Walker, D. J., and Perrin, P. J., 1990, The regulation of resistance to *Schistosoma mansoni* by auto-anti-idiotypic immunity. III. An analysis of effects on epitopic recognition, idiotypic expression, and anti-idiotypic reactivity at the clonal level, *J. Immunol.* **145**:2272–2280.
220. Dorf, M. E., and Benacerraf, B., 1984, Suppressor cells and immunoregulation, *Ann. Rev. Immunol.* **2**:127–158.
221. Pappagianis, D., and Zimmer, B. L., 1990, Serology of coccidioidomycosis, *Clin. Microbiol. Rev.* **3**:247–268.
222. Huppert, M., and Bailey, J. W., 1965, Immunodiffusion as a screening test for coccidioidomycosis serology, *Sabouraudia* **2**:284–291.
223. Huppert, M., and Bailey, J. W., 1965, The use of immunodiffusion tests in coccidioidomycosis. I. The accuracy and reproducibility of the immunodiffusion test which correlates with complement fixation, *Am. J. Clin. Pathol.* **44**:364–368.
224. Huppert, M., and Bailey, J. W., 1965, The use of immunodiffusion tests in coccidioidomycosis. II. An immunodiffusion test as a substitute for the tube precipitin test, *Am. J. Clin. Pathol.* **44**:369–373.
225. Pappagianis, D., Putman, E. W., and Kobayashi, G. S., 1961, Polysaccharide of *Coccidioides immitis*, *J. Bacteriol.* **82**:714–723.
226. Dolan, M. J., Cox, R. A., Williams, V., and Woolley, S., 1989, Development and characterization of a monoclonal antibody against the tube precipitin antigen of *Coccidioides immitis*, *Infect. Immun.* **57**:1035–1039.
227. Cole, G. T., Kruse, D., Zhu, S., Seshan, K. R., and Wheat, R. W., 1990, Composition, serologic reactivity, and immunolocalization of a 120-kilodalton tube precipitin antigen of *Coccidioides immitis*, *Infect. Immun.* **58**:179–188.
228. Catanzaro, A., and Flatauer, F., 1983, Detection of serum antibodies in coccidioidomycosis by solid-phase radioimmunoassay, *J. Infect. Dis.* **147**:32–39.
229. Weiner, M. H., 1983, Antigenemia detected in human coccidioidomycosis, *J. Clin. Microbiol.* **18**:136–142.
230. Galgiani, J. N., Dugger, K. O., Ito, J. I., and Wieden, M. A., 1984, Antigenemia in primary coccidioidomycosis, *Am. J. Trop. Med. Hyg.* **33**:645–649.
231. Rixford, E., and Gilchrist, T. C., 1896, Two cases of protozoan (coccidioidal) infection of the skin and other organs, *Johns Hopkins Hosp. Rep.* **1**:209–265.
232. Pappagianis, D., Miller, R. L., Smith, C. E., and Kobayashi, G. S., 1960, Response of monkeys to respiratory challenge following subcutaneous inoculation with *Coccidioides immitis*, *Am. Rev. Respir. Dis.* **82**:224–247.
233. Pappagianis, D., Levine, H. B., Smith, C. E., Berman, R. J., and Kobayashi, G. S., 1961, Immunization of mice with viable *Coccidioides immitis*, *J. Immunol.* **86**:28–34.
234. Converse, J. L., Castleberry, M. W., Besemer, A. R., and Synder, E. M., 1962, Immunization of mice against coccidioidomycosis, *J. Bacteriol.* **84**:46–52.
235. Converse, J. L., Castleberry, M. W., and Snyder, E. M., 1963, Experimental viable vaccine against pulmonary coccidioidomycosis in monkeys, *J. Bacteriol.* **86**:1041–1051.
236. Converse, J. L., Deauville, G. A., Snyder, E. M., Ray, J. G., Jr., and Seaquist, M. E., 1965, Control of tissue reactions in monkeys vaccinated with viable *Coccidioides immitis* by prevaccination with killed *Coccidioides immitis*, *J. Bacteriol.* **90**:783–788.
237. Converse, J., 1964, Experimental primary cutaneous coccidioidomycosis in the monkey, *J. Bacteriol.* **87**:81–85.
238. Pulliam, J. D., Converse, J. L., Snyder, E. M., Esterly, J. R., and Lowe, E. P., 1967, Experimental irradiated arthrospore vaccine against coccidioidomycosis in mice, *J. Bacteriol.* **94**:1394–1399.
239. Kong, Y. M., and Levine, H. B., 1967, Loss and recovery of virulence of arthrospores and spherule–endospores of *Coccidioides immitis* in live vaccines, in: *Proc. 2nd Coccidioidomycosis Symp.* (L. Ajello, ed.), University of Arizona Press, Tucson, pp. 189–195.

240. Kong, Y. M., and Levine, H. B., 1967, Experimentally induced immunity in the mycoses, *Bacteriol. Rev.* **31**:35–53.
241. Foley, J., Berman, R. J., and Smith, C. E., 1960, X-ray irradiation of *Coccidioides immitis* arthrospores: survival curves and avirulent mutants isolated, *J. Bacteriol.* **79**:480–487.
242. Castleberry, M. W., Converse, J. L., and Soto, P. J., Jr., 1964, Antibiotic control of tissue reactions in dogs vaccinated with viable cells of *Coccidioides immitis*, *J. Bacteriol.* **87**:1216–1220.
243. Friedman, L., and Smith, C. E., 1956, Vaccination of mice against *Coccidioides immitis*, *Am. Rev. Tuberc. Pul. Dis.* **74**:245–248.
244. Castleberry, M. W., Converse, J. L., Sinski, J. T., Lowe, E. P., Pakes, S. P., and Del Favero, J. E., 1965, Coccidioidomycosis: studies of canine vaccination and therapy, *J. Infect. Dis.* **115**: 41–48.
245. Levine, H. B., Kong, Y. M., and Smith, C. E., 1965, Immunization of mice to *Coccidioides immitis*: dose, regimen and spherulation stage of killed spherule vaccines, *J. Immunol.* **94**: 132–142.
246. Levine, H. B., and Smith, C. E., 1967, The reactions of eight volunteers injected with *Coccidioides immitis* spherule vaccine; first human trials, in: *Proc. 2nd Coccidioidomycosis Symp.* (L. Ajello, ed.), Symposium Specialists, Miami, Fla., pp. 197–200.
247. Pappagianis, D., Levine, H. B., and Smith, C. E., 1967, Further studies on vaccination of human volunteers with killed *Coccidioides immitis*, in: *Proc. 2nd Coccidioidomycosis Symp.* (L. Ajello, ed.), Symposium Specialists, Miami, FL, pp. 201–210.
248. Pappagianis, D., and Levine, H. B., 1975, The present status of vaccination against coccidioidomycosis in man, *Am. J. Epidemiol.* **102**:30–41.
249. Walker, J., 1985, Progress report on the coccidioidomycosis vaccine trail, in: *Proc. 4th Int. Conf. Coccidioidomycosis* (H. E. Einstein, and A. Catanzaro, eds.), National Foundation for Infectious Diseases, Washington, D.C., p. 347.
250. Pappagianis, D., 1986, Evaluation of the protective effect of killed *Coccidioides immitis* vaccine in man, *26th Int. Conf. Antimicrobial. Agents Chemotherapy*, American Society for Microbiology, Washington, D.C. (Abstr. 784).
251. Pappagianis, D., Hector, R., Levine, H. B., and Collins, M. S., 1979, Immunization of mice against coccidioidomycosis with a subcellular vaccine, *Infect. Immun.* **25**:440–445.
252. Kong, Y. M., Levine, H. B., and Smith, C. E., 1963, Immunogenic properties of non-disrupted and disrupted spherules of *Coccidioides immitis* in mice, *Sabouraudia* **2**:131–142.
253. Graybill, J. R., 1987, Chemotherapy of the systemic mycoses, *Seminars in Respiratory Medicine* **9**:207–216.
254. Ampel, N. M., Wieden, M. A., and Galgiani, J. N., 1989, Coccidioidomycosis: clinical update, *Rev. Infect. Dis.* **11**:897–911.
255. Ellis, W. G., Sobel, R. A., and Nielsen, S. L., 1982, Leukoencephalopathy in patients treated with amphotericin B methyl ester, *J. Infect. Dis.* **146**:125–137.
256. Dismukes, W. E., 1988, Azole antifungal drugs: old and new, *Ann. Intern. Med.* **109**: 177–179.
257. Galgiani, J. N., Stevens, D. A., Graybill, J. R., Dismukes, W. E., and Cloud, G. A., 1988, Ketoconazole therapy of progressive coccidioidomycosis. Comparison of 400- and 800-mg doses and observations at higher doses, *Am. J. Med.* **84**:603–610.
258. Tucker, R. M., Denning, D. W., Dupont, B., and Stevens, D. A., 1990, Itraconazole therapy for chronic coccidioidal meningitis, *Ann. Intern. Med.* **112**:108–112.
259. Tucker, R. M., Galgiani, J. N., Denning, D. W., Hanson, L. H., Graybill, J. R., Sharkey, K., Eckman, M. R., Salemi, C., Libke, R., Klein, R. A., and Stevens, D. A., 1990,Treatment of coccidioidal meningitis with fluconazole, *Rev. Infect. Dis.* **12**(suppl):S380–S389.
260. Lawrence, H. S., 1955, The transfer in humans of delayed skin sensitivity to streptococcal M substance and tuberculin with disrupted leukocytes, *J. Clin. Invest.* **34**:219–230.
261. Lawrence, H. S., 1969, Transfer factor, *Adv. Immunol.* **11**:195–266.

262. Graybill, J. R., Silva, J., Jr., Alford, R. H., and Thor, D. E., 1973, Immunologic and clinical improvement of progressive coccidioidomycosis following administration of transfer factor, *Cell. Immunol.* **8:**120–135.
263. Catanzaro, A., Spitler, L., and Moser, K. M., 1974, Immunotherapy of recurrent coccidioidomycosis, *J. Clin. Invest.* **54:**690–701.
264. Stevens, D. A., Pappagianis, D., Marinkovich, V. A., and Waddell, T. F., 1974, Immunotherapy in recurrent coccidioidomycosis, *Cell. Immunol.* **12:**37–48.
265. Catanzaro, A., and Spitler, L., 1976, Clinical and immunologic results of transfer factor therapy in coccidioidomycosis, in: *Transfer Factor: Basic Properties and Clinical Applications* (M. S. Ascher, A. A. Gottlieb, and C. H. Kirkpatrick, eds.), Academic Press, New York, pp. 477–491.
266. Graybill, J. R., 1977, The clinical course of coccidioidomycosis following transfer factor therapy, in: *Coccidioidomycosis. Current Clinical and Diagnostic Status* (L. Ajello, ed.), Symposia Specialists, Miami, FL, pp. 335–345.
267. Catanzaro, A., Friedman, P. J., Schillaci, R., Graybill, R., Einstein, H., and Spitler, L., 1985, Double-blind study of transfer factor treatment of coccidioidomycosis, in: *Coccidioidomycosis. Proc. 4th Int. Conf.* (H. E. Einstein, and A. Catanzaro, eds.), National Foundation for Infectious Diseases, Washington, D.C., pp. 330–338.
268. Kirchner, H., 1986, The interferon system as an integral part of the defense system against infections, *Antiviral Res.* **6:**1–17.
269. Hoeprich, P. D., and Merry, J. M., 1988, Effect of recombinant human interleukin 2 in experimental murine coccidioidomycosis, *Diagn. Microbiol. Infect. Dis.* **9:**115–118.
270. Farrar, W. L., Johnson, H. M., and Farrar, J. J., 1981, Regulation of the production of immune interferon and cytotoxic T lymphocytes by interleukin 2. *J. Immunol.* **126:**1120–1125.

9

Histoplasmosis

BETTY WU-HSIEH and D. H. HOWARD

1. INTRODUCTION

The zoopathogenic fungus *Histoplasma capsulatum* has three varieties: *H. capsulatum* var. *capsulatum*, *H. capsulatum* var. *duboisii*, and *H. capsulatum* var. *farciminosum*. The variety *capsulatum* is the cause of classic histoplasmosis; the variety *duboisii* causes the African variety of histoplasmosis; and the variety *farciminosum* causes epizootic lymphangitis in horses and mules.[1,2] This chapter will concentrate almost exclusively on the disease caused by the *capsulatum* variety because the amount of information on this variety far exceeds that on the other two and because, to most readers, this is the disease meant by the term "histoplasmosis." When interesting comparisons with the *duboisii* variety are instructive, they will be made, but consideration of the *farciminosum* variety is beyond the scope of this chapter. In order to avoid unnecessary reiteration the term *H. capsulatum*, unless other indicated, will mean the variety *capsulatum*.

2. DESCRIPTION OF DISEASE

2.1. Classical Histoplasmosis

The vast majority of infections with *H. capsulatum* var. *capsulatum* are completely asymptomatic or, at least, subclinical. Such infections will, at most, consist of mild, self-limited upper respiratory tract infections. Frequently the

---

BETTY WU-HSIEH and D. H. HOWARD • Department of Microbiology and Immunology, UCLA School of Medicine, Los Angeles, California 90024-1747.

Fungal Infections and Immune Responses, edited by Juneann W. Murphy *et al.* Plenum Press, New York, 1993.

only sign of encounter with the fungus is a conversion of delayed-type skin reactivity to histoplasmin from negative to positive.[3–6] Nevertheless, distinct diseases do result from infection by the fungus and these will be described briefly. (Recent discussions of this sort of information are contained in references 6–8.) The topical outline used here is adapted from that of Saag and Dismukes.[6,7] None of the forms of histoplasmosis is contagious.

2.1.1. Pulmonary Disease

2.1.1a. Acute. The common form of symptomatic illness is an acute pulmonary disease. The signs are an influenzalike illness composed of fever, chills, cough, malaise, weakness, or headache. The illness varies in severity from mild to severe, even requiring hospitalization. "The chest X ray reveals single or multiple areas of patchy pneumonitis with a predominant alveolar parenchymal pattern and hilar or paratracheal lymphadenopathy."[6,9,10] The disease resolves spontaneously over a period of 2–4 weeks. The roentgenographic signs resolve in 2–4 months with residual calcifications. These are more uncommon forms of acute disease (for a description, see references 6 and 7).

2.1.1b. Chronic (cavitary). This form of histoplasmosis has been studied extensively by Goodwin and his colleagues.[4] Chronic cavitary disease generally occurs in individuals with underlying chronic obstructive pulmonary disease. The course is indolent and accompanied by cough, weight loss, fever, night sweats, and malaise. Hemoptysis is uncommon.

2.1.2. Progressive Disseminated Disease

Primary pulmonary infections are often accompanied by spread of the organisms beyond the lungs and transient fungemia has been recorded.[10a] Organs rich in cells of the reticuloendothelial system (spleen, lymph nodes, and liver) are frequently involved. Usually these metastatic sites heal uneventfully with the advent of an acquired immunity. Rarely in normal hosts, but more commonly in immunocompromised individuals, the disseminated disease results in symptomatic progressive disease. There are three recognized forms of progressive disease: 1) acute (severe) disseminated histoplasmosis most commonly seen in infants and young children and in the immunosuppressed adult; 2) moderate disseminated histoplasmosis that occurs in adults and is less severe than the acute form; and 3) chronic disseminated histoplasmosis with focal involvement in any organ system. (For a discussion of various forms of disseminated histoplasmosis, see reference 6.)

2.1.3. Immune-Mediated Disease

There are late occurring immunologically mediated syndromes seen in histoplasmosis. These syndromes are considered to be caused by antigenic stimulation of a sensitized, hyperresponsive host. Three commonly encountered forms are: 1) mediastinal granuloma; 2) histoplasmomas; and 3) presumptive ocular histoplasmosis syndrome.[11–13]

2.2. African Histoplasmosis

This form of histoplasmosis is clinically distinct from the classical histoplasmosis discussed above. An acute form of pulmonary disease has not yet been described. Destructive granulomas and suppurative lesions of the skin, subcutaneous tissues, and bones are characteristic of the disease. A good summary of current information on African histoplasmosis is presented by Rippon in his textbook.[2]

3. ETIOLOGY

3.1. *Histoplasma capsulatum* var. *capsulatum*

In its anamorphic phase of growth, *H. capsulatum* occurs in two morphological forms: a yeast form composed of blastoconidia, and a mycelial form consisting of hyphae upon which are borne both micro- and macroconidia.[2,4,14–17] This dimorphism is associated with the parasitic and saprophytic phases of growth of the fungus, but some isolates will express the two distinct morphological aspects of the asexual life cycle in artificial culture media.[18,19] Thus, on common laboratory media incubated at 23° to 25°C, *H. capsulatum* grows as a mycelium, whereas on suitable media at 35° to 37°C, some isolates of the fungus convert to the blastoconidial form of growth.[18–21] In the tissues of a host, the commonly encountered form is the blastoconidium (yeast) which occurs in the cytoplasm of various mononuclear phagocytes.[9,10] *Histoplasma capsulatum* is, then, a facultative intracellular parasite,[2,9] and this fact figures prominently in discussions of the immune responses in histoplasmosis.

A teleomorphic phase of *H. capsulatum* is known. The name given to this form is *Ajellomyces capsulatus*.[21a] The mating type locus has two alleles and the fungus is heterothallic. Ascospores within asci contained in cleistothecia are the morphological features of this aspect of growth. Instability of tester strains has limited the use of this form of growth in genetic studies.

Berliner[22] described two colony variants of *H. capsulatum*, called brown (B) and albino (A). The B type was more virulent for rabbits and mice but the differences were not marked.[23,24] Other sorts of colony variants have been described and one must be aware of such occurrences in the pursuit of routine identification procedures in the clinical laboratory.[25] A comprehensive study of the cytological and ultrastructural aspects of the *H. capsulatum* has been published.[26,27] The physiological and genetic controls over dimorphism have been reviewed.[28] Differential hybridization techniques have been used to identify yeast phase specific DNA sequences in *H. capsulatum*.[29] One of these, yps-3, has been employed as a probe for the identification of *H. capsulatum*.[29,30] A study of the value of the probe in comparison to other techniques of identification has not been published.

Some preliminary work by Medoff and his colleagues[31] has implied that the transition to the yeast form of growth is mandatory for pathogenicity of *H. capsulatum* in mice. The transition of conidia of *H. capsulatum* involves a temperature shift of 25°C to 37°C.[32] Such a temperature shift is accompanied by the production of heat–shock proteins that may play a central role in temperature adaption and morphogenic change.[28,33] Moreover, the heat–shock response has some immunological implications as well.[33a,34,35]

Information of the cell-wall structure of *H. capsulatum* is meager. Kaufman and Blumer found several serotypes among strains of *H. capsulatum*, a finding that suggests differences in chemical composition of the cell walls.[36] Pine and Boone[37] and Domer and her colleagues[38,39] have described two chemotypes, I and II. The distribution of the glycosidic linkages of the glucans from yeast cells varied between the two chemotypes and was one basis for their distinction. Strains labelled chemotype II had large amounts of α-(1,3)-glucan whilst those designated type I had very low levels.[38,39] The α-(1,3)-glucan has been associated with virulence of strains of *Paracoccidioides brasiliensis*[40,41] but Domer[38,39] did not find a similar correlation in *H. capsulatum*. Klimpel and Goldman set about to isolate spontaneously occurring, isogenic, avirulent clones of *H. capsulatum* for study of virulence factors.[42] Several such clones were identified and each contained very much reduced α-(1,3)-glucan.[43] These results indicated a correlation of the presence of α-(1,3)-glucan with virulence, a correlation observed with *P. brasiliensis*.[40,41] Among strains labelled chemotype I by Domer (markedly reduced in α-(1,3)-glucan), however, were several virulent representatives. Klimpel and Goldman[43] have suggested that the apparent discrepancy between their results and those of Domer *et al.* might be related to a difference in regulation of α-(1,3)-glucan synthesis. Thus, both chemotype I and II strains produce α-(1,3)-glucan *in vivo* but chemotype I strains repress its expression *in vitro*. Derepression has not been accomplished, however, and the enigma remains unresolved. A much larger number of isolates and variants will need to be examined before conclusions or correlations of cell-wall structure and virulence can be reached.

Other approaches to strain separation among isolates of *H. capsulatum* have been made. A notable example is the recent report of Restriction Fragment Length Polymorphism (RFLP) of mitachondrial deoxyribonucleic acid (mit DNA) and ribosomal DNA (rDNA). Three classes of isolates could be recognized from these RFLP patterns: class 1 consisted of the Downs strain, class 2 consisted of North American and African isolates of the *duboisii* variety, and class 3 consisted of isolates from Central and South America.[44,44a] Other examples of class 1 have been identified among isolates from AIDS patients in the St. Louis area (W. E. Goldman, *personal communication*). Recently, an isolate from soil with a distinct RFLP pattern has been recovered from soil and set apart in a class 4.[45] Chromosomal mapping performed in agarose gels has revealed seven chromosome-sized molecules in the Downs strain of *H. capsulatum*.[46] When other strains were tested, considerable variation in band mobility was detected, a result comparable to that obtained with *Candida albicans*[47,48] and *Cryptococcus neoformans*.[49] The resolution of the chromosome number if *H. capsulatum* will require development of appropriate probes.

After the discovery of the teleomorphic form of *H. capsulatum* (*A. capsulatus*), it was considered that a direct genetic approach to such problems as dimorphism and pathogenicity of the fungus could be made.[50] The instability (sterility) of tester strains, however, has severely limited the use of the teleomorph form of growth. Alternative approaches have focused on the manipulation of the yeast cell form of growth. Alternative approaches to identifying sensory control systems in *H. capsulatum* have been initiated. Some genes have been cloned and a transformation system is under development.[28,51,52] A complete analysis of this sort of work is beyond the scope of this chapter (see references given).

3.2. *Histoplasma capsulatum* var. *duboisii*

The hyphal form of growth of this variety is identical to that of var. *capsulatum*.[2] The yeast form is, however, markedly different. The blastoconidia of var. *duboisii* are much larger (12–15 um) than are those of var. *capsulatum*. Thus, the organism is not an obvious intracellular parasite, though it may be found in the giant cells that are prominent components of the granulomata evoked by var. *duboisii*.[2] Cleistothecia result from pairing of tester strains *A. capsulatus* with *H. capsulatum* var. *duboisii*. This positive result was a major fact in the decision on the varietal rather than species status of *H. duboisii*.[2]

4. EPIDEMIOLOGY

Primary histoplasmosis acquired by inhalation of microconidia of *Histoplasma capsulatum* from natural sources.[2–5] The first three cases were uncovered in Panama in 1906 by Samuel Taylor Darling who concluded that the etiologic agent, observed within histiocytes, was a plasmodialike protozoan.[53] The first to culture *H. capsulatum* was W. A. DeMonbreun, who correctly described the etiologic agent to be a fungus with a diphasic life cycle.[14] Although the early reported cases of histoplasmosis were observed in patients with extensive disseminated disease, it is now recognized that primary histoplasmosis is customarily self-limited and benign.[6]

The infection is cosmopolitan.[2] Cases have been reported from every continent, save Antarctica.[1,2] The areas of highest incidence are the United States and Latin America. Histoplasmosis caused by *H. capsulatum* var. *duboisii* occurs almost exclusively in Africa.[2] It should be noted that cases of histoplasmosis caused by *H. capsulatum* var. *capsulatum* also occur in Africa[2,5] and that there are no serological or immunological methods for distinguishing the two forms of disease,[2] except for the rather different clinical presentation by patients with African histoplasmosis.[2,28]

Histoplasma capsulatum is a soil inhabitant. It is common in habitats enriched by bird droppings especially those of pigeons, grackles, oil birds, starlings, and chickens. Bats sustain a natural infection that involves the gastrointestinal tract and bat droppings may be heavily contaminated with the fungus.[1,2,15]

Histoplasma capsulatum is not a regular opportunist but is known to produce progressive infections in persons who are immunocompromised by hematologic malignancies,[2–4,54–57] cytotoxic therapy,[2,15,58–60] or persons with acquired immunodeficiency syndrome (AIDS).[61–69]

5. HOST DEFENSES

5.1. Innate Resistance

From the observation that infection by *H. capsulatum* is subclinical in approximately 95% of the cases[2–4,6,7] and is manifest only by a positive histoplasmin skin test or by residual calcifications, it is clear that humans have efficient ways of dealing with this potentially invasive and destructive fungus. The immunological bases for asymptomatic and subclinical infections may involve some of the same mechanisms as those responsible for recovery of individuals with symptomatic disease and for acquired resistance, but in this section we will consider only those systems operative without specific stimulation by the invading fungus, that is, soluble factors found in serum (or plasma *in vivo*), polymorphonuclear leukocytes (PMNL), unstimulated mononuclear phagocytes (MNP), and natural killer (NK) cells. The proper functioning of these humoral and cellular elements of innate resistance is likely to involve genetic control. Certainly, susceptibility to infection with *H. capsulatum* varies greatly among different genetic strains of mice.[70–73]

The sex and age of the host also influence susceptibility to experimental infection. For example, white Swiss mice up to 15 weeks of age are more susceptible to infection than are mice 17 weeks of age or older.[74] Salvin has attributed this age difference in susceptibility to the "development of sexual maturity." In his own studies, no difference in susceptibility was noted between the two sexes in immature mice (35 days of age).[75,76] In older mice

(8–14 weeks), however, females were more resistant than were males.[74] Similar results have been obtained with BALB/c mice.[77]

Recently, a model of opportunistic infection has been developed with *H. capsulatum* functioning as the opportunist. A variant of the lymphocytic choriomeningitis virus (LCMV) was isolated that causes chronic infection in adult mice. The mice are severely immunocompromised and die from an ordinarily sublethal dose of *H. capsulatum*.[78] The murine model of virus-induced immunosuppression will be useful in studying opportunistic fungal infections. A conceptually similar model employing *Candida albicans* in animals immunosuppressed by Freund leukemia virus infection has been reported.[79]

5.1.1. *Transferrin and Lactoferrin*

The iron binding activity of transferrin accounts in part for the inhibition of growth of many species of fungi in media containing animal serum.[80–83] The transferrin in serum or the conalbumin in egg white can bind enough iron to inhibit the growth of *H. capsulatum* in media containing those substances.[83–86]

H. capsulatum acquires iron by means of a hydroxamic type of siderophore[87–89] but unlike most gram-negative bacteria, this siderophore does not allow the fungus to remove iron *in vivo* from high affinity iron-binding compounds such as transferrin or lactoferrin.[90] The percent saturation of transferrin in the sera of normal subjects or of patients with histoplasmosis is in the range that would inhibit *H. capsulatum in vitro*.[86] These observations indicate that plasma transferrin could be important in restricting the extracellular replication of *H. capsulatum*; however, *H. capsulatum* is a facultative *intracellular* parasite. Mononuclear phagocytes have transferrin receptors.[91–93] Transferrin attached to the receptor is internalized by endocytosis. Within the acidified endocytic vacuole, iron is released and distributed to the labile iron pool of the cell.[94,95] The resulting apotranferrin bound to the receptor returns to the cell surface where the apotranferrin is released.[96] Thus, *H. capsulatum* in its customary intracellular location is not exposed to transferrin, and since it grows unimpeded in unstimulated mononuclear phagocytes, must obtain iron, presumably from the labile iron pool and probably via its own siderophore.

In a stimulated phagocyte a different sequence of events is possible. For example, *Legionella pneumophila* requires iron but does not have siderophores and growth of the bacterium is restricted to environments where iron is available. In a recent study Byrd and Horwitz[93] investigated the role of iron in the intracellular behavior of *L. pneumophila* in human monocytes. They showed that: 1) the intracellular replication of *L. pneumophila* was dependent on iron; 2) IFN-γ-activated monocytes inhibited *L. pneumophila* by limiting iron availability; and 3) the transferrin receptors of the cell were downregulated on IFN-treated cells.[93] Such limitation of the iron availability also inhibits the intracellular growth of *H. capsulatum* within IFN-γ-activated murine peritoneal macrophages.[93a]

The iron-binding protein lactoferrin that is found in the specific granules of PMNL also has fungistatic properties when not fully saturated with iron.[97] The activity of partially saturated lactoferrin against *H. capsulatum* has not been reported, but it theoretically could affect the fungus after release into the phagolysosome. Since lactoferrin is also released into the extracellular fluid during phagocytosis of particles by PMNL, it could augment the transferrin effect and, thus, further deplete the extracellular sources of iron. It is also known that the iron-lactoferrin complexes enter mononuclear phagocytes,[98] however, and might not come into direct contact with *H. capsulatum* in such cells.

5.1.2. Early Pathogenesis in Natural and Experimental Disease

The pathogenetic events that characterize subclinical or acute primary disease in humans are not precisely known because those forms of the disease are by definition mild, uncomplicated events whose histopathological appearance is not customarily studied.[2–4] The early pathogenesis recorded in experimental infections is usually inferred to be representative.[10] The cellular infiltrates noted in such infections consist primarily of lymphocytes, plasma cells, and mononuclear histiocytes.[99–103] Areas of intense PMNL infiltration with necrosis are also observed in early experimental lesions.[78,99,100,103] The cell types that comprise the inflammatory cell infiltrate have been examined for their ability to affect *H. capsulatum*, and the results of such studies have been reviewed previously and only a brief update of such literature will be given here.[83]

5.1.2a. Polymorphonuclear Leukocytes. In the early literature it was recorded that polymorphonuclear leukocytes (PMNL from humans,[104] guinea pigs,[105], and mice[106] killed blastoconidia of *H. capsulatum*. But more recent reports indicate that the viability of the fungus was not significantly reduced in the presence of PMNL from humans.[107]

A series of observations by Schnur and Newman[107a] has shown that: 1) human PMNL bind yeast cells through the CD 18 receptor, as do human cultured macrophages and human alveolar macrophages (see below); 2) unopsonized yeast cells were phagocytosed but phagocytosis was markedly enhanced by both heat-labile and heat-stable serum opsonins; and 3) the opsonins increased the attachment of yeast cells to the PMNL. The fate of *H. capsulatum* within the PMNL was not reported.

A number of antimicrobial systems operative within PMNL have been described.[81,108,109] It is convenient to divide such mechanisms into two types:

oxidative and nonoxidative.[109] The interaction of these antimicrobial systems with fungi in general[109] and with *H. capsulatum* in particular have been reviewed and will not be reconsidered in detail here.[83]

Both the microconidia and blastoconidia of *H. capsulatum* are sensitive to H_2O_2[110] and their sensitivity is increased by the presence of I^- or Cl^-.[110] Strains of the fungus vary in their sensitivity to H_2O_2, but the relative sensitivity is not correlated with the level of catalase in the different strains.[111] Hydrogen peroxide, halides, and myeloperoxidase (MPO) combine to form a potent antifungal system that is active against both microconidia and blastoconidia of *H. capsulatum*.[110]

It has been found[107a] that yeast cells of *H. capsulatum* stimulate a respiratory burst in PMNL as measured by nitroblue tetrazolium, oxygen consumption, luminol-enhanced and nonenhanced chemiluminescence, and H_2O_2 production. The PMNL that had phagocytized *H. capsulatum*, however, did not induce the release of superoxide anion. Understanding the mechanism of such entrapment will require further work.

There is only scant information on nonoxidative or microbicidal mechanisms of PMNL operative against *H. capsulatum*. What little information is available has been reviewed.[83] Certain of the cationic peptides from PMNL and from macrophages[109] have broad antimicrobial activities. At least two of these peptides, macrophage cationic peptides, MCP-1 and MCP-2[109] do kill *H. capsulatum*[106] under specified conditions of ionic strength and pH,[112] but do not affect intracellular growth of the fungus within mouse macrophages.[106]

5.1.2b. Mononuclear phagocytes (MNP). Three types of cells are used in studies on the interaction of fungi and MNP: circulating monocytes, alveolar macrophages, and peritoneal macrophages. An early report that *H. capsulatum* is killed by human circulating monocytes[104] was not repeated by another group of investigators.[113] *H. capsulatum* survives and multiples within rabbit and mouse alveolar macrophages[114,115] and mouse and guinea pig peritoneal macrophages.[115–120] Both yeast cells and microconidia of *H. capsulatum* are phagocytized by human cultured macrophages and alveolar macrophages.[113] The yeast cells survive the encounter and display normal intracellular generation times.[121] (Conidia may convert to yeast cells by direct budding so that the fate of such reproductive propagules is unlikely to be different from yeast cells).[121a] Cytokines do activate the macrophages to inhibit intracellular growth but opsonization doesn't augment the fungistatic effects.[121] Macrophage cell lines have been used in studies on the intracellular fate of *H. capsulatum*. Virulent but not avirulent strains of *H. capsulatum* can grow in P338D_1 macrophages.[122] The avirulent cells are lacking α-(1,3)-glucan in their cell walls and persist in distorted forms within the macrophages. A few such forms regain α-(1,3)-glucan in their cell walls. The P388D_1 macrophages cannot be stimulated to suppress the intracellular growth of *H. capsulatum* with IFN-γ.[123]

Nonprofessional phagocytes have also been used in studies on the intracellular fate of *H. capsulatum* (for the early literature, see reference 116). Human endothelial cells and hamster tracheal epithelial cells will phagocytize yeast cells of *H. capsulatum*.[124,125] As described above, strains of *H. capsulatum* that lack α-(1,3)-glucan in their cell walls are relatively avirulent for mice. Such strains grow well in hamster tracheal epithelial cells but poorly in P388D_1 macrophages. The virulent parents of these strains that do have α-(1,3)-glucan in their cell walls grow well in P388D_1 macrophages but are not taken up very well by the epithelial cells. Whether these differences in interaction with mammalian cells in culture are related to surface phenomenon in regard to the presence or absence of α-(1,3)-glucan will require further study.[122,125]

Recognition and binding of *H. capsulatum* of both yeast cells or microconidia to human cultured macrophages and alveolar macrophages is mediated by the macrophage CD 18 complex.[113,126] Ingestion of *H. capsulatum* yeast cells by human MNP-derived macrophages was followed by the generation of superoxide and hydrogen peroxide.[126] Thus, the behavior of *H. capsulatum* in human MNP-derived macrophages is different from that in human PMNL where superoxide anion is not released following phagocytosis of yeasts, though peroxide is formed by both types of phagocytes.[126,143a]

Facultative intracellular parasites must cope with the hostile environment of phagocytic cells. Moulder[127] has outlined three ways by which intracellular parasites avoid destruction by lysosomal substances: 1) escape from the phagosome; 2) evasion of phagosome–lysosome fusion (P-LF); or 3) resistance to antimicrobial substances in the lysosome. *Histoplasma capsulatum* yeast cells have not been observed outside a phagosome, but P-LF was not irrefutably demonstrated by electron microscopy.[128–131] More recent work has provided evidence that phagolysosomes are formed by phagocytes after ingestion of *H. capsulatum* and that the fungus resides and multiplies within them unharmed by their contents of potentially microbicidal substances.[132]

On the basis of current evidence frustration of P-LF by *H. capsulatum* is not the mechanism of intracellular survival. It appears that subsequent to P-LF, the fungus is able to suppress phagosome acidification.[132a] It is known that *H. capsulatum* alkalinizes instead of acidifying liquid shake cultures and that it accomplishes this by the production of large amounts of ammonia (Howard, *unpublished observations*). But whether or not such ammonia production occurs intracellularly has not been studied.

Some facultative intracellular parasites fail to trigger the oxidative burst that accompanies phagocytosis of most particles.[133–135] This method of avoiding microbicidal substances is apparently used by *H. capsulatum*.[136,137] The *in vivo* relevance of this observation, however, is somewhat modified by the observation that yeast cells opsonized with *Histoplasma*-specific antibody did evoke an oxidative burst.[137] Thus, Fc-mediated phagocytosis may evoke a different response by the phagocytic cells than that produced after CD 18 mediated phagocytosis. This same behavior is displayed by human PMNL phagocytizing *H. capsulatum*. From such observations one could speculate that phagocytosis of *H. capsulatum* early in infection before antibody formation would permit intracellular replication in the absence of a burst while subsequent phagocytosis of opsonized yeasts would lead to a burst and control of intracellular multiplication.[138,138a] Newman *et al.*,[113] however, have shown that opsonization with immune serum did not modify the intracellular generation time of *H. capsulatum* in human macrophages.

Antifungal mechanisms in MNP have not been studied as often as have those in PMNL. MPO is present in circulating monocytes but not in peritoneal macrophages.[81] When circulating monocytes are cultivated *in vitro*, they lose their MPO.[81] The loss of MPO during culture, however, does not affect the ability of stimulated macrophages to inhibit *H. capsulatum*.[113] Hydrogen peroxide is formed by peritoneal macrophages and has been indicated as a potential effector molecule. Lysozyme is very abundant in MNP and a lysosomal-rich extract from normal rabbit alveolar macrophages was shown to inhibit protein synthesis by *H. capsulatum* yeasts.[139] Cationic proteins with inhibitory effects against *H. capsulatum* have also been described.[109,113]

5.1.2c. Natural Killer (NK) Cells. The interaction of lymphocytes of the NK variety with *H. capsulatum* has been reported.[140,141] Antibody-dependent cellular cytotoxicity (ADCC) mediated by human peripheral blood natural cytotoxic lymphocytes has also been described in preliminary communications.[140,141]

The possibility that NK cells may contribute to natural cellular resistance against *H. capsulatum* is supported by the finding that beige (bg) mice, which are deficient in NK cell activity,[142] are more susceptible to intravenous challenge with yeast cells of the fungus than are bg/+ mice.[143]

5.1.3. Complement

Fresh serum is important for phagocytosis of *H. capsulatum* by mouse PMNL. Human PMNL recognize and bind *H. capsulatum* by CD 18 adherence receptors (see above) and unopsonized yeasts are phagocytized by the phagocytes. But phagocytosis by the cells is markedly augmented by both heat-labile and heat-stable opsonins.[143a] Although in an early study[144] fresh serum was shown to augment the uptake of the fungus by mouse peritoneal macrophages, a more recent investigation has provided no evidence that serum influences phagocytosis by peritoneal macrophages from mice[137] but opsonization does affect the oxidative burst.[138]

Viable or killed yeast cells of *H. capsulatum* have been shown to activate the alternate complement pathway,[145] and there is a report on the role of properdin in the phagocytosis and intracellular inhibition of the fungus in mouse peritoneal macrophages.[145a] Activation of the classical complement pathway by immune complexes or soluble products of fungal cells has been described as important in the phagocytosis and intracellular fate of *H. capsulatum* within mouse macrophages.[145a] Such complement activation may also generate chemotactic factors but this aspect has not been studied with *H. capsulatum*.

5.2. Immune Responses

5.2.1. Humoral

5.2.1a. General. High levels of serum antibodies are frequently observed in patients with progressive histoplasmosis but some B-cell response is seen in nearly all who are infected.[2–5] The antibody response is of value in establishing a diagnosis of the disease, but current evidence suggests only a minor role for antibodies in recovery from disease or the acquired resistance that is so prominent a feature of the disease.[146]

5.2.1b. Immunoglobulins. IgM is elevated in the acute phase of histoplasmosis but returns to normal values in the chronic, stabilized or recovery phases of the disease. Serum IgG levels are elevated in active histoplasmosis.[147–149] Serum IgA is elevated in histoplasmosis, particularly in chronic pulmonary disease.[147,149] There are conflicting reports on IgE levels in active histoplasmosis. In one study the levels were increased,[148] while, in another study, the concentrations were the same in disease as they were in normal subjects.[149] It should be noted that, in the latter study, the sera had been stored for 15 years and IgE is known to be labile.[150]

5.2.1c. Immune Complexes. Circulating immune complexes have been reported in a patient with disseminated histoplasmosis.[151] The presence of immune complexes was associated with peripheral blood eosinophilia and mesangiopathic glomerulonephritis. Direct immunofluorescent examination of a renal biopsy specimen showed the presence of IgA, IgM, and C3, but *H. capsulatum* antigen was not detected.

In a more comprehensive study, circulating immune complex levels were measured in sera of histoplasmosis patients with a Clq-binding assay.[152] Thirty-six percent of 55 patients with active disease had circulating immune complexes. But less than 5% of 21 healthy, histoplasmin skin-test positive persons and 7% of 82 healthy skin-test negative persons had such complexes.[83] The composition of immune complexes in terms of *H. capsulatum* antigen and anti-*Histoplasma* antibody was not determined.

5.2.1d. Role of Antibodies in Host Defense. Immune serum augments phagocytosis of the fungus (see above) but intracellular killing is not enhanced.[113,119,145a,153] Antibody-dependent cellular cytotoxicity (ADCC) has been reported with *H. capsulatum*; however, the work describing ADCC activity has been presented in abstract form only and details of the fate of *H. capsulatum* are not available.[140,141]

Adoptive transfer of spleen cells from mice immunized with the fungus results in immunity to challenge with *H. capsulatum* in the recipients.[154–156] Prior treatment of the transferred spleen cells with anti-Ig plus complement does not affect protective transfer, a fact that indicates the protection is not mediated by B cells. The assumption of nonparticipation by B cells is further strengthened by the observation that B-cell enriched spleen cells from immune donors does not protect against challenge.[157] Taken together, these results are in accord with the failure to transfer immunity with serum from *Histoplasma*-immunized donors.[154,158]

Progressive disseminated histoplasmosis is characterized by anergy and high levels of serum antibodies. These facts have led some investigators to suggest an immunosuppressive role for antibody either alone or in antigen-antibody complexes.[159–162] For example, Taylor *et al.*[161] reported that mice given serum from *H. capsulatum*-infected donors were more susceptible to intraperitoneal challenge with the fungus than were recipients of normal mouse serum. The work to date hasn't clarified whether these effects are those of free antibody or antibody complexed to antigen.

5.2.2. Cell-Mediated Immunity (CMI)

5.2.2a. General. Early work by Salvin[163–165] showed that organs rich in cells of the mononuclear phagocyte system such as the liver, spleen, and kidneys were important in clearing *H. capsulatum* from the tissues of animals given a sublethal infection. Injection of animals was followed by accumulations of the fungus occurred in Kupffer's cells of the liver and in the macrophages of other organs. The numbers of fungi were reduced over the course of several weeks, but viable organisms often persisted for long periods of time in host tissue.[75,76,164,166,167] Subsequent work has shown that procedures that alter the proper functioning of cells of the mononuclear phagocyte system have disastrous consequences in experimental murine histoplasmosis.[168–170] A cellular basis for acquired resistance in histoplasmosis in established by six features of natural or experimental disease: 1) the occurrence of delayed-type hypersensitivity (DTH) in healthy individuals exposed to the fungus;[2–4,171,172] 2) the adoptive transfer of specific DTH and resistance to normal recipients by lymphoid cells but not by serum from immunized animals;[154–156,158,173,174] 3) the *in vitro* transformation response of lymphocytes from skin-test positive individuals by *H. capsulatum* antigen;[159,175–181] 4) the activation of normal macrophages by lymphokines released from immune lymphocytes;[182–185] 5) the progressive disease that develops in experimental animals immunosuppressed by administration of lymphocytotoxic drugs,[72,186–191] or by thymectomy,[191] or that observed in athymic mice;[192,193] and 6) the increased susceptibility of persons who are immunosuppressed by hematological malignancies,[2–4,54,57,59] cytotoxic therapy,[2,54,58–60] or acquired immune deficiency syndrome.[61–68] Conclusions about the importance of cell-mediated versus humoral immunity in recovery from and resistance to histoplasmosis are reinforced by the report of an adequate immunologic defense against *H. capsulatum* by a patient with hypogammaglobulinemia.[194]

5.2.2b. Cutaneous Delayed-Type Hypersensitivity (DTH). Cutaneous reactivity of the delayed type consistently appears in individuals after they are exposed to *H. capsulatum*.[2–4,171,172] In 1945, Palmer published a study on subjects with nontuberculous pulmonary calcification and sensitivity to histoplasmin[195] that revealed a high incidence of histoplasmin reactivity among such subjects. It is now well known that a benign form of histoplasmosis is extraordinarily common in the Central Mississippi Valley, the Ohio River Valley, and along the Appalachian Mountains.[171]

Experimental infections in animals result in delayed-type sensitivity to histoplasmin,[75,76,89,101,196–204,205] and this sensitivity is transferable by means of T lymphocytes.[154–157,173,174,203] That acquired resistance also accompanies such transfers is one of several bases for the conclusion that DTH to histoplasmin in a normal individual signifies immunity to infection with *H. capsulatum*.

Depressed skin-test reactivity to histoplasmin is a common manifestation of progressive, disseminated histoplasmosis and denotes a poor prognosis.[3,4,55,206] The decreased DTH response is specific for *H. capsulatum* antigen in some patients but in others, particularly those with severe disease, a generalized cutaneous anergy occurs.[159,207] The mechanisms responsible for depressed DTH have not been characterized but may be related to the activation of suppressor cells. In patients with disease regression, the anergy is reversed.[207]

5.2.2c. Lymphocyte Transformation (LT). Chaparas *et al.*[208] showed that peripheral blood mononuclear cells from healthy, histoplasmin skin-test positive persons reacted to *H. capsulatum* antigens *in vitro*, a finding that has been confirmed by several investigators.[159,175–181] These findings have been developed into diagnostic tests of limited application (see below).

Evidence has been presented that heat-killed yeast cells of *H. capsulatum* and derived extracts are mitogenic to splenocytes from nonimmune mice.[209,210] Conversely, Yamamoto *et al.*[211] reported that a high molecular weight substance isolated from yeast cells of the fungus inhibits LT responses of nonimmune murine splenocytes to various mitogens.[211]

5.2.2d. Lymphokines.

Macrophage Migration Inhibitory Factor (MIF). Johnson and Scherago[212] showed significant retardation of the migration of peritoneal macrophages from guinea pigs experimentally infected with *H. capsulatum* by histoplasmin. The correlation between production of MIF by immune lymphocytes and the *in vivo* DTH response to histoplasmin was confirmed in studies of humans,[213,214] mice,[157,174,191,203,215] and guinea pigs.[200,216] Production of this lymphokine by patients with active histoplasmosis has not been evaluated.

Macrophage Activating Factor (MAF). H. capsulatum is a facultative intracellular parasite of the cells of the reticuloendothelial system.[2–4] The intracellular behavior of the fungus has been studied by a number of investigators.[116–119,147,153,217–220] Extension of such work to immunized animals showed that mononuclear phagocytes from mice immunized against *H. capsulatum* inhibit the intracellular growth of the fungus[120,153,221,222] and that growth of the fungus within normal macrophages is inhibited by lymphocytes freshly harvested from immunized mice.[221,222] Therefore, immune lymphocytes acting as mediator cells of acquired immunity can activate normal macrophages to inhibit the intracellular growth of *H. capsulatum.* Activation is mediated by lymphokines generated in cultures of immune splenocytes stimulated with *Histoplasma* antigen.[182] Supernatants from concanavalin A-stimulated T-cell hybridomas also activate macrophages to suppress the intracellular growth of *H. capsulatum.*[183,184]

The macrophage activating factor secreted by spleen cells from *Histoplasma*-immunized mice and that from concanavalin A-stimulated T-cell hybridomas has high antiviral activity, is inactivated by acid (pH 2.0), and is relatively heat stable (56°C, 30 min).[182,183] The current evidence supports the view that MAF is IFN-γ. Addition of murine recombinant IFN-γ endows macrophages with antihistoplasma activity.[184] Neither alpha interferon nor beta interferon activate macrophages for *Histoplasma* growth inhibitory activity.[183] Addition of rabbit polyclonal antibody to murine IFN-γ neutralizes the macrophage-activating activity of supernatants from immune splenocytes.[184,185] While the intracellular growth of *H. capsulatum* yeast cells is completely suppressed within macrophages, the fungus is not killed.[185]

Activated macrophages secrete a plasminogen activator.[223] Tiku *et al.*[224] report that splenic macrophages from *Histoplasma*-infected mice produce more of this fibrinolysin than do splenic macrophages from noninfected controls. Addition of splenic T cells (nylon wool nonadherent) from infected mice to normal peritoneal cells markedly enhanced fibrinolytic activity, suggesting that activation was mediated by *H. capsulatum* antigen-reactive T cells, either via a soluble lymphokine or by direct interaction of T cells with macrophages. Macrophage activation, as measured by the secretion of this enzyme, preceded clearance of *H. capsulatum* from spleens of infected mice.[224]

5.2.2e. The Role of CMI in Host Defense. The cumulative experience reported in the preceding sections provide evidence that CMI is crucial to host resistance to *H. capsulatum.* The L3T4+ T-cell subset is a mandatory component of CMI defense against *H. capsulatum.*[225] Activated macrophages *in vivo* are thought to mostly occur at the site(s) of infection,[226] thereby facilitating the interaction between lymphocytes, macrophages, and specific antigen. In *H. capsulatum* infection, the interaction would occur in granulomatous inflammatory sites,[3,227,229,230] since, in other biologic systems,[228–233] granuloma formation is mediated by immune T cells that release lymphokines to localize and activate macrophages. Thus, the T-lymphocyte–

macrophage interactions that have been demonstrated in *in vitro* studies of *H. capsulatum* are likely to occur *in vivo*.

Work on key cells in the CMI that develops in histoplasmosis has shown that the L3T4+ T-cell subset is an essential component of the immune response.[225] George Deepe and his colleagues have set about to develop and characterize murine T-cell lines and clones responsive to antigen of *H. capsulatum*.[234] The T cells and clones so far derived are of the helper/inducer phenotype (Thy-1.2+, Lyt-1+, L3T4+, Lyt-2$^-$). Supernatants from antigen-stimulated clones activated macrophages to suppress the intracellular growth of *H. capsulatum*.[234] A preliminary characterization of antigenic determinants in the complex filtrate product, histoplasmin that stimulate T cells has been carried out.[235] T-cell clones failed to transfer protection against *H. capsulatum* perhaps because of "defective" trafficking or poor survival (of cells) or both.[236]

The work of Deepe and his colleagues outlined has shown the central importance of CD4+ T cells in ridding a host's tissues of *H. capsulatum*. Further work from the group has defined some of the requirements for presentation of histoplasmin by accessory cells T-cell lines[237] and has begun to identify epitopes in yeast-phase antigens that stimulate T-cell hybridomas.[238]

5.2.3. *Regulation of Immune Response*

5.2.3a. Suppressor Cells and Factors. Disseminated forms of histoplasmosis most frequently occur in patients who are immunologically compromised.[2–4,54–60] In some patients, however, dissemination occurs without recognized predisposing factors and is accompanied by cellular anergy,[55,159,239,240] an observation that suggests that reduced CMI responses might result from immunosuppressive mechanisms. Stobo *et al.*[241,242] reported that macrophages from patients with disseminated histoplasmosis secreted a soluble suppressor factor(s) that mediated suppression of *in vitro* lymphocyte transformation (LT) responses by interacting with a population of short-lived, low-density regulatory T cells. Their interpretation was based on the following observations: 1) freshly isolated peripheral blood mononuclear cells from patients with disseminated disease showed diminished LT responses to mitogens and histoplasmin antigen; 2) depressed responses were not observed with mononuclear cells that had been maintained in culture for 7 days (i.e., after depletion of short-lived T cells); 3) addition of the soluble supernatant fraction obtained from patients' macrophages, but not from T cells, to freshly isolated peripheral blood cells of patients or normal subjects resulted in suppressed LT responses; 4) peripheral blood cells that had been maintained in culture for seven days were refractory to suppression by the macrophage-derived suppressor factor(s); and 5) addition of freshly isolated low-density, suppressor T cells to cultured peripheral blood cells rendered them susceptible to the suppressive effects of the soluble supernatant fraction from macrophages.

Immunoregulatory imbalances in disseminated histoplasmosis are further documented by a shift in the relative numbers of T-helper to T-suppressor cells[240,243] and by the demonstration of increased suppressor cell activity.[207,239] Comparisons were made in one study, of circulating T-lymphocyte subpopulations in patients with various manifestations of histoplasmosis.[243] Acute pulmonary, rheumatologic, disseminated, and chronic inflammatory forms of histoplasmosis were all associated with an increased number of OKT8+ (suppressor/cytotoxic) lymphocytes which resulted in a lower than normal OKT4+ (helper/inducer) to OKT8+ lymphocyte ratio. In contrast, cavitary disease was associated with an increase in OKT4+ cells, a decrease in OKT8+ cells and, accordingly, a higher than normal OKT4/OKT8 ratio. Recovery was accompanied by a return of the ratios to normal values.[243] Thus, alterations of T-cell populations with increases in suppressor/cytotoxic cells is seen in many clinical forms of histoplasmosis and not just with disseminated disease.

Artz and Bullock developed a highly reproducible animal model of disseminated histoplasmosis.[244] Intravenous injection of mice with certain doses of yeast cells produced progressive, disseminated disease during the first three weeks of infection, which resolved by eight weeks postinfection. During the height of infection (weeks one through three), the spleens were enlarged as a result of the influx of T lymphocytes,[245] and there was thymic involution.[244] Along with these histologic changes, the mice were markedly depressed in their 1) footpad DTH responses to histoplasmin and challenge with sheep erythrocytes; 2) *in vitro* blastogenic responses to concanavalin A and histoplasmin; 3) primary *in vitro* antibody responses to sheep erythrocytes; and 4) production of interleukin 1 and 2.[246–248] These depressed responses were attributed to suppressor-cell activity because splenocytes and lymph node cells from infected mice suppressed the primary antibody response to both T-cell dependent and T-cell independent antigens, and T-lymphocyte mediated cytotoxicity.[246,247] Two populations of suppressor cells, T lymphocytes and macrophagelike cells, were identified in the spleens and lymph nodes of infected animals.[247,248] The suppressor T cells were phenotypically Thy 1.2+, Lyt-2+, I-J+ cells.[248] The macrophagelike suppressor cells were poorly adherent to plastic but could be removed by passage over nylon wool or by treatment with carbonyl iron.[247,248] Administration of cyclophosphamide abolished the suppression that was mediated by T cells (nylon wool nonadherent) but did not affect the *in vitro* suppression mediated by macrophages (nylon wool adherent) or unfractionated spleen cells and *in vivo* DTH responses to histoplasmin or sheep erythrocytes.[249] The persistence of suppression after cyclophosphamide treatment indicates that cyclophosphamide-resistant macrophagelike cells are important mediators of suppression in *H. capsulatum*-infected mice. The mechanism by which these macrophagelike cells exert their immunosuppressive effects does not appear to be due to prostaglandin synthesis, since indomethacin does not reverse the *in vivo* suppression of DTH response to histoplasmin or sheep erythrocytes.[249] More recent work has indicated that arachidonic acid metabolism may be involved in abnormalities in CMI seen in disseminated histoplasmosis.[249a]

Suppressor cells were demonstrable in both C3H/HeJ and C57BL/6 mice following intravenous infection with *H. capsulatum.* At one week postinfection, a suppressor factor was produced by spleen cells from C3H/HeJ mice, whereas a helper factor was released by spleen cells from C57BL/6 mice.[250] Thus, even though splenocytes from infected C3H/HeJ and C57BL/6 mice are suppressive, the populations of their spleen cells actually produce disparate immunoregulatory factors with different functions and properties.[250] The surface phenotypes of spleen cells that produce suppressor factors were Thy-1.2+, Lyt-2+, I-J+, while that of the cells releasing helper factors were Thy-1.2+, Lyt-1+.[250] Both cell populations required the participation of accessory cells having characteristics of macrophages.[251,252]

The route of infection with *H. capsulatum* is important in the induction of suppressor cells in experimental animals. Nickerson and Faircloug[204] reported that C57BL/6 mice infected by an intratracheal inoculation gave strong blastogenic responses to heat-killed *H. capsulatum* yeasts. Although there was an early and transient antigen-nonspecific suppression of immune responses, it was not mediated by suppressor cells but rather was correlated with the size of the inoculum, which at high levels may have caused a blockade of antigen-processing macrophages. In contrast to the results after intratracheal infection, intravenous injection of an equivalent number of yeast cells resulted in the generation of potent suppressor cells.[204]

5.2.3b. Serum Factors. Sera from patients with active histoplasmosis or from those who have recovered from the disease suppress the LT responses to histoplasmin of healthy, histoplasmin positive donors.[159,162,175,176] It has been suggested, but not proven, that the suppressed responses are attributed to antibody or to antibody–antigen complexes.[162]

Work on experimental infections in mice has shown that subcutaneous injection of normal animals with sera from *H. capsulatum*-infected animals adversely affects the response of recipients to challenge with the fungus.[161] This same procedure also resulted in alteration of certain cellular responses.[160] The prevalence of antibodies in experimental and natural histoplasmosis recommends further study, both of complexes and of direct effects of antibodies on T-cell function.[161]

6. SERODIAGNOSIS AND IMMUNODIAGNOSIS OF HISTOPLASMOSIS

6.1. Serodiagnosis of Histoplasmosis

The salient features of serodiagnostic responses in histoplasmosis and the antigens used to detect them have been reviewed in detail a number of times.[34,172,206,253–260]

Serologic tests for histoplasmosis include complement fixation (CF), immunodiffusion (ID), and latex agglutination (LA). The CF test is performed with both whole yeast cells and a mycelial-phase culture filtrate antigen, called histoplasmin. Demonstrable CF titers to the yeast cell antigen are often found in the first three to four weeks of infection.[206,253–258,261] CF antibodies to histoplasmin develop later in the course of infection and are generally lower in titer than those obtained with yeast cells, and may occur in the absence of CF antibody reactivity to yeast cells, particularly in patients with chronic histoplasmosis.[253] The sensitivity of the CF assay when both yeast- and mycelial-phase antigens are used is reported to be as high as 96%.[256,257,261–264] But in some studies, one-third to one-half of patients with chronic pulmonary or progressive, disseminated histoplasmosis are negative for CF antibodies. False-positive reactions are well documented in patients with other mycotic diseases, in *Histoplasma*-sensitized persons who have been skin tested with histoplasmin, and in persons with nonmycotic diseases.[206,258,265,266] The magnitude of CF titers does not assist in distinguishing the pulmonary from extrapulmonary forms of the disease or establish the extent of disease severity.[253,254,261] But increasing titers (fourfold or greater) denote an unfavorable clinical response and, conversely, decreasing titers denote a favorable prognosis for recovery.

The specificity of antibody responses to *H. capsulatum* can be established by use of the ID test in which reference *Histoplasma* antigen and antiserum are used. Of the multiple antigens present in histoplasmin, the two that have been designated H and M are specific for *H. capsulatum*.[267] The occurrence of precipitin antibody to the H antigen, either alone or in combination with antibody to the M antigen, provides strong evidence of histoplasmosis but occurs in less than 27% of patients.[261,264,268,269] Reactivity to the M antigen is demonstrable in approximately 75% of patients with histoplasmosis but occurs in both active or inactive disease and may result from histoplasmin skin testing in sensitized persons.[268–270]

The LA test detects antibodies that are produced within two or three weeks after primary infection.[271] Titers of ≥1:32 are usually considered positive, but there are a number of clinical situations that give false-positive reactions (*e.g.*, rheumatoid arthritis, tuberculosis, bronchogenic carcinoma, mycoplasmal pneumonia, among others).[271–273]

More sensitive serologic procedures for detecting antibodies have been sought by using enzyme-linked immunosorbent assays (ELISA)[274] or radioimmunoassays.[275,276]

Antibodies detected by the various means just described have been found within the IgM, IgG, and IgA classes.[147,149,150,277–279] Latex agglutination antibodies occur in the IgM class, while precipitating antibodies are in the IgG class.[147] Complement-fixing antibodies mostly occur in the IgG class though some also are present in the IgM class.[147] Anti-*Histoplasma* IgA antibody was not demonstrated in histoplasmosis sera when evaluated by immunoelectrophoresis[147] but was detected by radioimmunoassays.[149,277,278] Anti-*Histoplasma*

IgA was associated predominantly with chronic pulmonary histoplasmosis.[149] A similar profile of humoral antibody responses has been demonstrated in bats (*Artibeus lituratus*) experimentally infected with this fungus.[280]

6.2. Immunodiagnosis of Histoplasmosis

Newberry *et al.*[175] and Kirkpatrick *et al.*[176] established that LT responses were depressed in patients with active histoplasmosis, the extent of which correlated with disease severity. Similarly depressed LT responses in active histoplasmosis patients were reported by Cox.[159] Transformation responses to *Candida albicans* antigen and to the mitogens phytohemagglutinin and concanavalin A were unimpaired in most patients. Contrary to these results,[159,175,176] Alford and Goodwin[177] reported that most patients with chronic pulmonary histoplasmosis exhibited a vigorous histoplasmin-induced blastogenesis *in vitro*. These divergent results may be due to differences in patient populations or possibly to the fact that Alford and Goodwin[177] assayed mononuclear cells that had been passed through a nylon wool column that, as suggested by Cox,[162] may have removed an adherent suppressor cell population(s).

In a subsequent study, Alford and Goodwin[178] monitored LT responses in three patients with chronic pulmonary histoplasmosis and found that disease exacerbation occurred with depressed LT responses in two of the three patients, one of whom showed a generalized state of suppressed reactivity, as evidenced by reduced responses to *Candida albicans* extract and to phytohemagglutinin. A similar transient suppression of the cutaneous DTH response of patients with exacerbated disease has been noted.[207]

The *in vitro* LT assay has been employed in studies to evaluate immune responsiveness during the course of disease in experimentally infected or immunized animals.[157,173,200,204,222,246,281] As with human histoplasmosis, immune animals and those with self-limited infection manifest strong LT responses to *H. capsulatum* antigens and, conversely, animals with progressive, disseminated disease are markedly suppressed in their response.

Watson *et al.* have reported that the depression of cellular immune responses in mice with experimental histoplasmosis is correlated with a diminution of interleukin 1 (IL-1) production by their spleen.[252] Similarly, Deepe and Bullock have shown that the impaired proliferative activity *in vitro* of peripheral blood mononuclear cells is associated with IL-1 secretion.[282]

7. PREVENTION AND THERAPY

7.1. Immunization

Experimental infections have been induced in a variety of laboratory animals, including monkeys,[14] chickens,[283–285] dogs,[286] frogs,[287] mice,[16,70,71,168,190,196,288–291] guinea pigs,[200,216] hamsters,[292–295] rats,[2,296] and rabbits.[297] Salvin conducted the first experiments on induction of acquired resistance in mice.[163–165,197] These early observations were confirmed and extended during the next two decades.[75,76,153,158,163,285,287,291,298–306]

The results from experiments on induced immunity to histoplasmosis have been summarized before.[75,76,205,260,285,307–309] Briefly, the salient features of these experiments are: 1) many different sorts of animals may serve as experimental models; 2) resistance can be induced by sublethal infection with either morphological form (yeast or mycelial) of the fungus or by vaccination with killed yeast cells or fractions thereof;[310–312] 3) sublethal infections are more immunogenic than nonliving cells or extracts; 4) multiplication of the challenge inoculum is suppressed in immunized animals but often is not eliminated altogether; 5) the immunogenic moieties occur predominantly in the cell walls of the fungus; 6) significant protection can be engendered by immunogens administered subcutaneously, intramuscularly, intravenously, and intraperitoneally, but not orally; 7) a certain amount of cross-protection has been observed in some systems, but a high degree of specificity accompanies some preparations, *e.g.*, ribosomes;[306,312] and 8) the protection engendered by immunization is T-cell dependent.[33a]

It has been estimated that 30 to 40 million persons within the United States have been infected with *H. capsulatum*,[313,314] and this number increases annually by 150,000 to 500,000.[314–316] Although most of these people deal effectively with primary infection, the potential progressive nature and serious prognostic outlook for chronic pulmonary or disseminated disease[317] warrants consideration of a vaccine for humans, in particular, those who, by reason of their occupations, are at risk of exposure to large inocula. In spite of this need, there appears to be no serious consideration given to development of a product for use in humans.[308]

7.2. Therapy

7.2.1. *Chemotherapy*

In its most common form, histoplasmosis is an acute, self-limited infection of the respiratory tract that does not customarily require therapy. Unusually severe cases may require a short course of therapy. In some, but not all, cases of chronic cavitary disease, therapy is recommended. Disseminated disease is ultimately fatal; with treatment mortality ranges from 10–30%.[318] Immune-mediated forms of disease do not customarily require antifungal therapy.[318]

From the time of its introduction until about 1980, amphotericin B (AMB) was the most effective antifungal agent available.[318,319] With the advent of the azoles alternative antifungals became available. Two of these, miconazole and ketoconazole (KTZ), have been used clinically.[318–320] Currently, KTZ has received the greatest amount of attention. Disseminated histoplasmosis is being reported with regularity among patients with AIDS. In such patients AMB remains the drug of choice.[318]

7.2.2. *Immunotherapy*

The depressed CMI in patients with chronic pulmonary or disseminated histoplasmosis and the ability to adoptively transfer immunity in experimental animal models provides a rationale for immunotherapeutic intervention. The efficacy of transfer factor, the dialyzable fraction of leukocytes from immune donors, in the transfer of DTH responses to specific antigens[321] prompted evaluation of its therapeutic effects in histoplasmosis.[322,323] The published results, though limited to two patients, indicated that T-lymphocyte reactivity was augmented during immunotherapy, but the effect was transient.

The availability of recombinant lymphokines offers a new and exciting approach to immunotherapeutic intervention (reviewed in references 324–326). Immunotherapy with recombinant interleukin-2 (IL-2), a lymphokine that promotes the expansion of antigen-reactive T cells[327,328] and B cells[329,330] and activates natural killer and cytotoxic T cells,[330–333] has been evaluated in experimental histoplasmosis.[247] Doses of 200 to 20,000 units, administered daily in mice infected with *H. capsulatum*, did not augment *in vivo* DTH responses or modify the course of disease. Recombinant IL-2 was effective, however, in augmenting *in vitro* T-cell responses of infected mice.[252,334] Recombinant IFN-γ augments the host response to some intracellular pathogens.[335,336] The therapeutic use of rMuIFN-γ in mice, however, did not alter the course of the infection.[73,185] Yet removal of endogenous IFN-γ by anti-IFN-γ antibody augments multiplication of *H. capsulatum* in tissues of infected mice. A possible explanation for this difference could be that the *in vivo* half-life of rMuIFN-γ is 10 min, whereas that of monoclonal antibodies against endogenously produced IFN-γ is 14 days.[185] Other cytokines have been shown to be important to the outcome of experimental histoplasmosis.[338]

ACKNOWLEDGMENTS. During the construction of this chapter the research activities of the authors were sponsored in part by Public Health Service Grants AI-22963 and AI-25134 from The National Institutes of Allergy and Infectious Diseases. We are grateful to those investigators who sent us reprints and preprints of their work, and we apologize to any whose work may not have come to our attention.

We thank Lois F. Howard for her help in preparation of the manuscript.

REFERENCES

1. Ajello, L., 1983, Histoplasmosis—a dual entity: Histoplasmosis capsulati and histoplasmosis duboisii, *L'Igiene Moderna* **79:**3–30.

2. Rippon, J. W., 1988, *Medical Mycology*, chapters 15 and 16, W. B. Saunders Company, Philadelphia.
3. Domer, J. E., and Moser, S. A., 1980, Histoplasmosis, a review, *Rev. Med. Vet. Mycol.* **15:** 159–164.
4. Goodwin, R. A., and DesPrez, R. M., 1978, Histoplasmosis. State of the art, *Am. Rev. Respir. Dis.* **117:**929–956.
5. Schwarz, J., 1981, *Histoplasmosis*, Praeger, New York.
6. Alsip, S. G., and Dismukes, W. E., 1986, Approach to the patient with suspected histoplasmosis, in: *Current Clinical Topics in Infectious Diseases* (J. S. Remington and M. N. Swantz, eds.), McGraw Hill, New York, pp. 254–296.
7. Saag, M. S., and Dismukes, W. E., 1988, Histoplasmosis, in: *Medicine for the Practicing Physician* (J. W. Hurst, ed.), Butterworths, Stoneham, MA, pp. 370–373.
8. Threlkeld, M. G., and Dismukes, W. E., 1989, Endemic mycoses, in: *Opportunistic Infections in Patients with the Acquired Immunodeficiency Syndrome* (G. L. Leong and J. Mills, eds.), Marcel Dekker, Inc., New York, pp. 285–313.
9. Schwarz, J., 1971, Histoplasmosis, in: *Human Infections with Fungi, Actinomycetes, and Algae*, (R. D. Baker, ed.), Springer–Verlag, New York, pp. 67–130.
10. Schwarz, J., 1960, The primary lesion in histoplasmosis, in: *Histoplasmosis* (H. C. Sweany, ed.), C. C. Thomas, Springfield, pp. 292–317.
10a. Paya, C. V., Roberts, G. D., and Cockerill III, F. R., 1987, Transient fungemia in acute pulmonary histoplasmosis detection by a new blood-culturing technique, *J. Infect. Dis.* **156:** 313–315.
11. Goodwin, R. A., Nickell, J. A., and DesPrez, R. M., 1972, Mediastinal fibrosis complicating healed primary histoplasmosis and tuberculosis, *Medicine* **51:**227–246.
12. Goodwin, R. A., and Snell, J. D., 1969, The enlarging histoplasmoma. Concept of a tumor-like phenomenon incompassing the tuberculoma and coccidioidoma, *Am. Rev. Respir. Dis.* **100:**1–18.
13. Smith, R. E., and Ganley, J. P., 1971, An epidemic of presumed ocular histoplasmosis, *Trans. Am. Acad. Ophthalmol. Otol.* **75:**994–1005.
14. DeMonbruen, W. A., 1934, The cultivation and characteristics of Darling's *Histoplasma capsulatum*, *Am. J. Trop. Med.* **14:**93–135.
15. Howard, D. H., 1984, The epidemiology and ecology of blastomycosis, coccidioidomycosis, and histoplasmosis, *Zbl. Bakt. Hyg.* **A257:**219–227.
16. Hansmann, G. H., and Schenken, J. R., 1934, A unique infection in man caused by a new yeast-like organism, a pathogenic member of the genus *Sepedonium*, *Am. J. Pathol.* **10:** 731–738.
17. Howell, A., 1939, Studies on *Histoplasma capsulatum* and similar form species. I. Morphology and development, *Mycologia* **31:**191–216.
18. Pine, L., 1960, Morphological and physiological characteristics of *Histoplasma capsulatum*, in: *Histoplasmosis* (H. C. Sweaney, ed.), C. C. Thomas, Springfield, pp. 40–75.
19. Pine, L., 1962, Nutritional determinants of fungus morphology, in: *Fungi and Fungous Diseases*, (G. Dalldorf, ed.), C. C. Thomas, Springfield, pp. 84–101.
20. Salvin, S. B., 1947, Cultural studies on the yeast-like phase of *Histoplasma capsulatum* Darling, *J. Bacteriol.* **54:**655–660.
21. Yen, C. M., and Howard, D. H., 1970, Germination of blastospores of *Histoplasma capsulatum*, *Sabouraudia* **8:**242–252.
21a. McGinnis, M. R., and Katz, B., 1979. *Ajellomyces* and its synonym *Emmonsiella*, *Mycotaxon* **8**:157–164.
22. Berliner, M. D., 1971, Biological implications of morphological variants in *Histoplasma capsulatum* primary isolates, in: *Histoplasmosis: Proceedings of the Second National Conference* (L. Ajello, E. W. Chick, and M. L. Furcolow, eds.), C.C. Thomas Publishers, Springfield, IL.
23. Daniels, L. S., Berliner, M. D., and Campbell, C. C., 1968, Varying virulence in rabbits infected with different filamentous types of *Histoplasma capsulatum*, *J. Bacteriol.* **96:**1535–1539.

24. Tewari, R. P., and Berkhout, F. J., 1972, Comparative pathogenicity of albino and brown types of *Histoplasma capsulatum* for mice, *J. Infect. Dis.* **125**:504–508.
25. Morris, P. R., Tecreni, A. A., and DiSalvo, A. F., 1986, Red pigmented *Histoplasma capsulatum*—an unusual variant, *J. Med. Vet. Mycol.* **24**:231–233.
26. Garrison, R. G., 1983, Ultrastructural cytology of pathogenic fungi, in: *Fungi Pathogenic for Humans and Animals*, Part A (D. H. Howard, ed.), Marcel Dekker, Inc., New York, pp. 229–321.
27. Garrison, R. G., 1985, Cytological and ultrastructural aspects of dimorphism, in: *Fungal Dimorphism* (P. J. Szaniszlo, ed.), Plenum Press, New York, pp. 15–47.
28. Maresca, B., and Kobayashi, G. S., 1989, Dimorphism in *Histoplasma capsulatum*: A model for the study of cell differentiation in pathogenic fungi, *Microbiol. Rev.* **53**:186–209.
29. Keath, E. J., Painter, A. A., Kobayashi, G. S., and Medoff, G., 1989, Variable expression of a yeast-phase-specific gene in *Histoplasma capsulatum* strains differing in thermotolerance and virulence, *Infect. Immun.* **57**:1384–1390.
30. Keath, E. J., Spitzer, E. D., Painter, A. A., Travis, S. J., Kobayashi, G. S., and Medoff, G., 1989, DNA probe for the identification of *Histoplasma capsulatum*, *J. Clin. Microbiol.* **27**: 2369–2372.
31. Medoff, G., Kobayashi, G. S., Painter, A., and Travis, S., 1987, Morphogenesis and pathogenicity of *Histoplasma capsulatum*, *Infect. Immun.* **55**:1355–1358.
32. Caruso, M., Sacco, M, Medoff, G., and Maresca, B., 1987, Heat shock 70 gene is differentially expressed in *Histoplasma capsulatum* strains with different levels of thermotolerance and pathogenicity, *Mol. Microbiol.* **1**:151–158.
33. Howard, D. H., Zeuthen, M. L., and Dabrowa, N., 1991, The stress in response in *Candida albicans*, in: *Cellular and Molecular Biology of Candida albicans*, (C. R. Prasad, ed.), Springer–Verlag, Heidelberg, pp. 62–71.
33a. Gomez, F. J., Gomez, A. M., and Deepe, G. S., Jr., 1992, An 80-kilodalton antigen from *Histoplasma capsulatum* that has homology to heat shock protein 70 induces cell-mediated immune responses and protection of mice. *Infect. Immun.* **60**:2565–2571.
34. Zeuthen, M. L., and Howard, D. H., 1989, Thermotolerance and the heat-shock response in *Candida albicans*, *J. Gen. Microbiol.* **135**:2509–2518.
35. Born, W., Happ, M. P., Dallas, A., Reardon, C., Kubo, R., Shinnick, T., Brennan, P., and O'Brien, R., 1990, Recognition of heat shock proteins and 28 cell function, *Immunol. Today* **11**:40–43.
36. Kaufman, L., and Blumer, S., 1966, Occurrence of serotypes among *Histoplasma capsulatum* strains, *J. Bacteriol.* **91**:1434–1439.
37. Pine, L., and Boone, C. J., 1968, Cell wall composition and serological reactivity of *Histoplasma capsulatum* serotypes and related species, *J. Bacteriol.* **96**:789–798.
38. Domer, J. E., Hamilton, J. G., and Harkin, J. C., 1967, Comparative study of the cell walls of the yeast-like and mycelial phases of *Histoplasma capsulatum*, *J. Bacteriol.* **94**:466–474.
39. Domer, J. E., 1971, Monosaccharide and chitin content of cell walls of *Histoplasma capsulatum* and *Blastomyces dermatitidis*, *J. Bacteriol.* **107**:870–877.
40. San-Blas, F. San-Blas, G., and Cova, L. J., 1976, A morphological mutant of *Paracoccidioides brasiliensis* strain IVIC PG9. Isolation and cell wall characterization, *J. Gen. Microbiol.* **93**: 209–218.
41. San-Blas, G., San-Blas, F., and Serrano, L. E., 1977, Host–parasite relationships in the yeastlike form of *Paracoccidioides brasiliensis* strain IVIC PG9, *Infect. Immun.* **15**:343–346.
42. Klimpel, K. R., and Goldman, W. E., 1987, Isolation and characterization of spontaneous avirulent variants of *Histoplasma capsulatum*, *Infect. Immun.* **55**:528–533.
43. Klimpel, K. R., and Goldman, W. E., 1988, Cell walls from avirulent variants of *Histoplasma capsulatum* lack α-(1,3)-glucan, *Infect. Immun.* **56**:2997–3000.
44. Vincent, R. D., Goewert, R., Goldman, W. E., Kobayashi, G. S., and Lambowitz, A. M., 1986, Classification of *Histoplasma capsulatum* isolates by restriction fragment polymorphisms, *J. Bacteriol.* **165**:813–818.

44a. Spitzer, E. D., Keath, E. J., Travis, S. J., Painter, A. A., Kobayashi, G. S., and Medoff, G., 1990, Temperature-sensitive variants of *Histoplasma capsulatum* isolated from patients with acquired immunodeficiency syndrome, *J. Inf. Dis.* **162**:258–261.
45. Spitzer, E. D., Kasker, B. A., Travis, S. J., Kobayashi, G. S., and Medoff, G., 1989, Use of mitochondrial and ribosomal DNA polymorphisms to classify clinical and soil isolates of *Histoplasma capsulatum, Infect. Immun.* **57**:1409–1412.
46. Steele, P. E., Carle, G. F., Kobayashi, G. S., and Medoff, G., 1989, Electrophoretic analysis of *Histoplasma capsulatum* chromosomal DNA, *Mol. Cell. Biol.* **9**:983–987.
47. Magee, B. B., Koltin, Y., Gorman, J. A., and Magee, P. T., 1988, Assignment of cloned genes to the seven electrophoretically separated *Candida albicans* chromosomes, *Mol. Cell. Biol.* **8**: 4721–4726.
48. Rustschenko-Bulgac, E., Sherman, F., and Hicks, J. B., 1990 Chromosomal rearrangements associated with morphological mutants provide a means for genetic variation of *Candida albicans, J. Bacteriol.* **172**:1276–1283.
49. Perfect, J. R., Magee, B. B., and Magee, P. T., 1989, Separation of chromosomes of *Cryptococcus neoformans* by pulsed field gel electrophoresis, *Infect. Immun.* **57**:2624–2627.
50. Kwon-Chung, K. J., 1972, *Emmonsiella capsulata*: Perfect state of *Histoplasma capsulatum, Science* **177**:368–369.
51. Worsham, P. L., and Goldman, W. E., 1988, Selection and characterization of *ura 5* mutants of *Histoplasma capsulatum, Mol. Gen. Genet.* **214**:348–352.
52. Worsham, P. L., and Goldman, W. E., 1990, Development of a genetic transformation system for *Histoplasma capsulatum*: Complementation of uracil auxotrophy, *Mol. Gen. Genet.* **221**:358–362.
53. Darling, S. T., 1906, A protozoan general infection producing pseudotubercules in the lungs and focal necrosis of the liver, spleen and lymph nodes, *J.A.M.A.* **46**:1283–1285.
54. Reddy, P., Gorelick, D. F., Brasher, C. A., and Larsh, H., 1970, Progressive disseminated histoplasmosis as seen in adults, *Am. J. Med.* **48**:629–636.
55. Smith, J. W., and Utz, J. P., 1972, Progressive disseminated histoplasmosis. A prospective study of 26 patients, *Ann. Intern. Med.* **76**:557–565.
56. Davies, S. F., Khan, M., and Sarosi, G. A., 1978, Disseminated histoplasmosis in immunologically suppressed patients. Occurrence in a nonendemic area, *Am. J. Med.* **64**:94–100.
57. Bryan, C. S., and DiSalvo, A. F., 1979, Overwhelming opportunistic histoplasmosis, *Sabouraudia* **17**:209–212.
58. Kauffman, C. A., Israel, K. S, Smith, J. W., White, A. C., Schwarz, J., and Brooks, G. F., 1978, Histoplasmosis in immunosuppressed patients, *Am. J. Med.* **64**:923–932.
59. Wheat, L. J., Slama, T. G., Norton, J. A., Kohler, R. B., Eitzen, H. E., French, M. L. V., and Sathapatayavongs, B., 1982, Risk factors for disseminated or fatal histoplasmosis, *Ann. Intern. Med.* **96**:159–163.
60. Wheat, L. J., Smith, E. J., Sathapatayavongs, B., Batteiger, B., Filo, R. S., Leapman, S. B., and French, M. V., 1983, Histoplasmosis in renal allograft recipients, two large urban outbreaks, *Arch. Intern. Med.* **143**:703–707.
61. Jones, P. G., Cohen, R. L., Batts, D. H., and Silva, J., Jr., 1983, Disseminated histoplasmosis, invasive pulmonary aspergillosis, and other opportunistic infections in a homosexual patient with acquired immune deficiency syndrome, *Sex. Transm. Dis.* **10**:202–204.
62. Pasternak, J., and Bolivar, R., 1983, Histoplasmosis in acquired immunodeficiency syndrome—diagnosis by bone marrow examination, *Arch. Intern. Med.* **143**:1495.
63. Wheat, L. J., and Small, C. B., 1984, Disseminated histoplasmosis in the acquired immune deficiency syndrome, *Arch. Intern. Med.* **144**:2147–2149.
64. Bonner, J. R., Alexander, J., Dismukes, W. E., App, W., Griffin, F. M., Little, R., and Shin, M. S., 1984, Disseminated histoplasmosis in patients with the acquired immune deficiency syndrome, *Arch. Intern. Med.* **144**:2178–2181.
65. Taylor, M. N., Baddour, L. M., and Alexander, J. R., 1984, Disseminated histoplasmosis

associated with the acquired immune deficiency syndrome, *Am. J. Med.* **77**:579–580.
66. Hazelhurst, J. A., and Vismer, H. F., 1985, Histoplasmosis presenting with unusual skin lesions in acquired immunodeficiency syndrome (AIDS), *Br. J. Dermatol.* **113**:345–348.
67. Wheat, L. J., Slama, T. G., and Zeckel, M. L., 1985, Histoplasmosis in the acquired immune deficiency syndrome, *Am. J. Med.* **78**:203–210.
68. Huang, C. T., Mcgarry, T., Cooper, S., and Saunders, R., 1987, Disseminated histoplasmosis in the acquired immunodeficiency syndrome—report of 5 cases from a nonendemic area, *Arch. Intern. Med.* **147**:1181–1184.
69. Bonner, J. R., Alexander, J., Dismukes, W. E., App, W., Griffin, F. M., Little, R., and Shin, M. S., 1984, Disseminated histoplasmosis in patients with the acquired immune deficiency syndrome, *Arch. Intern. Med.* **144**:2178–2181.
70. Howell, A., Jr., and Kipkie, G. F., 1951, A comparison of the susceptibility by intracerebral inoculation of six strains of mice with male dba line 1 mice, *Am. J. Trop. Med.* **31**:33–41.
71. Chick, E. W., and Roberts, G. D., 1974, The varying susceptibility of different genetic strains of laboratory mice to *Histoplasma capsulatum*, *Mycopathol. Mycol. Appl.* **52**:251–253.
72. Patton, R. M., Riggs, A. R., Compton, S. B., and Chick, E. W., 1976, Histoplasmosis in purebred mice: Influence of genetic susceptibility and immune depression on treatment, *Mycopathologica* **60**:39–43.
73. Wu-Hsieh, B., 1989, Relative susceptibilities of inbred mouse strains C57BL/6 and A/J to infection with *Histoplasma capsulatum*, *Infect. Immun.* **57**:3788–3792.
74. Saslow, S., and Schaefer, J., 1955, Relation of sex and age to resistance of mice to experimental *Histoplasma* infection, *Proc. Soc. Exp. Biol. Med.* **90**:400–402.
75. Salvin, S. B., 1956, Acquired resistance in experimental histoplasmosis, *Trans. N.Y. Acad. Sci.*, Ser. II, 18, **5**:462–468.
76. Salvin, S. B., 1960, Resistance of animals and man to histoplasmosis, in: *Histoplasmosis* (H. C. Sweany, ed.), C. C. Thomas, Springfield, pp. 99–112.
77. Taylor, M. L., Reyes-Montes, M. R., Garcia-Gonzalez, R,. Casasola, J., and Hernandez-Ramirez, A., 1982, Immune response changes with age and sex as factors of variation in resistance to *Histoplasma* infection, in: *Proceedings VIII Congress ISHAM*, Massey University (M. Baxter, ed.), Palmerston North, New Zealand, pp. 260–264.
78. Wu-Hsieh, B., Howard, D. H., and Ahmed, R., 1988, Virus-induced immunosuppression: A murine model of susceptibility to opportunistic infection, *J. Infect. Dis.* **158**:232–235.
79. Moors, M. A., Jones, S. M., Klycyck, K. K., Rogers, T. J., Buckley, H. R., and Blank, K. J., 1990, Effect of Friend leukemia virus infection on susceptibility to *Candida albicans*, *Infect. Immun.* **58**:1796–1801.
80. Weinberg, E. D., 1978, Iron and infection, *Microbiol. Rev.* **42**:45–66.
81. Howard, D. H., 1981, Mechanisms of resistance in the systemic mycoses, in: *Immunology of Human Infection* (A. J. Nahmias and R. J. O'Reilly, eds.), Plenum Publishing Corporation, New York, pp. 294–475.
82. Garraway, M. O., and Evans, R. C., 1984, *Fungal Nutrition and Physiology*, John Wiley and Sons, New York, pp. 146–151.
83. Wu-Hsieh, B, and Howard, D. H., 1989, Histoplasmosis, chapter 7, in: *Immunology of Fungal Diseases* (R. A. Cox, ed.), CRC Press, Inc., Boca Raton, FL, pp. 199–225.
84. Silva, M., and Buckley, H. R., 1972, Activity of egg white against fungi pathogenic to man, in: *Fungi and Fungous Diseases* (G. Dalldorf, ed.), C. C. Thomas, Springfield, pp. 277–291.
85. Sutcliffe, M. C., Savage, A. M., and Alford, R. H., 1980, Transferrin-dependent growth inhibition of yeast-phase *Histoplasma capsulatum* by human serum and lymph, *J. Infect. Dis.* **142**:209–219.
86. Caldwell, C. W., and Sprouse, R. F., 1982, Iron and host resistance to histoplasmosis, *Am. Rev. Respir. Dis.* **125**:674–677.
87. Burt, W. R., 1982, Identification of coprogen B and its breakdown products from *Histoplasma capsulatum*, *Infect. Immun.* **35**:990–996.

88. Burt, W. R., Underwood, A. L., and Appleton, G. L., 1981, Hydroxamic acid from *Histoplasma capsulatum* that displays growth factor activity, *Appl. Environ. Microbiol.* **42:**560–563.
89. Burt, W. R., and Smith, R. A., 1983, Studies on experimental murine histoplasmosis: Host protection and cellular immunity, *Can. J. Microbiol.* **29:**102–107.
90. Bullen, J. J., Rogers, J. J., and Griffiths, E., 1974, Bacterial iron metabolism in infection and immunity, in: *Microbial Iron Metabolism* (J. B. Neilands, ed.), Academic Press, Inc., New York, pp. 517–551.
91. Hirata, T., Bitterman, P. B., Mornex, J., and Crystal, R. G., 1986, Expression of the transferrin receptor gene during the process of mononuclear phagocyte maturation, *J. Immunol.* **136:**1339–1345.
92. Andreesen, R., Osterholz, H., Bodemann, K. J., Brass, K. J., Costabel, U., and Lohr, G. W., 1984, Expression of transferrin receptors and intracellular ferritin during terminal differentiation of human monocytes, *Blut* **49:**195–202.
93. Byrd, T. F., and Horwitz, M. A., 1989, Interferon gamma-activated human monocytes downregulate tranferrin receptors and inhibit the intracellular multiplication of *Legionella pneumophila* by limiting the availability of iron, *J. Clin. Invest.* **83:**1457–1465.
93a. Lane, T. E., Wu-Hsieh, B. A., and Howard, D. H., 1991, Iron limitation and the gamma-interferon-mediated antihistoplasma state of murine macrophages. *Infect. Immun.* **59:** 2274–2278.
94. Newman, R., Schneider, C., Sutherland, R., Vodinelech, L., and Greaves, M., 1982, The transferrin receptor, *Trends Biochem. Sci.* **7:**397–400.
95. Jacob, A., 1977, Low molecular weight intracellular iron transport compounds, *Blood* **50:** 433–439.
96. Dautry-Varsat, A., Ciechanover, A., and Lodish, H. F., 1983, pH and recycling of transferrin during receptor mediated endocytosis, *Proc. Natl. Acad. Sci. USA* **80:**2258–2262.
97. Kirkpatrick, C. H., Green, I., and Rich, R. R., 1971, Inhibition of growth of *Candida albicans* by iron-unsaturated lactoferrin: Relation to host-defense mechanisms in chronic mucocutaneous candidiasis, *J. Infect. Dis.* **124:**539–544.
98. Van Snick, J. L., Masson, P. L., and Heremans, J. F., 1974, The involvement of lactoferrin in the hyposideremia of acute inflammation, *J. Exp. Med.* **140:**1068–1084.
99. Procknow, J. J, Page, M. I., and Loosli, C. G., 1960, Early pathogenesis of experimental histoplasmosis, *Arch. Pathol.* **69:**413–426.
100. Kaplan, W., Kaufman, L., and McClure, H. M., 1972, Pathogenesis and immunological aspects of experimental histoplasmosis in cynomologus monkeys (*Macaca fascicularis*), *Infect. Immun.* **5:**847–853.
101. Schlitzer, R. L., Chandler, F. W., and Larsh, H. W., 1981, Primary acute histoplasmosis in guinea pigs exposed to aerosolized *Histoplasma capsulatum*, *Infect. Immun.* **33:**575–582.
102. Sumner, R. M., Bergholm, E., and Nickerson, D. A., 1985, Characterization of cellular infiltrates during murine pulmonary histoplasmosis, *Abstr. Annu. Meet. ASM*, F19, p. 367.
103. Baughman, R. P., Kimm, C. K., Vinegar, A., Hendricks, D. E., Schmidt, D. J., and Bullock, W. E., 1986, The pathogenesis of experimental histoplasmosis. Correlative studies of histopathology, bronchoalveolar lavage, and respiratory function, *Am. Rev. Respir. Dis.* **134:** 771–776.
104. Holland, P., 1971, Circulating human phagocytes and *Histoplasma capsulatum*, in: *Histoplasmosis, Proceedings of the 2nd National Conference* (L. Ajello, E. W. Chick, and M. L. Furcolow, eds.), C. C. Thomas, Springfield, pp. 380–383.
105. Howard, D. H., 1973, Fate of *Histoplasma capsulatum* in guinea pig polymorphonuclear leukocytes, *Infect. Immun.* **8:**412–419.
106. Howard, D. H., unpublished data, 1985.
107. Schaffner, A., Davis, C. E., Schaffner, T., Markert, M., Douglas, H., and Braude, A. I., 1986, *In vitro* susceptibility of fungi to killing by neutrophil granulocytes discriminates between primary pathogenicity and opportunism, *J. Clin. Invest.* **78:**511–524.

107a. Schnur, R. A., and Newman, S. L., 1990, The respiratory burst response to *Histoplasma capsulatum* by human neutrophils. Evidence for intracellular trapping of superoxide anion, *J. Immunol.* **144**:4765–4772.
108. Howard, D. H., 1980, How phagocytes kill fungi, *Dermatologia Venezolana* **18**:127–142.
109. Fleischmann, J., and Lehrer, R. I., 1985, Phagocytic mechanisms in host response, in: *Fungi Pathogenic for Humans and Animals*, Part B2 (D. H. Howard, ed.), Marcel Dekker, Inc., New York, pp. 123–149.
110. Howard, D. H., 1981, Comparative sensitivity of *Histoplasma capsulatum* conidiospores and blastospores to oxidative antifungal systems, *Infect. Immun.* **32**:381–387.
111. Howard, D. H., 1983, Studies on the catalase of *Histoplasma capsulatum*, *Infect. Immun.* **39**: 1161–1166.
112. Selsted, M. E., Szklarek, D., and Lehrer, R. I., 1984, Purification of antibacterial activity of antibacterial peptides of rabbit granulocytes, *Infect. Immun.* **45**:150–154.
113. Newman, S. L., Bucher, C., Rhodes, J., and Bullock, W. E., 1990, Phagocytosis of *Histoplasma capsulatum* yeasts and microconidia by human cultured macrophages and alveolar macrophages, *J. Clin. Invest.* **85**:223–230.
114. Sanchez, S., and Carbonell, L. M., 1975, Immunological studies on *Histoplasma capsulatum*, *Infect. Immun.* **11**:387–394.
115. Kimberlin, C. L., Hariri, A. R., Hempel, H. O., and Goodman, N. L., 1981, Interaction between *Histoplasma capsulatum* and macrophages from normal and treated mice: Comparison of the mycelial and yeast phases in alveolar and peritoneal macrophages, *Infect. Immun.* **34**:6–10.
116. Howard, D. H., 1959, Observation on tissue cultures of mouse peritoneal exudates inoculated with *Histoplasma capsulatum*, *J. Bacteriol.* **78**:69–78.
117. Howard, D. H., 1960, Effect of mycostatin and fungizone on the growth of *Histoplasma capsulatum* in tissue culture, *J. Bacteriol.* **79**:442–449.
118. Howard, D. H., 1964, Intracellular behavior of *Histoplasma capsulatum*, *J. Bacteriol.* **87**:33–38.
119. Howard, D. H., 1965, Intracellular growth of *Histoplasma capsulatum*, *J. Bacteriol.* **89**: 518–523.
120. Howard, D. H., 1973, Further studies on the inhibition of *Histoplasma capsulatum* within macrophages from immunized animals, *Infect. Immun.* **8**:577–581.
121. Newman, S., Bucher, C. ,and Gootee, L., 1990, Intracellular growth of *Histoplasma capsulatum* yeasts in human monocytes, cultured macrophages, and alveolar macrophages, *Abstr. Annu. Meet. ASM*, F98, p. 425.
121a. Szaniszlo, P., Jacob, C. W., and Geis, P. A., 1983, Dimorphism: morphological and biochemical aspects, in: *Fungi Pathogenic for Humans and Animals*, Part A (D. H. Howard, ed.), Marcel Dekker, Inc., New York, pp. 323–436.
122. Eissenberg, L. G., Poirier, S., and Goldman, W. E., 1990, Persistence of avirulent *Histoplasma capsulatum* yeasts in an altered morphology within P388D$_1$ macrophages, *Abstr. Annu. Meet. ASM*, F95, p. 424.
123. Wu-Hsieh, B., and Howard, D. H., 1989, Macrophage cell lines P388D$_1$ and IC-21 stimulated with gamma interferon fail to inhibit the intracellular growth of *Histoplasma capsulatum*, *Infect. Immun.* **57**:2903–2905.
124. Gatchel, S. L., Shigematsu, M. L., Maciag, T., and Tewari, R. P., 1986, Interactions of *Histoplasma capsulatum* with human endothelial cells in culture, *Abstr. Annu. Meet. ASM*, F16, p. 400.
125. Eissenberg, L. G., Klimpel, K. R., and Goldman, W. E., 1989, Invasion of respiratory epithelial cells by *Histoplasma capsulatum*: An alternative strategy for intracellular parasitism, *Abstr. Annu. Meet. ASM*, F122, p. 478.
126. Bullock, W. E., and Wright, S. D., 1987, Role of the adherence-promoting receptors, CR3, LFA-1, and p150,95, in binding of *Histoplasma capsulatum* by human macrophages, *J. Exp. Med.* **165**:195–210.

127. Moulder, J. W., 1985, Comparative biology of intracellular parasitism, *Microbiol. Rev.* **49**:298.
128. Dumont, A., and Robert, A., 1970, Electron microscopic study of phagocytosis of *Histoplasma capsulatum* by hamster peritoneal macrophages, *Lab. Invest.* **23**:278–286.
129. Dumont, A., and Piche, C., 1969, Electron microscopic study of human histoplasmosis, *Arch. Pathol.* **87**:168–178.
129a. Dumont, A., 1972, Ultrastructural aspects of phagocytosis of facultative intracellular parasites by hamster peritoneal macrophages, *J. Reticuloendothel. Soc.* **11**:469.
130. Von Behren, L. A., Chaudhary, S., Rabinovich, S, and Tewari, R. P., 1983, Electron microscopic study of phagosome–lysosome (P–L) fusion in normal mouse peritoneal macrophages after exposure to yeast cells of *Histoplasma capsulatum*, *Abstr. Annu. Meet. ASM*, F34, p. 388.
131. Edwards, M. R., 1982, Intracellular fate of *Histoplasma capsulatum* phagocytosed by murine peritoneal macrophages, *Fed. Proc.* **41**:963.
132. Eissenberg, L. G., Schlesinger, P. H., and Goldman, W. E., 1988, Phagosome-lysosome fusion in P388D$_1$ macrophages infected with *Histoplasma capsulatum*, *J. Leuk. Biol.* **43**: 483–491.
132a. Eissenberg, L. G., Goldman, B., and Schlessinger, P. H., 1992, Phagolysosomes containing *Histoplasma capsulatum* fail to acidify normally, *Abstr. Annu. Meet. ASM*, F-75, p. 511.
133. Anderson, S. E., and Remington, J. S., 1974, Effect of normal and activated human macrophages on *Toxoplasma gondii*, *J. Exp. Med.* **139**:1154–1174.
134. Wilson, C. B., Tsai, V., and Remington, J. S., 1980, Failure to trigger the oxidative burst by normal macrophages. Possible mechanism for survival of intracellular pathogens, *J. Exp. Med.* **151**:328–346.
135. Holzer, T. J., Nelson, K. E., Schauf, V., Crispen, R. G., and Anderson, B. R., 1986, *Mycobacterium leprae* fails to stimulate phagocytic cell superoxide anion generation, *Infect. Immun.* **51**:514–520.
136. Eissenberg, L. G., and Goldman, W. E., 1987, *Histoplasma capsulatum* fails to trigger release of superoxide from macrophages, *Infect. Immun.* **55**:29–34.
137. Wolf, J. E., Kerchberger, V., Kobayashi, G. S., and Little, J. R., 1987, Modulation of the macrophage oxidative burst by *Histoplasma capsulatum*, *J. Immunol.* **138**:582–558.
138. Wolf, J. E., Abegg, A. L., Travis, S. J., Kobayashi, G. S., and Little, J. R., 1989, Effects of *Histoplasma capsulatum* on murine macrophage functions: Inhibition of macrophage priming, oxidative burst, and antifungal activities, *Infect. Immun.* **57**:513–519.
138a. Wolf, J. E., Massof, S. E., Sherwin, J. R., and Considine, R. V., 1992, Inhibition of murine macrophage protein kinase C activity by nonviable *Histoplasma capsulatum*, *Infect. Immun.* **60**:2683–2687.
139. Calderone, R. A., and Peterson, E., 1979, Inhibition of amino acid uptake and incorporation into *Histoplasma capsulatum* by a lysosomal extract from rabbit alveolar macrophages, *J. Reticuloendothel. Soc.* **26**:11–19.
140. Tewari, R. P., Mkwanazi, J. B., McConnachie, P., Von Behren, L. A., Eagleton, L., Kulkarni, P., and Bartlett, P. C., 1981, Natural and antibody-dependent cellular cytotoxicity (ADCC) of human peripheral blood mononuclear cells (PBMC) to yeast cells of *Histoplasma capsulatum* (HC), *Abstr. Annu. Meet. ASM*, F66, p. 324.
141. Yamada, T., Khardori, N., and Tewari, R. P., 1982, Natural and antibody-dependent cellular cytotoxicity (ADCC) of macrophages and lymphocytes from normal and immune mice for yeast cells of *Histoplasma capsulatum* (HC), *Abstr. Annu. Meet. ASM*, F9, p. 328.
142. Roder, J. C., 1979, The beige mutation in the mouse. I. A stem cell predetermined impairment of natural killer cell function, *J. Immunol.* **123**:2174–2181.
143. Patino, M. M., Williams, D., Ahrens, J., and Graybill, J. R., 1987, Experimental histoplasmosis in the beige mouse, *J. Leukocyte Biol.* **41**:228–235.
143a. Newman, S., and Gootee, L., 1992, Fungistatic activity of human neutrophiles against *H.*

capsulatum yeasts is mediated by an oxygen-independent mechanism(s), *Abstr. Annu. Meet. ASM*, F-40, p. 505.

144. Miya, F., and Marcus, S., 1961, Effect of humoral factors on *in vitro* phagocytic and cytopeptic activities of normal and "immune" phagocytes, *J. Immunol.* **86**:652–668.
145. Ratnoff, W. D., Pepple, J. M., and Winkelstein, J. A., 1980, Activation of the alternate complement pathway by *Histoplasma capsulatum*, *Infect. Immun.* **30**:147–149.
145a. Wu, W. G., and Marcus, S., 1964, Humoral factors in cellular resistance. II. The role of complement and properdin in phagocytosis and cytopepsis by normal and "immune" macrophages, *J. Immunol.* **92**:397–443.
146. Schwarz, J., and Baum, G. L., 1963, Reinfection in histoplasmosis, *Arch. Pathol.* **75**:475–479.
147. Chandler, J. W., Jr., Smith, T. K., Newberry, W. M., Jr., Chin, T. D. Y., and Kirkpatrick, C. H., 1969, Immunology of the mycoses. II. Characterization of the immunoglobulin and antibody responses in histoplasmosis, *J. Infect. Dis.* **119**:247–254.
148. Cox, R. A., and Arnold, D. R., 1980, Immunoglobulin E in histoplasmosis, *Infect. Immun.* **29**:290–293.
149. Caldwell, C. W., Sprouse, R. F., and Everett, E. D., 1982, Antigen-specific immunoglobulins in patients with acute pulmonary, chronic pulmonary, or disseminated histoplasmosis, *Diagn. Microbiol. Infect. Dis.* **1**:49–57.
150. Ovary, Z., Itaya, T., Santomauro, J., and Richmond, G., 1978, On the stability of IgE antibodies, *J. Immunol. Methods* **24**:193–194.
151. Bullock, W. E., Artz, R. P., Bhathena, D., and Tung, K. S. K., 1979, Histoplasmosis—association with circulating immune complexes, eosinophilia, and mesangiopathic glomerulonephritis, *Arch. Intern. Med.* **139**:700–702.
152. Cox, R. A., Tewari, R. P., and Wheat, L. J., unpublished data, 1986 (see reference 83).
153. Hill, G. A., and Marcus, S., 1960, Study of cellular mechanisms in resistance to systemic *Histoplasma capsulatum* infection, *J. Immunol.* **85**:6–13.
154. Tewari, R. P., Sharma, D., Solotorovsky, M., Lafemina, R., and Balint, J., 1977, Adoptive transfer of immunity from mice immunized with ribosomes or live cells of *Histoplasma capsulatum*, *Infect. Immun.* **15**:789–795.
155. Tewari, R. P., Sharma, D. K., and Mathur, A., 1978, Significance of thymus-derived lymphocytes in immunity elicited by immunization with ribosomes or live yeast cells of *Histoplasma capsulatum*, *J. Infect. Dis.* **138**:605–613.
156. Williams, D. M., Graybill, J. R., and Drutz, D. J., 1981, Adoptive transfer of immunity to *Histoplasma capsulatum* in athymic nude mice, *Sabouraudia* **19**:39–48.
157. Khardori, N., Chaudhary, S., McConnachie, P., and Tewari, R. P., 1983, Characterization of lymphocytes responsible for protective immunity to histoplasmosis in mice, *Mykosen* **26**: 523–532.
158. Rowley, D. A., and Huber, M., 1956, Growth of *Histoplasma capsulatum* in normal, superinfected and immunized mice, *J. Immunol.* **77**:15–23.
159. Cox, R. A., 1979, Immunologic studies of patients with histoplasmosis, *Am. Rev. Respir. Dis.* **120**:143–149.
160. Taylor, M. L., Eslava-Campos, C., Garcia-Gonzales, R., and Hernandez-Ramirez, A., 1980, Relation between cellular and humoral immune response in mice immunized with *Histoplasma capsulatum*, in: *Human and Animal Mycology* (E. S. Kuttin and G. L. Baum, eds.), *Excerpta Medica*, Amsterdam, pp. 97–101.
161. Taylor, M. L., Diaz, S., Gonzalez, P. A., Sosa, A. C., and Toriello, C., 1984, Relationship between pathogenesis and immune regulation mechanisms in histoplasmosis: A hypothetical approach, *Rev. Infect. Dis.* **6**:775–782.
162. Cox, R. A., 1983, Cell-mediated immunity, in: *Fungi Pathogenic for Humans and Animals*, Part B1 (D. H. Howard, ed.), Marcel Dekker, Inc., New York, chapter 2.
163. Salvin, S. B., 1953, Immunization of mice against *Histoplasma capsulatum*, *J. Immunol.* **70**: 267–270.

164. Salvin, S. B., 1954, Cultural and serologic studies on nonfatal histoplasmosis in mice, hamsters, and guinea pigs, *J. Infect. Dis.* **94**:22–29.
165. Salvin, S. B., 1955, Further studies on immunization of mice against *Histoplasma capsulatum*, *Am. J. Hyg.* **61**:72–81.
166. Salvin, S. B., 1972, Immunity in the systemic mycoses, in: *Immunogenicity* (F. Borek, ed.), North Holland Publishing Co., Amsterdam, pp. 225–259.
167. Louria, D. B., Feder, N., Mitchell, W., and Emmons, C. W., 1959, Influence of fungus strain and lapse of time in experimental histoplasmosis and of volume of inoculum in cryptococcosis upon recovery of the fungi, *J. Lab. Clin. Med.* **53**:311–317.
168. Berry, C. L., 1969, The production of disseminated histoplasmosis in the mouse: The effects of changes in reticulo-endothelial function, *J. Pathol.* **97**:441–457.
169. Von Behren, L. A., Chaudhary, S., Khardori, N., Rabinovich, S., Shu, M.-D., and Tewari, R. P., 1983, Effect of silica on the susceptibility of mice to experimental histoplasmosis, *J. Reticuloendothel. Soc.* **34**:99–111.
170. Von Behren, L. A., Chaudhary, S., Rabinovich, S., Shu, M-D., and Tewari, R. P., 1983, Protective effect of poly-2-vinylpyridine-N-oxide on susceptibility of silica-treated mice to experimental histoplasmosis, *Infect. Immun.* **42**:818–823.
171. Edwards, L. B., Acquavira, F. A., Livesay, V. T., Cross, F. W., and Palmer, C. E., 1969, An atlas of sensitivity to tuberculin, PPD-B, and histoplasmin in the United States, *Am. Rev. Respir. Dis.* **99**:1–132.
172. Sathapatayavongs, B., Batteiger, B. E., Wheat, J. L., Slama, T. G., and Wass, J. L., 1983, Clinical and laboratory features of disseminated histoplasmosis during two large urban outbreaks, *Medicine* **62**:263–270.
173. Khardori, N., Von Behren, L., Chaudhary, S., McConnachie, P., Strano, A., and Tewari, R. P., 1986, Cellular mediators of anti-*Histoplasma* immunity. I. Protective immunity and cellular changes in spleens of mice immunized by sublethal infection with yeast cells of *Histoplasma capsulatum*, *Mykosen* **29**:103–115.
174. Sharma, D., Khardori, N., Chaudhary, S., Von Behren, L., McConnachie, P., and Tewari, R. P., 1986, Cellular mediators of anti-*Histoplasma* immunity. II. Protective immunity and delayed hypersensitivity in mice immunized by sublethal infection with yeast cells of *Histoplasma capsulatum*, *Mykosen* **29**:116–126.
175. Newberry, W. M., Chandler, J. W., Chin, T. D. Y., and Kirkpatrick, C. H., 1968, Immunobiology of the mycosis. I. Depressed lymphocyte transformation in chronic histoplasmosis, *J. Immunol.* **100**:436–444.
176. Kirkpatrick, C. H., Chandler, J. W., Jr., Smith, T. K., and Newberry, W. M., Jr., 1971, Cellular immunologic studies in histoplasmosis, in: *Histoplasmosis* (L. Ajello, E. W. Chick, and M. L. Furcolow, eds.), C. C. Thomas, Springfield, pp. 371–379.
177. Alford, R. H., and Goodwin, R. A., 1972, Patterns of immune responses in chronic pulmonary histoplasmosis, *J. Infect. Dis.* **125**:269–275.
178. Alford, R. H., and Goodwin, R. A., 1973, Variation in lymphocyte reactivity to histoplasmin during the course of chronic pulmonary histoplasmosis, *Am. Rev. Respir. Dis.* **198**:85–92.
179. Cox, R. A., 1979, Cross-reactions between antigens of *Coccidioides immitis*, *Histoplasma capsulatum*, and *Blastomyces dermatitidis* in lymphocyte transformation assays, *Infect. Immun.* **25**:932–938.
180. Alford, R. H., and Cartwright, B. B., 1981, Human lymphocyte blastogenesis induced by living and dead *Histoplasma capsulatum* yeasts and soluble yeast autolysates, *Sabouraudia* **19**: 85–95.
181. Alford, R. H., and Cartwright, B. B., 1983, Comparison of ketoconazole and amphotericin B in interference with thymidine uptake by and blastogenesis of lymphocytes stimulated with *Histoplasma capsulatum* antigens, *Antimicrob. Agents Chemother.* **24**:575–578.
182. Wu-Hsieh, B., and Howard, D. H., 1984, Inhibition of growth of *Histoplasma capsulatum* by lymphokine-stimulated macrophages, *J. Immunol.* **132**:2593–2597.

183. Wu-Hsieh, B., Zlotnik, A., and Howard, D. H., 1984, T-cell hybridoma-produced lymphokine that activates macrophages to suppress intracellular growth of *Histoplasma capsulatum*, *Infect. Immun.* **43**:380–385.
184. Wu-Hsieh, B., and Howard, D. H., 1987, Inhibition of the intracellular growth of *Histoplasma capsulatum* by recombinant murine gamma interferon, *Infect. Immun.* **55**:1014–1016.
185. Wu-Hsieh, B. A., and Howard, D. H., 1990, Gamma interferon and experimental murine histoplasmosis, in: *Microbial Determinants of Virulence and Host Response* (E. M. Ayoub, G. H. Cassell, W. C. Branch, Jr., and T. J. Henry, eds.), American Society for Microbiology, Washington, D.C., pp. 133–146.
186. Baum, G. L., Adriano, S. M., and Schwarz, J., 1954, Effect of cortisone on experimental histoplasmosis in mice, *Am. J. Clin. Pathol.* **24**:903–909.
187. Vogel, R. A., Michael, M., and Timpe, A., 1955, Cortisone in experimental histoplasmosis, *Am. J. Pathol.* **31**:535–543.
188. Berry, C. L., 1969, Modification of the host response in experimental histoplasmosis, *J. Pathol.* **97**:653–664.
189. Adamson, D. M., and Cozad, G. C., 1969, Effect of antilymphocyte serum on animals experimentally infected with *Histoplasma capsulatum* or *Cryptococcus neoformans*, *J. Bacteriol.* **100**:1271–1276.
190. Cozad, G. C., and Lindsey, T. J., 1974, Effect of cyclophosphamide on *Histoplasma capsulatum* infections in mice, *Infect. Immun.* **9**:261–265.
191. Anderson, K. L., and Ross, W. D., 1981, Susceptibility of immunodeficient mice to challenge infection with *Histoplasma capsulatum*, *Mycopathologia* **73**:73–78.
192. Williams, D. M., Graybill, J. R., and Drutz, D. J., 1978, *Histoplasma capsulatum* infection in nude mice, *Infect. Immun.* **21**:973–977.
193. Miyaji, M., Chandler, F. W., and Ajello, L., 1981, Experimental histoplasmosis capsulati in athymic nude mice, *Mycopathologia* **75**:139–148.
194. Biggar, W. D., Meuwissen, H. J., and Good, R. A., 1971, Successful defense against *Histoplasma capsulatum* in hypogammaglobulinemia, *Arch. Intern. Med.* **128**:585–587.
195. Palmer, C. E., 1945, Nontuberculous pulmonary calcification and sensitivity to histoplasmin, *Pub. Health Repts.* **60**:513–520.
196. Salvin, S. B., 1955, Hypersensitivity in mice with experimental histoplasmosis, *J. Immunol.* **75**:1–6.
197. Salvin, S. B., 1955, Resistance to reinfection in experimental histoplasmosis, *J. Immunol.* **74**: 214–221.
198. Box, E. D., and Briggs, N. T., 1961, Endotoxin susceptibility and delayed hypersensitivity in experimental histoplasmosis, *J. Immunol.* **87**:485–491.
199. Bauman, D. S., and Chick, E. W., 1973, Acute cavitary histoplasmosis in rhesus monkeys: Influence of immunological status, *Infect. Immun.* **8**:245–248.
200. Domer, J. E., 1976, *In vivo* and *in vitro* cellular responses to cytoplasmic and cell wall antigens of *Histoplasma capsulatum* in artificially immunized or infected guinea pigs, *Infect. Immun.* **13**:790–799.
201. McMurray, D. N., Thomas, M. E., Greer, D. L., and Tolentino, N. L., 1978, Humoral and cell-mediated immunity to *Histoplasma capsulatum* during experimental infection in neotropical bats (*Artibeus lituratus*), *Am. J. Trop. Med. Hyg.* **27**:815–821.
202. McMurray, D. N., and Greer, D. L., 1979, Immune responses in bats following intranasal infection with *Histoplasma capsulatum*, *Am. J. Trop. Med. Hyg.* **28**:1036–1044.
203. Sharma, D., Chaudhary, S., Burt, W. R., and Tewari, R. P., 1980, Characteristics of delayed hypersensitivity and macrophage migration inhibition factor in mice immunized with ribosomes or live yeast cells of *Histoplasma capsulatum*, in: *Human and Animal Mycology* (E. S. Kuttin and G. L. Baum, eds.), *Excerpta Medica*, Amsterdam, pp. 89–92.
204. Nickerson, D. A., and Fairclough, P., 1984, Immune responsiveness following intratracheal inoculation with *Histoplasma capsulatum* yeast cells, *Clin. Exp. Immunol.* **56**:337–344.

205. Salvin, S. B., 1963, Immunological aspects of the mycoses, *Progr. Allergy* **7**:213–328.
206. Furcolow, M. L., 1963, Tests of immunity in histoplasmosis, *N. Engl. J. Med.* **268**:357–361.
207. Artz, R. P., Jacobson, R. R., and Bullock, W. E., 1980, Decreased suppressor cell activity in disseminated granulomatous infections, *Clin. Exp. Immunol.* **41**:343–352.
208. Chaparas, S. D., Sheagren, J. N., DeMeo, A. L., and Hedrick, S., 1970, Correlation of human skin reactivity with lymphocyte transformation induced by mycobacterial antigens and histoplasmin, *Am. Rev. Respir. Dis.* **101**:67–73.
209. Ruiz, B. H., and Carvajal, R. E., 1986, Immunomodulation by *Histoplasma capsulatum* products: Polyclonal activation and mitogenic effects, *Mycopathologia* **93**:113–119.
210. Nickerson, D., and Friedrich, L. M., 1987, Stimulation of murine lymphocyte blastogenesis by mitogens in heat-killed *Histoplasma capsulatum* yeast cells, *Infect. Immun.* **55**:937–941.
211. Yamamoto, Y., Jingu, S., and Iwata, K., 1986, Active substance of *Histoplasma capsulatum* that inhibits blastogenic response of lymphocytes, *Microbiol. Immunol.* **30**:1–11.
212. Johnson, R. W., and Scherago, M., 1960, The effect of histoplasmin *in vitro* on the migration of leukocytes from guinea pigs experimentally infected with *Histoplasma capsulatum*, *Am. Rev. Respir. Dis.* **81**:96–99.
213. Thor, D. E., and Dray, S., 1968, A correlate of human delayed hypersensitivity: Specific inhibition of capillary tube migration of sensitized human lymph node cells by tuberculin and histoplasmin, *J. Immunol.* **101**:51–61.
214. Thor, D. E., and Dray, S., 1968, The cell migration-inhibition correlate of delayed hypersensitivity, *J. Immunol.* **101**:469–480.
215. Anderson, K. L., and Sakaguchi, A. Y., 1978, Migration inhibition factor study in *Histoplasma capsulatum*, *Mycopathologia* **64**:3–7.
216. Domer, J. E., and Ichinose, H., 1977, Cellular immune responses in guinea pigs immunized with cell walls of *Histoplasma capsulatum* prepared by several different procedures, *Infect. Immun.* **16**:293–301.
217. Larsh, H. W., and Shepard, C. C., 1958, HeLa cells and *Histoplasma capsulatum*. Phagocytosis and subsequent intracellular growth, *J. Bacteriol.* **76**:557–563.
218. Hill, G. A., and Marcus, S., 1959, Nature of resistance in mouse histoplasmosis, *Tuberculogy* **18**:33–35.
219. Hill, G. A., and Marcus, S., 1959, Resistance induced against *Histoplasma capsulatum*: Quantitative aspects, *J. Infect. Dis.* **105**:26–30.
220. Howard, D. H., and Otto, V., 1973, Protein synthesis by phagocytized yeast cells of *Histoplasma capsulatum*, *Sabouraudia* **7**:186–194.
221. Howard, D. H., and Otto, V., 1977, Experiments on lymphocyte-mediated cellular immunity in murine histoplasmosis, *Infect. Immun.* **16**:226–231.
222. Howard, D. H., Otto, V., and Gupta, R. K., 1971, Lymphocyte-mediated cellular immunity in histoplasmosis, *Infect. Immun.* **4**:605–610.
223. Unkeless, J. G., Gordon, S., and Reich, E., 1974, Secretion of plasminogen activation by stimulated macrophages, *J. Exp. Med.* **146**:157–171.
224. Tiku, M. L., McNabb, P. C., and Tomasi, T. B., 1985, Macrophage-related fibrinolysis in experimental disseminated histoplasmosis, *Infect. Immun.* **49**:641–646.
225. Gomez, A. M., Bullock, W. E., Taylor, C. L., and Deepe, G. S., Jr., 1988, Role of LeT4+ T cells in host defense against *Histoplasma capsulatum*, *Infect. Immun.* **56**(7):1685–1691.
226. Karnovsky, M. L., and Lazdins, J. K., 1978, Biochemical criteria for activated macrophages, *J. Immunol.* **121**:809–813.
227. Parsons, R. J., and Zarafonetis, C. J. D., 1945, Histoplasmosis in man. Report of seven cases and a review of seventy-one cases, *Arch. Intern. Med.* **75**:1–23.
228. Deepe, G. S., Jr., Taylor, C. L., Srivastava, L., and Bullock, W. E., 1985, Impairment of granulomatous inflammatory response to *Histoplasma capsulatum* by inhibitors of angiotensin-converting enzyme, *Infect. Immun.* **48**:395–401.
229. Buchanan, R. D., Fine, D. P., and Colley, D. G., 1973, *Schistosoma mansoni* infection in mice

depleted of thymus-dependent lymphocytes. II. Pathology and altered pathogenesis, *Am. J. Pathol.* **71**:207–218.

230. Deepe, G. S., Jr., and Bullock, W. E., 1988, Histoplasmosis: A granulomatous inflammatory response, in: *Inflammation: Basic Principles and Clinical Correlates* (J. I. Gallin, I. M., Goldstein, and R. Snyderman, eds.), Raven Press, Inc., New York, pp. 733–749.
231. Wellhausen, S. R., Chensue, S. W., and Boros, D. L., 1980, Modulation of granulomatous hypersensitivity: Analysis by adoptive transfer of effector and suppressor T lymphocytes involved in granulomatous inflammation in murine schistosomiasis, in: *Basic and Clinical Aspects of Granulomatous Diseases* (D. L. Boros and T. Yoshida, eds.), Elsevier/North-Holland, Amsterdam.
232. Mathew, R. C., and Boros, D. L., 1986, Anti-L3T4 antibody treatment suppresses hepatic granuloma formation and abrogates antigen-induced interleukin 2 production in *Schistosoma mansoni* infection, *Infect. Immun.* **54**:820–826.
233. Chensue, S. W., Boros, D. L., and David, C. S., 1986, Regulation of granulomatous inflammation in murine schistosomiasis. *In vitro* characterization of T lymphocyte subsets involved in the production and suppression of migration inhibition factor, *J. Exp. Med.* **151**: 1398–1412.
234. Deepe, G. S., Jr., Smith, J. G., Sonnenfeld, G., Denman, D., and Bullock W. E., 1986, Development and characterization of *Histoplasma capsulatum*-reactive murine T-cell lines and clones, *Infect. Immun.* **54**(3):714–722.
235. Harris, J. E., and Deepe, G. S., Jr., 1988, Characterization of antigenic determinants in histoplasmin that stimulate *Histoplasma capsulatum*-reactive T cells in vitro, *Infect. Immun.* **56**(9):2343–2349.
236. Deepe, G. S., Jr., 1988, Protective immunity in murine histoplasmosis: Functional comparison of adoptively transferred T-cell clones and splenic T cells, *Infect. Immun.* **56**(9):2350–2355.
237. Harris, J. E., and Deepe, G. S., Jr., 1989, Requirements for histoplasmin presentation by accessory cells to a *Histoplasma capsulatum*-reactive T-cell line, *J. Leukocyte Biol.* **45**:105–113.
238. Deepe, G. S., Jr., and Brunner, G. D., 1990, Functional analysis of *Histoplasma capsulatum*-reactive T-cell hybridomas, *Infect. Immun.* **58**(6):5014–5058.
239. Couch, J. R., Abdou, N. I., and Sagawa, A., 1978, Histoplasma meningitis with hyperactive suppressor T cells in cerebrospinal fluid, *Neurology* **28**:119–123.
240. Lehmann, P. F., Gibbons, J., Senitzer, D., Ribner, B. S., and Freimer, E. H., 1983, T lymphocyte abnormalities in disseminated histoplasmosis, *Am. J. Med.* **75**:790–794.
241. Stobo, J. D., Sigrun, P., Van Scoy, R. E., and Hermans, P. E., 1976, Suppressor thymus-derived lymphocytes in fungal infections, *J. Clin. Invest.* **57**:319–328.
242. Stobo, J. D., 1977, Immunosuppression in man: Suppression by macrophages can be mediated by interactions with regulatory T cells, *J. Immunol.* **119**:918–924.
243. Payan, D. G., Wheat, L. J., Brahmi, Z., Ip, S., Hansen, W. P., Hoffman, R. A., Healey, K., and Rubin, R. H., 1984, Changes in immunoregulatory lymphocyte populations in patients with histoplasmosis, *J. Clin. Immunol.* **4**:98–108.
244. Artz, R. P., and Bullock, W. E., 1979, Immunoregulatory responses in experimental disseminated histoplasmosis: Lymphoid organ histopathology and serological studies, *Infect. Immun.* **23**:884–892.
245. Watson, S. R., Miller, T. B., Redington, T. J., and Bullock, W. E., 1983, Immunoregulation in experimental disseminated histoplasmosis: Flow microfluorometry (FMF) studies of the thy and lyt phenotypes of T lymphocytes from infected mice, *J. Immunol.* **131**:984–990.
246. Artz, R. P., and Bullock, W. E., 1979, Immunoregulatory responses in experimental disseminated histoplasmosis: Depression of T-cell-dependent and T-effector responses by activation of splenic suppressor cells, *Infect. Immun.* **23**:893–902.
247. Nickerson, D. A., Havens, R. A., and Bullock, W. E., 1981, Immunoregulation in disseminated histoplasmosis: Characterization of splenic suppressor cell populations, *Cell. Immunol.* **60**:287–297.
248. Watson, S. R., and Bullock, W. E., 1982, Immunoregulation in disseminated histoplasmosis:

Characterization of the surface phenotype of splenic suppressor T lymphocytes, *Infect. Immun.* **37:**940–945.

249. Deepe, G. S., Jr., Kravitz, G. R., and Bullock, W. E., 1983, Pharmacological modulation of suppressor cell activity in mice with disseminated histoplasmosis, *Infect. Immun.* **41:** 114–120.

249a. Wolf, J. E., Massof, S.E., and Peters, S. P., 1992, Alterations in murine macrophage arachidonic acid metabolism following ingestion of nonviable *Histoplasma capsulatum*. *Infect. Immun.* **60:**2559–2564.

250. Deepe, G. S., Jr., Watson, S. R., and Bullock, W. E., 1982, Generation of disparate immunoregulatory factors in two inbred strains of mice with disseminated histoplasmosis, *J. Immunol.* **129:**2186–2191.

251. Deepe, G. S., Jr., Watson, S. R., and Bullock, W. E., 1984, Cellular origins and target cells of immunoregulatory factors in mice with disseminated histoplasmosis, *J. Immunol.* **132:** 2064–2071.

252. Watson, S. R., Schmitt, S. K., Hendricks, D. E., and Bullock, W. E., 1985, Immunoregulation in disseminated murine histoplasmosis: Disturbances in the production of interleukins 1 and 2, *J. Immunol.* **135:**3487–3489.

253. Campbell, C. C., 1967, Serology of the respiratory mycoses, *Sabouraudia* **5:**240–259.

254. Buechner, H. A., Seabury, J. H., Campbell, C. C., Georg, L. K., Kaufman, L., and Kaplan, W., 1973, The current status of serologic, immunologic and skin tests in the diagnosis of pulmonary mycoses, *Chest* **63:**259–270.

255. Palmer, D. P., Kaufman, L., Kaplan, W., and Cavallaro, J. J., 1977, *Serodiagnosis of Mycotic Diseases*, C. C. Thomas, Springfield, IL.

256. Kaufman, L., 1980, Serodiagnosis of fungal diseases, in: *Manual of Clinical Immunology* (N. R. Rose and H. Friedman, eds.), American Society for Microbiology, Washington, D.C.

257. Seeliger, H. P. R., 1962, Serology of fungi and deep fungous infections, in: *Fungi and Fungous Diseases* (G. Dalldorf, ed.), C. C. Thomas, Springfield, IL, pp. 158–186.

258. Mackenzie, D. W. R., 1983, Serodiagnosis, in: *Fungi Pathogenic for Humans and Animals*, Part B1 (D. H. Howard, ed.), Marcel Dekker, Inc., New York, chapter 4.

259. Huppert, M., 1983, Antigens for measuring immunological reactivity, in: *Fungi Pathogenic for Humans and Animals*, Part B1 (D. H. Howard, ed.), Marcel Dekker, Inc., New York, chapter 5.

260. Reiss, E., 1986, *Histoplasma capsulatum*, in: *Molecular Immunology of Mycotic and Actinomycotic Infections* (E. Reiss, ed.), Elsevier, New York, chapter 5.

261. Wheat, J., French, M. L. V., Kohler, R. B., Zimmerman, S. E., Smith, W. R., Norton, J. A., Eitzen, H. E., Smith, C. D., and Slama, T. G., 1982, The diagnostic laboratory tests for histoplasmosis. Analysis of experience in a large urban outbreak, *Ann. Intern. Med.* **97:** 680–685.

262. Newberry, W. M., Tosh, F. E., and Doto, I. L., 1967, The complement fixation test in the diagnosis of chronic pulmonary histoplasmosis and blastomycosis, *J. Chronic Dis.* **20:**303–309.

263. Lowell, J. R., and Shuford, E. H., 1976, The value of the skin test and complement fixation test in the diagnosis of chronic pulmonary histoplasmosis, *Am. Rev. Respir. Dis.* **114:**1069–1075.

264. Wheat, L. J., Wass, J., Norton, J., Kohler, R. B., and French, M. L. V., 1984, Cavitary histoplasmosis occurring during two large urban outbreaks. Analysis of clinical, epidemiologic, roentgenographic, and laboratory features, *Medicine* **63:**201–209.

265. Kaufman, L., Terry, T., Schubert, J. H., and McLaughlin, D., 1967, Effects of a single histoplasmin skin test on the serological diagnosis of histoplasmosis, *J. Bacteriol.* **94:**798–803.

266. Terry, P. B., Rosenow, E. C., and Roberts, G. D., 1978, False-positive complement fixation serology in histoplasmosis, *JAMA* **239:**2453–2456.

267. Heiner, D. C., 1958, Diagnosis of histoplasmosis using precipitin reactions in agar gel, *Pediatrics* **20:**616–629.

268. Kleger, B., and Kaufman, L., 1973, Detection and identification of diagnostic *Histoplasma capsulatum* precipitates by counterelectrophoresis, *Appl. Microbiol.* **26:**231–238.

269. Bauman, D. S., and Smith, C. D., 1975, Comparison of immunodiffusion and complement fixation tests in the diagnosis of histoplasmosis, *J. Clin. Microbiol.* **2:**77–80.
270. Campbell, C. C., and Hill, G. B., 1964, Further studies on the development of complement fixation antibodies and precipitins in healthy histoplasmin-sensitive persons following a single histoplasmin skin test, *Am. Rev. Respir. Dis.* **90:**927–934.
271. Bennett, D. E., 1966, The histoplasmin latex agglutination test: Clinical evaluation and a review of the literature, *Am. J. Med. Sci.* **251:**175–183.
272. DiSalvo, A. F., and Corbett, D. S., 1976, Apparent false-positive histoplasmin latex agglutination tests in patients with tuberculosis, *J. Clin. Microbiol.* **3:**306–308.
273. Oxenhandler, R. W., Adelstein, E. H., and Rogers, W. A., 1977, Rheumatoid factor: A cause of false-positive histoplasmin latex agglutination, *J. Clin. Microbiol.* **5:**31–33.
274. Sharma, A., Johri, B. N., and Shriniwas, 1982, An enzyme immunoassay for yeast and mycelial phase-specific antibodies in histoplasmosis, *J. Immunol. Methods* **50:**115–121.
275. Reiss, E., Hutchinson, H., Pine, L., Ziegler, D. W., and Kaufman, L., 1977, Solid-phase competitive-binding radioimmunoassay for detecting antibody to the M antigen of histoplasmin, *J. Clin. Microbiol.* **6:**598–604.
276. George, R. B., Lambert, R. S., Bruce, M. J., Pickering, J. W., and Wolcott, R. M., 1981, Radioimmunoassay: A sensitive screening test for histoplasmosis and blastomycosis, *Am. Rev. Respir. Dis.* **124:**407–410.
277. McKelvey, E. M., and Fahey, J. L., 1965, Immunoglobulin changes in disease: Quantitation on the basis of heavy polypeptide chains, IgG (G), IgA (A), and IgM (M), and of eight polypeptide chains, type K (I) and L (II), *J. Clin. Invest.* **44:**1778–1787.
278. Sprouse, R. F., Caldwell, C. W., and Everett, E. D., 1981, Radioimmunoassays that use ^{125}I-labeled protein A for determination of *Histoplasma capsulatum*-specific immunoglobulin G, A, and M class antibodies in histoplasmosis, *J. Clin. Microbiol.* **13:**66–72.
279. Wheat, L. J., Kohler, R. B., French, M. L. V., Garten, M., Kleiman, M., Zimmerman, S. E., Schlech, W., Ho, J., White, A., and Brahmi, Z., 1983, Immunoglobulin M and G histoplasmal antibody response in histoplasmosis, *Am. Rev. Respir. Dis.* **128:**65–70.
280. McMurray, D. N., Stroud, J., Murphy, J. J., Carlomagno, M. A., and Greer, D. L., 1982, Role of immunoglobulin classes in experimental histoplasmosis in bats, *Dev. Comp. Immunol.* **6:** 557–567.
281. Tewari, R. P., Khardori, N., McConnachie, P., Von Behren, L. A., and Yamada, T., 1982, Blastogenic responses of lymphocytes from mice immunized by sublethal infection with yeast cells of *Histoplasma capsulatum*, *Infect. Immun.* **36:**1013–1018.
282. Deepe, G. S., Jr., and Bullock, W. E., 1989, Correlative studies of blastogenic responses and interleukin 1 production by mononuclear cells from patients with zoopathogenic fungal infections, *Sabouraudia* **27:**159–168.
283. Schwarz, J., Baum, G. L., Wang, J. K., Binham, E. L., and Rubel, H., 1957, Successful infection of pigeons and chickens with *Histoplasma capsulatum*, *Mycopathologia* **8:**189–193.
284. Tewari, R. P., and Campbell, C. C., 1965, Isolation of *Histoplasma capsulatum* from feathers of chickens inoculated intravenously and subcutaneously with the yeast phase of the organism, *Sabouraudia* **4:**17–22.
285. Salvin, S. B., and Neta, R., 1982, Immunopathology of the mycotic infections, in: *The Reticuloendothelial System—Immunopathology*, Volume 4 (N. R. Rose and B. V. Siegel, eds.), Academic Press, New York, pp. 145–199.
286. Farrell, R. L., Cole, C. R., Prior, J. A., and Saslaw, S., 1953, Experimental histoplasmosis. I. Methods for production of histoplasmosis in dogs (20538), *Proc. Soc. Exp. Biol. Med.* **84:** 51–54.
287. Scherr, G. H., and Rippon, J. W., 1959, Experimental histoplasmosis in cold-blooded animals, *Mycopathologia* **11:**241–249.
288. Parsons, R. J., 1942, Experimental histoplasmosis in mice, *Arch. Pathol.* **34:**229–239.
289. Tager, M., and Liebow, A. A., 1942, Observations on histoplasmosis induced infections in the mouse, *Yale J. Biol. Med.* **14:**469–488.

290. Schaefer, J., and Saslaw, S., 1954, Some factors affecting resistance of mice to experimental histoplasmosis, *Proc. Soc. Exp. Biol. Med.* **85:**223–225.
291. Anderson, K. L., and Marcus, S., 1970, Immunity to histoplasmosis induced in mice by components of *Histoplasma capsulatum*, *Am. Rev. Respir. Dis.* **102:**614–622.
292. O'Hern, E. M., 1961, Resistance of hamsters in infections with *Histoplasma capsulatum*, *J. Immunol.* **87:**728–736.
293. O'Hern, E. M., and Fishbein, L., 1968, Studies on histoplasmosis. IV. Elution characteristics of the histo-inhibitory factor (HIF), *Mycopathol. Mycol. Appl.* **36:**33–41.
294. O'Hern, E. M., Fishbein, L. F., and Turner, P. H., 1969, Studies on histoplasmosis, V. Protective effects of HIF-coated histoplasma against challenge, *Mycopathol. Mycol. Appl.* **37:** 57–64.
295. Bauman, D. S., and Chick, E. W., 1969, An experimental model for studying extrapulmonary dissemination of *Histoplasma capsulatum* in hamsters, *Am. Rev. Respir. Dis.* **100:** 79–81.
296. Middleton, J. G., McVickar, D. L., and Peterson, J. C., 1950, Experimental histoplasmosis in the white rat, *Proc. Soc. Exp. Biol. Med.* **75:**164–166.
297. Day, R., 1949, Experimental ocular histoplasmosis, *Am. J. Ophthalmol.* **32:**1317–1330.
298. Salvin, S. B., and Bell, E. J., 1955, Resistance of mice with experimental histoplasmosis to infection with *Rickettsia typhi*, *J. Immunol.* **75:**57–62.
299. Grayston, J. T., and Salvin, S. B., 1956, Experimental histoplasmosis in immunized and nonimmunized mice, *Arch. Pathol.* **61:**422–444.
300. Knight, R. A., and Marcus, S., 1958, Polysaccharide skin test antigens derived from *Histoplasma capsulatum* and *Blastomyces dermatitidis*, *Am. Rev. Tuberc. Pulm. Dis.* **77:**983–989.
301. Knight, R. A., Hill, G., and Marcus, S., 1959, Immunization of mice with polysaccharides of *Histoplasma capsulatum* (24625), *Proc. Soc. Exp. Biol. Med.* **100:**356–358.
302. Salvin, S. B., 1965, Constituents of the cell wall of the yeast phase of *Histoplasma capsulatum*, *Am. Rev. Respir. Dis.* **92:**119–125.
303. Bauman, D. S., and Chick, E. W., 1969, Immunoprotection against extrapulmonary histoplasmosis in hamsters, *Am. Rev. Respir. Dis.* **100:**82–85.
304. Garcia, J. P., and Howard, D. H., 1971, Characterization of antigens from the yeast phase of *Histoplasma capsulatum*, *Infect. Immun.* **4:**116–125.
305. Feit, C., and Tewari, R. P., 1974, Immunogenicity of ribosomal preparations from yeast cells of *Histoplasma capsulatum*, *Infect. Immun.* **10:**1091–1097.
306. Tewari, R. P., 1975, Immunization against histoplasmosis, in: *The Immune System and Infectious Diseases* (E. Neter and F. Milgrom, eds.), S. Karger, Basel, 4th International Convocation on Immunology, Buffalo, New York, 1974, pp. 441–452.
307. Kong, Y.-C. M., and Levine, H. B., 1967, Experimentally induced immunity in the mycoses, *Bacteriol. Rev.* **31:**35–53.
308. Galgiani, J. N., 1985, Fungal vaccines, in: *Fungi Pathogenic for Humans and Animals*, Part B2 (D. H. Howard, ed.), Marcel Dekker, Inc., New York, chapter 9.
309. Reiss, E., 1985, Cell wall composition, in: *Fungi Pathogenic for Humans and Animals*, Part B2 (D. H. Howard, ed.), Marcel Dekker, Inc., New York, chapter 2.
310. Tewari, R. P., and LaFemina, R., 1983, Immunogenicity of subcellular fractions from yeast cells of *Histoplasma capsulatum*, *Jpn. J. Med. Mycol.* **24:**179.
311. Salvin, S. B., and Ribi, E., 1955, Antigens from yeast phase of *Histoplasma capsulatum*. II. Immunologic properties of protoplasm vs. cell walls, *Proc. Soc. Exp. Biol., N.Y.* **90:**287–294.
312. Segal, E., 1989, Fungal ribosomal vaccines, I. *Histoplasma* and *Candida* vaccines, *Mycopathologia* **105:**45–48.
313. Loosli, C. G., 1957, Histoplasmosis, *J. Chronic Dis.* **5:**473–488.
314. Emmons, C. W., Binford, C. H., Utz, J. P., and Kwon-Chung, K. J., 1977, Histoplasmosis, in: *Medical Mycology*, 3rd ed. (C. W. Emmons, C. H. Binford, J. P. Utz, and K. J. Kwon-Chung, eds.), Lea and Febiger, Philadelphia, PA.
315. Ajello, L., 1977, Systemic mycoses in modern medicine, *Contrib. Microbiol. Immunol.* **3:**2.

316. Bullock, W. E., and Deepe, G. S., Jr., 1983, Medical mycology in crisis, *J. Lab. Clin. Med.* **102:** 685–693.
317. Sarosi, G. A., Voth, D. V., Dahl, B. A., Doto, I. L., and Tosh, F. E., 1971, Disseminated histoplasmosis: Results of long-term follow up. A Center for Disease Control cooperative mycoses study, *Ann. Intern. Med.* **75:**511–516.
318. Saag, M. S., and Dismukes, W. E., 1988, Treatment of histoplasmosis and blastomycosis, *Chest* **93:**848–851.
319. National Institute of Allergy and Infectious Disease Mycoses Study Group, 1985, Treatment of blastomycosis and histoplasmosis with ketoconazole, *Ann. Int. Med.* **103:**861–872.
320. Paya, C. V., Hermans, P. E., Van Scoy, R. E., Ritts, R. E., Jr., and Hornburger, H. A., 1987, Repeatedly relapsing disseminated histoplasmosis: Clinical observations during long-term follow-up, *J. Infect. Dis.* **156:**308–312.
321. Ascher, M. S., Gottlieb, A. A., and Kirkpatrick, C. H., eds., 1976, *Transfer Factor*, Academic Press, Inc., New York.
322. Graybill, J. R., Ellenbogen, C., Drossman, D., Kaplan, P., and Thor, D. E., 1976, Transfer factor therapy of disseminated histoplasmosis, in: *Transfer Factor* (M. S. Ascher, A. A. Gottlieb, and C. H. Kirkpatrick, eds.), Academic Press, New York, p. 509–514.
323. Smith, C. R., Griffin, D. E., and Graybill, J. R., 1976, Chronic pulmonary histoplasmosis: Improved lymphocyte response with transfer factor, *Ann. Intern. Med.* **85:**708–709.
324. Maeyer, E. E., 1985, Immunomodulation by interferons: An appraisal, *Ann. Immunol. (Inst. Pasteur)*, **136D:**67.
325. Moller, G., ed., 1982, Interleukins and lymphocyte activation, *Immunol. Rev.* **65:**5–209.
326. Jacques, Y., and Soulliou, J. P., 1985, Third French workshop on interleukin 2: Joint report, *Lymphokine Res.* **4:**159–167.
327. Morgan, D. A., Fuscetti, F. W., and Gallo, R., 1976, Selective *in vitro* growth of T lymphocytes from normal human bone marrow, *Science* **193:**1007–1011.
328. Gillis, S., and Smith, K. A., 1977, Long term culture of tumor specific cytotoxic T cells, *Nature* (Lond.) **268:**154–156.
329. Marrack, P., Graham, S. D., Jr., Kushnir, E., Leibson, H. J., Roehm, N., and Kappler, J. W., 1982, Nonspecific factors in B cell responses, *Immunol. Rev.* **63:**33–49.
330. Farrar, J. J., Benjamin, W. R., Hilfiker, M. L., Howard, M., Farrar, W. L., and Fuller-Farrar, J., 1982, The biochemistry, biology, and role of interleukin 2 in the induction of cytotoxic T cell and antibody forming B cell responses, *Immunol. Rev.* **63:**129–166.
331. Henney, C. S., Kuribayashi, K., Kern, D. E., and Gillis, S., 1981, Interleukin 2 augments natural killer cell activity, *Nature* (Lond.) **291:**335–338.
332. Kern, D. E., Gillis, S., Okada, M., and Henney, C. S., 1981, The role of interleukin-2 (IL-2) in the differentiation of cytotoxic T cells: The effect of monoclonal anti-IL-2 antibody and absorption with IL-2 dependent T cell lines, *J. Immunol.* **127:**1323–1328.
333. Lutz, C. T., Glasebrook, A. L., and Fitch, F. W., 1981, Alloreactive cloned T cell lines. III. Accessory cell requirements for the growth of cloned cytolytic T-lymphocytes, *J. Immunol.* **126:**1404–1408.
334. Deepe, G. S., Jr., Taylor, C. L., Harris, J. E., and Bullock, W. E., 1986, Modulation of cellular responses in mice with disseminated histoplasmosis by recombinant interleukin-2, *Infect. Immun.* **53:**6–12.
335. Kiderlen, A. F., Kaufman, S. H. E., and Lohmann-Matthes, M. L., 1984, Protection of mice against the intracellular bacterium *Listeria monocytogenes* by recombinant immune interferon, *Eur. J. Immunol.* **14:**964–967.
336. Murray, H. W., Stern, J. J., Welte, K., Rubin, B. Y., Carriero, S. M., and Nathan, C. F., 1987, Experimental visceral leishmaniasis: Production of interleukin 2 and interferon-γ, tissue immune reactions, and response to treatment with interleukin 2 and interferon-γ, *J. Immunol.* **138:**2290–2297.
337. Wu-Hsieh, B. A., Lee G-S., Franco, M., and Hofman, F., 1992, Early activation of splenic macrophages by tumor necrosis factor alpha is important in determining the outcome of experimental histoplasmosis in mice, *Infect. Immun*, **60**:4230–4238.

10

Paracoccidioidomycosis

ANGELA RESTREPO-MORENO

1. DESCRIPTION OF THE DISEASE

Paracoccidioidomycosis (PCM) is a subacute or chronic disease caused by the dimorphic fungus *Paracoccidioides brasiliensis*. The primary infection occurs in the lungs, but dissemination to the mucous membranes, skin, lymph nodes, adrenals, and other organs occurs frequently. Inapparent and subclinical forms are also observed. This mycosis is mostly diagnosed in adult males and is restricted to Latin American countries.[1–4]

1.1. Portal of Entry

It is generally accepted that the fungus is exogenous to humans, but its precise niche in nature has not been discovered.[5] As a consequence, the portal of entry is still a matter of debate.

Originally, the frequency with which mucocutaneous lesions were observed in patients, led to the belief that trauma with contaminated plant debris resulted in infection. Presently, however, the portal of entry is considered to be inhalatory;[2,6,7] as evidence for the latter route, we have the results of aerosol-infected animals that developed a progressive disseminated disease,[8] histopathological data showing that nearly all the autopsy cases exhibit pulmonary lesions,[1] and clinical observations showing that 30% of the patients exhibit lung involvement as the only manifestation of the mycosis.[9] Furthermore, the fact that some patients present mucocutaneous lesions years

---

ANGELA RESTREPO-MORENO • Mycology Section, Corporación para Investigaciones Biologicas (CIB), Hospital Pablo Tobon Uribe, Medellin, Colombia.

Fungal Infections and Immune Responses, edited by Juneann W. Murphy *et al.* Plenum Press, New York, 1993.

after the diagnosis of the pulmonary problem also suggests that the primary infection occurs in the lungs and that the extrapulmonary manifestations are secondary lesions.[9,10]

1.2. Clinical Manifestations

Because of the fact that certain clinical manifestations were observed more frequently than others, four PCM forms were originally established, namely, tegumentary, lymphatic, visceral, and mixed. It was soon noticed that simultaneous involvement of two or more organs was common, a fact that invalidated the older topographic classification.[9,10] It would be more accurate to classify the clinical forms on the basis of physiopathogenic events; unluckily, however, the natural history of PCM can be reconstructed only partially because of our ignorance of the fungus's habitat. The following course of events has been proposed.

The primary infection (Fig. 1) occurs in the lungs and remains asymptomatic; in most cases, nonetheless, early dissemination to other organs is known to occur. The interplay between the host and the parasite is pivotal in determining the outcome of the infection. If the establishment of the fungus in its initial focus is controlled by the host's defenses, the primary infection is converted into an asymptomatic regressive (latent) form, which has been demonstrated by the presence in normal persons of either solitary nodules containing the fungus or by healed, primary lung complexes. When the hosts is unable to control the initial infection or if, later on, the original control exerted by such a host is lost, the fungus reproduces in tissues, an event that leads to the appearance of symptoms and to the characteristic progressive forms of PCM. In young adults and, especially, in children, the disease runs an acute or subacute course and is characterized by marked dissemination. Contrariwise, in adults, the course is chronic and dissemination is somewhat more restricted. If therapy is timely instaurated, most patients (young and adult), respond but develop residual lesions that may contain viable fungi; the latter are the source of future relapses. Most residual lesions produce incapacitating sequelae.[9,11]

1.3. Clinical Forms

Two different forms of PCM are recognized, the infection and the disease.[9,11] The infection (Table I) is documented in healthy paracoccidioidin-positive persons who live (or have lived) in the endemic areas. Also included in this form are those persons with inactive residual lesions that contain *P. brasiliensis*.

The disease is overtly symptomatic and, depending on its course, divided into two groups the acute (subacute), juvenile form and the chronic, adult form. Such forms are further subdivided on the basis of their severity.

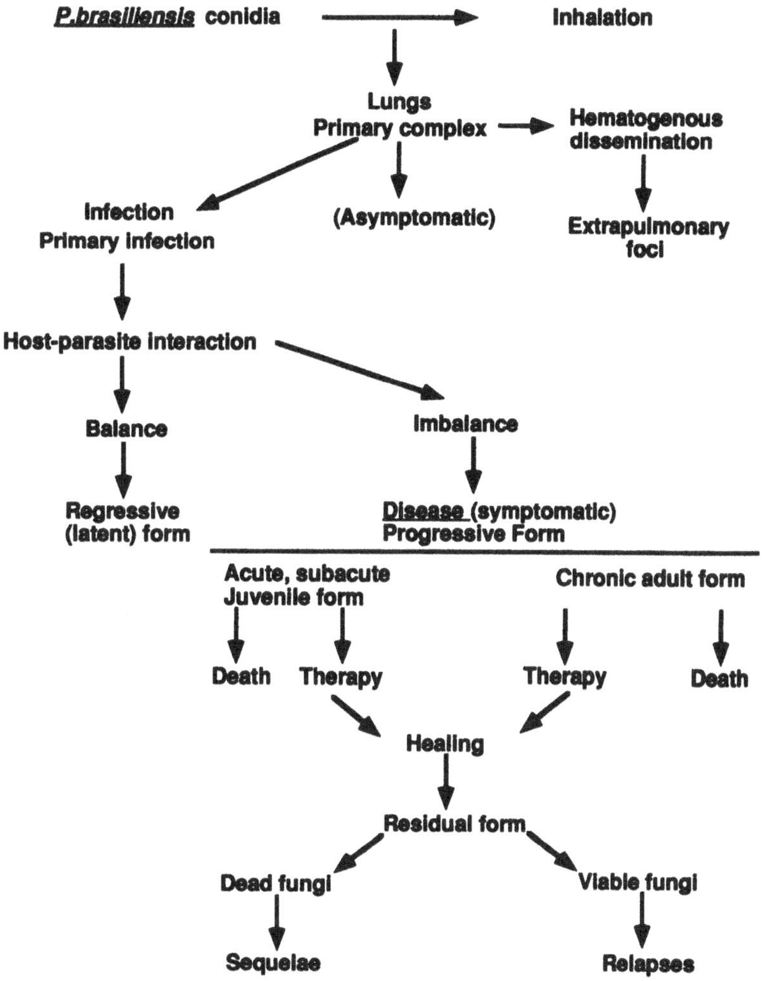

FIGURE 1. A model for the pathogenesis of paracoccidioidomycosis.

1.3.1. The Juvenile Progressive Form

This form is characterized by relatively short periods of incubation (several weeks to a few months); patients are children or young adults (up to 25 years of age) of both sexes. The disease is characterized by marked involvement of the reticuloendothelial system. Symptoms vary according to the type of organic involvement. One group of patients seeks medical attention because of enlarged lymph nodes, mostly cervical ones, spleno- and/or hepato-megaly. Patients in the remaining group present overt dissemination, are in poor condition and present multiple lesions. In such a group, the

TABLE I
Classification of Paracoccidioidomycosis

Paracoccidioidomycosis infection			
Paracoccidioidomycosis disease			
	Acute or subacute form (juvenile form)		
			moderate
			severe
	Chronic form (adult type)		
		unifocal	
			mild
			moderate
			severe
		multifocal	
			mild
			moderate
			severe
	Residual forms (sequelae)		

disease's course is more acute and the prognosis is more severe. Pulmonary involvement is not always apparent, since it is less severe than the remaining manifestations of the process. The lung roentgenologic study often reveals hypertrophy of hilar lymph nodes and/or milliary infiltrates. As a rule, the patients classified in the juvenile form exhibit impaired cellular immunity and present high anti-*P. brasiliensis* antibody titers.

Symptoms depend largely on the extent of fungal invasion; the affected lymph nodes not only enlarge in size and break open, but they also increase in number. If located in the abdominal cavity they may coalesce and form large pseudotumoral masses that exert pression over the adjoining organs. In such a case, the signs and symptoms are those of an acute abdominal process or of an abdominal obstruction. In the severely disseminated forms, the disease is devastating and is accompanied by generalized symptoms such as malaise, fever, anemia, and weight loss. Depending on the sites reached by the septic thrombii during dissemination, one observes lesions of the skin, the bones, the kidney, the brain, etc. The diagnosis of the juvenile form is frequently accomplished through biopsy, since the fungal nature of the disease is not suspected; differential diagnoses include lymphoma, leukemia, and tuberculosis. Histologically, there are variations, depending on the type of the disease; compact granulomas and a few *P. brasiliensis* cells are observed in the moderate forms, whereas, in the severe variety, there are few loose granulomas and abundant, actively reproducing fungal forms. The juvenile progressive form represents 3–5% of the total number of cases diagnosed.[2,9–13]

1.3.2. The Chronic Progressive, Adult-Type Form

This form is the predominant presentation of PCM and is observed in over 90% of patients. The disease course is slow and varies from a few months to several years; most patients are males over 30 years of age. In certain cases, the mycosis is unifocal with apparent involvement of only one organ (or system) and with the corresponding signs and symptoms. Other patients present alterations of more than one organ (or system) (multifocal (PCM).[11] The lungs are frequently involved but the corresponding manifestations (cough, expectoration, malaise) are inspecific; the pulmonary lesions may progress locally or disseminate to other organs and systems. PCM patients with lesions restricted to the lungs were not properly diagnosed in the past, since the "classical" manifestations of the mycosis (e.g., mucosal and/or cutaneous lesions) were absent; in such patients, the diagnosis was tuberculosis. The symptoms of pulmonary PCM are nonspecific and include cough, pleuritic pain, expectoration—at times blood tinged—fever, weight loss, asthenia, and generalized malaise; usually, there is some degree of dyspnea. Depending on the duration of the disease, the x-ray films reveal small, bilateral, cottony infiltrates located towards the central and basal portions of the lung; the upper fields are usually clear. Infiltrates may coalesce forming large, dense nodules (Fig. 2). Among the infiltrated parenchyma, one perceives the presence of fibrotic lines; the latter tend to increase during the course of the disease, even after instauration of appropriate therapy, and become established as permanent sequelae. Fibrosis may be severe enough to cause cardiorespiratory problems, including cor-pulmonale. Infiltrates and dense nodules are of the alveolar type while fibrotic lines correspond to the interstitial type. Besides these lesions, one may also find small cavities.[1,4,6,9,10,14,15]

The use of Gallium imaging does reveal the presence of radiologically and clinically unsuspected lesions.[16] Calcifications are the exception rather than the rule; pleural involvement and hypertrophy of the hilar lymph nodes are rare in adults. Contrariwise, emphisematous lesions and bullae are observed repeatedly. Auscultatory findings (rales, hypoventilation) are relatively minor in comparison with the severity of the x-ray images. Unifocal pulmonary PCM accounts for 30% of all cases; its diagnosis can be established by direct microscopy of sputum samples. Most patients with this type of disease have normal to slightly subnormal immune responses; antibody production is also normal.[4,6,9–11]

In those patients with chronic multifocal PCM, signs and symptoms vary according to the site involved. This form accounts for 65% of all cases. Because the external lesions, which are amenable to laboratory examination, are frequently present, the diagnosis is rapidly established by microscopy. Secondary lesions are observed most often in the mucous membranes, skin, lymph nodes, and adrenals; however, any other organ may be involved.

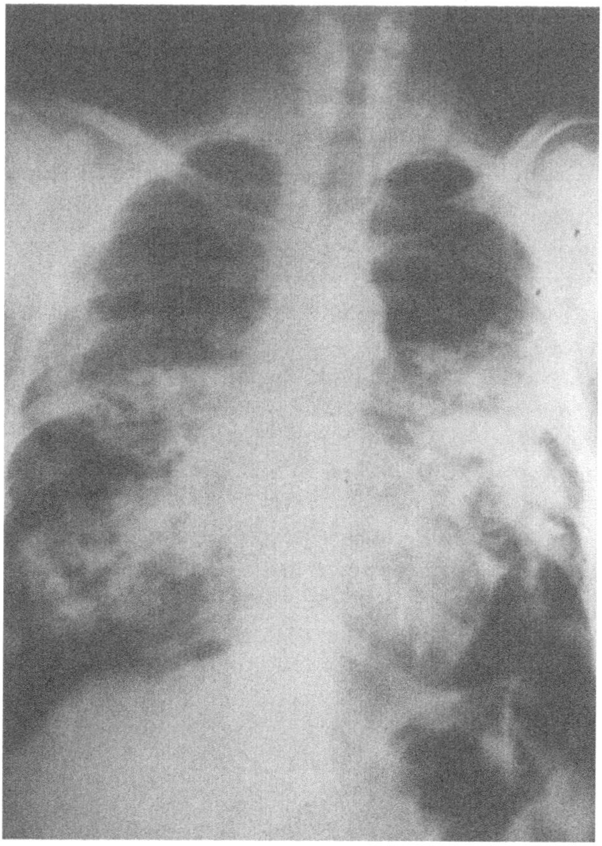

FIGURE 2. Chest X ray: bilateral infiltrates located in the central and lower fields. Upper fields appear free. Some bullous areas are apparent.

Patients are symptomatic and, besides the constitutional symptoms (weight loss, general malaise), one finds dysphonia, syalorrheae, skin hyperpigmentation, hypertrophied lymph nodes, mucosal and skin ulcerations, pleuritic pain, productive cough, and dyspnea. Chest x-rays usually reveal marked alterations similar to those described for unifocal pulmonary PCM.[1–4,6,9–11]

Most patients in this group seek medical advice because of the presence of oral mucosal lesions that are the cause of dysphonia, odinophagia, dysphagia, syalorrhea, and burning sensation. The lesions are ulcerative, infiltrated and with ill-defined borders (Fig. 3). Any site of the oral cavity may be affected (palate, gums, lips, tongue, uvula, epiglotis), as well as the larynx, pharynx, and vocal cords. When located in the lips, an edematous component is frequently present; the tonsils and the uvula exhibit destructive lesions. In

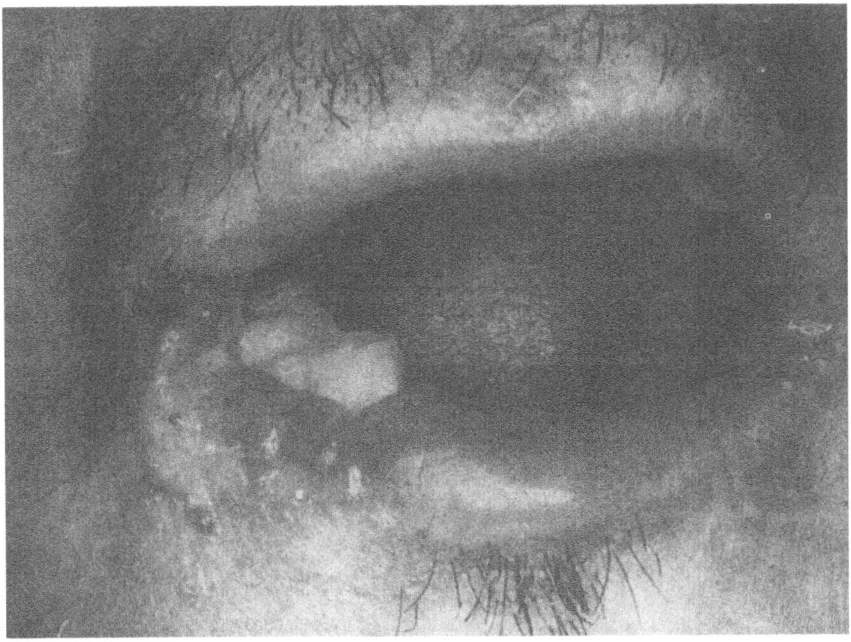

FIGURE 3. Mucocutaneous lesion in the lips' border. Note ulceration with granulomatous appearance, infiltration, and edema of surrounding area.

tissues, a granulomatous response is observed frequently. Skin lesions are usually multiple and appear at various sites; the face, however, is more often involved. Skin lesions are extremely polymorphic (e.g., ulcerative, nodular, verrucoid, or scaly); most lesions take a granulomatous appearance and usually cause infiltration of the surrounding tissues (Fig. 4). Histologically, skin lesions are characterized by pseudoepitheliomatous hyperplasia.[1,2,4,6,9–11]

Lymph node lesions are the predominant manifestation in children and young adults; however, older patients may also exhibit adenopathies. These lesions are preferentially, but not exclusively, located in the cervical, preauricular and submandibular areas; in juvenile patients, involvement of the mesenteric lymph nodes can occur, bringing forth a series of secondary problems. Adenopathies may be single or multiple and usually break open with formation of abscesses and fistular tracts. In some patients, there are compact granulomas with sparse fungal cells, whereas, in severe cases, tissue reaction is inadequate and there is abundance of yeast cells.[1,2,4,9,10,12–14]

The adrenals are target organs, with involvement noticed in over 50% of patients; there is a variety of lesions, from hypofunction to fully expressed Addison syndrome.[1,14,17] With lesser frequency, dissemination may also occur

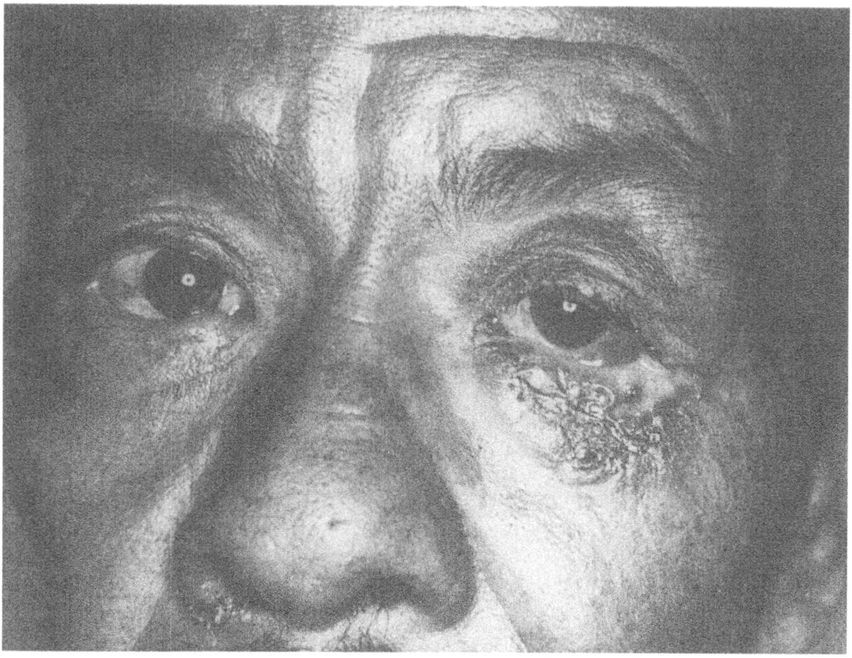

FIGURE 4. Skin lesions (palpebral and nasal) with ulceration, irregular borders, scaling, and retraction.

in the spleen, liver, intestines, central nervous system, vascular system, and bones. Very rarely, one finds lesions of the genitourinary tract.[1–4,10,12,14]

Paracoccidioidomycosis is diagnosed infrequently in patients with the acquired immunodeficiency syndrome (AIDS), but the disease is not especially aggressive.[18,19] Additionally, the mycosis is not common in other immunodepressive states; only 6 patients with immunodepression problems and PCM have been reported.[20,21]

1.3.3. Residual Forms

There are a number of residual lesions (sequelae) that may permanently hamper the well-being of PCM patients. Among these, fibrosis is important, especially in the lungs; as a result, approximately 80% of patients exhibit dyspnea of varying degrees of severity. Dysphonia and microstomy are also incapacitating sequelae of the infection. Stenosis of the glottis and trachea may appear when healing of the active lesions takes place by retraction. Unluckily, treatment halts the progress of the infection but does not hinder the formation of fibrosis. Destruction of the adrenals by fungal invasion may permanently impair the normal activity of such glands.[1,14,15,17]

2. ETIOLOGY

The dimorphic fungus *P. brasiliensis* is the only etiologic agent of the mycosis. In laboratory tests, and probably in nature as well, at temperatures of 17–28°C the fungus grows as a mold (M) while at higher test temperatures (30–37°C), and in the human body, it adopts a yeast form (Y).[15,22] The M form colonies grow slowly (15–25 days), are white to tan, have short aerial mycelium and, at times, produce a brownish tinge. Microscopically, the M colony is formed by thin, septate hyphae with practically no sporulation; special conidia (arthroconia and single-celled microconidia) can, however, be observed in carbohydrate-deficient culture media.[23] In experimental animals, these conidia have been shown to be infective by the respiratory route.[8] At 35–37°C, the Y colony is cerebriform, soft, and grows more quickly than the M form; microscopically, one observes oval to round yeast cells, 2–30 microns in size, some of which exhibit multiple sporulation with the "pilot's wheel" appearance, this is, a mother cell externally surrounded by multiple smaller buds (Fig. 5). Some yeast cells have only one or two buds or appear single. The cell wall is thick and refringent and the cytoplasm contains

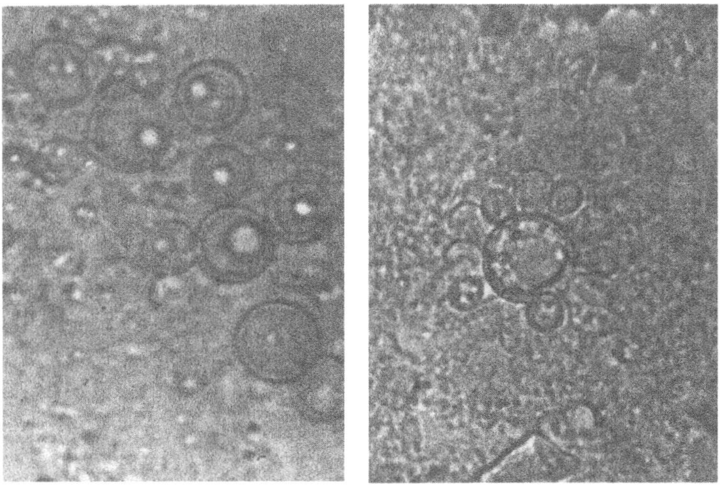

FIGURE 5. *P. brasiliensis* yeast cells (fresh preparation from pathological samples). There are large round-to-oval cells, with thick cell walls, lipid vacuoles, and multiple budding. Note cell with the "pilot's wheel" appearance (40 ×).

lipid vacuoles. In tissues *P. brasiliensis* exhibits the same type of Y morphology.[1,3,14,22] The yeast cell wall is mainly composed of α-1-3 glucan, while in the mycelial cell wall, β-1-3 glucan predominates. The former component appears to correlate with fungal virulence.[24]

3. EPIDEMIOLOGY

One of the most peculiar characteristics of paracoccidioidomycosis is its restricted geographic distribution. The disease is known to occur only in Latin America from Mexico (23N) to Argentina (34S); the disease, however, is not reported in all the countries within these latitudes; for instance, the Caribbean Islands, Chile, and Nicaragua have no reported cases of paracoccidioidomycosis. Within the recognized endemic countries, cases are not homogeneously distributed but are concentrated in certain ecological areas, such as the humid and very humid tropical and subtropical forests.[5,15] Brazil is the heart of the endemic area with 80% of all reported cases coming from this country; Colombia, Venezuela, and Argentina follow Brazil. Nonautochthonous cases have been reported in North America, Europe, and Japan[25,26] but in every instance, the patient has had a history of residence in one of the recognized endemic areas. It is worth mentioning that in some of the latter patients overt paracoccidioidomycosis was noticed 30–40 years after abandoning the endemic region. This indicates that PCM has extremely long periods of latency. It may be pertinent to point out that *P. brasiliensis* withstands the seemingly anaerobic conditions existing in partially healed nodular lesions[27] and as such, appears capable of dormancy in such locations.

PCM is rare in children, with only 3% of cases occurring in this group. Over 75% of patients are over 30 years of age.[15,26] Overt PCM exhibits another peculiarity, namely, its sex distribution. The disease is observed more frequently in men than in women, with a mean ratio of 15:1 or even higher; 84:1, 100:1 in certain areas.[15,26,28] On the other hand, the frequency of subclinical infections as measured by skin-test reactivity to paracoccidioidin, is similar for both sexes, indicating that the infection occurs in equal proportions in both men and women. Furthermore, when the disease occurs in children before puberty, there are no sex differences.[12,26] It has been shown that estrogens inhibit *P. brasiliensis* mycelial to yeast transformation, a step necessary for the microorganism's survival in host tissues.[29] When the initial contact with the fungus takes place, the hormonal levels may protect the female by hindering the expression of fungal dimorphism.

As it concerns occupation, 75% of patients are agricultural workers; nonetheless, PCM is diagnosed in urban settings in persons who have no agricultural pursuits. The disease has no special predilection for a particular race, but immigrants to the endemic areas tend to develop severe PCM.[3,4,15,26]

Paracoccidioidin skin-test surveys have been conducted in the endemic areas but the results have not been clear cut and have not led to the discovery of highly endemic areas.[5,6]

Man is *P. brasiliensis*' predilect host; there have been very few reports of the disease in animals; recently, however, a series of armadillos (*Dasypus novemcinctus*) have been shown to harbor the fungus in their internal organs.[5,30] PCM is not contagious from person to person, but a higher frequency of reactive paracoccidioidin skin tests has been recorded in the wives of patients with the disease.[26] Epidemic outbreaks have not been confirmed, probably because of the long latency periods characteristic of the disease. Incidence and prevalence are difficult to calculate as the disorder is not reportable and is easily misdiagnosed; an estimate indicates that there are four cases per one million inhabitants.[26]

It is generally accepted that the fungus' natural habitat is exogenous to man; however, its precise microniche remains undetermined. In spite of the many efforts undertaken by various researchers, *P. brasiliensis* has been isolated from Nature (soil) only twice and its isolation from bat guano has been discredited.[5] In the laboratory, the fungus requires a humid environment to grow and is capable of developing in water. These circumstances, plus the humid atmosphere of the endemic areas, which are characterized by abundant forests and water courses, lead to the belief that *P. brasiliensis* should have a microniche in or around water collections.[5]

4. HOST DEFENSES

In PCM the host–parasite interactions are complex; the former employs both arms of the immune system, as well as his innate resistance, to combat the microorganism, while the latter makes the most of its virulence powers to overcome such defenses. The fact that not every infected, paracoccidioidin-positive individual develops overt PCM indicates that humans are quite capable of restraining the fungus from becoming established in their tissues.[31–33] The following mechanisms of defense merit special comments.

4.1. Innate Resistance

Phagocytosis fulfills an important role in host defenses, as demonstrated by macrophage blockade: mice that receive blockers have more extensive lesions than control animals.[31] Peripheral blood PMNs from PCM patients phagocytose the fungus normally but are impaired in their digestive capacity.[34,35] Recent experiments with mouse PMNs revealed that death of the fungus does not occur unless the phagocytes had been activated previously with lymphokines.[35a] *P. brasiliensis* propagules—yeast or conidia—are susceptible to the products of oxidative metabolism generated by the H_2O_2-peroxidase-halide system; furthermore, the more virulent the isolate, the lower its susceptibility to neutrophil digestion.[36,37] *P. brasiliensis* activates the complement system through the alternative pathway but PMN's chemotaxis is not evoked directly. Complement, however, seems to play a role in phagocytosis, especially if accompanied by specific antibodies.[32,38]

In vitro natural killer cells have been shown to inhibit the growth of yeast targets; such cells may interact with the fungus in the lungs, during the initial phase of the infection and hinder its progress.[39]

It is well known that host resistance (or susceptibility) is controlled genetically. Certain HLA phenotypes have been detected with higher frequencies among patients with the disease.[32,40] By the same token, mice susceptibility to experimental infection varies with the strain; for instance, strain A/SN is resistant, while B10, D2/nSn, and B10A, are susceptible.[31] Further studies have revealed that mouse resistance to the fungus depends on a single autosomal gene.[41]

P. brasiliensis produces an array of antigenic moities, both cellular and metabolic, that are liberated (or produced) during fungal growth. Such antigens interact with the host's immune system evoking humoral and cell-mediated responses.[42]

4.2. Humoral Immunity

Antibodies of multiple specificities are produced regularly by PCM patients and *P. brasiliensis* experimentally infected animals. Four immunoglobulin classes (IgG, IgM, IgE and IgA) are detectable in patients; the predominant specific anti-*P. brasiliensis* antibodies are of the IgG class; however, IgE and IgA antibodies are also found.[33,43] Characteristically, PCM results in a polyclonal activation of the humoral system.[44] At time of diagnosis and during the early period of therapy, IgG levels are high, with their concentration paralleling the degree of invasion. These antibodies decrease later on with clinical improvement but may remain at low, stationary levels for 1–2 years posttherapy.[43] In contrast, IgA and IgM levels are usually normal at diagnosis.[33] High levels of IgE occur in active cases and their titer correlates with severity of the disease process; however, less than 1.0% of such IgE represents anti-*P. brasiliensis* antibodies.[45,46]

Antibodies are potent opsonizers and, in conjunction with the complement system, are deleterious to the fungus.[33,47] Experimental infections in mice that had been selected for their ability to produce high or low antibody levels suggested that such antibodies participate in host defense against the fungus.[48]

Circulating immune complexes have been detected in PCM patients. In a number of patients, a positive correlation among levels of the immune complexes, activity of the disease, and depression of cellular immunity has been shown.[49,50] To date, however, there are no indications that immune complexes cause direct deleterious effects to the host.[31] Circulating *P. brasiliensis* antigens can be detected in some patients with the disease.[51]

4.3. Cell-Mediated Immunity

It is well known that in PCM there is marked depression of cell-mediated immunity (CMI), the intensity of which parallels the severity of the infectious process.[31–33] CMI has been measured by skin testing, lymphoblastogenic assays, lymphokines production, T cell, and determination of total T lymphocytes and subsets. T-cell-mediated toxicity and T-cell-mediated suppression have also been used to further characterize CMI in this disease.[31–33]

Judging by skin testing, approximately half the patients with active disease fail to react to intradermal testing with homologous and heterologous antigens.[31–33] Furthermore, 30% of these patients are refractory to sensitization with 2-4-dinitrochlorobenzene.[31] Of importance, depression of delayed type hypersensitivity is a reversible phenomenon in that clinical remission is often associated with reacquisition of skin-test reactivity.[43]

Lymphocyte transformation tests have been used in an effort to better define the patient's CMI responses. The results indicate that there are significant differences between control subjects and patients with active disease, the latter exhibiting diminished responses to paracoccidioidin, heterologous antigens and mitogens.[43,52] It has been reported that the patient's own serum is capable of suppressing the blastogenic response of his or her lymphocytes and/or those obtained for normal controls. Furthermore, when the patient's lymphocytes are cultured in the presence of sera from normal donors, no inhibition occurs.[52] Remission of the disease is often accompanied by a decrease in the inhibitory effects exerted by sera. Thus, certain patients have suppressive factors that are capable of inhibiting the normal lymphocyte blastogenic responses.[31–33] The factor responsible for the suppressive effect has not been identified but excesses of antibody, immune complexes, and fungal products may be implicated.[31] Concerning the latter possibility, it has been demonstrated that *P. brasiliensis* produces a soluble antigen that, when injected intravenously, is capable of down-regulating the cellular immune response.[53,54]

In search for an explanation for the immune depression exhibited by PCM patients, the composition of the T-cell population has been studied. In untreated patients, characterization of the mononuclear cell subsets has revealed the presence of elevated numbers of monocyte/null cells, a low helper-to-suppressor ratio, and a reduction of both the percentage of total T cells and of the helper–inducer subset. Patients with severe disease had high levels of suppressor–cytotoxic T cells.[55] In biopsies, the granulomas have shown that the T-cell mantle that surrounds the fungal cells is mostly composed of T-helper cells. The histiocyte population in such granulomas is lys+ (phagocytic) with a few S-110+ (antigen presenter cells). It is possible that the interaction of the cell subsets in the granuloma itself results in liberation of stimulating factors, such as interleukin 2, which could activate the macrophages at the site of infection.[2,56] The presence of abundant

T-helper cells in the granulomas and the fact that the number of such cells in peripheral blood is low, suggest that such cells are trapped in the lesion; consequently, a process of recirculation of T lymphocytes must be operating.[56]

4.4. Histopathology

From experimental animal studies, early tissue interactions have been shown to consist of a nonspecific inflammation with a large accumulation of PMNs. Mononuclear cells, a large proportion of which are macrophages, start to appear a few days after inoculation. Later on, giant cells are formed, and in 5–7 days, epithelioid cells become recognizable with formation of granulomas.[31] Compact epithelioid granulomas with sparse fungal cells are formed in patients who have functional cellular immune systems. In patients who are deficient in their cell-mediated responses, granulomas tend to be less compact, and there are abundant, actively proliferating yeast cells.[31,55] Experiments with *P. brasiliensis* antigen coated particles have revealed that tissue reactions are similar to those obtained in response to whole yeast cells. These results strengthen the importance of CMI in the formation of ephitelioid granuloma.[2,31]

4.5. Regulation of the Immune Response

The studies by Jimenez and Murphy[53,54] have explored the characteristics of the system that suppresses cell-mediated immune responses. These authors found that a soluble antigen, when injected intravenously into mice, induced a subpopulation of suppressor cells capable of modulating the CMI responses. After one week, the lymph nodes and the spleen of these animals showed cells that, upon adoptive transfer, suppressed the afferent (inductor) limb of delayed type hypersensitivity (DTH). These suppressor cells were classified as T lymphocytes and the phenotype corresponded to L3T4 + Lyt+-1+2-I-J+; they were considered to be TS1 cells in the cascade of suppressor T cells. Although such cells inhibited the afferent limb, they did not interfere with the efferent (expression) limb. Nonetheless, if cells were taken 14 days after injection of the antigen, they were then capable of hindering the efferent limb. It was later demonstrated that these second order cells were T lymphocytes with a different phenotype (L3T4-Lyt-2+IJ+). The suppressor circuit probably starts with the induction of TS1 cells by the fungal antigen; these cells would liberate a factor capable of stimulating the production of the second order cells (Ts2), in charge of suppressing the efferent limb for DTH. If one were to extend the results of animal experimentation to humans, one could consider that a circuit of suppressor cells may be operative patients with the disease, as they are subjected to the action of circulating *P. brasiliensis* antigens. The T-cell-mediated immune depression that characterizes the disease could be ascribed to such a circuit.

By the same token, it has been shown that, in the mouse, the progression of the chronic disseminated form of PCM was associated with suppressor T cells; such cells hindered the normal response of circulating lymphocytes. It was further demonstrated that when mouse sera, rich in anti-*P. brasiliensis* antibodies, were passively transferred to normal animals before inoculation they were unable to exhibit DTH.[57]

In vitro experiments have shown that when resident peritoneal or pulmonary macrophages were co-cultured with *P. brasiliensis* yeast cells, there was a dramatic increase in fungal multiplication as compared with complete culture medium alone.[58] In sharp contrast to resident macrophages, lymphokine-activated peritoneal or pulmonary macrophages not only restricted the growth of the fungus, but killed a large proportion of the yeast cells in the inoculum.[58] Unpublished experiments by the author have shown that this is also the case with *P. brasiliensis* conidia, the presumed naturally infecting propagule; resident macrophages allow transformation of the conidia into yeast and fostered their multiplication; on the other hand, lymphokine-activated macrophages not only restricted the conidia-to-yeast conversion, but also killed a high proportion of the viable conidia present in the inoculum (Table II). These experiments reveal another facet of the host's multifactorial immunomodulating system.

TABLE II
Critical Role of Activation of Murine Phagocytes for Effective Killing of *P. brasiliensis* Propagules

Type of phagocyte	Activation of phagocytes	Percent of reduction in the viability of the inoculum: Yeast cells[a]	Conidia[b]	Ref. no.
Peritoneal polymorphonuclear leukocytes	Normal lymphokines[c]		3.8	Personal data (unpublished)
	Immune lymphokines[d]		48.1	
Peritoneal macrophages	None[e]	2.0		85
	Gamma interferon (100 U/ml)	24.0		
Pulmonary macrophages	None	18.5		86
	Gamma interferon (1000 U/ml)	45.5		
	Normal lymphokynes[c]		28.3	Personal data (unpublished)
	Immune lymphokines[d]		73.3	

[a]Viability determined by CFU.
[b]Viability determined by production of germ tubes in a microculture system.
[c]Supernatant of spleen cells from normal mice challenged with *P. brasiliensis* antigen.
[d]Supernatant of spleen cells from immune mice challenged with *P. brasiliensis* antigen.
[e]uCulture medium alone.

On the part of the parasite, its interaction with the host is also complex. It has been known that *P. brasiliensis* isolates vary in their virulence powers. The absence of α-(1,3)-glucan in an avirulent isolate led to studies tending to demonstrate that such α-(1,3)-glucan was directly related with the fungus virulence.[1] Cell-surface-associated alpha glucan may act as a shield against host defense mechanisms because their digestion by phagocytes is difficult to achieve.[59] Attenuation of virulence has been observed in isolates that have been subcultured repeatedly.[60] Furthermore, a variant of the fungus that reproduces as a yeast at room temperature (± 25°C) was found to be more virulent for mice than were the yeast (cultured at 36°C) or the conidia obtained from the same isolate.[61] It is apparent then, that multiple factors influence the isolate's capacity to grow in the host's tissues.

5. IMMUNODIAGNOSIS AND PROGNOSIS

PCM can be diagnosed by direct microscopic (KOH) or biopsy of properly collected samples. In over 90% of cases, the relative large size of the fungus (average 8–16 microns), its thick cell wall, refractoriness, and the characteristic multiple budding yeast cells allow proper identification. Histological examination is a very reliable procedure, especially if the silver methenamine stain is used. Phagocytosis of the fungus by both macrophages and giant cells is common. In most patients, there is granuloma formation.[1,2] Cultures are required to ascertain the activity of the mycotic process but fungal isolation is not always possible; nonetheless, with appropriate techniques and culture media, 60–80% of the samples yield the microorganism. Samples not heavily contaminated with bacteria can be cultured at 37°C and, in this case, the colonies will appear in 10–15 days; when such colonies are examined microscopically, the characteristic yeast cells become apparent. If the specimen contains bacteria and yeast flora—such as in sputum samples—the culture media should contain antibiotics and antimycotics; prolonged incubation (approximately 30 days) at 21–24°C renders better results in this case; colonies are white, compact, and have short aerial mycelia. In the regular mycological media, the mycelial phase is noncharacteristic, since sporulation is rare. In lean media, conidia of various sorts, mostly bulging arthroconidia, can be observed. It is recommended, however, that the mold be converted to the yeast form by subculturing it at 35–37°C.[3,15]

Immunodiagnosis of PCM can be accomplished by various means. Serologic tests designed to measure antibodies have been used for many years. The agar gel immunodiffusion technique (AGID) is simple to use and has a high predictive value.[32,62] The use of culture filtrate antigens derived from either the yeast or the mycelium forms allow recognition of 3 bands of precipitate; band 1 is present in 93% of cases at diagnosis and is specific for paracoccidioidomycosis; band 2 is detected with lesser frequency; band 3 is of rare occurrence and may cross-react with histoplasmin.[32,63,64] During the course of therapy, band 1 tends to disappear, but as much time as 3–5 years may elapse before the test becomes negative.[32] The counterimmunoelectrophoresis technique is also highly sensitive and specific but requires special equipment. When *P. brasiliensis* antigens are subjected to electrophoresis and are then allowed to interact with patients' sera, an arc of precipitate, termed E2, is formed; such an arc occurs in almost all active patients and is specific for PCM.[65]

Another very useful technique for the serological diagnosis of PCM is complement fixation (CF); the relative complexity of the test hinders its wider use; however, CF is a quantitative test and allows a more precise measurement of the antibody contents in serum samples than the agar gel procedures. Such measurements are important in the follow-up study of the patients because the results bear certain prognostic value. At diagnosis, CF titers tend to be high but, if the patient's response to therapy is adequate, titers decrease consistently. An increase in antibody contents, during or after termination of treatment, points towards an exacerbation of the mycotic process or to a relapse.[32,33,46] The CF is less specific than the AGID tests; 80% of the sera from PCM patients cross-react with *Histoplasma capsulatum* antigens, especially with the whole yeast reagent. The heterologous titers tend to be lower than the homologous ones. The simultaneous use of AGID and CF tests, however, allows proper interpretation of the results.[32,33]

Newer diagnostic procedures, such as various modalities of the enzyme-linked immunoabsorbent assays (ELISA), are being employed currently.[64,66–68] They are extremely sensitive but not very specific and consequently, interpretation of the results in a nonconfirmed patient represents a problem. Immunofluorescent techniques have also been used successfully by certain groups.[69] The indirect test has the advantage of allowing determination of the immunoglobulin types.

In general, antibody determination is important in diagnosis and follow-up of patients with PCM. When there are no external lesions amenable to microscopic examination, these tests often furnish the first clue to the correct diagnosis. Because PCM is a chronic disease requiring prolonged therapy, serologic tests constitute an adequate monitoring system to determine the patient's improvement.[62,69–71] Posttreatment follow-up observations also benefit from antibody level determination.[69,72] Serologic tests correlate with the extent of the mycosis; the higher the antibody titers (or number of precipitin bands), the more severe the disease. Serologic testing has only one drawback, namely, lack of antigen standardization. There are various types of antigens in use (culture filtrates, cell-derived products) and, although they appear adequate, comparisons between laboratories constitute a problem. At present, efforts are being made to characterize specific *P. brasiliensis* antigens; one of these specific antigens is a 43 KDA glycoprotein that is regularly recognized in PCM patients sera.[73,74] The production of this purified antigen is not too cumbersome and the activity of the various lots is reproducible.[75] This antigen has now been cloned and, consequently, mass production seems feasible.[76]

Tests have been used also to determine the presence of circulating antigens.[51,77] Antigenemia may become important for monitoring response to therapy.

Techniques aimed at determining the status of cell-mediated immunity in PCM patients have certain prognostic values but have not been used as extensively as those that measure antibody response. Skin tests with paracoccidioidin were much used in the past to attempt a diagnosis rather than to learn about the patient's DTH. It was soon learned that over 50% of active, untreated patients failed to recognize the antigens present in paracoccidioidin, a culture filtrate antigen, indicating that the intradermal test had no diagnostic value.[32,33,43] Furthermore, a certain proportion of healthy individuals from the endemic areas are reactive to paracoccidioidin as a result of previous contact with the fungus.[32,33] Nonetheless, certain patients who are unreactive at diagnosis, convert to paracoccidioidin positive during therapy, indicating restoration of their cell-mediated immunity; this change indicates improved prognosis.[32,33]

Tests have been conducted to determine the characteristics of the T-cell populations in patients with PCM. Generally, there is depression of cell-mediated immunity, the degree of which correlates with the severity of the mycosis.[2,31,52,55,56] Studies with monoclonal antibodies have revealed that there is an imbalance of the mononuclear cell subsets with a low helper-to-suppressor ratio.[55] Successful treatment brings forth not only clinical improvement but also a restoration of the above balance.[43] In most patients, B-cell function appears normal but the humoral response is rather hyperactive with polyclonal activation of the system (increased IgG, IgE, and IgA levels).[44] The levels of the former two immunoglobulins increase with dissemination of the mycosis and with the severity of its clinical presentation.[45]

6. PREVENTION AND THERAPY

At present, prevention of PCM is not feasible. The lack of precise knowledge concerning the fungus habitat and the circumstances conducive to encountering *P. brasiliensis*, hinders any attempt at preventing the infection.

On the other hand, when PCM becomes manifested as a disease entity, there are a number of therapies that can be instaurated with relative success;[78] nonetheless, healing of the mycosis brings forth incapacitating sequelae.

6.1. Chemotherapy

Most of the antifungal agents used in the treatment of PCM are fungistatic rather than fungicidal; consequently, for an effective response to occur, the patient's immune response must cooperate. As already discussed, cellular immunity is impaired in an important proportion of the patients. It is mandatory, therefore, to offer supportive therapeutic measurements aimed at restoration of the general well-being of the patient. There are various categories of drugs that can be used for treatment of PCM, as follows in the next three sections (6.1.1., 6.1.2., and 6.1.3.).

6.1.1. Sulfonamides

PCM is the only mycosis amenable to sulfonamide therapy. First used by Ribero,[78] in 1940, it is still employed by some because of facility for oral administration, low cost, and low levels of toxicity. In adults, sulfadiazine is given at a maximum dose of 6 g/day and in children, it is administered in divided doses of 60 to 100 mg/kg/day. After achieving clinical and serologic improvement, the dosage can be diminished. If the slow-acting sulfonamides (sulfamethoxypyridazine) are preferred, maximum daily dose is 1 to 2 g/day. The main disadvantage of sulfonamide therapy is its duration, as the patient must remain under continuous treatment for three to five years; otherwise, relapses occur in 25–30% of patients, with half of them becoming resistant to the drug. Sulfonamides may be used in ambulatory patients with mild to moderate disease; these drugs can also be employed as maintenance therapy after completion of a course of amphotericin B.[78–80] The patient should be instructed to increase water intake when taking sulfonamides to avoid crystalluria. Improvement should be seen within the first 3 months of therapy.[78–80] The combination trimethoprim-sulfamethoxazol has been found effective in patients who relapse.[80] Recently, a combination of sulfadiazine and trimethoprim (cotrimazine) has been tested with apparently good results.[81]

6.1.2. Amphotericin B

This polyene is highly effective in PCM and until 1976, it was the only drug available for the treatment of seriously ill patients or for those who were resistant to sulfonamides. The need for intravenous administration requiring a hospital setting, however, as well as its serious side effects and relative high cost, limit its application. A test dose of 1.0 mg of amphotericin B should be administered initially to determine potential adverse effects; this done, the dose should be adjusted to 0.5 mg/kg/day, with periodical increases up to 0.75 mg/kg/day; the intravenous infusion is to be given during a 4–6-hour period. Alternate day therapy can be given in patients who are not seriously ill. The duration of treatment depends on the patient's status, seriousness of the mycotic process, and tolerance. Serologic tests are useful adjuncts in determining the length of treatment. The patient's condition should improve after 3–4 weeks of therapy; 70% of patients improve with this drug. Some authors suggest giving a total dose of 1.0 g and then continuing therapy under an ambulatory basis, with sulfonamides or imidazole derivatives. Relapses are not uncommon (15%); there have been no reports on acquired resistance of *P. brasiliensis* to amphotericin B.[78,80] Amphotericin side effects are numerous and some are severe enough to hinder completion of therapy; particularly serious are those occurring in the kidneys. From the point of view of the patient's response, the combination amphotericin B-sulfonamide therapy offers advantages to single drug regimen.[69] The possibility of using liposomal amphotericin B is attractive but no specific data are available.

6.1.3. Imidazole Derivatives

The new orally administered imidazoles (ketoconazole, itraconazole, fluconazole) are effective in the treatment of PCM. The experience with ketoconazole is now sufficiently large to allow a fair judgment. Ketoconazole given at daily doses of 200 mg for 6–12 months produces sustained clinical improvement; such improvement becomes apparent shortly after (2–3 weeks) initiation of therapy. Careful observations have revealed that at the end of a 6-month course, 18% of patients had resolved all their pretherapy abnormalities, 70% have experienced major improvement, while 8% showed only minor improvement and required a more prolonged course of therapy. Only 2% of all patients appeared refractory to ketoconazole treatment. Relapse rate is lower (9%) than for other therapies. Ketoconazole can also be given to children at a dose of 50 to 200 mg/day (50 mg/kg/day) depending on the patient's weight. Most investigators in the endemic area believe that ketoconazole is the best therapeutic option available.[70,78–80] Ketoconazole's most important side effect, symptomatic hepatitis, is seen rarely (1 in 10,000 cases); the drug effects on testosterone synthesis, which occur when large doses (over 400 mg/) are given daily, is reversible. Due to interference with cytochrome P450-lycase, corticosteroid synthesis is decreased but hypoadrenalism symptoms have not been reported. Ketoconazole levels are decreased when rifampin is administered simultaneously; consequently, if the patient has concomitant tuberculosis, both drugs should be given separately. Ketoconazole requires an acid pH for proper absorption and, consequently, antacids and cimetidine-related compounds should not be ingested together.[77–82]

A newer imidazole derivative itraconazole, has now been tried in over 100 PCM patients. Itraconazole is more potent (it is administered at 100 mg dose) than ketoconazole, hepatic toxicity has not been demonstrated and there is no interference with testosterone or corticosteroid synthesis. Itraconazole shares with ketoconazole its oral absorption (under acidic pH) but produces higher plasma levels and is selectively bound to tissues. The duration of therapy varies from 3 to 12 months depending on the patient's condition and the promptness of response. Improvement begins to be noticed 1–2 weeks after initiation of therapy. In 93% of patients, 6 months of therapy appear adequate to control the mycotic process. The drug has been used in a few patients with the juvenile type of PCM with good results. As for relapses, the information is still preliminary; however, in a series of 30 itraconazole-treated patients who have been followed-up for 1–2 years post-therapy, no relapses have been documented;[71] in a different series, 7% of the patients relapsed.[83] If, on further trials and lengthier posttherapy follow-ups, itraconazole continues to show such a high effectivity, it would replace ketoconazole as the drug of choice since it has fewer side effects and the corresponding treatment periods appear shorter.

The newer of the imidazole derivatives that has reached the stage of clinical trials is fluconazole. It differs from the other compounds in that it is water soluble and is available for both oral and parenteral administration. Preliminary trials indicate that fluconazole is also an effective therapeutical agent; however, more patients and more prolonged observation periods are deemed necessary for an appropriate evaluation.[84]

In PCM, the therapeutic measures just discussed require careful monitoring of the patient's condition during therapy; clinical, radiological and mycological examinations should be performed on a regular basis. Because this mycosis is not diagnosed as promptly as would be desirable, there will be additional problems to consider, among them emaciation, fibrotic sequelae, adrenal hypofunction, and others. Consequently, even the best of therapies may be ineffective; cumulative mortality ranges from 17 to 25%, with higher figures being recorded in juvenile patients.[78]

6.1.4. *Immunotherapy*

Immune regulation measures have not been tried in humans. Experimental animal data shows that levamizole helps in restoring cellular immunity, with prompter clearing of infection in the animals receiving ketoconazole and levamizole simultaneously.[85] *In vitro* studies have shown that γ-interferon significantly promotes killing of *P. brasiliensis* yeast cells by pulmonary and peritoneal macrophages.[86–87] Because the fungus immunosuppresses the host, antimycotic therapy tilts the balance in the host's favor by decreasing fungal multiplication and allowing the host to increase expression of immune defenses. It has been repeatedly shown that the patient's cellular immunity improves with antifungal therapy with no additional measures being required to restore the immune imbalance.[31,32] Such restoration may take some time, however, and in severely ill patients, a form of immunotherapy would be highly desirable.

REFERENCES

1. Angulo, A., and Pollak, L., 1971, Paracoccidioidomycosis, in: *The Pathologic Anatomy of the Mycoses. Human Infections with Fungi, Actinomycetes and Algae* (R. H. Baker, ed.), Springer–Verlag, Berlin, pp. 507–576.
2. Franco, M., Mendes, R. P., Moscardi-Bacchi, M. M., Rezkallah-Iwasso, M., and Montenegro, M. R., 1989, Paracoccidioidomycosis, *Bailliere's Clin. Trop. Med. Comm. Dis.* **4:**185–220.
3. Lacaz, C. S., Porto, E., and Martins, J. E. C., 1984, *Micologia Medica*, 7th Ed., Sarvier Publishers, Sao Paulo, pp. 190–216.
4. Londero, A. T., and Melo, I. S., 1988, Paracoccidioidomicose, *J. Brazil Med. (JBM)* **55:** 96–115.
5. Restrepo, A., 1985, The ecology of *P. brasiliensis*: A puzzle still unsolved, *J. Med. Vet. Mycol.* **28:**323–334.
6. Giraldo, R., Restrepo, A., Gutierrez, F., Robledo, M., Londono, F., Hernandez, H., Sierra, F. and Calle, G., 1976, Pathogenesis of paracoccidioidomycosis. A model based on the study of 46 patients, *Mycopathologia* **58:**63–70.
7. Restrepo, A., Trujillo, M., and Gomez, I., 1989, Innaparent lung involvement in patients with the subacute juvenile type of paracoccidioidomycosis, *Rev. Inst. Med. Trop. Sao Paulo* **31:** 18–22.
8. McEwen, J. G., Bedoya, V., Patiño, M. M., Salazar, M. E., and Restrepo, A., 1987, Experimental paracoccidioidomycosis induced by the inhalation of conidia, *J. Med. Vet. Mycol.* **25:** 165–175.
9. Restrepo, A., Robledo, M., Giraldo, R., Hernandez, H., Sierra, F., Gutierrez, F., Londono, F., Lopez, R., and Calle, G., 1976, The gamut of paracoccidioidomycosis, *Am. J. Med.* **61:**33–42.
10. Londero, A. T., 1986, Paracoccidioidomicose. Patogenia, formas clinicas, manifestacoes pulmonares e diagnostico, *J. Pneumol. (Brazil)* **12:**41–60.
11. Franco, M., Montenegro, M. R., Mendes, R. P., Marques, S. A., Dillon, N. L., and Mota, N. G. S., 1987, Paracoccidioidomycosis. A recently proposed classification of its clinical forms, *Rev. Soc. Brasil. Med. Trop.* **20:**129–132.
12. Londero, A. T., and Melo, I. S., 1983, Paracoccidioidomycosis in Childhood. A critical review, *Mycopathologia* **82:**49–55.
13. Londero, A. T., Gonzalez, A. J. R., Cruz, M. L. S., Rozembaum, R., Cunha, R. Q., Machado, E. S., Vieira, A. R., Carvalho, F. G., Braga, M. P., Azevedo, E. C. L., Wanke, B., Cruz, M. F., and Menezes, J. A., 1987, Paracoccidioidomicose disseminada "infanto-juvenil" em adolescentes, *Arq. Bras. Med.* **61:**5–12.
14. Salfelder, K., Doehnert, G., and Doehnert, H. R., 1969, Paracoccidioidomycosis: Anatomic study with complete autopsies, *Virchows Arch. Pathol. Anat. Physiol.* **348:**51–56.
15. Del Negro, G., Lacaz, C. S., and Fiorillo, A. M. (eds.), 1982, *Paracoccidioidomicose (Blastomicose Sudamericana)*, Sarvier, Sao Paulo, Brazil.
16. Giorgi, M. C. R., Camargo, E. E., Pinto, W. P., and del Negro, G., 1987, Gallium-67 imaging in the diagnosis of blastomycosis, *Eur. J. Nucl. Med.* **13:**300–304.
17. Abad, A., Gomez, I., Velez, P., and Restrepo, A., 1986, Adrenal function in paracoccidioidomycosis. A prospective study in patients before and after ketoconazole therapy, *Infection* **14:**22–26.
18. Goldani, L. Z., Maritnez, R., Landell, G. A. M., Machado, A. A., and Coutinho, V., 1989, Paracoccidioidomycosis in a patient with the acquired immunodeficiency syndrome, *Mycopathol.* **105:**71–74.
19. Pedro, R., de J., Aoki, F. H., Bricato, R. S. B. S., Branchini, M. L. M., and Goncalves, F. L., 1989, Para-coccidioidomycosis and HIV infection, *Rev. Inst. Med. Trop.* **31:**119–125.
20. Sugar, A. M., Restrepo, A., and Stevens, D. A., 1984, Paracoccidioidomycosis in the immunosuppressed host, *Am. Rev. Respir. Dis.* **129:**340–342.
21. Londero, A. T., Santos, W., da Silva, L. A., and Ramos, C. D., 1987, Paracoccidioidomicose

associada a droga imunosupressora en paciente con lupus eritematoso sistemico, *J. Pneumol.* (Brazil) **13**:224–226.
22. Lacaz, C. S., 1975, Aspectos micologicos da paracoccidioidomicose, *Ars Curandi. Rev. Terap. Med. (Brazil)* **7**:6–13.
23. Bustamante, B., McEwen, J., Tabares, A. M., and Restrepo, A., 1985, Characteristics of the conidia produced by the mycelial form of *P. brasiliensis*. *Sabouraudia (J. Med. Vet. Mycol.)* **23**: 407–414.
24. San Blas, G., and San Blas, F., 1972, *P. brasiliensis*: cell wall structure and virulence. A review, *Mycopathol.* **62**:77–86.
25. Ajello, L., and Polonelli, L., 1985, Imported paracoccidioidomycosis: A public health problem in non-endemic areas, *Eur. J. Epidemiol.* **1**:160–165.
26. Greer, D. L., and Restrepo, A., 1975, *Epidemiology of paracoccidioidomycosis*, in: *The Epidemiology of Human Mycotic Diseases* (Y. Al-Doory, ed.), Charles C. Thomas, Springfield, IL, pp. 117–141.
27. Restrepo, A., de Bedout, C., Cano, L. E., Arango, M. D., and Bedoya, B., 1981, Recovery of *P. brasiliensis* from a partially calcified lymph node lesion by microaerophilic incubation in liquid media, *Sabouraudia* **19**:295–300.
28. Marquez, S. A., Franco, M. F., Mendes, R. P., Silva, N. C. A., Baccili, C., Curcelli, E. D., Feracin, A. C. M., Oliveira, C. S., Tagliarini, J. V., and Dillon, N. L., 1983, Some epidemiological aspects of paracoccidioidomycosis in Botucatu endemic area, State of Sao Paulo, Brazil, *Rev. Inst. Med. Trop. Sao Paulo* **25**:87–92.
29. Restrepo, A., Salazar, M. E., Cano, L. E., Stover, E. P., Feldman, D., and Stevens, D. A., 1984, Estrogens inhibit mycelium to yeast transformation in the fungus *P. brasiliensis*. Implications for resistance of females to paracoccidioidomycosis, *Inf. Imm.* **46**:346–353.
30. Naiff, R. D., Ferreira, L. C. L., Barret, T. V., Naiff, M. F., and Arias, J. R., 1986, Enzootic paracoccidioidomycosis in armadillos (*D. novemcinctus*) in Para State, Brazil, *Rev. Inst. Trop. Sao Paulo* **28**:19–27.
31. Franco, M. F., 1987, Host–parasite relationship in paracoccidioidomycosis, *J. Med. Vet. Mycol.* **25**:5–18.
32. Restrepo, A., 1987, Immune response to *P. brasiliensis* in human and animal hosts, in: *Current Topics in Medical Mycology*, Vol. 2 (M. R. McGinnis, ed.), Springer–Verlag, London, pp. 239–277.
33. Jimenez-Finkel, B., and Restrepo, A., 1989, Paracoccidioidomycosis, in: *Immunology of Fungal Diseases* (R. A. Cox, ed.), CRC Press, Boca Raton, FL, pp. 227–247.
34. Schaffner, A., Davis, C. E., Schaffner, T., and Markert, M., 1986, *In vitro* susceptibility of fungi to killing by neutrophil granulocytes discriminates between primary pathogenicity and opportunism, *J. Clin. Invest.* **78**:511–524.
35. Goihman-Yahr, M., Rothenberg, A., Bretana, A., Isturiz, G., and Rosquete, R., 1989, Digestion of killed *P. brasiliensis* by neutrophils, *Mycopathol.* **106**:53–58.
35a. Cano, L. E., Salazar, M. E., and Restrepo, A., 1989, Phagocytosis and killing of *P. brasiliensis* conidia by PMNs. Abstr. 73, American Society for Microbiology, p. 470.
36. McEwen, J., Brummer, E., Stevens, D. A., and Restrepo, A., 1987, Effect of PMN lymphocytes on the yeast form of *P. brasiliensis*, *J. Trop. Med. Hyg.* **36**:603–608.
37. McEwen, J. G., Salazar, M. E., and Restrepo, A., 1987, Susceptibility of *P. brasiliensis* conidia to products of oxidative metabolism, *Exp. Mycol.* **11**:241–244.
38. Calich, V. L. G., Kipnis, T. L., Mariano, M., Fava Netto, C., and Dias, W. D., 1979, The activation of the complement system of *P. brasiliensis* in vitro: its opsonic effect and possible significance for an in vivo model of infection, *Clin. Immunol.* **12**:20–30.
39. Jimenez, B. E., and Murphy, J. W., 1984, In vitro effects of natural killer cells against *P. brasiliensis* yeast phase, *Inf. Imm.* **46**:552–558.
40. Lacerda, G. B., Arce-Gomez, B., and Queiroz-Tellez, F., 1988, Increased frequency of HLA-B40 in patients with paracoccidioidomycosis, *J. Med. Vet. Mycol.* **26**:253–256.
41. Calich, V. L. G., Burger, E., Kashino, S. S., Fazioli, R. A., and Singer-Vermes, L. M., 1987,

Resistance to *P. brasiliensis* in mice is controlled by a single autosomal gene, *Inf. Imm.* **55:** 1919–1923.
42. San Blas, G., and San Blas, F., 1989, Antigenic structure of *P. brasiliensis*, in: *Immunology of Fungal Diseases* (E. Kurstak, ed.), Marcel Dekker, Inc., New York, pp. 171–192.
43. Restrepo, A., Restrepo, M., Restrepo, F., Moncada, L. H., Aristizabal, L., and Velez, H., 1978, Immune responses in paracoccidioidomycosis. A controlled study of 16 patients before and after treatment, *Sabouraudia* **16:**151–163.
44. Chequer-Bou-Habib, P., Daniel Ribeiro, C., Banic, D. M., Franciscone do Valle, A. C., and Galvao-Castro, B., 1989, Polyclonal activation in paracoccidioidomycosis, *Mycopathol.* **108:** 89–94.
45. Arango, M., and Yarzabal, L, 1982, T cell dysfunction and hyperimmunoglobulinemia E in paracoccidioidomycosis, *Mycopathol.* **79:**153–158.
46. Biagioni, L., Souza, M. J., Chamma, L. G., Mendes, R. P., Marques, S. A., Mota, N. G. S., and Franco, M., 1984, Serology of paracoccidioidomycosis. Correlation between class specific antibodies and clinical forms of the disease, *Transact. Royal Soc. Trop. Med. Hyg.* **78:**617–621.
47. Nogueira, M. E. S., Mendes, P. P., Marquez, S. A., and Franco, M., 1986, Detection of antibodies in paracoccidioidomycosis by the technique of complement-mediated lysis, *Braz. J. Med. Biol. Res.* **19:**241–247.
48. Carvalhaes, M. S., Silva, W. D., Girman, E. G., Sant'Anna, O. A., Abrahamsohn, P., and Kipnis, T. L., 1986, Experimental paracoccidioidomycosis in high and low antibody-producer mice, *Ann. Inst. Pasteur/Immunol.* **137:**127–141.
49. Arango, M., Oropeza, F., Anderson, D., Contreras, C., Bianco, M., and Yarzabal, L., 1982, Circulating immune complexes and in vitro reactivity in paracoccidioidomycosis, *Mycopathol.* **79:**153–158.
50. Chequer-Bou-Habib, D., Oliveira-Neto, M. P., Ferreira-da-Cruz, M. F., and Galvao-Castro, B., 1989, The possible role of circulating immune complexes in the deficiency of cell-mediated immunity in paracoccidioidomycosis, *Brasilian J. Med. Biol. Res.* **22:**205–212.
51. Garcia, N. M., Del Negro, G. M. B., Martins, H. P., and Lacaz, C. S., 1987, Detection of paracoccidioidomycosis circulating antigen by the immunoelectrophoresis immunodiffusion technique, *Rev. Inst. Med. Trop. Sao Paulo* **29:**327–328.
52. Costa, J. C., Pagnano, P. M. G., Bechelli, L. M., Fiorillo, A. M., and Lima Filho, E. C., 1983, Lymphocyte transformation test in patients with paracoccidioidomycosis, *Mycopathol.* **84:** 55–63.
53. Jimenez-Finkel, B. E., and Murphy, J. W., 1988, Characterization of efferent T suppressor cells induced by *P. brasiliensis*-specific afferent T suppressor cells, *Inf. Imm.* **56:**744–750.
54. Jimenez-Finkel, B. E., and Murphy, J. W., 1988, Induction of antigen specific T suppressor cells by soluble *P. brasiliensis* antigen, *Inf. Imm.* **56:**734–743.
55. Mota, N. G. S., Rezkallah-Iwasso, M. T., Peracoli, M. T. S., Andi, R. C., Mendes, R. P., Marcondes, J., Marques, S. A., Dillon, N. L., and Franco, M., 1985, Correlation between cell-mediated immunity and clinical forms of paracoccidioidomycosis, *Trans. Royal Soc. Trop. Med. Hyg.* **79:**765–772.
56. Moscardi-Bacchi, M., Soares, A., Mendes, R, Marques, S., and Franco, M., 1989, In situ localization of T lymphocyte subsets in human paracoccidioidomycosis, *J. Med. Vet. Mycol.* **27:**149–158.
57. Castaneda, E., Brummer, E., Pappagianis, D., and Stevens, D. A., 1988, Regulation of immune response by T suppressor cells and by serum in chronic paracoccidioidomycosis, *Cell Immunol.* **117:**1–11.
58. Brummer, E., Hanson, L. H., Restrepo, A., and Stevens, D. A., 1989, Intracellular multiplication of *P. brasiliensis* in macrophages: killing and restriction of multiplication of activated macrophages, *Inf. Imm.* **57:**2289–2294.
59. San Blas, G., 1985, *P. brasiliensis* cell wall glucan, pathogenicity and dimorphism, in: *Current Topics in Medical Mycology* (M. R. McGinnis, ed.), Springer–Verlag, New York, pp. 235–257.

60. Brummer, E., Restrepo, A., Hanson, L. H., and Stevens, D. A., 1990, Virulence of *P. brasiliensis*. The influence of in vitro passage and storage, *Mycopathol.* **109:**13–17.
61. Villar, L. A., and Restrepo, A., 1989, Virulence of a variant of *P. brasiliensis* that exists in the yeast form at room temperature, *J. Med. Vet. Mycol.* **27:**141–148.
62. Cano, L. E., and Restrepo, A., 1987, Predictive value of serologic tests in the diagnosis and follow-up of patients with paracoccidioidomycosis, *Rev. Inst. Med. Trop. Sao Paulo* **29:** 276–283.
63. Restrepo, A., and Moncada, L. H., 1974, Characterization of the precipitin bands detected in the immunodiffusion test for paracoccidioidomycosis, *Appl. Microbiol.* **28:**138–142.
64. McGowan, K. L., and Buckely, H. R., 1985, Preparation and use of cytoplasmic antigens for the serodiagnosis of paracoccidioidomycosis, *J. Clin. Microbiol.* **22:**39–43.
65. Yarzabal, L., 1971, Anticuerpos precipitantes especificos de la blastomicosis sudamericana revelados por immunoelectroforesis, *Rev. Inst. Med. Trop. Sao Paulo* **13:**320–327.
66. Camargo, Z. P., Guerch, J. L., Drouhet, E., and Improvisi, L., 1984, Enzyme-linked immunosorbent assay (ELISA) in the paracoccidioidomycosis, *Mycopathol.* **88:**31–35.
67. Mendes-Giannini, M. J. S., Camargo, M. E., Lacaz, C. S., and Ferreira, A. W., 1984, Immunoenzymatic absorption test for serodiagnosis of paracoccidioidomycosis, *J. Clin. Microbiol.* **20:**103–108.
68. Cano, L. E., Brummer, E., Stevens, D. A., and Restrepo, A., 1986, An evaluation of the enzyme-linked immunosorbent assay (ELISA) for quantitation of antibodies to *P. brasiliensis*, *J. Med. Vet. Mycol.* **24:**467-476.
69. Marques, S. A., Dillon, N. L., Franco, M. F., Habermann, M. C., Lastoria, J. C., Stolf, H. O., Marcondes, J., Silva, N. C. S., Caraviani, M. R., and Curi, P., 1985, Paracoccidioidomycosis: A comparative study of the evolutionary serologic, clinical and radiologic results for patients treated with ketoconazole or amphotericin B plus sulfonamides, *Mycopathol.* **89:**19–23.
70. Restrepo, A., Gomez, I., Cano, L. E., Arango, M. D., Gutierrez, F., and Sanin, A., 1983, Treatment of paracoccidioidomycosis with ketoconazole, A 3 year experience, *Am. J. Med.* **74** (Suppl B):48–52.
71. Naranjo, M. S., Trujillo, M., Munera, M. I., Gomez, I., and Restrepo, A., 1990, Treatment of paracoccidioidomycosis with itraconazole, *J. Med. Vet. Mycol.* **28:**71–80.
72. Restrepo, A., Gomez, I., Cano, L. E., Arango, M. D., Gutierrez, F., and Sanin, A., 1983, Post-therapy status of paracoccidioidomycosis patients treated with ketoconazole, *Am. J. Med.* **74**(1B):53–57.
73. Stambuk, B. U., Puccia, R., de Almeida, M. L. C., Travassos, L. R., and Schenkman, S., 1988, Secretion of the 43 kDa glycoproteins antigen by *P. brasiliensis*, *J. Med. Vet. Mycol.* **26:** 367–374.
74. Camargo, Z. P., Unterkircher, C., and Travassos, L. R., 1989, Identification of antigenic polypeptides of *P. brasiliensis* by immunoblotting, *J. Med. Vet. Mycol.* **27:**407–412.
75. Camargo, Z. P., Unterkiercher, C., Campoy, S. P., and Travassos, L. R., 1988, Production of *P. brasiliensis* exo-antigens for immunodiffusion tests, *J. Clin. Microbiol.* **26:**2147–2151.
76. Taba, M. R. M., Silveira, J. F., Travassos, L. R., and Schenkman, S., 1989, Expression in *E. coli* of a gene coding for epitopes of a diagnostic antigen of *P. brasiliensis*, *Exp. Mycol.* **13:** 223–230.
77. Mendes-Giannini, M. J. S., Bueno, J. P., Shikanai, M. A., Ferreira, A. W., and Masuda, A., 1989, Detection of the 43000 molecular-weight glycoprotein in sera of patients with paracoccidioidomycosis, *J. Clin. Microbiol.* **27:**2842.
78. Restrepo, A., 1990, Paracoccidioidomycosis, in: *Antifungal Drug Therapy*: A complete guide for the practitioner (P. H. Jacobs and L. Nall, eds.), Marcel Dekker, New York, p. 181–205.
79. Borelli, D., 1987, Terapia de la paracoccidioidomicosis. Valor actual de los antiguos tratamientos, *Rev. Argentina Micol.* Suppl., 13–20.
80. Del Negro, G., 1987, Avances terapeuticos em micoses profundas, com enfase a paracoccidioidomicose, *Ann. Brasil Dermatol.* **62:**209–212.

81. Barraviera, B., Mendes, R. P., Machado, J. M., Pereira, P. C. M., Souza, M. J., and Neira, D. A., 1989, Evaluation of treatment of paracoccidioidomycosis with cotrimazine, *Rev. Inst. Med. Trop. Sao Paulo* **31**:53–55.
82. Negroni, R., 1987, Estado actual del empleo del ketoconazol en paracoccidioidomicosis (ketoconazol: 6 años despues), *Rev. Argentina Micol.* Suppl., 21–26.
83. Negroni, R., Palmieri, O., Karen, F., Tiraboschi, I. N., and Galimberti, R. L., 1987, Oral treatment of paracoccidioidomycosis and histoplasmosis with Itraconazole, *Rev. Infect. Dis.* **9**(Suppl. 1):S47–S50.
84. Diaz, M., Negroni, R., Montero-Gei, F., Castro, L. G. M., Sampaio Borelli, D., Restrepo, A., Franco, L., Bran, J., Arathoon, E. G., and Stevens, D. A., A Pan-American 5 year study of fluconazole therapy for deep mycoses in the immunocompetent host, *Clin. Infect. Dis.* **14**(Suppl 1):S68–S76.
85. Rezkallah-Iwasso, M. T., Mota, N. G. S., Gomez, M. E., and Montenegro, M. R., 1984, The effect of levamisole on experimental paracoccidioidomycosis in the siriam hamster: Immunologic and histopathologic correlation, *Mycopathol.* **84**:171–180.
86. Brummer, E., Hanson, L. H., and Stevens, D. A., 1988, Gamma-interferon activation of macrophages for killing of *P. brasiliensis* and evidence for non-oxidative mechanisms, *Int. J. Immunol.* **10**:945–952.
87. Brummer, E., Hanson, L. H., Restrepo, A., and Stevens, D. A., 1988, In vivo and in vitro activation of pulmonary macrophages by IFN-gamma for enhanced killing of *P. brasiliensis* or *B. dermatitidis*, *J. Immunol.* **140**:2786–2789.

11

Cryptococcosis

THOMAS R. KOZEL

1. CRYPTOCOCCOSIS

Cryptococcus neoformans is an encapsulated yeast that may produce localized or disseminated infections in humans and a variety of animals. Although dissemination may occur in otherwise healthy individuals, its primary importance is that of an opportunistic fungal pathogen. It is the most common life-threatening fungal infection in patients with the acquired immunodeficiency disease (AIDS).

1.1. Portal of Entry

Infection by *C. neoformans* probably occurs by inhalation of dust and debris that are contaminated with the yeast. Viable cryptococci have been found in a variety of natural sources, particularly sites heavily contaminated with pigeon and other bird droppings.[1] Samples of infectious particles obtained from pigeon excreta have shown that some of the particles are between 1 and 5 μm in diameter, a size that is well suited for deposition in the alveoli.[2,3]

1.2. Clinical Manifestations of Cryptococcosis

1.2.1. Pulmonary Cryptococcosis

Infection of the respiratory system by *C. neoformans* has been categorized on the basis of radiologic–pathologic correlates.[4] The first form is sapro-

THOMAS R. KOZEL • Department of Microbiology and the Cellular and Molecular Biology Program, University of Nevada School of Medicine, Reno, Nevada 89557-0046.

Fungal Infections and Immune Responses, edited by Juneann W. Murphy *et al.* Plenum Press, New York, 1993.

phytic colonization of the airways, and possibly the air spaces, without significant clinical disease. The yeast can be cultured from sputum of such asymptomatic individuals, but there are no specific radiographic findings and there is no histopathologic evidence of pulmonary parenchymal involvement.[4–6] In the second form, cryptococci grow within the pulmonary parenchyma with little or no inflammatory reaction. Lesions are visible radiographically as sharply circumscribed masses and consist of a gelatinous solid fungal form. Histopathologic examination shows the presence of typical budding yeast forms with abundant capsule formation, suggesting that the gelatinous character of the lesion is due to the presence of the capsular polysaccharide. The third form is a typical mononuclear inflammatory reaction, which usually appears as a granuloma. The radiographic manifestations of this inflammatory form consist of three primary types: 1) segmental consolidation, 2) poorly defined masses, or 3) irregular "masslike" infiltrates.[4]

Cultures of sputum and lung biopsy specimens are frequently negative in patients with pulmonary cryptococcosis. For example, lung biopsy specimens from 5 of 9 patients with pulmonary cryptococcosis were negative when cultured.[7] Similarly, a review of the published literature found that only 19 of 101 patients with proven invasive pulmonary disease had positive sputum cultures.[8] Despite the high frequency of negative sputum cultures in pulmonary cryptococcosis, positive cultures must be interpreted with caution. The presence of the yeast in sputum in the absence of radiographic abnormalities is most likely due to simple colonization.[4–6] As a consequence, most authors recommend that antifungal therapy not be initiated in an otherwise normal patient on the basis of a positive sputum culture alone.[5–7,9] However, there is a need for aggressiveness in establishing a diagnosis of pulmonary cryptococcosis in compromised patients, particularly patients with AIDS. Dissemination is common in such patients, consequently there is general agreement that antifungal therapy should be given if a diagnosis of pulmonary cryptococcosis is made in a severely immunocompromised patient.[7,10,11] A prospective study is needed to determine whether AIDS patients who are colonized with *C. neoformans* are at increased risk for development of pulmonary and disseminated cryptococcosis.

1.2.2. Disseminated Cryptococcosis

Systemic cryptococcosis is most likely a consequence of dissemination from a pulmonary focus. Several authors have argued that disseminated cryptococcosis is due to reactivation of a prior pulmonary infection.[4,12,13] Most normal individuals are able to contain the infection within the lungs, but dissemination will occur in an occasional patient with no known immunologic abnormality. In contrast, dissemination is part of the natural history of untreated pulmonary cryptococcosis in the immunocompromised host.[7,10]

Most patients with disseminated cryptococcosis have meningeal involvement.[14] The onset of cryptococcal meningitis is usually insidious, although the course of infection can be acute, particularly in patients who are severely immunocompromised.[15] The most common signs and symptoms in cryptococcal meningitis are headache (87%), fever (60%), nausea and vomiting (53%), changes in mental status (52%), and signs of meningitis such as neck stiffness (50%).[14] Approximately 10% of patients with cryptococcal meningitis may be asymptomatic.[16,17]

The reasons for the predilection of *C. neoformans* for the central nervous system are not known. Incubation of encapsulated cryptococci in normal human serum leads to deposition of opsonic fragments of C3 on the capsular surface.[18–20] In contrast, cryptococci within a patient's spinal fluid show no deposition of C3 fragments in the capsule.[18] These data suggest that the central nervous system may be a favorable niche for avoidance of the host's natural anti-cryptococcal defenses. Polacheck et al.[21] suggested that the brain may provide a favorable nutritional environment for the yeast. Genetic studies have shown that phenoloxidase activity is associated with the virulence of *C. neoformans*.[22] Polacheck et al.[21] demonstrated that glucose represses the synthesis of phenoloxidase, leading to the hypothesis that the low steady-state concentration of glucose in the brain would support derepression of the synthesis of phenoloxidase. This increased synthesis of phenoloxidase would, in turn, allow the yeast to utilize catecholamines in the brain more efficiently than in other organs. Other explanations evoked to explain the predilection for the central nervous system include the presence of optimal concentrations of essential nutrients in the CNS[23] and the absence of putative inhibitory factors in CSF that are normally present in serum.[24]

Cryptococcemia is the most common extraneural manifestation of systemic cryptococcosis. Patients with cryptococcemia usually have concomitant CNS involvement, but in one study as many as 29% of patients with cryptococcemia had no evidence of meningitis.[12] Cryptococcosis may involve other extraneural sites, including the skin, kidney, heart, spleen, pancreas, adrenals, ovaries, lymph nodes, skeletal muscle, liver, and gastrointestinal tract.[25]

Cryptococcosis in the patient with AIDS is similar in many respects to cryptococcosis in patients who are severely immunocompromised for other reasons. There are, however, several unusual aspects to cryptococcosis in the AIDS patient. The initial clinical presentation can be quite subtle. As a consequence, the presence of persistent fever and headache in an individual with AIDS is a basis for evaluation for CNS cryptococcosis.[26,27] Extraneural infection concomitant with meningitis is relatively common, occurring in up to 50% of patients with cryptococcosis complicating AIDS.[27] Many patients do not have the classic signs and symptoms of meningitis. For example, in two independent studies less than half had nausea, vomiting, or altered mental status and only 20–31% had a stiff neck.[28,29] Relapse after apparently successful treatment of cryptococcal meningitis is reported in approximately half of those who complete therapy.[30,31]

2. ETIOLOGY OF CRYPTOCOCCOSIS

Cryptococcus neoformans is the etiological agent of cryptococcosis. It is an encapsulated yeast that reproduces by budding. Kwon-Chung found that *C. neoformans* is a Basidiomycete with the perfect state *Filobasidiella neoformans*.[32] Two varieties of the yeast have been identified. In both cases, the variety is closely related to the serotype that is produced by the yeast. *C. neoformans* var. *neoformans* is serotype A or D, whereas *C. neoformans* var. *gattii* is serotype B or C. Cross-hybridization between the two varieties shows only 55–63% homology,[33] and numerous biochemical differences have been demonstrated between var. *neoformans* and var. *gattii*. The corresponding perfect states produced by compatible mating types are *F. neoformans* var. *neoformans* and *F. neoformans* var. *bacillisporus*.[34]

3. EPIDEMIOLOGY OF CRYPTOCOCCOSIS AND *C. NEOFORMANS*

3.1. Occurrence of Cryptococcosis among Various Risk Groups

Cryptococcal infections are acquired from environmental sources. There are no known instances of human to human transmission. The frequency of subclinical infection by *C. neoformans* is not known, but some information can be deduced from skin test surveys. Studies using soluble extracts of cryptococcal cells have shown positive skin test results in 4.2%,[35] 5%,[36] and 8%[37] of those tested. The specificity of such skin tests cannot be assured, but it is attractive to speculate that these results represent the frequency of subclinical infection in a normal population. Perhaps it is more than a coincidence that the frequency of cryptococcosis among patients with AIDS is also 5–10%.[28,38,39] The frequency of positive skin test results increases among groups with a presumed higher rate of exposure to the yeast. Laboratory personnel who work with *C. neoformans* had positive skin test frequencies of 81%;[36] 32% of a group of pigeon breeders had positive skin tests; and positive skin tests were observed in 26 of 82 healthy individuals in a small Oklahoma community where three cases of cryptococcal meningitis occurred within 1 year.[40]

Host factors have a great influence on the epidemiology of cryptococcosis. The skin test surveys noted above suggest that infection of normal individuals may be relatively common. Presumably, the high level of innate resistance to the yeast restricts the infection. However, on some occasions otherwise normal individuals will develop disseminated disease. The factors that allow disseminated disease to occur in normal patients are not known but may be related to the inoculum size, the virulence of the infecting strain, or the presence of subtle immune deficiencies in otherwise normal individuals.[37,41] Disseminated cryptococcosis is most commonly associated with use of adrenal corticosteroids or with underlying illness that predisposes to disease. Patients with neoplastic disease, particularly chronic lymphatic leukemia, Hodgkin's disease, chronic myelogenous leukemia, and multiple myeloma, are at particular risk.[42] Other risk factors include collagen diseases, sarcoidosis, and immunosuppression associated with organ transplantation.

The importance of AIDS as a predisposing condition cannot be overemphasized. Prior to the advent of AIDS, there were an estimated 200–400 cases of cerebral meningitis in the United States per year.[43] A 6.8% occurrence of extrapulmonary cryptococcosis in patients with AIDS,[38] coupled with an estimated 50,000 AIDS cases in the United States in 1990,[44] suggests that the annual number of cases of extrapulmonary cryptococcosis may be 10 times higher than the number observed prior to AIDS. The occurrence of cryptococcosis in AIDS patients varies markedly with the patient risk category.[38] The percentage of AIDS patients with extrapulmonary cryptococcosis is lowest among homosexual men who are not known to be intravenous drug abusers (5.2%). Cryptococcosis occurs among 11.5% of AIDS patients who are heterosexual intravenous drug abusers. The occurrence of cryptococcosis among AIDS patients in central Africa is much higher, with reported frequencies in excess of 30%.[45,46] An explanation for this differential occurrence could be the presence of cryptococcal strains of more or less virulence in each patient group or subtle differences in the immune deficiency in each group. Perhaps the most likely explanation is a variation in exposure to cryptococci in the environment of each group.

3.2. Epidemiologic Differences between Serotypes

The A and D serotypes (*C. neoformans* var. *neoformans*) differ markedly from the B and C serotypes (*C. neoformans* var. *gattii*) with regard to their natural habitat,[47] geographic distribution,[47,48] and occurrence in patients with AIDS. *C. neoformans* var. *neoformans* has a worldwide distribution. Variety *neoformans* accounts for almost all isolates from Europe, Canada, Japan, and the United States (excluding southern California). Serotype A is the most commonly isolated serotype in North America. Serotype D is isolated rarely in North America, but it has been relatively common in European countries, such as Italy, Denmark, Germany, and Switzerland. The distribution of *C. neoformans* var. *gattii* differs markedly from the distribution of *C. neoformans* var. *neoformans*, having an unusually high prevalence in Australia, Brazil, Hawaii, southern California, southeast Asia, and central Africa. These results indicate that *C. neoformans* var. *gattii* has an environmental niche that is limited to tropical and subtropical regions. Among the strains of *C. neoformans* var. *gattii*, serotype B is much more prevalent than serotype C. Most isolates of serotype C have been from southern California.

Early epidemiological studies found that *C. neoformans* could be isolated from pigeon excreta or from soil contaminated with droppings from pigeons and other birds. Pigeon droppings have also been shown to contain bacteria that have anti-cryptococcal activity, amoeba that ingest cryptococci, and fungivorous mites and sow bugs.[49] These findings indicate that a complex of biological and nutritional factors influences the growth, persistence, and distribution of *C. neoformans* in nature. The high frequency of cryptococcosis in AIDS patients in central Africa led Swinne et al. to examine soil, droppings, and dust samples near the homes of 20 patients with AIDS-associated cryptococcosis.[50,51] The results of the survey showed that cryptococci were not only associated with pigeon and chicken droppings, the yeast was also isolated indoors in the domestic environment from dust and air, as well as from the intestinal contents of cockroaches. These results emphasize the ubiquitous presence of cryptococci and suggest that contact with the yeast may be more common in central Africa than has been generally appreciated.

Recent studies conducted in Australia by Ellis and Pfeiffer[159,160] found that *C. neoformans* var. *gattii* has a specific association with *Eucalyptus camaldulensis*, the river red gum tree. It has been suggested that the infectious fungal propagules are released into the environment when the trees flower.[160] *C. neoformans* var. *gattii* has also been isolated from *E. camaldulensis* trees near San Francisco.[161] If *E. camaldulensis* proves to be the sole natural habitat for var. *gattii*, this would explain the occurrence of *C. neoformans* var. *gattii* in tropical and subtropical regions as well as the infrequency with which *C. neoformans* var. *gattii* is isolated from patients with AIDS who, presumably, would have limited contact with *E. camaldulensis*.[52–56]

4. HOST DEFENSES AGAINST CRYPTOCOCCOSIS

Host resistance to cryptococcosis is quite high. The factors that account for this resistance are doubtless a combination of innate resistance and the ability of the immune system to control the yeast. The extraordinary sensitivity of patients with AIDS to disseminated cryptococcosis strongly implicates cellular immunity as a critical factor in host resistance. Conversely, the relative infrequency of cryptococcosis in the neutropenic patient suggests that neutrophils may be less important in host resistance than the role played by neutrophils in other systemic fungal infections, such as candidiasis or aspergillosis.

4.1. Innate Resistance

4.1.1. Serum Sensitivity of C. neoformans

Human serum contains one or more components that inhibit the growth of *C. neoformans*.[24,57–59] Inhibition by serum is not dependent on the complement system because the inhibitory activity is not inactivated by heating the serum at 56°C.[58,59] The inhibitory factor does not appear to be antibody, because the inhibitory activity is not removed by adsorption with cryptococcal cells and globulin purified from serum is not inhibitory.[58] The inhibitory factor(s) is trypsin sensitive, is nondialyzable, and is not inhibited by EDTA.[24] The significance of the serum inhibitory factor in resistance to cryptococcosis is not known.

4.1.2. *Phagocytosis of C. neoformans*

Studies with a murine model of cryptococcosis indicate that phagocytosis of *C. neoformans* is an important factor in innate resistance to cryptococcosis. Treatment of mice with silica to destroy host macrophages markedly decreases the resistance of mice to cryptococcal infection.[60] In contrast, activation of host macrophages by immunization with BCG enhances resistance to cryptococcosis.[60]

The cryptococcal capsule is a potent inhibitor of phagocytosis. Nonencapsulated variants of the yeast are quite sensitive to phagocytosis, whereas encapsulated isolates are engulfed poorly.[61,62] Studies of encapsulated cryptococci have noted a direct relationship between capsule size and resistance to phagocytosis.[63–65] The precise mechanism for inhibition of phagocytosis by the cryptococcal capsule is not known. Any model for inhibition of phagocytosis must explain why small-capsule strains are phagocytosis sensitive while large-capsule strains are not. Studies to date have shown that inhibition is due to a recognition failure rather than any direct effect by the capsular polysaccharide on phagocytes.[66–68] Inhibition only occurs when the polysaccharide is bound to the yeast in the form of a capsule.[66,68] Finally, inhibition is not associated with simple physicochemical properties of the capsule, such as negative surface charge or hydrophilicity.[69,70] These pieces of information suggest that inhibition of phagocytosis by large capsules may be due to a failure in opsonization of the yeast that would normally occur with nonencapsulated or small-capsule isolates.

Incubation of encapsulated cryptococci in normal human serum leads to the binding of IgG[18,19,71] and fragments of complement component C3[18,19,72] to the yeast. Diamond et al. first provided indirect evidence that an antibody in normal human serum is opsonic for encapsulated cryptococci.[18] The opsonic activity of human serum was markedly reduced by adsorption at 0°C with cryptococcal cells. The opsonic activity of the adsorbed serum was restored by the addition of purified human IgG, thus identifying IgG as the essential opsonin that was removed by adsorption. Ultrastructural studies demonstrated that incubation of encapsulated cryptococci in normal human serum leads to binding of IgG within the capsule, thus providing direct evidence that normal serum contains potentially opsonic antibody that is reactive with the cryptococcal capsule.[19,71] The anti-cryptococcal antibody in normal human serum is not opsonic unless complement activation has also occurred.[18] Thus there is an apparent synergy between IgG and complement fragments in opsonization of encapsulated cryptococci.

The cryptococcal capsule is a potent activator of the complement system. Incubation of encapsulated cryptococci in normal human serum leads to the binding of approximately 1–4 × 10^7 molecules of C3 to each yeast cell.[20,64] Ultrastructural studies identified the capsule itself as the primary binding site for the C3 fragments.[19,64] Elution of the fragments and analysis by SDS-PAGE identified iC3b as the most abundant fragment bound to the yeast.[20] Incubation of encapsulated cryptococci with an alternative complement pathway that is reconstituted from the six proteins of the alternative pathway produces a pattern of activation and binding of C3 fragments that is qualitatively and quantitatively identical to the pattern observed when the yeast is incubated in normal serum.[73] These data indicate that the classical complement pathway has little or no role in opsonization of the yeast.

4.1.3. Growth Inhibition of C. neoformans by Phagocytes

Studies designed to assess the anti-cryptococcal activities of phagocytic cells have produced highly variable results. Discrepancies between results may be due to differences in 1) cell type, 2) manner of opsonization, 3) level of activation of the phagocytes, and 4) the species used as a source of phagocytic cells. Antibody-dependent anti-cryptococcal activity has been reported for mouse resident peritoneal macrophages, human neutrophils, and human nonadherent mononuclear cells.[74–76] In contrast, opsonization of encapsulated cryptococci by anticapsular antibody does not facilitate the anti-cryptococcal activity of cultured human monocytes; indeed, the presence of anticapsular antibody significantly enhances fungal growth.[57] Gamma interferon increases the anti-cryptococcal activity of murine macrophages[76,77] but suppresses the *in vitro* anti-cryptococcal activity of cultured human monocytes.[57]

There is increasing evidence that ingestion is not essential for the anti-cryptococcal action of phagocytes.[74,76–78] These studies suggest that host defenses may have found a way to overcome the phagocytosis-inhibiting action of the capsular polysaccharide. Murine bone marrow-derived macrophages stimulated with gamma interferon suppress the growth of *C. neoformans* by a mechanism that is extracellular and independent from phagocytosis.[77] The stimulatory effect of gamma interferon is augmented by the presence of bacterial lipopolysaccharide.[77] This latter observation is consistent with studies in several systems which found that gamma interferon requires a synergistic signal which can be provided by lipopolysaccharide.[79] Granger et al.[78] observed that murine peritoneal macrophages also require the presence of lipopolysaccharide for optimal extracellular anti-cryptococcal activity.

Several mechanisms have been proposed to account for the anti-cryptococcal activity of phagocytes. These include the myeloperoxidase-mediated fungicidal system,[65] lysosomal granules,[80] and an as yet unidentified product of the L-arginine-dependent, nitrate-generating pathway.[81] The L-arginine system appears to be particularly important because killing of cryptococci by murine resident peritoneal cells is correlated with stimulation of nitrate synthesis but not with stimulation of the respiratory burst or lysosomal enzyme release.[76]

4.1.4. *Natural Killer Cells*

Natural killer (NK) cells may contribute to innate resistance to murine cryptococcosis. NK cells bind to encapsulated cryptococci, forming "conjugates" analogous to NK-tumor cell conjugates.[82] Murine NK cells inhibit *in vitro* growth of *C. neoformans*.[83] Moreover, the requirements for cations in binding and growth inhibition of cryptococci are similar to the requirements for binding and killing of tumor cell targets.[84] *In vivo* results indicate that anti-cryptococcal NK activity may have biological relevance. Mice with high NK cell activity have an increased ability to clear cryptococci early in the course of an experimental infection.[85] Conversely, mice homozygous for the beige mutation, a mutation that leads to impaired NK cell activity, have a reduced ability to clear cryptococci from lung and spleen when compared to heterozygous bg/+ mice.[86–88] Studies with beige mice should be interpreted with some caution because beige mice may have other abnormalities that would influence innate resistance to cryptococcosis.

Patients with AIDS have impaired NK-cell function; moreover, the loss of NK-cell activity has been suggested as a marker for the progression of HIV infection to overt AIDS.[89,90] A deficiency in NK-cell function is one possible link between the immune deficiencies that occur in the AIDS patient and a putative mechanism for resistance to cryptococcosis. Nevertheless, the importance of NK cells in human cryptococcosis remains unknown. A recent study reported that neither resting human NK cells nor human lymphokine-activated killer cells had any effect on the *in vitro* growth of *C. neoformans*.[91] Human NK cells, however, have a marked ability to inhibit the growth of cryptococci that are opsonized with anti-cryptococcal antibody,[91] suggesting a possible function for naturally occurring anti-cryptococcal antibody.

4.2. Immune Responses

4.2.1. *Antigenic Composition of C. neoformans*

Antigens used in most studies of the immune response to *C. neoformans* have been either whole yeast cells or extracts from culture filtrates of the yeast. These are complex antigenic mixtures that may have multiple properties. Studies by Cherniak and his collaborators[92–94] have attempted to identify the critical antigenic components in these preparations. To date, three antigenic fractions have been isolated and characterized in some detail: the capsular polysaccharide, a second polysaccharide termed *galactoxylomannan*, and a mannoprotein.

The capsular polysaccharide of *C. neoformans* occurs in four serotypes, types A, B, C, and D. The polysaccharides are polymers of glucuronic acid, xylose, and mannose, and are referred to as *glucuronoxylomannans* (GXM).[92] The GXM is the major serotypic polysaccharide produced by the yeast. It is the primary constituent of the polysaccharide capsule, and it is found in copious amounts in culture filtrates of the yeast. There is also increasing evidence that there are variant forms of the polysaccharide that are intermediate between the four serotypes.[95,96] Polysaccharides of all four serotypes contain a backbone of α-(1→3)-linked D-mannopyranosyl residues. Side chains of single xylopyranosyl and glucuronopyranosyl residues are attached to the mannan backbone. Regardless of serotype, the glucuronosyl residue is linked β-(1→2) to the backbone. Xylose residues are linked β-(1→2) to the mannan backbone, and, depending on the serotype, also may be linked β-(1→4) to the backbone. Serotypes D and A are the simplest of the polysaccharides, with the mannosyl residues substituted only at 0–2. Serotypes B and C are most complex because the mannopyranosyl residues are also substituted with xylopyranosyl residues at 0–4 as well as 0–2. The four serotypes also differ with regard to the extent of O-acetylation. Serotype D is the most heavily O-acetylated, while serotype C is the least O-acetylated.

The galactoxylomannan (GalXM) and mannoprotein are minor polysaccharides found in culture filtrates of the yeast. The mannoprotein is much more active serologically than the GalXM.[97] The GalXM consists of xylose: mannose:galactose in a molar ratio of 1:1.5:1.8. The GalXM lacks the glucuronosyl residue found in the GXM and is, therefore, relatively uncharged. Cryptococcal mannoprotein consists of xylose, mannose, and galactose in molar ratios of 1:7.6:0.8 and contains approximately 20% peptide.[93] The mannoprotein is adherent to concanavalin A, whereas the GalXM is not.[93] Both the GalXM and the mannoprotein are produced by acapsular mutants of *C. neoformans*. The cell wall mannoprotein of many nonpathogenic yeasts is surface exposed. In contrast, most mannoprotein in the cryptococcal cell wall is located in the inner two thirds of the cell wall. Available evidence indicates that the cryptococcal mannoprotein is a nonstructural element of the cell wall.[98]

4.2.2. *Humoral Immunity*

The capsular polysaccharide is the primary antigen that stimulates the production of antibodies. There is increasing evidence that normal individuals have circulating antibodies reactive with the GXM. Henderson et al.[99]

found that sera from 89% of normal subjects have low levels of antibody to cryptococcal polysaccharide. Similarly, Dromer et al.[100] found that 69% of serum samples from healthy subjects had anticapsular antibodies. In contrast, only 20% of patients with AIDS had anticapsular antibodies. This latter observation is consistent with the observation that patients with AIDS lose their ability to produce IgG antibodies to T-independent antigens.[101] The antigenic stimulus for the anticapsular antibodies in normal subjects is not known. The antibodies could be a response to exposure to *C. neoformans* in nature. Alternatively, the antibodies could be cross-reactive antibodies stimulated by the numerous polysaccharides that are ubiquitous in the environment.

Some cryptococcosis patients develop modest titers of anticapsular antibody during the course of an infection. The reported incidence of positive tests for serum antibody in cryptococcal infections has varied from 45 to 80%.[102–104] On occasion, circulating antibody has been detected early in infection, throughout the infection, or concurrent with circulating antigen. Conversion of antibody tests in the cryptococcosis patient from negative to positive appears to be a rare occurrence.[102] There are differing reports on the immunoglobulin class of anti-cryptococcal antibody found in human sera. Antibodies detected in sera from patients with cryptococcosis are reported to be predominantly of the IgG class.[105,106] whereas antibody produced in response to immunization of volunteers with cryptococcal polysaccharide is predominantly IgM.[99]

Cryptococcal polysaccharide has many of the physical, chemical, and immunological properties of T-independent antigens. T-independent antigens are so defined because they elicit an antibody response in both normal and athymic mice.[107] The response of mice to immunization with cryptococcal polysaccharide is markedly enhanced when functional T lymphocytes are abrogated by treatment of the mice with anti-lymphocyte serum.[108] Like other T-independent antigens,[109] it appears that suppressor T cells regulate the immune response to cryptococcal polysaccharide. Other similarities to T-independent antigens, such as pneumococcal polysaccharide, polyvinylpyrrolidone, and D-amino acid polymers,[107] include a high molecular weight, poor *in vivo* degradation, and a linear structure with one or more repeating determinants.

Exposure to doses of GXM that are greater than the optimal immunizing dose renders recipients immunologically unresponsive to subsequent immunization by an optimal immunizing dose. Immunologic unresponsiveness is expressed as a reduced ability to produce circulating antibody.[104,110] Studies at the cellular level demonstrated that the unresponsiveness is due to a suppression of antibody synthesis rather than neutralization of circulating antibody by excess antigen.[110] Studies of patients with cured cryptococcal meningitis have shown that systemic cryptococcosis is often associated with prolonged specific immunologic unresponsiveness.[99] Cured patients have a reduced ability to produce antibody in response to immunization with cryptococcal polysaccharide, yet their response to immunization with pneumococcal polysaccharide does not differ from normal volunteers. The clinical importance of this apparently acquired immune unresponsiveness is not known.

The role of antibody in resistance to cryptococcosis is controversial. Evidence that anticapsular antibody enhances resistance is derived from several lines of evidence. First, the presence of circulating anticapsular antibody in serum from cryptococcosis patients is associated with a favorable prognosis.[111,112] Production of anti-cryptococcal antibody appears to be deficient in some patients with cryptococcosis, suggesting that a specific immunologic defect may contribute to the pathogenesis of disseminated disease.[99] The second line of evidence for the role of antibody in resistance is the abundance of *in vitro* data demonstrating antibody-dependent cellular effector mechanisms. Anticapsular antibody facilitates killing of the yeast by mouse resident peritoneal macrophages,[76] human peripheral blood leukocytes,[74] and human neutrophils.[75] It should be emphasized, however, that antibody enhancement of killing is not observed with all potential effector cells. Killing of the yeast by cultured human monocytes is not enhanced by opsonization with anticapsular antibody.[57]

Passive immunization studies provide both the most persuasive and most controversial evidence for a role for antibodies in resistance to cryptococcosis. Both Gadebusch[113] and Graybill et al.[114] reported that mice could be protected against experimental cryptococcosis by passive immunization with rabbit anti-cryptococcal antibody. A key feature of both reports was the need for antibody to be present at the time and site of challenge. In contrast, Louria and Kaminski found that passive immunization has no effect on survival time or numbers of cryptococci in brain tissue if the mice are treated 24 hours after challenge at a site that differs from the challenge route.[115] Without doubt, the most impressive results have been the use of monoclonal antibodies to protect DBA/2 mice from the fatal cryptococcal pneumonia that occurs in these C5-deficient mice.[116,117] Monoclonal antibodies have been markedly less effective when used in C5-sufficient Swiss Webster or BALB/c mice.[117,118] The antibodies effect a significant reduction in numbers of yeasts in lung and spleen, but they have no effect on the numbers of cryptococci in brain tissue, and they are unable to prolong the life of infected mice.[118]

4.2.3. *Cell-Mediated Immunity*

There is abundant evidence that cell-mediated immunity is important in resistance to cryptococcosis. Skin test and lymphocyte transformation surveys of individuals working in laboratories with *C. neoformans* show a high percentage of reactivity to cryptococcin,[37,41] yet significant laboratory infections by the yeast are virtually unknown. Similarly, pigeon breeders who have an increased exposure to the yeast have a high frequency of delayed cutaneous hypersensitivity to cryptococcin, yet this subject group does not have an increased frequency of symptomatic cryptococcosis.[35] This indicates that normal subjects readily develop delayed hypersensitivity responses to the yeast, and natural resistance is quite high. The most serious infections occur in patients with cellular immune deficiencies, such as those associated with corticosteroid therapy, reticuloendothelial malignancy, organ transplantation, sarcoidosis, and AIDS. Even patients with cryptococcosis who are otherwise normal appear to have subtle defects in cellular immunity that may predispose to cryptococcosis.[37,41] These latter immunologic abnormalities were apparent at least one year after successful antifungal therapy, suggesting that the defect was not caused by the cryptococcal infection itself but may have been present prior to infection. Taken together, these data strongly suggest that *C. neoformans* most commonly requires a host who is deficient in some aspect of cell-mediated immunity before dissemination can occur.

Passive immunization studies support the argument that cell-mediated immunity is a potent factor in resistance to cryptococcosis. Studies in guinea pig and murine models of cryptococcosis have shown that resistance to *C. neoformans* is conferred on recipient animals by passive transfer of lymphoid cells.[119,120] T lymphocytes in the lymphoid cells from the immune donors are the cells responsible for transfer of resistance. *In vitro* studies have shown inhibition of the growth of *C. neoformans* by coincubation of the yeast with T lymphocytes from immunized mice.[121] Thus, there is direct evidence that T lymphocytes can act as anti-cryptococcal effector cells.

4.2.4. *Immunoregulation by Cryptococcal Antigens*

Cryptococcal antigens induce the formation of a regulatory network of lymphoid cells that can amplify or suppress the delayed-type hypersensitivity response to *C. neoformans*. These regulatory networks have been characterized in a murine model of cryptococcosis.[122–129] Induction of the suppressor network follows intravenous injection of soluble cryptococcal antigen at concentrations that mimic the antigenemia that might occur in patients with cryptococcosis. The suppressor network induced by cryptococcal antigens suppresses the delayed-type hypersensitivity than mice normally develop in response to 1) immunization with soluble cryptococcal antigens or 2) infection with small numbers of viable cryptococci.

Two distinct preparations of cryptococcal antigen induce the suppressor network. The first antigenic preparation is a concentrated culture filtrate of the yeast. As noted previously, culture filtrates of *C. neoformans* contain numerous components, such as the major serotypic polysaccharide (GXM), the galactoxylomannan, and the mannoprotein. The active component in this antigenic mixture that is responsible for the immune suppression is not known. The second antigenic preparation that has been studied is serum from mice infected with *C. neoformans*. Serum from infected mice will produce suppression of delayed-type hypersensitivity in a manner similar to the suppression observed in response to the culture filtrate antigen.[123] Transfer by serum from infected mice demonstrates that the cryptococcal components that induce suppression are produced *in vivo*. The suppressive component in sera from infected mice was shown to be a cryptococcal antigen. Passage of the suppressive sera over columns containing anti-cryptococcal antibody removes the suppressive component. Moreover, the desorbed cryptococcal antigen is suppressive when transferred to normal mice.[129]

The suppressor cell network includes at least three distinct populations of suppressor T lymphocytes. A suppressive dose of cryptococcal antigen induces cyclophosphamide-sensitive T cells to develop into first-order (Ts1) suppressor cells.[128] These first-order suppressor cells or a soluble factor (TsF1) produced by Ts1 cells can have two functions. Ts1 cells can 1) suppress the induction of lymphocytes responsible for the delayed-hypersensitivity response (T_{DH})[125] and 2) induce second-order suppressor T cells (Ts2).[128] Ts2 cells or the soluble factors produced by Ts2 cells function with third-order suppressor T cells (Ts3) to suppress the expression of T_{DH} cells. Ts3 develop from cyclophosphamide-sensitive precursors following immunization with cryptococcal antigens in complete Freund's adjuvant.[122]

Amplification of the anti-cryptococcal delayed-type hypersensitivity response involves a second regulatory cell population. Immunization of mice with cryptococcal antigen leads to induction of cells responsible for delayed-type hypersensitivity (T_{DH}). These cells will adoptively transfer the delayed-type hypersensitivity to naive recipients.[125] Amplifier cells (T_{AMP}) are induced in mice given a primary immunizing dose of cryptococcal antigen in complete Freund's adjuvant.[130] T_{AMP} cells amplify the anti-cryptococcal delayed-type hypersensitivity response in recipient mice when they are transferred at the time of immunization of the recipient. T_{AMP} cells are distinct from T_{DH} cells and appear to have many properties similar to helper T cells.[126]

The cryptococcal antigen(s) responsible for elicitation and regulation of cell-mediated responses have not been determined with certainty. The most carefully characterized antigenic preparation has been CneF, a concentrated culture filtrate of *C. neoformans*. Depending on the dose and route of administration, CneF will 1) induce delayed-type hypersensitivity in naive mice,[123] 2) induce the formation of splenic T lymphocytes responsible for *in vitro* growth inhibition of *C. neoformans*,[123] 3) downregulate delayed-type hypersensitivity by induction of Ts1, Ts2, and Ts3 lymphocytes,[110,123,127] 4) induce an amplified delayed-type hypersensitivity response by induction of T_{AMP} cells in naive mice, and 5) elicit a delayed-type hypersensitivity response in the footpads of previously immunized mice.[131] Cnef is a complex mixture and it is attractive to speculate that all of the above responses are elicited or induced by the same component. In this regard, the mannoprotein, a component of

Cnef, elicits the delayed-type hypersensitivity responses in mice previously immunized with either Cnef or whole heat-killed cryptococci.[132]

5. IMMUNODIAGNOSIS

5.1. Serology

5.1.1. Assays for Antibody

Several assays, including whole cell agglutination, passive hemagglutination, ELISA, complement fixation, and indirect immunofluorescence, have been utilized to assess the production of antibody to *C. neoformans*. Almost all assays are designed to detect antibody specific for the major serotypic polysaccharide (GXM). Despite the wide spectrum of available assays, tests for anti-cryptococcal antibody are not a part of the normal workup of a patient with suspected cryptococcosis. Tests for antibody are often negative; only 40% of sera from patients with cryptococcosis have antibody detectable by tube agglutination or immunofluorescence.[103,112] Tests for antibody appear to have some prognostic value because a positive test for antibody is associated with a favorable prognosis, and a negative test correlates with treatment failure.[112]

5.1.2. Assays for Antigen

Cryptococcal polysaccharide (GXM) is shed in large amounts into the serum and CSF of patients with systemic cryptococcosis. This polysaccharide is readily detected by the latex agglutination assay. In this assay, latex particles are coated with the globulin portion of rabbit antiserum to *C. neoformans*.[111] The assay is quite sensitive, detecting polysaccharide concentrations as low as 30–60 ng/ml.[133,134] Recent studies of cryptococcosis in patients with AIDS have shown that the latex agglutination assay is positive in 98% of sera[28] and in 90–100% of CSF samples[26,28,29] from patients with cryptococcal meningitis.

There are several possible sources of false-positive reactions in the latex agglutination assay. The latex particles are coated with rabbit globulins. As a consequence, false-positive reactions are produced by the IgM anti-IgG antibody present in the serum of most patients who have rheumatoid arthritis.[133] Consequently, a control for the presence of rheumatoid factor is an essential part of latex agglutination assays for cryptococcal polysaccharide. Rheumatoid factor and other "interfering factors" can be eliminated by the treatment of samples with dithiothreitol,[135] EDTA-heat extraction,[136] or by treatment with pronase.[137] Positive results in the cryptococcal latex agglutination assay have also been reported in sera from two patients with disseminated infection by *Trichosporon beigelii*[138] and in serum from a patient with septicemia due to the Gram-negative bacterium designated DF-2.[139]

The latex agglutination assay has prognostic value in following the response of patients to antifungal therapy. An extensive study of prognostic factors in cryptococcosis showed that patients who died with active cryptococcosis had higher antigen titers than patients who were cured.[112] A high serum or CSF titer at the end of treatment correlated with treatment failure. Similarly, the failure of a positive antigen titer to decrease by at least fourfold or become negative signaled a treatment failure.

Despite the high sensitivity and specificity of the latex agglutination assay, several anomalies have become apparent in studies of sera and CSF from AIDS patients with cryptococcal infections. First, the extraordinarily high antigen titers (1/1,000,000) reported in some patients[39,140] raise questions as to the ability of LA to accurately reflect antigen levels, since such titers would normally require polysaccharide concentrations of at least 10 mg/ml. Second, there are numerous anecdotal reports of apparent "inhibitory factors" that may produce artificially low LA titers in sera from patients with AIDS. Finally, several reports indicate that, in contrast to cryptococcosis in patients who do not have AIDS,[112] serum antigen levels do not reflect either the severity of disease or the clinical response to antifungal therapy in AIDS patients with cryptococcosis.[28,30,39]

An enzyme immunoassay (EIA) has been recently described which uses a polyclonal antibody capture system and a monoclonal antibody indicator system for detection of cryptococcal polysaccharide.[162] The EIA had a 12-fold sensitivity advantage over latex agglutination for detection of serotypes A and B polysaccharides and an equivalent sensitivity for serotypes C and D polysaccharides. The EIA did not show false positive reactions with sera containing rheumatoid factor; it allowed for direct specimen detection without pronase or other treatment of serum. As with latex agglutination, *T. beigelii* antigens produced a positive reaction with EIA.

5.2. Cellular Aspects of Immunodiagnosis

Delayed hypersensitivity skin tests, lymphocyte transformation, and leukocyte migration inhibition have been used to evaluate cell-mediated immunity of patients to *C. neoformans*. Antigens used in these assays include killed cryptococcal cells and cryptococcin, a soluble extract of cryptococcal cells. In general, assays of cellular immunity in human cryptococcosis have received little emphasis and have little diagnostic value. Skin testing may have epidemiologic value, since several studies have shown that skin testing with cryptococcin can distinguish between individuals exposed or not exposed to *C. neoformans*.[35,36,41]

6. PREVENTION AND TREATMENT

6.1. Immunization and Vaccines

Attempts at active immunization against cryptococcal infection have met with very limited success. Several studies attempted to induce protective immunity by immunization with the capsular polysaccharide.[141–143] The results were uniformly negative. The failure of polysaccharide vaccines examined to date may be due to 1) an inability of antibody to protect, 2) a failure to elicit levels of antibody sufficient for protection, or 3) use of inappropriate animal models or experimental design for evaluation of efficacy. Recent studies with a tetanus toxoid conjugate of cryptococcal polysaccharide indicate that induction of high levels of antibody is possible.[163] If conjugate vaccines exhibit protective efficacy, there is a potential role for immunization in the protection of individuals at high risk for cryptococcosis.

In contrast to the poor results observed to date with polysaccharide vaccines, mice have increased survival rates and prolongation of survival time after immunization with killed whole-cell vaccines.[142,144–148] Small capsule strains or decapsulated cells induce higher levels of protective immunity than large-capsule strains.[142,147,148] The efficacy of the vaccine is improved by the use of adjuvants such as *Bordetella pertussis* vaccine.[146,149] Protective immunity also is produced by immunization with viable cryptococci using 1) small numbers of highly virulent cryptococci,[150] 2) comparatively large numbers of less virulent strains,[150] or 3) virulent cryptococci administered by routes that do not lead to dissemination.[149,151–153] Finally, almost all studies have shown that, like other infectious diseases that are controlled primarily by cellular immunity, immunity is relative rather than absolute. It is not difficult to overwhelm the fragile protective immunity with larger challenge innocula.

6.2. Treatment

6.2.1. *Chemotherapy*

Prior to the occurrence of AIDS, amphotericin B used alone or in combination with flucytosine was the cornerstone of therapy.[154] An extensive clinical study has shown that a combination of amphotericin B and flucytosine is at least as effective as amphotericin B alone.[155] The combination cured or improved more patients, produced fewer failures or relapses, effected a more rapid sterilization of the CSF, and was less toxic than amphotericin B alone. Moreover, the combination permitted a lower dose of amphotericin B. The duration of combination therapy is guided by the immune status or underlying disease of the patient, the severity of the disease, and several serum and CSF variables, including cryptococcal antigen levels.[156]

6.2.2. *Therapy in the Patient with AIDS*

Treatment of cryptococcosis in the patient with AIDS differs in several respects from treatment of patients who do not have AIDS. First, therapy for acute or first-episode cryptococcal meningitis in patients with AIDS is associated with high failure and relapse rates. As a consequence, an induction course of antifungal therapy is followed by maintenance therapy.[26,28,29] Second, many patients with AIDS cannot tolerate flucytosine, and amphotericin B is used in somewhat higher doses than would normally be used in the patient who does not have AIDS or in the combination amphotericin B/flucytosine regimen.[26,31] At least one study reported that the addition of flucytosine offers no advantage over treatment with amphotericin B alone.[28] If flucytosine is used in patients with AIDS, the dosage is usually reduced.[26,31] Patients receiving flucytosine should be monitored for hematologic signs of toxicity,[31] and serum concentrations of flucytosine should be monitored for levels predictive of toxicity.[26,31,154] Finally, the disappearance of cryptococcal antigen in serum and CSF that ideally occurs in the treatment of cryptococcosis does not necessarily occur in the patient with AIDS.[31] Serum titers may remain unchanged, rise or fall, and are not consistent prognostic indicators.[27] As a consequence, decisions regarding continuation of induction therapy must be based on a stabilization or decrease in antigen levels, culture for viable cryptococci, and clinical signs of response to treatment.[26,31]

Fluconazole is a promising alternative or adjunct to amphotericin B in the treatment of cryptococcosis in the patient with AIDS. Fluconazole is water soluble and is well adsorbed after oral administration. Moreover, studies in a subhuman primate model have shown that the drug readily crosses the blood–brain barrier, producing concentrations in the CSF that are >80% of serum concentrations.[157] Clinical studies have shown that fluconazole is effective suppressive therapy that is well tolerated for long periods of time.[140] Itraconazole, like fluconazole, is a triazole that can be administered orally. A recent study of 33 patients with cryptococcosis found a response rate of 100% in patients with cryptococcemia and 65% in patients with cryptococcal meningitis.

6.2.3. *Immunotherapy*

The potential value of passive immunization with monoclonal antibodies in treatment of cryptococcosis is uncertain. The inability of passive immunization to prolong life and to effect clearance of the yeast from the brain of Swiss Webster and BALB/c mice suggests that it may be difficult to treat a CNS infection by passive immunization alone. However, studies by Sanford et al.[118] have shown that monoclonal antibodies of the IgG2a and IgG2b isotypes will reduce the fungal burden in lung and spleen. Thus, there is a possibility that passive immunization could be used in conjunction with more conventional antibiotic therapy either to increase the efficacy of a treatment protocol or to reduce the levels of antibiotic needed for treatment of cryptococcosis. The viability of this approach is suggested by an earlier study which found that passive immunization with anti-cryptococcal antibody markedly improved the ability of amphotericin B to prolong the survival of mice with experimental cryptococcosis.[158]

Finally, several studies have shown that treatment of mice with monoclonal antibodies specific for interferon gamma,[164] tumor necrosis factor alpha,[165] and granulocyte-macrophage colony stimulating factor[165] renders the mice more susceptible to experimental cryptococcosis. These results indicate that cytokines are essential mediators of resistance to cryptococcosis. It is likely that susceptibility in many risk groups is related to deficiencies in production of one or more essential cytokines. If this proves to be the case, therapeutic use of cytokines has the potential to reverse predisposition to cryptococcosis.

REFERENCES

1. Emmons, C. W., 1955, Saprophytic sources of *Cryptococcus neoformans* associated with the pigeon (*Columba livia*), *Am. J. Hyg.* **62:**227–232.
2. Powell, K. E., Dahl, B. A., Weeks, R. J., and Tosh, F. E., 1972, Airborne *Cryptococcus neoformans*: Particles from pigeon excreta compatible with alveolar deposition, *J. Infect. Dis.* **125:**412–415.
3. Ruiz, A., and Bulmer, G. S., 1981, Particle size of airborne *Cryptococcus neoformans* in a tower, *Appl. Environ. Microbiol.* **41:**1225–1229.
4. Feigin, D. S., 1983, Pulmonary cryptococcosis: Radiologic-pathologic correlates of its three forms, *Am. J. Roentgenol.* **141:**1263–1272.
5. Tynes, B., Mason, K. N., Jennings, A. E., and Bennett, J. E., 1968, Variant forms of pulmonary cryptococcosis, *Ann. Intern. Med.* **69:**1117–1125.
6. Warr, W., Bates, J. H., and Stone, A., 1968, The spectrum of pulmonary cryptococcosis, *Ann. Intern. Med.* **69:**1109–1116.
7. Kerkering, T. M., Duma, R. J., and Shadomy, S., 1981, The evolution of pulmonary cryptococcosis, *Ann. Intern. Med.* **94:**611–616.
8. Campbell, G. D., 1966, Primary pulmonary cryptococcosis, *Am. Rev. Respir. Dis.* **94:** 236–243.
9. Houk, V. N., and Moser, K. M., 1965, Pulmonary cryptococcosis: Must all receive amphotericin B?, *Ann. Intern. Med.* **63:**583–596.
10. Wasser, L., and Talavera, W., 1987, Pulmonary cryptococcosis in AIDS, *Chest* **92:**692–695.
11. Williams, D. W., Krick, J. A., and Remington, J. S., 1976, Pulmonary infection in the compromised host. Part 1, *Am. Rev. Respir. Dis.* **114:**359–394.
12. Perfect, J. R., Durack, D. T., and Gallis, H. A., Cryptococcemia, *Medicine* **62:**98–109.
13. Bulmer, G. S., 1990, Twenty-five years with *Cryptococcus neoformans*, *Mycopathologia* **109:** 111–122.
14. Sabetta, J. R., and Andriole, V. T., 1985, Cryptococcal infection of the central nervous system, *Med. Clin. North Am.* **69:**333–344.
15. Diamond, R. D., 1985, *Cryptococcus neoformans*, in: *Principles and Practice of Infectious Diseases*, 2nd ed. (G. L. Mandell, R. G. Douglas, Jr., and J. E. Bennett, eds.), John Wiley & Sons, New York, pp. 1460–1468.

16. Patterson, T. F., and Andriole, V. T., 1989, Current concepts in cryptococcosis, *Eur. J. Clin. Microbiol. Infect. Dis.* **8**:457–465.
17. Roux, P., Touboul, J.-L., Feuilhade de Chauvin, M., Delacour, T., Revuz, J., Basset, D., Mayaud, C., and Lancastre, F., 1986, Disseminated cryptococcosis diagnosed in AIDS patients by screening for soluble serum antigens, *Lancet* **1**:1154.
18. Diamond, R. D., May, J. E., Kane, M. A., Frank, M. M., and Bennett, J. E., 1974, The role of the classical and alternative complement pathways in host defenses against *Cryptococcus neoformans* infection, *J. Immunol.* **112**:2260–2270.
19. Kozel, T. R., Highison, B., and Stratton, C. H., 1984, Localization on encapsulated *Cryptococcus neoformans* of serum components opsonic for phagocytosis by macrophages and neutrophils, *Infect. Immun.* **43**:574–579.
20. Kozel, T. R., and Pfrommer, G. S. T., 1986, Activation of the complement system by *Cryptococcus neoformans* leads to binding of iC3b to the yeast, *Infect. Immun.* **52**:1–5.
21. Polacheck, I., Hearing, V. J., and Kwon-Chung, K. J., 1982, Biochemical studies of phenoloxidase and utilization of catecholamines in *Cryptococcus neoformans*, *J. Bacteriol.* **150**: 1212–1220.
22. Kwon-Chung, K. J., and Rhodes, J. C., 1986, Encapsulation and melanin formation as indicators of virulence in *Cryptococcus neoformans*, *Infect. Immun.* **51**:218–223.
23. Littman, M. L., 1958, Capsule synthesis by *Cryptococcus neoformans*, *Trans. N.Y. Acad. Sci.* **20**: 623–648.
24. Igel, H. J., and Bolande, R. P., 1966, Humoral defense mechanisms in cryptococcosis: Substances in normal human serum, saliva, and cerebrospinal fluid affecting the growth of *Cryptococcus neoformans*, *J. Infect. Dis.* **116**:75–83.
25. Lewis, J. L., and Rabinovich, S., 1972, The wide spectrum of cryptococcal infections, *Am. J. Med.* **53**:315–322.
26. Dismukes, W. E., 1988, Cryptococcal meningitis in patients with AIDS, *J. Infect. Dis.* **157**: 624–628.
27. Grant, I. H., and Armstrong, D., 1988, Fungal infections in AIDS: cryptococcosis, *Infect. Dis. Clin. North Am.* **2**:457–464.
28. Chuck, S. L., and Sande, M. A., 1989, Infections with *Cryptococcus neoformans* in the acquired immunodeficiency syndrome, *N. Engl. J. Med.* **321**:794–799.
29. Zugar, A., Louie, E., Holzman, R. S., Simberkoff, M. S., and Rahal, J. J., 1986, Cryptococcal disease in patients with the acquired immunodeficiency syndrome: diagnostic features and outcome of treatment, *Ann. Intern. Med.* **104**:234–240.
30. Kovacs, J. A., Kovacs, A. A., Polis, M., Wright, W. C., Gill, V. J., Tuazon, C. U., Gelmann, E. P., Lane, H. C., Longfield, R., Overturf, G., Macher, A. M., Fauci, A. S., Parrillo, J. E., Bennett, J. E., and Masur, H., 1985, Cryptococcosis in the acquired immunodeficiency syndrome, *Ann. Intern. Med.* **103**:533–538.
31. Armstrong, D., 1988, Life-threatening opportunistic fungal infection in patients with the acquired immunodeficiency syndrome, *Ann. N.Y. Acad. Sci.* **544**:443–450.
32. Kwon-Chung, K. J., 1975, A new genus, *Filobasidiella*, the perfect state of *Cryptococcus neoformans*, *Mycologia* **67**:1197–1200.
33. Aulakh, H. S., Straus, S. E., and Kwon-Chung, K. J., 1981, Genetic relatedness of *Filobasidiella neoformans* (*Cryptococcus neoformans*) and *Filobasidiella bacillispora* (*Cryptococcus bacillisporus*) as determined by deoxyribonucleic acid base composition and sequence homology studies, *Int. J. Syst. Bacteriol.* **31**:97–103.
34. Kwon-Chung, K. J., Polacheck, I., and Bennett, J. E., 1982, Improved diagnostic medium for separation of *Cryptococcus neoformans* var. *neoformans* (serotypes A and D) and *Cryptococcus neoformans* var. *gattii* (serotypes B and C), *J. Clin. Microbiol.* **15**:535–537.
35. Newberry, W. M., Jr., Walter, J. E., Chandler, J. W., Jr., and Tosh, F. E., 1967, Epidemiologic study of *Cryptococcus neoformans*, *Ann. Intern. Med.* **67**:724–732.
36. Atkinson, A. J., Jr., and Bennett, J. E., 1968, Experience with a new skin test antigen prepared from *Cryptococcus neoformans*, *Am. Rev. Respir. Dis.* **97**:637–643.

37. Graybill, J. R., and Alford, R. H., 1974, Cell-mediated immunity in cryptococcosis, *Cell. Immunol.* **14:**12–21.
38. Selik, R. M., Starcher, E. T., and Curran, J. W., 1987, Opportunistic diseases reported in AIDS patients: frequencies, associations, and trends, *AIDS* **1:**175–182.
39. Eng, R. H. K., Bishburg, E., and Smith, S. M., 1986, Cryptococcal infections in patients with acquired immune deficiency syndrome, *Am. J. Med.* **81:**19–23.
40. Muchmore, H. G., Felton, F. G., Salvin, S. B., and Rhoades, E. R., 1969, Delayed hypersensitivity to cryptococcin in man, *Sabouraudia* **6:**285–288.
41. Schimpff, S. C., and Bennett, J. E., 1975, Abnormalities in cell-mediated immunity in patients with *Cryptococcus neoformans* infections, *J. Allergy Clin. Immunol.* **55:**430–441.
42. Kaplan, M. H., Rosen, P. P., and Armstrong, D., 1977, Cryptococcosis in a cancer hospital: Clinical and pathological correlates in forty-six patients, *Cancer* **39:**2265–2274.
43. Rippon, J. W., 1988, *Medical Mycology—The Pathogenic Fungi and the Pathogenic Actinomycetes*, 3rd ed., W. B. Saunders Company, Philadelphia.
44. Statistics from the World Health Organization and the Centers for Disease Control, 1990, *AIDS 1990* **4:**375–379.
45. Clumeck, N., Sonnet, J., Taelman, H., Mascart-Lemone, F., de Bruyere, M., Vandeperre, P., Dasnoy, J., Marcelis, L., Lamy, M., Jonas, C., Eyckmans, L., Noel, H., Vanhaeverbeek, M., and Butzler, J.-P., 1984, Acquired immunodeficiency syndrome in African patients, *N. Engl. J. Med.* **310:**492–497.
46. Katama, C., 1984, Acquired immunodeficiency syndrome (AIDS) in Africans, *Ann. Soc. Belg. Med. Trop.* **64:**379–389.
47. Bennett, J. E., Kwon-Chung, K. J., and Howard, D. H., Epidemiologic differences among serotypes of *Cryptococcus neoformans*, *Am. J. Epidemiol.* **105:**582–586.
48. Kwon-Chung, K. J., and Bennett, J. E., 1984, Epidemiologic differences between the two varieties of *Cryptococcus neoformans*, *Am. J. Epidemiol.* **120:**123–130.
49. Ruiz, A., Neilson, J. B., and Bulmer, G. S., 1982, Control of *Cryptococcus neoformans* in nature by biotic factors, *Sabouraudia* **20:**21–29.
50. Swinne, D., Kayembe, K., and Niyimi, M., 1986, Isolation of saprophytic *Cryptococcus neoformans* var. *neoformans* in Kinshasa, Zaire, *Ann. Soc. Belg. Med. Trop.* **66:**57–61.
51. Swinne, D., Deppner, M., Laroche, R., Floch, J. J., and Kadende, P., 1989, Isolation of *Cryptococcus neoformans* from houses of AIDS-associated cryptococcosis patients in Bujumbura (Burundi), *AIDS* **3:**389–390.
52. Bottone, E. J., Salkin, I. F., Hurd, N. J., and Wormser, G. P., 1987, Serogroup distribution of *Cryptococcus neoformans* in patients with AIDS, *J. Infect. Dis.* **156:**242.
53. Rinaldi, M. G., Drutz, D. J., Howell, A., Sande, M. A., Wofsy, C. B., and Hadley, W. K., 1986, Serotypes of *Cryptococcus neoformans* in patients with AIDS, *J. Infect. Dis.* **153:**642.
54. Clancy, M. N., Fleischmann, J., Howard, D. H., Kwon-Chung, K. J., and Shimizu, R. Y., 1990, Isolation of *Cryptococcus neoformans gatti* from a patient with AIDS in southern California, *J. Infect. Dis.* **161:**809.
55. Shimizu, R. Y., Howard, D. H., and Clancy, M. N., 1986, The variety of *Cryptococcus neoformans* in patients with AIDS, *J. Infect. Dis.* **154:**1042.
56. Swinne, D., Nkurikiyinfura, J. B., and Muyembe, T. L., 1986, Clinical isolates of *Cryptococcus neoformans* from Zaire, *Eur. J. Clin. Microbiol.* **5:**50–51.
57. Levitz, S. M., and Farrell, T. P., 1990, Growth inhibition of *Cryptococcus neoformans* by cultured human monocytes: role of the capsule, opsonins, the culture surface, and cytokines, *Infect. Immun.* **58:**1201–1209.
58. Howard, D. H., 1961, Some factors which effect the initiation of growth of *Cryptococcus neoformans*, *J. Bacteriol.* **82:**430–435.
59. Baum, G. L., and Artis, D., 1963, Characterization of the growth inhibition factor for *Cryptococcus neoformans* (GIFc) in human serum, *Am. J. Med. Sci.* **246:**87–91.
60. Monga, D. P., 1981, Role of macrophages in resistance of mice to experimental cryptococcosis, *Infect. Immun.* **32:**975–978.

61. Bulmer, G. S., and Sans, M. D., 1967, *Cryptococcus neoformans*. II. Phagocytosis by human leukocytes, *J. Bacteriol.* **94:**1480–1483.
62. Mitchell, T. G., and Friedman, L., 1972, In vitro phagocytosis and intracellular fate of variously encapsulated strains of *Cryptococcus neoformans*, *Infect. Immun.* **5:**491– 498.
63. Hasenclever, H. F., and Mitchell, W. O., 1960, Virulence and growth rates of *Cryptococcus neoformans* in mice, *Ann. N.Y. Acad. Sci.* **89:**156–162.
64. Kozel, T. R., Pfrommer, G. S. T., Guerlain, A. S., Highison, B. A., and Highison, G. J., 1988, Strain variation in phagocytosis of *Cryptococcus neoformans*: dissociation of susceptibility to phagocytosis from activation and binding of opsonic fragments of C3, *Infect. Immun.* **56:** 2794–2800.
65. Diamond, R. D., Root, R. K., and Bennett, J. E., 1972, Factors influencing killing of *Cryptococcus neoformans* by human leukocytes in vitro, *J. Infect. Dis.* **125:**367–376.
66. Kozel, T. R., 1977, Non-encapsulated variant of *Cryptococcus neoformans*. II. Surface receptors for cryptococcal polysaccharide and their role in inhibition of phagocytosis by polysaccharide, *Infect. Immun.* **16:**99–106.
67. Kozel, T. R., and Gotschlich, E. C., 1982, The capsule of *Cryptococcus neoformans* passively inhibits phagocytosis of the yeast by macrophages, *J. Immunol.* **129:**1675–1680.
68. Bulmer, G. S., and Sans, M. D., 1968, *Cryptococcus neoformans*. III. Inhibition of phagocytosis, *J. Bacteriol.* **95:**5–8.
69. Kozel, T. R., Reiss, E., and Cherniak, R., 1980, Concomitant but not causal association between surface charge and inhibition of phagocytosis by cryptococcal polysaccharide, *Infect Immun.* **29:**295–300.
70. Kozel, T. R., 1983, Dissociation of a hydrophobic surface from phagocytosis of encapsulated and non-encapsulated *Cryptococcus neoformans*, *Infect. Immun.* **39:**1214–1219.
71. Kozel, T. R., Pfrommer, G. S. T., Guerlain, A. S., Highison, B. A., and Highison, G. J., 1988, Role of the capsule in phagocytosis of *Cryptococcus neoformans*, *Rev. Infect. Dis.* **10:**S436–S439.
72. Goren, M. B., and Warren, J., 1968, Immunofluorescence studies of reactions at the cryptococcal capsule, *J. Infect. Dis.* **118:**215–229.
73. Kozel, T. R., Wilson, M. A., Pfrommer, G. S. T., and Schlageter, A. M., 1989, Activation and binding of opsonic fragments of C3 on encapsulated *Cryptococcus neoformans* by using an alternative complement pathway reconstituted from six isolated proteins, *Infect. Immun.* **57:** 1922–1927.
74. Diamond, R. D., and Allison, A. C., 1976, Nature of the effector cells responsible for antibody-dependent cell-mediated killing of *Cryptococcus neoformans*, *Infect. Immun.* **14:** 716–720.
75. Miller, G. P. G., and Kohl, S., 1983, Antibody-dependent leukocyte killing of *Cryptococcus neoformans*, *J. Immunol.* **131:**1455–1459.
76. Levitz, S. M., and DiBenedetto, D. J., 1988, Differential stimulation of murine resident peritoneal cells by selectively opsonized encapsulated and acapsular *Cryptococcus neoformans*, *Infect. Immun.* **56:**2544–2551.
77. Flesch, I. E. A., Schwamberger, G., and Kaufmann, S. H. E., 1989, Fungicidal activity of IFN-τ-activated macrophages: extracellular killing of *Cryptococcus neoformans*, *J. Immunol.* **142:**3219–3224.
78. Granger, D. L., Perfect, J. R., and Durack, D. T., 1986, Macrophage-mediated fungistasis in vitro: requirements for intracellular and extracellular cytotoxicity, *J. Immunol.* **136:** 672–680.
79. Adams, D. O., and Hamilton, T. A., 1984, The cell biology of macrophage activation, *Ann. Rev. Immunol.* **2:**283–318.
80. Lehrer, R. I., and Ladra, K. M., 1977, Fungicidal components of mammalian granulocytes active against *Cryptococcus neoformans*, *J. Infect. Dis.* **136:**96–99.
81. Granger, D. L., Hibbs, J. B., Jr., Perfect, J. R., and Durack, D. T., 1988, Specific amino acid

(L-arginine) requirement for the microbiostatic activity of murine macrophages, *J. Clin. Invest.* **81:**1129–1136.
82. Nabavi, N., and Murphy, J. W., 1985, In vitro binding of natural killer cells to *Cryptococcus neoformans* targets, *Infect. Immun.* **50:**50–57.
83. Murphy, J. W., and McDaniel, D. O., 1982, In vitro reactivity of natural killer (NK) cells against *Cryptococcus neoformans, J. Immunol.* **128:**1577–1583.
84. Hidore, M. R., and Murphy, J. W., 1989, Murine natural killer cell interactions with a fungal target, *Cryptococcus neoformans, Infect. Immun.* **57:**1990–1997.
85. Cauley, L. K., and Murphy, J. W., 1979, Response of congenitally athymic (nude) and phenotypically normal mice to *Cryptococcus neoformans* infection, *Infect. Immun.* **23:**644–651.
86. Murphy, J. W., 1984, Natural killer cells against a mycotic pathogen, *Microbiology* **1984:** 328–332.
87. Marquis, G., Montplaisir, S., Pelletier, M., Mousseau, S., and Auger, P., 1985, Genetic resistance to murine cryptococcosis: The beige mutation (Chediak-Higashi Syndrome) in mice, *Infect. Immun.* **47:**288–293.
88. Hidore, M. R., and Murphy, J. W., 1986, Natural cellular resistance of beige mice against *Cryptococcus neoformans, J. Immunol.* **137:**3624–3631.
89. Voth, R., Rossol, S., Graff, E., Laubenstein, H. P., Schroder, H. C., Muller, W. E. G., Meyer Zum Buschenfelde, K. H., and Hess, G., 1988, Natural killer cell activity as a prognostic parameter in the progression to AIDS, *.J. Infect. Dis.* **157:**851–852.
90. Brenner, G. B., Dascal, A., Margolese, R. G., and Wainberg, M. A., 1989, Natural killer cell function in patients with acquired immunodeficiency syndrome and related diseases, *J. Leukoc. Biol.* **46:**75–83.
91. Miller, M. F., Mitchell, T. G., Storkus, W. J., and Dawson, J. R., 1990, Human natural killer cells do not inhibit growth of *Cryptococcus neoformans* in the absence of antibody, *Infect. Immun.* **58:**639–645.
92. Cherniak, R., Reiss, E., Slodki, M. E., Plattner, R. D., and Blumer, S. O., 1980, Structure and antigenic activity of the capsular polysaccharide of *Cryptococcus neoformans, Mol. Immunol.* **17:**1025–1032.
93. Turner, S. H., Cherniak, R., and Reiss, E., 1984, Fractionation and characterization of galactoxylomannan from *Cryptococcus neoformans, Carbohydr. Res.* **125:**343–349.
94. Cherniak, R., Reiss, E., and Turner, S. H., 1982, A galactoxylomannan antigen of *Cryptococcus neoformans* serotype A, *Carbohydr. Res.* **103:**239–250.
95. Spiropulu, C., Eppard, R. A., Otteson, E., and Kozel, T. R., 1989, Antigenic variation within serotypes of *Cryptococcus neoformans* detected by monoclonal antibodies specific for the capsular polysaccharide, *Infect. Immun.* **57:**3240–3242.
96. Bhattacharjee, A. K., Kwon-Chung, K. J., and Glaudemans, C. P. J., 1981, Capsular polysaccharide from a parent strain and from a possible, mutant strain of *Cryptococcus neoformans* serotype A, *Carbohydr. Res.* **95:**237–248.
97. Reiss, E., Cherniak, R., Eby, R., and Kaufman, L., 1984, Enzyme immunoassay detection of IgM to galactoxylomannan of *Cryptococcus neoformans, Diagn. Immunol.* **2:**109–115.
98. Vartivarian, S. E., Reyes, G. H., Jacobson, E. S., James, P. G., Cherniak, R., Mumaw, V. R., and Tingler, M. J., 1989, Localization of mannoprotein in *Cryptococcus neoformans, J. Bacteriol.* **171:**6850–6852.
99. Henderson, D. K., Bennett, J. E., and Huber, M. A., 1982, Long-lasting, specific immunologic unresponsiveness associated with cryptococcal meningitis, *J. Clin. Invest.* **69:**1185–1190.
100. Dromer, F., Aucouturier, P., Clauvel, J.-P., Saimot, G., and Yeni, P., 1988, *Cryptococcus neoformans* antibody levels in patients with AIDS, *Scand J. Infect. Dis.* **20:**283–285.
101. Amman, A. J., Schiffman, G. S., Abrams, D., Volberding, P., Ziegler, J., and Conant, M., 1984, B-cell immunodeficiency in acquired immunodeficiency syndrome, *JAMA* **251:**1447–1449.

102. Bindschadler, D. D., and Bennett, J. E., 1968, Serology of human cryptococcosis, *Ann. Intern. Med.* **69:**45–52.
103. Kaufman, L., and Blumer, S., 1968, Value and interpretation of serological tests for the diagnosis of cryptococcosis, *Applied Microbiology* **16:**1907–1912.
104. Kozel, T. R., Gulley, W. F., and Cazin, J., Jr., 1977, Immune response to *Cryptococcus neoformans* soluble polysaccharide: immunological unresponsiveness, *Infect. Immun.* **18:**701–707.
105. Blumer, S. O., and Kaufman, L., 1977, Characterization of immunoglobulin classes of human antibodies to *Cryptococcus neoformans*, *Mycopathologia* **61:**55–60.
106. Vogel, R. A., 1966, The indirect fluorescent antibody test for the detection of antibody in human cryptococcal disease, *J. Infect. Dis.* **116:**573–580.
107. Basten, A., and Howard, J. G., 1973, Thymus independence, *Contemp. Top. Immunobiol.* **2:** 265–291.
108. Breen, J. F., Lee, I. C., Vogel, F. R., and Friedman, H., 1982, Cryptococcal capsular polysaccharide-induced modulation of murine immune responses, *Infect. Immun.* **36:**47–51.
109. Baker, P. J., Stashak, P. W., Amsbaugh, D. ., Prescott, B., and Barth, R. F., 1970, Evidence for the existence of two functionally distinct types of cells which regulate the antibody response to type III pneumococcal polysaccharide, *J. Immunol.* **105:**1581–1583.
110. Murphy, J. W., and Cozad, G. C., 1972, Immunological unresponsiveness induced by cryptococcal capsular polysaccharide assayed by the hemolytic plaque technique, *Infect. Immun.* **5:**896–901.
111. Gordon, M. A., and Vedder, D. K., 1966, Serologic tests in diagnosis and prognosis of cryptococcosis, *Am. Med. Assoc.* **197:**961–967.
112. Diamond, R. D., and Bennett, J. E., 1974, Prognostic factors in cryptococcal meningitis: A study of 111 cases, *Ann. Intern. Med.* **80:**176–181.
113. Gadebusch, H. H., 1958, Passive immunization against *Cryptococcus neoformans*, *Proc. Soc. Exp. Biol. Med.* **98:**611–614.
114. Graybill, J. R., Hague, M., and Drutz, D. J., 1981, Passive immunization in murine cryptococcosis, *Sabouraudia* **19:**237–244.
115. Louria, D. B., and Kaminski, T., 1965, Passively-acquired immunity in experimental cryptococcosis, *Sabouraudia* **40:**80–84.
116. Dromer, F., Charreire, J., Contrepois, A., Carbon, C., and Yeni, P., 1987, Protection of mice against experimental cryptococcosis by anti-*Cryptococcus neoformans*. monoclonal antibody, *Infect. Immun.* **55:**749–752.
117. Dromer, F., Perronne, C., Barge, J., Vilde, J. L., and Yeni, P., 1989, Role of IgG and complement component C5 in the initial course of experimental cryptococcosis, *Clin. Exp. Immunol.* **78:**412–417.
118. Sanford, J. E., Lupan, D. M., Schlageter, A. M., and Kozel, T. R., 1990, Passive immunization against *Cryptococcus neoformans* with an isotype-switch family of monoclonal antibodies reactive with cryptococcal polysaccharide, *Infect. Immun.* **58:**1919–1923.
119. Diamond, R. D., 1977, Effects of stimulation and suppression of cell-mediated immunity on experimental cryptococcosis, *Infect. Immun.* **17:**187–194.
120. Lim, T. S., and Murphy, J. W., 1980, Transfer of immunity to cryptococcosis by T-enriched splenic lymphocytes from *Cryptococcus neoformans*-sensitized mice, *Infect. Immun.* **30:**5–11.
121. Fung, P. Y. S., and Murphy, J. W., 1982, In vitro interactions of immune lymphocytes and *Cryptococcus neoformans*, *Infect. Immun.* **36:**1128–1138.
122. Khakpour, F. R., and Murphy, J. W., 1987, Characterization of a third-over suppressor T cell (Ts3) induced by cryptococcal antigen(s), *Infect. Immun.* **55:**1657–1662.
123. Murphy, J. W., and Moorhead, J. W., 1982, Regulation of cell-mediated immunity in cryptococcosis. I. Induction of specific afferent T suppressor cells by cryptococcal antigen, *J. Immunol.* **128:**276–283.
124. Fidel, P. L., Jr., and Murphy, J. W., 1988, Characterization of an in vitro-stimulated, *Cryptococcus neoformans*-specific second-order suppressor T cell and its precursor, *Infect. Immun.* **56:**1267–1272.

125. Murphy, J. W., 1985, Effects of first-order *Cryptococcus*-specific T-suppressor cells on induction of cells responsible for delayed-type hypersensitivity, *Infect. Immun.* **48**:439–445.
126. Fidel, P. L., Jr., and Murphy, J. W., 1990, Characterization of a cell population with amplifies the anticryptcococcal delayed-type hypersensitivity response, *Infect. Immun.* **58**:393–398.
127. Murphy, J. W., and Mosley, R. L., 1985, Regulation of cell-mediated immunity in cryptococcosis. III. Characterization of second-order T-suppressor cells (Ts2), *J. Immunol.* **134**: 577–584.
128. Murphy, J. W., Mosley, R. L., and Moorhead, J. W., 1983, Regulation of cell-mediated immunity in cryptococcosis. II. Characterization of first-order T-suppressor cells (Ts1) and induction of second-order suppressor cells, *J. Immunol.* **130**:2876–2881.
129. Murphy, J. W., and Cox, R. A., 1988, Induction of antigen-specific suppression by circulating *Cryptococcus neoformans* antigen, *Clin. Exp. Immunol.* **73**: 174–180.
130. Fidel, P. L., Jr., and Murphy, J. W., 1989, Effects of cyclosporin A on the cells responsible for the anticryptococcal cell-mediated immune response and its regulation, *Infect. Immun.* **57**: 1158–1164.
131. Murphy, J. W., and Pahlavan, N., 1979, Cryptococcal culture filtrate antigen for detection of delayed-type hypersensitivity in cryptococcosis, *Infect. Immun.* **25**:284–292.
132. Murphy, J. W., Mosley, R. L., Cherniak, R., Reyes, G. H., Kozel, T. R., and Reiss, E., 1988, Serological, electrophoretic, and biological properties of *Cryptococcus neoformans* antigens, *Infect. Immun.* **56**:424–431.
133. Bennett, J. E., and Bailey, J. W., 1971, Control for rheumatoid factor in the latex test for cryptococcosis, *Am. J. Clin. Pathol.* **56**:360–365.
134. Prevost, E., and Newell, R., 1978, Commercial cryptococcal latex kit: Clinical evaluation in a medical center hospital, *J. Clin. Microbiol.* **8**:529–533.
135. Gordon, M. A., and Lapa, E. W., 1974, Elimination of rheumatoid factor in the latex test for cryptococcosis, *Am. J. Clin. Pathol.* **61**:488–494.
136. Eng, R. and Person, A., 1981, Serum cryptococcal antigen determination in the presence of rheumatoid factor, *J. Clin. Microbiol.* **14**:700–702.
137. Stockman, L., and Roberts, G. D., 1982, Specificity of the latex test for cryptococcal antigen: a rapid, simple method for eliminating interference factors, *J. Clin. Microbiol.* **16**: 965–967.
138. McManus, E. J., Bozdech, M. J., and Jones, J. M., 1985, Role of the latex agglutination test for cryptococcal antigen in diagnosing disseminated infections with *Trichosporon beigelii*, *J. Infect. Dis.* **151**:1167–1169.
139. Westerink, M. A. J., Amsterdam, D., Petell, R. J., Stram, M. N., and Apicella, M. A., 1987, Septicemia due to DF-2: cause of a false-positive cryptococcal latex agglutination result, *Am. J. Med.* **83**:155–158.
140. Sugar, A. M., 1988, Oral fluconazole as suppressive therapy of disseminated cryptococcosis in patients with acquired immunodeficiency syndrome, *Am. J. Med.* **85**:481–489.
141. Kligman, A. M., 1947, Studies of the capsular substance of *Torula histolytica* and the immunologic properties of Torula cells, *J. Immunol.* **57**:395–401.
142. Goren, M. B., 1967, Experimental murine cryptococcosis: effect of hyperimmunization to capsular polysaccharide, *J. Immunol.* **98**:914–922.
143. Gadebusch, H. H., 1963, Immunization against *Cryptococcus neoformans* by capsular polysaccharide, *Nature* **199**:710.
144. Hoff, C. L., 1941, Immunity studies of *Cryptococcus hominis* (*Torula histolytica*) in mice, *J. Lab. Clin. Med.* **27**:751–754.
145. Abrahams, I., and Gilleran, T. G., 1960, Studies on actively acquired resistance to experimental cryptococcosis in mice, *J. Immunol.* **85**:629–635.
146. Abrahams, I., 1966, Further studies on acquired resistance to murine cryptococcosis: enhancing effect of *bordetella pertussis*, *J. Immunol.* **96**:525–529.
147. Gadebusch, H. H., 1960, Specific degradation of *Cryptococcus neoformans* 3723 capsular

polysaccharide by a microbial enzyme. III. Antibody stimulation by partially decapsulated cells, *J. Infect. Dis.* **107**:406–409.

148. Kase, A., and Metzger, J. F., 1962, Isolation of a dry variant from a mucoid strain of *Cryptococcus neoformans* and preliminary comparative studies, *J. Bacteriol.* **83**:926–927.
149. Dykstra, M. A., and Friedman, L., 1978, Pathogenesis, lethality, and immunizing effect of experimental cutaneous cryptococcosis, *Infect. Immun.* **20**:446–455.
150. Louria, D. B., 1960, Specific and non-specific immunity in experimental cryptococcosis in mice, *J. Exp. Med.* **111**:643–665.
151. Moser, S. A., Lyon, F. L., Domer, J. E., and Williams, J. E., 1982, Immunization of mice by intracutaneous inoculation with viable virulent *Cryptococcus neoformans*: immunologic and histopathological parameters, *Infect. Immun.* **35**:685–696.
152. Perceval, A. K., 1965, Experimental cryptococcosis: hypersensitivity and immunity, *J. Pathol. Bacteriol.* **89**:645–655.
153. Vanbreuseghem, R., 1967, White mice sensitivity to *Cryptococcus neoformans*. Production of resistance against a cerebral inoculation, *Ann. Soc. Belg. Med. Trop.* **47**: 281–294.
154. Sugar, A. M., 1990, Overview: treatment of cryptococcal meningitis, *Rev. Infect. Dis.* **12**: S338–S348.
155. Bennett, J. E., Dismukes, W. E., Duma, R. J., Medoff, G., Sande, M. A., Gallis, H., Leonard, J., Fields, B. T., Bradshaw, M., Haywood, H., McGee, Z. A., Cate, T. R., Cobbs, C. G., Warner, J. F., and Alling, D. W., 1979, A comparison of amphotericin B alone and combined with flucytosine in the treatment of cryptococcal meningitis, *N. Engl. J. Med.* **301**:126–131.
156. Dismukes, W. E., Cloud, G., Gallis, H. A., Kerkering, T. M., Medoff, G., craven, P. C., Kaplowtiz, L. G., Fisher, J. F., Gregg, C. R., Bowles, C. A., Shadomy, S., Stamm, A. M., Diasio, R. B., Kaufman, L., Soong, S.-J., and Blackwelder, W. C., 1987, Treatment of cryptococcal meningitis with combination amphotericin B and flucytosine for four as compared with six weeks, *N. Engl. J. Med.* **317**:334–341.
157. Arndt, C. A. S., Walsh, T. J., McCully, C. L., Balis, F. M., Pizzo, P. A., and Poplack, D. G., 1988, Fluconazole penetration into cerebrospinal fluid: implications for testing fungal infections of the central nervous system, *J. Infect. Dis.* **157**:178–180.
158. Gordon, M. A., and Lapa, E., 1964, Serum protein enhancement of antibiotic therapy in cryptococcosis, *J. Infect. Dis.* **114**:373–377.
159. Ellis, D. H., and Pfeiffer, T. J., 1990, Natural habitat of *Cryptococcus neoformans* var. *gattii*, *Infect. Immun.* **28**:1642–1644.
160. Ellis, D. H., and Pfeiffer, T. J., 1990, Ecology, life cycle, and infectious propagule of *Cryptococcus neoformans*, *Lancet* **336**:923–925.
161. Pfeiffer, T., and Ellis, D., 1991, Environmental isolation of *Cryptococcus neoformans gattii* from California, *J. Infect. Dis.* **163**:929–930.
162. Gade, W., Hinnefeld, S. W., Babcock, L. S., Gilligan, P., Kelly, W., Wait, K., Greer, D., Pinilla, M., and Kaplan, R. L., 1991, Comparison of the PREMIER cryptococcal antigen enzyme immunoassay and the latex agglutination assay for detection of cryptococcal antigens, *J. Clin. Microbiol.* **29**:1616–1619.
163. Devi, S. J. N., Schneerson, R., Egan, W., Ulrich, T. J., Bryla, D., Robbins, J. B., and Bennett, J. E., 1991, *Cryptococcus neoformans* serotype A glucuronoxylomannan-protein conjugate vaccines: synthesis, characterization, and imunogenicity, *Infect. Immun.* **59**:3700–3707.
164. Salkowski, C. A., and Balish, E., 1991, A monoclonal antibody to gamma interferon blocks augmentation of natural killer cell activity induced during systemic cryptococcosis, *Infect. Immun.* **59**:486–493.
165. Collins, H. L., and Bancroft, G. J., 1992, Cytokine enhancement of complement-dependent phagocytosis by macrophages: synergy of tumor necrosis factor-α and granulocyte-macrophage colony-stimulating factor for phagocytosis of *Cryptococcus neoformans*, *Eur. J. Immunol.* **22**:1447–1454.

12

Chromoblastomycosis

GARY E. HOLLICK

1. INTRODUCTION

Chromoblastomycosis is a chronic, slowly progressive infection of the cutaneous and subcutaneous tissue. The disease is found primarily in tropical and subtropical climates but can occur in more temperate parts of the world. The first reported case was by Rudolph[1] who published the description of a disease he called figueira. One year later, Lane[2] and Medlar[3] described a patient from New England who had verrucous lesions on the foot. This represented the first reported case of chromoblastomycosis from the United States and also established *Phialophora verrucosa* as the first recognized etiologic agent. In 1920, Pedroso and Gomes[4] reported a patient whom they had followed since 1911 with a nodular and ulcerated disease of the leg and foot that they called blastomycose negra. These diseases were ultimately shown to be the same. Subsequently, many additional cases have been reported to occur in the United States and world wide. A summary of these reports was compiled and published by Al-Doory[5] and Carrion.[6] Additionally, several excellent reviews of chromoblastomycosis have been recently published.[7,8]

The disease has become known by a variety of different names. The more common synonyms include figueira, formigueiro, chromomycosis, phaeosporotrichosis, verrucous dermititis, Pedroso's disease, Fonseca's disease, and Carrion's disease. The name chromoblastomycosis was first proposed by Terra *et al.*[9] to describe the disease caused by the colored yeastlike fungus that they had seen in tissue. Moore and Almeida,[10] however, felt that this name was somewhat misleading since it implied that the organism grew

GARY E. HOLLICK • Department of Pathology and Laboratory Medicine, Microbiology Laboratory, Rochester General Hospital, Rochester, New York 14621.

Fungal Infections and Immune Responses, edited by Juneann W. Murphy *et al.* Plenum Press, New York, 1993.

in tissue as a yeast. They felt that a more generic name should be given and proposed that the name chromomycosis replace chromoblastomycosis. After Moore and Almeida's[10] proposal, many different diseases, caused by a variety of different dematiaceous fungi, have been included under the umbrella of chromomycosis. This has caused much confusion in the literature and has led to the proposal of several different disease classification systems.[11–14]

In an attempt to clarify the different disease entities, Ajello *et al.*[14] created the name phaeohyphomycosis for those infections that were clinically, pathologically and mycologically distinct from the classic description of chromoblastomycosis. Phaeohyphomycosis would encompass those diseases in which any combination of dematiaceous yeastlike cells, hyphae, or pseudohyphae was present in tissue. Chromoblastomycosis would be restricted to the original description of chronic verrucous cutaneous lesions as described by Terra *et al.*[9] Also, as pointed out by McGinnis,[15,16] this nomenclature can accommodate the isolation of new etiologic agents or the taxonomic reclassification of existing fungi capable of causing chromoblastomycosis without a corresponding change in the disease name. For the purposes of this chapter, the term chromoblastomycosis will be used.

2. DESCRIPTION OF THE DISEASE

2.1. Clinical Features

Chromoblastomycosis occurs primarily in the tropical and subtropical climates but cases have been reported from North, South, and Central America, Europe, Asia, Africa, and Australia. The infection is acquired following trauma and is seen most often in individuals engaged in some type of agricultural activity. The infection usually begins on the distal part of an extremity such as the leg, foot, or hand. Other body areas such as the trunk, buttocks, shoulders, face, and ears can also be infected. In some countries, the number of cases involving the shoulder, chest or trunk exceeds those involving the leg. Dissemination to other parts of the body or involvement of other organs is rarely reported.[17] If dissemination does occur it is generally to the brain. The major complications are secondary bacterial infections that can lead to lymphedema and elephantiasis.[18] The disease is very slow to progress and as many as twenty-years can elapse before medical attention is sought. Lesions are painless unless secondary bacterial infection leads to suppuration. At this time, medical attention is usually sought.

The primary lesion begins as a small, nonpurulent, pink or violaceous scaly papule. These small papules may persist for many months or years, after which a new crop of satellite lesions appears in the same or an adjacent area. These occur as a result of lymphatic spread or by autoinoculation. The lesions ultimately become raised 1–3 mm above the surface of the skin to form an irregular plaque. Lesions at this stage tend to enlarge and coalesce. The plaques may continue to rise and develop into larger pedunculated and verrucous lesions that resemble a cauliflower head (Fig. 1A–C). These growths may reach 1–2 cm above the surface of the skin. The surface of the lesion may be studded with small ulcerations that resemble black dots. These dots are composed of clotted blood and fungi that are being eliminated from the lesion. When palpated, the lesions may discharge a whitish, caseous material. Ulceration of the lesion is not an inherent part of the disease progression but can occur in some patients because of secondary bacterial infection.

A less common form of the disease is described as having flat, annular, papular lesions with a raised active border. These lesions do not get the typical cauliflowerlike appearance (Fig. 1B). Healing involves extensive keloid formation resulting in a fibromatous appearance. There is usually not any invasion of the bone or muscle or formation of sinus tracts, as seen with mycetomas. Active lesions must be differentiated from blastomycosis, yaws, tertiary syphilis, mycetoma, leishmaniasis, sporotrichosis, and cutaneous tuberculosis. Al-Doory[5] has suggested that there are two features that characterize chromoblastomycosis. These features are the presence of the distinctive margins of the lesion and the development of satellite lesions.

In a summary of 159 cases, Carrion[6] described five types of lesions that he had seen on patients. He categorized the lesions, based on appearance, as nodular, tumorous, verrucous, plaque, or cicatrical. These types were not mutually exclusive and more than one type could often occur at any one time on the same patient. Nodular lesions are moderately elevated, soft, dull pink to violaceous growths that can be smooth, verrucous, or scaly. With time, these lesions become tumorous in appearance. These tumorous lesions are larger, papillomatous, lobulated and covered with dirty gray epidermal debris. These lesions become enormous in size and resemble a cauliflower head. Verrucous lesions are characterized by extensive hyperkeratosis and growths with a wartlike appearance. Carrion[6] noted that this type of lesion is frequently seen along the border of the foot. In contrast to the other lesions, plaque lesions tend to be flat or slightly elevated. They are variable in size and shape and may have a reddish to violaceous color. Cicatricial lesions are characterized as growths that enlarge by expanding centrifugally. Healing, with scarring, takes place in the center of the lesion. These lesions often cover extensive areas of the body.

2.2. Histopathological Features

The major histopathological feature of chromoblastomycosis is the presence of a chestnut brown yeastlike cell called a sclerotic body. The sclerotic bodies appear to represent the tissue vegetative form of the fungus that arrested between the yeast and hyphal stage in response to environmental conditions in the host.[19] The sclerotic bodies are round to polyhedral,

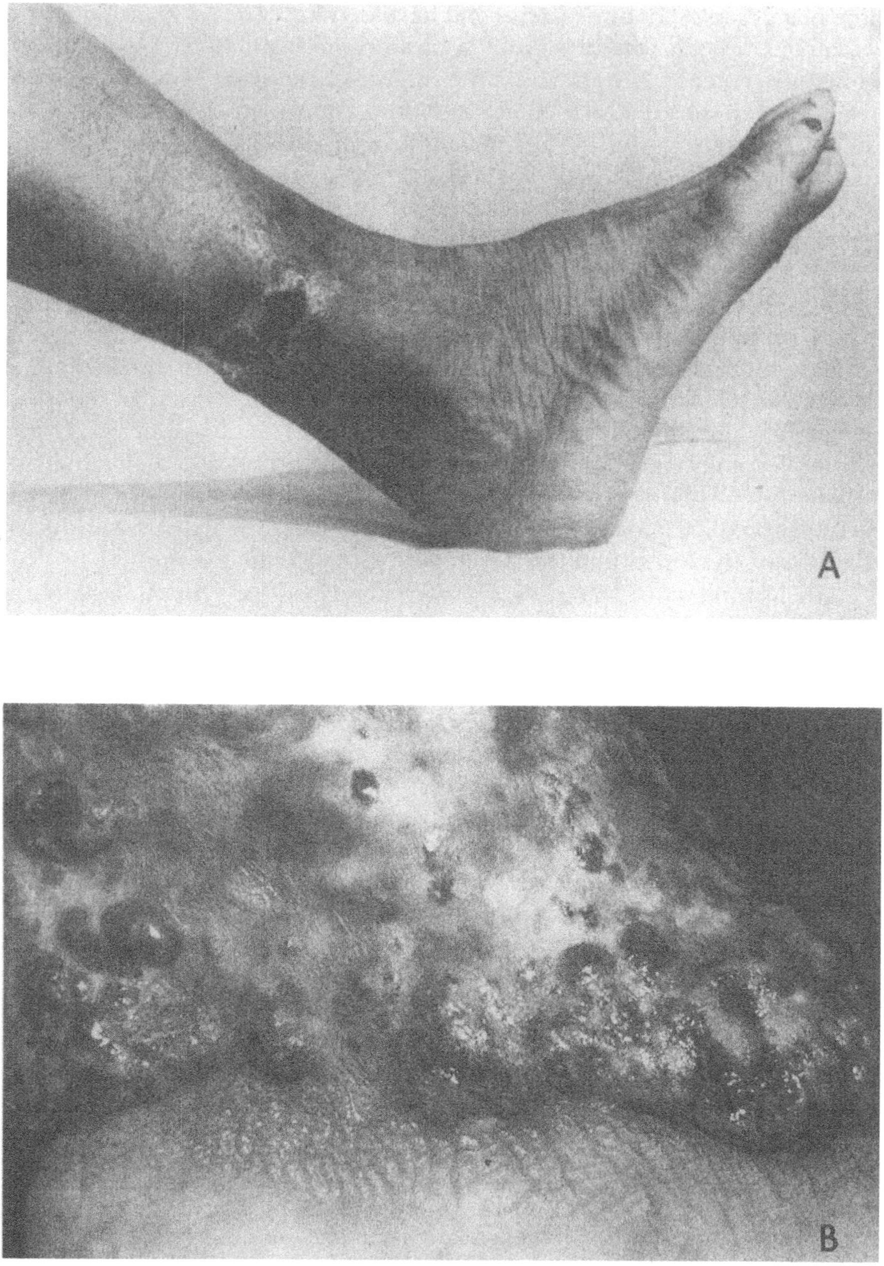

FIGURE 1. A) Early lesion of chromoblastomycosis, B) chronic verrucous lesions of chromoblastomycosis, and C) chronic verrucous chromoblastomycosis with ulcerated and satellite lesions. (A courtesy of Lea & Febiger, B courtesy of M. Rinaldi, C courtesy of ASCP Press.)

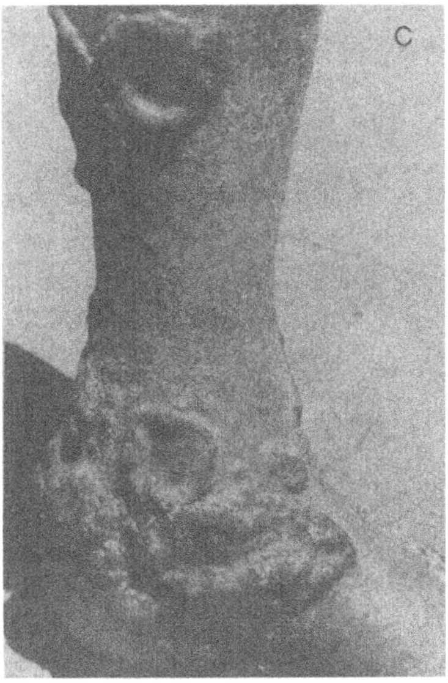

FIGURE 1. *(Cont.)*

muriform, and approximately 5–12 μm in diameter (Fig. 2). They may occur singly, in groups, or in short chains.

Rosen *et al.*[20] studied sclerotic bodies with the electron microscope. They reported finding three varieties of fungal bodies. They described sclerotic cells with thick walls and a well-developed infrastructure that included the presence of mitochondria, endoplasmic reticulum, and nuclear material. These cells probably represent the actively growing tissue form of the organism. Additionally, they noted cells laden with storage substances, and partially or completely degenerated cells. These latter forms were most often seen in association with macrophages and inflammatory cells and may represent cells being eliminated by the host. They also reported the presence of sclerotic bodies that had germ tubes and hyphal fragments. Using the scanning electron microscope, Harada and Kusunoki[21] confirmed this observation.

The tissue response seen in almost all cases involves extensive pseudoepitheliomatous hyperplasia. The epidermis is generally thickened and shows irregular growths into the dermis. The stratum corneum shows hyperkeratosis and is often distorted. The epidermal layers may be infiltrated with neutrophils, lymphocytes, eosinophils, and macrophages with the presence of giant cells. The dermis is characterized by the presence of a mixed mycotic granuloma consisting of a foreign body granuloma modified by the presence of lymphocytes, giant cells, and polymorphonuclear neutrophils.[22,23] Almost all cases show neutrophils, lymphocytes and macrophages organized concentrically around the sclerotic bodies.

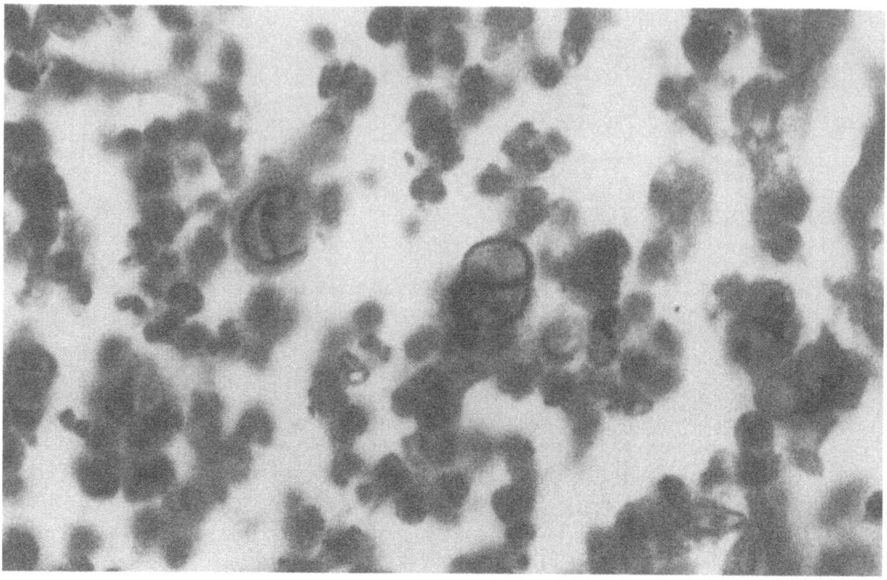

FIGURE 2. Group of sclerotic bodies showing the muriform septations (×175). (Figure courtesy of ASCP Press.)

Uribe *et al.*[24] studied the biopsies obtained from 26 patients. They postulated the following series of events for elimination of the organism after infection. The fungus comes into contact with the pseudoepithelial hyperplasia and moves toward the epidermis in association with other cellular elements. Microabscesses are formed from the influx of polymorphonuclear neutrophils. The sclerotic bodies enter the stratum spinosum where they reside in the microabscesses. The cells and sclerotic bodies then ascend to the horny layer through the normal upward current of cells of the stratified epithelia. Once in the horny layer, the fungus is eliminated. This accounts for the clotted blood and sclerotic bodies that appear as black dots on the verrucous lesions. Healing is by fibrosis and scarring. This series of events is consistent with Batres *et al.*[22] proposal of transepithelial elimination as the mechanism of organism removal from tissue.

3. ETIOLOGY AND LABORATORY DIAGNOSIS

3.1. Etiology

The organisms responsible for causing chromoblastomycosis belong to a group of five closely related species of dematiaceous fungi. In the past, much confusion has surrounded the exact taxonomic position of these fungi. The basis for this confusion stems from the polymorphic nature of conidia formation. These fungi display one or more of three divergent types of sporulation that depend(s) on physiological growth conditions. These three morphologic types are usually referred to as *Rhinocladiella*-like, *Cladosporium*-like, or *Phialophora*-like. *Cladosporium carrionii, Phialophora verrucosa,* and *Rhinocladiella aquasperosa* each typically produce one type of sporulation, however, under appropriate environmental conditions and a careful microscopic search, other types of sporulation can be seen.[15] More problematic is the exact taxonomic position of *Fonsecaea pedrosoi* and *Fonsecaea compacta.* In culture, each of these fungi can display more than one type of sporulation at a time. For that reason, these fungi have been placed at one time or another into the genera *Phialophora, Hormodendrum, Acrotheca, Phialoconidiophora, Rhinocladiella,* or *Trichosporium.* The genus *Fonsecaea,* as used by McGinnis and Schell,[25] will be used here.

3.1.1. P. verrucosa Medlar, 1915

Growth is slow and, at first, olive-gray and dome shaped. With age, the colonies become flattened and covered with short mouse-gray aerial hyphae. The conidiophores are simple or branched and pale brown in color. The phialides are a well-formed flask or vase shape, 4–7 u in length, with a pronounced, flared, cup-shaped collarette at the apex (Fig. 3A). Conidia are borne singly in succession and held together in a ball at the apex of the phialide in a gelatinous material. Spores are single celled, ellipsoidal to oblong, pale brown, smooth, and 1.5×3.4 u in size. A thorough search of this organism growing on cornmeal agar can reveal the *Rhinocladiella* and *Cladosporium* type of sporulation.

3.1.2. C. carrionii Trejos, 1954

Growth is very slow and, at first, gray-green and velvety with a raised center. The edges of the colonies may be surrounded by black submerged hyphae. Conidiophores are erect, branched, and pale brown in color. The conidia are one celled, 2.5×5 u in size, oval, and may appear in branched chains (Fig. 3B). Shield cells and spores with terminal dysjunctors can be seen. Phialides with collarettes may occur with some isolates.

3.1.3. F. pedrosoi (Brumpt) Negroni, 1936

Colonies are slow growing, velvety or wooly, and gray-black to olive-gray in color. Four types of sporulation are seen; however, depending on environmental conditions, one type often predominates. The predominate type of sporulation is usually single-celled conidia produced on swollen denticles on

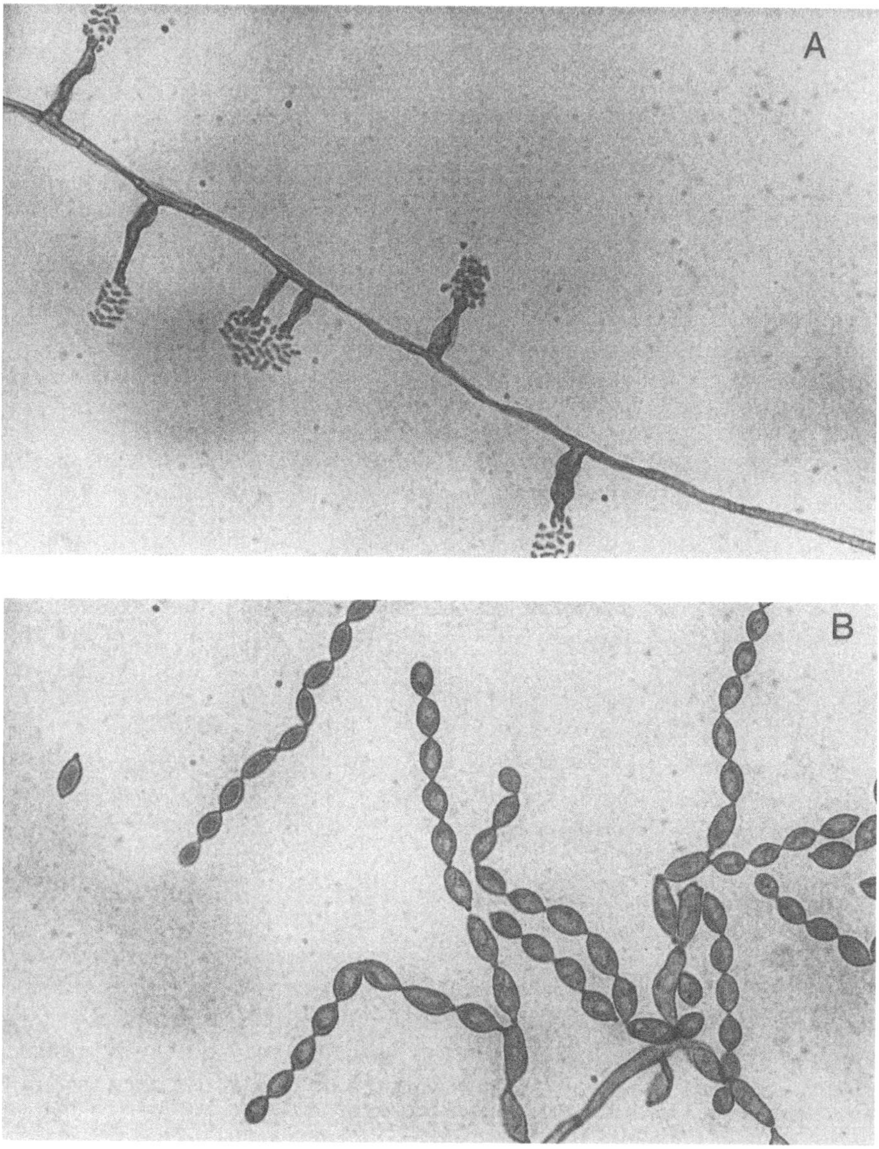

FIGURE 3. A) *P. verrucosa*. Vase shaped phialide with accumulation of conidia around the flared lip of the phialide, B) *C. carrionii*. Long chains of branched, elliptical conidia, C) *F. pedrosoi*. *Rhinocladiella* type of sporulation, D) *F. compacta*, and compact head of conidia with *Cladosporium* type of arrangement. E) *R. aquasperosa*. Typical sympodial arrangement of spores (×700). (E, courtesy of M. Rinaldi.)

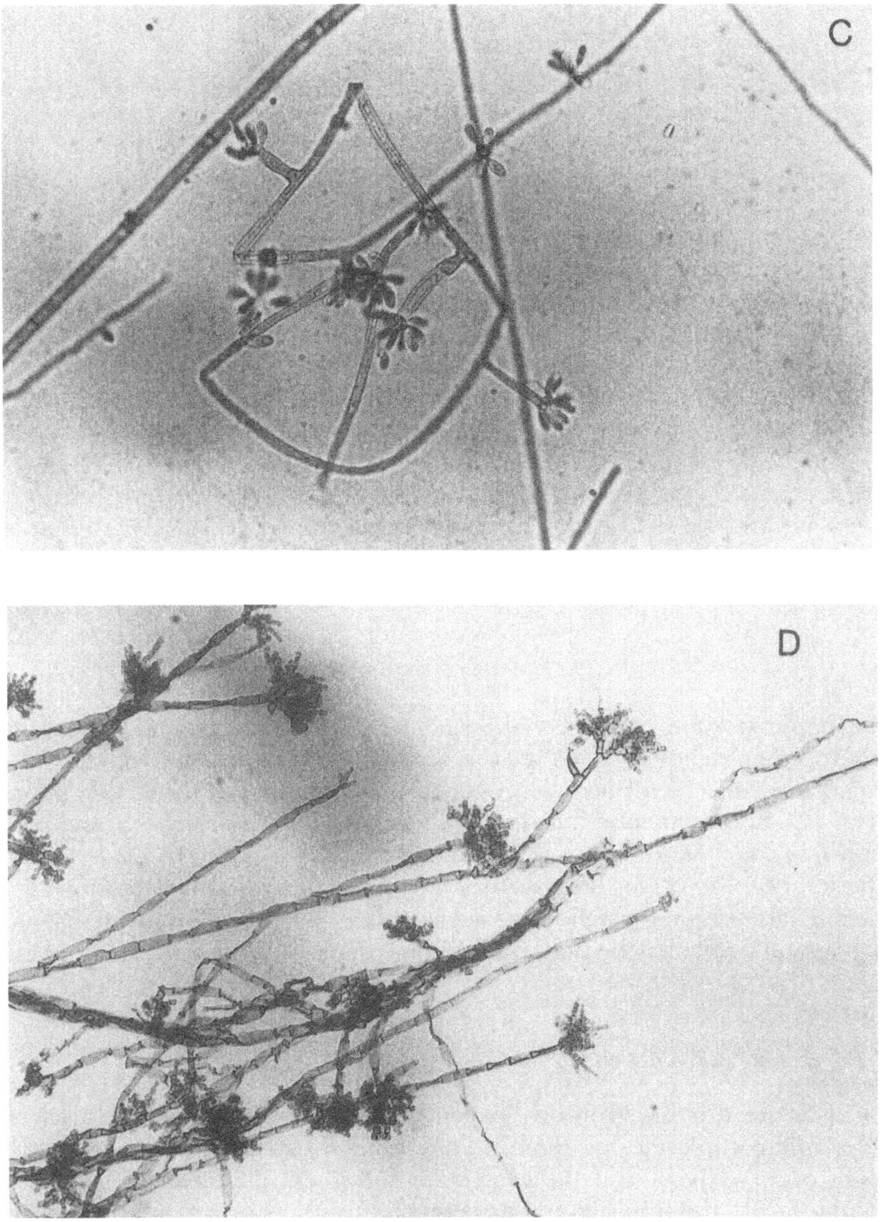

FIGURE 3. *(Cont.)*

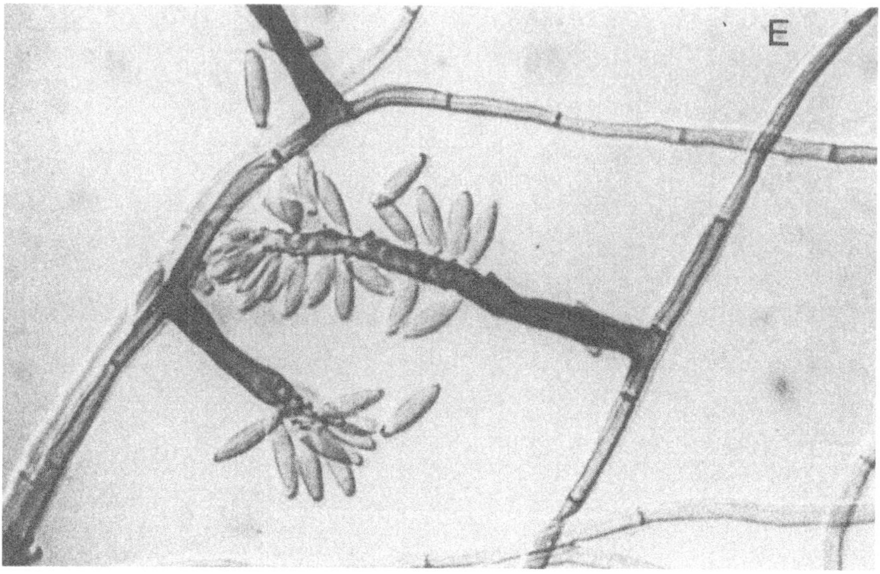

FIGURE 3. (*Cont.*)

the upper portion of the conidiogenous cells. Secondary conidia may be produced on dendicles from the distal portion of the primary conidia. The conidia can also be formed pleurogenously on the sides of the conidiophore (Fig. 3C). Spores are one celled, 4 × 8 u in size, ellipsoid, smooth, and pale brown in color. Short chains of conidia, like those produced by *Cladosporium* species, can also occur. Rare isolates have been described that produce the typical flask-shaped phialide, as seen with *P. verrucosa.*[15] Sporulation on Sabourauds agar may be poor but can be enhanced by growth on cornmeal agar.

3.1.4. F. compacta Carrion, 1940

Colonies are slow growing, heaped, and dark olive to olivaceous-black in color. Microscopically, this organism resembles *F. pedrosoi* except that it has a more compact head and the spores are subspherical instead of ellipsoid. Conidia are 4 × 8 u in size and develop on the apices of irregularly swollen conidiophores (Fig. 3D). The primary conidia may give rise to small subhyaline secondary conidia. Chains of conidia, like those produced by *Cladosporium,* and flask-shaped phialides, like those produced by *Phialophora,* also may be present.

3.1.5. R. aquasperosa (Borelli) Schell, McGinnis and Borelli, 1983

The organism grows initially as a wooly, dark olivaceous-gray colony. Conidiophores are erect and cylindrical with conidia arranged around the apex (Fig. 3E). The conidia are one celled, clavate and 2 × 5 u in size with a conspicuous basal scar. Sporulation similar to that seen with *Exophiala* and *Wangiella* can also be seen.

3.2. Laboratory Diagnosis

Crusts, exudates, or biopsy materials are appropriate specimens for both culture and direct microscopic examination. Zaias and Rebell[26] recommend that the black dots present on the verrucous lesion be selected as a site for culture and direct examination. Direct examination is performed by placing a portion of the specimen onto a microscope slide and adding several drops of 10–20% potassium hydroxide. Gentle heating of the slide over an open flame will help to clarify the tissue and allow clear visualization of the sclerotic bodies. Microscopic examination will reveal the characteristic chestnut-brown muriform sclerotic bodies. McGinnis[15] states that the sclerotic bodies can be better visualized in verrucous lesions than in the flat annular type of lesions.

Specimens should be plated onto Sabourauds agar and Sabourauds agar with antibiotics and cycloheximide. The use of the latter medium will help prevent overgrowth with contaminating bacteria and fungi. Media should be incubated in a moist atmosphere at 25–30°C and held four to six weeks before being discarded as negative. Incubation at 35°C may inhibit some strains of *C. carrionii.*

Identification of isolates is made by morphologic and microscopic observation. The agents causing chromoblastomycosis can be confused microscopically with the agents causing phaeohyphomycosis and, additionally, with some nonpathogenic dematiaceous fungi. Hydrolysis of starch and gelatin has been advocated for separating *C. carrionii* and *X. bantiana* from the nonpathogenic members of *Cladosporium.*[27] Other physiological tests such as urea hydrolysis and decomposition of casein, xanthine, hypoxanthine, and tyrosine are of little value in identifying this group of organisms. Espinel-Ingroff *et al.*[28] have recently published data to suggest that, while the pathogenic *Cladosporium* species fail to hydolyze gelatin, so do many saprophytic *Cladosporium* species. The ability to hydrolyze gelatin appeared to be medium dependent. McGinnis *et al.*[29] have reported that growth at 40°C would clearly differentiate *X. bantiana* from *C. carrionii.* More recently, *X. bantiana* was found to assimilate inositol while *C. carrionii* was negative for this characteristic.[30]

The identification of *Fonsecaea* species and *Phialophora* species is based primarily on morphological characteristics. Espinel-Ingroff[28] showed that, despite previous recommendations, proteolytic activity cannot reliably separate pathogenic from nonpathogenic species of dematiaceous fungi. Additionally, because of the variable results obtained, even known pathogenic dematiaceous fungi could not be separated from each other. On the other hand, Espinel-Ingroff *et al.*[30] were able to use carbohydrate assimilation tests from the API 20C system to identify *Phialophora* species. Using this system, one could easily separate *P. verrucosa* from *P. parasitica*, *P. repens*, and *P. richardsiae*. *F. pedrosoi*, however, could not be differentiated from *F. compacta* by using this system. An exoantigen test has been described that was found to be useful in identifying *F. pedrosoi* and *P. verrucosa* and separating these organisms from similar dematiaceous fungi.[31] The lack of commercial availability precludes the use of this test for diagnostic purposes.

4. EPIDEMIOLOGY

Chromoblastomycosis has a worldwide distribution and is seen primarily in countries that have tropical or subtropical climates. The disease is common in Mexico, Cuba, the Dominican Republic and has been reported from several countries of Central and South America. Al-Doory[5] and Londero and Ramos[32] have summarized several hundred cases that occurred in Brazil. Cases have also been reported from Venezuela but rarely from Peru, Argentina or Uruguay.[33] Scattered reports have appeared from Martinique,[34] Russia,[35] Finland,[36] Czechoslovakia, Rumania, Algeria, Italy,[37] Australia,[38] and South Africa.[39] Three reports have summarized a large number of cases from Madagascar[40] and Japan.[41,42] A comprehensive review of the epidemiology and geographic distribution of the disease can be found in a report published by Brygoo and Destombes.[43]

Most cases have been described in males, although one report found a higher than expected incidence in females[44] (quoted in 15). All ages appear to be susceptible, with most patients falling between 30 and 50 years of age. For some reason, children in areas of high prevalence do not appear to get the infection. Human to human transmission has not been reported.

The most frequently isolated organism from cases of chromoblastomycosis in many countries is *F. pedrosoi* followed by *P. verrucosa* and *C. carrionii*.[5,6,43,44] In certain sections of Venezuela,[44] Australia[38] and South Africa,[39] however, the most commonly isolated agent is *C. carrionii*. Fewer than a dozen cases have been attributed to *F. compacta*,[37] and the rarest agent of chromoblastomycosis is *R. aquasperosa*.

All the agents of chromoblastomycosis are thought to be saprophytic fungi living on wood,[45–47] wood pulp,[48] soil[46] or plant debris.[46] Gezuele *et al.*[45] reported finding *F. pedrosoi* in 34 of 328 samples of wood, soil and plant debris. *P. verrucosa* was isolated from 31 of these samples. Similarly, Iwatsu *et al.*[46] were able to isolate *P. verrucosa* from 16 of 120 wood samples. Unlike

Gezuele *et al.*[45] they were unable to isolate *P. verrucosa* from soil. A total of 177 samples of wood and soil yielded only one isolate of *F. pedrosoi*. Dixon *et al.*[47] were also unable to isolate *F. pedrosoi* or *P. verrucosa* from soil. Five of 39 samples of woody plant material yielded *P. verrucosa*. *C. carrionii* has been isolated from wooden fence posts in Australia.[49]

5. HOST DEFENSES

5.1. Innate Resistance

The intact skin forms an effective mechanical barrier against the agents of chromoblastomycosis. In areas where these fungi are commonly isolated from soil, infection is always secondary to some type of trauma. Even after appropriate exposure, the host probably can ward off the infection. This may be partially due to slow growth of the organisms or the lack of virulence factors produced by this group of fungi. Alternatively, infection may be averted by the activation of the alternative complement pathway with the generation of factors responsible for the influx of polymorphonuclear leukocytes and macrophages. In this regard, Torinuki *et al.*[50] have demonstrated that *F. pedrosoi* can activate the complement system via the alternative pathway with the production of C5a. With the influx of inflammatory cells, the host could remove the fungus by transepithelial elimination, as discussed previously.

The pathology and clinical appearance of the disease may be caused by the host's immune response rather than the destructive effects of the fungus. The chronic and long-term nature of the disease is probably caused by a complex interplay between the inability of the inflammatory cells to effectively kill and digest the sclerotic bodies, and the hosts natural response to remove the foreign material by transepithelial elimination.

5.2. Immune Responses

5.2.1. Humoral

Antibody to *F. pedrosoi* was first reported by Martin *et al.*[51] in patients with chromoblastomycosis. Since then, precipitating antibody has been demonstrated in patients with chromoblastomycosis caused by *F. pedrosoi*[52,53] and *C. carrionii*.[54,55] Iwatsu *et al.*[52] were able to demonstrate low levels of antibody in eight patients. Healthy subjects and patients with sporotrichosis gave negative results in their precipitin assays. Similarly, Buckley and Murray[53] demonstrated antibody in 12 of 13 patients. Furthermore, they found that eight sera reacted with antigens from *F. compacta* and two sera reacted with antigens from *P. verrucosa*. The precipitin bands obtained did not show lines of identity with *F. pedrosoi* antigens. Antigens isolated from *F. pedrosoi*, *F. compacta* and *P. verrucosa* have been shown in other studies[56,57] to cross react. In two separate studies, Villalba[54,55] found antibody in eight of 13 and in 13 of 15 patients with infection caused by *C. carrionii*. All the cited studies found that the antibody levels detected in patients were extremely low. The role, if any, played by antibody in the host response to infection is unclear at this time. Martin *et al.*[51] did suggest that the antibody level falls during successful therapy. The lack of a standardized and commercially available antigen precludes the use of serological testing as an adjunct to diagnosis.

Infection in experimental animals has not shed any further light on the role of the humoral immune response in chromoblastomycosis. Kurita[58] found that antibody was detectable after 16 days in animals infected with *F. pedrosoi*. The titers were extremely low and antibody disappeared as the infection resolved. Ahrens *et al.*[59] also demonstrated low levels of antibody in experimentally infected nude mice. They did not, however, perform any transfer studies to ascertain the role of antibody in their animal model.

5.2.2. Cell-mediated Immunity

Very few studies have attempted to examine the role of cell-mediated immunity in chromoblastomycosis. In this regard, studies in humans are limited to the use of skin tests to elicit a delayed type hypersensitivity reaction. Iwatsu *et al.*[52] were able to obtain positive skin tests in a small number of patients infected with *F. pedrosoi*.

Studies done in animals have shown that positive skin tests,[60] production of macrophage inhibitory factor, and lymphocyte transformation[59] can be demonstrated after infection with *F. pedrosoi*. Additionally, Nishimura and Miyaji[61] were able to transfer cells passively into nude (nu/nu) mice before infection and prevent the fungus from growing freely in tissue. These animal studies do not reflect the infection as seen in humans since the animals were infected intravenously, and the disease either resolved on its own[58] or seeded to other organs and then resolved.[61] These studies proved the presence, but not necessarily a role, for cell-mediated immunity in chromoblastomycosis.

Polak[62] and Ahrens *et al.*[59] have recently described an animal model of chromoblastomycosis that resembles the disease as seen in humans. Only Ahrens *et al.*[59] studied the protective role of cell-mediated immunity. Infecting nu/nu mice subcutaneously led to the development of local lesions two weeks postinjection. These lesions evolved slowly for six months. After four months *F. pedrosoi* also could be found in the lungs, kidneys, and other parts of the skin. Histopathologic sections of lesions showed a mixed granulomatous reaction with many dark brown sclerotic bodies. This type of tissue response is seen in patients with chromoblastomycosis. Infection of nu/+ mice resulted in disease that resolved after four to six weeks. Reconstitution of the nu/nu mice with splenic T cells from nu/+ mice allowed the animals to ward off the infection. In mice defective in NK cell and macrophage function, infection behaved similar to that which occurred in the nu/+ animals. It would appear from these data that lymphocytes play a major protective role against the development of chromoblastomycosis.

6. PROGNOSIS AND THERAPY

6.1. Prognosis

Chromoblastomycosis typically runs a very chronic and benign course. Death is seldom the result of disease unless complications arise because the lesions become secondarily infected with bacteria or, in those rare instances, because dissemination to the brain occurs. Since most cases occur in rural, poor areas medical attention is not sought until bacterial infection occurs or the lesions become cosmetically displeasing. Therefore, it is not uncommon for patients to wait one to 20 years before medical attention is sought. This makes the treatment of chromoblastomycosis extremely difficult.

6.2. Therapy

In general, small localized lesions are more easily treated than are the generalized verrucous lesions. The use of iodides, iontophoresis of copper sulfate, vitamin D injections, and Amphotericin B injected intralesionally, topically, or intravenously has been used with limited success.[37] The treatment of choice is wide and deep excision followed by grafting. Success is contingent upon treating the disease in the early stages when the lesions are small. Unfortunately, this is not when most individuals seek medical attention. Curettage and cautery should not be performed since this could result in dissemination of the fungus through the lymphatics.[15]

Table I summarizes the surgical and nonsurgical methods used to treat the disease. Many of these studies are based on relatively few cases and a short follow-up period; in all instances, the studies are uncontrolled. Therefore, caution should be exercised when interpreting the results. Local heat treatment with a pocket warmer has been advocated for many years.[63–66] The duration of therapy has been reported to vary from one to 12 months before a cure can be achieved. Except in one instance, all the lesions treated were small and localized. Tagami *et al.*[64] were able to treat an extensive lesion on the buttock with an electric bed warmer. The patient had to spend 19 hours per day on the warmer. After three months of therapy the lesions had healed and cultures were negative. It has been speculated that the reason heat treatment is successful is that the fungi are killed at the temperatures achieved by the bed warmers (42–46°C).[65,66] This claim, however, could not be substantiated by Tagami *et al.*[66]

TABLE I
Summary of Treatment Methods for Chromoblastomycosis

Treatment	Duration	Total no. patients	No. cured (improved)	Organism	Reference
Local heat	1–12 months	7	7	*F. pedrosoi*	62, 64, 65, 66
Excision		2	2	*F. pedrosoi* *P. verrucosa*	67, 68
+ketoconazole	4 months	2	1 (1)	*C. carrionii*	67, 68, 69
+thiobendazole	6–22 months	3	1 (2)	*F. pedrosoi* (2)	70
Cryosurgery	30 sec–4 min 1–40 cycles	11	8 (3)	*F. pedrosoi*	71, 72
+5-fluorocytosine		1	1	*F. pedrosoi*	72
Mohs surgery		1	1		73
CO_2 laser		1	1		74
Thiabendazole	6–22 months	11	5 (6)	*F. pedrosoi* (4) *C. carrionii* (1)	70
5-fluorocytosine	2–12 months	23	16 (7)	*F. pedrosoi* (13) *Cladosporium* sp. (2) *R. aquasperosa* (1)	75, 76, 77
Ketoconazole	1–4 months	13	2 (8)	*F. pedrosoi* (1)	78, 79
Itraconazole	1–6 months	28	13 (14)	*F. pedrosoi* (13) *C. carrionii* (11)	80, 81, 82, 83
5-fluorocytosine					
+Amphotericin B	not given	5	4 (1)	not given	77
+ketoconazole	12 months	1	[1]	*F. pedrosoi*	84

Excision, coupled with treatment with ketoconazole[67–69] or thiabendazole,[70] has met with limited success. Half the patients treated with ketoconazole and two thirds of the patients treated with thiabendazole either were not cured or only improved on therapy.

Surgical techniques used for removal of the lesions have included cryosurgery,[71,72] Mohs micrographic surgery,[73] and removal of lesions with a carbon dioxide laser.[74] Pimentel *et al.*[71] treated 11 patients using cryosurgery with liquid nitrogen. All five patients with localized lesions were cured with no relapses for the follow-up period of 53 months. In six patients with generalized lesions, half attained a clinical and mycological remission, while the remaining patients had significant improvement without cure. Cure was achieved using freezing times of 30 seconds to 4 minutes with the number of treatment cycles varying from one to more than 40. Lesions healed forming flat, smooth, and shiny cicatrical tissue three to six weeks posttreatment.

Antifungal therapy has included thiabendazole, 5-fluorocytosine, the imidazoles, and Amphotericin B. Bayles[70] treated 11 patients with thiabendazole for periods ranging from six to 22 months. She was able to cure five patients, while six patients showed significant improvements of the lesions. Lopes *et al.*[75–77] have reported on the successful use of 5-fluorocytosine in the treatment of both localized and generalized lesions. Cure was achieved in 16 of 23 patients using a dose of 200 mg/kg/day every 6 hours for periods of up to 67 months. In seven patients, improvement was seen; however, organisms isolated from the lesions were resistant to 5-fluorocytosine.

Ketoconazole alone has not been found to be useful for the treatment of either localized or generalized lesions.[78,79] In a total of 10 treated patients, only two achieved a cure, while improvement of lesions was seen in the remaining eight patients.

Four studies have reported on the efficacy of using itraconazole in doses of 100–200 mg/day for up to two years.[80–83] Cure was achieved in 13 of 28 patients, while 14 patients showed significant improvement. Interestingly, one study showed that patients with infection caused by *C. carrionii* may be more amenable to treatment with itraconazole. Borelli[83] found that eight of nine patients with chromoblastomycosis caused by *C. carrionii* responded to therapy, whereas only two of five patients with disease caused by *F. pedrosoi* were cured.

Combination therapy using 5-fluorocytosine plus ketoconazole[84] or Amphotericin B[77] has been reported. Although the number of patients treated was small, the combination of 25 mg of Amphotericin B every other day plus 75 mg/kg/day of 5-fluorocytosine cured most patients. Follow up of four patients for two to three years showed no sign of recurrence.

To date, it would appear that the treatment of choice for small lesions would be surgical excision or cryosurgical removal. Antifungal therapy of more generalized lesions caused by *F. pedrosoi* would be 5-fluorocytosine or 5-fluorocytosine plus Amphotericin B to prevent the emergence of resistance. Disease caused by *C. carrionii* should be treated with itraconazole.

REFERENCES

1. Rudolph, M., 1914, Uber die brasilianische "Figueira," *Arch. Schiffs Tropen-Hyg.* **18:**498.
2. Lane, C. G., 1915, A cutaneous lesion caused by a new fungus (*Phialophora verrucosa*), *J. Cutan. Dis.* **33:**840–846.
3. Medlar, E. M., 1915, A cutaneous lesion caused by a new fungus, *Phialophora verrucosa*, with a study of the fungus, *J. Med. Res.* **32:**507–521.
4. Pedroso, A., and Gomes, J. M., 1920, Sobre quatro casos de dermatite verrucosa produzida pela *Phialophora verrucosa*, *Ann. Paulist. Med. Cir.* **11:**53–61.
5. Al-Doory, Y., 1972, *Chromomycosis*, Mountain Press Publishing Company, Missoula, Montana.
6. Carrion, A. L., 1950, Chromoblastomycosis, *Ann. N.Y. Acad. Sci.* **50:**1255–1282.
7. Milam, C. P., and Fenske, N. A., 1989, Chromoblastomycosis, *Dermatologic Clinics* **7:**219–225.
8. Fader, R. C., and McGinnis, M. R., 1988, Infections caused by dematiaceous fungi: Chromoblastomycosis and phaeohyphomycosis, *Infect. Dis. Clinics of North America* **2:**925–938.
9. Terra, F., Torres, M., de Fonseca, O., and Area Leao, A. E., 1922, Novo typo de dermatite verrucosa mycose por *Acrotheca* com associacao de leishmaniosa, *Brasil-Med.* **2:**363–368.
10. Moore, M., and Almeida, F., 1935, Etiologic agents of chromomycosis (chromoblastomycosis of Terra, Torres, Fonseca and Leao, 1922) of North and South America, *Rev. Biol. Hyg.* **6:** 94–97.
11. Zaias, N., 1978, Chromomycosis, *J. Cutan. Pathol.* **5:**155–164.
12. Weedon, D., and Ritchie, G., Cystic chromomycosis of the skin, *Pathol.* **11:**389–392.
13. McGinnis, M. R., Ajello, L., and Schell, W. A., 1985, Mycotic diseases. A proposed nomenclature, *Inter. J. Derm.* **24:**9–15.
14. Ajello, L., Georg, L. K., Steigbeigel, R. T., and Wang, C. J. K., 1974, A case of phaeohyphomycosis caused by a new species of *Phialophora*, *Mycologia* **66:**490–498.
15. McGinnis, M. R., 1983, Chromoblastomycosis and phaeohyphomycosis: New concepts, diagnosis, and mycology, *J. Amer. Acad. Dermatol.* **8:**1–16.
16. McGinnis, M. R., 1987, Infections caused by black fungi, *Arch. Dermatol.* **123:**1300–1302.
17. Carrion, A. L., 1975, Chromoblastomycosis and related infections. New concepts, differential diagnosis, and nomenclatural implications, *Inter. J. Dermatol.* **14:**27–32.
18. Vollum, D., 1977, Chromomycosis: a review, *Br. J. Dermatol.* **96:**454–458.
19. Szaniszlo, P. J., Hsieh, P. H., and Marlowe, J. D., 1976, Induction and ultrastructure of the multicellular (sclerotic) morphology in *Phialophora dermatitidis*, *Mycologia* **68:**117–130.
20. Rosen T., Gyorkey, F., Joseph, L. M., and Batres, E., 1980, Ultrastructural features of chromoblastomycosis, *Inter. J. Dermatol.* **19:**461–468.
21. Harada, S., and Kusunoki, T., 1983, Scanning electron microscopic observation of the parasitic forms of *Fonsecaea pedrosoi* in human skin lesion, *Mycopathologia* **82:**33–37.
22. Batres, E., Wolf, J. E., Jr., Rudolph, A. H., and Knox, J. M., 1978, Transepithelial elimination of cutaneous chromomycosis, *Arch. Dermatol.* **114:**1231–1232.
23. Walter, P., Garin, Y., and Richard-Lenoble, D., 1982, Chromoblastomycosis. A morphological investigation of the host–parasite interaction, *Virchows Arch. (Pathol. Anat.)* **397:**203–214.
24. Uribe, J. F., Zuluaga, A. I., Leon, W., and Restrepo, A., 1989, Histopathology of chromoblastomycosis, *Mycopathologia* **105:**1–6.
25. McGinnis, M. R., and Schell, W. A., 1981, The genus Fonsecaea and its relationship to the genera Cladosporium, Phialophora, Rhamichloridium, and Rhinocladiella, Pan Am. Health Organ Publ. No. 396, Washington, D.C., pp. 215–234.
26. Zaias, N., and Rebell, G., 1973, A simple and accurate diagnostic method in chromoblastomycosis, *Arch. Dermatol.* **108:**545–546.
27. Fuentes, C. A., and Bosch, Z. E., 1960, Biochemical differentiation of the etiological agents of chromoblastomycosis from non-pathogenic *Cladosporium* species, *J. Invest. Dermatol.* **34:** 419–421.
28. Espinel-Ingroff, A., Goldson, P. R., McGinnis, M. R., and Kerkering, T. M., 1988, Evaluation

of proteolytic activity to differentiate some dematiaceous fungi, *J. Clin. Microbiol.* **26**:301–307.
29. McGinnis, M. R., Borelli, D., Padhye, A. A., and Ajello, L., 1986, Reclassification of *Cladosporium bantianum* in the genus *Xylohypha*, *J. Clin. Microbiol.* **23**:1148–1151.
30. Espinel-Ingroff, A., McGinnis, M. R., Pincus, D. H., Goldson, P. R., and Kerkering, T. M., 1989, Evaluation of the API 20C yeast identification system for the differentiation of some dematiaceous fungi, *J. Clin. Microbiol.* **27**:2565–2569.
31. Espinel-Ingroff, A., Shadomy, S., Dixon, D., and Goldson, P., 1986, Exoantigen test for *Cladosporium bantianum*, *Fonsecaea pedrosoi*, and *Phialophora verrucosa*, *J. Clin. Microbiol.* **23**: 305–310.
32. Londero, A. T., and Ramos, C. D., 1976, Chromomycosis: a clinical and mycological study of thirty-five cases observed in the hinterland of Rio Grande du Sul, Brazil, *Am. J. Trop. Med. Hyg.* **25**:132–135.
33. Mackinnon, J. E., 1972, Regional peculiarities of some deep mycoses, *Mycopath. Mycol. Appl.* **46**:249–265.
34. Cales, D., and Helenon, R., 1976, La chromomycose a la Martinque, *Afr. Med.* **15**:457–466.
35. Malkina, A. Y., and N. N. Darchenkova, 1977, Distribution of chromomycosis in the world, *Vestu. Dermatol. Veneral.* **1**:41–45.
36. Sonck, C. E., 1975, Chromomycosis in Finland, *Dermatologia* **19**:189–193.
37. Rippon, J. W., 1988, *Medical Mycology. The Pathogenic Fungi and the Pathogenic Actinomycetes*, 3rd ed., W. B. Saunders Company, Philadelphia.
38. Leslie, D. F., and Beardmore, G. L., 1979, Chromoblastomycosis in Queensland: a retrospective study of 13 cases at the Royal Brisbane Hospital, *Australian J. Dermatol.* **20**:23–30.
39. Simson, F. W., 1946, Chromoblastomycosis. Some observations on the type of the disease in South Africa, *Mycologia* **38**:432–449.
40. Coulanges, P., and Locheron, P., 1981, La chromomycose a Madagascar, *Arch. Inst. Pasteur Madagascar* **48**:69–95.
41. Fukushiro, R., 1977, Some considerations on infections by dematiaceous fungi; with special regard to chromomycosis. *Jpn. J. Med. Mycol.* **18**:398–421.
42. Fukushiro, R., 1983, Chromomycosis in Japan, *Int. J. Dermatol.* **22**:221–229.
43. Brygoo, E. R., and Destombes, P., 1975, Geographical distribution of chromoblastomycosis and its various pathogenic agents, *Bull. Soc. Fr. Mycol. Med.* **4**:181–183.
44. Campins, H., and Scharyj, M., 1953, Chromoblastomicosis comentarios sobre 24 casos, con estudio clinico, histologico y micologico, *Gac. Med. Caracas* **61**:127–151.
45. Gezuele, E., Mackinnon, J. E., and Conti-Diaz, I. A., 1972, The frequent isolation of *Phialophora verrucosa* and *Phialophora pedrosoi* from natural sources, *Sabouraudia* **10**:266–273.
46. Iwatsu, T., Miyaji, M., and Okamoto, S., 1981, Isolation of *Phialophora verrucosa* and *Fonsecaea pedrosoi* from nature in Japan, *Mycopath.* **75**:149–158.
47. Dixon, D. M., Shadomy, H. J., and Shadomy, S., 1980, Dematiaceous fungal pathogens isolated from nature, *Mycopath.* **70**:153–161.
48. Wang, C. J. K., 1965, Fungi of pulp and paper of New York, State Univer. Coll. Forrestry Tech. Pub. 87, p. 76.
49. Ridley, M. F., 1957, The natural habitat of *Cladosporium carrionii*, a cause of chromoblastomycosis in man, *Australian J. Dermatol.* **4**:23–27.
50. Torinuki, W., Okohchi, K., Takematsu, H., and Tagami, H., 1984, Activation of the alternative complement pathway by *Fonsecaea Pedrosoi*, *J. Invest. Dermatol.* **83**:308–310.
51. Martin, D. S., Baker, R. D., and Conant, N. F., 1936, A case of verrucous-dermatitis caused by *Hormodendrum pedrosoi* (chromoblastomycosis) in North Carolina, *Amer. J. Trop. Med.* **16**: 593–609.
52. Iwatsu, T., Miyaji, M., Taguchi, H., Okamoto, S., and Kurita, N., 1979, Skin test-active substance prepared from culture filtrate of *Fonsecaea pedrosoi*, *Mycopath.* **67**:101–105.
53. Buckley, H. R., and Murray, I. G., 1966, Precipitating antibodies in chromomycosis, *Sabouraudia* **5**:78–80.

54. Villalba, E., 1988, Detection of antibodies in the sera of patients with chromoblastomycosis by counter immunoelectrophoresis I. Preliminary results, *J. Med. Vet. Mycol.* **26:**73–74.
55. Villalba, E., and Yegres, J. F., 1988, Detection of circulating antibodies in patients affected by chromoblastomycosis by *Cladosporium carrionii* using double immunodiffusion, *Mycopath.* **102:**17–19.
56. Cooper, B. H., and Schneidau, J. D., 1970, A serological comparison of *Phialophora verrucosa, Hormodendrum pedrosoi* and *Cladosporium carrionii* using immunodiffusion and immunoelectrophoresis, *Sabouraudia* **8:**217–226.
57. Gordon, M. A., and Al-Doory, Y., 1965, Application of fluorescent antibody procedures to the study of pathogenic dematiaceous fungi, *J. Bacteriol.* **89:**551–556.
58. Kurita, N., 1979, Cell-mediated immune responses in mice infected with *Fonsecaea pedrosoi, Mycopath.* **68:**9–15.
59. Ahrens, J., Graybill, J. R., Abishawl, A., Tio, F. O., and Rinaldi, M. G., 1989, Experimental murine chromomycosis mimicking chronic progressive human disease, *Am. J. Trop. Med. Hyg.* **40:**651–658.
60. Iwatsu, T., Miyaji, M., Taguchi, H., and Okamoto, S., 1982, Evaluation of skin test for chromoblastomycosis using antigens prepared from culture filtrates of *Fonsecaea pedrosoi, Phialophora verrucosa, Wangella dermatidis,* and *Exophiala ieanselmei, Mycopath.* **77:**59–64.
61. Nishimura, K., and Mayaji, M., 1981, Defense mechanisms of mice against *Fonsecaea pedrosoi* infection, *Mycopath.* **76:**155–166.
62. Polak, A., 1984, Experimental infection of mice by *Fonsecaea pedrosoi* and *Wangiella dermatitidis, J. Med. Vet. Mycol.* **22:**167–169.
63. Yanase, K., and Yamada, M., 1978, 'Pocket-warmer' therapy of chromomycosis, *Arch. Dermatol.* **114:**1095.
64. Tagami, H., Ohi, M., Aoshima, T., Moriguchi, M., Suzuki, N., and Yamada, M., 1979, Topical heat therapy for cutaneous chromomycosis, *Arch. Dermatol.* **115:**740–741.
65. Kinbara, T., Fukushiro, R., and Eryu, Y., 1982, Chromomycosis-Report of two cases successfully treated with local heat therapy, *Mykosen* **25:**689–694.
66. Tagami, H., Ginoza, M., Imaizumi, S., and Urano-Suchisa, S., 1984, Successful treatment of chromomycosis with topical heat therapy, *J. Amer. Acad. Dermatol.* **10:**615–619.
67. Wackym, P. A., Gray, G. F., Jr., Richie, R. E., and Gregg, C. R., 1985, Cutaneous chromomycosis in renal transplant recipients. Successful management in two cases, *Arch. Intern. Med.* **145:**1036–1037.
68. Gruber, B., Rippon, J. W., and Dayal, V. S., 1988, Phaeomycotic cyst (chromoblastomycosis) of the neck, *Arch. Otolaryngol. Head Neck Surg.* **114:**1031–1032.
69. Greene, J. N., Foulis, P. R., and Yangro, B. G., 1990, Case report: Chromomycosis in a steroid-dependent patient with chronic obstructive pulmonary disease, *Amer. J. Med. Sci.* **299:**54–57.
70. Bayles, M. A. H., 1971, Chromomycosis. Treatment with thiabendazole, *Arch. Dermatol.* **104:** 476–485.
71. Pimentel, E. R., Castro, L. G. M., Cuce, L. C., and Sampaoi, S. A., 1989, Treatment of chromomycosis by cryosurgery with liquid nitrogen: A report on eleven cases, *J. Dermatol. Surg. Oncol.* **15:**72–77.
72. Nobre, G., Oliveira, A. D. S., Verde, S. F., Martins, O., and Picoto, A. D. S., 1980, Chromomycosis: Report of a case and management by cryosurgery, topical chemotherapy, and conventional surgery, *J. Dermatol. Surg. Oncol.* **6:**576–578.
73. Pavlidakey, G. P., Snow, S. N., and Mohs, F. F., 1986, Chromoblastomycosis treated by Mohs micrographic surgery, *J. Dermatol. Surg. Oncol.* **12:**.1073–1075.
74. Kuttner, B. J., and Siegle, R. J., 1986, Treatment of chromomycosis with a CO2 laser, *J. Dermatol. Surg. Oncol.* **12:**965–968.
75. Lopes, C. F., Cisalpino, E. O., Alvarenga, R. J., Armond, S., Porto, R. V., Maia, F. A., and

Peixoto, Y., 1971, Treatment of chromomycosis with 5-fluorocytosine, *Inter. J. Dermatol.* **10:** 182–191.
76. Lopes, C. F., Alvarenga, R. J., Cisalpino, E. O., Resende, M. A., and Oliveira, L. G., 1978, Six years' experience in treatment of chromomycosis with 5-fluorocytosine, *Inter. J. Dermatol.* **17:** 414–418.
77. Lopes, C. F., 1981, Recent developments in the therapy of chromoblastomycosis, *Bull. Pan Am. Health Organ.* **15:**58–64.
78. McBurney, E. I., 1982, Chromoblastomycosis treatment with ketoconazole, *Cutis* **30:** 746–748.
79. Symoens, J., Moens, M., Dom, J., Scheijgrond, H., Dony, J., Schuermans, V., Legendre, R., and Finestine, N., 1980, An evaluation of two years of clinical experience with ketoconazole, *Rev. Infect. Diseases* **2:**674–687.
80. Heyl, T., 1985, Treatment of chromomycosis with itraconazole, *Br. J. Dermatol.* **112:**728–729.
81. Restrepo, A., Gonzalez, A., Gomez, I., Arango, M., and Bedout, C. D., 1988, Treatment of chromoblastomycosis with itraconazole, *Ann. N.Y. Acad. Sci.* **544:**504–516.
82. Lavalle, P., Suchil, P., De Ovando, F., Reynoso, S., 1987, Itraconazole for deep mycoses: Preliminary experience in Mexico, *Rev. Infect. Diseases* **9**(S)**:**564–570.
83. Borelli, D., 1987, A clinical trial of itraconazole in the treatment of deep mycoses and leishmaniasis, *Rev. Infect. Diseases* **9**(S)**:**557–563.
84. Silber, J. G., Gambert, M. E., Green, K. M., and Shalata, A. R., 1983, Treatment of chromomycosis with ketoconazole and 5-fluorocytosine, *J. Am. Acad. Dermatol.* **8:**236–238.

13

Mycotic Mycetoma

MOHAMAD M. EL-ZAATARI
and MICHAEL R. McGINNIS

1. INTRODUCTION

The disease mycetoma has been known since ancient times in India, where it was described in the religious book *Atharva Veda* as "padaval mika," which means foot anthill. Gill in 1842 and Colebrook in 1848, as part of their *Madura Dispensary Reports* were the first investigators to clearly describe the disease.[1] Additionally, Colebrook was the first to publish the term "madura foot", the common name used in the Madura District, to describe this clinical entity. He believed that the infection developed following a puncture wound by a thorn but did not realize that the etiologic agents of the disease were microorganisms. The term mycetoma (pl. mycetomata), which means fungus tumor, was first introduced in 1861 by Carter.[2] He also classified mycetomata into two varieties based on the color of the granules; the black variety he described was probably caused by *Madurella mycetomatis*, whereas the light-colored variety was caused by an actinomycete.

Historically, all of the actinomycetes known to cause mycetoma were classified in the single genus *Actinomyces*. Because they were classified as species of *Actinomyces*, Pinoy[3] proposed the name actinomycoses for these bacterial infections. He believed that the term mycetoma should be restricted to infections that were caused by true fungi. Chalmers and Archibald[4,5] attempted to further clarify mycetomata by proposing that the term maduromycoses should be used for "true mycetoma" caused by fungi and actinomycoses for diseases caused by actinomycetes. The granules formed in

MOHAMAD M. EL-ZAATARI and MICHAEL R. McGINNIS • Department of Pathology, University of Texas Medical Branch, Galveston, Texas 77555.

Fungal Infections and Immune Responses, edited by Juneann W. Murphy *et al.* Plenum Press, New York, 1993.

cases described as actinomycosis were always composed of fine, nonsegmented filaments in contrast to true hyphae and vesicles composing granules of fungal origin. They further divided maduromycoses into three types based on the color of the granules that were formed in tissue. In contemporary mycology, mycetoma is divided into two principle types depending on the classification of the etiologic agents. Actinomycotic (also termed actinomycetoma) refers to those infections caused by an aerobic actinomycete (see Chapter 24), whereas eumycotic is used to designate mycetoma caused by a fungus. This basic differentiation is extremely important for proper management of the disease. In this chapter, only mycetomata caused by fungi will be considered.

2. DESCRIPTION OF THE DISEASE

2.1. Portal of Entry

Mycetoma refers to a localized, noncontagious, slowly progressing, chronic infection that involves the cutaneous and subcutaneous tissues, fascia, and bone. Following the traumatic implantation of an etiologic agent that is associated with soil or plant debris into the subcutaneous tissue of a healthy person, the fungus begins to grow in the damaged tissue. It may develop at any body site,[6] but approximately 70 percent of cases involve the foot, 12 percent the hands, and 18 percent other sites such as legs, knees, thighs, arms, head and neck, chest wall, and the back. The lesion usually starts as a small papule or nodule that continues to increase in size. As the infection progresses, it becomes characterized by tumefaction, draining sinuses, and hyphal aggregates of the fungus known as granules (syn. grains).

2.2. Clinical Manifestations

Regardless of the etiologic agent, the clinical picture of the disease is essentially the same. The patient often recalls a history of trauma, such as a thorn prick or splinters. The male to female ratio is 5:1 and the age of most patients is between 20 and 40 years. The exact incubation period is unknown, but it is believed to vary between several weeks or months to several years depending on the patient and the etiologic agent. Infections caused by fungi progress at a much slower rate than those caused by aerobic actinomycetes. Patients often do not seek medical help early in the disease process because of the painless nature of mycetomata.

The earliest manifestation of the disease is a small, firm, painless subcutaneous papule or nodule that is attached to the skin. As it gradually increases in size, sinuses form that communicate to the surface of the skin through which pus and granules are discharged to the surface. The granules are hyphal aggregates of the fungus that are organized into a distinctive structure that varies with the specific etiologic agents. As the infection spreads directly along the fascial planes to adjacent tissue, new sinuses form as the old ones heal. Following healing of the involved tissue, massive fibrosis occurs, which has a tumorlike appearance (Fig. 1). Hematogenous spread of the etiologic agent is unknown. In 1 percent of cases,[6] regional lymph nodes are involved. When the lesions extend to bone, which often takes months or years, bone destruction occurs with small cavities filling with granules. The granules provide support for the bone,[6] which prevents spontaneous fractures. Complete remodeling often occurs. Osteoporosis and subcutaneous granulomata that are sometimes calcified are seen during radiographic

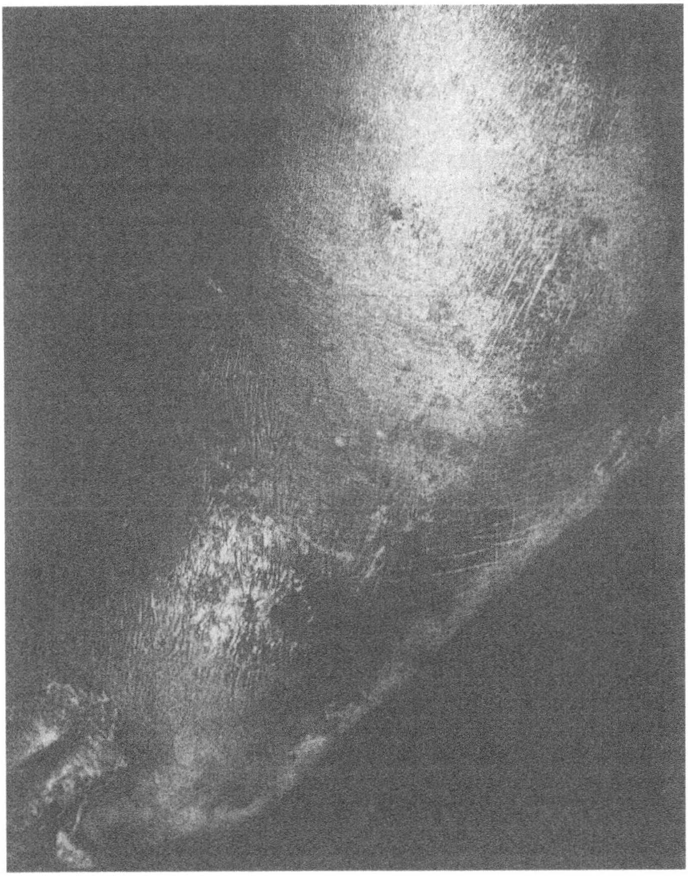

FIGURE 1. Mycetoma of foot illustrating tumefaction and sinuses caused by *Madurella mycetomatis*.

examination of the infected areas. Early osteolytic damage includes loss of cortical margin and external erosion of the bone. As the disease progresses, local damage of the blood and lymphatic vessels and nerves occurs. Frequently, secondary bacterial infections and osteomyelitis producing total bone destruction occurs. Pain is associated with the disease when multiple fistula are present, secondary infections occur, and osteitis and arthritis have developed.[7]

3. ETIOLOGY

The fungi listed in Table I are the currently known etiologic agents of mycotic mycetoma. Even though dermatophytes have been reported as agents of mycetoma, they do not cause this disease.[8]

4. EPIDEMIOLOGY

The disease occurs worldwide, with most patients living in tropical and temperate zones located between the Tropics of Cancer and Capricorn. The disease is frequently reported in countries such as India, Mexico, Senegal, and Sudan where it has an extremely high endemicity. *Madurella mycetomatis* occurs all over the world (Table II),[8] where it is the most commonly reported fungal etiology of mycetoma in the world. In the United States, *Pseudallescheria boydii* is the most commonly reported agent of mycotic mycetoma.[9]

5. HOST DEFENSES

The host response in mycetomata is granulomatous, as seen in tissue sections. The dermis and subcutaneous tissues have localized abscesses con-

TABLE I
Etiologic Agents of Mycotic Mycetoma

Acremonium falciforme	*Exophiala jeanselmei*	*Neotestudina rosatii*
A. kiliense	*Fusarium moniliforme*	*Phialophora cyanescens*
A. recifei	*F. solani var. coeruleum*	*Plenodomus avramii*
Aspergillus flavus	*F. solani var. minus*	*Pseudallescheria boydii*
A. nidulans	*Leptosphaeria senegalensis*	*Pseudochaetosphaeronema larense*
Corynespora cassiicola	*L. tompkinsii*	*Pyrenochaeta mackinnonii*
Curvularia geniculata	*Madurella grisea*	*P. romeroi*
C. lunata	*M. mycetomatis*	

TABLE II
Frequency of Major Fungi Causing Mycetoma

Principal species	Percentage of 943 reported cases tabulated from the literature
Madurella mycetomatis	69.2
Pseudallescheria boydii	10.6
Leptosphaeria senegalensis	10.5
M. grisea	6.0
Acremonium spp.[a]	2.7
Pyrenochaeta romeroi	1.0

[a]Includes *Acremonium* isolates not identified to species, as well as *A. falciforme*, *A. kiliense*, and *A. recifei*.

taining solitary granules, or clusters of granules surrounded by neutrophils. As the disease becomes more chronic, there is a chronic inflammatory reaction around the abscesses with highly vascular granulation tissue infiltrated by epithelioid cells, lymphocytes, plasma cells, and multinucleated giant cells. An area of extensive fibrosis surrounds the granules, which results in tumefaction and deformity.[10]

The granules are approximately 0.5–2 mm or larger in diameter, varying in shape, color, and texture. These characteristics, as well as their internal structure (Fig. 2 and Fig. 3), can be used to identify the major agents involved. Demonstration of granules is essential to make a histopathologic diagnosis of mycetoma. For example, *M. mycetomatis* (Fig. 4) and several other fungi form black granules. Melanin, host protein, and tissue debris cause the dark color associated with these granules.[11] In a study using electron microscopy, the granules produced by *Acremonium kiliense*, *P. boydii*, and *M. mycetomatis* were found to be essentially the same except that *M. mycetomatis* has a dark homogeneous coating of a cementlike substance associated with the fungal cells composing the granules. The cell walls of the hyphal cells composing the granules were greatly thickened, some of which appeared to have fused with each other, and many cells appeared dead.[12] The granules of *M. mycetomatis*[13] have numerous polysaccharide microfibrillar structures that are externally contiguous to the fungus and extend into the cementlike material (granule matrix). There was some host material seen, especially at the periphery of the granules, which could provide a protective barrier for the fungus against antibodies and antifungal agents. In contrast, this material might be the remnants of decomposed fungal cells. The granules present an important antigenic challenge to the host immune system.

Antibodies formed against fungi causing mycetoma have been demonstrated in patient sera by immunodiffusion (ID) and counter-immunoelectrophoresis (CIE). Using an enzyme-linked immunosorbent assay (ELISA),[14]

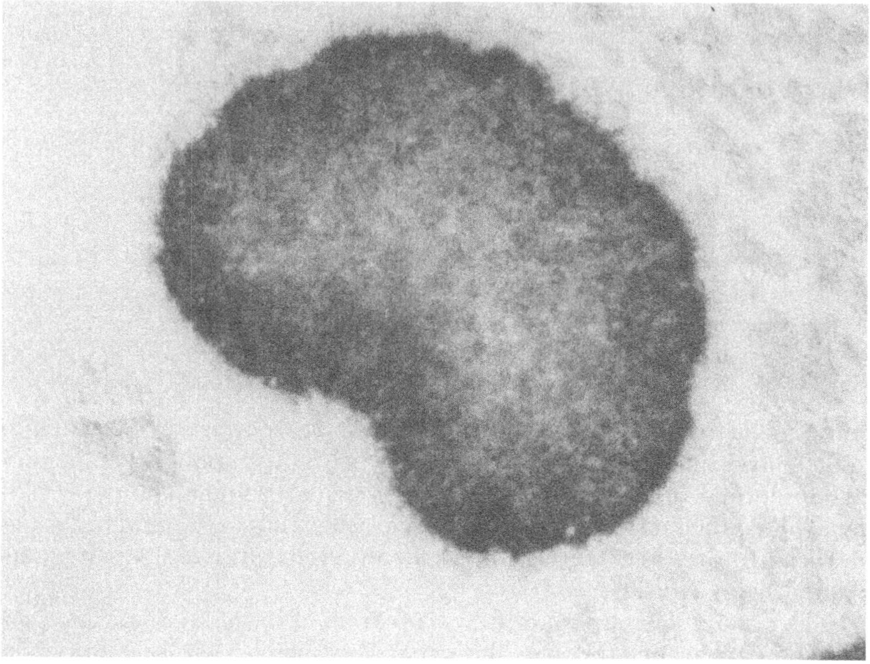

FIGURE 2. Granule of *Pseudallescheria boydii* in tissue. (160×, Gridley fungal stain.)

antibodies to the etiologic agents have been found in both uninfected individuals and patients living in the same geographic area. In a recent study, a specific antibody response was clearly demonstrated in patient sera obtained from patients having well-established disease caused by *M. mycetomatis*.[15] The dominant immunoglobulin class was IgM, which was present in 12 of 13 patients. It has been suggested that the presence of polysaccharide antigens, which are common components of the fungal wall, may be largely responsible for the stimulation of IgM class antibody response. Only 1 of 13 patients had a barely detectable IgM level. IgG has been detected in low levels in some patients. Homologous and heterologous antigens were detected by immunoblotting in patient sera in the same study.

Some individuals are more susceptible to developing mycetoma, whereas others are not, despite the existence of the etiologic agents commonly occurring in the environment. In another study of mycetoma patients,[16] it was found that such patients had a partially compromised cell-mediated immunity (CMI) response. The underlying cause was related to malnutrition with low protein intake, multiple bacterial, parasitic, and viral infections in childhood.

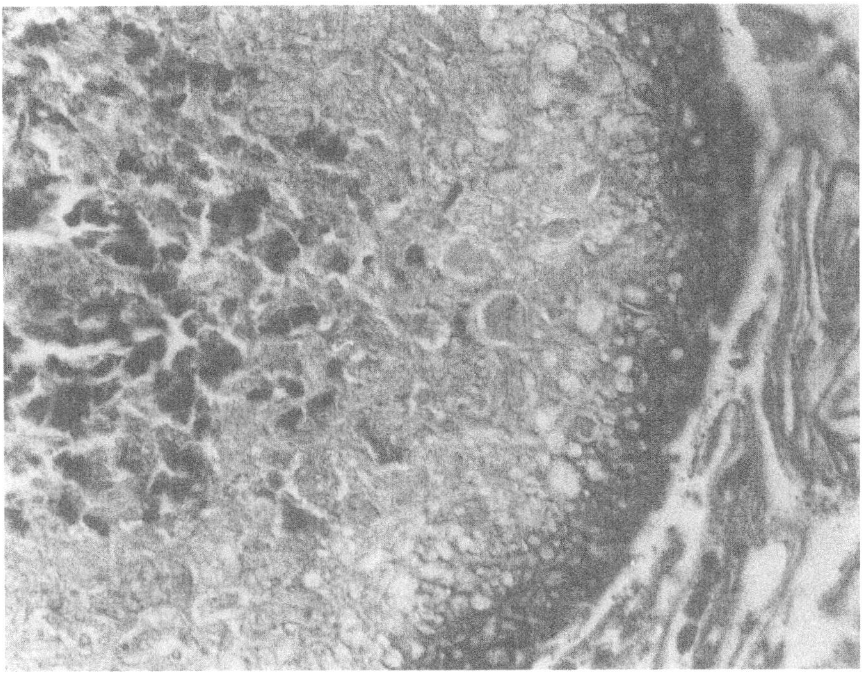

FIGURE 3. Granule of *Madurella grisea* in tissue. (400×, H and E.)

6. IMMUNODIAGNOSIS AND PROGNOSIS

Serologic diagnosis, at present, is limited to techniques such as CIE, ID, ELISA, and indirect fluorescent antibody testing. Most of these methods are sensitive and specific, although false negative results may occur when using ID, especially during the early stages of the disease. ELISA is extremely sensitive, but cross reactions between antibodies raised to different species of fungi complicates the serodiagnosis of the disease. CIE is sensitive, reliable, and used mostly as a means to follow the patient's progress during therapy.[6] Indirect fluorescent antibody testing is highly specific and sensitive[17] but has limited practical application.

A definite diagnosis of the etiologic agent can be established only by culture of the causative organism. Pus, exudate, biopsy materials, and the gauze bandage are examined for the presence of granules. If none are readily visible, granules usually can be expressed from the sinuses by applying localized pressure. Granules are washed in sterile saline to remove debris, crushed, and then examined and inoculated into proper isolation media. The granules are composed of hyphae that are approximately 2–5 μm in diame-

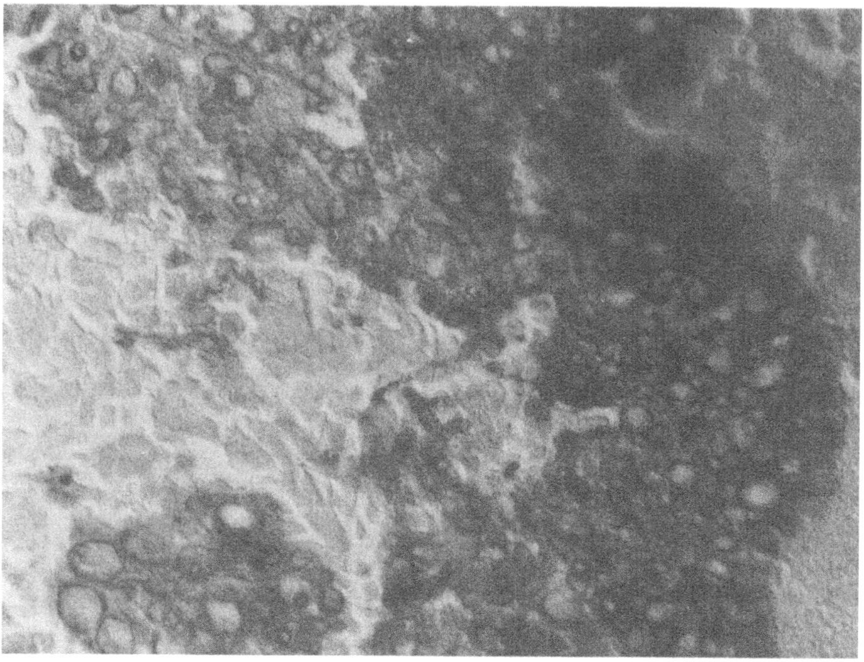

FIGURE 4. Granule of *Madurella mycetomatis* in tissue. (400×, Gridley fungal stain.)

ter, often with swollen cells that are 12–30 μm in diameter. Mold inhibitory agar or Sabouraud glucose agar without cyclohexamide should be inoculated with the clinical material and then incubated at 30°C for 4 weeks.

Prognosis can be good, but this depends on several factors. The most important include early diagnosis by the physician and the site of infection. Mycetoma involving the head and neck usually responds poorly to treatment.[18] Identification of the causal organism, successful treatment, and the rate of recurrence are extremely important.

7. PREVENTION AND THERAPY

There is no acquired or active immunity to the disease. Health, education, appropriate medical care, and improvement of the socioeconomic stature of the population at risk may help prevent some cases of mycetoma.

Unfortunately, many clinicians elect to manage mycetoma with aggressive surgery, such as amputation, before using chemotherapy. There is ample time to initiate proper management because mycetoma is a chronic disease that continues for years. There is ample time to evaluate the disease and isolate the etiologic agent prior to the initiation of management. Mahgoub[6] recommends that a combination of surgical and medical treatment be used to treat mycetomata. There are reports of ketoconazole being effective against fungi such as *M. mycetomatis* and *P boydii*[6,19,20] and griseofulvin, to a lesser degree, against *M. mycetomatis*.[21] Chemotherapy is usually required for up to 10 months and should be discontinued once the CIE serology yields normal results. Unfortunately, many cases do not respond to medical treatment. Surgical management may have a recurrence rate as high as 80 percent,[6,22] especially if the surgeons do not have adequate experience with the disease. Surgical management should include excision of tissue so that a clear margin is seen.

REFERENCES

1. Carter, H. V., 1860, On a new striking form of fungus diseases, principally affecting the foot, and prevailing endemically in many parts of India. *Trans. Med. Phys. Soc. Bombay* **6**:104–142.
2. Carter, H. V., 1861, On "mycetoma" or the fungus-disease of India: Including notes of recent cases and new observations on the structure, and c. of the entophytic growth. *Trans. Med. Phys. Soc. Bombay* **7**:206–221.
3. Pinoy, E., 1913, Actinomycoses et mycetomes. *Bull. Inst. Pasteur* **11**:929–938, **11**:977–984.
4. Chalmers, A. J., and Archibald, R. G., 1916, A Sudanese maduromycosis. *Ann. Trop. Med. Parasitol.* **10**:169–223.
5. Chalmers, A. J., and Archibald, R. G., 1918, The classification of the mycetomas, *J. Trop. Med. Hyg.* **21**:121–123.
6. Mahgoub, E.S., 1985, Mycetoma, *Sem. Dermatol.* **4**:230–239.
7. Mariat, F., Destombes, P., and Segretain, G., 1977, The mycetomas: Clinical features, pathology, etiology and epidemiology. *Contr. Microbiol. Immunol.* **4**:1–39.
8. McGinnis, M. R., and Fader, R. C., 1988, Mycetoma: A contemporary concept, *Infect. Dis. Clinics. North Am.* **2**:939–954.
9. Green, W. O., and Adams, T. E., 1964, Mycetoma in the United States. A review and report of seven additional cases, *Am. J. Clin. Pathol.* **42**:75–91.
10. Chandler, F. W., Kaplan, W., and Ajello, L., 1980, A Colour Atlas and Textbook of the Histopathology of Mycotic Diseases. Wolfe Medical Publications, London.
11. Wheeler, M. H., and Bell, A. A., 1988, Melanins and their importance in pathogenic fungi, in: *Current Topics in Medical Mycology*, (M. R. McGinnis, ed.), Springer–Verlag, New York, p. 338.
12. Hay, R. J., and Collins, M. J., 1983, An ultrastructural study of pale eumycetoma grains, *Sabouraudia* **21**:.261–269.
13. Wethered, D. B., Markey, M. A., Hay, R. J., Mahgoub, E. S., and Gumaa, S. A., 1987, Ultrastructural and immunogenic changes in the formation of mycetoma grains, *J. Med. Vet. Mycol.* **25**:39–46.
14. McLaren, M. L., and Mahgoub, E. S., Georgakopoulos, E., 1978, Preliminary investigation on the use of the enzyme-linked immunosorbent assay (ELISA) in the serodiagnosis of mycetoma, *Sabouraudia* **16**:225–228.
15. Wethered, D. B., Markey, M. A., Hay, R. J., Mahgoub, E. S., and Gumaa, S. A., 1988, Humoral immune responses to mycetoma organisms: Characterization of specific antibodies by the use of enzyme-linked immunosorbent assay and immunoblotting, *Trans. Roy. Soc. Trop. Med. Hyg.* **82**:918–923.

16. Mahgoub, E. S., Gumaa, S. A., and El Hassan, M. A., 1977, Immunological status of mycetoma patients, *Bull. Soc. Pathol. Exot. Filiales* **70**:48–54.
17. Joshi, K. R., and Singhvi, S., 1988, Serodiagnosis of mycetoma, *Trans. Roy. Soc. Trop. Med. Hyg.* **82**:334.
18. Gumaa, S. A., Mahgoub, E. S., and El Sid, M. A., 1986, Mycetoma of the head and neck, *Am. J. Trop. Med. Hyg.* **35**:594–600.
19. Ramos, L., Luque, A., and Alvarez, D., 1986, Aislamiento e identificaion de *Madurell mycetomatis* de un micetoma podal, *Bol. Micol.* **3**:5–8.
20. Stierstorfer, M. B., Schwartz, B. K., McGuire, J. B., and Miller, A. C., 1988, *Pseudallescheria boydii* mycetoma in northern England, *Int. J. Dermatol.* **27**:383–387.
21. Mahgoub, E. S., 1976, Medical management of mycetoma, *Bull. World Health Organ.* **54**: 303–310.
22. Visvanathan, R., 1989, Surgical treatment of fungal mycetoma, *Can. J. Surgery* **32**:74–76.

14

Pseudallescheriasis

CHESTER R. COOPER, JR. and IRA F. SALKIN

1. GENERAL DESCRIPTION

Pseudallescheria boydii is a ubiquitous mold that can be readily isolated from both soil and water.[1–3] Although best known as an etiological agent of eumycotic mycetoma, this fungus can also cause a wide spectrum of clinical manifestations collectively termed pseudallescheriasis.* *Pseudallescheria boydii* is generally regarded to have low inherent virulence, with most nonmycetoma infections having been reported in immunocompromised or otherwise debilitated hosts. Thus, symptoms in such individuals may mimic many aspects of aspergillosis and other opportunistic fungal infections. This, in turn, may present the clinician with difficult choices in disease management. Complicating these factors is the relative paucity of chemotherapeutic agents that can readily and successfully be used in treating the disease, especially in cases of invasive infections.

This chapter will focus primarily on the descriptions of the most common types of pseudallescheriasis in humans and the etiology, immunology, diagnosis, and clinical treatment of this disease. Although reports of animal infections exist,[1] they are beyond the scope of this chapter. Information on the clinical aspects of human cases will be discussed only in general terms and the

*Eumycotic mycetomas caused by *P. boydii* are considered in detail in a separate chapter of this volume.

CHESTER R. COOPER, JR., and IRA F. SALKIN • Laboratories for Mycology, Wadsworth Center for Laboratories and Research, New York State Department of Health, Albany, New York 12201-0509.

Fungal Infections and Immune Responses, edited by Juneann W. Murphy *et al.* Plenum Press, New York, 1993.

reader is referred to the published reports for particular details. The cases enumerated in the text and summarized in Tables I and II were obtained from references cited in reviews by Rippon[1] and by Elder and Roberts,[2] as well as from a literature search of documented cases published in English-language journals from 1980 through early 1990. Several other reports from foreign language journals have not been included, and other documented incidents probably have been inadvertently omitted. These additional reports undoubtedly reflect many of the aspects of pseudallescheriasis represented by those cases that are noted in this review. The authors apologize to those individuals whose case reports were not cited.

TABLE I
Reported Infections Caused by *Pseudallescheria boydii* (*Scedosporium apiospermum*)

Type of infection	Number of cases reported	Outcome of infection			
		Resolved	Death[a]	Persistent	Unknown
Pulmonary					
colonization	51	26	1[b] (2.0)	1	23
fungoma	21	14	5[c] (23.8)	—	2
invasive/pneumonia					
without dissemination	6	5	—	1	—
with dissemination	18	2	16 (88.9)	—	—
unspecified[d]	19	3	6 (31.6)	2	8
Central nervous system					
meningitis	7	1	5 (71.4)	—	1
brain abscess	10	5	5 (50.0)	—	—
Sinusitis	13	8	4 (30.8)	1	—
Endocarditis	5	1	4 (80.0)	—	—
Miscellaneous					
arthritis/osteomyelitis	20	17	—	1	2
keratitis	14	12	—	—	2
wound/soft tissue	13[e]	10	—	1	2
endophthalmitis	6[f]	3	2 (50.0)	—	—
optical orbit	2	2	—	—	—
Totals	**205**	**109**	**48 (23.4)**	**7**	**40**

Note: The data in this table were obtained from references cited in Rippon[1] and in Elder and Roberts[2] as well as from other documented case reports published in English language journals.[4–14,16–20,22] Data from diverse reports of otomycosis are not included.
[a]The number in parentheses represents the number of deaths as a percentage of the number of infections reported.
[b]Nine deaths unrelated to colonization by *P. boydii* are not included.
[c]Two additional deaths caused by subsequent *Aspergillus* infection are not included.
[d]This category includes those case reports of colonization or fungoma in which data regarding the type of lesion is unclear or missing.
[e]This sum includes a single case of chronic prostatitis.
[f]One patient developed endophthalmitis subsequent to successful recovery from *Pseudallescheria* pneumonia. This individual is counted twice in this table.

TABLE II
General Profiles of Clinical Cases of Infection by *Scedosporium inflatum*

Case number	History and predisposing factors	Clinical presentation	Therapy employed[a]	Outcome of infection
1	Parenteral trauma	Septic arthritis (knee)	Mic, Keto, Flu, AmB	Amputation; cleared
2	Parenteral truama	Septic arthritis (knee)	Mic, Keto, AmB, and debridement	Cleared
3	Tibial fracture	Septic arthritis (knee)	Debridement, AmB	Cleared
4	Parenteral trauma	Septic arthritis (knee)	Debridement, AmB, Keto, and Mic	Cleared
5	Parenteral trauma	Osteomyelitis	Debridement	Cleared
6	Parenteral trauma	Osteomyelitis	Debridement, AmB, Keto	Cleared
7	Multiple fractures and trauma	Osteomyelitis and cellulitis	Debridement, AmB, Keto	Cleared
8	Scleroderma/skin ulcers	Gangrenous skin ulcers	Debridement	Cleared
9	IV drug abuse	Septic arthritis (hip)	Drainage, AmB, 5-FC	Cleared
10	Fever, chronic cough	Lung infection	Bronchoscopy (2×)	Cleared
11	Kidney transplant	Peritonitis	AmB, Mic	Died

Adapted from Wilson *et al.*[25]
[a]Mic, miconazole; Keto, ketoconazole; Flu, fluconazole; AmB, amphotericin B.

2. CLINICAL MANIFESTATIONS OF PSEUDALLESCHERIASIS

The vast majority of infections caused by *P. boydii* are mycetomas, which are rarely life threatening.[1] Yet, there is an increasing number of infections by *P. boydii* or its anamorph, *Scedosporium apiospermum*, which have gained access to debilitated individuals through pulmonary or parenteral routes and have proved to be fatal. Surprisingly, excluding incidents of mycetoma and otomycosis, the collective mortality rate for all reported cases of pseudallescheriasis is greater than 23% (Table I).

The following sections describe those human infections caused by *P. boydii* and *S. apiospermum*. We have made no attempt to distinguish between infections caused by the teleomorph or the anamorph for reasons to be discussed below (see Section 4). Finally, a separate section is devoted to similar infections caused by *Scedosporium inflatum*.

2.1. Pulmonary Infection

Other than pedal mycetoma, the pulmonary system is the most common site of infection by *P. boydii*.[1,2] The clinical manifestations can be differentiated into several categories including colonization, fungoma (fungus ball), and invasive disease. With regard to colonization and fungoma, the infections tend to remain localized, often occurring in patients having prior lung cavitation due to various etiological agents. Affected individuals frequently exhibit a chronic cough and hemoptysis. The latter can range from the presence of blood in the sputum to fatal hemorrhage.

Histologically, pulmonary disease by *P. boydii* is quite similar to aspergillosis.[1–3] Fungal abscesses are partly enclosed in granulated tissue with marked fibrosis and necrosis of the parenchyma. Additionally, the tubercles become chronically inflamed and giant cells are present. Sputum from colonized patients will yield *P. boydii* when cultured on appropriate media. Similarly, the fungus can be isolated from tissue specimens obtained following surgical resection of fungomas.

Examination of stained tissue sections from patients colonized with *P. boydii* generally reveals loose, hematoxylinophilic mycelial strands or yellow-brown hyphal aggregates, which may occasionally bear conidia and dichotomous branches. Although similar to that of *Aspergillus* in tissue, the branching pattern of *P. boydii* is more haphazard and less progressive. Small grains, similar in appearance and composition to those formed in mycetomas, are sometimes found as well. Conversely, fungomas consist of concentric rings of septated hyphae that usually bear conidia. The hyphae of *P. boydii* seen in these types of infection are narrower (2–5 μm) than those found in cases of aspergillosis. Also, when produced *in vivo*, the brown and ovoid conidia of *P. boydii* measure approximately 5 by 10 μm and are borne terminally or laterally on short conidiophores, quite different from those found in aspergillosis. Finally, both *P. boydii* and *Aspergillus* may coexist in the same infection, making identification of the etiological agents by culture and serological techniques of great importance.

Through 1989, more than 70 cases of *P. boydii* colonization and fungoma have been reported (Table I).[1,4–6] One markedly consistent characteristic of most of these cases is that they occur in patients having underlying pulmonary disease or other predisposing conditions. Moreover, in those reports that included follow-up examinations, more than half represented simple transient colonization of the lung. These two themes were dramatically demonstrated in a classic study by Travis *et al.*[4] in which 36 isolates of *P. boydii* were obtained from 32 patients with respiratory tract symptoms over a 10-year period. Of these patients, only two exhibited signs of clinically significant infection. Twenty-six of the 30 (87%) colonized patients had some form of underlying pulmonary disease and five (17%) were receiving corticosteroids. In follow-up studies of 22 of the 30 patients, 13 remained free of pulmonary disease, whereas nine died from circumstances unrelated to *P. boydii* infection. Similarly, in Rippon's[1] summary of 54 cases of pulmonary colonization and fungoma, 48 (89%) patients had some form of underlying disease prior to or concomitant with infection due to *P. boydii*. While most survived following surgery and/or chemotherapy, 11 (23%) patients died as a result of pulmonary pseudallescheriasis.

In contrast, the above conditions are relatively benign when compared to invasive pulmonary infection. As noted by Elder and Roberts,[2] such infections by *P. boydii* can be divided into four categories: 1) normal hosts with invasive pulmonary disease; 2) severe lung infections in compromised hosts with or without extrapulmonary dissemination; 3) subsequent extrapulmonary dissemination following resolution of an initial lung infection in a compromised host; and 4) compromised patients with disseminated disease but no evidence of lung involvement. The latter category may include a number of the case reports described below placed within other categories of clinical manifestations. In those instances where such cases have no parenteral route of entry into the body, the reader should be aware that dissemination probably occurred following inhalation of infectious propagules and pulmonary infection.

By far the most severe instances of invasive pulmonary infection are those that precede dissemination to the visceral organs.[1–4,6–9] Such infected individuals usually develop pneumonia or pneumonitis and are typically immunocompromised or have some type of underlying disease unrelated to psuedallescheriasis. Unchecked, invasive pulmonary pseudallescheriasis results in vascular involvement, formation of nodular infarcts, and tissue necrosis. The organism typically spreads to the brain, thyroid, kidneys, and heart. Aspergillosis can also accompany or follow the initial infection of *P. boydii*. Mortality rates are quite high; two-thirds of patients having invasive disease died (Table I). Most individuals who did recover were apparently diagnosed early in the disease process and received appropriate treatment before dissemination of the infection. Once dissemination occurs following invasive pulmonary pseudallescheriasis, however, all cases proved to be lethal. Overall, there is approximately a 24% mortality rate from all forms of pulmonary involvement (28 deaths out of 115 reported cases; Table I).

2.2. Central Nervous System Involvement

Other than pulmonary disease, the second largest cause of death due to *P. boydii* is infection of the central nervous system (CNS) through the formation of brain abscesses or the involvement of the meninges (Table I). For purposes of this review, we define CNS infections as those arising from a break in the venal or membranous barriers that separate the brain from the rest of the body and not by dissemination from other *active* foci of infection. Hence, this definition excludes many cases resulting from dissemination of invasive pulmonary disease, sinusitis, etc. but not those in which no evidence of prior or concomitant lung infection exists. In fact, of the 17 cases that we recognized, 12 involve individuals with no obvious lung infection and who presumably inhaled the etiological agent. Four of these patients were apparently normal hosts and eight were immunocompromised. In five other cases, a parenteral means of infection was noted.

A recent review by Berenguer *et al.*[10] cited 21 cases of CNS infection by *P. boydii*. Criteria for diagnosis of CNS pseudallescheriasis were established by the authors and included 1) clinical symptoms consistent with pachymeningitis or leptomeningitis, or demonstration of intracranial abscess, and 2) isolation of *P. boydii* from CNS tissue or fluid, or visualization of septate hyphae in the meninges or brain with concurrent isolation of *P. boydii* from sites outside the CNS. After review of the 21 cases, however, we can only agree that 15 cases fit the above criteria as well as our definition of CNS pseudallescheriasis. We judge the remaining six cases to result from dissemination of pulmonary or sinus infections. Of the 15 cases, having a collective mortality rate of 60% (Table I), eight involve brain abscesses (50% mortality) and seven are meningial infections (71% mortality). All four individuals (one abscess and three meningitis patients), who were not clinically treated, died. Of the remaining 11 patients treated either surgically, with drugs, or both, five died.

In the reported cases of brain abscesses,[9–11] lesions are generally confined to the brain in patients who tend to have underlying disease or other predisposing conditions. Histologically, large areas of necrosis are typically found. The organism, as in pulmonary infections, is generally seen as mycelial strands that exhibit dichotomous branching and occasionally bear conidia (Fig. 1). Meningal infections, by contrast, usually result from traumatic implantation of infectious propagules into immunocompetent individuals,[1,9–11] but demonstration of the organism in the meninges is not always

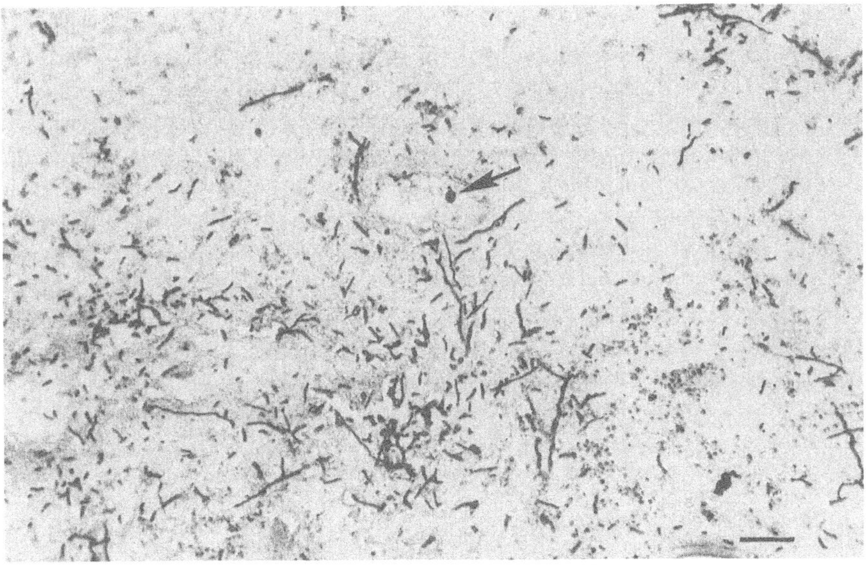

FIGURE 1. Photomicrograph of a thin section of human brain tissue infected with *Pseudallescheria boydii.* Note the dichotomously branching hyphae and the large, sporelike structure (arrow) present in this section. Gomori methenamine silver stain. Bar represents 40 μm.

possible. Dissemination from the CNS to other tissues can occur in both types of infection.

2.3. Sinusitis

Thirteen cases of sinusitis[1,4,12,13] due to *P. boydii* have been documented (Table I). Of these, seven patients (54%) had no underlying disease or known predisposing conditions. Samples from the sinus region of 10 patients were culture-positive for *P. boydii*; another was positive by immunofluorescent methods. Infection typically involves the maxillary sinus, sphenoid sinus, or nasal septum, and symptoms exhibited by patients can be mistaken for rhinocerebral zygomycosis. Most cases are resolved by surgical debridement, but antifungal agents have also been used, though their value in treatment is unclear. Perhaps the most important aspect of this type of infection is its proximity to and possible extension to the brain. Four deaths (31% of the reported incidents of sinusitis) have been associated with initial *P. boydii* sinusitis. Postmortem examination of two cases documented brain lesions due to *P. boydii.* No autopsy was performed on the other two individuals, but other evidence strongly indicated brain involvement due to *P. boydii.*

2.4. Endocarditis

To date, five cases of endocarditis[1,14] caused by *P. boydii* have been reported (Table I). Four involved heart valve transplants and the remaining case was associated with implantation of a pacemaker. Only one individual, a valve transplant recipient, survived following surgery and treatment with miconazole (80% mortality). One of the transplant patients was an intravenous drug abuser, whereas none of the others had any predisposing factors prior to surgery. Cultures of *P. boydii* were recovered from the blood or transplanted tissue of each patient. The exact source of the infectious agent is not known, but contamination of the prosthetic devices is a logical assumption. Rippon[1] notes that tissue-source replacements for heart valves are generally soaked in solutions containing amphotericin B and nystatin. Most isolates of *P. boydii*, however, are resistant to these two drugs (see Section 6.2).[15]

2.5. Miscellaneous Infections

The categories of disease described above account for nearly all the reported deaths caused by pseudallescheriasis (Table I). More than 30% of those individuals contracting either pulmonary, CNS, sinus, or endocardial infections due to *P. boydii* died. Yet, infection by this fungus is not limited to these organ systems. For instance, *P. boydii* has been identified as the causative agent of lesions of the soft tissues,[1,6,16,17] the eyes (keratitis, endophthalmitis, and optical orbit),[1,18,19] the ears,[1,2] and joints and bones (arthritis/osteomyelitis).[1,20] Such infections can be caused by dissemination of the etiological agent from the lung following inhalation, but most often the disease results from introduction of the organism through the protective layers of the skin or eye.

Implantation of *P. boydii* by dermal or subcutaneous trauma usually results in mycetoma, but alternative pathological events can occur. Thirteen cases of nonmycetoma pseudallescheriasis of the soft tissues have been recorded[1,6,16] including a case that resembled lymphocutaneous sporotrichosis.[21] Improvement in the latter did not occur upon treatment with iodides, but the infection did respond when the antifungal regimen was changed to sulfadiazine and levamisol. Additionally, 19 collective cases of infection of the eye and surrounding tissues have been noted.[1,18,19] Some of these infections led to blindness, yet only one progressed to involve the CNS resulting in patient death. An increase in the frequency of infections leading to arthritis and osteomyelitis has also been noted.[1,20] In most instances, the lesions have been limited to the hands, feet, synovial fluids, or knees. Finally, a number of cases of otomycosis (data not shown; see ref. 1) and one instance of chronic prostatitis[22] due to *P. boydii* infection have been reported. No severe complications from these infections were noted.

In summary, the significance of these miscellaneous infections is that most are readily resolved (>83%) by appropriate surgical intervention, chemotherapy, or a combination of both. Even when instances of otomycosis are excluded, the mortality rate from these infections is extremely low (<2%).

2.6. Infections Caused by *S. inflatum*

In 1984, Malloch and Salkin[23] described a new species of *Scedosporium* isolated from a bone biopsy of a 6-year-old boy. The new species, *S. inflatum*, differs from *S. apiospermum* in its mode of conidia formation and with respect to several physiological characteristics (see Section 4.3 below). Since the initial isolation of *S. inflatum* from the case of osteomyelitis, 14 other incidents of infection in humans (11 cases) and animals (3 cases) have been reported (Table II).[24] The case reports demonstrated the association of *S. inflatum* with focally invasive infections, in general, and suggested a narrow spectrum of clinical disease in contrast to *S. apiospermum*.[25] Like other opportunistic pathogens, however, *S. inflatum* is now being recognized more and more in the clinical environment as is its potential to cause serious disseminated infections. In the last several years, the authors' laboratories have received no less than six reference cultures for identification that have been determined to be *S. inflatum* (unpublished results). These include specimens originally isolated from blood and spinal fluid, sites not described in the case reports presented by Wilson *et al.*[25]

As noted, the published case histories of the 11 human infections by *S. inflatum* depicted localized disease. Five patients exhibited septic arthritis of the knee or hip, three more were afflicted with osteomyelitis, and the remaining three individuals had skin ulcers, pulmonary colonization, or a surgical wound infection (Table II). Histologically, the *S. inflatum* infections were very similar to those produced by *P. boydii*.[24,25] Progressive tissue necrosis occurred in two cases and the etiological agent appeared as hyphal fragments which formed conidialike structures. Treatment of each case generally included antifungal agents, debridement, drainage of the lesion, or a combination of these. In several instances, antifungal treatment was unsuccessful owing to the apparent resistance of the fungus[25] (also, see Section 6 below). Except for two instances, all infections cleared and remained so in eight of the patients during follow-up examinations (average of >2 years). One patient was lost to follow-up studies. Of the two exceptions, amputation was required to resolve the infection in one individual and the other patient died, apparently from peritonitis caused by *S. inflatum*.

3. ETIOLOGY AND EPIDEMIOLOGY

The etiologic agent of pseudallescheriasis is ubiquitous, having been isolated from a variety of ecological niches, including soils, potted plants, sewage, river water, and livestock manure.[1,2,26] Although distributed worldwide, *P. boydii* accounts for relatively few cases of eumycotic mycetoma outside the United States. Perhaps this is because of differing pathogenic potentials according to geographical regions.

Like most disease-causing fungi, *P. boydii* exists as a saprobe until given the opportunity to penetrate host defense mechanisms. Hence, it is host factors that generally determine the presence and extent of infection. For instance, of 110 pulmonary cases of pseudallescheriasis reported between 1955 and 1987 in which adequate records were kept, underlying disease was noted for 94 (>85%) patients.[1,4–6] Additionally, more than 29% of those individuals acquiring pulmonary, cardiac, CNS, or sinus infections did not survive, whereas less than 2% of infections in other types of tissues resulted in death (Table I). With regard to distribution of human infections, Rippon notes that mycetoma caused by *P. boydii* is more frequent in men than in women (3:1 to 5:1).[1] In contrast, our evaluation of nonmycetoma infections caused by *P. boydii* shows that, although disease is more common in males overall, the frequency is not as disparate (3:2; data not shown).

4. MYCOLOGY

It is significant to note that other than the superficial mycosis agent, *Piedraia hortae*, *P. boydii* is the only fungal pathogen of humans that is routinely isolated from clinical material as a teleomorph.[1] While teleomorphic forms of other fungal pathogens are known (e.g., some members of the genus *Aspergillus* are anamorphic forms of the genus *Eurotium*) they are not recovered upon direct isolation from specimens. *Scedosporium apiospermum*, the anamorph of *P. boydii*, is regularly obtained in culture, either prior to or concomitant with, the formation of cleistothecia or the fruiting structures that characterize *P. boydii*. In the majority of case reports of pseudallescheriasis, however, few authors have described the causative agent as *S. apiospermum*. Rather, they use the teleomorph designation, despite the fact that the anamorph was probably the actual entity isolated. In many instances, the teleomorph may never be recovered or develop in culture.

This does not infer that the disease should be known as scedosporiosis as opposed to pseudallescheriasis. First, the latter term is ingrained in the lexicon of medical mycology. Second, since Emmons[27] demonstrated that the anamorphic state of *P. boydii* was *S. apiospermum*, we now realize that these two binomials simply apply to different stages in the life cycle of the same organism. Consequently, we are dealing with the same etiologic agent and there is no need to change the name of the disease. Although we have made no attempt to separate case reports of pseudallescheriasis based upon the particular state of the fungus that was recovered in culture, we suggest that future reports clearly indicate which of the two forms or both was obtained upon initial isolation.

4.1. *Pseudallescheria boydii*

Until recently, the identification and classification of *P. boydii* was complicated because of various taxonomic interpretations provided by mycologists. Since Shear's original description in 1922, the teleomorphic stage of *P. boydii* has been known by several binomials, including *Allescheria boydii*, *Pseudallescheria sheari*, and *Petriellidium boydii*.* In 1982, McGinnis *et al.*[29] provided the correct taxonomic interpretation in establishing *Pseudallescheria boydii*. Since ascocarps formed by this fungus are cleistothecia, *P. boydii* is classified within the family Microascaceae.

Pseudallescheria boydii grows rapidly on most common mycological media forming abundant aerial hyphae. In general, it grows as a white colony eventually becoming brownish gray in color, with a gray to black reverse. It usually forms globose, light brown to black cleistothecia beneath the agar surface (Fig. 2). Induction of ascocarp formation may require repeated passage on nutrient-deficient media (e.g., water agar). Within the cleistothecia develop subglobose to globose asci containing eight ovoid to ellipsoidal ascospores. The ascospores are smooth walled, yellow-brown to copper in color, and possess two germ pores. Usually, the asci are evanescent and rupture of a mature cleistothecium releases a mass of ascospores.

4.2. *Scedosporium apiospermum*

Two anamorphic states of *P. boydii* are described by Shear.[30] One forms synnemata and is now considered to be in the genus *Graphium* (Fig. 3). Although this anamorph has not been reported to have been isolated from a human infection, a case of an animal infection caused by this anamorph has been described;[31] however, it is usually recovered associated with the second anamorph, *S. apiospermum*, or it may not be isolated at all from clinical specimens. *Scedosporium apiospermum*, the more common of the two anamorphs, forms terminal or lateral uninflated conidiogenous cells bearing one to several clavate conidia (Fig. 4).[32] Terminal conidia are pushed from the tips of conidiogenous cells by subsequent formation of new conidia. Clusters of conidia may form by remaining next to the conidiogenous cells upon separation.

4.3. *Scedosporium inflatum*

It has been noted that another species of *Scedosporium*, *S. inflatum*, is an increasing cause of human and animal mycoses.[23–25] This fungus is morphologically similar to *S. apiospermum* and *Scopulariopsis brumptii* in that all three

*For a historical summary of the derivation of these epithets, see the discussion in Rippon[1] and McGinnis.[28]

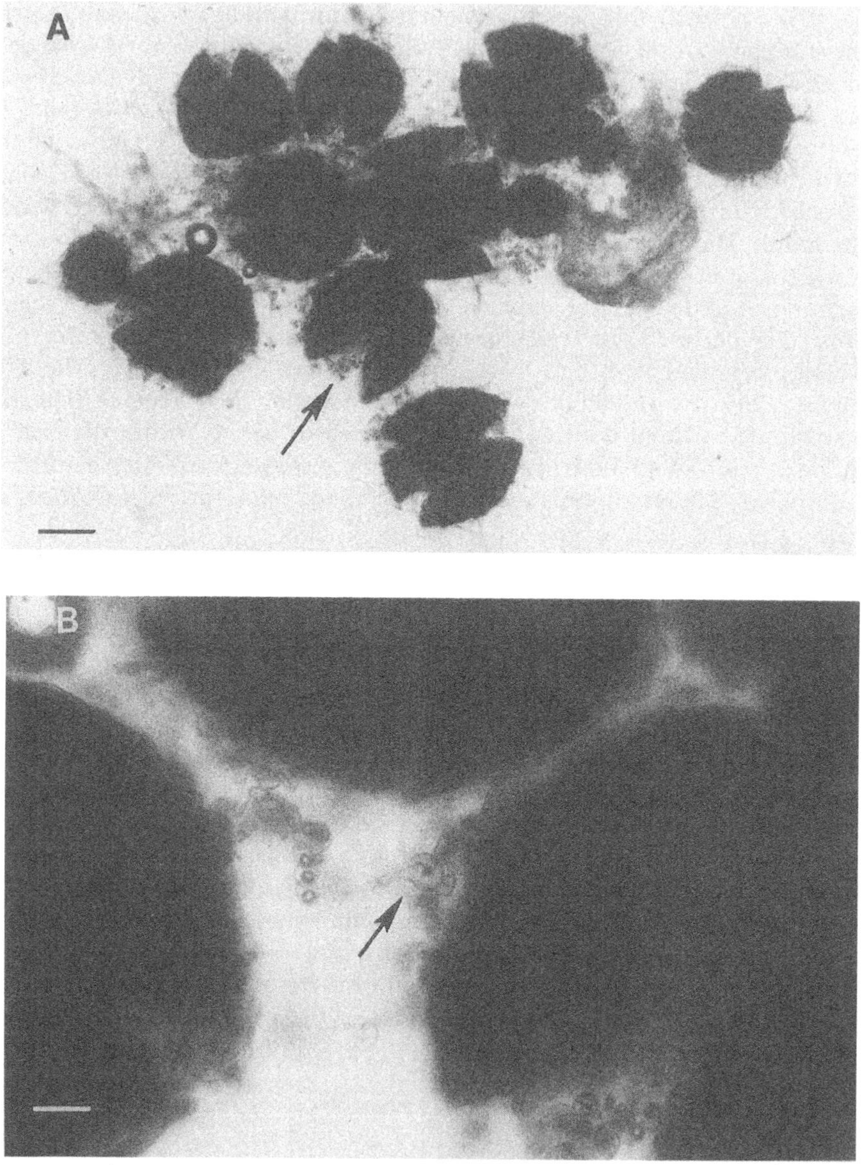

FIGURE 2. Photomicrographs of unstained cleistothecia of *Pseudallescheria boydii* that have been crushed in water. A) Note ascospores (arrow) being released from the cleistothecium. Bar represents 40 μm. B) A more highly magnified photomicrograph of the same preparation reveals round to oval ascospores containing a prominent inclusion (lipid body?). An ascogenous hypha (?) is also observable (arrow). Bar represents 10 μm.

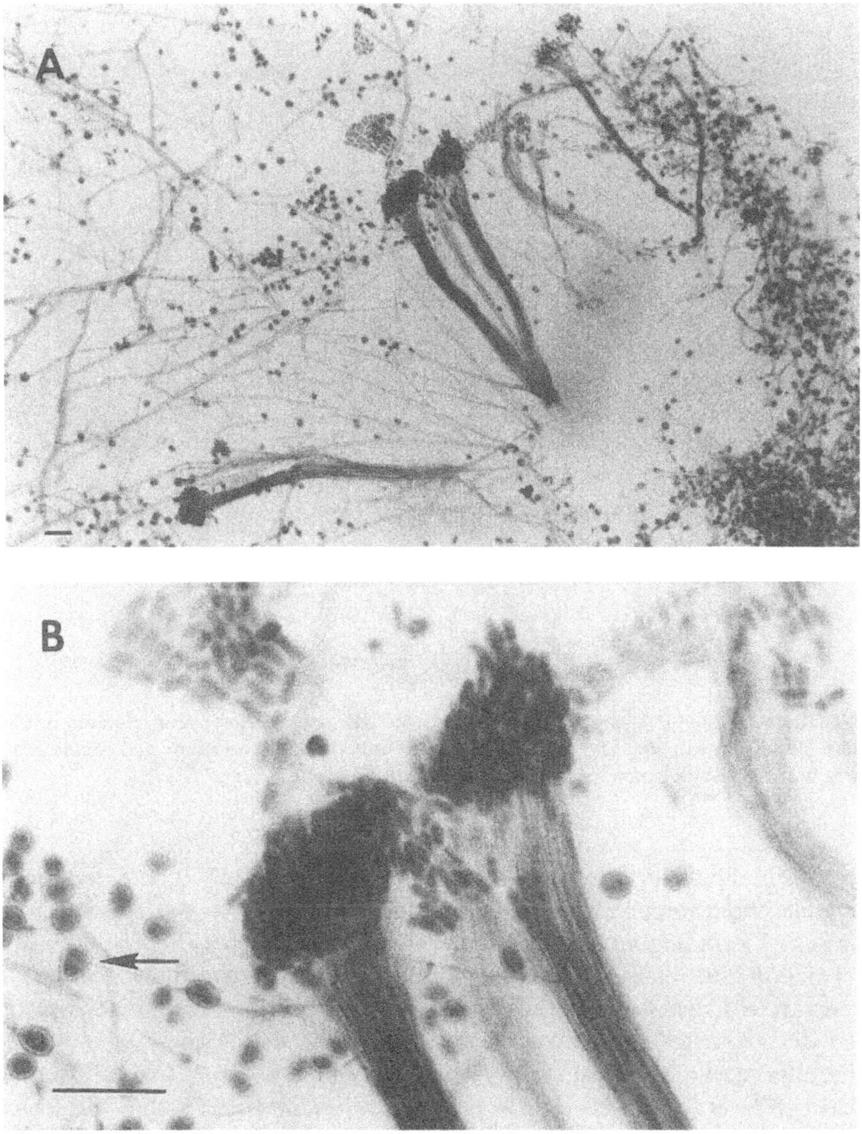

FIGURE 3. Photomicrographs of synnemata (*Graphium* state) of *Pseudallescheria boydii.* Several synnemata at low magnification (A) and (B) at higher magnification showing the terminal ends of three such structures from the same photomicrograph. Note the cylindrical nature of the conidia produced from the synnemata as compared to the clavate conidia (arrow) typical of the *Scedosporium* anamorph of *P. boydii.* Lactophenol blue stain. Bars in each figure represent 20 μm.

FIGURE 4. Photomicrograph of conidiophores of *Scedosporium apiospermum* observed by Hoffman phase contrast. Note the presence of both terminal and lateral uninflated conidiogenous cells bearing several clavate conidia. Bar represents 10 μm.

form annelloconidia through percurrent proliferation.[32] The conidia of *S. inflatum*, however, are flasked-shaped and accumulate as wet clusters at the apices of conidiogenous cells with widely-inflated bases (Fig. 5). Repeated formation and releases of conidia results in variably spaced annellations. In contrast to *S. apiospermum*, the conidia of *S. inflatum* do not always completely secede but instead remain attached to the side of the conidiogenous cell, thereby creating the illusion of sympodial conidiogenesis. A major difference between *S. brumptii* and the two species of *Scedosporium* is that the former produces conidia arranged in a catenulate fashion (Fig. 6).

Additionally, there are physiological differences that may be used to separate *S. apiospermum* from *S. inflatum*. Results of preliminary studies indicate that *S. apiospermum* will grow on cycloheximide-containing media (Mycosel), but does not produce urease nor does it grow at 45°C (see ref. 24; S. Rosenthal, I. Salkin, I. Weitzman, and M. Kemna, unpublished results). Conversely, *S. inflatum* will not grow on Mycosel but does produce urease and does grow as reduced colonies at 45°C.

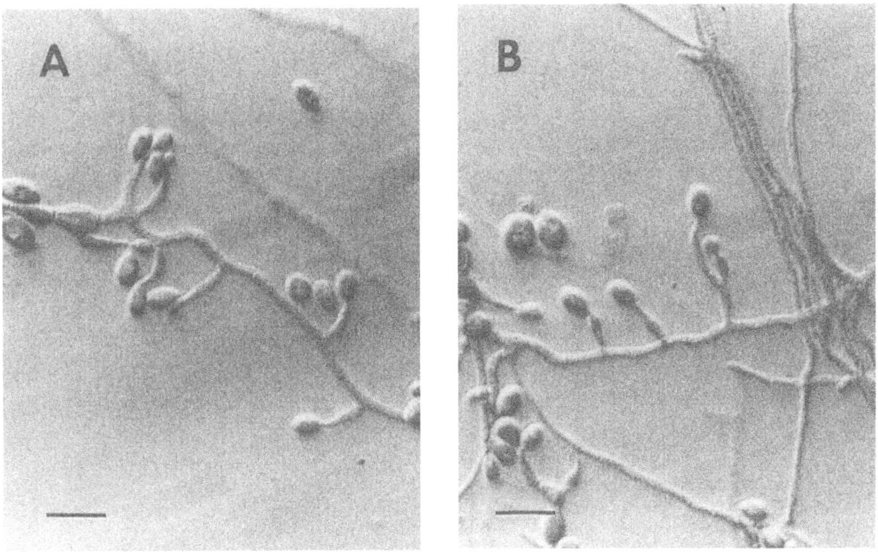

FIGURE 5. Photomicrographs of conidiophores of *Scedosporium inflatum* observed by Hoffman phase contrast (A and B). Note that the ovoid conidia are borne at the apices of conidiogenous cells with widely-inflated bases. Bars in each figure represent 10 μm.

5. BIOLOGICAL STUDIES

Despite the frequent isolation of *P. boydii* in cases of eumycotic mycetoma and its increasing role in opportunistic infections, few biological studies involving this fungus have been reported. Most of the published reports have focused on nutritional studies, animal models of disease, immunology/serology, and antifungal susceptibility. These topics are discussed briefly below.

5.1. Nutritional Studies

The nutritional studies by Wolf *et al.*,[33] Cazin and Decker, [34] and Gordon[35] have provided most of the basic physiological information about *P. boydii*. To summarize the findings of these investigators, the majority of *P. boydii* strains appeared capable of assimilating various carbohydrates, a wide variety of organic or inorganic nitrogen sources, and most were autotrophic with respect to their growth factor requirements. Growth was not stimulated by the addition of water- or fat-soluble vitamins. Interestingly, Mohr and Muchmore[36] have shown that estrogen inhibits growth of *P. boydii in vitro* and have proposed a possible hormonal influence on infection. In support of their hypothesis, the authors subsequently reported the case of eumycotic mycetoma by *P. boydii* in a female patient that would temporarily resolve when she

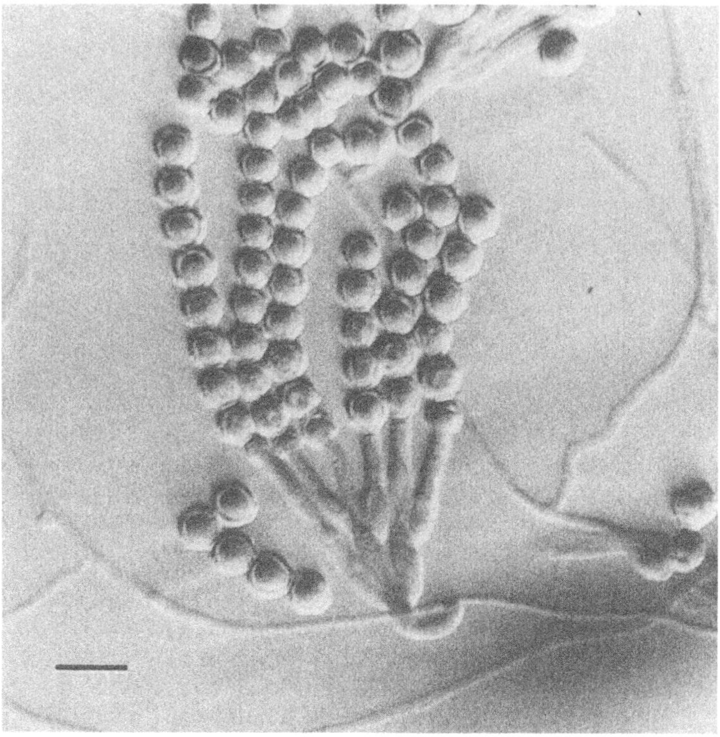

FIGURE 6. Photomicrograph of multibranched annellides of *Scopulariopsis brumptii* observed by Hoffman phase contrast. Note the round, thick-walled conidia arranged in a catenulate fashion. Bar represents 10 μm.

became pregnant.[37] The infection was eventually resolved with hormonal treatments.

Ascogonial development appeared nutritionally related as well as strain dependent. Strains of *P. boydii* that consistently formed cleistothecia did so only on nutritionally deficient media. Conversely, conidial production was enhanced on media containing a rich source of organic nitrogen.

5.2. Experimental Animal Infections

Studies employing mice as model hosts for experimental infections with *P. boydii* have been conducted by a number of researchers. Most, however, have been directed at producing localized mycetomas and not systemic infection. Because the investigations were generally unsuccessful, it was concluded that *P. boydii* was relatively avirulent for laboratory animals. This has been contradicted by studies involving the production of systemic disease using environmental and clinical isolates of *P. boydii*. In fact, wide variations in the ability of *P. boydii* strains to cause disease has been routinely noted. Lupan and Cazin[38] investigated the virulence of *P. boydii* in mice using isolates from human infections. Although the LD_{50} (i.e., dosage of inoculum that is lethal for 50% of the mice) differed for individual isolates, most mice consistently acquired fulminant infections when inoculated by the intravenous route. In contrast, other routes did not lead to fatal infections unless corticosteroid-treated mice were employed. Similarly, Bell[39] showed that human isolates of *P. boydii* inoculated intraperitonally into mice were significantly more virulent than those strains isolated from soil or manure (95% mortality versus 28% mortality within 21 days). A recent study in our laboratories also indicated that virulence was strain-dependent in mice infected via tail vein injections (D. M. Dixon, B. Breslin, and C. R. Cooper, unpublished data). In both our study and in Bell's,[39] progressive torticollis caused by CNS involvement was a common symptom of infected mice. This was confirmed on autopsy. Lesions containing conidia and hyphae were also noted in most of the other major internal organs.

5.3. Humoral and Cellular Immunity

Previous immunological studies of *P. boydii* infections were generally limited to the comparison of antigenic relationships among etiologic agents of eumycotic mycetoma and the development of methods for serological diagnosis of mycetoma (see ref. 40 and 41). Few investigations have been specifically directed towards studying the immune response involved in nonmycetoma pseudallescheriasis, although one might logically expect that they would be similar to those exhibited in cases of mycetoma. Since this chapter is concerned with *P. boydii* infections other than mycetoma, the following discussion will emphasize those few experimental studies of non-mycetoma pseudallescheriasis. For more information regarding the immune response to mycetoma caused by *P. boydii*, the reader is referred to the chapter in this volume concerning eumycotic mycetomas.

Of the few studies on nonmycetoma infection, perhaps the best noted is that by Lupan and Cazin[42] in which mice were used as models to investigate the humoral immune response in pseudallescheriasis. Since previous studies had established strain-dependent virulence, the humoral response to six strains of *P. boydii* was studied by producing systemic disease in mice through intravenous injection of conidia. The mortality rates in these experiments ranged from 49% to 84% (within 28 days). Most survivors produced significant levels of antibodies to *P. boydii*, but there was no statistical correlation between virulence and ability to produce antibodies nor between particular strains and the levels of antibody generated. Interestingly, however, not all survivors within each experimental group produced measurable levels of antibodies. The authors proposed that seroconversion in these mice depended upon substantial proliferation of the fungus *in vivo*, such as in a chronic infection. In the same study, Lupan and Cazin[42] injected mice with viable and nonviable conidia either intramuscularly or subcutaneously. In contrast to mice injected intravenously, none of these animals died. Mice injected with viable and nonviable conidia produced significant levels of antibody, but higher titers were found in those infected with viable propagules. Additionally, higher antibody levels were found in mice injected intramuscularly than in those infected by the subcutaneous route. From their results, the authors suggested that strain variation, as identified by differences in LD_{50}s and humoral response, might dictate whether infections became chronic or systemic in nature.

Although the role of cellular immunity has been widely studied in a number of fungal infections, it has yet to be well characterized with regard to nonmycetoma pseudallescheriasis. In a broad-based study of fungal infections and cellular immunity, Schaffner *et al.*[43] compared the susceptibility of primary pathogens (e.g., *Coccidioides*) and opportunists (e.g., *Pseudallescheria*) to killing by neutrophil granulocytes. Their investigations showed that all fungi studied were ingested by neutrophils, thereby triggering the release of hydrogen peroxide. The primary pathogens, however, were less susceptible to killing by the peroxide. The authors suggested that the differences in the susceptibility of the fungi studied may explain in part the difference in virulence between primary and opportunistic pathogens.

6. DIAGNOSIS AND TREATMENT OF PSEUDALLESCHERIASIS

The relative virulent nature of *P. boydii*, as well as its resistance to various chemotherapeutic agents, makes the rapid and accurate diagnosis of nonmycetoma infections imperative. Although standard culture techniques are helpful, it can take weeks before an isolate is identified. During this time, unless the appropriate antifungal therapy is empirically applied, the etiologic agent can disseminate to other tissues and further complicate the patient's recovery. Hence, there is a need for more expeditious and specific methods to differentiate *P. boydii* infections from those caused by other fungi. Described below are some of the more common nonculture-related methods employed in the clinical mycology laboratory to diagnose nonmycetoma pseudallescheriasis, as well as information pertaining to the chemotherapeutic treatment of such infections.

6.1. Serological and Histopathological Diagnosis

Procedures involving the serodiagnosis and prognostic evaluation of various forms of pseudallescheriasis are standard in many laboratories. These methods were initially developed to clinically study eumycotic mycetoma, but they are equally valuable for studying nonmycetoma forms of pseudallescheriasis. Much of the early serologic work with *P. boydii* was performed by Lupan and Cazin[42,44,45] who isolated several antigenic fractions from different isolates of both *P. boydii* and its anamorph *S. apiospermum*. They were able to show that the teleomorph and the anamorph were immunologically equivalent. They also demonstrated that sera from culturally proven cases of pseudallescheriasis contained high levels of antibody to antigen derived from *P. boydii* or *S. apiospermum*. Sera from patients having other types of fungal infections did not react with their antigen preparations. Using this as a basis for study, immunodiffusion tests have been developed as reliable indicators of disease. The appearance of precipitin bands in human cases of pulmonary colonization and fungoma by *P. boydii* can be detected readily, as can their disappearance on remission of the infection.[1,46–49] Similar results were obtained in experimental mice infections.[42] Potentially, immunodiffusion tests can be useful in distinguishing transient pulmonary involvement from chronic or disseminated infections.

Similar methodology has also been applied to questions regarding etiology when cultures were not available and to histopathological analysis of tissue sections. Morace and Polonelli[50] developed an exoantigen test to serologically identify *P. boydii* from other morphologically similar hyphomycetous fungi. The methodology is simple, specific, and allows accurate identification of *P. boydii* isolates within three days after sufficient growth of pure cultures has been obtained. Fluorescent antibody reagents have also been developed and used to identify *P. boydii in situ* (Fig. 7).[51] They can be produced from hyphal fragments, conidia, soluble antigens, and even whole or fractionated serum, although the best preparations were made from hyphal cell walls. One disadvantage to these reagents is that to enhance specificity they often must be absorbed against species of *Aspergillus*, *Fusarium*, and *Scopulariopsis*.

6.2. Antifungal Susceptibility

Perhaps the area of greatest clinical significance with regard to *P. boydii* or *S. inflatum* infections is the identification of efficacious antifungal agents. Few studies have been centrally focused towards this goal. Most of the data concerning the antifungal susceptibilities of these fungi have been through case studies. The following discussion broadly summarizes current knowledge and trends in treating diseases caused by *P. boydii* and *S. inflatum*.

One of the earliest agents reported to have fungistatic activity against *P. boydii* was estrogen (see Section 5.1).[36,37] Subsequent studies, however, have been more conventional in that they evaluated the susceptibility of *P. boydii* to currently available antifungal agents. Of various published reports, those by Nielsen, Lutwick *et al.*, and Galgiani *et al.* have provided substantial information concerning the inhibitory effects of amphotericin B (AmB), 5-fluoro-

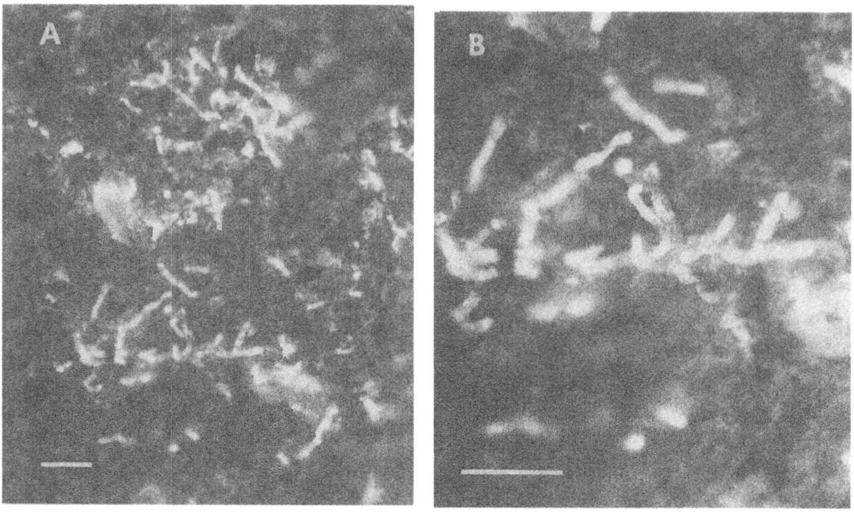

FIGURE 7. Low (A) and high (B) magnification photomicrographs of a thin section of human brain tissue infected with *Pseudallescheria boydii*. The hyphae have been stained with an FITC-conjugated antibody specific for *P. boydii*. Compare the dichotomously branching hyphae with those depicted in Fig. 1. Bars in each figure represent 20 μm.

cytosine (5-FC), miconazole (MCZ), and ketoconazole (KTZ) on both human and environmental isolates of *P. boydii*.[15,52,53] Using *in vitro* assays, these studies have collectively shown *P. boydii* to be resistant to AmB, 5-FC, and KTZ, but generally susceptible to the antifungal activities of MCZ. Galgiani *et al.* did attempt to demonstrate that KTZ was effective against pulmonary and skeletal pseudallescheriasis in that five of seven patients did improve with KTZ treatment.[53] It was clear from their studies of 22 clinical isolates, however, that MCZ was superior to KTZ with respect to antifungal activity. The minimum inhibitory concentration (MIC) of MCZ at which 50% of the isolates were susceptible was 8 to 40 times lower than that of KTZ.

Despite these data, it is also clear that data from *in vitro* susceptibility studies do not necessarily correlate with *in vivo* treatment results. In the investigations by Galgiani *et al.* and Lutwick *et al.*, it was noted that some patients treated with MCZ suffered a relapse on cessation of therapy.[52,53] Furthermore, although the efficacy of MCZ is relatively high in the treatment of many types of pseudallescheriasis,[1] CNS involvement by *P. boydii* is somewhat recalcitrant to MCZ treatment.[10] Yet, no other antifungal agents have been shown to be any more effective in such infections. Moreover, the American Thoracic Society recommends the use of MCZ for pulmonary pseudallescheriasis and also recognizes that KTZ may also be an effective chemotherapeutic agent for some patients.[54]

Little information exists with regard to other antifungal drugs. Itraconazole (ITRA) has not been shown to be particularly effective[55] and almost no information has been reported as to the usefulness of fluconazole (FLU). Preliminary experiments using a mouse model of infection to assess the efficacy of currently available and experimental antifungal agents, including FLU, are in progress, however (D. M. Dixon, B. Breslin, and C. R. Cooper, unpublished data).

A generally similar spectrum of antifungal resistance has been shown with *S. inflatum*.[24,25] These reports showed *S. inflatum* to be resistant to AmB, MCZ, KTZ, and FLU (MICs >18, 20, 12, and 80 μg/ml, respectively). Additionally, 10 of 14 isolates were resistant to ITRA (MIC > 10 μg/ml), but the remaining strains were susceptible to this antimycotic at MICs ≤ 0.018 μg/ml. It is very important to note, though, that these *in vitro* assays were performed as serial broth dilutions of the drugs. In the present authors' laboratories, different results were obtained for MCZ and KTZ when isolates of *S. inflatum* were assayed using a semisolid agar-based medium to dilute the drugs. It appears that the differences in the susceptibility of a fungus are associated with the assay method. Consequently, caution should be taken when trying to relate *in vitro* results to clinical treatment of infection.

7. SUMMARY

Nonmycetoma infections by *P. boydii* and those caused by the emerging pathogen *S. inflatum* should be regarded as potential life-threatening conditions, even in those patients having no predisposing conditions. Typically, though, most nonmycetoma infections caused by *P. boydii* are pulmonary in origin and frequently occur in immunosuppressed individuals or in those having prior lung disease. The mortality rate of *P. boydii* infections in humans, excluding mycetoma and otomycosis, is greater than 23%. Histopathologically, the fungus appears quite similar to *Aspergillus*. A positive, differential diagnosis can be made using fluorescent antibody staining of tissue sections. Although isolation of *P. boydii* from clinical specimens other than pulmonary sources would also indicate an active disease process, recovery of the fungus from sputum or lung tissue does not necessarily denote disease but may only be indicative of transient colonization that may or may not resolve without further treatment. In the latter case, prognostic evaluations can be made using serological techniques. Finally, infections caused by either *P. boydii* or *S. inflatum* appear most responsive to miconazole administered intravenously, surgical resection, or both.

The literature search for this contribution was completed in early 1990. Since that time, a large number of clinical and scientific reports have appeared in both English- and foreign-language journals concerning pseudallescheriasis, its biology, and its epidemiology. A summary of all these reports is not practical here. However, as an example of the rapidly growing recognition of this disease and its etiological agents, a brief literature search of only English-language journals from early 1990 through mid-1992 revealed at least 32 new cases of disease caused by *P. boydii* and at least 19 by *S. inflatum*. Two of the new cases attributed to *P. boydii* occurred in AIDS patients. Undoubtedly, more cases similar to these will be identified in the near future. In addition, a recent study by Guého and de Hoog (*J. Mycol. Med.* **118**:3–9, 1991) suggests that there may be two pathogenic variants of *P. boydii*: one responsible for pulmonary and disseminated disease, and another having a predilection for the nose and ear. Based upon DNA homology studies, these authors also suggest that *S. inflatum* is identical to *Lomentospora prolificans* and therefore should be properly described as *Scedosporium prolificans*. However, preliminary experiments conducted in our laboratory do not necessarily support this conclusion. Using cultures derived from the same type strains of *L. prolificans* and *S. inflatum* studied by Guého and de Hoog, whole-cell DNAs isolated from these two organisms possess different restriction fragment length polymorphisms and thereby indicate that the type strains are not identical.

ACKNOWLEDGMENTS. The authors wish to thank our colleagues in the Laboratories for Mycology and elsewhere for providing some of the unpublished results of studies mentioned in this work.

REFERENCES

1. Rippon, J. W., 1988, *Medical Mycology. The Pathogenic Fungi and the Pathogenic Actinomycetes*, 3rd ed., W. B. Saunders Co., Philadelphia.
2. Elder, B. L., and Roberts, G. D., 1986, Pseudallescheriasis, in: *Fungal Diseases of the Lung* (G. A. Sarosi and S. F. Davies, eds.), Grune and Stratton, Inc., Orlando, pp. 205–215.
3. Chandler, F. W., and Watts, J. C., 1987, *Pathologic Diagnosis of Fungal Infections*, ASCP Press, Chicago.
4. Travis, L. B., Roberts, G. D., and Wilson, W. R., 1985, Clinical significance of *Pseudallescheria boydii*: A review of 10 years' experience, *Mayo Clinic Proc.* **60**;531–537.
5. Pluss, J. L., and Opal, S. M., 1985, An additional case of pulmonary *Pseudallescheria boydii* improved with ketoconazole therapy, *Chest* **87**:843.
6. Anaissie, E., Bodey, G. P., Kantarjian, H., Ro, J., Vartivarian, S. E., Hopfer, R., Hoy, J., and Rolston, K., 1989, New spectrum of fungal infections in patients with cancer, *Rev. Infect. Dis.* **11**:369–378.
7. Schwartz, D. A., Amenta, P. S., and Finkelstein, S. D., 1989, Cerebral *Pseudallescheria boydii* infection: unique occurrence of fungus ball formation in the brain, *Clin. Neurol. Neurosurg.* **91**:79–84.
8. Guyotat, D., Piens, M. A., Bouvier, R., and Fiere, D., 1987, A case of disseminated *Scedosporium apiospermum* infection after bone marrow transplantation, *Mykosen* **30**:151–154.
9. Dworzack, D. L., Clark, R. B., Borkowski, W. J., Jr., Smith, D. L., Dykstra, M., Pugsley, M. P., Horowitz, E. A., Connolly, T. L., McKinney, D. L., Hostetler, M. K., Fitzgibbons, J. F., and Galant, M., 1989, *Pseudallescheria boydii* brain abscess: association with near drowning and

efficacy of high-dose, prolonged miconazole therapy in patients with multiple abscesses, *Medicine* **68:**218–224.
10. Berenguer, J., Diaz-Mediavilla, J., Urra, D., and Muñoz, P., 1990, Central nervous system infection caused by *Pseudallescheria boydii*: Case report and review, *Rev. Infect. Dis.* **11:**890–896.
11. Perez, R. E., Smith, M., McClendon, J., Kim, J., and Eugenio, N., 1988, *Pseudallescheria boydii* brain abscess. Complication of an intravenous catheter, *Amer. J. Med.* **84:**359–362.
12. Morgan, M. A., Wilson, W. R., Neel, H. B., III, and Roberts, G. D., 1984, Fungal sinusitis in healthy and immunocompromised individuals, *Amer. J. Clin. Pathol.* **82:**597–601.
13. Fessler, R. G., and Brown, F. D., 1989, Superior sagittal sinus infection with *Petriellidium boydii*: case report, *Neurosurgery* **24:**604–607.
14. Armin, A. R., Reddy, V. B., and Orfei, E., 1987, Fungal endocarditis caused by *Pseudallescheria (Petriellidium) boydii* in an intravenous drug user, *Texas Heart Inst. J.* **14:**321–324.
15. Nielsen, H. S., 1967, Effects of amphotericin B in vitro on perfect and imperfect strains of *Allescheria boydii*, *Appl. Microbiol.* **15:**86–91.
16. Kemp, H. B., Bedford, A. F., and Fincham, W. J., 1982, *Pseudallescheria boydii* infection of the knee: a case report, *Skel. Radiol.* **9:**114–117.
17. Sheftel, T. G., Mader, J. T., and Cierny, G., 1987, *Pseudallescheria boydii* soft tissue abscess, *Clin. Orthop.* **215:**212–216.
18. Stern, R. M., Zakov, Z. N., Meisler, D. M., Hall, G. S., and Martin, A., 1986, Endogenous *Pseudallescheria boydii* endophthalmitis, *Cleveland Clinic Quart.* **53:**197–203.
19. Caya, J. G., Farmer, S. G., Williams, G. A., Franson, T. R., Komorowski, R. A., and Kies, J. C., 1988, Bilateral *Pseudallescheria boydii* endophthalmitis in an immunocompromised patient, *Wis. Med. J.* **87:**11–14.
20. Fernandez-Guerrero, M. L., Ruiz-Barnes, P., and Ales, J. M., 1987, Postcraniotomy mycetoma of the scalp and osteomyelitis due to *Pseudallescheria boydii*, *J. Infect. Dis.* **156:**855.
21. Conti-Diaz, A., 1980, Micetomas y procesos premicetomatoses en el Uruguay, *Mycopathologia* **72:**59–64.
22. Meyer, E., and Herrold, R. D., 1961, *Allescheria boydii* isolated from a patient with chronic prostatitis, *Amer. J. Clin. Pathol.* **35:**155–159.
23. Malloch, D., and Salkin, I. F., 1984, A new species of *Scedosporium* associated with osteomyelitis in humans, *Mycotaxon* **21:**247–255.
24. Salkin, I. F., McGinnis, M. R., Dykstra, M. J., and Rinaldi, M. G., 1988, *Scedosporium inflatum*, an emerging pathogen, *J. Clin. Microbiol.* **26:**498–503.
25. Wilson, C. M., O'Rourke, E. J., McGinnis, M. R., and Salkin, I. F., 1990, *Scedosporium inflatum*: Clinical spectrum of a newly recognized pathogen, *J. Infect. Dis.* **161:**102–107.
26. Summerbell, R. C., Krajden, S., and Kane, J., 1989, Potted plants in hospitals as reservoirs of pathogenic fungi, *Mycopathologia* **106:**13–22.
27. Emmons, C. W., 1944, *Allescheria boydii* and *Monosporium apiospermum*, *Mycologia* **36:**188–193.
28. McGinnis, M. R., 1980, *Laboratory Handbook of Medical Mycology*, Academic Press, New York.
29. McGinnis, M. R., Padhye, A., and Ajello, L., 1982, *Pseudallescheria* Negroni et Fischer, 1943 and its later synonym *Petriellidium* Mallock, 1970, *Mycotaxon* **14:**94–103.
30. Shear, C. L., 1922, Life history of an undescribed ascomycete isolated from granular mycetoma of man, *Mycologia* **14:**239–243.
31. Käufer, I., and Weber, A., 1977, *Graphium fructicola* als Ursache einer Systemmykose beim Hund, *Mykosen* **20:**39–46.
32. Dykstra, M. J., Salkin, I. F., and McGinnis, M. R., 1989, An ultrastructural comparison of conidiogenesis in *Scedosporium apiospermum*, *Scedosporium inflatum* and *Scopulariopsis brumptii*, *Mycologia* **81:**896–904.
33. Wolf, F. T., Bryden, R. R., and MacLaren, J. A., 1950, The nutrition of *Monosporium apiospermum*, *Mycologia* **42:**233–241.
34. Cazin, J., and Decker, D. W., 1964, Carbohydrate nutrition and sporulation of *Allescheria boydii*, *J. Bacteriol.* **88:**1624–1628.

35. Gordon, M. A., 1957, Nutrition and sporulation of *Allescheria boydii*, *J. Bacteriol.* **73:**199–205.
36. Mohr, J. A., and Muchmore, H. G., 1968, Inhibition of *Allescheria boydii* by naturally occurring estrogen, in: *Antimicrobial Agents Chemotherapy* (G. L. Hobby, ed.), American Society for Microbiology, Bethesda, MD, pp. 423–424.
37. Mohr, J. A., and Muchmore, H. G., 1968, Maduromycosis due to *Allescheria boydii*, *J. Amer. Med. Assoc.* **204:**335–336.
38. Lupan, D. M., and Cazin, J., 1973, Pathogenicity of *Allescheria boydii* for mice, *Infect. Immun.* **8:**743–751.
39. Bell, R. G., 1978, Comparative virulence and immunodiffusion analysis of *Petriellidium boydii* (Shear) Malloch strains isolated from feedlot manure and a human mycetoma, *Can. J. Microbiol.* **24:**856–886.
40. Mackenzie, D. W. R., 1983, Serodiagnosis, in: *Fungi Pathogenic for Humans and Animals, Part B, Pathogenicity and Detection: I* (D. H. Howard, ed.), Marcel Dekker, Inc., New York, pp. 121–218.
41. Huppert, M., 1983, Antigens used for measuring immunological reactivity, in: *Fungi Pathogenic for Humans and Animals, Part B, Pathogenicity and Detection: I* (D. H. Howard, ed.), Marcel Dekker, Inc., New York, pp. 218–302.
42. Lupan, D. M., and Cazin, J., Jr., 1979, Humoral response to experimental petriellidiosis, *Infect. Immun.* **24:**843–850.
43. Schaffner, A., Davis, C. E., Schaffner, T., Markert, M., Douglass, H., and Braude, A. I., 1986, In-vitro susceptibility of fungi to killing by neutrophil granulocytes discriminates between primary pathogenicity and opportunism, *J. Clin. Invest.* **78:**511–524.
44. Lupan, D. M., and Cazin, J., Jr., 1976, Serological diagnosis of petriellidiosis (allescheriosis). I. Isolation and characterization of soluble antigens from *Allescheria boydii* and *Monosporium apiospermum*, *Mycopathologia* **58:**31–38.
45. Lupan, D. M., and Cazin, J., Jr., 1977, Serological diagnosis of petriellidiosis (allescheriosis). II. Indirect (passive) hemagglutination assay for antibody to polysaccharide antigens of *Petriellidium (Allescheria) boydii* and *Monosporium apiospermum*, *Mycopathologia* **58:**31–38.
46. Seale, J. P., and Hudson, J. A., 1985, Successful medical treatment of pulmonary petriellidiosis, *South. Med. J.* **78:**473–476.
47. Hainer, J. W., Ostrow, J. H., and Mackenzie, D. W. R., 1974, Pulmonary monosporosis: report of a case with precipitating antibody, *Chest* **66:**601–603.
48. Kathuria, S. K., and Rippon, Jr., 1982, Non-aspergillus aspergilloma, *Amer. J. Clin. Pathol.* **78:** 870–873.
49. Rippon, J. W., and Cramichael, J. W., 1976, Petriellidiosis (allescheriasis): four unusual cases and review of literature, *Mycopathologia* **58:**117–124.
50. Morace, G., and Polonelli, L., 1981, Exoantigen test for identification of *Petriellidium boydii* cultures, *J. Clin. Microbiol.* **14:**237–240.
51. Jackson, J. A., Kaplan, W., Kaufman, L., and Standard, P., 1983, Development of fluorescent-antibody reagents for demonstration of *Pseudallescheria boydii* in tissues, *J. Clin. Microbiol.* **18:** 668–673.
52. Lutwick, L. I., Galgiani, J. N., Johnson, R. H., and Stevens, D. A., 1976, Visceral fungal infection due to *Petriellidium boydii* (*Allescheria boydii*). In vitro drug sensitivity studies, *Amer. J. Med.* **61:**632–640.
53. Galgiani, J. N., Stevens, D. A., Graybill, J. R., Stevens, D. L., Tillinghast, A. J., and Levine, H. B., 1984, *Pseudallescheria boydii* infections treated with ketoconazole. Clinical evaluations of seven patients and in vitro susceptibility results, *Chest* **86:**219–224.
54. Sarosi, G. A., Bates, J. H., Bradsher, R. W., Davis, S. F., Dismukes, W. E., Geroge, R. B., Graybill, J. R., and Utz, J. P., 1988, Chemotherapy of pulmonary mycoses, *Amer. Rev. Resp. Dis.* **134:**1078–1081.
55. Viviani, M. A., Tortorano, A. M., Woenstenborghs, R., and Cauwenbergh, G., 1987, Experience with itraconazole in deep mycoses in Northern Italy, *Mykosen* **30:**233–234, 237–244.

15

Aspergillosis

JUDITH C. RHODES

1. DESCRIPTION OF THE DISEASE

Aspergillosis has been used to describe a group of disparate diseases that runs the gamut of clinical presentations from allergy to colonization to invasive disease to intoxication and that has in common being caused by members of the genus *Aspergillus* or their metabolites.

1.1. Portal of Entry

In allergic and invasive aspergillosis and in aspergilloma, the organism enters the body by inhalation of conidia of *Aspergillus*, an ubiquitous fungus. The conidia are easily airborne,[1] round to oval, less than 8 μm in diameter,[2] and, thereby, able to reach to alveoli of the lungs. Rarely, invasive infection is acquired by direct implantation of the fungus.[3,4] In the case of intoxication, ingestion of foodstuffs contaminated with preformed toxic metabolites is the mode of acquisition.

1.2. Clinical Manifestations

Although the clinical manifestations of allergic aspergillosis, aspergilloma, and invasive aspergillosis are described in separate sections, there is

JUDITH C. RHODES • Department of Pathology and Laboratory Medicine, University of Cincinnati College of Medicine, Cincinnati, Ohio 45267-0529.

Fungal Infections and Immune Responses, edited by Juneann W. Murphy *et al.* Plenum Press, New York, 1993.

often considerable overlap in the classification.[5,6] For example, some patients with aspergilloma may also be allergic to *Aspergillus* antigens, lesions in some cases of invasive aspergillosis may cavitate and form an aspergilloma, or a patient with allergic bronchopulmonary aspergillosis may develop disseminated disease.[7] This overlap in clinical presentation seen in patients with aspergillosis is predicated on the spectrum of response of the host's defense system.

1.2.1. Clinical Manifestations of Allergic Aspergillosis

Allergic aspergillosis may manifest itself as extrinsic allergic alveolitis or as allergic bronchopulmonary aspergillosis (ABPA). In the former case, the affected individual is nonatopic and responds to repeated exposure to organic dust containing high concentrations of conidia and hyphal fragments of *Aspergillus* with the development of hypersensitivity pneumonitis, as in farmer's lung. Symptoms of cough, fever, chills, malaise, and dyspnea usually occur four or more hours after exposure.[8] ABPA is a syndrome that is primarily diagnosed in young (less than 35 years),[9] atopic individuals who display the following constellation of signs and symptoms: asthma with current or history of pulmonary infiltrates, eosinophilia, *Aspergillus* serum precipitins, elevated serum IgE and IgE antibodies to *Aspergillus*, type I skin reaction to *Aspergillus* antigens, and central bronchiectasis.[10] Expectoration of brownish mucous plugs filled with fungal hyphae is common (Fig. 1). Late stages of ABPA are characterized by irreversible obstructive lung damage due to fibrosis. Staging of ABPA into acute, remission, recurrent exacerbation, corticosteroid-dependent asthma, and fibrotic stages has been proposed as an aid to managing these patients.[11] Recently, allergic *Aspergillus* sinusitis, a form of chronic sinusitis in which thick, green inspissated mucus filled with eosinophils and Charcot-Leyden crystals is seen, has been recognized as a separate clinical entity.[12]

1.2.2. Clinical Manifestations of Aspergilloma

Aspergilloma, or fungus ball, most commonly represents colonization of a cavity by a member of the genus *Aspergillus*. Preexisting pulmonary cavities of diverse origin, frequently tuberculosis, may be colonized by the fungus.[13] Also, cavitation may occur as a late sequela of necrosis or infarction in invasive aspergillosis, often in conjunction with the return of granulocytopoiesis.[14] The most common symptom of aspergilloma is intermittent hemoptysis; although the hemoptysis is usually not extensive, fatal episodes have been reported. Additionally, some patients may experience fever, malaise, and weight loss. Although the lung apices have been reported as the most common site for aspergillomas,[13] this may simply reflect that old tuberculous cavities are most commonly colonized. Aspergillomas have also been reported in the sinuses and nose.[15]

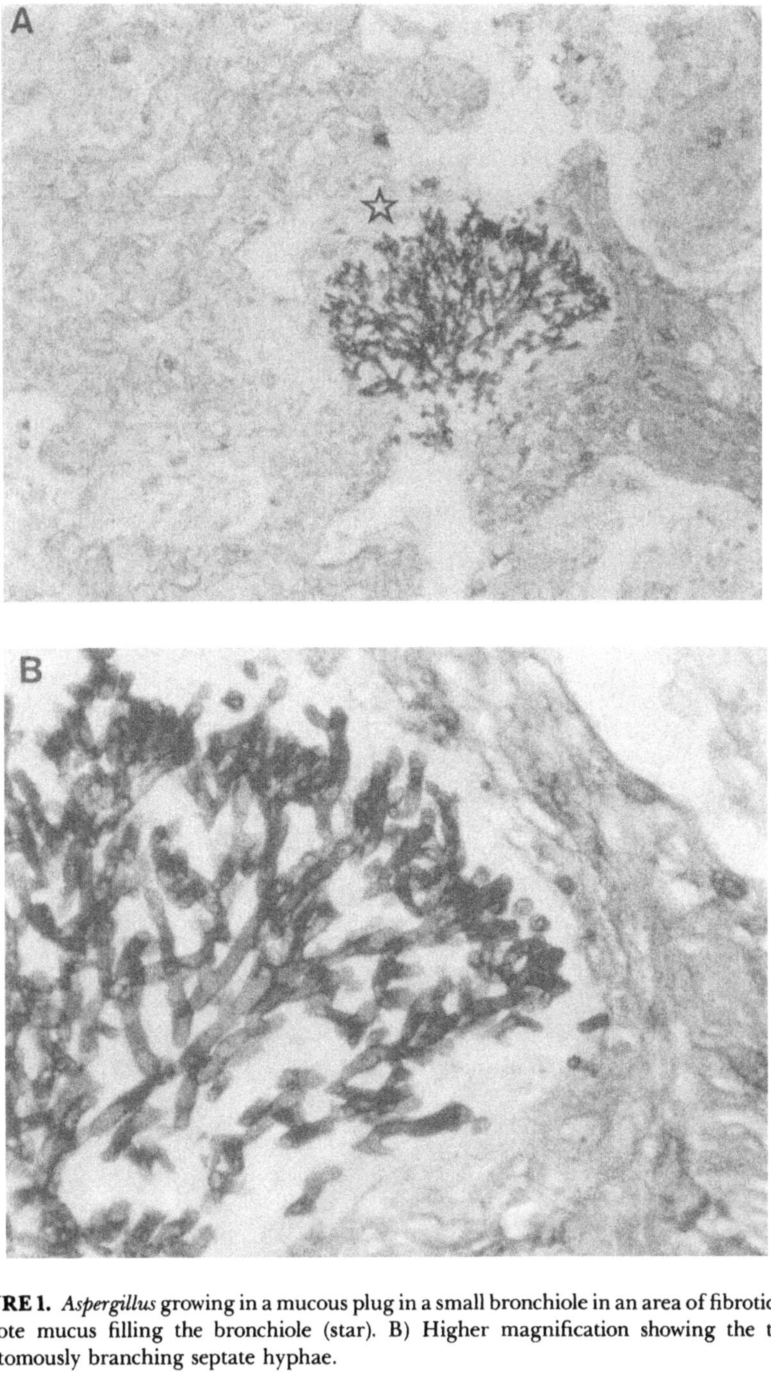

FIGURE 1. *Aspergillus* growing in a mucous plug in a small bronchiole in an area of fibrotic lung. A) Note mucus filling the bronchiole (star). B) Higher magnification showing the typical dichotomously branching septate hyphae.

1.2.3. Clinical Manifestations of Invasive Aspergillosis

Invasive aspergillosis is an aggressive, life-threatening disease of immunocompromised patients. Granulocytopenia and administration of broad spectrum antibiotics and corticosteroids have been shown to be major predisposing conditions in this form of aspergillosis.[16–18] Clinically, patients have bronchopneumonia, often accompanied by pleuritic pain and persistent fever, which progresses through consolidation or cavitation.[6] The histopathologic hallmark of invasive aspergillosis is angioinvasion (Fig. 2); the septate, dichotomously branching hyphae penetrate through the tissue without regard for anatomic barriers and enter vessels. This penetration of blood vessels may lead to infarction, as well as being a mode of dissemination for the fungus.[19] Organs from the brain to the kidney to the bowel may be targets of dissemination,[6] and symptoms of dissemination are usually referable to ischemic necrosis of the target organ. In some neutropenic patients, an invasive form of *Aspergillus* sinusitis has been reported, which clinically may mimic rhinocerebral mucormycosis.[20] Finally, chronic necrotizing or "semi-invasive" aspergillosis is a recently recognized syndrome seen in normal to

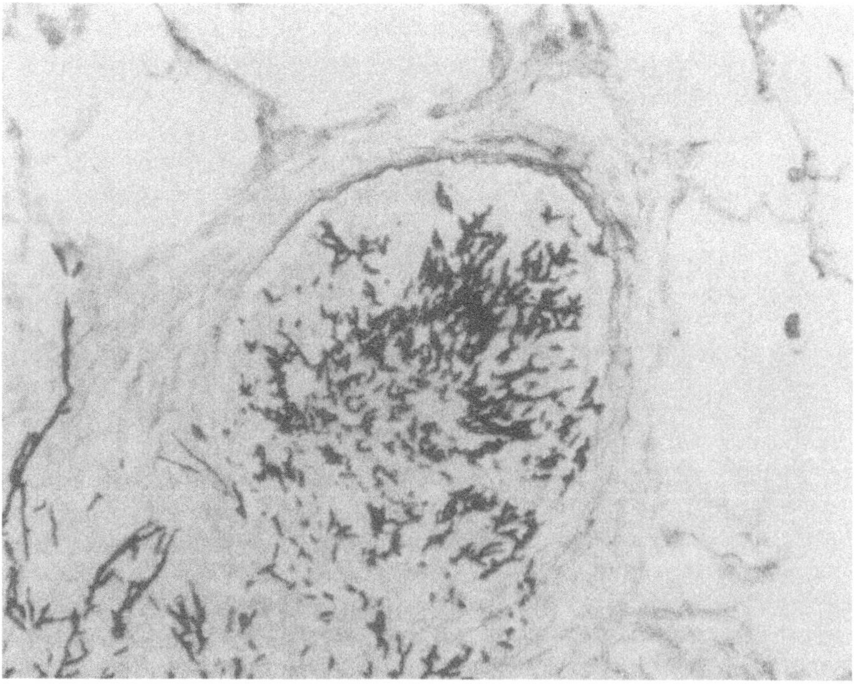

FIGURE 2. Small vessel in the lung showing invasion by hyphae of *A. fumigatus*.

moderately immunocompromised patients.[21,22] The clinical features of this form of aspergillosis are those of a chronic respiratory illness accompanied by fever and weight loss. Long-standing pulmonary infiltrates may cavitate over a period of months and form an aspergilloma.

1.2.4. *Clinical Manifestations of Aspergillus-Related Intoxication*

Intoxication by metabolites of *Aspergillus* is most commonly associated with the ingestion of grain contaminated with aflatoxin, even though many other mycotoxins are produced by members of this genus. Aflatoxicosis is well described in animal models, in which the toxin is one of the most potent hepatic carcinogens known. The occurrence of acute aflatoxicosis in man is less well documented. In a 1974 outbreak in India, almost 400 people were affected and 106 of those died. Jaundice was the presenting feature; in some patients, the course was mild and self-limited, but in others ascites appeared, followed by lower extremity edema and sudden death secondary to massive gastrointestinal hemorrhage.[23] Evidence for chronic effects of aflatoxin ingestion have been reported more commonly. For example, in Transkei a positive correlation between aflatoxin ingestion and development of primary hepatic carcinoma has been reported in goldminers,[24] and there is evidence for antibody to aflatoxin epitopes in exposed populations in Kenya.[25] Finally, evidence that dietary exposure to mycotoxins may alter immune function suggests that the potential for these compounds to alter human health may be far greater than recognized previously.[26]

2. ETIOLOGY

Aspergillus is a very large genus containing over 200 species.[1] Only a handful of these species, however, have been associated with disease, and, of those, over 95% of all infections can be accounted for by three species.

2.1. Description of the Genus

The genus is characterized by septate hyphae from which arise nonseptate conidiophores; although these conidiophores may be smooth or rough, hyaline or pigmented, they all terminate in a swollen vesicle. One or two rows of sterigmata are formed on the surface of the vesicle. Chains of round to oval, variously pigmented conidia are produced from the tips of the sterigmata. The shape of the vesicle and the arrangement of the sterigmata thereon determine the shape of the conidial head, from columnar to radiate, a key feature in speciating members of the genus (Fig. 3). Finally, the color of the conidia is the determining factor for the color of the colony. Following growth on appropriate medium, observation of these macroscopic and mi-

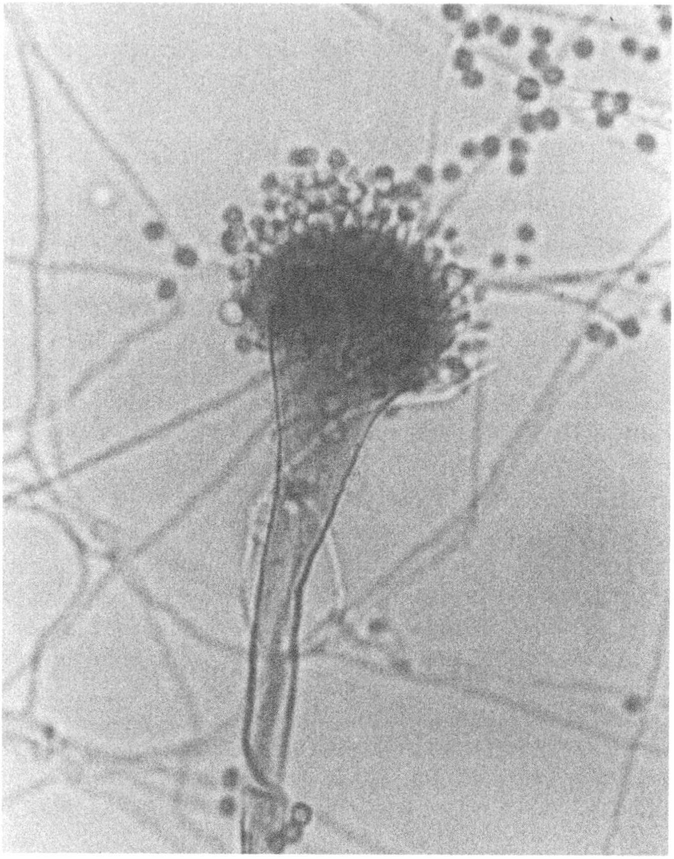

FIGURE 3. Conidial head of *A. fumigatus* illustrating the clavate vesicle with a single row of sterigmata on the upper half resulting in a columnar arrangement of conidial chains.

croscopic features is usually sufficient to identify the organism, using one of several published keys to the genus.[27,28]

2.2. Description of Species Associated with the Disease

Of the documented species of *Aspergillus*, *A. fumigatus* causes the large majority of cases of both invasive and noninvasive aspergillosis. Indeed, the allergic forms of disease appear to be caused almost exclusively by this organism.[9] Both aspergilloma and invasive aspergillosis are also caused by *A. flavus* and *A. niger*.[29] Other organisms that have been reported from cases of aspergillosis include *A. nidulans*, *A. terreus*, *A. carneus*, *A. candidus*, *A. oryzae*, *A. ustus*, *A. avenaceus*, *A. versicolor*, and *A. deflectus*.[1,30]

The restricted list of pathogenic species from the genus suggests that these organisms may share some properties not found in other species that confer on them an intrinsic pathogenic advantage. One obvious feature that these human pathogens must share is the ability to grow efficiently at 37°C.[31] The production of proteinolytic enzymes is another property or virulence mechanism that has been investigated in the pathogenesis of invasive disease and allergic disease. The ability to produce elastase has been correlated with mouse virulence in strains of *A. fumigatus*.[32] Likewise, isolates from cases of invasive aspergillosis in humans all showed the ability to digest elastin, regardless of the species.[33] Finally, antigens with proteinase activity have been isolated from *Aspergillus* and have been shown to react with the sera from patients with allergic aspergillosis and aspergilloma.[34,35]

Intoxications caused by aflatoxin, are due most commonly to *Aspergillus parasiticus*, *A. flavus*, and other toxigenic members of the *flavus–oryzae* group.

3. EPIDEMIOLOGY

Members of the genus *Aspergillus* are ubiquitous fungi found throughout the world. It has been estimated that one mature conidiophore of *A. fumigatus* can produce 50,000 conidia.[36] It is perhaps not surprising that these fungi are among the most frequently recovered in surveys of airborne fungi, both indoors and outdoors.

3.1. General Epidemiology

Aspergillus species grow well on virtually any organic debris. Therefore, they are found naturally in leaves, grains, soil, and composts.[37] Compost is a particularly rich source for the thermotolerant/thermophilic species; *A. fumigatus* is often the predominant fungal member of the compost microbiota, with more than 10^8 conidia per gram of compost material.[2] Likewise, numbers of airborne conidia in such places as hay barns have been estimated to be as high as 10^9 per cubic meter.[36] These numbers are exceptional, however, and the number of conidia in outdoor air probably rarely exceeds 200 per cubic meter.[38] Interestingly, indoor air usually contained larger numbers of *Aspergillus* conidia than outdoor air sampled over the same period.[39] This has led hospital epidemiologists to advocate the use of high efficiency filtration units to reduce the numbers of airborne particles in the rooms of patients at risk to undetectable levels. In general, however, with the exception of those individuals with occupational exposure (e.g., sanitation workers involved in the composting of sewage sludge), everyday exposure to the conidia of *Aspergillus* may be of little concern to the majority of the population.

3.2. Hospital Epidemiology

Over the past two decades, there have been several reports of outbreaks of nosocomial aspergillosis.[39] In most of these, there was an association with new construction, reconstruction, or renovation of the hospital during or immediately preceding the outbreak. Under normal circumstances, numbers of *Aspergillus* conidia in hospital air are quite low.[40] These incidents have served as one of the driving forces for the design, installation, and use of high efficiency air filtration of rooms or wards for those patients who must undergo prolonged periods of severe neutropenia. The utility of these protective environments in preventing nosocomial aspergillosis, however, is not accepted universally.[41]

4. HOST DEFENSE

Host defense to invasive aspergillosis is virtually exclusively the province of innate mechanisms. In the allergic forms of aspergillosis, the acquired responses appear to contribute more to the pathology than to the control of disease.

4.1. Innate Mechanisms

Numerous retrospective and prospective reviews of invasive aspergillosis were able to analyze common factors that were responsible for placing patients at risk for these infections. Granulocytopenia and administration of broad spectrum antibiotics and corticosteroids were consistently identified as key predisposing factors.[16–18] Likewise, the recognition that patients with chronic granulomatous disease, in which their professional phagocytic cells fail to undergo the respiratory burst, are susceptible to invasive aspergillosis added further weight to the concept that the presence of normally functioning granulocytes was the key factor in defense against invasive disease.[42] It was subsequently shown in a murine model that both the granulocyte and monocyte function had to be ablated for invasive infection to develop.[43] Currently, bronchoalveolar macrophages are believed to serve on the first line of defense and to be responsible for the ingestion and killing of inhaled conidia.[44,45] Those conidia escaping the macrophages and going on to germinate and form hyphae are the target of attack by the granulocytes, which adhere to the hyphae, degranulate, and kill the fungus.[46] Only in individuals in which this dyad of innate defense is perturbed is acquisition of the invasive form of aspergillosis a threat.

4.2. Immune Response

In colonizing and in invasive forms of aspergillosis, study of the immune response has been targeted toward detecting that response as an aid to determining the diagnosis or prognosis of patients. In allergic aspergillosis, studies have sought to explain the pathology of the disease, which appears to be caused primarily by that very response. In all studies on the immune response to *Aspergillus* antigens, progress has been hampered by the lack of common, reference antigens that are capable of detecting responses to all common species and by which results obtained in multiple centers may be compared.

4.2.1. Humoral Immune Response in Aspergillosis

4.2.1a. Humoral Response in Allergic Aspergillosis. Patients with extrinsic allergic alveolitis or hypersensitivity pneumonitis develop IgG precipitin antibodies, and the pulmonary pathology appears to be mediated by immune complex deposition, perhaps in conjunction with cell-mediated mechanisms (see Section 4.2.2).[9] In contrast, IgE appears to play the major role in the pathophysiology of ABPA. The patients have elevated total IgE, which rises and falls with exacerbations and remission of clinical symptoms.[11] Specific anti-*Aspergillus* IgE is also demonstrable in ABPA patients. Type I immediate hypersensitivity responses, the IgE-mediated wheal and flare, are present in virtually all ABPA patients challenged intradermally with *Aspergillus* antigens. This type I response is believed to be responsible for the bronchospasm component of the disease. IgG anti-*Aspergillus* precipitin antibodies are also produced by ABPA patients, with almost 100% having positive serological tests for these antibodies. Approximately 50% of patients in one study had positive type III skin responses, the IgG-mediated hemorrhagic or Arthus type reaction, when tested.[10] Pulmonary immune complex deposition, secondary to elevated IgG antibodies, may account for much of the inflammatory component seen during active phases of ABPA.

4.2.1b. Humoral Response in Aspergilloma. Patients with aspergilloma, or fungus ball, usually mount a brisk IgG response to *Aspergillus* antigens, judging by the multiple precipitin bands formed on immunodiffusion to extracts of the infecting fungus.[47] The response does not appear to provide any protection to the patient, however. In cases of aspergilloma in which the fungus ball has formed secondarily to cavitation of an invasive lesion, the ability of patients to produce antibodies is governed more by their degree of immunosuppression than by the classification of disease. Additionally, some patients with aspergillomas that have formed in preexisting cavities, show a marked allergic component in their illnesses.[48]

4.2.1c. Humoral Response in Invasive Aspergillosis. Because of their profound degrees of immunosuppression, patients with invasive aspergillosis mount little to no humoral response to the fungus when the standard serological tests are used.[47] When a sensitive immunoblotting procedure was used, however, sera from all 38 patients tested showed antibodies to an immunodominant 58 kilodalton antigen from *Aspergillus fumigatus*.[49] Antibody production demonstrated by this or other techniques, however, has not been shown to offer any real benefit to the patient. In animal models, there is no evidence for protection by antibodies in invasive aspergillosis.[50]

4.2.2. *Cellular Response in Aspergillosis*

4.2.2a. Cellular Response in Allergic Aspergillosis. The presence of bronchocentric granulomatosis as a pathologic hallmark of ABPA argues that there must be a cell-mediated component in the responses of these patients to *Aspergillus*.[5] Lymphocyte proliferation in response to antigens of *A. fumigatus* has been demonstrated in some ABPA patients.[9] Likewise, fractionation of antigenic extracts of *A. fumigatus* yielded components that elicited positive delayed type (type IV) skin reactivity in, and proliferation of, lymphocytes from sensitized guinea pigs.[51] One fraction actually inhibited lymphocyte proliferation not only to the *Aspergillus* antigen but also to subsequent stimulation by concanavalin A.[52] A human correlate to this has not been described.

4.2.2b. Cellular Response in Aspergilloma. With the exception of those aspergilloma patients who have an allergic response to their infecting organism, there is little evidence to support the presence of a cell-mediated response in these patients. In general, the pathology of aspergillomas is one in which inflammatory response, either acute or chronic, is minimal. The host defense against this form of infection has not been well studied.

4.2.2c. Cellular Response in Invasive Aspergillosis. The role of T cells in invasive disease has been investigated in experimental systems, with little consensus as to its importance. For example, nu/nu mice fare no worse than their nu/+ littermates when infected with conidia of *A. fumigatus*. Nu/+ mice can be immunized protectively against the fungus, however, and the protective immunity can be transferred to nu/nu recipients with adherent peritoneal cells from the immunized nu/+ mice.[53] This suggests a role for T cells and macrophages in the acquisition of immunity to aspergillosis, but the issue is still open to investigation. In man, reviews of cases of invasive aspergillosis have not suggested that T-cell defects were primary predisposing factors for disease but rather than granulocytopenia is the risk factor most commonly associated with infection.[18] Additional recent evidence on the incidence of invasive aspergillosis in patients with the acquired immunodeficiency syndrome also suggests that, even in the face of profound loss of CD4+ T cells, this fungal infection rarely occurs, except when patients become neutropenic, secondary to azidothymidine therapy.[54]

4.2.3. Immune Regulation in Aspergillosis

Regulation of the immune response to aspergillosis has been little studied. Of interest are the description of metabolic products of the fungus that are immunomodulatory *in vitro*. One of the best characterized of these is gliotoxin, produced *in vitro* by *A. fumigatus*.[55] Gliotoxin is a potent inhibitor of macrophage phagocytosis and also appears able to inhibit the ability of spleen cells to induce major histocompatibility complex restricted cytotoxic cells. Additionally, *A. fumigatus* allergen I, which is involved in IgE binding and elicitation of the IgE-mediated response, is also a cytotoxin of the mitogillin family.[56] Finally, *A. fumigatus* also produces phospholipid compounds that are active in inhibiting the activation of complement via the alternative pathway.[57,58] All these factors have the potential to influence both the innate and immune responses that provide adequate host defense to *Aspergillus* in the normal host. It is not currently known whether gliotoxin or the complement inhibitor, however, is produced *in vivo* in biologically significant quantities.

5. IMMUNODIAGNOSIS AND PROGNOSIS

Many potential systems for the immunodiagnosis of aspergillosis have been described in the literature. Those based on detection of antibody to the organism have been very successful in allergic aspergillosis and aspergilloma, and those used for the detection of fungal antigen have great potential for the diagnosis of invasive aspergillosis. Assays for aspects of cell-mediated immune function have been little used in these infections.

5.1. Serologic Tests for Antibodies to *Aspergillus*

Detection of specific precipitating antibodies to *Aspergillus* by double immunodiffusion has been the most frequently used serologic test for the diagnosis of aspergillosis.[59] The test is available commercially from a number of sources and is technically simple to perform and to interpret.[47] The presence of one or more precipitin band is one of the diagnostic criteria outlined for the diagnosis of allergic bronchopulmonary aspergillosis. Published studies would suggest that 70–100% of patients with ABPA are positive for IgG precipitating antibodies directed toward *Aspergillus*, depending on the antigen used and whether or not serum was concentrated prior to testing.[9] Additionally, rates of positivity in patients with extrinsic allergic alveolitis are also very high. Precipitins are present in 98–99% of aspergilloma patients, and multiple bands, often three or more, are routinely present.[13] In patients with invasive aspergillosis, however, the experience with detecting precipitating antibody response has been disappointing.[59] Immunoblotting is a sensitive technique that has been used to detect antibodies in sera from patients with invasive aspergillosis and patients with ABPA.[49,60] Although the correlation between the sizes of the recognized antigens has not been exact, it has great research potential for determining which antigens may be the most appropriate ones for subsequent testing. This information, coupled with a format such as ELISA, may allow wider availability of a more sensitive and specific test for aspergillosis.[61]

5.2. Serologic Tests for Detection of *Aspergillus* Antigen

Antibody production in the immunocompromised host with invasive aspergillosis may be difficult to detect. Therefore, methods have been sought for diagnosing this infection that would rely on the measurement of a product of fungal origin and thereby be independent of the host's ability to respond. *Aspergillus* galactomannan circulates during infection and appears in the urine, presumably after clearance by a receptor-mediated process by Kupffer cells in the liver.[62,63] In addition to galactomannan, at least seven other *Aspergillus*-related antigens have been detected in the urine of patients with invasive disease by immunoblotting.[64] Investigators have developed radioimmunoassay, ELISA, and immunoblot procedures to detect antigen, either free or in immune complexes in serum, bronchoalveolar fluid, or urine.[64–70] In two of the largest studies on patient samples, a radioimmunoassay used to measure antigenemia showed 74% sensitivity and 90% specificity, and either of two ELISA procedures used to measure antigenemia or antigenuria showed greater than 95% sensitivity and specificity.[64,71] In the latter study, significantly more urine samples (20% and 44%) than serum samples (14% and 18%) were positive; this corroborates previous results obtained in experimentally infected animals.[62,64,70] Although antigen detection for the early diagnosis of invasive aspergillosis holds great promise for the future, determination of their full diagnostic potential awaits their widespread availability.

Another approach to the diagnosis of aspergillosis that is independent of host immunological response is the detection of fungal metabolites in the body fluids of patients. The test for D-arabinitol in candidiasis is an example of such a test. In an analogous fashion, the presence of high levels of D-mannitol in the serum and tissues of experimentally infected animals has been shown to correlate with the presence and extent of invasive aspergillosis.[72] Likewise, the presence of oxalic acid in bronchoalveolar lavage fluid has been proposed as a presumptive marker for aspergillosis.[73]

5.3. Prognostic Markers in Aspergillosis

In allergic forms of aspergillosis, antibody titers may be of some prognostic value. It has been suggested that precipitins may decrease with successful corticosteroid therapy. IgE levels, however, are the most useful parameter to follow in the treatment of ABPA patients.[74] The total IgE often rises prior to a clinical relapse, and duration of therapy may be judged based on the fall in IgE level.

Levels of antigenemia and antigenuria may correlate with the clinical course in invasive aspergillosis. Data in animal studies and some from patient studies would suggest that antigen levels rise as the clinical condition worsens.[62,70,71] Also, some studies have shown that efficacious antifungal therapy causes antigen levels to fall;[75] however, this has not been confirmed in all studies.[64] At this time, it is probably premature to derive any universal correlates for antigen levels in this disease.

6. PREVENTION AND THERAPY

Prevention of aspergillosis involves, for the most part, avoidance of the etiological agent. Therapy of allergic forms of disease depends on inhibiting the immune response responsible for the pathogenesis, whereas treating invasive disease requires aggressive antifungal therapy.

6.1. Prevention of Aspergillosis

Aspergillosis is caused by a ubiquitous fungus; therefore, prevention of disease is not accomplished easily. Because of the forms of the disease and the way in which the normal host defends against them, vaccination is not a logical or practical approach. Therefore, what can be done to intervene in the cycle of infection?

Avoidance of the etiological agents, the pathogenic species of *Aspergillus*, is probably the major factor in preventing disease. Patients with allergic aspergillosis should not visit or work in hay barns, sludge composting plants, or other places known to harbor high numbers of conidia.[2,37] Indeed, those places should be off limits to patients with other conditions that might predispose them to infection, such as pulmonary cavitary disease or a condition that could cause immunosuppression. For those who must work in known risk areas, use of an appropriate respirator can help to markedly decrease the number of conidia respired.[2]

Patients who are extremely granulocytopenic, secondary to chemotherapy or bone marrow transplantation, are profoundly susceptible to invasive aspergillosis. As discussed in Section 3.2, this fact has prompted hospital epidemiologists and others to suggest that special rooms or wards with high efficiency air filtration units be constructed for use with these patients. Although the institution of such measures has been reported to be successful in reducing the incidence of invasive disease in high-risk patients, data from other centers has shown no change in incidence since the provision of a protective environment.[39,4,1]

6.2. Therapy of Aspergillosis

Therapeutic approaches to the clinical forms of aspergillosis vary according to our understanding of their pathogenesis.

6.2.1. *Therapy of Allergic Aspergillosis*

Treatment of this form of aspergillosis is targeted toward suppression of the immunological response to the fungus rather than to the fungus itself. Attempts to eradicate the organism, usually through the aerosolization of an antifungal agent, have been largely abandoned because of their poor initial efficacy and high relapse rates.[9] Corticosteroids are now the therapeutic choice for these patients. Criteria for the administration of corticosteroids in ABPA have been well described and are largely based on the stage of the patient's illness and the total serum IgE levels.[74] Whether early diagnosis and treatment of these patients may prevent progression to central bronchiectasis is not known, but that hypothesis is suggested by initial patient data on prevention of fibrosis.

6.2.2. *Therapy of Aspergilloma*

Aspergilloma is the form of aspergillosis in which the therapeutic choices are the widest and the guidelines the least well defined.[13] The options range from conservative medical management without direct intervention to surgical resection with or without concomitant antifungal therapy. The primary difficulty in administering antifungal agents is obtaining adequate levels of the agent in the fungus ball. Systemic therapy with amphotericin B or itraconazole may be warranted in the subset of patients with aspergilloma with chronic necrotizing aspergillosis.[76] In others, local treatment via the bronchial tree of percutaneous instillation has been recommended as the best way to get adequate drug levels at the site.[13] Theoretically, surgery is the best treatment, especially in cases complicated by significant hemoptysis, but it is often not feasible. Additionally, the mortality rate and complication rate may be unacceptably high. Medical management without direct intervention may prove successful in that approximately 10% of aspergilloma patients experience spontaneous cure.

6.2.3. *Therapy of Invasive Aspergillosis*

Successful therapy of invasive aspergillosis in the granulocytopenic patient is critically dependent on the return of the granulocytes themselves. Antifungal therapy may function only to extend the time available to the patient for bone marrow regeneration with subsequent production of circulating granulocytes. The primary weapon in the physician's arsenal to treat this entity is amphotericin B.[77] Although highly toxic, it is still the best antifungal available for fungicidal activity. There are many new options on the horizon for treatment of these patients. One of these is liposome encapsulated amphotericin B. This preparation causes significantly less toxicity than the parent formulation while achieving superior tissue levels of drug. An early report on the response of invasive aspergillosis to liposome encapsulated amphotericin B reported five complete responses out of seven patients treated.[77] The azole antifungal itraconazole has been reported to be both inhibitory and fungicidal *in vitro* against *Aspergillus* species. As experience accumulates on the use of this agent in patients with invasive disease, as well as other forms of aspergillosis, the data appear quite promising. Response rates of 12 of 15 patients, including 10 of 12 immunocompromised patients in one study and 15 of 21 in another, are representative of the promise that this drug, and other of the new azoles, may hold for treating this problematic infection.[75,78,79]

ACKNOWLEDGMENTS. The technical assistance of J. Card in preparing the illustrations is gratefully acknowledged. This work was supported in part by grants AI26145 and AI82392 from National Institutes of Health.

REFERENCES

1. Kwon-Chung, K. J., 1988, *Aspergillus*: Diagnosis and description of the genus, in: *Aspergillus and Aspergillosis* (H. V. Bossche, D. W. R. Mackenzie, and G. Cauwenbergh, eds.), Plenum Press, New York, pp. 11–21.
2. Joseph, J. M., 1983, Aspergillosis in composting, in: *Occupational Mycoses* (A. F. DiSalvo, ed.), Lea and Febiger, Philadelphia, pp. 123–141.
3. Allo, M. D., Miller, J., Townsend, T., and Tan, C., 1987, Primary cutaneous aspergillosis associated with Hickman intravenous catheters, *N. Engl. J. Med.* **317:**1105–1108.
4. Magid, M. L., Prendiville, J. S., and Esterly, N. B., 1988, Violaceous nodules on the arm of a child with acute lymphocytic leukemia: Primary cutaneous aspergillosis (*Aspergillus flavus*), *Arch. Dermatol.* **124:**122–126.
5. Hay, R. J., 1988, Pulmonary aspergillosis: The clinical spectrum, in: *Aspergillus and Aspergillosis* (H. V. Bossche, D. W. R. Mackenzie, and G. Cauwenbergh, eds.), Plenum Press, New York, pp. 97–105.
6. Young, R. C., Bennett, J. B., Vogel, C. L., Carbone, P. P., and DeVita, V. T., 1970, Aspergillosis: The spectrum of the disease in 98 patients, *Medicine* **49:**147–173.
7. Starke, I. D., and Keal, E. E., 1980, Cerebral aspergilloma in a patient with allergic bronchopulmonary aspergillosis, *Br. J. Dis. Chest.* **74:**301–305.
8. Kurup, V. P., Barboriak, J. J., and Fink, J. N., 1984, Hypersensitivity pneumonitis, in: *Mould Allergy* (Y. Al-Doory and J. F. Domson, eds.), Lea and Febiger, Philadelphia, pp. 216–243.
9. Glimp, R. A., and Bayer, A. S., 1981, Fungal pneumonias (part 3): Allergic bronchopulmonary aspergillosis, *Chest* **80:**85–94.
10. Hoehne, J. H., Reed, C. E., and Dickie, H. A., 1973, Allergic bronchopulmonary aspergillosis is not rare: With a note on preparation of antigen for immunologic tests, *Chest* **63:**177–181.
11. Patterson, R., Greenberger, P. A., Radin, R. C., and Roberts, M., 1982, Allergic bronchopulmonary aspergillosis: Staging as an aid to management, *Arch. Intern. Med.* **96:**286–291.

12. Jonathan, D., Lund, V., and Milroy, C., 1989, Allergic aspergillus sinusitis: An overlooked diagnosis? *J. Laryngol. Otol.* **103**:1181–1183.
13. De Coster, A., Dierckx, P., and Grivignee, A., 1988, Aspergilloma, in: *Aspergillus and Aspergillosis* (H. V. Bossche, D. W. R. Mackenzie, and G. Cauwenbergh, eds.), Plenum Press, New York, pp. 107–114.
14. Albelda, S. M., Talbot, G. H., Gerson, S. L., Miller, W. T., and Cassileth, P. A., 1985, Pulmonary cavitation and massive hemoptysis in invasive pulmonary aspergillosis: influence of bone marrow recovery in patients with acute leukemia, *Am. Rev. Respir. Dis.* **131**:115–120.
15. Pena, C. E., 1975, Aspergillus intranasal fungus ball, *Am. J. Clin. Pathol.* **64**:343–344.
16. Aisner, J., Murillo, J., Schmipff, S. C., and Steere, A. C., 1979, Invasive aspergillosis in acute leukemia: Correlation with nose cultures and antibiotic use, *Ann. Intern. Med.* **90**:4–9.
17. Gustafson, T. L., Schaffner, W., Lavely, G. B., Stratton, C. W., Johnson, H. K., and Hutcheson, R. H., Jr., 1983, Invasive aspergillosis in renal transplant recipients: Correlation with corticosteroid therapy, *J. Infect. Dis.* **148**:230–238.
18. Gerson, S. L., Talbot, G. H., Hurwitz, S., Strom, B. L., Lusk, E. J., and Cassileth, P. A., 1984, Prolonged granulocytopenia: The major risk factor for invasive pulmonary aspergillosis in patients with acute leukemia, *Ann, Intern. Med.* **100**:345–351.
19. Rhodes, J. C., and Koelliker, D. D., 1993, Pathologic and laboratory aspects of aspergillosis, in: *Progress in Infectious Diseases* (R. Genta, ed.), Field and Wood Medical Publishers, Philadelphia. In press.
20. Peterson, D. E., and Schmipff, S. C., 1989, *Aspergillus* sinusitis in neutropenic patients with cancer: A review, *Biomed. Pharmacother.* **43**:307–312.
21. Gefter, W. B., Weingrad, T. R., Epstein, D. M., Ochs, R. H., and Miller, W. T., 1981, "Semi-invasive" pulmonary aspergillosis: A new look at the spectrum of *Aspergillus* infections of the lung, *Radiology* **140**:313–321.
22. Binder, R. E., Faling, L. J., Pugatch, R. D., Mahasaen, C., and Snider, G. L., 1982, Chronic necrotizing pulmonary aspergillosis: A discrete clinical entity, *Medicine* **61**:109–124.
23. Krishnamachari, K. A. V. R., Bhat, R. V., Nagarajan, V., and Tilak, T. B. G., 1975, Hepatitis due to aflatoxicosis: An outbreak in western India, *Lancet* **1**(May 10):1061–1063.
24. van Rensburg, S. J., van Schalkwyk, G. C., and van Schalkwyk, D. J., 1990, Primary liver cancer and aflatoxin intake in Transkei, *J. Environ. Pathol. Toxicol. Oncol.* **10**:11–16.
25. Autrup, H., Seremet, T., and Wakhisi, J., 1990, Evidence for human antibodies that recognize an aflatoxin epitope in groups with high and low exposure to aflatoxins, *Arch. Environ. Health.* **45**:31–34.
26. Pestka, J. J., and Bondy, G. S., 1990, Alteration of immune function following dietary mycotoxin exposure, *Can. J. Physiol. Pharmacol.* **68**:1009–1016.
27. Raper, K. B., and Fennell, D. I., 1965, *The Genus Aspergillus*, Williams and Wilkins Company, Baltimore.
28. Rhodes, J. C., and Kwon-Chung, K. J., 1989, Identification of agents of systemic mycoses, in: *Medical Mycology: A Practical Approach* (E. G. V. Evans and M. D. Richardson, eds.), IRL Press, Oxford, pp. 141–170.
29. Herbert, P. A., and Bayer, A. S., 1981, Fungal pneumonia (part 4): Invasive pulmonary aspergillosis, *Chest* **80**:220–225.
30. Rinaldi, M. G., 1983, Invasive aspergillosis, *Rev. Infect. Dis.* **5**:1061–1077.
31. Rhodes, J. C., 1988, Virulence factors in fungal pathogens, *Microbiolog. Sci.* **5**:252–254.
32. Kothary, M. H., Chase, T., Jr., and Macmillan, J. D., 1984, Correlation of elastase production by some strains of *Aspergillus fumigatus* with ability to cause pulmonary invasive aspergillosis in mice, *Infect. Immun.* **43**:320–325.
33. Rhodes, J. C., Bode, R. B., and McCuan-Kirsch, C. M., 1988, Elastase production by clinical isolates of *Aspergillus*, *Diag. Microbiol. Infect. Dis.* **10**:165–170.
34. Robinson, B. W. S., Venaille, T. J., Mendis, A. H. W., and McAleer, R., 1990, Allergens as

proteases: An *Aspergillus fumigatus* proteinase directly induces human epithelial cell detachment, *J. Allergy Clin. Immunol.* **86**:726–731.
35. Piechura, J. E., Kurup, V. P., and Daft, L. J., 1990, Isolation and immunochemical characterization of fractions from membranes of *Aspergillus fumigatus* with protease activity, *Can. J. Microbiol.* **36**:33–41.
36. Mackenzie, D. W. R., 1988, *Aspergillus* in man, in: *Aspergillus and Aspergillosis* (H. V. Bossche, D. W. R. Mackenzie, and G. Cauwenbergh, eds.), Plenum Press, New York, pp. 1–8.
37. Sinski, J. T., 1975, The epidemiology of aspergillosis, in: *The Epidemiology of Human Mycotic Diseases* (Y. Al-Doory, ed.), Charles C. Thomas, Publisher, Springfield, pp. 210–226.
38. Nolard, N., Detandt, M., and Beguin, H., 1988, Ecology of *Aspergillus* species in the human environment, in: *Aspergillus and Aspergillosis* (H. V. Bossche, D. W. R. Mackenzie, and G. Cauwenbergh, eds.), Plenum Press, New York, pp. 35–41.
39. Walsh, T. J., and Dixon, D. M., 1989, Nosocomial aspergillosis: Environmental microbiology, hospital epidemiology, diagnosis and treatment, *Eur. J. Epidemiol.* **5**:131–142.
40. Herman, L. G., 1980, *Aspergillus* in patient case areas, *Ann. N.Y. Acad. Sci.* **353**:140–146.
41. Meyers, J. D., 1990, Fungal infections in bone marrow transplant patients, *Semin. Oncol.* **17** (Suppl. 6):10–13.
42. Cohen, M. S., Isturiz, R. E., Malech, H. L., Root, R. K., Wilfert, C. M., Gutman, L., and Buckley, R. H., 1981, Fungal infection in chronic granulomatous disease: The importance of the phagocyte in defense against fungi, *Am. J. Med.* **71**:59–66.
43. Schaffner, A., Douglas, H., and Braude, A., 1982, Selective protection against conidia by mononuclear and against mycelia by polymorphonuclear phagocytes in resistance to *Aspergillus*: Observations on these two lines of defense *in vivo* and *in vitro* with human and mouse phagocytes, *J. Clin. Invest.* **69**:617–631.
44. Merkow, L. P., Epstein, S. M., Sidransky, H., Verney, E., and Pardo, M., 1971, The pathogenesis of experimental pulmonary aspergillosis: An ultrastructural study of alveolar macrophages after phagocytosis of *A. flavus* spores *in vivo*, *Am. J. Pathol.* **62**:58–74.
45. Levitz, S. M., Selsted, M. E., Ganz, T., Lehrer, R. I., and Diamond, R. D., 1986, *In vitro* killing of spores and hyphae of *Aspergillus fumigatus* and *Rhizopus oryzae* by rabbit neutrophil cationic peptides and bronchoalveolar macrophages, *J. Infect. Dis.* **154**:483–489.
46. Diamond, R. D., Huber, E., and Haudenschild, C. C., 1983, Mechanisms of destruction of *Aspergillus fumigatus* hyphae mediated by human monocytes, *J. Infect. Dis.* **147**:474–483.
47. Goodman, N. L., 1982, Aspergillosis, *Clin. Microbiol. Newsl.* **4**:9–11.
48. Batten, J., 1974, Clinical aspects of *Aspergillus* infections in man, in: *Aspergillosis and Farmer's Lung in Man and Animal* (R. de Haller and F. Suter, eds.), Hans Huber Publishers, Bern, pp. 61–68.
49. Fratamico, P. M., and Buckley, H. R., 1991, Identification and characterization of an immunodominant 58-kilodalton antigen of *Aspergillus fumigatus* recognized by sera of patients with invasive aspergillosis, *Infect. Immun.* **59**:309–315.
50. Lehman, P. F., and White, L. O., 1976, Acquired immunity to *Aspergillus fumigatus*, *Infect. Immun.* **13**:1296–1298.
51. Kim, S. J., and Chaparas, S. D., 1978, Characterization of antigens from *Aspergillus fumigatus*: Preparation of antigens from organisms grow in a completely synthetic medium, *Am. Rev. Respir. Dis.* **118**:547–551.
52. Chaparas, S. D., Morgan, P. A., Holobaugh, P., and Kim, S. J., 1986, Inhibition of cellular immunity by products of *Aspergillus fumigatus*, *J. Med. Vet. Mycol.* **24**:67–76.
53. Williams, D. W., Weiner, M. H., and Drutz, D. J., 1981, Immunologic studies of disseminated infection with *Aspergillus fumigatus* in the nude mouse, *J. Infect. Dis.* **143**:726–733.
54. Denning, D. W., Follansbee, S. E., Scolaro, M., Norris, S., Edelstein, H., and Stevens, D. A., 1991, Pulmonary aspergillosis in the acquired immunodeficiency syndrome, *N. Engl. J. Med.* **324**:654–662.

55. Mullbacher, A., Waring, P., and Eichner, R. D., 1985, Identification of an agent in cultures of *Aspergillus fumigatus* displaying anti-phagocytic and immunomodulating activity *in vitro*, *J. Gen. Microbiol.* **131:**1251–1258.
56. Arruda, L. K., Platts-Mills, T. A., Fox, J. W., and Chapman, M. D., 1990, *Aspergillus fumigatus* allergen I, a major IgE-binding protein, is a member of the mitogillin family of cytotoxins, *J. Exp. Med.* **172:**1529–1532.
57. Washburn, R. G., Hammer, C. H., and Bennett, J. E., 1986, Inhibition of complement by culture supernatants of *Aspergillus fumigatus*, *J. Infect. Dis.* **154:**944–951.
58. Washburn, R. G., DeHart, D. J., Agwu, D. E., Bryant-Varela, B. J., and Julian, N. C., 1990, *Aspergillus fumigatus* complement inhibitor: Production, characterization, and purification by hydrophobic interaction and thin-layer chromatography, *Infect. Immun.* **58:**3508–3515.
59. de Repentigny, L., and Reiss, E., 1984, Current trends in immunodiagnosis of candidiasis and aspergillosis, *Rev. Infect. Dis.* **6:**301–312.
60. Bernstein, J. A., Zeiss, C. R., Greenberger, P. A., Patterson, R., Marhoul, J. F., and Smith, L. L., 1990, Immunoblot analysis of sera from patients with allergic bronchopulmonary aspergillosis: Correlation with disease activity, *J. Allergy Clin. Immunol.* **86:**532–539.
61. Harvey, C., Shaw, R. J., and Longbottom, J. L., 1987, Diagnostic specificity of a sandwich ELISA for *Aspergillus*-related diseases, *J. Allergy Clin. Immunol.* **79:**324–330.
62. Dupont, B., Huber, M., Kim, S. J., and Bennett, J. E., 1987, Galactomannan antigenemia and antigenuria in aspergillosis: Studies in patients and experimentally infected rabbits, *J. Infect. Dis.* **155:**1–11.
63. Bennett, J. E., Friedman, M. M., and Dupont, B., 1987, Receptor-mediated clearance of *Aspergillus* galactomannan, *J. Infect. Dis.* **155:**1005–1010.
64. Rogers, T. R., Haynes, K. A., and Barnes, R. A., 1990, Value of antigen detection in predicting invasive pulmonary aspergillosis, *Lancet* **336:**1210–1213.
65. Schaffer, P. J., Medoff, G., and Kobayashi, G. S., 1979, Demonstration of antigenemia by radioimmunoassay in rabbits experimentally infected with *Aspergillus*, *J. Infect. Dis.* **139:** 313–319.
66. Weiner, M. H., 1980, Antigenemia detected by radioimmunoassay in systemic aspergillosis, *Ann. Intern. Med.* **92:**793–796.
67. Sabetta, J. R., Miniter, P., and Andriole, V. T., 1985, The diagnosis of invasive aspergillosis by an enzyme-linked immunosorbent assay for circulating antigen, *J. Infect. Dis.* **152:**946–953.
68. Wilson, E. V., Hearn, V. M., and Mackenzie, D. W. R., 1987, Evaluation of a test to detect circulating *Aspergillus fumigatus* antigen in a survey of immunocompromised patients with proven or suspected disease, *J. Med. Vet. Mycol.* **25:**365–375.
69. Haynes, K. A., Latge, J. P., and Rogers, T. R., 1990, Detection of *Aspergillus* antigens associated with invasive infection, *J. Clin. Microbiol.* **28:**2040–2044.
70. Yu, B., Niki, Y., and Armstrong, D., 1990, Use of immunoblotting to detect *Aspergillus fumigatus* antigen in sera and urine of rats with experimental aspergillosis, *J. Clin. Microbiol.* **28:**1575–1579.
71. Talbot, G. H., Weiner, M. H., Gerson, S. L., Provencher, M., and Hurwitz, S., 1987, Serodiagnosis of invasive aspergillosis in patients with hematologic malignancy: Validation of the *Aspergillus fumigatus* antigen radioimmunoassay, *J. Infect. Dis.* **155:**12–27.
72. Wong, B., Brauer, K., Tsai, R. R., and Jayasimhulu, K., 1989, Increased amounts of the *Aspergillus* metabolite D-mannitol in tissue and serum of rats with experimental aspergillosis, *J. Infect. Dis.* **160:**95–103.
73. Benoit, G., De Chauvin, M. F., Cordonnier, C., Astier, A., and Bernaudin, J.-F., 1985, Oxalic acid level in bronchoalveolar lavage fluid from patients with invasive pulmonary *aspergillosis*, *Am. Rev. Respir. Dis.* **132:**748–751.
74. Patterson, R., Greenberger, P. A., Halwig, J. M., Liotta, J. L., and Roberts, M., 1986, Allergic bronchopulmonary aspergillosis: Natural history and classification of early disease by serologic and roentgenographic studies, *Arch. Intern. Med.* **146:**916–918.

75. Patterson, T. F., Miniter, P., and Andriole, V. T., 1990, Efficacy of fluconazole in experimental invasive aspergillosis, *Rev. Infect. Dis.* **12**(Suppl. 3):S281–S285.
76. Dupont, B., 1990, Itraconazole therapy in aspergillosis: Study in 49 patients, *J. Am. Acad. Dermatol.* **23**:607–614.
77. Lopez-Berestein, G., 1989, Liposomal-amphotericin B in the treatment of systemic mycoses in patients with cancer, in: *Diagnosis and Therapy of Systemic Fungal Infections* (K. Holmberg and R. D. Meyer, eds.), Raven Press, New York, pp. 159–166.
78. Denning, D. W., Tucker, R. M., Hanson, L. H., and Stevens, D. A., 1990, Itraconazole in opportunistic mycoses: Cryptococcosis and aspergillosis, *J. Am. Acad. Dermatol.* **23**:602–607.
79. Denning, D. W., Hanson, L. H., and Stevens, D. A., 1990, In vitro activity of saperconazole (R66 905) compared with amphotericin B and itraconazole against *Aspergillus species*, *Eur. J. Clin. Microbiol. Infect. Dis.* **9**:693–697.

16

Phaeohyphomycosis

MOHAMAD M. EL-ZAATARI
and MICHAEL R. McGINNIS

1. INTRODUCTION

In 1974, Ajello and coworkers[1] created the term *phaeohyphomycosis* (*phaeo*, brown; *hypho*, hyphal) to encompass a group of fungal infections distinct from chromoblastomycosis. Historically, phaeohyphomycosis has been referred as chromomycosis (in part), cerebral chromomycosis, cladosporiosis, phaeomycotic cyst, phaeosporotrichosis, and subcutaneous phaeomycotic cyst. McGinnis and colleagues[2,3] have recently refined the concept of phaeohyphomycosis by recognizing several clinically distinct presentations. Phaeohyphomycosis consists of a distinct, heterogeneous, and important group of mycotic infections in which the etiologic agents occur in tissue as either dematiaceous yeastlike cells, pseudohyphaelike elements, septate hyphae, or any combination of these forms.

Phaeohyphomycosis is a general term referring to infections caused by dematiaceous fungi, which vary from hyaline to pale brown to black in tissue, due to the presence of melanin in the fungal cell walls. Melanin contributes to their darkly colored colonies on artificial media. The fungal elements may be short or elongated, regular or distorted to swollen in shape, yeastlike, or any combination of these forms may be present in the same lesion. In some cases, examination of several tissue sections may be required to clearly visualize the fungal elements. If the dematiaceous nature of the fungal elements in tissue are not apparent, a Fontana-Masson stain can be used to confirm the

MOHAMAD M. EL-ZAATARI and MICHAEL R. McGINNIS • Department of Pathology, University of Texas Medical Branch, Galveston, Texas 77555.

Fungal Infections and Immune Responses, edited by Juneann W. Murphy *et al.* Plenum Press, New York, 1993.

presence of melanin in the fungal cell walls. Melanin appears to have an association with virulence. Dixon and coworkers[4] found that melanin-deficient mutants of *Wangiella dermatitidis* are less virulent in mice than the dematiaceous wild type.

Phaeohyphomycosis includes a broad spectrum of opportunistic infections caused by a number of different dematiaceous fungi, regardless of their taxonomic classification. Common and widely used terms, such as black piedra, tinea nigra, or mycotic keratitis, refer to very traditional clinical forms of phaeohyphomycosis. Mycetoma, chromoblastomycosis, and phaeohyphomycosis represent three reference points for a continuum of infections caused by the dematiaceous fungi. In tissue, the etiologic agents of mycetoma produce granules that are well-organized mycelial structures. The sclerotic bodies or muriform cells that are characteristic of chromoblastomycosis represent structures that are phenotypically intermediate between a yeast and hyphal form. Fungi such as *Exophiala jeanselmei* may cause mycetoma, subcutaneous phaeohyphomycosis,[5] and chromoblastomycosis.[6] In another example, the classic agent of chromoblastomycosis, *Phialophora verrucosa*, has been recovered from cases of subcutaneous phaeohyphomycosis.[7] Diseases caused by dematiaceous fungi are a reflection of the dynamic interactions between the host and the fungus.

2. ETIOLOGY AND EPIDEMIOLOGY

There are more than 30 known fungi capable of causing phaeohyphomycosis (Tables I–III). Fungi like *Phaeoannellomyces werneckii*, a salt-tolerant black yeast, usually causes infections in patients living in tropical regions and other areas surrounded by salt water. *Piedraia hortae*, the cause of black piedra, has been associated with specific Indian tribes in South America, as well as with other general populations of people. It is not clearly understood whether there are genetic factors, environment conditions, or cultural factors involved in black piedra. Probably all of these influence the attack rate of *P. hortae* in some areas.

One of the most common etiologic agents of subcutaneous phaeohyphomycosis is *Exophiala jeanselmei*. The fungus frequently causes infections in patients having diabetes following the traumatic introduction of a splinter harboring the fungus. *Wangiella dermatitidis* has been frequently reported from Japan and Asian countries, and a spectrum of other species have been isolated from human and animal infections. Some of these fungi are rare and only have been seen causing disease in severely immunosuppressed patients. Medically important species of *Bipolaris* and *Exserohilum*, which were considered to be species of *Helminthosporium* or *Drechslera*, are important opportunistic pathogens.[8] The identification of these fungi is often difficult because they have not been previously clearly defined.

TABLE I
Clinical Classification and Common Etiologic Agents of Superficial and Cutaneous Phaeohyphomycosis

Superficial phaeohyphomycosis: Infections confined to the stratum corneum with little or no tissue response. If hair is involved, the fungi grow superficially around the hair shaft with little damage.
  Black piedra
    Piedraia hortae
  Skin infections
    Phaeoannellomyces werneckii (syn. *Exophiala werneckii*)
    Stenella araguata
Cutaneous and corneal phaeohyphomycosis: Keratinized tissues are invaded. Nonliving layers are generally involved, often exhibiting extensive tissue destruction.
  Dermatomycosis
    Alternaria spp.
    Nattrassia mangiferae (syn. *Hendersonula toruloidea*)
  Mycotic keratitis
    Botryodiplodia theobromae
    Curvularia geniculata
    C. lunata
    C. senegalensis
    Exophiala jeanselmei
    Exserohilum rostratum (syn. *Drechslera rostrata*)
  Onychomycosis
  Botryodiplodia theobromae
  Nattrassia mangiferae
  Phyllosticta sp.

Only cultivation can be used to determine the identification of the various etiologic agents because these fungi cannot be distinguished from one another on the basis of tissue response nor tissue morphology. Tissue morphology can be of help at times to recognize some groups of fungi. For example, the presence of annellated yeast cells suggest *Exophiala* or *Phaeoannellomyces* rather than *Bipolaris* or *Wangiella*. The various agents of phaeohyphomycosis are worldwide in distribution, common in nature, and can be isolated from plant material, wood, soil, and other environments.

TABLE II
Some Etiologic Agents of Subcutaneous Phaeohyphomycosis

Alternaria alternata	*Phialophora bubakii*	*Phoma* sp.
Bipolaris spicifera (syn. *Drechslera spicifera*)	*P. parasitica*	*Phoma hibernica*
Exophiala jeanselmei	*P. repens*	*Scytalidium dimidiatum*
E. moniliae	*P. richardsiae*	*Wangiella dermatitidis*
E. spinifera	*P. verrucosa*	*Xylohypha emmonsii*

TABLE III
Some Etiologic Agents of Systemic Phaeohyphomycosis

Bipolaris hawaiiensis (syn. *Drechslera hawaiiensis*)
B. spicifera
Cladosporium cladosporioides
Curvularia geniculata
C. pallescens
Dissitimurus exedrus
Exophiala jeanselemei
Wangiella dermatitidis
Xylohypha bantiana (syn. *Cladosporium bantianum, C. trichoides*)

3. CLINICAL FEATURES, HISTOPATHOLOGY, AND THERAPY

Based upon tissue involvement, phaeohyphomycosis can be divided into five types of infectious disease: superficial, cutaneous and corneal, subcutaneous, and systemic and allergic.

3.1. Superficial Phaeohyphomycosis

This is a benign superficial fungal infection caused by fungi such as *Phaeoannellomyces werneckii* (syn., *Exophiala werneckii*), commonly referred to as tinea nigra. The disease is rarely caused by *Stenella araguata*.

These fungi infect the stratum corneum, probably following mild traumatic inoculation of heavily keratinized skin, especially on the palms of the hands and the soles of the feet. Little or no tissue response to the presence of the fungus occurs. The lesions are typically solitary, darkly pigmented macules with an irregular and well-defined margin (Fig. 1). If the hair is involved, such as black piedra caused by *Piedraia hortae*, the fungus grows superficially around the hair shaft, with little damage to the hair. The black, hard, firmly attached nodules are ascomycete fruiting bodies that contain ascospores.

Tinea nigra, which is prevalent in tropical and subtropical regions, involves children and young adults more commonly than adults. The infection is caused by a salt-tolerant black yeast that is not contagious. The disease is mainly of cosmetic importance and has little or no discomfort, nor host reaction. The fungus grows in the outer layers of the stratum corneum.

A diagnosis of superficial phaeohyphomycosis caused by fungi such as *P. werneckii* can be made by examining skin scrapings mounted in 10% potassium hydroxide and observing dematiaceous yeasts and branched, septate, dematiaceous hyphae that are typically 1.5–3 μm wide; the yeast cells are 1.5–5 μm in diameter. Skin scrapings should be plated onto a culture medium such as mold inhibitory agar without antibiotics. Normally, *P. wer-*

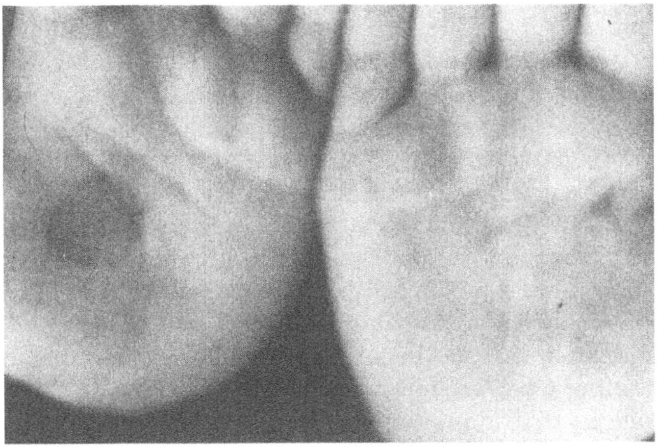

FIGURE 1. Superficial phaeohyphomycosis involving the palm of the hand caused by *Phaeoannellomyces werneckii.*

neckii will grow within a few days when incubated at 25–30°C. The exoantigen test is capable of specifically identifying *P. werneckii.*[9] In tissue sections stained with hematoxylin and eosin, the fungi can be observed as narrow, frequently branched hyphae in the mild to moderate hyperkeratotic layers of the epidermis. The pigmented nature of the cell wall should be evident.

Tinea nigra is a curable infection when treated with topical agents such as thiabendazole, keratolytic agents such as 4% salicylic acid, or mechanical removal of the infected layer of the stratum corneum. Ketoconazole and other imazoles are very effective.

3.2. Cutaneous Phaeohyphomycosis

This is a form of a phaeohyphomycosis in which the fungi invade keratinized tissue i.e., the stratum corneum, hair, and nails. It differs in the degree of tissue damage and host immune response in comparison to superficial phaeohyphomycosis. Dermatomycosis, mycotic keratitis, and onychomycosis (Table I) are examples of this disease. Demonstration and isolation of pigmented fungi in clinical material obtained from lesions will help establish the diagnosis.

Dermatomycosis differs from dermatophytosis in being caused by dematiaceous fungi rather than dermatophytes. However, regardless of whether the etiologic agent is dermatophyte or a dematiaceous fungus, the clinical appearance, host response, and diagnosis are essentially the same. Infections caused by *Nattrassia mangiferae* (syn. *Hendersonula toruloidea*) are resistant to griseofulvin treatment.[10]

Mycotic keratitis is an infection that occurs following trauma to the corneum and the subsequent implantation of the fungus. A fluffy, gray-white nodule starts at the site of infection, after which a small ulcer forms, resulting from liquefaction of the superficial layers of the adjacent cornea. The center of the ulcer frequently has a wide halo around it with radial lines that may be observed in the halo. An inflammatory response occurs, but vascularization is frequently absent. Fungal penetration into the anterior chamber of the eye can be enhanced if antimicrobial agents or steroids are used to prevent bacterial infection and to inhibit the inflammatory response. Mycotic keratitis is known to be caused by a number of dematiaceous fungi (see Table I).

Swabs are not recommended for culture nor the collection of material for direct examination. It is necessary to see dematiaceous fungal elements in corneal scraping taken from the ulcer edge or deep within the base of the ulcer as a means to distinguish between a contaminant and a pathogen. For definitive identification, isolation of the fungus from corneal scrapings is required.

The drug of choice for treatment is natamycin (pimaricin). It should be continued for several weeks after the occurrence of reepithelialization of the ulcer. To monitor the healing process, biweekly corneal scrapings have been shown to be helpful.

Onychomycosis, or nail infections, can be caused by a number of dematiaceous fungi (see Table I). The infections are difficult to distinguish on clinical grounds from other fungal infections, because nail plate thickening, opacification, and onycholysis are common features. The nail tends to be darkly pigmented, which reflects, in part, the growth of the dematiaceous fungus throughout the nail. Infection caused by *N. mangiferae* is often evident as paronychia resembling an infection caused by a *Candida* spp.[11]

Diagnosis can be made by direct examination of nail scrapings in 30% KOH containing 40% (wt/vol.) DMSO. Fungal hyphae may be difficult to see at times. If hyaline hyphae are present, they are morphologically indistinguishable from dermatophytes. Nail scrapings should be plated onto a medium without cycloheximide because *N. mangiferae* is inhibited by this antimicrobial agent.

The use of oral ketoconazole is the treatment of choice. Topical application of antifungal agents is not effective. Unfortunately, both choices are often ineffective in eliminating this fungal disease.

3.3. Subcutaneous Phaeohyphomycosis

This is a heterogeneous group of diseases generally resulting from the traumatic implantation of the fungus into the subcutaneous tissue, resulting in a cystlike localized lesion with abscess formation. Often, the patient can recall injury to the area by objects such as splinters, which are presumed to be the vehicle of transmission. Splinters are occasionally observed in histologic tissue sections. In most instances, the patient presents with a solitary, discrete, asymptomatic, well-encapsulated subcutaneous nodule (Fig. 2) that may be on the foot, leg, hand, arm, back, or other body site. Occasionally, such lesions have been misdiagnosed as cynovial cyst, epidermoid inclusion cysts, or foreign body granulomata. Most of the immune competent patients, especially those having subcutaneous phaeohyphomycosis caused by *E. jeanselmei*, are rural workers having daily trauma to the skin from thorns, sticks, or other objects. McGinnis noted in a survey of 36 cases that the male to female ratio was 23:13. In immunosuppressed and diabetic patients, nodules have developed after a minor trauma with contaminated objects.

The lesion slowly develops at the site of entry of the fungus to form a nodule that is slightly elevated and is palpable under the skin. The process is chronic and forms a nodule in the subcutaneous tissue and dermis. The overlying epidermis is not usually affected. The central area of the nodule becomes necrotic, resulting in the transformation of the nodule into an encapsulated abscess that may be 2 cm or more in diameter, which often has a liquified center. In well-developed abscesses, yellowish-gray viscid, purulent fluid can be aspirated. Draining sinuses from the abscesses have not been observed. Local dissemination to regional lymph nodes or to distant sites is extremely rare, but in the immunosuppressed patients the lesions may

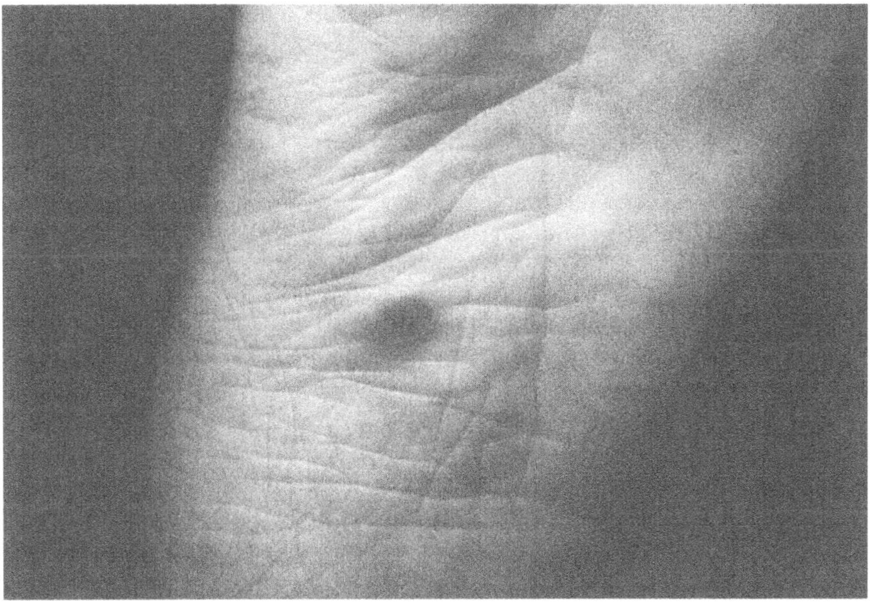

FIGURE 2. Subcutaneous phaeohyphomycosis involving the sole of the foot caused by *Exophiala jeanselmei.*

enlarge with the tissue invasion or disseminate, which results in a fatal outcome. Many fungi have been reported to cause subcutaneous phaeohyphomycosis (see Table II). The most commonly isolated causative agents are *E. jeanselmei*, *W. dematitidis*, *Bipolaris* spp.,[8] and *Phialophora* spp.

Skin scrapings, purulent aspirates, and excised abscess material are good specimens for direct examination and culture. Specimens are examined in 10–20% KOH after gentle heating to accelerate the clearing process. Tissue should be homogenized before direct examination and plating. The presence of pale to brownish, irregularly swollen, sepatate, branched or unbranched hyphae, and dematiaceous yeastlike cells, found in conjunction with hyphal elements within the clinical specimens, is diagnostic. The clinical specimens should be plated on selective and nonselective media, since some of the etiologic agents of phaeohyphomycosis may be sensitive to cycloheximide. The cultures are incubated at 25–30°C for up to 4 weeks before being reported as negative. Most dematiaceous fungi will form visible colonies within 1–2 weeks.

The histopathologic examination of the infected tissue specimens shows dense collagenous connective tissue surrounding the abscess, with compact epithelioid histocytes, multinucleated giant cells, neutrophils, occasional eosinophils, and lymphocytes at the inner margin of the abscess. At the center of the process, necrotic debris mixed with degenerated neutrophils, fibrin, and vegetative matter are often seen. The individual and aggregates of dematiaceous fungal elements are usually clearly evident in the wall of the abscess, as well as in the center (Figs. 3 and 4).

The treatment of choice is surgical excision of the subcutaneous cyst. Antifungal agents such as ketoconazole, 5-fluorocytosine, and recently itraconazole[12] have been used as adjunct therapy to prevent the recurrence or the spread of infection if the excision is performed improperly.

3.4. Systemic Phaeohyphomycosis

These are rare infections, but they are the most serious forms of phaeohyphomycosis. The infections occur primarily in immunosuppressed persons, but also have been reported in previously healthy individuals. In cerebral phaeohyphomycosis caused by *Xylohypha bantiana*, *Fonsecaea pedrosoi*, *Wangiella dermatitidis*, and other fungi, Bennett and colleagues[13] reviewed 35 cases. The male to female ratio in that review was 24:11. Some patients were immunosuppressed, and others have no apparent predisposing illness. Dixon and coworkers[14] recently reviewed 30 culture-documented cases of infection caused by *X. bantiana*; 26 cases involved the central nervous system. There was young male predomination, and pharmacologic immunosuppression was not found to be an important predisposing factor. In most systemic cases, the lungs were the primary site of infection after inhalation of fungal elements from the environment. The primary pulmonary infection is usually inapparent and often goes unnoticed, with later dissemination to other internal

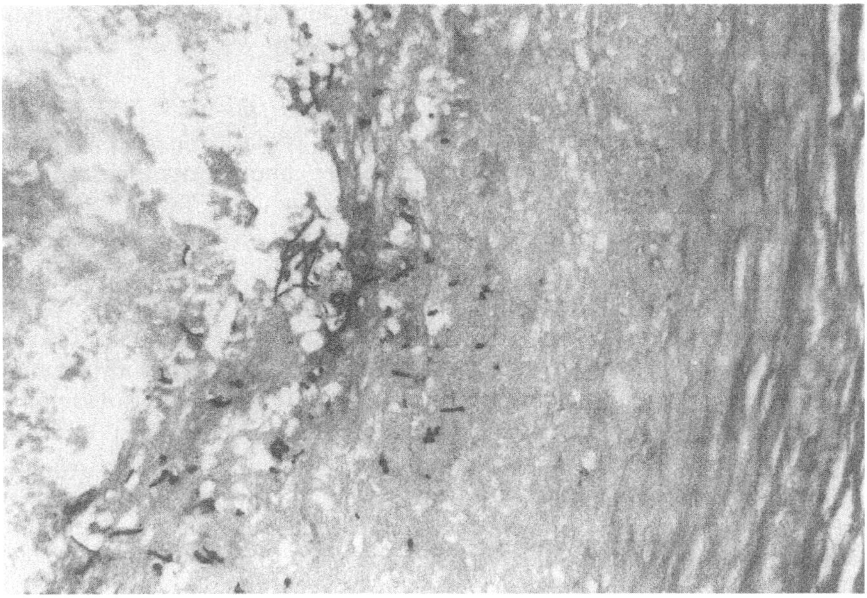

FIGURE 3. Fungal elements of *Exophiala jeanselmei* in a subcutaneous lesion surrounded by dense collagenous connective tissue. 100×, Gomori methenamine sliver stain.

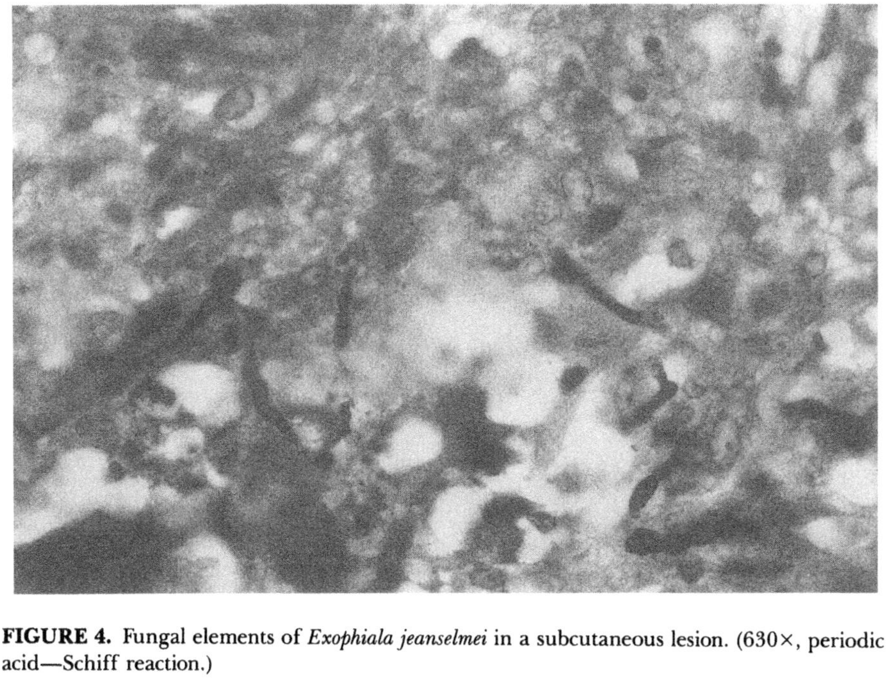

FIGURE 4. Fungal elements of *Exophiala jeanselmei* in a subcutaneous lesion. (630×, periodic acid—Schiff reaction.)

organs. Sometimes infection of the paranasal sinuses may result in dissemination of the fungus to the central nervous system, especially in severally debilitated patients.

One of the etiologic agents listed in Table III, *Xylohypha bantiana* (syn. *Cladosporium bantianum, C. trichoides*),[15,16] shows a remarkable predilection for the brain, particularly the cerebral hemispheres and contiguous meninges. The clinical picture includes headache, nausea, vomiting, fever, nuchal pain and rigidity, and mental confusion. Cerebral lesions, as seen in CT scan of the brain or at autopsy (Fig. 5), are either solitary or multiple and closely resemble the abscess seen in subcutaneous phaeohyphomycosis. The lesion can also appear as a generalized inflammatory infiltrate with necrosis. Dissemination to other organs does not occur as often as cerebral involvement.

There are no immunodiagnostic test procedures for the diagnosis of either systemic or subcutaneous phaeohyphomycosis. Histologic examination of the lesion should reveal dematiaceous, branched, septate hyphae in unstained or in hematoxylin and eosin stained sections (Fig. 6). If no pigmentation is detected, the Fontana-Masson melanin stain can be used to confirm the dematiaceous nature of the cell walls and to accentuate the fungi in tissue.

Treatment of systemic phaeohyphomycosis is by surgical excision of the encapsulated lesions, combined with administration of intravenous ampho-

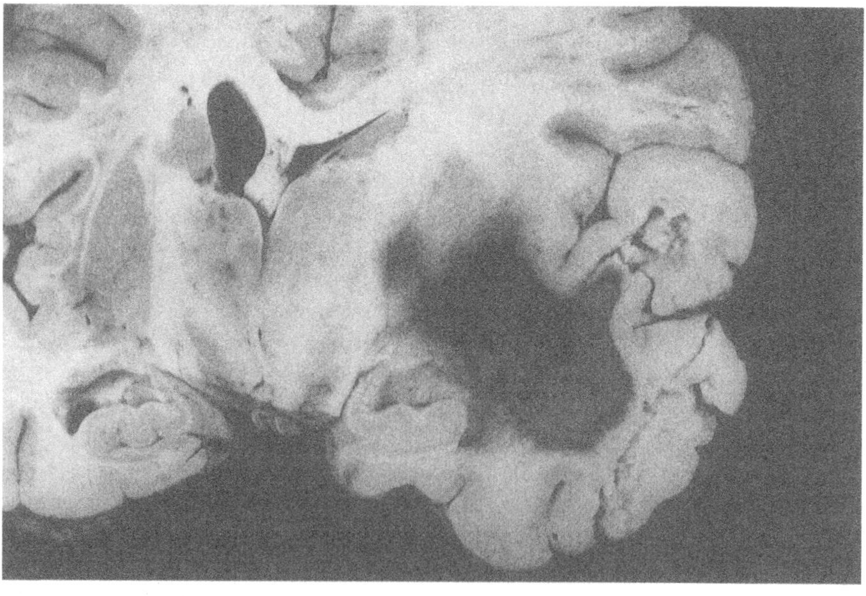

FIGURE 5. Abscess in a brain caused by *Xylohypha bantiana*.

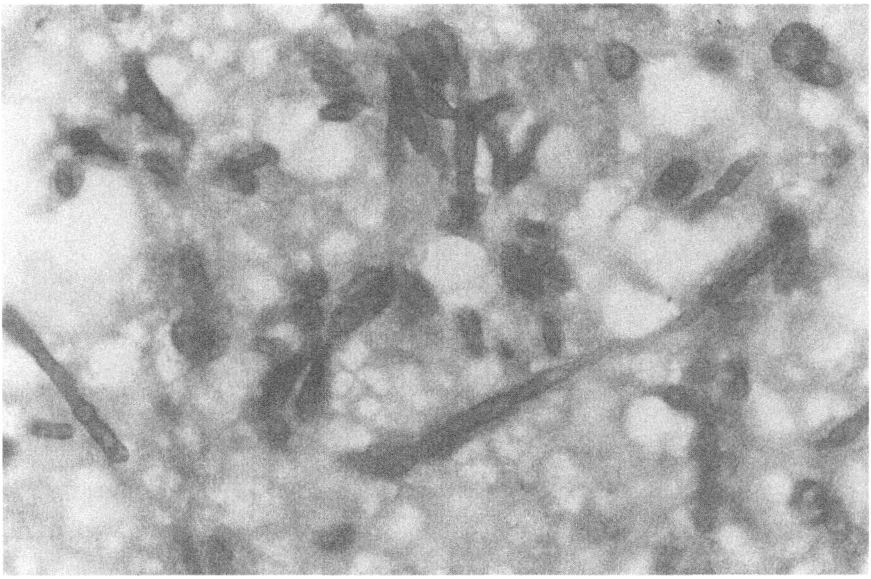

FIGURE 6. *Xylohypha bantiana* in brain tissue. (1000×, H&E.)

tericin B. Cerebral phaeohyphomycosis has a grave prognosis, even with aggressive therapy. The most important factor is surgical management, as systemic antifungal chemotherapy apparently did not influence the outcome.[13]

Up to 12% of patients on peritoneal dialysis will develop peritonitis caused by fungi.[17] In addition to *Candida* species, infections have also been attributable to dematiaceous fungi, such as *Bipolaris*, *Exophiala*, and *Exserohilum* species. Cure usually requires removal of the dialysis catheter and systemic antifungal chemotherapy.

3.5. Allergic Phaeohyphomycosis

Dematiaceous fungi, such as *Alternaria*, *Bipolaris*, *Cladosporium*, *Curvularia*, and *Exserohilum* species, are capable of producing localized colonization and infection of the paranasal sinuses. Patients usually suffer from recurrent attacks of sinusitis. Nasal polyps or paranasal sinus masses are typically detected on examination of the patient. Patients may or may not have an underlying disease, but nasal deformaties may exist. Histologic examination (Fig. 7) and culture of the mucous plugs are important as a means to establish a diagnosis of allergic fungal rhinosinusitis cause by dematiaceous fungus. Most patients respond to surgical debridement alone, although amphotericin

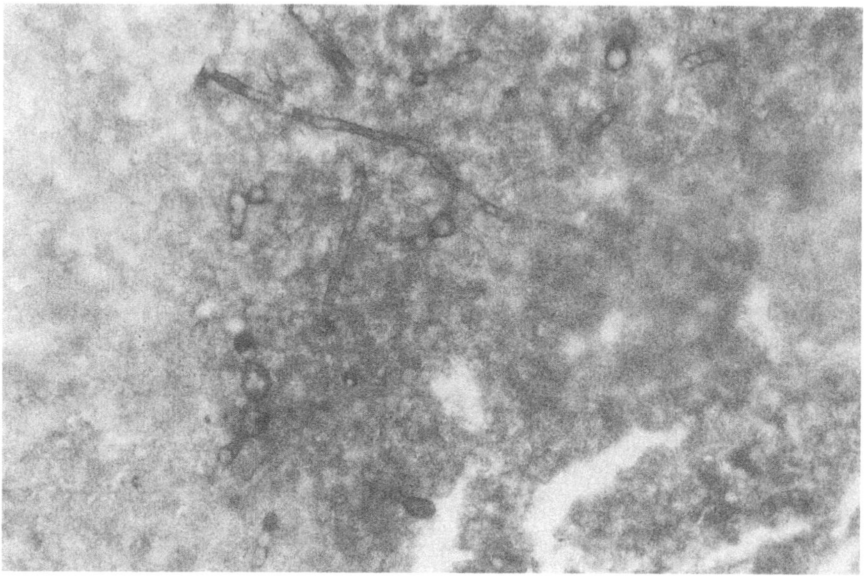

FIGURE 7. Fungal elements in tissue from nasal sinus of *Curvularia verruculosa*. (630×, Gomori methenamine sliver stain.)

B may also be required in the presence of bone invasion or immunosuppression. In a recent case report, itraconazole was shown to be an effective means to prevent recurrence after surgical debridement in an immunocompetent host.[18] Also, ketoconazole was effective to prevent relapse after surgical debridement.

REFERENCES

1. Ajello, L., Georg, L. K., Steigbigel, R. T., and Wang, C. J. K., 1974, A case of phaeohyphomycosis caused by a new species of *Phialophora*, *Mycologia* **66**:490–498.
2. McGinnis, M. R., 1983, Chromoblastomycosis and phaeohyphomycosis: New concepts, diagnosis, and mycology, *J. Am. Acad. Dermatol.* **8**:1–16.
3. McGinnis, M. R., Ajello, L., and Schell, W. A., 1985, Mycotic diseases: A proposed nomenclature, *Int. J. Dermatol.* **24**:9–15.
4. Dixon, D. M., Polak, A., and Conner, G. W., 1989, Mel^- mutants of *Wangiella dermatitidis* in mice: Evaluation of multiple mouse and fungal strains, *J. Med. Vet. Mycol.* **27**:335–341.
5. Sindhuphak, W., MacDonald, E., Head, E., and Hudson, R. D., 1985, *Exophiala jeanselmei* infection in a postrenal transplant patient, *J. Am. Acad. Dermatol.* **13**:877–881.
6. Naka, W., Harada, T., Nishikawa, T., and Fukushiro, R., 1986, A case of chromoblastomycosis: With special reference to the mycology of the isolated *Exophiala jeanselmei*, *Mykosen* **29**:445–452.
7. Schnadig, V. J., Long, E. G., Washington, J. M., McNeely, M. C., and Troum, B. A., 1986,

Phialophora verrucosa-induced subcutaneous phaeohyphomycosis. Fine needle aspiration finding, *Acta. Cytol.* **30:**425–429.
8. McGinnis, M. R., Rinaldi, M. G., and Winn, R., 1986, Emerging agents of phaeohyphomycosis: Pathogenic species *Bipolaris* and *Exserohilum, J. Clin. Microbiol.* **24:**250–259.
9. Standard, P. G., Kaufman, L., and Padhye, A. A., 1988, An immunological procedure for the rapid identification of *Phaeoannellomyces werneckii, Diag. Clin. Immunol.* **5:**367–370.
10. Gentles, J. C., and Evans, E. G. V., 1970, Infection of the feet and hands with *Hendersonula toruloidea, Sabouraudia* **8:**72–75.
11. Hay, R. J., and Moore, M. K., 1984, Clinical features of superficial fungal infections caused by *Hendersonula toruloidea* and *Scyalidium hyalinum, Br. J. Dermatol.* **110:**677–683.
12. Kotylo, P. K., Israel, K. S., Cohen, J. S., and Bartlett, M. S., 1989, Subcutaneous phaeohyphomycosis of the finger caused by *Exophiala spinifera, Am. J. Clin. Pathol.* **91:**624–627.
13. Bennett, J. E., Bonner, H., Jennings, A. E., and Lopez, R. I., 1973, Chronic meningitis caused by *Cladosporium trichoides, Am. J. Clin. Pathol.* **59:**398–407.
14. Dixon, D. M., Walsh, T. J., Merz, W. G., and McGinnis, M. R., 1989, Infections due to *Xylohypha bantiana (Cladosporium trichoides), Rev. Infect. Dis.* **11:**515–525.
15. McGinnis, M. R., Borelli, D., Padhye, A. A., and Ajello, L., 1986, Reclassification of *Cladosporium bantianum* in the genus *Xylohypha, J. Clin. Microbiol.* **23:**1148–1151.
16. McGinnis, M. R., Padhye, A. A., and Ajello, L., 1987, *Cladosporium trichoides* var. *chlamydosporum*: An invalidly published name, *Mycologia* **79:**803–805.
17. Eisenberg, E. S., Leviton, I., and Soeiro, R., 1986, Fungal peritonitis inpatients receiving peritoneal dialysis: Experience with 11 patients and review of the literature, *Rev. Infect. Dis.* **8:**309–321.
18. Maskin, S. L., Fetchick, R. J., Leone, C. R., Jr., Sharkey, P. K., and Rinaldi, M. G., 1989, *Bipolaris hawaiiensis* caused phaeohyphomycotic orbitopathy. A devastating fungal sinusitis in an apparently immunocompetent host, *Ophtalmology* **96:**175–179.

17

Fungal Allergy

VISWANATH P. KURUP and JORDAN N. FINK

1. INTRODUCTION

Allergy is one of the commonest acute and chronic diseases encountered in humans. The term *allergy* was coined by Von Pirquet in 1906 to designate an altered state in humans caused by the introduction of a foreign substance.[1] After the first encounter with a foreign substance (antigen), changes occur in specific responsiveness and, as a result, the reaction to the foreign substance is either increased (hypersensitivity) or decreased (immunity). Thus, an allergic reaction occurs as a result of adverse effects of a specific immune interaction, such as antibody and/or lymphocytes, and specific substances, such as allergens/antigens. Among the allergens associated with allergic diseases in humans, those from molds represent a major group.

Fungi have been associated with a number of allergic disorders in humans. The major allergic diseases caused by fungi are allergic asthma, allergic rhinitis, allergic sinusitis, allergic bronchopulmonary mycoses, and hypersensitivity pneumonitis.[2–6] The pathogenesis of these diseases may vary considerably and may represent only one type of immunologic mechanism or may constitute a combination of several mechanisms acting simultaneously.

Although symptoms in some patients with allergic rhinitis, conjunctivitis, and asthma have been attributed to exposure to fungal spores, it is extremely difficult to demonstrate immediate wheal and flare of IgE-mast-cell-mediated skin test reactivity in such patients to a single antigen alone.

VISWANATH P. KURUP and JORDAN N. FINK • Research Service, Zablocki Veterans Affairs Medical Center and Allergy–Immunology Division, Medical College of Wisconsin, Milwaukee, Wisconsin 53295.

Fungal Infections and Immune Responses, edited by Juneann W. Murphy *et al.* Plenum Press, New York, 1993.

The majority demonstrate reactivity to several allergens from different sources and with varied chemical characteristics. Additionally, many of these patients also demonstrate bronchial hypersensitivity and, therefore, there is further confusion in establishing the etiological role of fungal antigens in their symptoms.

2. ALLERGIC RHINITIS

After exposure to mold spores or mycelial particles, susceptible individuals may develop classical forms of nasal allergy, "hay fever" or allergic rhinitis.[2] The frequently encountered fungi associated with this disease are listed in Table I. The symptoms and signs of fungal-induced allergic rhinitis are usually indistinguishable from those caused by the inhalation of pollen, dust, animal danders, and insect allergens. Allergic rhinitis can be present as a seasonal or nonseasonal disease. Allergic rhinitis is due to the interaction of allergens and IgE fixed to specific mast cells in the epithelial surface of the nasal mucosa. This results in the release of a number of potent chemical mediators, including histamine, eosinophil and other chemotactic factors, leukotrienes, and prostaglandins, which then contribute to the pathogenesis of the disease.[3–7] In seasonal allergic rhinitis, largely attributed to pollen sensitivity, the significance of exposure to fungal spores cannot be underestimated. Persistent or intermittent nasal obstruction and nasal discharge are common but cannot be differentiated from other forms of rhinitis. Eye irritation with lacrimation, occasional redness, and itching are also commonly

TABLE I
Fungi Reported as Allergens

Zygomycetes	Deuteromycetes	Deuteromycetes (*Cont.*)
Mucor	*Acremonium*	*Paecilomyces*
Rhizopus	*Alternaria*	*Penicillium*
Ascomycete	*Aspergillus*	*Phoma*
Chaetomium	*Aureobasidium*	*Sporotrichum*
Claviceps	*Botrytis*	*Stemphylium*
Daldinia	*Candida*	*Torula*
Didymella	*Cladosporium*	*Trichoderma*
Erysiphe	*Coniosporium*	*Trichophyton*
Eurotium	*Curvularia*	*Trichothecium*
Microsphaera	*Drechslera*	Mushrooms (Basidiomycetes)
Zylaria	*Epicoccum*	*Agaricus*
Basidiomycetes	*Fusarium*	*Cantharellus*
Puccinia	*Gliocladium*	*Ganoderma*
Tilletia	*Helminthosporium*	*Polyporus*
Urocystis	*Monilia*	*Xylobolus*
Ustilago	*Nigrospora*	

seen in these patients. The clinical features may also include sneezing, clear watery rhinorrhea, nasal congestion, and pruritis. Sneezing and nocturnal cough have also been reported subsequent to exposure to fungal spores. The symptoms of allergic rhinitis cause marked discomfort that may interfere with the life-style of the patients. Complications of allergic rhinitis include sinusitis, nasal polyps, craniofacial abnormalities, and behavioral problems.

The diagnosis of allergic rhinitis is based on the clinical history, the demonstration of IgE-mediated skin reactivity to appropriate allergens, and environmental analysis for aeroallergens. Skin test reactivity to exposed allergen is important but must be interpreted properly.[8] *In vitro* tests, such as the radioallergosorbent test (RAST) and the paper radioimmunosorbent test (PRIST), are useful in the diagnosis.[1–9] However, reliable standardized allergens/antigens for these tests are not universally available; in addition, the *in vitro* tests are expensive.[10]

The management of the disease involves symptomatic relief through treatment with antihistamines, oral decongestants, and intranasal or rarely systemic corticosteroids. Avoiding environmental exposure to possible allergens is the best course of action. Immunotherapy may be of benefit in some individuals but must be administered cautiously.

3. ALLERGIC ASTHMA

Asthma develops following exposure to specific allergens. Although a number of fungi have been implicated in allergic asthma, the most frequently recognized ones are few. These fungi are listed in Table I.

Classic asthma due to molds may be relatively rare; it can occur in conjunction with dust- or pollen-induced asthma. Mold-associated asthma may be more prevalent in children than in adults in some geographical areas. On exposure to mold spores, nasal symptoms that suggest a cold may occur, followed by coughing and wheezing. Usually no fever will be present unless there is an associated respiratory infection. The symptoms of mold-induced asthma may be differentiated from those due to other inhalant allergens by history of specific exposure, skin test reactivity, or *in vitro* demonstration of specific IgE antibodies to specific antigens. The lack of sufficiently pure and reliable allergenic extracts, however, poses a major problem in the development of *in vitro* and *in vivo* tests.[10] *In vitro* tests are expensive and less sensitive than dermal hypersensitivity tests.

As reactions to mold may coexist with other conditions, such as pollen, dust, and food allergy, it is essential that mold-induced asthma be specifically differentiated. In children careful clinical evaluations are needed to differentiate the disease from croup and cystic fibrosis. In adults chronic obstructive pulmonary diseases and chronic bronchitis need to be differentiated from mold asthma.

4. ALLERGIC FUNGAL SINUSITIS

Sinus disease due to fungi constitute several clinicopathological syndromes, which include invasive, mycetoma production and allergic fungal sinusitis.[5,11,12] The invasive fungal infection as well as fungal mycetoma differ considerably from the allergic sinusitis in their pathophysiology. Allergic fungal sinusitis occurs in atopic individuals, is chronic in nature, and affects multiple sinuses without tissue invasion.

Several fungi have been implicated in allergic fungal sinusitis, the most important of which are *Aspergillus*,[5] *Curvularia*,[11,12] *Alternaria*,[13] and *Bipolaris*.[13] The majority of the patients are young, and atopic and nasal polyposis, asthma, and rhinitis are frequently noted.[13] Mucoid impaction, "allergic mucin"-containing fragments of the fungi, similar to those seen in the sputum plugs of allergic bronchopulmonary aspergillosis (ABPA), is consistently noticed in allergic sinusitis.[5] The sinus exudate may contain eosinophils and Charcot-Leyden crystals, in addition to epithelial cells and cellular debris. Fungus hyphae may be abundant and can be demonstrated by staining with Gomori's silver methanamine stain. No histological evidence of tissue invasion can be detected. Sinus tissue usually shows a thickened basement membrane and edematous stroma containing eosinophils, neutrophils, mast cells, plasma cells, and lymphocytes. Progressive thickening and clouding of several or all paranasal sinuses may be seen on radiological examination.

Patients with allergic fungal sinusitis demonstrate immediate and late cutaneous reactivity to the appropriate fungal antigen. Total and specific serum IgE and IgG antibodies have shown marked elevations in these patients.[11]

Since there is no invasion of the fungus, systemic treatment is not warranted. However, debridement of allergic mucin and restoration of sinus mucus drainage are appropriate therapy. This will help to decrease the fungal antigen load and theoretically can decrease the stimulus for immunologically mediated inflammation. Because of the similarity of fungal sinusitis with ABPA, treatment with corticosteroids has also been suggested.

5. ALLERGIC BRONCHOPULMONARY MYCOSES

Allergic bronchopulmonary mycoses (ABPM) are allergic lung diseases characterized by chest roentgenographic infiltrates, peripheral blood eosinophilia, and immunological pulmonary inflammation with parenchymal lung destruction. Other common features include proximal bronchiectasis, the demonstration of serum precipitins to the offending antigen, elevated total and specific serum IgE, and immediate and late cutaneous hypersensitivity to fungal antigens. The majority of the bronchopulmonary mycoses are reported to be caused by *Aspergillus fumigatus* (Af) or another species of *Aspergillus*; less frequently other species of fungi have also been implicated[3] (Table II). Because of the similarities of the different bronchopulmonary mycoses and the universal presence of ABPA, the latter disease will be dealt with in more detail.

The first report of ABPA came from England in 1952.[14] Ever since then, large number of case reports have appeared from all parts of the world. ABPA has been reported to occur in 20% of asthmatic patients admitted to hospitals and in 5% of all rhinitis cases, while the incidence of ABPA in cystic fibrosis may vary from 10% to 25%. Thus the actual incidence of ABPA may be much higher than is indicated from the number of reported cases.

5.1. Clinical Diagnosis

The diagnosis of ABPA is usually not difficult if all the criteria such as asthma, peripheral blood eosinophilia, roentgenographic infiltrates, sputum-containing plugs of *Aspergillus* hyphae, elevated total serum IgE, and specific antibodies to Af or other fungal antigens; immediate and late skin test reactivity to specific antigens; positive sputum culture; and proximal bronchiectasis are present.[3,4,15] Characteristic radiographic features include upper lobe infiltrates, which may shift from one location to another as the episode takes its course. Repeated episodes may lead to edema, dilatation, and destruction of bronchial walls, resulting in cavitation, atelectasis, and fibrosis. Pulmonary function studies may show diminished diffusion and reduction in vital and total lung capacities. Reversibility of obstruction may be documented after administration of bronchodilators, depending on the extent of damage to the pulmonary parenchyma and bronchial walls. Pulmonary function and chest x-rays may return to normal between episodes.[3,4]

TABLE II
Fungi Reported from Allergic Bronchopulmonary Mycoses

Fungi	Reference	Aspergillus	Reference
Candida	35	*A. fumigatus*	14
Curvularia	36, 37	*A. terreus*	39
Dreschleria	37	*A. oryzae*	40
Stemphyllum	38	*A. ochraceans*	41
		Helminthosporium	42

5.2. Immunology

The immunology of ABPA is not completely understood.[15–17] All patients with ABPA have elevated IgG and IgE antibodies specific to Af or to other fungi involved in the disease.[3,18,19] Brummund et al. reported that although IgG and IgE are consistently elevated in all ABPA sera studied, there was evidence of polyclonal elevation of all isotypes.[20] Because of the high levels of precipitating antibodies and the presence of a late intradermal reaction to *Aspergillus* antigens, an immune complex mechanism has also been implicated in ABPA. Additionally, complement components and antigen–antibody complexes have also been detected in pulmonary lesions of ABPA patients.

Although IgG and IgE antibodies to the involved antigens are essential in the development of ABPM, the lymphocytic infiltration and granuloma formation in the lung implicates a cell-mediated immune response in these patients as well. Antigen-induced lymphocyte transformation has been shown to occur in ABPA by several investigators.[16,17] However, the lymphoproliferation decreased during the acute exacerbation and lymphocyte phenotype assays demonstrated that T-helper inducer cells showed a significant increase, while in the peripheral circulation the suppressor cell population was normal. Interleukin-2 positive T-cells also have been detected from both peripheral and lung lymphocytes, indicating that helper T-cells are activated by *Aspergillus*. No difference was noted between the peripheral and lung lymphocytes.[21]

5.3. Laboratory Diagnosis

All patients exhibit immediate cutaneous reactivity to Af antigens. An Arthus-type reaction will usually follow and occasionally a delayed reaction may also be produced.[15] Total serum IgE is of major value in the diagnosis and management of the disease.[19] At the time of diagnosis, patients demonstrate elevated total serum IgE and a moderate increase in the levels of total serum IgG and IgA.[22] *A. fumigatus*-specific antibodies can be detected by an agar gel double-diffusion test. The specific IgG, IgA, IgE, and subclasses of IgG can be estimated by ELISA. Although standardized antigens are not currently available, with careful use of appropriate controls ELISA can be conveniently used to demonstrate specific antibodies in the patient sera.[23]

5.4. Patient Management

The management of patients with ABPA includes control of both asthma and ABPA. The source of antigen should be determined and exposure should be avoided. Prednisone is the drug of choice for the treatment of ABPM. Other agents, such as antifungal agents, cromolyn, bronchodilators, and inhaled corticosteroids, have been tried but with only limited success. Repeated evaluation of patients is necessary for successful control.

6. HYPERSENSITIVITY PNEUMONITIS

Hypersensitivity pneumonitis (HP) or extrinsic allergic alveolitis is caused by exposure to a number of organisms in susceptible individuals.[6] Included among them are farmer's lung, bagassosis, and mushroom worker's lung, resulting from occupational exposure to thermophilic actinomycetes present in hay, bagasse, and mushroom compost, respectively, while ventilation-system-induced pneumonitis is a result of the contaminated heating or air-conditioning systems of residential and commercial buildings.[6,24] These diseases result from the inhalation of actinomycetes antigens belonging to the genus *Faenia* and *Thermoactinomyces*. HP due to fungi have also been reported and the common organisms implicated are *Aspergillus* and *Penicillium* species. The organisms commonly associated with HP are shown in Table III.

The first report of a well-characterized example of hypersensitivity, namely, farmer's lung, came from Great Britain in 1932.[25] Subsequently other occupationally linked HP were recognized from several parts of the world.[26–29]

TABLE III
Hypersensitivity Pneumonitis Caused by Fungi and Actinomycetes

Disease	Source of environment	Antigen
Farmer's lung	Moldy hay, moldy corn	*Faenia rectivirgula* (*Micropolyspora faeni*) *Thermoactinomyces candidus*, *T. vulgaris*, *Saccharomonospora viridis*
Bagassosis	Moldy bagasse	*Thermoactinomyces sacchari*, *T. candidus*, *T. vulgaris*
Mushroom worker's lung	Moldy compost	*Thermoactinomyces vulgaris*, *T. candidus*, *Saccharomonospora viridis*, *Faenia rectivirgula*
Ventilation systems pneumonitis	Contaminated forces air systems	*Thermoactinomyces candidus*, *T. vulgaris*, *Penicillium* spp.
Maple bark disease	Contaminated bark	*Cryptostroma corticale*
Malt worker's lung	Moldy malt	*Aspergillus clavatus*, *A. fumigatus*
Sequoiosis	Moldy redwood dust	*Graphium* sp., *Aureobasidium pullulans*, *Trichoderma* sp.
Cheese worker's lung	Cheese mold	*Penicillium caesi*
Suberosis	Moldy cork	*Penicillium caesi*
Sauna taker's lung	Moldy tub	*Aureobasidium pullulans*

Mushroom worker's lung, caused by the inhalation of compost dust containing thermophilic actinomycetes, has also been reported.[24,26] Moldy bagasse was reported as the antigen responsible for bagassosis, while contaminated air-handling systems with thermophilic actinomycetes were attributed to the development of HP in exposed susceptible individuals.[27] The altered environment, particularly closed buildings and artificial air-handling systems, might have contributed to the selective growth and proliferation of the various organisms, resulting in sensitization and development of the disease in susceptible individuals.[29]

6.1. Clinical Diagnosis

The common and most distinguishable form of HP is the acute form resulting from intermittent exposure to an antigenic source. The patient becomes sensitized following inhalation of antigens and on subsequent exposure develops characteristic symptoms usually 4–8 hours after inhalation of the antigens. In the acute form the symptoms may persist up to 12 hours and the patient may recover spontaneously. The symptoms usually reappear each time the patient is exposed to the antigens. The patients remain asymptomatic in between the exposures.

In the chronic form of HP, irreversible lung damage may occur frequently. Continuous exposure to the antigens may result in an intense immunological inflammatory reaction in the lung. Fibrosis with irreversible pulmonary insufficiency can be seen in these patients. Lung biopsies may reveal interstitial fibrosis with granuloma formation and alveolar wall thickened by infiltration with lymphocytes, plasma cells, and eosinophils. The patient may develop ventilatory impairment and diffusion defects that do not respond to prolonged avoidance or corticosteroid therapy.[15] Radiological findings vary from mild interstitial infiltrates to abnormalities similar to those seen in acute pulmonary edema and interstitial pneumonitis. During acute attacks, soft patchy ill-defined parenchymal densities that tend to coalesce may be seen in all lung fields. In chronic phases of the disease the lung may present the appearance of a fibrotic process with diffuse interstitial disease. The x-ray findings are usually nonspecific.

Pulmonary function studies show considerable variation in HP caused by thermophilic actinomycetes. In the acute forms a primary restrictive abnormality is characteristic, although some degree of small-airways disease and airway obstruction may be present. The diffusion capacity is frequently abnormal in the early part of the disease and diminishes further with advancing severity. In the chronic form of the disease, severe fibrosis with irreversible pulmonary insufficiency can be seen. These patients may develop ventilatory impairment and diffusion defects that do not respond to prolonged avoidance or corticosteroid therapy. Lung biopsies from these patients show interstitial fibrosis with granuloma formation and alveolar wall thickening due to infiltration of lymphocytes, plasma cells, and eosinophils.

6.2. Immunology

The most characteristic feature of HP is the universal presence of serum antibodies against offending antigens.[6,24] The antigens used in detecting the antibodies are complex and have inconsistent reactivity with the antibodies. Moreover, cross-reactivities between different species of some antigens have been reported, however, in most instances, the antibody concentrations are high enough to be detected by the agar-gel double-diffusion method. Although antibodies against some antigens belonging to all immunoglobulin isotypes have been reported, most significantly high levels of IgG, IgM, and IgA have been detected in the sera of most patients with farmer's lung.[30] Antigen–antibody crossed electrophopresis (CIE) and Western blots of two-dimensional polyacrylamide gel electrophoresis can be used in the differentiation of patients from exposed individuals.[31]

The role of cell-mediated immunity in HP has been recognized because of the granulomatous inflammatory response in the lung, the delayed-type skin test reactivity, and the presence of antigen-sensitized lymphocytes in the peripheral circulation and the lungs. Cell-mediated immunity has also been demonstrated by the macrophage migration inhibition assay and by antigen-induced lymphocyte transformation.[6,24] Lymphocyte phenotype studies using monoclonal antibodies failed to show any differences between asymptomatic but exposed individuals with serum precipitins and patients.[32,33] Bronchoalveolar lavage studies of HP patients indicate marked lymphocytosis with a predominance of suppressor cells. The suppressor cells from patients were functionally defective, while those from asymptomatic exposed individuals were apparently normal.[32,33]

Because of the presence of circulating antibodies against antigens, the involvement of immune complexes in pathogenesis has been considered. This view, however, is questioned because of the presence of high levels of circulating antibodies in the sera of asymptomatic exposed individuals. The 4- to 6-hour delay in the onset of symptoms after exposure to the antigens and immune complex deposition in the lung tissue and Arthus reaction on skin testing with the specific antigens supports the immune-complex-induced pathogenesis. On the other hand, sensitized lymphocytes in the peripheral circulation and in the lung, typical lymphocytic granuloma, lack of neutrophil infiltration in the lung, and delayed-type skin test reactivity all support the presence of cell-mediated immunity. These findings indicate that all types of immune mechanisms may be involved in HP. Various animal model studies also support the fact that activated T cells and immune complexes may be involved in the pathogenesis of HP.[34]

6.3. Laboratory Diagnosis

Hypersensitivity pneumonitis is characterized by the presence of circulating antibodies against the offending antigens. Antigens extracted from the organisms may be used in agar-gel double diffusion or in ELISA for detecting circulating antibodies.[6,30] Knowledge of the organisms present in the environment may also help to determine the antigen involved in the disease.[6] Hence, careful culture and identification methods may be used to determine the organisms contaminating the environment. Patients may frequently show antigen-induced lymphocyte transformation and a delayed hypersensitivity skin reaction on testing with appropriate antigens. However, standardized antigens are not available, although the currently available characterized antigens may give reasonably good results if used with appropriate controls and careful techniques.[6,25,30]

6.4. Patient Management

The treatment of HP due to organic dust antigens depends on recognition of the antigen and contaminated environment involved in the disease. The most important therapeutic measure in HP should be avoidance of offending antigens. Patients should use masks with filters capable of removing the dust. The moisture content of the hay and other materials has to be controlled to minimize the growth of the thermophiles. In addition, frequent cleaning of air exchangers and appropriate ventilation in the work area are essential for controlling exposure to offending agents. Corticosteroids are frequently useful in acute cases, while appropriate therapy for progressive pulmonary insufficiency may be necessary in patients with irreversible lung damage.

7. CONCLUSIONS

Allergic diseases due to fungi are important because of their widespread occurrence and disabling feature; however, not much progress has been made in the specific diagnosis and successful management of the disease. Antigens available currently for *in vitro* or *in vivo* diagnostic procedures are crude antigens and may show considerable variations in their reactivity and dependability. Although attempts have been made to obtain pure and standardized antigens for immunodiagnosis, results to date are not very encouraging. Our understanding of the immune mechanisms has also not progressed due to the nonavailability of standardized reagents. Appropriate control measures and patient management can be directed to understanding immune regulation in these diseases.

ACKNOWLEDGMENTS. This study was supported by the Veterans Affairs Medical Research Service and National Institutes of Health (AI-23071). The editorial assistance of Mavis Fuderer is gratefully acknowledged.

REFERENCES

1. Bierman, W. C., and Van Arsdel, P. P., Jr., 1987, Clinical evaluation of the patient with allergic and immunologic disease, in: *Principles of Immunology and Allergy* (R. F. Lockey and S. C. Bukantz, eds.), W. B. Saunders, Philadelphia, pp. 1–26.
2. Burge, H. A., 1989, Airborne—allergenic fungi, *Immunol. Allergy Clin. North. Am.* **9:** 307–319.
3. Greenberger, P. A., 1988, Allergic bronchopulmonary aspergillosis and fungoses, *Clin. Chest Med.* **9:**599–608.
4. Fink, J. N., 1988, Allergic bronchopulmonary aspergillosis, *Hosp. Pract.* **23:**105–128.
5. Katzenstein, A. A., Sale, S. R., and Greenberger, P. A., 1983, Allergic Aspergillus sinusitis: A newly recognized form of sinusitis, *J. Allergy Clin. Immunol.* **72:**89–98.
6. Kurup, V. P., 1989, Hypersensitivity pneumonitis due to sensitization with thermophilic actinomycetes, *Immunol. Allergy Clin. North. Am.* **9:**285–306.
7. Slavin, R. G., 1987, Rhinitis, sinusitis, otitis and oral lesions, in: *Principles of Immunology and Allergy* (R. F. Lockey and S. C. Bukantz, eds.), W. B. Saunders, Philadelphia, pp. 27–44.
8. Tarlo, S. M., Fradkin, A., and Tobin, R. S., 1988, Skin testing with extracts of fungal species derived from homes of allergy clinic patients in Toronto, Canada, *Clin. Allergy* **18:**45–52.
9. Kozak, P. P., Jr., and Hoffman, D. R., 1984, Critical review of diagnostic procedures for mould allergy, in: *Mould Allergy* (Y. Al-Doory and J. F. Davison, eds.), Lea and Febiger, Philadelphia, pp. 157–186.
10. Yunginger, J. W., 1988, Allergenic extracts: Characterization, standardization and prospects for the future. *Pediatr. Clin. North Am.* **30:**795–805.
11. Brummund, W., Kurup, V. P., Harris, C. J., Duncavage, J. A., and Arkins, J. A., 1986, Allergic sinorbital mycoses, *JAMA* **256:**3249–3253.
12. Macmillan, R. H., Cooper, P. H., Brody, B. A., and Mills, A. S., 1987, Allergic fungal sinusitis due to *Curvularia lunata*, *Human Pathol.* **18:**960–964.
13. Gourley, D. S., 1989, Allergic fungal sinusitis, *Insights Allergy* **4:**1–7.
14. Hinson, K. F. W., Moon, A. J., and Plummer, N., 1952, Bronchopulmonary aspergillosis, *Thorax* **7:**317–333.
15. Bardana, E. J., 1981, The clinical spectrum of aspergillosis. Part 2. Classification and description of saprophytic, allergic and invasive variants of human disease, *CRC Crit. Rev. Clin. Lab. Sci.* **13:**85–159.
16. Knutsen, A. P., and Slavin, R. G., 1989, In-vitro T-cell responses in patients with cystic fibrosis and allergic bronchopulmonary aspergillosis, *J. Lab. Clin. Med.* **113:**428–435.
17. Walker, C. A., Fitzharris, P., Longbottom, J. L., and Taylor, J. N., 1989, Lymphocyte sensitization to *Aspergillus fumigatus* in allergic bronchopulmonary aspergillosis, *Clin. Exp. Immunol.* **76:**34–40.
18. Patterson, R., Roberts, M., Ghory, A. C., and Greenberger, P. A., 1980, IgA antibody against *Aspergillus fumigatus* antigen in patients with allergic bronchopulmonary aspergillosis, *Clin. Exp. Immunol.* **42:**395–398.
19. Patterson, R., Rosenburg, M., and Roberts, M., 1977, Evidence that *Aspergillus fumigatus* growing in the airway of man can be a potent stimulus of specific and non-specific IgE formation. *Am. J. Med.* **63:**257–262.
20. Brummund, W., Resnick, A., Fink, J. N., and Kurup, V. P., 1987, *Aspergillus fumigatus*-specific antibodies in allergic bronchopulmonary aspergillosis and aspergilloma: Evidence for a polyclonal antibody response. *J. Clin. Microbiol.* **25:**5–9.
21. Jarmoc, L. M., Maton, M. C., Hutcheson, P. S., Bedrossian, C. W., and Slavin, R. G., 1988, The immunopathology of cystic fibrosis with and without allergic bronchopulmonary aspergillosis. *J. Allergy Clin. Immunol.* **81:**185 (abstract).
22. Kurup, V. P., Resnick, A., Kalbfleisch, J., and Fink, J. N., 1990, Antibody isotype responses in Aspergillus induced diseases, *J. Lab. Clin. Med.* **115:**298–303.

23. Kurup, V. P., Resnick, A., Scribner, G. H., Gunasekaran, M., and Fink, J. N., 1986, Enzyme profile and immunochemical characterization of *Aspergillus fumigatus* antigens, *J. Allergy Clin. Immunol.* **78:**1166–1173.
24. Kurup, V. P., Barboriak, J. J., and Fink, J. N., 1984, Hypersensitivity pneumonitis, in: *Mould Allergy* (Al-Doory, Y., and Domson, J. F.), Lea and Febiger, Philadelphia, pp. 216–243.
25. Campbell, J. M., 1932, Acute symptoms following work with hay, *Br. Med. J.* **II:**1143–1144.
26. Bringhurst, L. S., Byrne, R. N., and Gershon-Cohen, J., 1959, Respiratory disease of mushroom workers. Farmer's lung, *JAMA* **171:**15–18.
27. Buechner, H. A., 1960, Bagassosis: Peculiarities of its geographical pattern and report of the first case from Peru and Puerto Rico, *JAMA* **174:**1237–1241.
28. Dickie, H. A., and Rankin, J., 1958, Farmer's lung and acute granulomatous interstitial pneumonitis occurring in agricultural workers, *JAMA* **167:**1069–1076.
29. Banaszak, E. F., Thiede, W. H., and Fink, J. N., 1970, Hypersensitivity pneumonitis due to contamination of an air conditioner, *N. Engl. J. Med.* **283:**271–276.
30. Brummund, W., Kurup, V. P., Resnick, A., Milson, T. J., Jr., and Fink, J. N., 1988, Immunological response to *Faenia rectivirgula* (*Micropolyspora faeni*) in a dairy farm family, *J. Allergy Clin. Immunol.* **82:**190–195.
31. Kurup, V. P., 1987, Detection of relevant farmer's lung antigens by using immunoblots of two-dimensional electrophoresis. *Diagn. Immunol.* **5:**25–29.
32. Leatherman, J. W., Michael, A. F., Schwartz, B. A., and Hoidal, J. R., 1984, Lung T-cells in hypersensitivity pneumonitis, *Ann. Intern. Med.* **100:**390–392.
33. Cormier, Y., Belanger, J., Beaudoin, M., Laviolette, R., Beaudoin, R., and Herbert, T., 1984, Abnormal bronchoalveolar lavage in asymptomatic dairy farmers: A study of lymphocytes, *Am. Rev. Respir. Dis.* **130:**1046–1049.
34. Keller, R. H., Calvanico, N. J., and Stevens, J. O., 1982, Hypersensitivity pneumonitis in nonhuman primates, *J. Immunol.* **128:**116–122.
35. Akiyama, K., Mathison, D. A., Riker, J. B., Greenberger, P. A., and Patterson, R., 1984, Allergic bronchopulmonary candidiasis, *Chest* **85:**699–670.
36. Halwig, J. M., and Brueske, D. A., Greenberger, P. A., Dreisin, R. B., and Sommers, H. M., 1985, Allergic bronchopulmonary curvulariopsis, *Am. Rev. Respir. Dis.* **132:**186–188.
37. McAleer, R., Kroenert, D. B., Elder, J. L., and Foudist, J. H., 1981, Allergic bronchopulmonary disease caused by *Curvalaria lunate* and *Drechslera hawaiiensis*, *Thorax* **36:**338–344.
38. Benatar, S. R., Allan, B., Hewitson, R. P., and Don, P. A., 1980, Allergic bronchopulmonary stemphyliosis, *Thorax* **35:**515–518.
39. Laham, M. N., Allen, R. C., and Greene, J., 1981, Allergic bronchopulmonary aspergillosis (ABPA) caused by *Aspergillus terreus*: Specific lymphocyte sensitization and antigen directed serum opsonic activity, *Ann. Allergy* **46:**74–80.
40. Akiyama, K., Takizawa, H., Suzuki, M., Miyachi, S., Ichinoke, M., and Yanagihara, Y., 1987, Allergic bronchopulmonary aspergillosis due to *Aspergillus oriyzae*, *Chest* **91:**285–286.
41. Novey, H. S., and Wells, I. D., 1978, Allergic bronchopulmonary aspergillosis caused by *Aspergillus ochraceus*, *Am. J. Clin. Pathol.* **70:**840–843.
42. Hendrick, D. J., Ellithorpe, D. B., Lyon, F., Hattier, P., and Salvaggio, J. E., 1982, Allergic bronchopulmonary helminthosporiosis, *Am. Rev. Respir. Dis.* **126:**935–938.

18

Fungal Infections in the Immunocompromised Host

Clinical Syndromes and Diagnosis

HERMAN CHMEL

1. INTRODUCTION

Fungal infections are a major cause of morbidity and mortality in the immunocompromised host. In the last 25 years, the frequency of invasive fungal infections has increased remarkably. Unbiased data concerning the true incidence and prevalence of invasive fungal infections in different patient populations is limited. Hart *et al.*[1] reported 132 invasive fungal infections in normal and immunocompromised patients. Seventy infections occurred in immunocompromised patients, 81% caused by *Candida* spp., *Aspergillus* spp., and *Zygomycete* spp. and 12% by *Cryptococcus* spp. In contrast, 62 invasive fungal infections occurred in normal patients, and none were caused by *Candida, Aspergillus,* or *Zygomycete* spp. Muller *et al.*[2] reported the following incidence of opportunistic deep-seated mycosis, candidiasis 92.6%; aspergillosis 6.7%; cryptococcosis 0.35%, and zygomycosis 0.35%. Rose and Varkey,[3] in a comprehensive review of invasive fungal infections, reported per 10,000 hospital discharges the following rates: *Candida* spp. 7.08, *Aspergillus* spp. 1.4, and *Zygomycete* spp. 0.23. Data from the National Nosocomial Infection Surveillance Program (CDC)[4] representing 1984 rates showed *Can-*

HERMAN CHMEL • Division of Infectious and Tropical Diseases, Department of Internal Medicine, College of Medicine, University of South Florida, Tampa, Florida 33612. *Present address*: Department of Medicine, St. Francis Medical Center, Trenton, New Jersey 08629.

Fungal Infections and Immune Responses, edited by Juneann W. Murphy *et al.* Plenum Press, New York, 1993.

dida spp. accounted for 5.5% of all isolates and was the eighth most common nosocomial pathogen. The category listed as "other fungi" accounted for 1.7% of all isolates and ranked 11th. Although data from autopsies are limited and selective, the relative frequency of invasive fungal infections in transplant and oncologic patients can be inferred by their review. The observed frequency of invasive fungal infections is 20–30% in patients with acute leukemia, 10–15% in patients with lymphoma, and 5% in patients with other malignancies.[5] The frequency in organ transplant recipients is 28–42% for liver, 10–35% for heart, 2–30% for bone marrow, and 0–20% for kidney.[6] Overall invasive candidiasis accounted for death in 10–40% of patients and invasive aspergillosis in 5–15%.[7]

In patients with human immunodeficiency virus (HIV) infection, the frequency of invasive fungal infections is difficult to ascertain, although as many as 95% of patients will manifest evidence of oropharyngeal candidiasis sometime during their illness.[8] Cryptococcosis develops in 6–13% of patients with acquired immunodeficiency syndrome (AIDS),[9] while disseminated infections due to *Histoplasma* and *Coccidioides* number less than 100 reported cases.[10]

The agents responsible for systemic fungal infection can be separated into two groups, true pathogenic and opportunistic fungi (Table I). The true pathogenic fungi cause infection in normal hosts that are self-limited but in compromised hosts can be devastating. Once acquired these fungi can remain in a pseudolatent state, recrudesce, and disseminate when the patient becomes immunosuppressed.

A variety of factors have been identified that compromise the patients' immune defenses, predisposing them to serious and fatal fungal infections (Table II).[11–16] Table III outlines additional factors that influence the type of invasive fungal infection seen in the compromised host.[17,18] These risk factors are discussed in more detail in other sections of this book.

Table IV lists some of the diagnostic problems encountered in trying to establish the presence or absence of an invasive fungal infection. The most definitive diagnostic procedure is to demonstrate the fungus in tissue by

TABLE I
Some Agents Responsible for Systemic Fungal Infections

True pathogenic	Opportunistic
Crypotococcus neoformans	*Candida* spp.
Histoplasma capsulatum	*Aspergillus* spp.
Coccidioides immitis	*Zygomycete* spp.
Blastomyces dermatiditis	*Trichosporon* spp.
Paracoccidioides braziliensis	*Fusarium* spp.
	Pseudoallescheria boydii

TABLE II
Host Factors Influencing the Development of Invasive Fungal Infections

	Altered host factors			
Organism	Mucocutaneous	Neutrophile	T lymphocyte/ monocyte	Lymphocyte
Candida spp.	+++	+++	+++	(?)
Cryptococcus neoformans	(?)	(?)	+++	(?)
Aspergillus spp.	+	+++	(?)	(?)
Zygomycete spp.	+	+++	(?)	(?)
Histoplasma capsulatum	(?)	—	+++	(?)
Coccidioides immitis	(?)	—	+++	(?)
Trichosporon spp.	++	+++	(?)	(?)
Fusarium spp.	+++	+++	(?)	(?)
Pseudoallescheria boydii	+	+++	(?)	(?)

Modified after Armstrong *et al.*[11]
+ = slight association.
++ = moderate association.
+++ = definite association.
(?) = questionable association.

biopsy or to isolate the putative fungus in a normally sterile body fluid (cerebrospinal fluid or blood). Many times the former procedure cannot be done because the patient is quite ill or has potentially serious bleeding problems related to severe thrombocytopenia. These problems are most often seen in patients suspect for invasive aspergillosis and zygomycosis.

Definitive diagnostic tests for disseminated candidiasis are lacking. Blood cultures are usually not helpful. Fifty percent of patients with autopsy-documented disseminated candidiasis have not had a positive blood culture prior to death for *Candida* spp. from blood.[19–22] Serologic tests, including antigen and metabolic detection, have been pursued in an effort to better diagnose disseminated candidiasis.[22–24] However, at this writing there is no commercial test available to reliably diagnose disseminated candidiasis.

Concerning aspergillosis and zygomycosis, definitive diagnosis is by demonstrating the organism in tissue biopsies. Serologic tests for zygomycotic infections are lacking. Serologic tests for aspergillus antigen are only available in research laboratories and look promising.[24]

TABLE III
Some Predisposing Factors to Fungal Infection

Broad spectrum antibiotic therapy	Metabolic acidosis and hyperglycemia
Immunosuppressive chemotherapy	Gastrointestinal tract surgery
Corticosteroid therapy	Hematologic malignancy
Debilitation and malnutrition	Intravascular catheters
Parenteral hyperalimentation	

TABLE IV
Diagnostic Dilemmas in Invasive Fungal Infections

Fungi	Clinical features	Cultures[a]	Serology: Antibody	Serology: Antigen	Serology: Metabolite
Candida spp.	Nonspecific <10% retinal and/or skin lesions	Blood: < 50% + late Urine: false + or − Sputum: false + or −	False +/−	Infrequently + universally unavailable	Arabinitol, mannose not commercially available, experimental
Cryptococcus neoformans	Lung: nonspecific CNS: signs meningitis Disseminated: nonspecific may see skin lesions	Blood: occasionally + Urine: occasionally + CSF: + most cases Skin lesions: +	May be +	Frequently + even in culture—disease	—
Aspergillus spp.	Lung: symptoms of infarct with/without cavity CNS: symptoms of mass lesion or stroke Heart: infarct symptoms Disseminated: infarct/ nonspecific Skin: necrosis	Biopsy specimens: often + Blood: very rare + CSF: usually − Sputum: may be + or false +	False negatives, not useful usually	Galactomannan/ other may be +, not readily available	—
Zygomycete spp.	Rhinocerebral: characteristics necrosis/ cranial nerve Lung: Infarction ± cavitation Disseminated: nonspecific Skin: necrosis	Biopsy: often + Blood: usually − CSF: usually − Skin: +	Experimental	—	—

Fusarium spp.	Lung: nonspecific, mimic Aspergillus Disseminated: Skin lesions	Blood: + Skin biopsy: +	—	—	—
Pseudoallescheria boydii	Lung: infarction ± cavitation Disseminated: nonspecific	Biopsy: often +	—	—	—
Trichosporin spp.	Mimics disseminated candidiasis Skin: purpura progressing to necrosis	Blood: + Biopsy: +	—	False + Cryptococcal Antigen test	—
Histoplasma capsulatum	Lung: nonspecific cavitary to diffuse interstitial infiltrates Disseminated: mucosal ulcers skin lesions, fever, lymphadenopathy, hepatosplenomegaly, pneumonia	Sputum: infrequent + Blood: may be + Bone marrow: frequent + in disseminated disease Biopsy: +	CF > 1:8 +[b] ID H band: recent infection[b] M band: old, recent skin test false + not helpful in compromised host	Experimental (+ in urine/serum)	—
Coccidioides immitis	Lung: nonspecific Disseminated: nonspecific	Sputum: ±, highly infectious Blood: infrequent + CSF: ±	CF ≥ 1:16 + ID: + recent infection Other: various tests available	—	—

[a] +: positive, −: negative
[b] CF = complement fixation; ID = immunodiffusion; H = histoplasmin antigen; M = mycelial antigen.

The other fungi that are discussed in this paper have more reliable methods for diagnosis and are discussed elsewhere.

The list of fungi that can cause invasive infection in the immunocompromised patient continues to grow.[25] A brief discussion concerning the diagnosis and clinical manifestations of the more frequently encountered fungal infections is presented.

2. CANDIDIASIS

Candida spp. are responsible for most fungal infections in the immunocompromised host. Although there are over 80 species, *C. albicans*, *C. tropicalis*, *C. parapsilosis*, *C. krusei*, and *C. glabrata* account for most infections. *Candida* spp. are normal commensals of humans, inhabiting the oropharynx, gastrointestinal tract, and vagina.[26] Most *Candida* spp. infections are endogenously acquired and occur when a defect in host immune defense allows invasion of the body. Recovery of *Candida* spp. from a normally sterile body site or recovery of the same species from multiple sites may indicate disseminated infection.

2.1. Clinical Syndromes and Diagnosis

2.1.1. Oropharyngeal Candidiasis

Oral candidiasis (thrush) classically presents as creamy-white raised patches adherent to the mucosa of the oropharynx that bleed easily with scraping. Infection occurs in patients treated with broad-spectrum antibiotics, chemotherapy, and steroids; patients using steroid aerosol inhalers; patients with HIV infection; and other cellular immunodeficiency. Four distinct clinical variants of oral candidiasis are seen in HIV infection.[27] Erythematous (atrophic) candidiasis affects 50% of HIV-infected patients, pseudomembraneous and angular candidiasis are next in frequency, while hyperplastic candidiasis is less often encountered. Oral candidiasis may also predict the development of other opportunistic infections and full-blown AIDS in adults.[28–30] Diagnosis is made by demonstrating pseudophypha or yeast forms on Gram stain and growing *Candida* spp. in cultures from scrapings of the mucosal lesions.

2.1.2. Candida Epiglottitis

Candida spp. have been reported to cause epiglottitis in immunocompromised hosts.[31–33] Patients complain of difficulty swallowing, severe pain, and odynophagia. Diagnosis is accomplished by visualizing thrushlike lesions involving the epiglottitis by laryngoscopy and by recovering material for Gram stain and culture.

2.1.3. *Candida Esophagitis*

Esophagitis occurs in patients who receive cytotoxic chemotherapy or corticosteroids, and in those who have hematologic malignancies, diabetes mellitus, esophageal motility problems, and HIV infection.[34–37] Patients complain of dysphagia or odynophagia. Half of the patients may not have symptoms. Barium swallow may show irregular and edematous folds and ulcerations, which are suggestive of the diagnosis. Definitive diagnosis is made by esophagoscopy, demonstrating a friable mucosa with an overlying yellow-white pseudomembrane, and biopsy or brushings. Endoscopy should probably not be performed in severely immunocompromised patients with neutropenia and thrombocytopenia because of the possibility of bacteremia and bleeding.

2.1.4. *Hepatosplenic Candidiasis*

Hepatosplenic candidiasis has become a significant problem in immunocompromised patients because of the increased use of intensive-dose chemotherapy. Hepatic candidiasis occurs in 5–59% of patients with disseminated infection, but occasionally the infection is confined to the liver and/or spleen.[38–40] Tables V and VI describe the clinical and laboratory characteristics in hepatosplenic candidiasis. Diagnosis is made by biopsy, demonstrating yeasts or pseudohypha in hepatic tissue with a surrounding granulomatous

TABLE V
Hepatosplenic Candidiasis: Clinical and Laboratory Features

	N (73)	%		N (73)	%
Sex			Laboratory anomalies		
Male	30	(41)	Leukocytosis > 10,000	21	(30)
Female	32	(44)	Elevated alkaline phosphatase	45	(62)
Unknown	11	(15)	Elevated bilirubin	17	(23)
Age			Elevated transaminases	14	(20)
Range (yrs)	1–64		Noninvasive imaging		
Median	26		CT	53/58	(91)
Underlying disease			US	31/45	(68)
Acute leukemia	64	(87)	Tc liver scan	12/32	(40)
Lymphoma	4	(6)	Sites affected (65)		
Sarcoma	2	(3)	Liver	28	(38.5)
Aplastic anemia	3	(4)	Spleen	5	(7.0)
Clinical presentation			Liver + spleen	39	(54.5)
Fever	63	(86)	Evidence of dissemination	18	(24.6)
Abdominal pain	40	(55)	Causative species	29	(40)
Hepatomegaly	33	(45)	*C. albicans*	23	(80)
Splenomegaly	30	(41)	*C. tropicalis*	5	(17)
			C. albicans/tropicalis	1	(3)
			Outcome (68) death	30	(44)

TABLE VI
Hepatosplenic Candidiasis: Characteristics of Presentation

Patients with acute leukemia with prolonged neutropenia following chemotherapy who achieve remission
Develops fever unresponsive to antibiotic therapy
Anorexia, fever, weight loss
Abdominal pain, right upper quadrant
Hepatosplenomegaly
Markedly elevated alkaline phosphatase

reaction. Occasionally patients will require exploratory laporatomy to establish the diagnosis. It is also important to note that even after presumed successful therapy, patients at autopsy may have residual evidence of yeast in the liver.[40]

2.1.5. Candida Urinary Tract Infection

Candida spp. have been isolated from the urine of approximately 10% of normal patients. Asymptomatic candiduria, cystitis, pyelonephritis, papillary necrosis, and fungus ball formation have been caused by *Candida* spp.[41] *C. albicans* and *C. glabrata* account for the majority of infections. Risk factors for the development of candida urinary tract infection include an indwelling (Foley) catheter, the use of broad-spectrum antibiotics, and diabetes mellitus. Clinical and laboratory criteria do not reliably distinguish between colonization and the various types of infection. Invasive procedures, such as cystoscopy and contrast studies of the urinary tract, may be helpful with diagnosis, especially for fungus ball formation.

2.1.6. Candidemia and Disseminated Candidiasis

Candida fungemia has become an important infection in the hospital because of its increasing incidence and the excess mortality associated with it.[42,43] Candidemia can cause dissemination of the yeast leading to systemic (invasive) candidiasis. Risk factors for fungemia have included diabetes mellitus, malnutrition, renal failure, gastrointestinal tract surgery, burns, trauma, transplantation, neutropenia, leukemia, the use of broad-spectrum antibiotics and corticosteroids, hyperalimentation, the use of central intravenous catheters, and chemotherapy.[42,43] Wey *et al.*[43] recently reported a study describing the risk factors for nosocomial candidemia. They concluded that the use of multiple antibiotics, prior hemodialysis, the use of a Hickman catheter, and multiple sites other than blood positive for *Candida* spp. had the highest probability of developing nosocomial candidemia.

Vascular catheters, especially central lines used for parenteral nutrition, contribute substantially to candidemia. Colonization of intravascular catheters by *Candida* spp. occurs in 2–8% of patients, with as many as 20% developing candidemia.[44–46] The ultimate complication of colonization of intravascular catheters is disseminated infection, which in the immunocompromised patient carries a high mortality rate. Another source of fungemia is colonization or infection in the gastrointestinal tract.

Disseminated candidasis is most frequently reported in patients with acute leukemia receiving cytoreductive chemotherapy. In this setting, the most important risk factors are profound and persistent neutropenia, and the use of vascular catheters.[47] The source of *Candida* spp. is either the skin or the gastrointestinal tract. Clinically, there are very few findings suggestive of disseminated disease. These include skin and eye lesions in less than 10% of patients. Diagnosis is by biopsy, blood cultures, or cultures of various clinical specimens. Blood cultures are positive in less than 50% of patients with autopsy-proven candidiasis. Serological tests have included tests for antibody, antigen, and metabolic byproducts of *Candida*. Tests for antibody lack sensitivity and specificity, and are therefore not reliable for diagnosing disseminated candidiasis.[22,24,48] Currently available tests for circulating *Candida* antigen are useless.[22–24,48] Detection of *Candida* metabolites, such as D-arabinitol and mannose, have proven useful but are only available in research laboratories.[22–24] Currently the diagnosis of disseminated candidiasis is difficult; there are no useful clinical or serological methods readily available to clinicians.

3. ASPERGILLOSIS

Aspergillus species are ubiquitous organisms found in soil, water, decaying vegetation, and debri. Their conidia are easily dislodged and dispersed in the air, allowing susceptible hosts to become infected. *A. fumigatus* is responsible for most infections, while *A. flavus*, *A. niger*, and *A. terreus* are infrequently isolated.[48] *Apsergillus* spp. are primarily respiratory pathogens. Nosocomial outbreaks have occurred in hospitals during construction or renovation or when air-conditioning systems are heavily contaminated.[49,50] In the setting of prolonged neutropenia, nasal cultures positive for *Aspergillus* spp. may serve as evidence of infection.[51]

3.1. Clinical Syndromes and Diagnosis

Aspergillus spp. are primarily respiratory tract pathogens, causing mostly pulmonary infections. Various clinical syndromes have been described in the immunocompromised host (Table VII).[52–54] Risk factors predisposing to invasive disease include acute leukemia, the use of high-dose steroids or

TABLE VII
Clinical Syndromes Caused by Aspergillus

Hemorrhagic necrotic bronchopneumonia
Pulmonary infarction
Invasive mycetoma
Rhinocerebral infection
Cerebral infarction
Disseminated infection
Myocardial infarction
Gastrointestinal infarction
Necrotic skin lesions

cytoreductive therapy, and bone marrow transplant patients. The common denominator in these patients is severe and protracted neutropenia. Occasionally locally invasive disease may be seen in the immunosuppressed host where prior tissue damage had occurred, for example, in the skin.[55]

In the immunocompromised host, the spores of *Aspergillus* spp. are inhaled and can produce an acute necrotizing bronchopneumonia, hemorrhagic pulmonary infarct, or fungus balls with cavitation.[52] The infarction and necrosis are related to the propensity of aspergillus hypha to invade blood vessels. Most patients with acute pneumonia develop shortness of breath, a nonproductive cough, occasionally hemoptysis, and have rales on physical examination. If infarction occurs, the patient has the classic signs and symptoms of pulmonary embolus: pleuritic chest pain, fever, hemoptysis, pleural rub, and a wedge-shaped infiltrate.[53] With progression, the lesion cavitates and on radiographs one can appreciate fungus balls in the cavities (invasive mycetomas). The prognosis is usually poor.

Disseminated aspergillosis is the second most common form of invasive aspergillosis, occurring in approximately one quarter of immunocompromised patients.[52,53] Most patients have protracted granulocytopenia.[56] Any organ may be involved. The infections may present as acute vascular events. An early and rapid diagnosis of invasive aspergillosis is difficult to make.

Diagnosis is achieved by demonstrating the septate hypha branching at acute angles in tissue or by culture of infected tissue. Respiratory tract cultures may reveal *Aspergillus* spp., and their clinical significance has to be interpreted in light of the patient's clinical presentation. Aspergillus antibody detection in immunocompromised patients is usually not helpful.[22–24] Methods for the detection of aspergillus antigens, including galactomannon, in various body fluids, including urine, have been developed and appear promising but are only available in research laboratories.[22–24] Currently the diagnosis of invasive aspergillosis is difficult; in the proper setting, the clinical manifestations are suggestive and supporting laboratory tests other than culture are not universally available.

4. ZYGOMYCOSIS

In the immunocompromised host, zygomycotic infections are most often caused by *Mucor* spp. and *Rhizopus* spp., although *Absidia*, *Cunninghamella*, and *Rhizomucor* have also been reported to cause infection.[57] *Zygomycete* spp. account for as much as 15% of fungal infections in the compromised host, making them the third most common fungus responsible for infection in this patient population.[1,58] Zygomycetes are ubiquitous in nature and are found in soil, on fruit and bread, and in decaying debri. Infection results most often from inhaling the spores, although ingestion and traumatic innoculation into the skin have been reported.[57] Risk factors include diabetes mellitus with ketoacidosis, hematologic malignancies, patients receiving cytoreductive chemotherapy resulting in neutropenia, corticosteroid therapy, skin damage (severe burns), organ transplants, illicit intravenous drug use, contaminated air in the hospital environment, and patients receiving deferoxamine therapy.[59] The clinical manifestations of disease are protean and are very similar to invasive *Aspergillus* spp. infections (vascular invasion, tissue infarction, and necrosis). Zygomycetes can be differentiated from aspergillus in that zygomycetes are broad, nonseptate hypha branching irregularly at right angles. Most forms of zygomycosis are infrequently diagnosed antemortem, except the rhinocerebral form.[59]

4.1. Clinical Syndromes and Diagnosis

4.1.1. Rhinocerebral Infection

Rhinocerebral infection is the most common clinical form of zygomycosis.[57,60] Most patients have poorly controlled diabetes with ketoacidosis, the remainder of infections occur in patients with metabolic acidosis (chronic renal failure), acute leukemia, and patients receiving cytoreductive chemotherapy or corticosteroids. Patients complain of headache, facial and eye pain, nasal discharge, fever, and facial swelling. Infection begins in the nasopharynx, spreads to the nasal and paranasal sinuses, orbit, and brain. Examination may reveal a black necrotic eschar in the nares or on the hard palate. The pathologic features are those of vascular invasion, infarction, and hemorrhagic necrosis. Computerized axial tomography is helpful in defining the extent of the disease. Scrapings or biopsy of involved tissue (eschar) may reveal large nonseptate hypha with right angle branching and may provide a diagnosis rapidly. Serological tests for diagnosis are currently not available.[59]

4.1.2. Pulmonary Infection

Pulmonary infection is the second most common from of zygomycosis seen in immunocompromised patients. It occurs in patients with neutropenia related to cytoreductive therapy, acute and uncontrolled leukemia, lymphoma, and diabetes mellitus. Disseminatin occurs frequently. Clinical findings are nonspecific. Infection is characterized by necrotizing hemorrhagic pneumonia and pulmonary infarction. Diagnosis is by biopsy of involved tissue.

4.1.3. Disseminated Zygomycosis

Disseminated zygomycosis occurs in severely immunocompromised patients. Recent reports suggest that prior splenectomy or deferoxamine therapy may be predisposing factors to disseminated disease.[59] Although rarely described in patients with cell-mediated immunodeficiency, disseminated zygomycosis has been reported in patients with AIDS who use intravenous heroin.[59] Clinical manifestations reflect vascular invasion and organ infarction. Antemortem diagnosis is infrequent. Diagnosis requires a high index of suspicion and biopsy of the affected area to demonstrate the broad nonseptate right-angle branching hypha.

5. CRYPTOCOCCOSIS

Cryptococcosis, an infection occurring in either normal or immunocompromised patients, is caused by *Cryptococcus neoformans*. Two varieties of the organism exist; *C. neoformans* var. *neoformans* (serotypes A/D), accounting for over 90% of isolates in the United States, and *C. neoformans* var. *gattii* (serotype B/C).[61,62] The organism is distributed widely in nature, being associated with pigeon excreta. Most cases of cryptococcosis are reported in patients with cell-mediated immune defects, including patients with HIV infections. Patients acquired the organism and infection by inhaling basidiospores of *C. neoformans*. Diagnosis is accomplished by demonstrating and culturing the encapsulated yeast from clinical specimens and serologically by the detection of cryptococcal polysaccharide antigen in various body fluids.

5.1. Clinical Syndromes and Diagnosis

5.1.1 Central Nervous System Infection

The most common clinical form of cryptococcosis is central nervous system infection. *C. neoformans*, one of the more common causes of chronic meningitis, is the most common cause of fungal central nervous system infection in the compromised patient.[63,64] The organisms are inhaled, produce subclinical pneumonia, and spread hematogenously to the central nervous system. The onset is usually insidious and variable. Fever, headache, and changes in mental status are seen in many patients. In patients with AIDS, many organisms may be present in the cerebrospinal fluid, with a minimal inflammatory response and very little in the way of symptoms. Six to 13% of patients with AIDS develop a cryptococcal infection.[65] Approximately 75% will present with signs of meningoencephalitis.

Cryptococcal meningoencephalitis is the second most common opportunistic central nervous system infection in patients with AIDS. Table VIII summarizes some of the findings of cryptococcosis in AIDS patients. Laboratory diagnosis is made by observing the encapsulated yeast in clinical specimens; by detecting cryptococcal antigen in serum, urine, or spinal fluid; or by culture. India ink preparations have a low sensitivity and are positive in 40% or less of culture-proven cases in non-AIDS patients. In AIDS patients the India ink preparation is positive in over 80% of cases due to the unusually large number of organisms in spinal fluid. The latex agglutination test for cryptococcal antigen in serum and spinal fluid is sensitive, detecting 95–99% of culture-proven cases of meningitis.[66] The test can be falsely positive in patients with disseminated trichosporonosis, DF-2 sepsis, and rarely zygomycosis.[59]

5.1.2. Pulmonary Infection

Pulmonary infection is the second most common clinical form of cryptococcosis. Asymptomatic or symptomatic pulmonary lesions with or without disseminated disease may be seen. The presentation is variable. Chest radio-

TABLE VIII
Cryptococcosis in AIDS Patients

Third most frequent infectious cause of neurologic disease in AIDS patients
Fourth most frequent opportunistic infection
6–13% incidence
Central nervous system infection > pulmonary/disseminated disease < other extraneurological
Meningoencephalitis occurs in 66–84% AIDS patients
Disseminated infection in 50%
CNS manifestations: fever/headache 85%; meningismus 30%; altered mental status 20%; nonspecific 50%
Symptom duration 31 days (CNS disease)
Diagnosis: CT scan normal or nonspecific; CSF minimal abnormalities (16% glucose, leukocytes and protein normal), India ink, culture and antigen positive in most patients; serum for antigen usually positive
Overall outcome poor

graphs may show various types of lesions from nodules to a diffuse interstitial pattern. The latter pattern is seen more commonly in AIDS patients. Diagnosis is by biopsy of pulmonary lesions. The latex agglutination test for cryptococcal antigen is positive in 10% of patients. Sputum and pleural cultures may be negative in 50% of patients.

5.1.3. Disseminated Infection

In the immunocompromised patient, pulmonary disease is associated with a high likelihood of dissemination.[67] Any organ may be involved, the liver and spleen being most frequent. Skin lesions may be seen in as many as 15% of patients. Cryptococcal antigen can be detected in the serum of approximately 60% of patients with disseminated disease.

6. TRICHOSPORONOSIS

6.1. Clinical Syndromes and Diagnosis

Trichosporon spp. are widely distributed in nature, being found in soil, and may colonize or be part of the human gastrointestinal tract flora. *T. beigelii* is the most common cause of disseminated trichosporonosis in patients with hematologic malignancies and granulocytopenia.[68–70] The lungs, kidney, and eyes are the major organ systems involved.[70]

The clinical manifestations are diverse and suggest disseminated candidiasis. Most infections occur in neutropenic patients. The portal of entry is felt to be the gastrointestinal tract. Purpuric skin lesions progressing to large necrotic areas are common.[70]

Diagnosis is made by recovering the organism from blood, skin, or tissue at autopsy. The latex agglutination test for cryptococcal antigen can be falsely positive in the serum of patients with disseminated trichosporonosis.

7. FUSARIOSIS

Fusarium species are ubiquitous in nature, being found in soil, and are important plant pathogens. Pulmonary and disseminated infection are seen in neutropenic patients, including bone marrow transplants.[71] Disseminated infection may be associated with erythematous maculopapular to nodular skin lesions. *F. solarri*, *F. moniliforme*, and *F. proliferatum* are most commonly isolated. Most infections are fatal. Diagnosis is by culturing the organism from tissue specimens. In tissue sections *Fusarium* can be confused with *Aspergillus* spp. and *Pseudoallescheria*.

8. PSEUDALLESCHERIASIS

8.1. Clinical Syndromes and Diagnosis

Pseudoallescheria boydii is ubiquitous in nature. Infections are being described with increasing frequency in immunocompromised patients, including patients with neutropenia.[72] The organism may be inhaled or innoculated into the body. Infections mimic those of aspergillus and zygomycetes. Pulmonary infection with massive hemorrhage and mycotic thrombi has been seen. Disseminated infections have been reported. Diagnosis is by culturing the organism from meaningful clinical specimens. Microscopically in tissue sections, it can be confused with *Apsergillus* species.

9. HISTOPLASMOSIS

9.1. Clinical Syndromes and Diagnosis

Histoplasma capsulatum, a true dimorphic fungus, is found in soil and is endemic in river valleys in certain areas of North, Central, and South America. Infection is usually acquired by inhaling the organism (conidia). Most pulmonary infections are asymptomatic, self-limiting, and become dormant (latent). Occasionally disseminated or chronic cavitary histoplasmosis is seen in normal hosts and can be associated with significant morbidity and mortality. In patients with problems in cell-mediated immunity, reactivation of a latent focus of histoplasmosis can occur, resulting in a confined pulmonary or disseminated infection. Risk factors for disseminated disease include cell-mediated immune defects and lymphocytopenia.[73] The major clinical syndromes seen in the immunocompromised patient include pulmonary and disseminated infection.[74] Clinical manifestations are variable and may include fever, lymphadenopathy, hepatosplenomegaly, pneumonia, mucosal ulcers, and/or skin lesions.

In patients with AIDS, disseminated histoplasmosis is being recognized more frequently in certain geographic areas. Historically, most patients have traveled or resided in an endemic area. The infection usually represents reactivation of a latent focus and can be very severe. Clinical manifestations vary but include weight loss and fever.

Diagnosis is best accomplished by demonstrating the organism in clinical specimens, including tissue biopsies. Bone marrow aspiration and biopsy with culture offers the best results.[75] Blood cultures in patients with AIDS are frequently positive. Serologic tests for antibody are not reliable in many immunocompromised patients, including AIDS. Antigen detection in serum and urine in disseminated infection may be useful but are only available in research laboratories.[76]

10. COCCIDIOIDOMYCOSIS

10.1. Clinical Syndromes and Diagnosis

Coccidioides immitis is found in soil in geographically restricted areas (southwestern United States, Mexico, Central and South America). Infection is acquired by inhaling the organism (arthrospores). Coccidioidomycosis, usually asymptomatic, is self-limiting to mild subclinical disease. Disseminated infection occurs in immunocompromised patients. Once acquired by residing or traveling in an endemic area, the risk for dissemination increases in patients with cell-mediated immune defects and patients treated with chemotherapy.[77–79]

Pulmonary and disseminated diseases are the most frequent clinical syndromes seen in the immunocompromised host. In patients with AIDS disseminated coccidioidomycosis is seen with increasing frequency in endemic areas. Although most infections represent reactivation of a dormant focus, AIDS patients moving to an endemic area may experience primary disease with dissemination.[80]

Diagnosis is accomplished by demonstrating the fungus in biopsy specimens or by culture. Occasionally the organism can be observed on direct smear of sputum. Coccidioidal serology is often reactive, especially the complement fixation test for antibody.[66,81]

11. OTHER FUNGI

A number of other fungi are appearing in immunocompromised patients, causing invasive disease. For example, phaeohyphomycosis and hyalohyphomycosis are the emerging opportunistic diseases in immunocompromised patients.[25] We should be prepared to recognize these infections clinically and then to isolate and identify the responsible fungus so that appropriate therapy can be initiated in the infected patient.

REFERENCES

1. Hart, P. D., Russell, E., Jr., and Remington, J. S., 1969, The compromised host and infection, II, Deep fungal infection, *J. Infect. Dis.* **120:**169–191.
2. Miller, J., Kappe, R., Kubitza, D., Fessler, R., and Scheidecker, I., 1988, The incidence of deep seated mycoses in Freiburg, *Mycoses* **31(Suppl. 1):**9–28.
3. Rose, H. D., and Varkey, B., 1975, Deep mycotic infection in the hospitalized adult: A study of 123 patients, *Medicine* **54:**499–507.
4. Centers for Disease Control, 1986, Nosocomial infection surveillance, in: *CDC Surveillance Summaries, MMWR* **35:**17–29SS.
5. Bodey, J. P., 1984, Candidiasis in cancer patients, *Am. J. Med.* **77:**13–19.

6. Dummer, J. S., Hardy, A., Poorsatter, A., and Ho, M., 1983, Early infections in kidney, heart and liver transplant recipients on cyclosporin, *Transplantation* **36:**259–267.
7. Weber, D. J., and Rutala, W. A., 1989, Epidemiology of hospital-acquired fungal infections, in: *Diagnosis and Therapy of Systemic Fungal Infections* (K. Holmberg and R. D. Meyer, eds.), Raven Press, New York, p. 2.
8. Anonymous, 1988, Orofacial manifestations of HIV infection, *Lancet* **1:**976–977.
9. Grant, I. H., and Armstrong, D., 1988, Fungal infections in AIDS: Cryptococcosis, in: *Infectious Disease Clinics of North America*, Vol. 2 (M. A. Sande and P. A. Volberding, eds.), W. B. Saunders, Philadelphia, pp. 827–839.
10. Macher, A. M., DeVinatea, M. L., Tuur, S. M., and Angritt, P., AIDS and the mycoses, in: *Infectious Disease Clinics of North America*, Vol. 2 (D. Drutz, ed.), W. B. Saunders, Philadelphia, pp. 821–839.
11. Armstrong, D., Chmel, H., Singer, C., Tapper, M., and Rosen, P. P., 1975, Nonbacterial infections associated with neoplastic disease, *Eur. J. Cancer* **11(Suppl):**79–94.
12. Bodey, G. P., Buckley, M., Sathe, Y. S., and Freireich, E. J., 1966, Quantitative relationships between circulating leukocytes and infections in patients with acute leukocytes and infections in patients with acute leukemia, *Ann. Intern. Med.* **64:**328–340.
13. Kirkpatrick, C. H., 1984, Host factors in defense against fungal infections, *Am. J. Med.* **77:** 1–12.
14. Newman, A. K., and Schimpff, S. C., 1987, Hospital hotel services as risk factors for infection among immunocompromised patients, *Rev. Infect. Dis.* **9:**206–213.
15. Armstrong, D., 1981, Fungal infections in the compromised host, in: *Clinical Approach to Infection in the Compromised Host* (R. H. Rubin and L. S. Young, eds.), Plenum Publishing, New York, pp. 195–228.
16. Gold, J. W. M., 1984, Opportunistic fungal infections in patients with neoplastic disease, *Am. J. Med.* **76:**458–463.
17. Weber, D. J., and Rutala, W. A., 1989, Epidemiology of hospital-acquired fungal infections, in: *Diagnosis and Therapy of Systemic Fungal Infections* (K. Holmberg and R. D. Meyers, eds.), Raven Press, New York, pp. 1–25.
18. Anaissie, E., and Bodey, G. P., 1989, Nosocomial fungal infections, in: *Infectious Disease Clinics of North America*, Vol. 3 (D. J. Weber and W. A. Rutala, eds.), W. B. Saunders, Philadelphia, pp. 867–882.
19. Edwards, J. E., Jr., Drutz, D. J., Bennett, J. E., and Remington, J. S., 1986, in: *Disseminated Candidiasis. A Major Problem in Cancer and Postoperative Patients*. Part 1, Academy Professional Information Sources, New York.
20. Body, B. A., Pfaller, M. A., Durrer, J., Koontz, F., and Groschel, D. H., 1988, Comparison of the lysis centrifugation and radiometric blood culture systems for recovery of yeasts, *Eur. J. Clin. Microbiol. Infect. Dis.* **7:**417–420.
21. Kiehn, T. E., Wong, B., Edwards, F. F., and Armstrong, D., 1983, Comparative recovery of bacteria and yeasts from lysis-centrifugation and a conventional blood culture system, *J. Clin. Microbiol.* **18:**300–304.
22. Bennett, J. E., 1987, Rapid diagnosis of candidiasis and aspergillosis, *Rev. Infect. Dis.* **9:** 398–402.
23. DeRepentigny, L., and Reiss, E., 1984, Current trends in immunodiagnosis of candidiasis and aspergillosis, *Rev. Infect. Dis.* **6:**301–312.
24. Barnes, R. A., 1990, Immunological diagnosis of fungal infection in the immunocompromised host, *Rev. Med. Microbiol.* **1:**58–65.
25. Rinaldi, M. G., 1989, Emerging Opportunists, in: *Infectious Disease Clinics of North America*, Vol. 3 (D. Drutz, ed.), W. B. Saunders, Philadelphia, pp. 65–76.
26. Odds, F. C., 1984, Ecology and epidemiology of *Candida* species, *Zentralbl. Bakteriol. Parasitenkd. Infektionskr. Hyg. Abt. 1 Reihe A* **257:**207–212.
27. Anonymous, 1989, Oral candidosis in HIV infection, *Lancet* **2:**1491–1492.

28. Klein, R. S., Harris, C. A., Small, C. B., Moll, B., Lasser, M., and Friedland, G., 1984, Oral candidiasis in high risk patients as the initial manifestation of the acquired immunodeficiency syndrome, *N. Engl. J. Med.* **311**:354–357.
29. Syrjanen, S., Valle, S. L., and Antonen, J., 1988, Oral candidal infections as a sign of HIV infection in homosexual men, *Oral Surg.***65**:36–40.
30. Schiodt, M., and Pindborg, J. J., 1987, Epidemiology and clinical oral manifestations of human immune deficiency virus infection: A review, *Int. J. Oral Maxillofac. Surg.* **16**:1–19.
31. Coleman, M. F., 1986, Epiglottitis in immunocompromised patients, *Head Neck Surg.* **8**: 466–468.
32. Walsh, T. J., and Gray, W. C., 1987, Candida epilottitis in immunocompromised patients, *Chest* **91**:482–485.
33. Cole, S., Zawin, M., Lungberg, B., Hoffman, J., Bailey, L., and Ernstoff, M. S., 1987, Candida epiglottitis in an adult with acute non-lymphocytic leukemia, *Am. J. Med.* **82**:662–664.
34. Nathieson, R., and Dutta, S. K., 1983, Candida esophagitis, *Digest. Dis. Sci.* **28**:365–371.
35. Welik, R. M., Starcher, E. T., and Curran, J. W., 1987, Opportunistic diseases reported in AIDS patients: Frequencies, associations, and trends, *AIDS* **1**:175–182.
36. Holt, H., 1986, Candida infections of the esophagus, *Gut* **9**:227–238.
37. Walsh, T. S., Hamilton, S. R., and Belitos, N., 1988, Esophageal candidiasis. Diagnoses and treatment of an increasingly recognized fungal infection, *Postgrad. Med.* **84**:193–205.
38. Tashjian, L. S., Abramson, J. S., and Peacock, J. E., Jr., 1983, Focal hepatic candidiasis: A distinct clinical variant of candidiasis in immunocompromised patients, *Rev. Infect. Dis.* **6**: 689–703.
39. Haron, E., Feld, R., Tuffnell, P., Patterson, B., Hasselback, R., and Matlow, A., 1987, Hepatic candidiasis: An increasing problem in immunocompromised patients, *Am. J. Med.* **83**:17–26.
40. Thaler, M., Pastakia, B., Shawker, T. H., O'Leary, T. O., and Pizzo, P. A., 1988, Hepatic candidiasis in immunocompromised patients: A new or evolving syndrome, *Ann. Intern. Med.* **108**:88–100.
41. Odds, F. C., 1988, *Candida and Candidiosis*, 2nd ed., W. B. Saunders, Philadelphia, pp. 169–174.
42. Bodey, G. P., and Fainstein, V., 1985, *Candidemia*, Raven Press, New York, pp. 135–154.
43. Wey, S. B., Mori, M., Pfaller, M. A., Woolson, R. F., and Wenzel, R. P., 1989, Risk factors for hospital-acquired candidemia, *Arch. Intern. Med.* **149**:2349–2353.
44. Collignon, P. J., Soni, N., and Pearson, I. Y., 1986, Is semi-quantitative culture of central vein catheter tips useful in the diagnosis of catheter-associated bacteremia, *J. Clin. Microbiol.* **24**: 532–535.
45. Torres-Rojas, J. R., Stratton, C. W., Sanders, C. V., Horsman, T. A., Hawley, H. B., Dascomb, H. E., and Vial, L. J., 1982, Candidal suppurative peripheral thrombophlebitis, *Ann. Intern. Med.* **96**:431–435.
46. Prager, R. L., and Silva, J., Jr., 1984, Colonization of central venous catheters, *South. Med. J.* **77**:458–461.
47. Odds, F. C., 1988, *Candida and Candidiosis*, 2nd ed., W. B. Saunders, Philadelphia, pp. 206–230.
48. Musial, C. E., Cockerill, F. R., and Roberts, G. D., 1988, Fungal infections of the immunocompromised host: Clinical and laboratory aspects, *Clin. Microbiol. Rev.* **1**:349–364.
49. Sarubbi, F. A., Jr., Kopf, H. B., Wilson, M. B., McGinnis, M. R., and Rutala, W. A., 1982, Increased recovery of *Aspergillus flavus* from respiratory specimens during hospital construction, *Am. Rev. Resp. Dis.* **125**:33–38.
50. Opal, S. M., Asp, A. A., Cannady, P. B., Jr., Morse, P. L., Burton, L. J., and Hammer, P. G., 1986, Efficacy of infection control measures during a nosocomial outbreak of disseminated aspergillosis associated with hospital construction, *J. Infect. Dis.* **153**:634–637.
51. Aisner, J., Murillo, J., Schimpff, S. C., and Steere, A. C., 1979, Invasive aspergillosis in acute leukemia: Correlation with nose cultures and antibiotic use, *Ann. Intern. Med.* **90**:4–9.
52. Rinaldi, M. G., 1983, Invasive aspergillosis, *Rev. Infect. Dis.* **5**:1061–1077.

53. Young, R. C., Bennett, J. E., Vogel, C. L., Carbone, P. P., and DeVita, V. T., 1970, Aspergillosis: The spectrum of the disease in 98 patients, *Medicine* **49:**147–173.
54. Levity, S. M., 1989, Aspergillosis, in: *Infectious Disease Clinics of North America*, Vol. 3 (D. J. Drutz, ed.), W. B. Saunders, Philadelphia, pp. 1–18.
55. McCarty, J. M., Flam, M. S., Pullen, G., 1986, Outbreak of primary cutaneous aspergillosis related to intravenous arm boards, *J. Pediatr.* **108:**721–724.
56. Gerson, S. L., Talbot, G. H., Hurwitz, S., Strom, B. L., Lusk, E. J., and Cassileth, P. A., 1984, Prolonged granulocytopenia: The major risk factor for invasive pulmonary aspergillosis in patients with acute leukemia, *Ann. Intern. Med.* **100:**345–351.
57. Rinaldi, M. G., 1989, Zygomycosis, in: *Infectious Disease Clinics of North America*, Vol. 3 (D. J. Drutz, ed.), W. B. Saunders, Philadelphia, pp. 19–41.
58. Cho, S. Y., and Choi, H. Y., 1979, Opportunistic fungal infection among cancer patients, *Am. J. Clin. Pathol.* **72:**617–621.
59. Ingram, C. W., Sennesh, J., Cooper, J. N., and Perfect, J. R., 1989, Disseminated zygomycosis: Report of four cases and review, *Rev. Infect. Dis.* **11:**741–754.
60. Karam, F., and Chmel, H., 1990, Rhino-orbito cerebral mucormycosis, *Ear, Nose Throat J.* **69:** 187–193.
61. Perfect, J. R., 1989, Cryptococcosis, in: *Infectious Disease Clinics of North America*, Vol. 3 (D. J. Drutz, ed.), W. B. Saunders, Philadelphia, pp. 77–102.
62. Bennett, J. E., Kwon-Chung, K. J., and Howard, D. H., 1977, Epidemiology differences among serotypes of *Cryptococcus neoformans*, *Am. J. Epidemiol.* **105:**582–590.
63. Sabetta, J. R., and Andriole, V. T., 1985, Cryptococcal infections of the central nervous system, *Med. Clin. North Am.* **69:**333–342.
64. Salaki, J. S., Louria, D. B., and Chmel, H., 1984, Fungal and yeast infections of the central nervous system. A clinical review, *Medicine* **63:**108–131.
65. Zuger, A., Louie, E., Holzman, R. S., Simberkoff, M. S., and Rahal, J. J., 1986, Cryptococcal disease in patients with the acquired immunodeficiency syndrome, *Ann. Intern. Med.* **104:** 234–240.
66. Kaufman, L., and Reiss, E., 1985, Serodiagnosis of fungal diseases, in: *Manual of Clinical Microbiology* (E. H. Lennette, A. Balows, W. J. Hausler, Jr., and H. J. Shadomy, eds.), American Society of Microbiology, Washington, D.C., pp. 924–944.
67. Kerkering, T. M., Duma, R. J., and Shadomy, S., 1981, The evolution of pulmonary cryptococcosis, *Ann. Intern. Med.* **94:**611–616.
68. Hoy, J., Hsu, K. C., Rolston, K., Hopfer, R. L., Lura, M., and Bodey, G. P., 1986, *Trichosporon beigelli* infection: A review, *Rev. Infect. Dis.* **8:**959–967.
69. Walsh, T. J., Newman, K. R., Moody, M., Wharton, R. C., and Wade, J. C., 1986, Trichosporonosis in patients with neoplastic disease, *Medicine* **65:**268–279.
70. Walsh, T. J., 1989, Trichosporonosis, in: *Infectious Disease Clinics of North America*, Vol. 3 (D. J. Drutz, ed.), W. B. Saunders, Philadelphia, pp. 43–52.
71. Anaissie, E., Kantarjian, H., Ro, J., Jones, P., and Bodey, G. P., 1988, The emerging role of Fusarium infections in patients with cancer, *Medicine* **67:**77–83.
72. Travis, L. B., Roberts, G. D., and Wilson, W. R., 1985, Clinical significance of *Pseudoallescheria boydii*: A review of 10 years experience, *Mayo Clin. Proc.* **60:**531–537.
73. Wheat, L. J., Slama, T. G., and Norton, J. A., 1982, Risk factors for disseminated or fatal histoplasmosis, *Ann. Intern. Med.* **96:**159–163.
74. Kaufman, C. A., Israel, K. S., and Smith, J. W., 1978, Histoplasmosis in immunosuppressed patients, *Am. J. Med.* **64:**923–932.
75. Wheat, L. J., Histoplasmosis, in: *Infectious Disease Clinics of North America*, Vol. 2 (D. J. Drutz, ed.), W. B. Saunders, Philadelphia, pp. 841–859.
76. Wheat, L. J., Kohler, R. B., and Tewari, R. P., Diagnosis of disseminated histoplasmosis by detection of histoplasma capsulation antigen in serum and urine specimens, *N. Engl. J. Med.* **314:**83–88.

77. Rutala, P. J., and Smith, J. W., 1978, Coccidioidomycosis in potentially compromised hosts: The effects of immunosuppressive therapy in dissemination, *Am. J. Med. Sci.* **275:**283–295.
78. Deresinski, S. C., and Stevens, D. A., 1975, Coccidioidomycosis in compromised hosts, *Medicine* **54:**377–395.
79. Ampel, N. M., Ryan, K. J., Carry, P. J., and Schifman, R. B., 1986, Fungemia due to *Coccidioides immitis*: An analysis of 16 episodes in 15 patients and a review of the literature, *Medicine* **65:**312–321.
80. Meyers, R. D., and Holmberg, K., 1989, Fungal infections in HIV patients, in: *Diagnosis and Therapy of Systemic Fungal Infections* (K. Holmberg and R. D. Meyers, eds.), Raven Press, New York, pp. 93–94.
81. Pappagianis, D., and Zimmer, B. L., 1990, Serology of coccidioidomycosis, *Clin. Microbiol. Rev.* **3:**247–268.

19

Immunochemistry of Fungal Antigens (Part A)

Primary Dimorphic Pathogens

ERROL REISS

1. INTRODUCTION

This chapter reviews some recent developments in the characterization of antigens that are potentially or actually useful as probes of the immune response in the primary systemic mycoses. Other sources should be consulted for a broader perspective.[1–4] A species-specific immunodominant protein exocellular antigen ("exoantigen") has been identified for each of the systemic fungi. Procedures for purifying these proteins have become available.

Abbreviations: (All carbohydrates are the D enantiomers, and are in the pyranose configuration, unless otherwise indicated) BGG, bovine gamma globulin; BSA, bovine serum albumin; C, common fungal antigen, a carbohydrate; C3, third component of complement; CDN, coccidioidin; CF, culture filtrate; Con A, concanavalin A; DTT, dithiothreitol; EIA, enzyme immunoassay; ELISA, enzyme-linked immunosorbent assay; EM, electron microscopy; endo H, endoglycosidase H, an N,N′-diacetyl chitobiase; g.l.c.-m.s., gas liquid chromatography-mass spectrometry; Glc*p*, glucose; Glc*p*A, glucosyluronic acid; GM, galactomannan, mAbs, monoclonal antibodies; Man*p*, mannose; MP, mannoprotein; n.m.r., nuclear magnetic resonance spectrometry; pAbs, polyclonal antibodies; pI, isoelectric point; RIA, radioimmunoassay; SDS-PAGE, sodium dodecyl sulfate-polyacrylamide gel electrophoresis; SPH, spherulin; TCA, trichloroacetic acid; TP, tube precipitin; Western blot, immunoelectro-transfer blot.

ERROL REISS • Molecular Mycology Section, Mycotic Diseases Branch, Division of Bacterial and Mycotic Diseases, National Center for Infectious Diseases, Centers for Disease Control and Prevention, Atlanta, Georgia 30333.

Fungal Infections and Immune Responses, edited by Juneann W. Murphy *et al.* Plenum Press, New York, 1993.

Most of these antigens have been adapted to primary binding immunoassays to detect antibodies in patients.

Admixed with the protein exoantigens in unpurified culture supernatants are galactomannan (GM)-based cell wall polysaccharides or glycoproteins that show extensive cross-reactivity among genera. These antigens are also useful because they can be directly detected in body fluids (antigenuria in histoplasmosis) or their respective antibody concentrations can provide early evidence of infection (coccidioidomycosis).

2. BLASTOMYCOSIS—*BLASTOMYCES DERMATITIDIS* ANTIGENS

Some immunodiagnostic methods for blastomycosis tend to lack specificity because of the presence of a common fungal antigen (galactomannan, GM) shared by the major genera of primary systemic, dimorphic fungal pathogens.[5] Whole yeast forms, homogenate supernatants, and culture filtrates (CF) all contain this antigen, contributing to cross-reactions in agglutinin, complement fixation, and immunofluorescence tests that attempt to discriminate *B. dermatitidis* from *Histoplasma capsulatum*. Nonetheless, diagnostic specificity of precipitin tests has not been compromised because of the presence of a species-specific exoantigen, A.[6]

This protein antigen, and a similar moiety, WI-1, are the subjects of research designed to replace former methods with primary binding enzyme immunoassays (EIAs) and radioimmunoassays (RIAs). As knowledge becomes available about the structure and antigenicity of the immunodominant proteins, studies to unravel their roles in the physiology of the fungus and in the host–fungus interaction can be anticipated.

2.1. WI-1

A protein antigen was released from frozen-thawed *B. dermatitidis* yeast forms.[7] The antigen, WI-1, had a molecular weight of 115–120 kDa, depending on the strain, and was electroeluted from SDS-PAGE gels. This purified antigen was radioiodinated and a radioimmunoprecipitation test was devised with staphylococcal protein-A as the precipitant. Residual free antigen in the supernatant was measured. This RIA detected antibodies in a panel of 68 blastomycosis patients, yielding 85% sensitivity and 98% specificity. PAbs against purified WI-1 localized the antigen on the *B. dermatitidis* yeast form cell surface by immunofluorescence and immunocolloidal gold electron microscopy (EM), but not on the surface of *H. capsulatum* or *Candida albicans*.

2.2. A Antigen

The A antigen is the exoantigen that forms an immunoprecipitate specific for *B. dermatitidis* and does not cross-react with exoantigens of other genera of primary, systemic, dimorphic fungal pathogens.[8] Anion exchange chromatography of the supernatant from yeast form cultures revealed up to 20 proteins by SDS-PAGE.[9] Of these, the most immunoreactive protein had a molecular weight of 98 kDa and is presumably the A antigen. Immunoblotting with sera from blastomycosis patients supported this view. When the semipurified A-antigen preparation was adapted to indirect ELISA, antibodies were detected more sensitively in patients, but the specificity was lower than in the western immunoblot test.

Although the exoantigen reaction has been validated as a means of unequivocal identification of *B. dermatitidis* cultures, this specificity is somewhat less than ideal in western immunoblots probed with sera of patients with histoplasmosis. In that case, sera from 3 of 14 histoplasmosis patients reacted with the 98-kDa antigen, indicating either dual exposure to the two fungi whose endemic areas overlap, or an epitope within the 98-kDa antigen that is common to both *B. dermatitidis* and *H. capsulatum*.[9] The similarity in molecular weights between the 98-kDa active moiety in the A-antigen-containing complex and the 115- to 120-kDa WI-1 antigen increases the need for side-by-side comparisons to determine if these antigens have the same specificity.

3. COCCIDIOIDOMYCOSIS—*COCCIDIOIDES IMMITIS* ANTIGENS

The major immunoprecipitating antigens of *C. immitis* present in coccidioidin (CDN) and spherulin (SPH) were systematically enumerated by two-dimensional crossed rocket immunoelectrophoresis.[10] Twenty-six antigens were identified in CDN and 12 in SPH. This nomenclature was confirmed and extended by Cole et al.[11] The antigens are numbered, with the exception of one, called CS, designated by Cole et al.,[11] which was not present in the system as originally defined. That antigen was present both in an extract of the arthroconidial wall and in CDN. The usefulness of such a scheme is that it provides a point of reference for studies by different groups, thus contributing to the standardization of antigens.

Only two antigens, F and TP, are of proven value for serodiagnostic assays. A third antigen, HS, has come to light that is useful mainly for immunoidentification of *C. immitis*. Progress is being made in determining their roles in the physiology of the fungus and their intracellular locations, but at present it is not clear whether they contribute to pathogenicity or are merely convenient marker molecules.

3.1. Coccidioidin (CDN) for Serological Tests

The toluene-induced autolysate method produces an antigen satisfactory for the immunodiffusion variation of the tube precipitin (TP) test.[12,13] Mycelial form cultures from 24 strains of *C. immitis* are individually grown on glucose–yeast extract dialyzate medium for 3 days. Although such a large number of strains are used to ensure adequate coverage of the antigenic mosaic, occasionally only a single strain is used, i.e., *C. immitis* Silveira (ATCC 28868). After autolysis and removal of cell particulates, the CDNs are pooled, ultrafiltered, and dialyzed versus buffer. This preparation resulted in a single line in immunodiffusion containing the heat-stable TP antigen, described in more detail below. Huppert and Bailey[14] also described a variation of this method to produce CDN containing the heat-labile F antigen, suitable for the quantitative immunodiffusion analogue of the complement fixation (CF) test. This variation avoided toluene autolysis and resulted in a filtrate antigen solution. Some workers prefer to combine the toluene-induced autolysate with the filtrate antigen solution to obtain the totipotent CDN.

3.2. Spherulin (SPH)

SPH is a distilled water lysate of spherules, the tissue form of *C. immitis*, that is prepared by first cultivating *C. immitis* for 5 days at 40°C in modified Converse medium, without Tamol N, in a 20% CO_2–air atmosphere.[15–17] The endosporulating spherules were harvested, resuspended in deionized water, and incubated at 35°C on a gyratory shaker for 20 days. After killing with Formalin, the partially autolyzed spherule culture was further disrupted in a Braun cell homogenizer. The soluble portion of this product was SPH. A supernatant of the spherule phase after 5 days' growth was produced for comparative purposes.

3.3. Tube Precipitin (TP) Antigen

The TP-antigen is a heat-stable (60°C, 30 minutes) constituent of CDN consisting of a monomer, molecular weight 110–120 kDa, that associates via disulfide bridge(s) to form a 240-kDa dimer.[18] TP antigen is a concanavalin A (Con A)-binding glycoprotein that contains 3-*O*-methyl Man*p*.[19] This form of the antigen was isolated by alkali extraction (1 M NaOH) from the inner wall of the infectious propagule, the arthroconidium of the saprophytic, mycelial form. Moreover, an antigen with similar molecular weight and antigenic characteristics was obtained from CDN.[18]

Although the antigen was extracted from the mycelial form, it was also located by immunocolloidal gold EM in the outer layer and segmentation apparatus of the spherule-form wall.[19] The relation of the TP antigen to Huppert's numerical scheme of two-dimensional crossed rocket immunoprecipitin arcs is not finally reconciled. Cox and Britt[20] provided evidence that the TP antigen forms an incomplete precipitin arc in Huppert's scheme, similar in migration but distinct from antigen 2.[10]

Kruse and Cole,[18] used a burro reference antiserum against CDN. The reference antigen was a composite of culture filtrate and toluene autolysates. In that case their purest TP antigen was a Con-A eluate that was subjected to SDS-PAGE. The electroeluted antigen, molecular weight range 97–116 kDa, was then fractionated on a high-pressure, gel-permeation column, yielding a 120-kDa glycoprotein that was identical to reference TP-antigen preparations in immunodiffusion and in the classical TP test. Moreover, this purified 120-kDa glycoprotein was identical to Huppert's antigen 2 in advancing line immunoelectrophoresis.

TP antigen was also isolated by pronase digestion of TP-anti-TP immunoprecipitates.[21] The pronase-resistant TP antigen was present in a neutral and a negatively charged form based on anion exchange chromatography. Gel permeation revealed heterogeneity in the molecular weights of these components. Neither form would enter PAGE gels, even at a low porosity comonomer concentration. The neutral TP antigen had an average molecular weight of 140 kDa and contained 3-*O*-methyl Man*p*:Man*p*:Glc*p* at 8:1.2:1 (molar ratio). The charged TP-antigen molecular weight was higher, average 225 kDa, and it had equimolar amounts of 3-*O*-methyl-Man*p*, Man*p*, Gal*p*, Glc*p*. The two forms of TP antigen could be separated by lectin affinity because neutral TP was more tightly bound to Con A.

In most respects the two studies that characterized column-purified TP-antigen agree that it is glycoprotein in nature, in which 3-*O*-methyl Man*p* is a key determinant; resistant to pronase; able to bind to anion exchange columns; and occurring in two principal molecular weight forms: ~120–140 kDa and possibly as two associated polypeptides with a combined weight of 220–240 kDa.

The TP antigen is not specific for *C. immitis* but produces cross-reactions with an antigen common to other primary, systemic, dimorphic fungi: *Blastomyces dermatitidis*, *Histoplasma capsulatum*, and *Paracoccidioides brasiliensis*. Antigens of gymnoascaceous saprophytes that produce arthroconidia resembling *C. immitis* also cross-react: *Malbranchea dendritica* and the *Malbranchea* state of *Uncinocarpus reesii* produce 3-*O*-methyl-Man*p*.[19] Other saprophytes that look like the mycelial form of *C. immitis*—*Arachniotus reticulatus* and *Auxarthron zuffianum* species—also produce the TP antigen.[22]

3.4. F Antigen

F antigen was so named either because of its usefulness in the CF test or because it is abundant in filtrates of mycelia or spherule forms, in contrast to its lower concentration in lysates. F antigen is a heat-labile (60°C, 30 minutes) protein, with a molecular weight estimated by gel-permeation chromatography of 110 kDa that becomes dissociated under conditions of SDS-PAGE to a 45- to 48-kDa monomer.[23,24] F antigen was purified by immunoaffinity chromatography using monospecific antibodies.[23] F antigen is specific for *C. immitis*.

3.5. HS Antigen

HS substance found in CDN is a heat-stable (HS, 60°C, 30 minutes) antigen that in immunodiffusion is not identical to the TP antigen.[22] Furthermore, HS antigen, unlike the TP antigen, is specific for *C. immitis* and is absent in fungi in other genera of primary, systemic, fungal pathogens and saprophytic, arthroconidia-producing look-alikes. HS antigen is released by the mold form of *C. immitis* during the early stages of growth. HS is recommended as an excellent antigen for the exoantigen test but is not recommended for testing human serum specimens *by immunodiffusion* because precipitins against HS are not frequently encountered in serum from coccidioidomycosis patients. The TP antigen, while recommended for testing serum specimens, is not recommended for the exoantigen test because of its cross-reactivity with other fungi.

Since its first description by Kaufman et al.,[22] the HS antigen has been further investigated.[25,26] The HS antigenic complex as originally described[22] is not a single immunoprecipitating antigen but contains at least six proteins visible in Coomassie blue-stained PAGE gels.[26] The number of components in silver-stained gels is unknown. A similar preparation, CDN-F, of Cox and Britt[25] contained six to seven immunoprecipitin arcs. The latter authors showed that CDN containing HS-exoantigen reactivity could be resolved by lectin affinity for Con A into an effluent fraction containing two precipitating antigens, one forming a line of identity in advancing line immunoelectrophoresis with a constituent in the HS reference antigen of Kaufman et al.[22] The same Con A-passaged semipurified HS antigen was reactive with a constituent of spherulin.[17]

3.5.1. Proteinase

A soluble fraction of the arthroconidial wall contained two proteinases, one of them with a molecular weight of 38 kDa (19 kDa under reducing conditions), corresponding to the well-characterized CS immunoprecipitate in the reference CDN-anti-CDN system.[26] This proteinase was electroeluted from PAGE gels and produced a line of identity when it was reacted in immunodiffusion against the HS antigen. At present it seems that the active, and perhaps immunodominant, antigenic moiety in the HS complex is a proteinase with a dimeric molecular weight of 39 kDa and a monomeric molecular weight of 19 kDa. Furthermore, Cole et al.[26] observed that the unfractionated HS antigen reacts very well in indirect ELISA with serum samples from coccidioidomycosis patients. This finding invites a reappraisal of the earlier belief that HS is of no diagnostic significance because of the paucity of HS-containing immunoprecipitin arcs with serum from coccidioidomycosis patients. Further studies are needed to establish if the heat-stable (56°C, 30 minutes) antigen that did not bind to Con A of Cox and

Britt[25] is the same as the proteolytic 19- or 38-kDa antigen of Cole et al.[26]

Thus far, the antigenic complex of *C. immitis*, for purposes of clinical immunoassays, consists of the TP and F antigens. The TP provides early, presumptive evidence of infection and the F antigen is of diagnostic and crucial prognostic value for monitoring patients. The HS antigen was described in the development of the "exoantigen" procedure for immunoidentification of *C. immitis*. Exoantigen methods avoid the necessity to experimentally infect animals to demonstrate endosporulating spherules.[22,27]

4. HISTOPLASMOSIS—*HISTOPLASMA CAPSULATUM* ANTIGENS

4.1. Histoplasmin

In its classical definition, histoplasmin is the cell-free supernatant fluid from static cultivation of a *H. capsulatum* mold-form human isolate for 3–6 months at ambient temperature on asparagine-glucose-glycerol salts (Smith's) medium.[28] This standard method is sanctioned by the U.S. Food and Drug Administration for the production of skin-test histoplasmin, which is standardized according to its ability to elicit positive delayed cutaneous hypersensitivity in infected persons. That is true, even if, as is most often the case, the infection is subclinical. Other variations include histoplasmin used for serology, which is "balanced" to exhibit the characteristic immunoprecipitin arcs, and yeast form histoplasmin. Histoplasmin is produced in our laboratory from *H. capsulatum* strain NIH 6623. Yeast-form growth from a single Pine's citrate agar slant is seeded into a heavy-duty 1-liter Erlenmeyer flask containing 300 ml of Smith's medium. The culture is incubated for 30 days at 25°C at 150 rpm on a gyratory shaker. Then Merthiolate is added to give a final concentration of 0.05%, and after an additional week flasks are assayed for purity and sterility. Sterility is validated by cultivation at 25°C on BHI + 5% sheep blood for 1 week. Cultures that pass the purity and sterility checks are centrifuged and the supernatant fluid is filter-sterilized and stored at 4°C.

Histoplasmin contains species-specific protein antigens and a common fungal antigen. Nomenclature for the immunoprecipitin arcs formed by two major histoplasmin proteins, H and M, was established many years ago by the pediatrician Douglas Heiner.[28] That nomenclature is still in use because the biological roles of the protein precipitinogens in the physiology of the fungus have not been elucidated. In addition to the protein antigens, the common fungal antigen, C, most probably a GM, cross-reacts with a similar antigen in the other systemic, dimorphic, fungal pathogens.[5] Interest in this latter antigen has been stimulated by reports, discussed below, that an antigen with similar characteristics may occur in the urine of infected patients, where it serves as an indicator of infection that does not depend on the immunocompetence of the human host.

The strong positive delayed cutaneous hypersensitivity evoked by histoplasmin in sensitized individuals is a correlate of the pulmonary granulomatous response evoked by *Histoplasma capsulatum* infection. The important role of T-cell-mediated immunity in histoplasmosis illustrated by these *in vivo* findings has received new *in vitro* support by the cloning of murine T_{helper} lymphocytes that are responsive to histoplasmin stimulation.[29]

4.2. H Antigen

H antigen is an acidic glycoprotein, molecular weight 108–110 kDa, pI 4.5, that is the first and most abundant antigen to be released from mycelial forms in the exoantigen method.[8,28] A somewhat different molecular weight, 120 kDa, was estimated by Harris and Deepe,[29] who also characterized a covalent glycosidic moiety in H antigen. Strong anodic mobility is displayed by the H antigen in immunoelectrophoresis. Anti-H precipitins are slow to appear in acute histoplasmosis, occurring in 32–39% of cases, usually in patients with active and progressive disease. Mouse pAbs against periodate-oxidized H localized the antigen within the cytoplasm of yeast forms.[30]

4.3. M Antigen

M antigen is a glycoprotein that is released into histoplasmin during growth and autolysis of mycelial forms. Only some *H. capsulatum* strains are good sources of M-rich histoplasmin. Precipitins against M-antigen commonly occur in acute histoplasmosis and may appear after a single positive skin test in a previously exposed person. Having a weak negative charge at pH 8.0, M antigen desorbs briskly from anion exchange columns. In immunoelectrophoresis the M antigen migrates only a short distance towards the anode. SDS-PAGE gels and western blots indicate that M antigen occurs as a weakly immunoreactive dimer of 150 kDa that dissociates into a 70- to 75-kDa antigenic subunit.[31] A role for dithiothreitol in promoting this dissociation has been inferred but further proof of the existence of intermolecular disulfide bonds is not available at present.[31] Harris and Deepe[29] have emphasized a 182-kDa moiety in histoplasmin as the active M antigen.

Anion exchange chromatography was the major first step taken in the purification of histoplasmin proteins. Upon application of the salt gradient M elutes first, followed by the more acidic H antigen. However, the heterogeneous and polydisperse nature of the C antigen causes it to elute over a wide range of ionic strength. Even though pH 3–3.5 is not the optimal pH for operating carboxymethyl-agarose cation exchange columns, Green and Pine[32] discovered that in that pH range the C antigen eluted in the column void volume, thus removing it from the protein antigens, which retained sufficient positive charge below their isoelectric pH to remain bound. This method is undergoing further development in the author's laboratory.

The intracellular locale in *H. capsulatum* of the M antigen is not known with certainty, but immunocolloidal gold EM using pAbs against purified M as a probe has shown selective deposition of immunogold in the plasmalemma.[30]

4.4. C Antigen, a Galactomannan

The common fungal antigen, C, present in histoplasmin is either a polysaccharide or a glycoprotein. The antigen activity is stable to boiling and labile to periodate oxidation. C antigen purified from histoplasmin by affinity for Con A-agarose had a molar ratio of Man*p*:Gal*p* = 4:1. There was a significant amino sugar content amounting to 6.7% of the dry weight.[28] mAbs against C antigen are readily produced in mice immunized with histoplasmin.[31] The mAb, CB4, also binds to a common epitope in coccidioidin, blastomycin, and paracoccidioidin antigens.[28] After SDS-PAGE the C-antigen activity is dispersed along the gel and is visible as a "smear" in the western blot probed with sera from mice and rabbits immunized with histoplasmin and, to a lesser extent, with sera from *H. capsulatum*-infected humans.

A heat-stable (100°C, 5 minutes) carbohydrate-containing antigen is detectable in the urine of patients with disseminated histoplasmosis.[33] Similar heat-stable antigen(s) were released into the bronchoalveolar lavage fluid of mice intranasally infected with *H. capsulatum*.[34] In both human and murine studies detection was accomplished by double-antibody sandwich immunoassay with either radioiodinated or enzyme-labeled indicator antibodies.[35] Capture and indicator antibodies were produced in rabbits by immunizing them with histoplasmin in complete Freund's adjuvant.

Histoplasma urine antigen was inactivated by periodate oxidation or by mixed glycosidases, and could be adsorbed from urine by Con A-agarose. Given the low molecular weight expected of urine antigen after glomerular filtration, the ability of the double antibody sandwich format to detect it implicates a polysaccharide, which would have high avidity due to a multiple, repeating epitope structure.

Antigenuria is usually present at the time of clinical presentation, whereas cultures may become positive 2–4 weeks later. In contrast to mannan antigenemia in candidiasis or GM antigenemia in aspergillosis, which may be transitory phenomena, antigenuria in histoplasmosis may persist for up to 2 months.[33] Perhaps because of shared epitopes, the test for *Histoplasma* antigenuria was positive in the urine of 4 of 25 blastomycosis patients and in the cerebrospinal fluid of a coccidioidomycosis patient.[36] Formal proof is lacking that the heat-stable antigen detected in urine is the same GM-containing C antigen characterized from histoplasmin, but the two activities share common properties.

5. PARACOCCIDIOIDOMYCOSIS—*PARACOCCIDIOIDES BRASILIENSIS* ANTIGENS

5.1. Exoantigen

The exoantigen, 1 or E, is the basis for a specific immunoidentification of *P. brasiliensis* cultures.[8] Agar gel immunodiffusion has also been used as a serodiagnostic test in paracoccidioidomycosis. The same antigen, on which the immunoidentification of *P. brasiliensis* cultures is based, is the major species-specific culture supernatant antigen, A, reactive with sera from patients with the disease.[37] That terminology is ambiguous because currently the term *A* is applied to the entirely distinct A antigen of *B. dermatitidis*.

A cell-wall-derived GM is common to the genera of primary systemic dimorphic fungi, including *P. brasiliensis*, and complicates the interpretation of sensitive antibody tests that use unpurified antigenic complexes.[5] A monomolecular and species-specific protein antigen with a molecular weight of 43 kDa has received recent emphasis.[38,39] Some evidence, summarized below, suggests that this 43-kDa protein may indeed be similar or identical to antigen 1 or E. If that is the case, then three different groups agree on the nature of the immunodominant secreted protein antigen of *P. brasiliensis*.[8,38,40]

5.2. 43-kDa Glycoprotein

A 43-kDa glycoprotein is continually produced and secreted into the medium during the growth of yeast forms, and its concentration declines in the stationary phase, possibly as a result of endogenous proteinases.[38] The antigen is optimally isolated at peak concentration from supernatants of yeast from cultures grown for 5 days on yeast extract–peptone–glucose cultures.[38] Partial purification was effected by its affinity for Con A and by immunoaffinity chromatography using the IgG from paracoccidioidomycosis patients as an adsorbent. Gel permeation chromatography was also a facile means of purifying this antigen away from a high molecular weight GM.[41] After SDS-PAGE, isolated antigen in gel was used to raise monospecific antiserum. This, in turn, was affinity-purified on a column containing solid-phase adsorbed exocellular *P. brasiliensis* antigen. Antibodies against the *P. brasiliensis* antigen E2[40] also reacted with the 43-kDa antigen, and the 43 kDa-containing immunoprecipitin arc that was formed in immunoelectrophoresis had a mobility similar to that obtained with antigen E2.

Virtually all paracoccidioidomycosis patients produce IgG against the 43-kDa glycoprotein, as shown in western immunoblots.[39,42] The patients typically react with a doublet of proteins in the 43- to 48-kDa region of blot strips, indicating either isozymes or different antigens. Weak IgG responses against this antigen occur in histoplasmosis patients.

The carbohydrate-containing portion of *P. brasiliensis* culture supernatants with affinity for Con A was analyzed.[41] One constituent was a high molecular weight peptido-GM, consisting of a (1→6) mannan linear portion extensively substituted with (1→2)-oligomannoside chains, possibly in α and β configurations. Single beta-Gal*p* nonreducing termini occurred, in contrast to the galactofuranosyl residues detected in mycelial form GM by Azuma et al.[5] Another *P. brasiliensis* preparation was enriched in a 55-kDa highly glycosylated protein that had a more complex L-fuco-galactomannan structure.[41]

5.3. Other Protein Antigens

As more sensitive western immunoblot methods are used to probe the humoral immune response to *P. brasiliensis*, other antigens have come to light that may be useful indicators of disease activity. IgG against a 70-kDa glycoprotein was frequently present before therapy of paracoccidioidomycosis. A favorable clinical response was heralded by a decline in those antibodies, followed by a more gradual decline in IgG against the major 43-kDa glycoprotein.[39] Homogenate-supernatant antigens derived from *P. brasiliensis* yeast forms present a different antigenic mosaic than was the case with culture filtrates of the mold forms.[42] Here too, a doublet in the 45- to 48-kDa range and a single 76-kDa antigen were reminiscent of antigens previously characterized from the mold-form culture filtrates. In addition, species-specific antigens occurred with molecular weights estimated at 27.5, 57, and 114 kDa. Such diversity in the number of antigens is expected from a eukaryotic organism, and as disease becomes more chronic or severe the multiplicity of antibodies would predictably increase.

6. SUBCUTANEOUS MYCOSES—SPOROTRICHOSIS—*SPOROTHRIX SCHENCKII* ANTIGENS

Sporotrichosis most often presents as a subcutaneous mycosis resulting from the traumatic implantation of thorns, wood splinters, hay, sphagnum moss, or soil contaminated with *S. schenckii*. Fixed cutaneous, lymphocutaneous, and extracutaneous forms of the disease occur.

6.1. Peptido-L-Rhamno-D-Mannan

The outer layer of the cell wall contains a species-specific peptido-L-rhamno-D-mannan, which accounts for the excellent specificity obtained in clinical assays for antibodies in sporotrichosis. Classical immunoassays measure the agglutination of whole yeast forms or of latex particles passively coated with supernatants of *S. schenckii* cultures.[1,43] Adaptation of the culture filtrate antigen to an indirect EIA was beneficial in increasing the sensitivity of antibody detection in serum and cerebrospinal fluid, obtaining more objective end points, so as to improve monitoring of the patient's response to therapy.[43]

6.2. Sporotrichin Proteins

There has been less impetus to search for protein antigens to improve the antigenic specificity of clinical assays because of satisfaction with the rhamnomannan. Nonetheless, analysis of proteins present in culture filtrate antigens (sporotrichin) has resulted in the discovery of some proteins that may have important implications for pathogenesis.[44] Western immunoblotting was conducted after SDS-PAGE of sporotrichin, using as probes serum specimens from patients with lymphocutaneous and extracutaneous forms of sporotrichosis. Not surprisingly, serum from patients with lymphocutaneous disease reacted with fewer *S. schenckii* proteins than those with extracutaneous disease. The immunodominant proteins in sporotrichin that evoked IgG in most patients had molecular weights estimated at 40 and 70 kDa. Patients with extracutaneous disease also reacted against a 32-kDa protein, and that antigen may be a very useful marker of deep infection.

6.3. Chymotryptic and Aspartyl Proteinases

S. schenckii produces two proteinases induced by growth on BSA-containing medium.[45,46] The two proteinases were separated by anion exchange chromatography using a continuous salt gradient; proteinase I eluted first, whereas desorption of proteinase II required 0.35 M NaCl. Repetitive cycles of gel permeation chromatography were required to remove extraneous pigment. The molecular weights estimated by gel permeation chromatography were consistent with the results of SDS-PAGE for 39-kDa proteinase II. Proteinase I had an apparent molecular weight of 36.5 kDa, measured by gel permeation, but SDS-PAGE showed the presence of five polypeptides, a 37-kDa component being most abundant. Thus two distinct proteinases with similar molecular weight are present. The pH optimum for proteinase I was pH 6.0, and it was inhibited by chymostatin. Proteinase II's pH optimum was pH 3.5 and pepstatin was inhibitory.

The manuscript for this chapter was submitted for publication in October of 1990.

REFERENCES

1. Reiss, E., 1986, *Molecular Immunology of Mycotic and Actinomycotic Infections*, Elsevier Science, New York, 423 pp.
2. Cox, R. A. (ed.), 1989, *Immunology of Fungus Infections*, CRC Press, Boca Raton, FL, 240 pp.
3. Drouhet, E., Cole, G. T., de Repentigny, L., Latge, J.- P., and Dupont, B. (eds.), 1988, *Fungal Antigens—Isolation, Purification, and Detection*, Plenum Press, New York, 469 pp.
4. Kurstak, E., Marquis, G., Auger, P., de Repentigny, L., and Montplaisir, S. (eds.), 1989, *Immunology of Fungal Diseases*, Marcel Dekker, New York, 494 pp.
5. Azuma, I., Kanetsuna, F., Tanaka, Y., Yamamura, Y., and Carbonell, L. M., 1974, Chemical and immunological properties of galactomannans obtained from *Histoplasma duboisii*, *Histoplasma capsulatum*, *Paracoccidioides brasiliensis* and *Blastomyces dermatitidis*, *Mycopathol. Mycol. Appl.* **54:**111–125.

6. Kaufman, L., and Reiss, E., 1986, Serodiagnosis of fungal diseases, in: *Manual of Clinical Immunology*, 3rd ed. (N. R. Rose, H. Friedman, and J. L. Fahey, eds.), American Society of Microbiology, Washington, D.C., pp. 446–466.
7. Klein, B. S., and Jones, J. M., 1990, Isolation, purification, and radiolabeling of a novel 120-kD surface protein on *Blastomyces dermatitidis* yeasts to detect antibody in infected patients, *J. Clin. Invest.* **85**:152–161.
8. Kaufman, L., and Standard, P. G., 1987, Specific and rapid identification of medically important fungi by exoantigen detection, *Ann. Rev. Microbiol.* **41**:209–225.
9. Hurst, S., Kaufman, L., and Jalbert, M., 1989, Western blot analysis and characterization of a purified *Blastomyces dermatitidis* culture filtrate antigen, *Prog. Abstr. Annu. Mtg. Amer. Soc. Microbiol.* (New Orleans) #F-31, p. 463.
10. Huppert, M., Spratt, N. S., Vukovich, K. R., Sun, S. H., and Rice, E. H., 1978, Antigenic analysis of coccidioidin and spherulin determined by two-dimensional immunoelectrophoresis, *Infect. Immun.* **20**:541–551.
11. Cole, G. T., Sun, S. H., Dominguez, J., Yuan, L., Franco, M., and Kirkland, T. N., 1988, Wall-associated antigens of *Coccidioides immitis*, in: *Fungal Antigens* (E. Drouhet, G. T. Cole, L. de Repentigny, J. -P. Latge, and B. Dupont, eds.), Plenum Press, New York, pp. 395–415.
12. Pappagianis, D., Smith, C. E., Kobayashi, G. S., and Saito, M. G., 1961, Studies of antigens from young mycelia of *Coccidioides immitis*, *J. Infect. Dis.* **208**:35–44.
13. Huppert, M., and Bailey, J. W., 1965, The use of immunodiffusion tests in coccidioidomycosis II. An immunodiffusion test as a substitute for the tube precipitin test, *Am. J. Clin. Pathol.* **35**:369–373.
14. Huppert, M., and Bailey, J. W., 1965, The use of immunodiffusion tests in coccidioidomycosis I. The accuracy and reproducibility of the immunodiffusion test which correlates with complement fixation, *Am. J. Clin. Pathol.* **35**:364–368.
15. Levine, H., Cobb, J. M., and Scalarone, G. M., 1969, Spherule coccidioidin in delayed dermal sensitivity reactions of experimental animals, *Sabouraudia* **7**:20–32.
16. Scalarone, G. M., Levine, H. B., Pappagianis, D., and Chaparas, S. D., 1974, Spherulin as a complement-fixing antigen in human coccidioidomycosis, *Am. Rev. Respir. Dis.* **110**:324–328.
17. Cox, R. A., and Britt, L. A., 1987, Antigenic identity of biologically active antigens in coccidioidin and spherulin, *Infect. Immun.* **55**:2590–2596.
18. Kruse, D., and Cole, G. T., 1990, Isolation of tube precipitin antibody-reactive fractions of *Coccidioides immitis*, *Infect. Immun.* **58**:169–178.
19. Cole, G. T., Kruse, D., Zhu, S., Seshan, K. R., and Wheat, R. W., 1990, Composition, serological activity, and immunolocalization of a 120-kilodalton tube precipitin antigen of *Coccidioides immitis*, *Infect. Immun.* **58**:179–188.
20. Cox, R. A., and Britt, L. A., 1986, Isolation of a coccidioidin component that reacts with immunoglobulin M precipitin antibody, *Infect. Immun.* **53**:449–453.
21. Zimmer, B. L., and Pappagianis, D., 1989, Immunoaffinity isolation and partial characterization of the *Coccidioides immitis* antigen detected by the tube precipitin and immunodiffusion-tube precipitin tests, *J. Clin. Microbiol.* **27**:1759–1766.
22. Kaufman, L., Standard, P. G., Huppert, M., and Pappagianis, D., 1985, Comparison and diagnostic value of the coccidioidin heat-stable (HS and tube precipitin) antigens in immunodiffusion, *J. Clin. Microbiol.* **22**:515–518.
23. Cox, R. A., Britt, L. A., and Michael, R. A., 1987, Isolation of *Coccidioides immitis* F antigen by immunoaffinity chromatography with monospecific antiserum, *Infect. Immun.* **55**:227–232.
24. Zimmer, B. L., and Pappagianis, D., 1988, Characterization of a soluble protein of *Coccidioides immitis* with activity as an immunodiffusion-complement fixation antigen, *J. Clin. Microbiol.* **26**:2250–2256.
25. Cox, R. A., and Britt, L. A., 1986, Isolation and identification of an exoantigen specific for *Coccidioides immitis*, *Infect. Immun.* **52**:138–143.
26. Cole, G. T., Zhu, S., Pan, S., Yuan, L., Kruse, D., and Sun, S. H., 1989, Isolation of antigens with proteolytic activity from *Coccidioides immitis*, *Infect. Immun.* **57**:1524–1534.
27. Standard, P. G., and Kaufman, L., 1977, Immunological procedures for the rapid and

specific identification of *Coccidioides immitis* cultures, *J. Clin. Microbiol.* **5**:149–153.

28. Reiss, E., and Bragg, S. L., 1988, Immunochemical analysis of histoplasmin proteins and polysaccharide, in: *Fungal Antigens—Isolation, Purification, and Detection* (E. Drouhet, G. T. Cole, L. de Repentigny, J. -P. Latge, and B. Dupont, eds.), Plenum Press, New York, pp. 431–444.
29. Harris, J. E., and Deepe, Jr., G. S., 1988, Characterization of antigenic determinants in histoplasmin that stimulate *Histoplasma capsulatum*-reactive T cells in vitro, *Infect. Immun.* **56**: 2343–2349.
30. Graybill, J. R., Patino, M., and Ahrens, J., 1988, In situ localization of antigens of *Histoplasma capsulatum* using colloidal gold immune electron microscopy, *Mycopathology* **104**:181–188.
31. Reiss, E., Knowles, J. B., Bragg, S. L., and Kaufman, L., 1986, Monoclonal antibodies against the M-protein and carbohydrate antigens of histoplasmin characterized by the enzyme-linked immunoelectrotransfer blot method, *Infect. Immun.* **53**:540–546.
32. Green, J. H., and Pine, L., 1985, Preparation of h and m antigens of *Histoplasma capsulatum* free of heterologous antigens, *Curr. Microbiol.* **12**:209–216.
33. Wheat, L. J., Kohler, R. B., and Tewari, R. P., 1986, Diagnosis of disseminated histoplasmosis by detection of *Histoplasma capsulatum* antigen in serum and urine specimens, *N. Engl. J. Med.* **314**:83–88.
34. Graybill, J. R., Patino, M. M., Gomez, A. M., and Ahrens, J., 1985, Detection of histoplasmal antigens in mice undergoing experimental pulmonary histoplasmosis, *Am. Rev. Respir. Dis.* **132**:752–756.
35. Zimmerman, S. E., Connolly-Stringfield, P., Wheat, L. J., French, M. L. V., and Kohler, R. B., 1989, Comparison of sandwich solid-phase radioimmunoassay and two enzyme-linked immunosorbent assays for detection of *Histoplasma capsulatum* polysaccharide antigen, *J. Infect. Dis.* **160**:678–685.
36. Wheat, L. J., Connolly-Stringfield, P., Kohler, R. B., Frame, P. T., and Gupta, M. R., 1989, *Histoplasma capsulatum* polysaccharide antigen detection in diagnosis and management of disseminated histoplasmosis in patients with acquired immunodeficiency syndrome, *Am. J. Med.* **87**:396–400.
37. Restrepo, A, and Drouhet, E., 1970, Etude des anticorps precipitants dans las blastomycose sudamericaine par l'analyse immunoelectrophoretique des antigenes de *Paracoccidioides brasiliensis*, *Ann. Inst. Pasteur (Paris)* **119**:338–346.
38. Stambuk, B. U., Puccia, R., de Almeida, M. L. C., Travassos, L. R., and Schenkman, S., 1988, Secretion of the 43 kDa glycoprotein antigen by *Paracoccidioides brasiliensis*, *J. Med. Vet. Mycol.* **26**:367–373.
39. Camargo, Z. P., Unterkircher, C., and Travassos, L. R., 1989, Identification of antigenic polypeptides of *Paracoccidioides brasiliensis* by immunoblotting, *J. Med. Vet. Mycol.* **27**:407–412.
40. Yarzabal, L. A., Bout, D., Naquira, R., Fruit, J., and Andrieu, S., 1977, Identification and purification of the specific antigen of *Paracoccidioides brasiliensis* responsible for immunoelectrophoretic band E, *Sabouraudia, J. Med. Vet. Mycol.* **15**:79–85.
41. Puccia, R., Schenkman, S., Gorin, P. A. J., and Travassos, L. R., 1986, Exocellular components of *Paracoccidioides brasiliensis*: Identification of a specific antigen, *Infect. Immun.* **53**: 199–206.
42. Casotto, M., 1990, Characterization of the cellular antigens of *Paracoccidioides brasiliensis* yeast form, *J. Clin. Microbiol.* **28**:1188–1193.
43. Scott, E. N., Kaufman, L., Brown, A. C., and Muchmore, H. G., 1987, Serologic studies in the diagnosis and management of meningitis due to *Sporothrix schenckii*, *N. Engl. J. Med.* **317**: 935–940.
44. Scott, E. N., and Muchmore, H. G., 1989, Immunoblot analysis of antibody responses to *Sporothrix schenckii*, *J. Clin. Microbiol.* **27**:300–304.
45. Tsuboi, R., Sanada, T., Takamori, K., and Ogawa, H., 1987, Isolation and properties of extracellular proteinases from *Sporothrix schenckii*, *J. Bacteriol.* **169**:4104–4109.
46. Tsuboi, R., Sanada, T., and Ogawa, H., 1988, Influence of culture medium pH and proteinase inhibitors on extracellular proteinase activity and cell growth of *Sporothrix schenckii*, *J. Clin. Microbiol.* **26**:1431–1433.

20

Immunochemistry of Fungal Antigens (Part B)

Opportunistic Pathogens

ERROL REISS

1. INTRODUCTION

Emphasis in this chapter is placed on recent advances in the immunochemistry of the three major opportunistic mycoses: candidiasis, cryptococcosis, and aspergillosis. Surely a broad range of other fungi can infect the immunocompromised host, but relatively little is known of their antigenic structure.

Abbreviations: (All carbohydrates are the D enantiomers, and are in the pyranose configuration, unless otherwise indicated) ABPA, allergic bronchopulmonary aspergillosis; BGG, bovine gamma globulin; BSA, bovine serum albumin; C3, third component of complement; CF, culture filtrate; Con A, concanavalin A; CR3, neutrophil receptor for C3; CWMP, cell wall mannoprotein; DTT, dithiothreitol; EAC, sheep erythrocyte-antibody-complement; EIA, enzyme immunoassay; ELISA, enzyme-linked immunosorbent assay; EM, electron microscopy; endo H, endoglycosidase H, an N,N′-diacetyl chitobiase; GalXM, galactoxylomannan; g.l.c.-m.s., gas liquid chromatography-mass spectrometry; Glc*p*, glucose; Glc*p*A, glucosyluronic acid; GM, galactomannan; GXM, glucuronoxylomannan; IA, invasive aspergillosis; IC, invasive candidiasis; MAbs, monoclonal antibodies; ME, mycelial extract; Man*p*, mannose; MP, mannoprotein; n.m.r., nuclear magnetic resonance spectrometry; pAbs, polyclonal antibodies; pI, isoelectric point; RIA, radioimmunoassay; SDS-PAGE, sodium dodecyl sulfate-polyacrylamide gel electrophoresis; TCA, trichloroacetic acid; western blot, immunoelectro-transfer blot.

ERROL REISS • Molecular Mycology Section, Mycotic Diseases Branch, Division of Bacterial and Mycotic Diseases, National Center for Infectious Diseases, Centers for Disease Control and Prevention, Atlanta, Georgia 30333.

Fungal Infections and Immune Responses, edited by Juneann W. Murphy *et al.* Plenum Press, New York, 1993.

Some recent general references cited in the previous chapter, "Immunochemistry of Fungal Antigens (Part A)" provide a broader perspective.

As seen in the systemic, dimorphic, fungal pathogens, a dichotomy exists in the antigenic structure between the major antigenic cell envelope polysaccharides or glycoproteins and, on the other hand, the protein antigens. Cell wall mannoproteins or heteromannans (whether in the wall or, in one case, the capsule) are capable of modulating immunity, including its downregulation. Detection of polysaccharides in body fluids: mannan of *Candida albicans*, glucuronoxylomannan of *Cryptococcus neoformans*, and galactomannan of *Aspergillus fumigatus* have contributed diagnostic methods independent of a functioning immune system in the host.[1–3]

Protein antigens have pathogenic potential according to their enzymatic activities, whether as proteinases capable of attacking the host, as catalase, or as yet uncharacterized functions. Because proteins are amenable to a molecular biological approach, such studies, should go far towards further defining the structure and function of fungal protein antigens. Successful detection of a *C. albicans* heat-labile protein in the blood during candidiasis, described below, has kindled interest in extending this approach to other fungi.

2. DISEASES CAUSED BY OPPORTUNISTIC YEASTS

2.1. Candidiasis—*Candida albicans* Antigens

2.1.1. *C. albicans Cell Wall Mannoprotein (CWMP)*

2.1.1a. Definition. CWMP is the readily soluble outer mural layer and is the immunodominant cell surface antigen. CWMP, sometimes called phosphomannoprotein, or simply mannan, is a homopolymer of Man*p* linked by *O*- and *N*-glycosidic bonds to structural protein. The protein is ~5% of the dry weight. The general organization of the mannan portion consists of an inner core linked by chitobiosyl aspartamido bonds to the protein and a contiguous outer chain region, containing the antigenic oligomannoside epitopes. The configuration of linkage and glycosidic bond arrangements in the outer chain region consist of a linear (1→6)-α-mannan backbone heavily substituted with oligomannosides containing (1→2)-α and sparse but important (1→3)-α mannosyl residues. The degree of polymerization (d.p.) of the antigenic side chains is 2–7. Base-labile oligomannosides are also attached directly to the protein moiety via *O*-glycosidic ester bonds to serine and threonine. The actual repeat unit structure of the outer chain region is not yet clear and two models have been proposed. Kobayashi et al.[4] proposed a comblike structure with linear oligomannosyl antigenic determinants, whereas Kagaya et al.[5] proposed a more highly branched arboreal structure.

Awareness has gradually emerged about the antigenic importance of a domain in the CWMP outer chain occupied by (1→2)-β oligomannosyl residues.[6,7] Acid-labile (pH 2, 1 hour, 100°C) phosphodiesters may connect (1→2)-β oligomannosides to the otherwise α-linked CWMP.[4] These β-linked oligomannosides comprise 9% of the total carbohydrate of the CWMP. Both serotypes A and B appear to contain these linear (1→2)-β mannosyl antigenic determinants (d.p. 2–7).[4,8]

2.1.1b. CWMP: Extraction. CWMP is extracted by stirring blastoconidia in hot neutral buffer. The cell-free supernatant then receives cetyltrimethyl ammonium bromide (CTAB). After adjustment to pH 8.8 with sodium borate, CWMP is recovered in the precipitate.[9] By compositional analysis, serotype B CWMP contained carbohydrate 90%, protein 3.1%, and phosphate 1.48%, and had an $[\alpha]_D$ of +63.1°.[4] CWMP produces broad and diffuse elution profiles on gel permeation columns, indicative of polydispersity in its synthesis and in the course of its extraction. Molecular weights (gel-permeation columns calibrated with dextrans) are in the range of 40–250 kDa. Bulk CWMP can be separated by anion exchange chromatography into fractions based on their net charge, probably resulting from the degree of phosphorylation.[10]

2.1.1c. CWMP: Range of Interactions with the Immune System. This topic, while peripheral to the scope of the present work, is noteworthy because these interactions create a rationale for CWMP structure elucidation. As the most abundant *C. albicans* surface antigen, one that activates complement by the alternative pathway, CWMP is a likely target for opsonization.[10,11] Domer systematically reviewed CWMP immunomodulation and only some recent highlights are presented here.

Blastogenesis of peripheral blood mononuclear cells is stimulated by CWMP, provided they are from *C. albicans*-sensitized persons.[12] Strong delayed-type hypersensitivity can be elicited by CWMP in mice.[10] Human B cells are stimulated by CWMP to go from the G_0 to G_1 phase, but T-helper cells or their factors are required for antibody production.[12] According to their net charge, CWMP subfractions can either stimulate or suppress unrelated T-dependent antibody responses.[10] Small oligomannoside catabolites of CWMP can exert suppressive activity on antigen-elicited lymphocyte blastogenesis.[13] Keeping these properties in mind, CWMP cannot be overlooked as a key antigen in the host–*C. albicans* interaction. During invasive candidiasis (IC) of the immunocompromised host, circulating CWMP–immune complexes can be detected by heat dissociation and either EIA or RIA as a specific diagnosis.[1,2]

2.1.1d. CWMP: Antigenicity. C. albicans serotype A was for a long time regarded as containing all the determinants present in serotype B and, in addition, the A-specific determinant.[14] Tsuchiya's agglutinin-adsorption scheme produced monospecific antiserum reactive with the serotype A-specific epitope, antigen 6.[15] MAbs have been produced that react with the

CWMP of serotype A.[9] A serotype B antigen, 13b, in Tsuchiya's[15] scheme was identified with monospecific antiserum, but no serotype-B-specific mAbs have been produced as yet. Linkages present in the acid-stable mannan of serotype B are all (1→2)-α, with penultimate and type-specific (1→3)-α mannosyl residues.[4] Terminal nonreducing (1→3)-α-mannosyl residues occur in both serotypes.

Kagaya et al.[5] proposed that the serotype-A-specific determinant consists of a hexasaccharide with a (1→3)-α mannosyl nonreducing terminus in an otherwise (1→2)-α mannotetraose. Further, they proposed a single (1→6)-α mannosyl side substituent at the penultimate mannosyl residue from the nonreducing terminus.

Although the (1→2)-β mannan moiety is a minor portion by weight, its removal, yielding acid-stable mannan, is accompanied by a major reduction in serological activity with rabbit anti-*C. albicans* serotype A or anti-serotype B.[4] Beta-linked oligomannosides probably occur in both serotypes but a definitive structural analysis of this domain within serotype A CWMP has not yet appeared.

MAbs with specificity for (1→2)-β-oligomannosides were produced from mice immunized with heat-killed blastoconidia of serotype B.[16] Most of the mAbs produced reacted against the intact CWMP but not against the acid-stable CWMP of serotype B, underlining the immunodominance of the (1→2)-β Man*p* linkages for that serotype. Surprisingly, acid-treated CWMP of serotype A reacted with a majority of the mAbs, indicating some acid-stable β-mannan in serotype A. One mAb, 18d7, was reactive only with the native CWMP of serotypes A and B. Presumably it binds to a unique acid-labile β-mannosyl determinant.

The morphogenetic transition from the yeast form to the mycelial form is accompanied by a marked reduction in the acid-labile and hence phosphate-bound (1→2)-β-mannan.[17] This finding is important because the wall of the germ tube and hyphal forms of *C. albicans* is thinner than that of the yeast forms. A considerable literature has accumulated *pro* and *con* the existence of germ-tube-specific antigens (see Section 2.1.5.). Clearly, serum produced against germ tubes and adsorbed with yeast forms is capable of selective immunofluorescent staining of the germ tube portion. This finding, however, may relate to the increased accessibility of proteins in the periplasmic space of germ tubes. In yeast forms periplasmic proteins may be occluded by a thicker wall structure, one that is shielded by an antigenic (1→2)-β-mannan.

2.1.2. The C. albicans Surface Receptors for iC3b and C3d

A fortuitous observation led to elucidating a novel *C. albicans* surface receptor for opsonic complement fragments iC3b and C3d. While studying the ability of a lymphocytic cell line to elicit rosetting by complement-coated sheep erythrocytes (EAC), Heidenreich and Dierich[18] observed a *C. albicans* pseudohyphal contaminant that avidly bound to EAC. Flow cytometry of *C. albicans* using mAbs against complement receptors as markers showed that only mAbs against the α chain of polymorphonuclear neutrophil CR3, the iC3b receptor, were present on *C. albicans*.[19] The iC3b receptor, extracted by disruption of surface-radioiodinated *C. albicans*, was immunoprecipitated with anti-CR3 mAbs and had a molecular weight of 130 kDa.[20] Morphogenesis to mycelial forms increased the expression of the iC3b receptors. *C. albicans* blastoconidia treated with anti-CR3 mAbs were more readily phagocytosed.

An mAb against *C. albicans* CWMP was used as an adsorbent to purify the candidal C3d receptor from a homogenate-supernatant of *C. albicans* blastoconidia.[21] The immunoaffinity-purified antigen was capable of blocking EAC3d rosette formation around *C. albicans*. The C3d receptor was a 60-kDa glycoprotein that could bind the lectin, concanavalin A (Con A). At this point the *C. albicans* receptors for C3d and iC3b appear to be distinct molecules.

2.1.3. C. albicans Major Cytoplasmic Protein Antigens

2.1.3a 47-kDa Protein. Western immunoblot was used to identify this protein as a clinically important antigen.[22–25] IgM antibodies against the 47-kDa glycoprotein were consistently produced by patients who survived invasive candidiasis (IC). Poor clinical outcome was correlated with a weak or absent response to this antigen. A glycosidic moiety was inferred by the affinity of the antigen for Con A. The antigen's native molecular weight, 74–79 kDa, was reduced upon heating (100°C, 5 minutes) in the dissociation buffer for SDS-PAGE. A 47-kDa *Candida* antigen was isolated from patients' serum by immunoaffinity chromatography using as the ligand either human convalescent IC serum, rabbit antiserum against a homogenate-supernatant, or rabbit antiserum produced against *Candida* antigen isolated by immunoaffinity from an IC patient. In each case the affinity-purified circulating antigen was identified by western immunoblot.

Matthews and Burnie[26] produced monospecific antiserum against the 47-kDa glycoprotein and applied it as an indicator antibody in a dot-blot immunobinding assay to detect the antigen circulating in the blood of 97% of 87 cases of IC. The serum antigen was heat labile.

2.1.3b. 48-kDa Enolase. This cytoplasmic protein did not bind to Con A and was purified by anion exchange chromatography.[27] Patients with IC contained antibodies that immunoprecipitated the radioiodinated antigen. MAbs were prepared against this antigen and then adapted to an EIA to detect the 48-kDa protein[28] in experimental murine IC.[29]

Proof that the 48-kDa Antigen is Enolase. Enolase is a yeast glycolytic enzyme that catalyzes the reaction of 2-phospho-D-glycerate to phosphoenolpyruvate.[30] The *Saccharomyces* enzyme consists of a dimer of two similar 40- to 50-kDa polypeptides. PAbs produced against purified *Saccharomyces cerevisiae* enolase precipitated a *C. albicans* 48-kDa protein and reduced the enolase activity of the cytoplasmic supernatant.[30] MAbs against the *C. albicans* antigen cross-reacted in western blot with a ~48-kDa protein in the *S. cerevisiae* homogenate. A 48-kDa protein was also translated *in vitro* from *C. albicans* mRNA in the rabbit reticulocyte expression system. The protein was then immunoprecipitated by the anti-*S. cerevisiae* enolase-IgG. Considering the functional and antigenic similarity between the *C. albicans* and *Saccharomyces* enolases, it would not be surprising if the *Candida* 48-kDa enzyme also occurred naturally as a higher molecular weight dimer.

2.1.4. 43-kDa Secreted Aspartyl Proteinase (EC3.4.23.6)

The production of an extracellular proteinase by *C. albicans* was observed by Staib in 1965.[31] Gradually its potential as a pathogenic factor has emerged. *C. albicans* serotype A, serotype B, and *C. tropicalis* produce proteinase and are the most pathogenic *Candida* species. The increased prevalence of serotype A in infections is arguably related to its greater proteinase expression than that of the B serotype. Of two isolates of serotype B studied, proteinase was expressed only on blastoconidia, but not on germ tubes.[32] Of the other less pathogenic *Candida* species, *C. parapsilosis* is only proteolytic under forced culture conditions and does not express proteinase during *in vitro* infection of phagocytes.[33]

A nitrous acid-induced proteinase-deficient mutant of reduced mouse virulence reverted to the proteinase-positive virulent form after mouse passage.[34] *C. albicans* releases the aspartyl proteinase into the culture medium during growth on proteinaceous substrates. The aspartyl proteinase is active in an acidic milieu that surrounds the blastoconidia, the result of glucose catabolism. Induction does not occur on Sabouraud agar because its high glucose content is repressive. Human or bovine serum albumin can serve as a nitrogen source. A simple basal salts medium with vitamins, containing 0.1% human serum albumin and adjusted to pH 5, induced proteinase secretion in 75 of 100 strains tested.[31] Significant accumulation of proteinase in liquid growth medium occurs after the primary growth phase, between the 8th and 11th days, but there is a peak of activity followed by its disappearance as a result of autolysis.[35]

2.1.4a. Purification. The enzyme is subject to alkaline degradation, thus purification is accomplished below pH 7. Above pH 8.4 it aggregates to a dimer and undergoes autolysis. Various combinations of anion exchange and gel permeation chromatography have been proposed.[36–38] The C.D.C. Molecular Immunology Laboratory method is summarized as follows. The cell-free culture supernatant is concentrated 10-fold by ultrafiltration and is dialyzed against 0.02 M citrate buffer, pH 6.3. The dialyzed retentate is applied to a DEAE-Sephadex A25 column and desorbed with 0.2 M citrate buffer. Fractions containing proteinase activity are concentrated by dialysis against polyethylene glycol 20,000 and then chromatographed on a Sephadex G 75 column. This column serves to separate the CWMP from proteinase. Proteinase-containing fractions are rechromatographed on DEAE-Sephadex, this time eluting with a continuous salt gradient, from 0.02 to 0.5 M citrate, pH 6.3. The product is free of CWMP, as detected in sandwich EIA, and consists of three isozymes in the molecular weight range of 41–48 kDa (Morrison and Reiss, unpublished observations).

Initially, Macdonald and Odds[37] indicated the aspartyl proteinase was a mannoenzyme, and development of an mAb[39] that reportedly binds to a carbohydrate site on the proteinase also suggests the possibility of glycosylation. In contrast to these observations, immunoblots probed with either antimannan antibody or with Con A-peroxidase failed to stain *C. albicans* aspartyl proteinase.[33] Purified aspartyl proteinase contains less than 1% mannan content and fails to stain on SDS-PAGE gels with the silver stain variation modified for carbohydrate detection.

2.1.4b. Spectrum of Activity. Candida proteinase was located by immunofluorescence within clogged blood vessels of a patient with acronecrosis of the fingers and toes as a consequence of IC.[40] Proteinases of *C. albicans* are implicated in coagulation and vasoconstriction. Reninlike activation of the angiotensin vasoconstriction pathway has been observed.[41] Aspartyl proteinase had 1/10 the activity of pure renin in cleaving angiotensinogen, the first step in the generation of the vasoconstrictor peptide, angiotensin II. The combined proteolytic effect of *C. albicans* can activate coagulation factors IX and X.

Aspartyl proteinase can degrade the stratum corneum,[42] and human sole skin was a suitable inducer substrate for the proteinase. *C. albicans* does not normally invade the living layer of the epidermis, nor can it digest whole human serum, implying the presence of a potent serum inhibitor, probably α_2-macroglobulin.

The enzyme was detected on *Candida* colonizing murine kidney[43] and decay of immunoglobulins in the kidney may be the direct result of proteinase.[44] Adequate specificity controls to ensure the absence of CWMP from such preparations are essential. *C. albicans* aspartyl proteinase degrades immunoglobulins present in normal saliva, including IgG, IgM, IgA_2, and secretory component.[45] Bacterial IgA-proteinases are not able to attack IgA_2 and are inhibited by antibodies in normal body fluids. Perhaps because the active site of aspartyl proteinases is conserved between microbes and humans, no such neutralizing antibodies were found in human saliva.[45]

2.1.4c. Cloning an Aspartyl Proteinase Gene of C. albicans. A genomic library of *C. albicans* CBS 2730 DNA was produced in bacteriophage λ, and the library was probed with the proA gene from *Saccharomyces cerevisiae*, encoding for a vacuolar proteinase. Twenty-two plaques hybridized with this probe and these were pooled and rescreened. The restriction map of one clone revealed two *Eco*RI fragments that hybridized to the *Saccharomyces* probe. They were subcloned into phage M13 and the amplified inserts were sequenced by dideoxy chain termination.[46]

The cloned *C. albicans* proteinase gene includes a total of 1705 base pairs. Starting at the 5′ end (relative to mRNA), a promotor region containing a TATA box regulatory sequence is found at position 360. The first methionine residue is 45 base pairs downstream from the TATA box at position 407 and predicts the synthesis of a 38 amino acid open reading frame, with a TAG amber termination codon. This 38 amino acid sequence contains hydrophobic amino acids and could serve as a signal peptide, providing extracellular targeting information. On the 3′ end of the signal peptide is an amber stop codon, followed by a second methionine residue at position 521. The initiation codon at bp 521 extended 1140 bp, giving an open reading frame encoding a 380 amino acid proteinase structural gene. The nucleotide structure suggests that a suppressor tRNA is required to overcome termination codon readthrough to produce the secreted form of aspartyl proteinase. This *C. albicans* proteinase is probably produced as a proenzyme that is activated by proteolysis to produce the mature enzyme. The amino-terminus sequence suggests that this cleavage site for the mature enzyme is at position 634, 12 amino acids before a homologous site in the *Saccharomyces* aspartyl proteinase. Given this maturation site, the functional enzyme should contain 340 amino acids, and have a molecular weight of 43 kDa. Overall there was 70% sequence homology between the *C. albicans* gene and the *Saccharomyces* pro A gene.

2.1.5. C. albicans Germ-Tube-Related Antigens

Interest in germ-tube-related antigens stems from the morphogenesis observed when *C. albicans* converts from the blastoconidial (yeast or Y) form through the production of germ tubes to the mycelial (M) form. Germ tube formation accompanies tissue invasion from mucosal foci and is a mechanism that enables *C. albicans* to evade or escape from endophagocytosis. Molecules selectively expressed during the Y→M transition may be involved in pathogenesis and/or can serve as markers more reflective of infection than those present in the yeast form.

Germ-tube-specific antibodies were produced by immunizing rabbits with germ tubes and adsorbing the serum with blastoconidia. The resulting antibodies stained only germ tubes by immunofluorescence.[47] Zymolyase, a bacterial enzyme complex containing (1→3)-β-glucanase and a proteinase, digested germ tube cell walls and released high molecular weight mannoproteins, as well as lower molecular weight proteins. Sundstrom et al.,[48] using this method and SDS-PAGE, described a 200-kDa mannoprotein present in both blastoconidia and germ tubes. Only the form of the antigen obtained from germ tubes was strongly reactive in western blot with germ tube antibodies. Strain–strain differences in germ tube mannoproteins were noted, *C. albicans* NIH B311 containing both 155-kDa and 200-kDa mannoproteins.

Simple dithiothreitol (DTT) extraction of freeze-thawed mycelia of the B311 strain released both low and high molecular weight antigens.[49] A 19-kDa antigenic protein was present only in mycelial extracts but was not expressed on the germ tube surface. As such the 19-kDa protein could not account for form-specific immunofluorescence. That function was performed by a mannoprotein both high in molecular weight and in the degree of glycosylation. Surface radioiodinated mycelia and blastoconidia were DTT-extracted and, when examined by SDS-PAGE, a 230- to 250-kDa mannoprotein was enriched in the mycelial extracts. Antiserum made mycelial-specific by adsorption with blastoconidia reacted in western blots with the 230- to 250-kDa antigen.

Treatment of the mannoproteins released from mycelia with endoglycosidase H caused disintegration of the 230- to 250-kDa mannoprotein.[49] This profound reduction in molecular weight was not observed by Sundstrom et al.,[48] who nevertheless showed that mannoprotein released by zymolyase from germ tubes was affected by *endo* H treatment, leading to modest but detectable molecular weight reduction of the antigen in a western blot. When material released during spheroplast formation was recovered and digested with pronase, the carbohydrate portion could be separated by gel permeation chromatography. This carbohydrate did not contain a germ-tube-specific antigen detectable by radioimmunoprecipitation.[48] The protein moiety was thus implicated as responsible for the germ tube specificity of the 200-kDa mannoprotein.

Isolated, detergent-washed cell walls from intrinsically radiolabeled *C. albicans* germ tubes were digested with zymolyase.[50] Fluorography indicated the presence of 180- and 260-kDa mannoproteins that were absent in similar digests from blastoconidia. Moreover, a mAb was produced against the purified 260-kDa mannoprotein that was specific for germ tubes and did not react with blastoconidia in immunofluorescence or in zymolyase digests of blastoconidia.

Thus, different laboratories agree that highly glycosylated, high molecular weight mannoproteins from germ tubes or from mycelia are qualitatively different from those of blastoconidia. These molecules have much in common with, and may be part of, the structural CWMP. As was discussed in the section on CWMP, morphogenesis may result in reduced expression of the $(1\rightarrow2)$-β-oligomannoside epitopes on the mycelial form surface. In one instance a mAb was produced that specified the germ tube surface.[50] Knowledge about its epitope specificity will be important in deciphering the structural changes induced by dimorphism. Another study in which mice were immunized with viable germ tubes resulted in the production of mAbs that bound to surface CWMP and to a 200-kDa mannan released by zymolyase from germ tubes or from blastoconidia.[51] None of these mAbs were specific for germ tubes, although the screening process was designed to detect such specificity.

Such research does not directly address the question about enzymes that may reside in the periplasmic space and as a result of topology of germ tubes may be accessible to the cell surface. For example, Sundstrom et al.[48] showed that approximately 12 protein antigens, intrinsically labeled with ^{35}S-methionine, were released during spheroplast production by zymolyase digestion of germ tubes. Eight of these proteins were immunoprecipitated by germ-tube-specific antiserum.

2.2. Cryptococcosis—*Cryptococcus neoformans* Antigens

2.2.1. *Antigenic Heteroglycans of the Cryptococcus neoformans Cell Envelope*

The structure of the acidic, viscous, capsular polysaccharide, glucuronoxylomannan (GXM), is known[52,53] but the epitopes encoding serotype specificity have not been clearly defined (Fig. 1). Most of the published structure of GXM is based on analysis of single isolates of each serotype. Not surprisingly, the picture emerging from analysis of several isolates reveals further complexity. Downregulation of immune responses at the cellular level induced by purified capsular material has been described.[54] Tolerogenic doses of cryptococcal capsular polysaccharide induce an idiotypic T-lymphocyte suppressor factor that binds to macrophages and inhibits their ability to phagocytose encapsulated cryptococci.[55] The galactoxylomannan (GalXM) complex has received only preliminary structure and functional characterization.[56] A preliminary report[57] indicates the GalXM can be detected in the cerebrospinal fluid during cryptococcal meningitis. The mannoprotein component of GalXM appears to be the active antigen present in CNEF, an antigenic complex present in the medium after the growth of *C. neoformans*. CNEF has been shown to modulate delayed-type hypersensitivity reactions via a complex network of afferent and efferent T-suppressor lymphocytes and their factors.[58]

Unlike the well-studied cell wall matrix of *C. albicans*,[7,14,59] the isolated

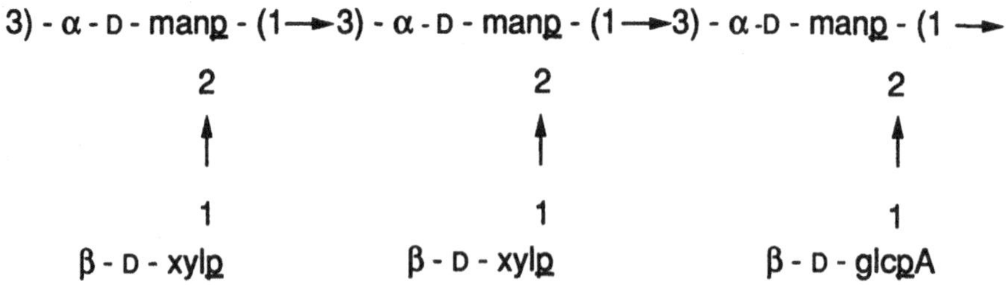

FIGURE 1. Typical structure of GXM from *Cryptococcus neoformans* serotype A. (Used with permission from Reference 67.)

wall of *C. neoformans* is the least well-known component of the cell envelope, although the nature of the glucans has been reported.[60] The interaction of *C. neoformans* glucan with the immune system is completely unknown and potentially important.

2.2.1a. Overview of Cell Envelope Structure. The cell envelope of *C. neoformans* consists of the glucuronoxylomannan (GXM) capsule, the fibrillar wall matrix of glucan-chitin, and a readily soluble galactoxylomannan (GalXM) complex. The GXM (M_r = ca. 8×10^5 Da)[61] has a linear (1→3)-α-mannan backbone substituted with single β-linked Xyl*p* and Glc*p*A nonreducing termini. *O*-Acetylation occurs at *O*-6 of unsubstituted mannosyl residues and varies according to the isolate from 2.8% to 10.5% of the GXM.[53] The molar ratio of Xyl*p*:Glc*p*A in the GXM is an approximate way to differentiate serotypes, D:A:B:C = 1:2:3:4. The Xyl*p* substitution is lower in the variety *C. neoformans* var. *neoformans* (serotypes A and D) than in *C. neoformans* var. *gattii* (serotypes B and C), but too few isolates have been analyzed to have confidence in an exact arithmetic progression in Xyl*p* molar ratios.

Uncertainty exists about the exact locale of the GalXM complex in the cell envelope. GalXM is released into the medium during growth of encapsulated and acapsular *C. neoformans*, and can be separated into a galacto-xylo polymer and a mannoprotein by the affinity of the latter for Con A.[56] The GalXM is not found in isolated cell walls of an acapsular mutant (J-67), which contain, instead, 88% Glc*p* as glucans and 7.3% N-acetylglucosamine as chitin.[60] A total of 15.1% of the wall dry weight was an insoluble (1→3)-α-glucan. An additional 33% of the wall was a (1→6)-β-glucan, also containing 6–19% (1→3)-α-Glc*p* residues. The (1→6)-β-glucan remained soluble after it was extracted from isolated cell walls.

2.2.2. *Glucuronoxylomannan (GXM) Capsule*

The capsule of *C. neoformans* is a high molecular weight, viscous, acidic heteropolysaccharide, glucuronoxylomannan (GXM). The linear portion of the molecule is a (1→3)-α-mannan. Single (1→2)-β-Glc*p*A residues are linked to the mannan backbone. Varying degrees of substitution of the mannan with single Xyl*p* residues occur. The predominant linkage is (1→2)-β, but there are also (1→4)-β linkages, such that in serotypes B and C some of the mannosyl residues in the backbone are disubstituted with single (1→4)-β-Xyl*p* and (1→2)-β-Glc*p*A. The extent of Xyl*p* substitution has been suggested as a rough guide to differentiate serotypes (Table I).

2.2.2a. C. neoformans Varieties neoformans and gattii: Serotypes. Four serotypes of *C. neformans*—A, B, C, and D—were described by Wilson et al.[62] by using agglutinin-adsorption tests. A fifth type—A,D—was reported later.[63] The status of surface antigenic factors of encapsulated cryptococci was reinvestigated by Ikeda et al.,[64] who devised a scheme of eight surface

TABLE I
***Cryptococcus neoformans* Cell Envelope GXM: Correlation between Variety, Serotype, and Molar Ratios of Sugar Components**

Variety	Serotype	Xyl*p*:Glc*p*A (molar ratio)[a]	Surface antigens[b]
neoformans	D	1	1, 2, 3, 8
	A, D	1	1, 2, 3, 7, 8[c]
	A	2	1, 2, 3, 7
gattii	B	3	1, 2, 4, 5
	C	4	1, 4, 6

[a]The molar ratio of Man*p* to other sugar constitutents is 3.
[b]Ref. 63.
[c]Ref. 62.

factors with respect to monospecific antisera. Four of these monospecific antisera defined the type-specific epitopes corresponding to the original Wilson et al.[62] scheme.

Serotypes are expected among encapsulated isolates of the same species, but the segregation of mating according to the serotypes was certainly an unexpected development. Kwon-Chung[65,66] reported that compatible mating types of *C. neoformans* serotypes A and D completed a basidiomycetous perfect (teleomorph) state, *Filobasidiella neoformans* var. *neoformans*. The teleomorph from *Filobasidiella neoformans* var. *bacillisporus* was described for compatible mating-type strains of serotypes B and C. Two varieties were defined for the asexual (anamorph) states of *C. neoformans*: var. *neoformans* constituted serotypes A and D, whereas var. *gattii* comprised serotypes B and C. The most basic difference in the capsular GXM of the two varieties is reduced xylosylation of the GXM in serotypes A and D, perhaps reflecting a difference in the concentration or activity of Xyl*p* transferase.

2.2.2b. GXM Epitopes: Poly- and Monoclonal Antibodies.

O-Acetyl Epitope. Cryptococcal GXMs are *O*-acetylated. Cherniak[53] summarized the degree of *O*-acetylation: serotype A, 6.5%; type B, 10.5%; type C, 3.0%, and finally type D, 2.8%. Those data were based on only one strain of each serotype. The site of *O*-acetylation was determined to be O-6 of Man*p* by comparing the ^{13}C-n.m.r. spectra of native GXM, *O*-deacetylated GXM, and periodate-oxidized GXM (the latter step removing all Xyl*p* and Glc*p*A). The degree of *O*-acetylation has been suggested as a pathogenic factor because avirulent species, *C. albidus* and *C. laurentii*, are not *O*-acetylated.[52] Complete *O*-deacetylation is accomplished by 24-hour incubation in pH 11 NH_4OH solution.[67] This condition is not stringent enough to disrupt other glycosidic bonds in GXM.

Various studies suggest that *O*-acetyl plays a role in the antigenic mosaic of GXM. Conjugation of the capsular polysaccharide to bovine gamma globulin (BGG) under conditions that preserved *O*-acetylation led to a potent antigen, whereas *O*-deacetylated polysaccharide conjugated to BGG was inert.[68] Rabbit-anti-*C. neoformans* serotype D serum reacted in immunodiffusion against either *O*-deacetylated or carboxyl-reduced GXM *but not* against GXM that was both carboxyl reduced *and* *O*-deacetylated. This observation can be explained by the presence of two different epitopes, one containing *O*-acetyl and one Glc*p*A,[61] but evidence for a complex epitope containing both *O*-acetyl and Glc*p*A was not presented.

PAbs against an *O*-acetyl-containing epitope were produced by adsorbing rabbit-anti-*C. neoformans* serum with cryptococci that had been *O*-deacetylated with 0.1 M NaOH.[69] Similarly, pAbs versus a Glc*p*A epitope were produced by adsorbing the same antiserum with cryptococci in which the Glc*p*A residues were reduced to Glc*p*. The anti-*O*-acetyl serum reacted with native GXM or carboxyl-reduced GXM but not with *O*-deacetylated GXM. Conversely, the anti-Glc*p*A serum reacted with native GXM or *O*-deacetylated GXM but not with carboxyl-reduced GXM.

Three mAbs were produced from mice immunized with GXM-ovine erythrocyte conjugates.[70] They were tested in immunodiffusion versus chemically modified GXM derivatives. Two mAbs reacted with native and *O*-deacetylated GXM, and one (mAb *302*) reacted with native and polyalcohol GXM (periodate-oxidized, $NaBH_4$-reduced) in which *O*-acetyl was preserved. MAb 302 appears to be specific for *O*-acetyl. This same mAb was reactive with the native GXM of serotypes A and D, but not with serotypes B and C. That observation has been confirmed with ELISA using 11 strains of serotypes B and C.[71]

The selective binding of this mAb to serotypes A and D can be explained by assuming that serotype D, the least substituted with Xyl*p*, has a more open linear mannan structure, one in which there might be an increased degree of *O*-acetylation. The lack of reactivity of mAb 302 with GXM of serotypes B and C could then be explained by the observed higher density of Xyl*p*, resulting in more limited access of the linear Man*p* residues to the antibody combining site. Finally, there may be a lower extent of *O*-acetylation in serotypes B and C.

Bhattacharjee et al.[52] proposed that serotypes A and D would be *O*-acetylated to a greater extent than serotypes B and C because of the dense Xyl*p* substitution of serotypes B and C. This proposal has received only modest support from observations on additional isolates. Ten strains of *C. neoformans* serotype A were analyzed, and the extent of GXM *O*-acetylation was 6.3–15.5% of the dry weight (mean *O*-acetyl content, 11 ± 3%).[72] Of six serotype B strains investigated by the same group, the degree of *O*-acetylation of GXM ranged between 6.3 and 10.4% (mean 8.5%). The possibility, however, remains that *O*-acetyl groups in serotypes A and D are more "solvent exposed" due to the more open structure of those polysaccharides and as such would be more available as epitopes.

MAbs have the potential to be serotype specific, and that was to be expected because of the ability of monospecific antisera produced by agglutinin adsorptions[64] to identify type-specific factors. MAbs against the GXM have been produced by different groups, and these have had restricted specificity, but none has achieved serotype specificity. Eckert and Kozel[70] produced three murine mAbs against polysaccharides of *C. neoformans* that were reactive in agglutinin and precipitin tests. One of them reacted with polysaccharides from all four serotypes, and a second mAb, discussed above in relation to its specificity for *O*-acetyl, reacted with serotypes A and D. Their third mAb was reactive with serotypes A, B, and D.

Dromer et al.[73] characterized mAb E, which in EIA and immunofluorescence reacted in decreasing order against *C. neoformans* serotypes A, B, and D. MAbs against GXM could bind, in a variety of EIAs and immunofluorescence, to either serotypes A and D (BD-1), or to A and B (BA-4), or to all four serotypes (CD-6).[74] Glc*p*A was the most important saccharide in the epitope for the binding of mAbs BD-1 or CD-6.

2.2.2c. Isolation, Purification, and Chromatography. Selective precipitation with cetyltrimethyl ammonium bromide (CTAB) is used to isolate the GXM from the culture medium after growth of the yeast.[75] The positively charged quaternary ammonium detergent complexes with the negatively charged GXM.

Precipitation of GXM with CTAB. Particle-free culture supernatant of *C. neoformans* is concentrated *in vacuo*, dialyzed versus deionized water, and then the polysaccharides are precipitated with 3 volumes of cold ethanol. The ethanol precipitate is dissolved in 0.2 M NaCl to give a carbohydrate concentration of 3–4 mg/ml. Then solid CTAB is added to give a final concentration of 3 mg/mg of ethanol precipitate. A 0.05% CTAB solution is added, and when NaCl reaches the critical electrolyte concentration the GXM forms an insoluble complex with CTAB. After removal of the precipitate and washing it with water, the complex is solubilized by adding 2 M NaCl until the complex is dissolved. The purified GXM can be dialyzed versus water and lyophilized, or characterized further by gel permeation and anion exchange chromatography.[76]

Molecular Weight. GXM purified from the growth medium of *C. neoformans* serotype D strain had a molecular weight of approximately 800 kDa estimated by gel permeation chromatography.[61] Turner[72] also used gel permeation to study the time course of depolymerization by sonication of GXM from serotype B. Their column was calibrated with polysaccharide molecular weight standards. Native molecular weight was estimated at 975 kDa. After 5 minutes of sonication the molecular weight was 536 kDa. After 15 minutes $\leq$20°C, there was a decline in molecular weight to 210 kDa. Cherniak et al.[77] sonicated the GXM of strain 371 for 15 minutes, resulting in a 10-fold reduction in viscosity with no detectable change in the composition and structure.

The results of two groups thus indicate the GXM recovered from the growth medium has a very high molecular weight, in the vicinity of 800–975 kDa. Sonication results in a controlled reduction in the viscosity and molecular weight. The chromatographic behavior and fine structure analysis are more convenient with sonicated GXM. A lower viscosity and molecular weight are desirable for conjugating GXM to proteins to immunize mice for mAb production.[67]

The salt concentration needed to desorb the GXM from diethylaminoethyl (DEAE) anion exchange resin was between 0.25 and 0.27 M. The molecular weights of two sonicated GXMs, measured by gel permeation, were 108.5 and 121.4 kDa. These polysaccharides contain less than 1% protein as measured by the Coomassie blue dye binding assay.

In summary, several properties of the GXM have been described, including monosaccharide composition and variation of molar ratios of Xyl*p*:Glc*p*A according to serotype; similarity between GXM released into the growth medium and cell-adherent GXM; controlled reduction in molecular weight as a result of sonication; anion exchange and gel permeation analysis of size, charge, and homogeneity; *O*-acetylation, and the effects of *O*-deacetylation.

2.2.2d. ^{13}C-Nuclear Magnetic Resonance (n.m.r.) Spectrometry. ^{13}C-n.m.r. spectrometry is a sensitive means of detecting the fingerprint of a given strain, of assigning signals in the spectrum to particular glycosidic bonds, and finally of deducing the structure, including minor variations. Further, sonicated, *O*-deacetylated GXM provide considerably sharper spectra than native GXM. ^{13}C-n.m.r. spectra were obtained from nine serotype A strains. The region due to the anomeric carbon atom allowed these strains to be characterized into groups.[72] Group 1 had the simplest pattern, with six major resonances (Fig. 2, Table II).

Examples of standard strains with this chemo group are CDC 9104 and NIH 201. The anomeric resonance assignments were made using the assignment of serotype B model GXM and by comparing the heights of each resonance with the abundance of methylated monosaccharide derivatives identified by methylation-fragmentation gas–liquid chromatography–mass spectrometry (g.l.c.-m.s.). Five major resonances in the anomeric carbon region have been assigned (Table II).

Serotype A chemo-group 1 strains consisted of *C. neoformans* serotype A 68, 196, 201, and 9104. They produced the simplest spectra. Chemo-group 2, illustrated by strains 271 and 355, gave the most complex spectra. They had three resonances for Xyl*p*-β-(1→2)- and a strong signal for Xyl*p*-β(1→4)-. Figure 2 shows the comparison of a typical C. neoformans serotype A-group 1 strain, 201, with a group 2 strain: 271. A further discussion of ^{13}C-n.m.r. chemotypes is outside the scope of this review.

In summary, serotype A strains can be grouped by the anomeric carbon regions of their ^{13}C-n.m.r. spectra. Some serotype A strains contain Xyl*p*-

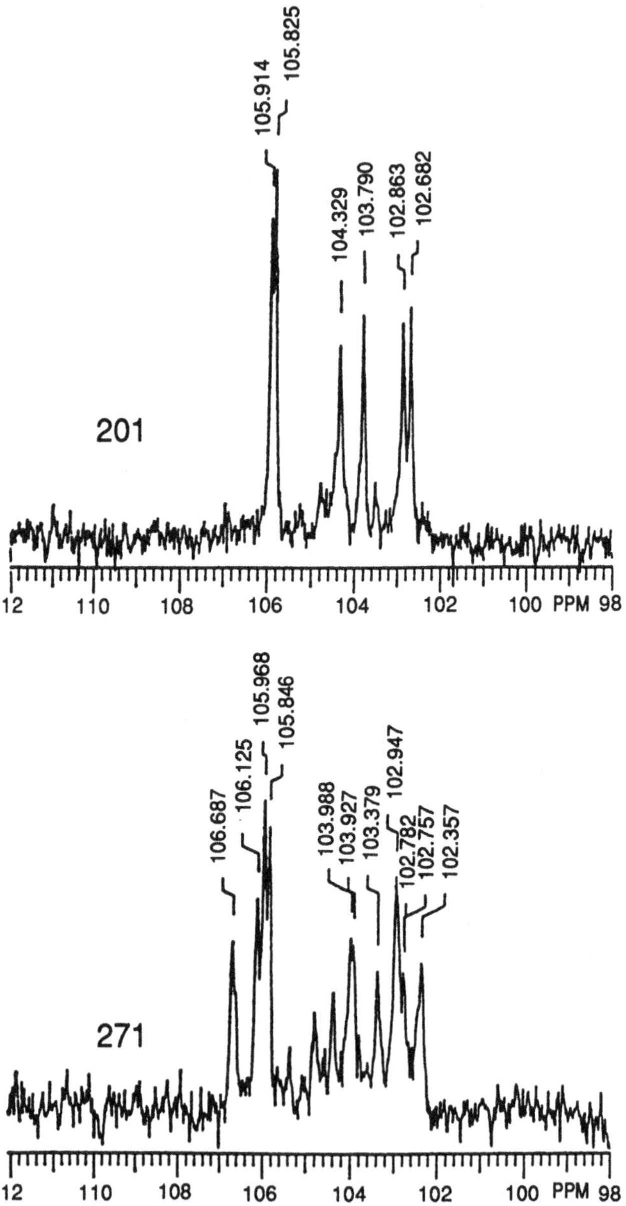

FIGURE 2. ^{13}C-n.m.r. spectra of anomeric carbon regions of GXM from *Cryptococcus neoformans* serotype A, chemogroups 1 (strain 201) and 2 (strain 271). (Used with permission from Reference 72.)

TABLE II
Major Anomeric Region ^{13}C-n.m.r Resonances of GXM from *Cryptococcus neoformans* Serotype A Strains (Chemo-Group 1)

p.p.m	Identity
105.97	(1 → 2)-β-Xyl*p*[a]
105.87	(1 → 2)-β-Xyl*p*[a]
104.38	Glc*p*A
103.89	Man*p* linked at 0–2 to Xyl*p*
102.88	Man*p* linked at 0–2 to Xyl*p*
102.73	Man*p* linked at 0–2 to Glc*p*A

[a]Typical resonances in Group 1 of serotype A., that is strain 201 (see Fig. 2); no (1 → 4)-β-D-Xyl*p* present. (Data modified from Reference 72.)

β-(1→4)-residues, and some Man*p* units in the linear portions are disubstituted with Xyl*p* and Glc*p*A.

2.2.3. *Galactoxylomannan (GalXM) Complex*

At least three polysaccharides are released in the broth cultures of *C. neoformans*.[56,76] These are resolved by CTAB into the precipitable GXM (88% of total carbohydrate) and the nonprecipitable GalXM complex (12%). Gal*p* is present only in the galactoxylomannan, thus explaining the sporadic reports of Gal*p* in "cryptococcal polysaccharide" going back to Rebers et al.[78] The galactoxylomannan contains 2% *O*-acetyl and molar ratios of Xyl*p*: Man*p*:Gal*p*:Glc*p*A = 1:1.8:1.9:0.2.

Lectin affinity chromatography on Con A-agarose[56] showed that the GalXM complex is composed of at least two polymers. Three peaks were obtained from the Con A column. The first peak, 70% of the applied sample, did not bind to the affinity matrix and consisted of a galactoxylomannan and only 2% peptide. The second peak, eluting with borate buffer, comprised only 3% of the applied sample and was enriched in mannan.

2.2.3a. Mannoprotein (MP). Peak III was eluted from the Con A column with 0.2 M methyl-α-D-mannopyranoside. This fraction, comprising 12% of the applied sample, was a mannoprotein (MP) with a 21% peptide moiety and 6% phosphate. Interestingly, of the total amino acid composition in the MP, fully 28.7% was serine and another 17% was threonine. These amino acids are known to participate in *O*-glycosidic ester bonds between the amino acids and mannosyl residues of the glycosidic portion of MPs. A high glucosamine content was also detected by the amino acid analyzer. N-glycosidic binds between N,N′-diacetylchitobiose and the peptide moiety are another form of glycoprotein linkage found in MPs.

The ^{13}C-n.m.r. spectrum of the MP was similar to that of *Saccharomyces cerevisiae*. The following assignments were made: 101.90 ppm, 2-*O*-Man*p*; 99.3 ppm, 2,6-di-*O*-substituted Man*p*.; 101.0 ppm. Thus an unsubstituted (1→6)-α-D-Man*p* side chain structure appeared feasible. Consistent with these structural assignments was a strong serological reaction with antiserum made against the hapten: α-Man*p*-(1→2)-[α-D-Man*p* (1→6)]-α-D-Man*p*.

Delayed cutaneous hypersensitivity responses in mice sensitized with killed cryptococci were elicited by the GalXM complex, and most of the reactivity resided in the MP.[79] The MP also was capable of binding to IgM of some cryptococcosis patients.[80]

3. DISEASES CAUSED BY OPPORTUNISTIC MOLDS

3.1. Aspergillosis—*Aspergillus fumigatus* Antigens

3.1.1. Allergic Bronchopulmonary Aspergillosis (APBA)

3.1.1a. Protein Antigens. The historical development of mycelial extract (ME) and culture filtrate (CF) antigens has been reviewed through 1985.[81] There is as yet no universally accepted antigen preparation for detecting IgG and IgE in patients with ABPA. Interlaboratory standardization is lacking. Some preparations have been characterized by two-dimensional immunoelectrophoresis and western immunoblot. In several instances the CF, ME, or other purified fractions were adapted to enzyme immunoassays (EIA) and evaluated with serum specimens from panels of ABPA patients and control subjects.

3.1.1b. Mycelial Extract (ME) Antigens.

Preparation of ME. Typically, liquid shake cultures were grown at 37°C on synthetic medium and the mycelial growth was harvested after 2–5 days. The washed mycelia was mechanically disrupted, and the cell-free supernatant (= ME) was either used directly or further fractionated by chromatography according to size, charge, affinity, or immunoaffinity. In *Aspergillus* antigen research, two steps are frequently used to discriminate nonglycosylated proteins from glycoproteins and from the omnipresent cell wall galactomannan (GM): 1) ammonium sulfate precipitation of the ME proteins[82] and 2) affinity chromatography on Con A-agarose to bind GM, soluble glucan, and glycoproteins.[83,84] In these cases Hearn et al.[84] further characterized the Con-A-bound fractions, whereas de Repentigny[83] was interested in the nonglycosylated proteins.

Hydrophobic interaction chromatography on phenyl-Sepharose columns has been used to fractionate an ME from *A. fumigatus*.[85] A carbohydrate-containing complex recovered by desorption with 4 M salt included a component with an estimated molecular weight >400 kDa. This fraction was designated the "470-kDa" antigenic complex. The molecular weight did not coincide with the pattern of proteins in the 50- to 77-kDa range obtained after SDS-PAGE.[85] Such a result could be explained by the occurrence of subunits that aggregate in the fungal culture and become dissociated under the conditions of SDS-PAGE. This phenomenon has been noted previously by Calvanico et al.[86]

Because of the differences observed in the SDS-PAGE profile and the estimated molecular weight of 470 kDa, other workers may be using similar antigens but with different nomenclature. Other protein-containing fractions desorbed from the phenyl Sepharose column[85] yielded catalase, molecular weight ~250 kDa, and a proteinase, molecular weight range 25–50 kDa [see Section].

Analysis of the column-purified "470-kDa" preparation indicated the carbohydrate: protein ratio was >1000. The presence of GM may be suspected but remains unproven. The 470-kDa antigenic complex[87] produced a characteristic immunoprecipitin arc in two-dimensional crossed rocket immunoelectrophoresis. The concentration of this antigen varied markedly from strain to strain. Antigens similar to the 470-kDa complex were also present in a TCA-soluble antigen previously used to detect IgG in normal humans and in those with various respiratory diseases, including ABPA patients.[88]

3.1.1c. Western Immunoblot Analysis. Substantial SDS-PAGE and western blotting has been done with *A. fumigatus* proteins and glycoproteins. An ME from *A. fumigatus* contained proteins in the 75- to 85-kDa range, another small cluster in the 30- to 40-kDa range, and several lower molecular weight proteins.[84] Chromatography of the ME on Con A-agarose resulted in a Con A-bound fraction with fewer and sharper proteins. Serum pooled from human aspergillosis cases reacted with Con A-bound and purified antigens in the 20-, 30- to 40-, and 80- to 85-kDa ranges. The immunodominant epitopes in the ME appeared to be in glycoproteins, because treatment with mixed glycosidases of *Cytophaga johnsonii* abolished all antigen–antibody binding. An ME was used in western immunoblot to screen a panel of confirmed APBA patients.[89] Sera from 21 of 22 APBA patients produced IgG and IgE that reacted with a protein antigen of about 70 kDa [see Ag 13, Section 3.1.1*e*].

An ME was passaged through a Con A-agarose column to *remove* glycoproteins and polysaccharides.[83] The *unbound* fraction contained nonglycosylated antigens of interest having molecular weights of 41, 54, and 71 kDa. These antigens elicited antibodies in rabbits who survived 10 days after experimental IA, and such antibodies were absent in normal rabbits. Two-dimensional immunoblotting indicated that 41-, 54-, and 71-kDa antigens originated in a single high molecular weight precursor protein. This phenomenon recalls the observations of Schønheyder and Andersen[85] and that of Calvanico et al.[86]

3.1.1d. Other Protein Antigens.

Catalase. Catalase and a proteinase are potent precipitating antigens of *A. fumigatus*, which may have a role in the pathogenesis of aspergillosis. Catalase can be detected in immunoprecipitin arcs by "bubblography."[87] Catalase-containing "~250-kDa" fractions, and proteinase-containing fractions "25–50 kDa" from a Sephadex G200 column were used to coat microtitration plates for EIA.[90] Two-dimensional crossed rocket immunoelectrophoresis and bubblography revealed two antigens with catalase activity in strains of *A. fumigatus* and *A. flavus*.[87]

C Antigen, Proteinase Activity. Proteinase activity was measured by zones of lysis in casein agar.[85] The proteinase-containing antigen was adapted to a diagnostic EIA and had sensitivity and specificity similar to the catalase antigen. Antigen 13 in Longbottom's scheme (see Section 3.1.1e.) was purified from a CF preparation of *A. fumigatus*[91] and was identical to the antigen originally described as a chymotrypsinlike proteinase (= C antigen) by an immunoenzymogram method of Tran van Ky et al.[92] The molecular weight of Ag 13 was estimated to be 70 kDa. The antigen was labile to 70°C for 5 minutes and bound to Con A. Purified Ag 13 was adapted to EIA. Ag 13 was not, however, a good elicitor of IgE using the *allergogram* method. Further studies are needed to reconcile whether or not C antigen, Ag-13, elastase, and the proteinase of Schønheyder and Andersen[85,90] are identical.

Elastase activity is separable from general proteinase activity found in the CF of *A. flavus* cultures grown in the presence of elastin and BSA.[93] Both activities were adsorbed from the CF by adding dry carboxymethyl cellulose. The elastase was purified by gel permeation and hydroxyapatite chromatography. *A. flavus* elastase has a molecular weight of 23 kDa and a pI of 7.6. The optimum pH if 7.5–8.2 and the optimum temperature range is 45–55°C. The enzymatic activity is inhibited by 1,10-phenanthroline, a known metalloproteinase inhibitor.

3.1.1e. Allergograms. Longbottom developed a numerical taxonomy of 15 *A. fumigatus* antigens/allergens based on findings with sera from ABPA patients.[94,95] Two-dimensional crossed, radio-immunoelectrophoresis was used to identify allergens by an ^{125}I-anti-IgE overlay technique in which rabbit antiserum provided a reference system of rocket immunoprecipitates. The electrophoretogram was overlaid with patient's serum serving as a source of IgE. The final incubation was with ^{125}I-anti IgE and then an autoradiogram was made, called an *allergogram.* Although present in minute amounts in serum, IgE can intercalate with the second-dimension immunoprecipitin arcs if there are vacant epitopes. Thus *A. fumigatus* antigens that are stimulants of IgE can be enumerated and their purification can be pursued.

Three antigens, numbered 2, 3, and 15 in the rabbit reference system, were major allergens in that they bound to IgE from greater than 50% of sera from ABPA patients. Antigen 3 (Ag 3) was precipitated with ammonium sulfate and was chromatographed on Sephacryl S200. It was a heat-labile protein of ~24 kDa molecular weight.

The allergogram system was used to identify Ag 7, which reacted with IgG from >70% of ABPA patients, but was only weakly reactive with IgE from those patients.[96] The antigen was detected in all CF and ME, but was enriched in ME preparations. The ME was chromatographed on a Sephacryl S200 column, and fractions were screened by rocket immunoelectrophoresis. The approximate molecular weight of Ag 7 was 150–200 kDa. SDS-PAGE revealed no high molecular weight antigen, only a fragment of ~36 kDa. The antigenic activity was stable to boiling and was bound by Con A. This material was similar to Schønheyder's "470 kDa"[85] antigenic complex. Antigens 7 and antigen 13 (= proteinase) were regarded as intracellular, since neither appeared in the CF before autolysis occurred.[97]

3.1.1f. Germling Antigens The antigenic composition of MEs prepared from young germling cultures fluctuates according to the age of the germlings. SDS-PAGE showed that the proteins extracted from germinating *A. fumigatus* conidia were distinct from those in both conidia and later mycelial growth.[98] The number of antigenic proteins in the ME increased and by 96-hours there was an enrichment in proteins in the 30- to 40-kDa range and another cluster in the 50- to 60-kDa range. These proteins elicited both IgG and IgE in serum from ABPA patients. Proteins in the 30- to 40-kDa range were very similar to those observed by Hearn et al.[84] Antiserum against nonviable 15-hour-old germinated conidia of *A. fumigatus* identified a major germling antigen as Ag 5 in the Longbottom[94] reference system.[97] Monospecific antiserum against Ag 5 tested positive for immunofluorescence on formalin-fixed germlings. Allergograms with sera pooled from ABPA patients indicated that both IgG and IgE were elicited by Ag 5. Ag 5 was first detectable in the CF from 24-hour-old *A. fumigatus* cultures. The molecular weight of Ag 5 was estimated as 35 kDa. It was labile to 56°C for 30 minutes and did not bind to Con A. Antigens with similar molecular weights in the 30- to 40-kDa range, including Ag 5,[97] the germling surface-accessible antigen of Hearn et al.,[84] and the antigens described by Piechura et al.,[98] are worthy of further attention.

3.1.1g. Other Subcellular Antigens. A membrane vesicle fraction was prepared from mechanically disrupted mycelia of *A. fumigatus*.[99] Antigens were solubilized with octyl-β-D-glucopyranoside. Two-dimensional crossed rocket immunoelectrophoresis with rabbit antiserum revealed five major precipitate peaks with the vesicle fraction as compared with over 20 major peaks with the unfractionated ME.

3.1.1h. Summary of A. fumigatus *Protein Antigens.* Despite the use of different *A. fumigatus* tester strains and different methods for antigen fractionation, a recurring theme is seen in gels and western blots of antigenic proteins with molecular weights of 30–40 kDa and 50–75 kDa. Such moieties should receive additional emphasis in the efforts aimed at further standardization. The exact molecular weight of the chymotrypsinlike proteinase of *A. fumigatus* is not clear. Existing estimates were made by gel permeation chromatography and are subject to confirmation by SDS-PAGE: Schønheyder and Andersen[85] have estimated the molecular weight as 25–50 kDa, whereas the "Ag 13" with proteinase activity characterized by Harvey and Longbottom[91] had a molecular weight of 70 kDa. The *A. flavus* elastase molecular weight was 23 kDa.[93] There are also significant antigens whose estimated molecular weights fall outside this range: the 250-kDa catalase of *A. fumigatus*. True comparisons can only be made when the antigens are run side by side and probed with sera from the a panel of human patients and/or a standard rabbit antiserum. A workshop to standardize the methodology and nomenclature for *A. fumigatus* antigens is needed.

3.1.2. Invasive Aspergillosis (IA) in the Immunocompromised Host

IA is a pulmonary-disseminated disease in the granulocytopenic host compromised by cancer, transplantation, or autoimmune disease. Precipitins may be present, but due to the host's immunosuppressed status they may be absent. Immunodiagnosis can be accomplished by using sensitive tests for antibodies or by the RIA detection of GM antigenemia.[100] An EIA-inhibition test also detected a similar antigen in the blood of rabbits with experimental IA[101] and in the blood and urine of IA patients.[102]

3.1.2a. Limitations of Antibody Measurements in IA. Western blotting suggests that some antigens are selectively expressed in IA, particularly a 40-kDa protein, and that patients' residual IgM- and IgG-forming capacity can advantageously be measured.[103] This observation received support from experimental IA in the cortisone-immunosuppressed rabbit.[83] Rabbits surviving until 10 days postinfection demonstrated antibodies to a trio of 41-, 54-, and 71-kDa antigens that were related to a single high molecular weight precursor. Indirect EIA was used to test for serum IgG concentrations in patients with histologically proven IA using commercial *A. fumigatus* antigen sources.[104] The patients were undergoing immunosuppressive therapy for management of a renal or hepatic allograft. All patients had significantly elevated IgG compared with control subjects. ME antigens tended to detect antibodies earlier in the illness, but CF antigens detected an overall higher number of positive responses. By the time of diagnosis of IA, antibodies in some patients fell to within a normal range. Four of the IA patients tested negative with respect to all four antigens within the period of 1 month before death. Each patient would have to be monitored prospectively from the onset of the immunosuppression so that the baseline antibody concentration could serve as the internal control.

3.1.2b. Circulating Galactomannan (GM) Antigen. The preliminary studies that led to the further refinements discussed here have been reviewed.[2,14,105,106] GM is the readily soluble outer antigenic layer of the cell wall of *A. fumigatus* and related species. The most thorough analysis of the carbohydrate chemistry of *Aspergillus* species GM was reported by Barreto-Bergter et al.[107,108] Structural features common to GMs of *A. fumigatus*, *A. flavus*, and *A niger* were a linear (1→6)-α-mannan backbone substituted with (1→2)-α-oligomannosyl chains having a degree of polymerization (d.p.) ≤3 and immunodominant determinants consisting of (1→5)-β-galactofuranosides (d.p. <4).[109] But *Candida albicans* mannan also has (1→2)-α-oligomannosyl chains, yet Weiner and Coats-Stephen (1979)[110] have shown that tests for antigenemia in rabbits infected with either *A. fumigatus* or with *C. albicans* are mutually exclusive. Thus other considerations, the d.p., conformation, or further substitution, distinguish the Man*p*-containing epitopes of these two fungi.

A competitive binding RIA has been validated in a large-scale examination of sera from leukemia patients.[100] In the RIA, the ^{125}I-labeled carbohydrate-containing antigen was combined with sufficient polyvalent rabbit antiserum to precipitate approximately 10,000 cpm in the presence of half-saturated ammonium sulfate. The presence of antigenemia results in inhibition of the radioimmunoprecipitation, and the percent inhibition is converted to antigen in ng/ml with respect to a standard curve. Greater than 10% inhibition is significant for a positive test. Serum samples are heated at acidic pH to dissociate antigen-serum complexes, presumably to residual amounts of antibodies in the otherwise immunosuppressed host.

Bennett[105] pointed out that *A. fumigatus* GM is subject to rapid removal from the blood by Man*p*-specific receptor-mediated endocytosis by Kupffer cells and other macrophages. In that case, much higher test sensitivities for antigenemia tests may be difficult to achieve. Other clinical fluids useful for the detection of soluble *A. fumigatus* GM are bronchoalveolar lavage fluid and urine. Andrews and Weiner[112] detected antigen in bronchoalveolar lavage fluid and cells. Dupont et al.[102] devised an indirect, competitive binding EIA (also known as inhibition EIA) to detect antigenemia and antigenuria in IA. Antiserum was produced by injecting rabbits intravenously with viable *A. fumigatus* conidia. GM antigen was isolated from an ME by sequential chromatography on anion exchange- and Con A-affinity columns. The presence of GM in the product was verified by gas-liquid chromatography. Purified GM was used to construct a standard curve from which the *in vivo*-produced antigen concentrations were interpolated. Serum samples were boiled for 10 minutes in PBS before use to dissociate serum-GM complexes. The lower limit of sensitivity for the EIA was 10–40 ng/ml.

Independently, de Repentigny et al.[101] detected GM-antigenemia by indirect inhibition EIA in cortisone-suppressed, experimentally infected rabbits. The antibody used in this EIA was produced against detergent-washed *A. fumigatus* cell walls. The serum-antigen was stable to boiling and to digestion with Pronase. It could be adsorbed by Con A-agarose and had a molecular weight estimated by ultrafiltration at 50–100 kDa. Based on these findings, and the use of purified GM to coat the microtitration plates for the inhibition EIA, the circulating antigen was confirmed to be GM. Serum-GM was detected in infected rabbits on days +2 to +4, with concentrations ranging from 95 to 350 ng/ml.

An inhibition EIA detected 10 ng of carbohydrate antigen per milliliter of serum.[112] Serum samples were pretreated to dissociate serum-antigen complexes by adding TCA, boiling, and centrifugation to remove denatured protein. Reference antiserum was obtained from rabbits injected subcutaneously with viable conidia.

An indirect double-antibody sandwich EIA with biotin-avidin amplification was used to screen sera from 45 leukemia patients being treated by bone marrow transplantation.[113] A heat-stable *A. fumigatus* antigen was detected in 3 of 4 patients with autopsy- or biopsy-proven IA. The geometry of this assay is more complicated than those previously used, but the utility of a double-antibody sandwich format appears to be valid.

A protein of approximately 80 kDa was detected by western immunoblot in serum from 6 of 15 immunosuppressed, infected rabbits.[114] The serum was first percolated through an immunoaffinity column to which rabbit anti-*A. fumigatus*-IgG was affixed. This capture antibody was raised in rabbits by immunizing them with a variety of preparations, including ME, whole sonicated mycelia, CF, and an alkaline extract of a hyphal mat. The immunoaffinity-purified serum-antigen was electrophoresed and then probed in a western blot test with the antibodies used for the immunoaffinity column. Although this is a research prototype, it illustrates the potential for detecting antigens other than the cell wall GM.

The manuscript for this chapter was submitted for publication in October of 1990.

REFERENCES

1. Jones, J. M., 1990, Laboratory diagnosis of invasive candidiasis, *Clin. Microbiol. Rev.* **3**:32–45.
2. de Repentigny, L., 1989, Serological techniques for diagnosis of fungal infection, *Eur. J. Clin. Microbiol. Infect. Dis.* **8**:362–375.
3. Odds, F. C., 1988, *Candida and Candidosis*, 2nd ed., Balliere Tindall, London, 468 pp.
4. Kobayashi, H., Shibata, N,. Nakada, M., Chaki, S., Mizugami, K., Ohkubo, Y., and Suzuki, S., 1990, Structural study of cell wall phosphomannan of *Candida albicans* NIH B-792 (serotype B) strain, with special reference to ^{1}H and ^{13}C NMR analyses of acid-labile oligomannosyl residues, *Arch. Biochem. Biophys.* **278**:195–204.
5. Kagaya, K., Miyakawa, Y., Fujihara, H., Suzuki, M., Soe, G., and Fukazawa, Y., 1989, Immunologic significance of diverse specificity of monoclonal antibodies against mannans of *Candida albicans*, *J. Immunol.* **143**:3353–3358.

6. Suzuki, M., and Fukazawa, Y., 1982, Immunochemical characterization of *Candida albicans* cell wall antigens: Specific determinant of *Candida albicans* serotype A mannan, *Microbiol. Immunol.* **26**:387–402.
7. Fukazawa, Y., 1989, Antigenic structure of *Candida albicans*—Immunochemical basis of serologic specificity of the mannans in yeasts, in: *Immunology of Fungal diseases* (E. Kurstak, G. Marquis, P. Auger, L. de Repentigny, and S. Montplaisir, eds.), Marcel Dekker, New York, pp. 37–62.
8. Shibata, N., Ichikawa, T., Tojo, M., Takahashi, M., Ito, N., Okubo, Y., and Suzuki, S., 1985, Immunochemical study of the mannans of *Candida albicans* NIH-207, NIH B-792, and J-1012 strains prepared by fractional precipitation with cetyltrimethylammonium bromide, *Arch. Biochem. Biophys.* **243**:338–348.
9. Reiss, E., de Repentigny, L., Kuykendall, R. J., Carter, A. W., Galindo, R., Auger, P., Bragg, S. L., and Kaufman, L., 1986, Monoclonal antibodies against *Candida tropicalis* mannan: Antigen detection by enzyme immunoassay and immunofluorescence, *J. Clin. Microbiol.* **24**: 796–802.
10. Domer, J. E., 1989, *Candida* cell wall mannan: A polysaccharide with diverse immunologic properties, *CRC Crit. Rev. Microbiol.* **17**:33–51.
11. Ray, T. L., and Wuepper, K. D., 1976, Activation of alternative (properdin) pathway of complement by *Candida albicans* and related species, *J. Investig. Dermatol.* **67**:700–703.
12. Manganey, M., Fischer, A., Le Deist, F., Latge, J.-P., and Durandy, A., 1989, Direct activation of human B-lymphocytes by *Candida albicans*-derived mannan antigen, *Cell Immunol.* **122**: 329–337.
13. Podzorski, R. P., Gray, G. R., and Nelson, R. D., 1990, Different effects of native *Candida albicans* mannan and mannan-derived oligosaccharides on antigen-stimulated lymphoproliferation *in vitro*, *J. Immunol.* **144**:707–716.
14. Reiss, E., 1986, *Molecular Immunology of Mycotic and Actinomycotic Infections*, Elsevier, New York, 423 pp.
15. Tsuchiya, T., Fukazawa, Y., Taguchi, M., Nakase, M., and Shinoda, T., 1974, Serologic aspects of yeast classification, *Mycopathology* **53**:77–91.
16. Tojo, M., Shibata, N., Kobayashi, M., Mikami, T., Suzuki, M., and Suzuki, S., 1988, Preparation of monoclonal antibodies reactive with β-1,2-linked oligomannosyl residues in the phosphomannan-protein complex of *Candida albicans* NIH B792 strain, *Clin. Chem.* **34**:539–543.
17. Shibata, N., Kobayashi, H., Tojo, M., and Suzuki, S., 1986, Characterization of phosphomannan-protein complexes isolated from viable cells of yeast and mycelial forms of *Candida albicans* NIH B-792 strain by the action of Zymolyase-100 T, *Arch. Biochem. Biophys.* **251**:697–708.
18. Heidenreich, F., and Dierich, M. P., 1985, *Candida albicans* and *Candida stellatoidea*, in contrast to other *Candida* species, bind iC3b and C3d but not C3b, *Infect. Immun.* **50**: 598–600.
19. Gilmore, B. J., Retsinas, E. M., Lorenz, J. S., and Hostetter, M. K., 1988, An iC3b receptor on *Candida albicans*: Structure, function, and correlates for pathogenicity, *J. Infect. Dis.* **157**: 38–46.
20. Eigentler, A., Schultz, T. F., Larcher, C., Breitwieser, E.-M., Myones, B. L., Petzer, A. L., and Dierich, M. P., 1989, C3bi-binding protein on *Candida albicans*: Temperature-dependent expression and relationship to human complement receptor type 3, *Infect. Immun.* **57**: 616–622.
21. Linehan, L., Wadsworth, E., and Calderone, R., 1988, *Candida albicans* C3d receptor, isolated by using a monoclonal antibody, *Infect. Immun.* **56**:1981–1986.
22. Matthews, R. C., Burnie, J. P., and Tabaqchali, S., 1984, Immunoblot analysis of the serological response in systemic candidiasis, *Lancet* **2**:1415–1418.
23. Matthews, R. C., Burnie, J. P., and Tabaqchali, S., 1987, Isolation of immunodominant

antigens from sera of patients with systemic candidiasis and characterization of the serological response to *Candida albicans*, *J. Clin. Microbiol.* **25**:230–237.

24. Burnie, J. P., Matthews, R. C., Fox, A., and Tabaqchali, S., 1985, Use of immunoblotting to identify antigenic differences between the yeast and mycelial phases of *Candida albicans*, *J. Clin. Pathol.* **38**:701–706.
25. Burnie, J. P., Matthews, R. C., Watson, J. G., Milligan, D. W., and Tabaqchali, S., 1986, Analysis of the antibody response developing in an infant with *Candida albicans* meningitis, *J. Infect.* **12**:79–83.
26. Matthews, R., and Burnie, J., 1988, Diagnosis of systemic candidiasis by an enzyme-linked dot immunobinding assay for a circulating immunodominant 47-kilodalton antigen, *J. Clin. Microbiol.* **26**:459–463.
27. Strockbine, N. A., Largen, M. T., Zweibel, S. M., and Buckley, H. R., 1984, Identification and molecular weight characterization of antigens from *Candida albicans* that are recognized by human sera, *Infect. Immun.* **43**:715–721.
28. Strockbine, N. A., Largen, M. T., and Buckley, H. R., 1984, Production and characterization of three monoclonal antibodies to *Candida albicans* proteins, *Infect. Immun.* **43**:1012–1018.
29. Walsh, T. J., Buckley, H. R., Hom-Eng, M., Johnson, D. E., Maret, M., and Gary, P., 1988, Analysis of a 48 kD cytoplasmic antigen of *Candida albicans* in experimental candidiasis, 10th Cong. Intern. Soc. Hum. Anim. Mycol., Barcelona, *Revista Iberica de Micologia* **5(Suppl)**:38 (1–111).
30. Mason, A. B., Brandt, M. E., and Buckley, H. R., 1989, Enolase activity associated with a *C. albicans* cytoplasmic antigen, Proc. 7th Intern. Symp. Yeasts, *Yeast* **5(Suppl.)**:S231–S239.
31. Staib, F., 1965, Serum-proteins as nitrogen source for yeast-like fungi, *Sabouraudia* **4**: 187–193.
32. Borg, M., and Rüchel, R., 1988, Expression of extracellular acid proteinase by proteolytic *Candida* spp. during experimental infection of oral mucosa, *Infect. Immun.* **56**:626–631.
33. Rüchel, R., Böning, B., and Borg, M., 1986, Characterization of a secretory proteinase of *Candida parapsilosis* and evidence for the absence of the enzyme under conditions of infection in vitro, *Infect. Immun.* **53**:411–419.
34. Kwon-Chung, K. J., Lehman, D., Good, C., and Magee, P. T., 1985, Genetic evidence for role of extracellular proteinase in virulence of *Candida albicans*, *Infect. Immun.* **49**:571–575.
35. Morrison, C. J., 1990, Morphogenetic/pathogenetic antigens of *Candida albicans*: Aspartyl proteinase, *Prog. Abstr. Annu. Mtg. Amer. Soc. Microbiol.*, Anaheim, CA, Session F.
36. Remold, H., Fasold, H., and Staib, F., 1968, Purification and characterization of a proteolytic enzyme from *Candida albicans*, *Biochim. Biophys. Acta* **167**:399–406.
37. Macdonald, F., and Odds, F. C., 1980, Inducible proteinase of *Candida albicans* in diagnostic serology and in the pathogenesis of systemic candidosis, *J. Med. Microbiol.* **13**:423–435.
38. Rüchel, R., 1981, Properties of a purified proteinase from the yeast *Candida albicans*, *Biochim. Biophys. Acta* **659**:99–113.
39. Borg, M., Watters, D., Reich, B., and Rüchel, R., 1988, Production and characterization of monoclonal antibodies against secretory proteinase of *Candida albicans* CBS 2730, *Zbl. Bakt. Hyg.* **A268**:62–73.
40. Rüchel, R., 1983, On the role of proteinases from *Candida albicans* in the pathogenesis of acronecrosis, *Zbl. Bakt. Mikr. Hyg.* **A255**:524–536.
41. Rüchel, R., 1983, On the renin-like activity of *Candida* proteinases and activation of blood coagulation in vitro, *Zbl. Bakt. Hyg.*, I. Abt. Orig. A **255**:368–379.
42. Negi, M., Tsuboi, R., Matsui, T., and Ogawa, H., 1984, Isolation and characterization of proteinase from *Candida albicans*: Substrate specificity, *J. Invest. Dermatol.* **83**:32–36.
43. Macdonald, F., and Odds, F. C., 1983, Virulence for mice of a proteinase-secreting strain of *Candida albicans* and a proteinase-deficient mutant, *J. Gen. Microbiol.* **129**:431–438.
44. Rüchel, R., 1984, A variety of *Candida* proteinases and their possible targets of proteolytic attack in the host, *Zbl. Bakt. Hyg.* **A 257**:266–274.

45. Rüchel, R., 1986, Cleavage of immunoglobulins by pathogenic yeasts of the genus *Candida*, *Microbiol. Sci.* **3**:316–319.
46. Lott, T. J., Page, L. S., Boiron, P., Benson, J., and Reiss, E., 1989, Nucleotide sequence of the *Candida albicans* aspartyl proteinase gene, *Nucleic Acids Res.* **17**:1779.
47. Sundstrom, P. M., and Kenny, G. E., 1985, Enzymatic release of germ tube-specific antigens from cell walls of *Candida albicans*, *Infect. Immun.* **49**:609–614.
48. Sundstrom, P. M., Nichols, E. J., and Kenny, G. E., 1987, Antigenic differences between mannoproteins of germ tubes and blastospores of *Candida albicans*, *Infect. Immun.* **55**: 616–621.
49. Ponton, J., and Jones, J. M., 1986, Identification of two germ-tube-specific cell wall antigens of *Candida albicans*, *Infect. Immun.* **54**:864–868.
50. Casanova, M., Gil, M. L., Cardenoso, L., Martinez, J. P., and Sentandreu, R., 1989, Identification of wall-specific antigens synthesized during germ tube formation by *Candida albicans*, *Infect. Immun.* **57**:262–271.
51. Sundstrom, P. M., Tam, M. R., Nichols, E. J., and Kenny, G. E., 1988, Antigenic differences in the surface mannoproteins of *Candida albicans* as revealed by monoclonal antibodies, *Infect. Immun.* **56**:601–606.
52. Bhattacharjee, A. K., Bennett, J. E., and Glaudemans, C. P. J., 1984, Capsular polysaccharides of *Cryptococcus neoformans*, *Rev. Infect. Dis.* **6**:619–624.
53. Cherniak, R., 1988, Soluble polysaccharides of *Cryptococcus neoformans*, *Curr. Top. Med. Mycol.* **2**:40–54.
54. Breen, J. F., Lee, I. C., Vogel, F. R., and Friedman, H., 1982, Cryptococcal capsular polysaccharide-induced modulation of murine immune responses, *Infect. Immun.* **36**:47–51.
55. Blackstock, R., Hall, N. K., and Hernandez, N. C., 1989, Characterization of a suppressor factor that regulates phagocytosis by macrophages in murine cryptococcosis, *Infect. Immun.* **57**:1773–1779.
56. Turner, S. H., Cherniak, R., and Reiss, E., 1984, Fractionation and characterization of galactoxylomannan from *Cryptococcus neoformans*, *Carbohydr. Res.* **125**:343–349.
57. van de Moer, A., Salhi, S. L., and Bastide, J.-M., 1989, IgM monoclonal antibody against the galactoxylomannan-mannoprotein complex of *Cryptococcus neoformans*, *Prog. Abstr. Intern. Conf. Cryptococcus, Cryptococcosis*, Jerusalem, #P-2.
58. Murphy, J. W., 1988, Influence of cryptococcal antigens on cell-mediated immunity, *Rev. Infect. Dis.* **10(Suppl. 2)**:S432–S435.
59. Shepherd, M. G., 1987, Cell envelope of *Candida albicans*, *CRC Crit. Rev. Microbiol.* **15**:7–25.
60. James, P. G., Cherniak, R., Jones, R. G., Stortz, C. A., and Reiss, E., 1990, Cell wall glucans of *Cryptococcus neoformans* cap 67, *Carbohydr. Res.* **198**:23–38.
61. Bhattacharjee, A. K., Kwon-Chung, K. J., and Glaudemans, C. P. J., 1979, The structure of the capsular polysaccharide from *Cryptococcus neoformans* serotype D, *Carbohydr. Res.* **73**: 183–192.
62. Wilson, D. E., Bennett, J. E., and Bailey, J. W., 1968, Serological grouping of *Cryptococcus neoformans*, *Proc. Soc. Exp. Biol. Med.* **127**:820–823.
63. Ikeda, R., Nishikawa, A., Shinoda, T., and Fukazawa, Y., 1985, Chemical characterization of capsular polysaccharide from *Cryptococcus neoformans* serotype A–D, *Microbiol. Immunol.* **29**:981–991.
64. Ikeda, R., Shinoda, T., Fukazawa, Y., and Kaufman, L., 1982, Antigenic characterization of *Cryptococcus neoformans* serotypes and its application to serotyping of clinical isolates, *J. Clin. Microbiol.* **16**:22–29.
65. Kwon-Chung, K. J., 1975, A new genus, *Filobasidiella*, the perfect state of *Cryptococcus neoformans*, *Mycologia* **67**:1197–1200.
66. Kwon-Chung, K. J., 1976, A new species of *Filobasidiella*, the sexual state of *Cryptococcus neoformans* B and C serotypes, *Mycologia* **68**:942–946.
67. Todaro-Luck, F., Reiss, E., Cherniak, R., and Kaufman, L., 1989, Characterization of

Cryptococcus neoformans capsular glucuronoxylomannan polysaccharide with monoclonal antibodies, *Infect. Immun.* **57:**3882–3887.
68. Goren, M. B., and Middlebrook, G. M., 1967, Protein conjugates of polysaccharide from *Cryptococcus neoformans*, *J. Immunol.* **98:**901–913.
69. Kozel, T. R., and Gotschlich, E. C., 1982, The capsule of *Cryptococcus neoformans* passively inhibits phagocytosis of the yeast by macrophages, *J. Immunol.* **129:**1675–1680.
70. Eckert, T. F., and Kozel, T. R., 1987, Production and characterization of monoclonal antibodies specific for *Cryptococcus neoformans* capsular polysaccharide, *Infect. Immun.* **55:** 1895–1899.
71. Spiropulu, C., Eppard, R. A., Otteson, E., and Kozel, T. R., 1989, Antigenic variation within serotypes of *Cryptococcus neoformans* detected by monoclonal antibodies specific for the capsular polysaccharide, *Infect. Immun.* **57:**3240–3242.
72. Turner, S. H., 1989, Strain to Strain Variation of the Glucuronoxylomannan Produced by *Cryptococcus neoformans*, Ph.D. dissertation, Georgia State University, Atlanta, Georgia.
73. Dromer, R., Salamero, J., Contrepois, A., Carbon, C., and Yeni, P., 1987, Production, characterization, and antibody specificity of a mouse monoclonal antibody reactive against *Cryptococcus neoformans* capsular polysaccharide, *Infect. Immun.* **55:**742–748.
74. Todaro-Luck, F., White, E. H., Reiss, E., and Cherniak, R., 1989, Immunoelectron-microscopic characterization of monoclonal antibodies (MAbs) against *Cryptococcus neoformans*, *Mol. Cell Probes* **3:**345–361.
75. Cherniak, R., Reiss, E., Slodki, M. E., Plattner, R. D., and Blumer, S. O., 1980, Structure and antigenic activity of the capsular polysaccharide from *Cryptococcus neoformans* serotype A, *Mol. Immunol.* **17:**1025–1032.
76. Cherniak, R., Reiss, E., and Turner, S. H., 1982, A galactoxylomannan antigen of *Cryptococcus neoformans* serotype A, *Carbohydr. Res.* **103:**239–250.
77. Cherniak, R., Jones, R. G., and Reiss, E., 1988, Structure determination of *Cryptococcus neoformans* serotype A- variant glucuronoxylomannan by ^{13}C-n.m.r. spectroscopy, *Carbohydr. Res.* **172:**113–138.
78. Rebers, P. A., Barker, S. A., Heidelberger, M., Dische, Z., and Evans, E. E., 1958, Precipitation of the specific polysaccharide of *Cryptococcus neoformans* A by types II and XIV antipneumococcal sera, *J. Am. Chem. Soc.* **80:**1135–1137.
79. Murphy, J. W., Mosley, R. L., Cherniak, R., Reyes, G. H., Kozel, T. R., and Reiss, E., 1988, Serological, electrophoretic, and biological properties of *Cryptococcus neoformans* antigens, *Infect. Immun.* **56:**424–431.
80. Reiss, E., Cherniak, R., Eby, R., and Kaufman, L., 1984, Enzyme immunoassay detection of IgM to galactoxylomannan of *Cryptococcus neoformans*, *Diagn. Immunol.* **2:**109–115.
81. Schønheyder, H., 1987, Pathogenetic and serological aspects of pulmonary aspergillosis, *Scand. J. Infect. Dis.* **(Suppl. 51):**1–62.
82. Harvey, C., and Longbottom, J. L., 1986, Characterization of a major antigenic component of *Aspergillus fumigatus*, *Clin. Exp. Immunol.* **65:**206–214.
83. de Repentigny, L., 1990, Antibody responses to *Aspergillus* antigens, *Prog. Abstr. Ann. Mtg. Amer. Soc. Microbiol.*, Anaheim, CA, session 64, p. 37.
84. Hearn, V. M., Wilson, E. V., and Mackenzie, D. W. R., 1988, Characterization of purified wall antigens obtained from *Aspergillus* species, in: *Fungal Antigens: Isolation, Purification, and Detection* (E. Drouhet, G. T. Cole, L. de Repentigny, J. -P. Latge, and B. Dupont, eds.), Plenum Press, New York, pp. 355–365.
85. Schonheyder, H., and Andersen, P., 1984a, Fractionation of *Aspergillus fumigatus* antigens by hydrophobic interaction chromatography and gel filtration, *Int. Arch. Allergy Appl. Immunol.* **73:**231–236.
86. Calvanico, N. J., Dupont, B. L., Huang, C. J., Patterson, R., Fink, J. N., and Kurup, V. P., 1981, Antigens of *Aspergillus fumigatus*. 1. Purification of a cytoplasmic antigen reactive with sera of patients with *Aspergillus*-related disease, *Clin. Exp. Immunol.* **45:**662–671.

87. Schønheyder, H., Storgaard, L., and Andersen, P., 1985, Variation of a 470,000 daltons antigen complex and catalase antigen in clinical isolates of *Aspergillus fumigatus*, *J. Med. Vet. Mycol.* **23**:339–348.
88. Bardana, E. J., Jr., 1980, The clinical spectrum of aspergillosis—part 1: Epidemiology, pathogenicity, infection in animals and immunology of *Aspergillus*, *CRC Crit. Rev. Clin. Lab. Sci.* **13**:21–83.
89. Leung, P. S. C., Gershwin, M. E., Coppel, R., Halpern, G., Novey, H., and Castles, J. J., 1988, Localization, molecular weight, and immunoglobulin subclass response to *Aspergillus fumigatus* allergens in acute bronchopulmonary aspergillosis, *Int. Arch. Allergy Appl. Immunol.* **85**:416–421.
90. Schønheyder, H., and Andersen, P., 1984b, IgG antibodies to purified *Aspergillus fumigatus* antigens determined by enzyme-linked immunosorbent assay, *Int. Arch. Allergy Appl. Immunol.* **74**:262–269.
91. Harvey, C., and Longbottom, J. L., 1987, Characterization of a second major antigen Ag 13 (Antigen C) of *Aspergillus fumigatus* and investigation of its immunological reactivity, *Clin. Exp. Immunol.* **70**:247–254.
92. Tran Van Ky, P., Torck, C., Vaucelle, T., and Floc'h, F., 1969, Etude comparee sur immunoelectrophoregramme des enzymes de l'extrait antigenique d'*Aspergillus fumigatus*, reveles par des serums experimentaux et des serums de malades atteints d'aspergillose, *Sabouraudia* **7**:73–84.
93. Rhodes, J. C., Amlung, T. W., and Miller, M. S., 1990, Isolation and characterization of an elastinolytic proteinase from *Aspergillus flavus*, *Infect. Immun.* **58**:2529–2534.
94. Longbottom, J. L., 1983, Allergic bronchopulmonary aspergillosis: Reactivity of IgE and IgG antibodies with antigenic components of *Aspergillus fumigatus* (IgE/IgG antigen complexes), *J. Allergy Clin. Immunol.* **72**:668–675.
95. Longbottom, J. L., 1986, Allergens of *Aspergillus fumigatus*: Identification and characterization of components combining with specific IgE antibody, *Zbl. Bakt, Hyg.* **A261**:503–508.
96. Harvey, C., and Longbottom, J. L., 1986, Characterization of a major antigenic component of *Aspergillus fumigatus*, *Clin. Exp. Immunol.* **65**:206–214.
97. Taylor, M. L., and Longbottom, J. L., 1988, Partial characterization of a rapidly released antigenic/allergenic component (Ag 5) of *Aspergillus fumigatus*, *J. Allergy Clin. Immunol.* **81**: 548–557.
98. Piechura, J. E., Riefel, R. S., and Daft, L. J., 1987a, Electrophoretic and serological analyses of cytoplasmic antigens from *Aspergillus fumigatus* during growth of conidia to mature mycelia, *J. Med. Vet. Mycol.* **27**:243–254.
99. Piechura, J. E., Riefel, R. S., and Daft, L. J., 1987b, Biochemical and immunochemical analyses of detergent solubilized antigens from membrane vesicles of *Aspergillus fumigatus*, *Can. J. Microbiol.* **33**:955–962.
100. Talbot, G. H., Weiner, M. H., Gerson, S. L., Provencher, M., and Hurwitz, S., 1987, Serodiagnosis of invasive aspergillosis in patients with hematologic malignancy: Validation of the *Aspergillus fumigatus* antigen radioimmunoassay, *J. Infect. Dis.* **155**:12–27.
101. de Repentigny, L., Boushira, M., Ste-Marie, L., and Bosisio, G., 1987, Detection of galactomannan antigenemia by enzyme immunoassay in experimental invasive aspergillosis, *J. Clin. Microbiol.* **25**:863–867.
102. Dupont, B., Huber, M., Kim, S. J., and Bennett, J. E., 1987, Galactomannan antigenemia and antigenuria in aspergillosis: Studies in patients and experimentally infected rabbits, *J. Infect. Dis.* **155**:1–11.
103. Matthews, R., Burnie, J. P., Fox, A. M., and Tabaqchali, S., 1985, Immunoblot analysis of serological responses in invasive aspergillosis, *J. Clin. Pathol.* **38**:1300–1303.
104. Trull, A. K., Parker, J., and Warren, R. E., 1985, IgG enzyme linked immunosorbent assay for diagnosis of invasive aspergillosis: Retrospective study over 15 years of transplant recipients, *J. Clin. Pathol.* **38**:1045–1051.

105. Bennett, J. E., 1987, Rapid diagnosis of candidiasis and aspergillosis, *Rev. Infect. Dis.* **9:** 398–402.
106. de Repentigny, L., and Reiss, E., 1984, Current trends in immunodiagnosis of candidiasis and aspergillosis, *Rev. Infect. Dis.* **6:**301–312.
107. Barreto-Bergter, E. M., Travassos, L. R., and Gorin, P. A. J., 1980, Chemical structure of the D-galacto-D-mannan component from hyphae of *Aspergillus niger* and other *Aspergillus* spp., *Carbohydr. Res.* **86:**273–285.
108. Barreto-Bergter, E. M., Gorin, P. A. J., and Travassos, L. R., 1981, Cell constituents of mycelia and conidia of *Aspergillus fumigatus*, *Carbohydr. Res.* **95:**205–218.
109. Bennett, J. E., Bhattacharjee, A. K., and Glaudemans, C. P. J., 1985, Galactofuranosyl groups are immunodominant in *Aspergillus fumigatus* galactomannan, *Mol. Immunol.* **22:** 251–254.
110. Weiner, M., and Coats-Stephen, M., 1979, Immunodiagnosis of systemic aspergillosis. I. Antigenemia detected by radioimmunoassay in experimental infection, *J. Lab. Clin. Med.* **93:**111–119.
111. Andrews, C. P., and Weiner, M. H., 1982, *Aspergillus* antigen detection in bronchoalveolar lavage fluid from patients with invasive aspergillosis and aspergillomas, *Am. J. Med.* **73:** 372–380.
112. Patterson, T. F., Miniter, P., Ryan, J. L., and Andriole, V. T., 1988, Effect of immunosuppression and amphotericin B on *Aspergillus* antigenemia in an experimental model, *J. Infect. Dis.* **158:**415–422.
113. Johnson, T. M., Kurup, V. P., Resnick, A., Ash, R. C., Fink, J. N., and Kalbfleisch, J., 1989, Detection of circulating *Aspergillus fumigatus* antigen in bone marrow transplant patients, *J. Lab. Clin. Med.* **114:**700–707.
114. Phillips, P., and Radigan, G., 1989, Antigenemia in a rabbit model of invasive aspergillosis, *J. Infect. Dis.* **159:**1147–1150.

21

Mycotoxins
Toxic Metabolites of Fungi

W. G. SORENSON

1. INTRODUCTION

Toxic metabolites are produced by a wide variety of fungi. The greatest impact on humans has come from those toxins that commonly contaminate foodstuffs or those fungi used directly as food, i.e., the mushrooms and other fleshy fungi. Mushroom poisoning is also called *mycetismus* and will not be discussed in the present chapter. Toxins produced in foodstuffs present a more insidious problem than mycetismus intoxication because the toxins may be present in basic staples, even in the absence of visible mold growth, and are normally detected only by chemical methods. Episodes of mycotoxicosis frequently arise in animals as unexplained illnesses in which the disease is not transmissible from animal to animal, i.e., it is not due to an infectious agent and drugs, or antibiotics have little effect on the course of the disease. Outbreaks of these diseases are often associated with particular climatic or seasonal conditions favoring fungal growth, the disease can often be traced to specific foodstuffs, and the foodstuffs may contain unusually large numbers of fungi.[1] Koch's postulates for mycotoxins involve finding the mycotoxin in substrate suspected to the cause of the episode of mycotoxicosis, isolating a fungus from the substrate capable of producing the toxin, inducing production of the mycotoxin *in vitro* by the isolate, and inducing the characteristic

W. G. SORENSON • National Institute for Occupational Safety and Health, Morgantown, West Virginia 26505 and Department of Microbiology and Immunology, West Virginia University, Morgantown, West Virginia 26506.

Fungal Infections and Immune Responses, edited by Juneann W. Murphy *et al.* Plenum Press, New York, 1993.

features of the toxicosis in experimental animals with materials from the *in vitro* culture.[2]

Mycotoxins may be produced in the field on live crop plants or are produced during storage. The serious study of mycotoxins has developed only since the early 1960s, when the aflatoxins were discovered to be the cause of massive numbers of deaths of turkey poults fed contaminated peanut meal in Britain. Since that discovery an enormous literature has developed.

Although we do not know the precise function of mycotoxins for the fungi involved, they are generally believed to be secondary metabolites, i.e., metabolites produced when the mycelium is subjected to some sort of nutrient limitation during growth. During the stationary phase, metabolism becomes "unbalanced" and pools of primary metabolites, such as various amino acids, acetyl CoA, and malonyl CoA, accumulate and become available for the biosynthesis of secondary metabolites, which are rarely produced when optimum concentrations of nutrients are available. Nearly all mycotoxins are derived from 1) amino acids, 2) the mevalonate pathway, 3) the acetate-malonate pathway, 4) the shikimic acid pathway, or 5) combinations of the above, e.g., lysergic acid and isolysergic acid are derived from tryptophan by condensation with an isoprene unit from mevalonic acid.[3] Secondary metabolism can lead to a tremendous variety of metabolites, often unique to individual species or to groups of closely related species. To illustrate this latter point, let us consider the 12,13-epoxytrichothecene mycotoxins. Trichothecin, produced by *Trichothecium roseum*, was discovered in 1949 and was the first of the group described. At the present time, more than sixty 12,13-epoxytrichothecene mycotoxins are known and new members of the group continue to be described. The trichothecenes are produced by several genera of Deuteromycetes, such as *Trichothecium*, *Fusarium*, *Myrothecium*, *Trichoderma*, *Cephalosporium*, and *Stachybotrys*. Production of specific mycotoxins has been shown to provide useful taxonomic data in the genus *Penicillium*.[4]

1.1. Detection and Analysis of Mycotoxins

1.1.1. Chromatography

In most instances, mycotoxins are found in complex organic mixtures and analysis must include extraction, cleanup, and/or partial purification. Common solvents for extraction of mycotoxins include chloroform, methylene chloride, ethyl acetate, acetonitrile, methanol, and acetone. Removal of unwanted contaminants is most often accomplished by column chromatography, liquid–liquid partition, or gel-permeation chromatography. Thin-layer chromatography (TLC) remains the preferred method of screening and analysis for the vast majority of mycotoxins.[5] TLC is relatively simple and quick, and, with modern instrumentation, yields reliable quantitative results.[6,7] Qualitative TLC is also useful, does not require expensive instrumentation, and is therefore inexpensive. In many cases confirmatory reactions can be preformed directly on the chromatoplate, extending the value of the procedure. Betina[5] has provided an excellent in-depth review of the use of TLC for the analysis of mycotoxins. High performance liquid chromatography (HPLC) has become increasingly popular and, in some cases, the two techniques have been used in combination. For example, Frisvad and Thrane[8] described a procedure for the analysis of 182 mycotoxins and other fungal secondary metabolites in a standardized system employing HPLC combined with TLC. All important groups of mycotoxins could be detected in the system.

1.1.2. Enzyme-linked Immunosorbent Assay (ELISA)

A number of poly- and monoclonal antibody preparations have been developed for use in immunoassays for mycotoxins.[9] These techniques offer speed, ease, reduced cost relative to conventional chromatographic methods of analysis, and are designed for use by personnel with limited technical expertise. Commercial kits now available involve competition for a solid-phase antibody between a mycotoxin–enzyme conjugate and an unconjugated mycotoxin in the sample extract. Such kits are available for aflatoxin B_1 and M_1, zearalenone, T-2 toxin, and deoxynivalenol. They are highly specific, rapid, and can be used directly for assaying methanol-water extracts of a wide range of foods.[9] Pestka[9] has published an extensive review of the use of immunoassays for mycotoxins. The small molecular size of mycotoxins precludes their direct use as antigens in immunoassays, i.e., the production of specific antibody against mycotoxins requires covalent binding to carrier molecules, e.g., proteins.

1.2. Toxicity

The relative toxicity of mycotoxins is usually expressed as the LD_{50} of the toxin, i.e., the dose that is lethal for 50% of the experimental animals exposed. Expressed in these terms, mycotoxins exhibit a wide range of toxicity, both when different mycotoxins are compared in the same way and when the effects of the same mycotoxin are compared in different animal species or by different routes of administration. For example, cyclochlorotine and satratoxin H have LD_{50} values below 1 mg/kg and can be considered "extremely toxic" whereas certain other compounds have LD_{50} values as high as 800 mg/kg.[10] Similar, different species of experimental animals vary in their susceptibility to acute toxicity, e.g., oral LD_{50} values range from 0.3 to 17.7 mg/kg for aflatoxin. Whereas the rat, mouse, and hamster are relatively resistant to aflatoxin (LD_{50} >10 mg/kg), cats, ducklings, pigs, and rabbits are much more susceptible (LD_{50} ≤1 mg/kg[10] The sex and strain of the experimental animal also may have a major effect on the degree of toxicity observed. In general, newborn and male animals are more susceptible than adults and females. Cyclochlorotine and citreoviridin are exceptional in that they are both more toxic to adult than newborn mice.[10] When lethal doses are compared with respect to the route of administration, the oral dose is generally the least toxic, followed by subcutaneous, intraperitoneal, and finally intravenous injection.[10]

1.3. Natural Exposure of Humans and Animals to Mycotoxins

Both humans and animals are typically exposed to mycotoxins via ingestion in the diet, and this can be considered the portal of entry into the body in most cases of natural intoxication due to mycotoxins. It must also be recognized, however, that exposure via inhalation is possible in some circumstances and emerging evidence suggests that inhalation exposure may produce more toxicity than by any other route.[11] Nearly all of the known mycotoxins are nonvolatile. Therefore inhalation of mycotoxins would require their absorption to particles of substrate or within airborne fungal spores.[12] The levels of mycotoxins in fungal spores can be very high, and levels as high as 200,000 ppb have been reported in the conidia of *Aspergillus flavus* and *Aspergillus parasiticus*.[13] The spore size of many of the species of toxigenic fungi are within the respirable range and, hence, they could be deposited in alveoli. Thus, high concentrations of mycotoxins adsorbed onto spores or particulates could reach the alveoli via inhalation. Mycotoxins present in the lung may interfere with immune function, and such mycotoxins (trichothecenes, gliotoxin, aflatoxins) are of special interest. A compromised immune system would affect the ability of the host to respond to allergens and infection, and could adversely affect the ability of the immune system to perform its surveillance function in the control of tumor development.

Dermal exposure to fungal propagules and toxins may result from direct contact with contaminated surfaces or particles in the air. Skin irritation and minor allergic reaction to dermally deposited allergens could occur but is likely to be less significant than the response of atopic individuals who inhale the offending material. When the contact material contains mycotoxins, however, the skin reaction may be quite significant. Workers cleaning up *Stachybotrys atra* contamination in a home reported that the dust and debris were highly irritating to the skin and respiratory system.[14] Approximately 50 mycotoxins are known to be produced by fungi common to house dust in Europe and North America.[15]

1.4. Etiology of the Mycotoxicoses

It is frequently difficult to establish etiology for specific episodes of intoxication by mycotoxins for several reasons. First, mycotoxin production in foodstuffs often occurs in "hot spots," i.e., localized areas within the commodity where conditions are favorable for the growth of the mycotoxigenic fungi. Other areas within the same sample or load of grain may contain very low levels or none of the toxin because of the uneven distribution of inoculum and/or because of localized high moisture content due to a leaky roof or other cause. Secondly, the foodstuff or feed may be consumed before mycotoxicosis is suspected. Thirdly, because individuals of the same species are variable in their ability to produce toxins and because the conditions favoring toxin production may not be known, it is not always possible to demonstrate that toxigenic fungi are present in the sample. For example, only a fraction of isolates of *A. flavus* tested produce aflatoxin in artificial culture.[16]

1.5. Immunomodulating Effects of Mycotoxins

The earliest indication of possible immunosuppressive or other immunomodulating effects due to mycotoxins were noted in connection with trichothecene intoxications.[17] These mycotoxins have a remarkable capacity to inhibit protein synthesis, to preferentially attack lymphoid tissues,[18] and to markedly increase susceptibility to infection.[18–22] Trichothecenes may inhibit both the humoral and cellular arms of the immune system. For example, Rosenstein *et al.*[23] reported that T-2 toxin decreased the humoral response to T-dependent antigens at the same time that the response to T-independent antigens was increased, Tryphonas *et al.*[24] found that repeated gavage with deoxynivalenol (DON) leads to reduced serum antibody titers and reduced plaque-forming response to sheep erythrocytes, and Pestka *et al.*[22] reported that the plaque-forming response was inhibited by dietary DON. Several investigators have reported inhibition of the mitogen response due to treatment with T-2 toxin,[25–28] DON,[20] roridin A,[29] and verrucarin A,[29] and inhibition of the delayed-type hypersensitivity response due to T-2 toxin[26] or DON.[22] Miller and Atkinson[30] reported that lymphocyte cultures exposed to low levels of DON produced enhanced stimulation of mitogen-induced blastogenesis and that the effect could also be demonstrated with medium conditioned by the exposure of macrophages to low levels of DON. These results suggest that interleukin synthesis by macrophages is stimulated by low levels of trichothecenes and that this activity results in immunomodulation. Although both B- and T-cell responses to mitogens are inhibited,[22,29] it appears that various subpopulations of lymphocytes have different susceptibilities to T-2 toxin[28] and, presumably, to other trichothecenes as well. T-2 toxin did not induce or interfere with the generation of suppressor cells by concanavalin A.[28] Tai and Pestka[31] made the interesting observation that the toxicity of T-2 toxin was markedly increased if coadministered with sublethal concentrations of lipopolysaccharide (LPS) from *Salmonella typhimurium*. Increased mortality was accompanied by an impaired splenic response to LPS, and it appears that T-2 toxin and LPS may act synergistically. This observation is highly significant when one considers the widespread occurrence of bacterial cell wall materials (and endotoxin) in nature and the fact that laboratory investigations of the immunotoxic effects of mycotoxins are often done with purified preparations of individual mycotoxins, whereas the natural exposure is much more complex.

Ochratoxin A has been shown to inhibit natural killer (NK) cell activity and to increase the growth of transplantable tumor cells,[32] and it has been implicated in the development of malignant disease,[33] possibly because of impaired tumor surveillance. Creppy *et al.*[34] reported that OA interferes with B-cell function, leading to reduced antibody synthesis, and treatment of T lymphocytes with OA almost completely abolished the proliferative response to PHA. Treatment of highly purified subpopulations of T lymphocytes (>98%) with OA indicated no selective inhibition of the different subpopulations.[33] These authors also demonstrated that OA inhibits the production of interleukin-2 (IL-2) and the expression of IL-2 receptors.

Patulin inhibits phagocytosis in rat alveolar macrophages.[35,36] Phagocytosis and phagosome–lysosome fusion were diminished in mouse peritoneal macrophages at >0.1 μg/ml by a mechanism independent of products of the oxidative burst.[37] Patulin diminished the mitogenic response to PHA, Con A, and PWM, accompanied by concomitant decreasing serum immunoglobulin levels in mice and rabbits.[38]

Gliotoxin and other epipolythiodioxopiperazines inhibit several leukocyte functions associated with the central role of the macrophage, including the concanavalin A-mediated proliferative response to splenocytes, the lipopolysaccharide-mediated proliferative response of splenocytes, target cell lysis by cytotoxic cells, interleukin-2 (IL-2) mediated T-cell proliferation, lymphokine release by activated T cells, and stimulatory cell function in mixed lymphocytes cultures.[39] Gliotoxin irreversibly inhibits murine T-cell proliferation to mitogens, and pretreatment of allogenic spleen cells permits their transfer into sublethally irradiated recipients without inducing a graft versus host reaction (GVH). Gliotoxin treatment of bone marrow cells allows the establishment of fully allogenic bone marrow chimeras free of GVH disease.[40] Sporidesmin is a related epipolythiodioxopiperazine and Mullbacher *et al.*[41] have shown that it is more active than gliotoxin in inhibition of phagocytosis, in its effect on stimulator cells in mixed lymphocyte cultures, and in its ability to inhibit mitogen stimulation of T lymphocytes.

Aflatoxin B_1 has been shown to inhibit DNA synthesis in lymphocytes, natural killer cell function, and generation of concanavalin A-induced suppressor cells in mice[42] and the cell-mediated immune response and phagocytic index in chickens.[43]

1.6. Control and Prevention

Obviously, the most direct means of preventing intoxication due to the presence of fungal metabolites in foods and feeds is to prevent the development of natural contamination of the commodity by fungi. Unfortunately, this is often difficult or impossible to accomplish since the fungi are ubiquitous and may infect the plant before harvest. Grain held at or below 13% moisture will not support the growth of fungi and, if the grain is not infected before harvest and if it can be dried as quickly as possible before storage, invasion by fungi and consequent mycotoxin production can be minimized or eliminated. In general, mycotoxins are quite stable and often can withstand cooking temperatures without degradation. Certain agricultural commodities (i.e., those known to be attacked by toxigenic fungi) are routinely screened for the presence of specific mycotoxins, e.g., peanuts and other commodities for the presence of aflatoxin. Since the aflatoxins are natural products of ubiquitous fungi, exposure in many parts of the world, particularly where food is in short supply, may be unavoidable. Most countries of the world have established control levels that limit the permissible amounts of specific mycotoxins in commodities sold for human consumption. The current level for aflatoxin in the United States is 20 ppb (20 μg/kg), which represents a compromise between the ideal and the reality that complete elimination of aflatoxin is not feasible. Some success has been achieved in decontamination of certain foodstuffs and extensive research over the past 20 years has provided overwhelming support for the use of ammoniation as a practical solution to aflatoxin detoxification in animal feeds.[44]

2. MYCOTOXICOSES

The remainder of this chapter will deal with representative mycotoxins from the perspective of specific mycotoxicoses, particularly those that have had the greatest impact on humans and their domestic animals.

2.1. Aflatoxicosis

2.1.1. Description of the Disease

Typically, the first signs of chronic aflatoxicosis in both avian and mammalian species are loss of appetite, weight loss, and general malaise.[45] Pathological studies of low-level intoxication reveal generalized jaundice and cirrhosis of the liver with bile duct proliferation and periportal fibrosis. Acute aflatoxicosis results in jaundice of the mucous membrane, widespread hemorrhages, and fatty accumulation in the liver. Liver pathology is the most characteristic and easily identifiable effect in most species. Although the liver is the primary target organ in most species, cancerous lesions have been observed in the kidney, colon, lung, and lacrimal gland of various animals challenged with aflatoxins. Cases of acute aflatoxicosis in humans have been reported from Taiwan, Canada, Uganda, West Germany, India, and Kenya, and chronic aflatoxicosis leading to elevated primary liver cancer has been reported from Uganda, Thailand, Kenya, Swaziland, Mozambique, and China.[46] Deaths due to occupationally related exposure to aflatoxin have been described among chemical engineers in Czechoslovakia,[47] biochemists working to purify aflatoxin by preparative thin-layer chromatography,[48] and peanut oil-press workers in the Netherlands.[49]

Reye's syndrome (RS) is a mysterious, virus-associated, children's disease of unknown etiology reported in the United States, the United Kingdom, Japan, Korea, Taiwan, Thailand, Malaysia, Singapore, Indonesia, and Czechoslovakia.[46] No constant etiologic agent has been found and Reye concluded in his original paper that etiology may not be constant in every case.[50] Dvorackova[46] has extensively studied RS in Czechoslovakia, and her studies suggest a real link between aflatoxin and at least some cases of RS. Her conclusion was based on the following facts: 1) RS is a children's disease and aflatoxin sensitivity is known to be greatest in the young; 2) clinicomorphological features, similar to those seen in RS, developed in young monkeys fed aflatoxin B_1; 3) metabolic abnormalities characteristic of RS and aflatoxicosis are similar; 4) aflatoxin B_1 was found in 60% of Czech study subjects and ranged from 5 to 1769 μg/kg (with an average level of approximately 200 μg/kg) of liver weight in those who died; and 5) no consistent viral association with RS has been demonstrated. It has been suggested that either both virus and toxin act directly or indirectly at a common cellular locus or that the toxin impairs immune defenses. Morphological changes in the lymphatic organs, ultrastructural alterations, and the known immunomodulating effects of aflatoxin would seem to support the latter hypothesis.[46]

2.1.2. *Toxigenic Fungi and Natural Occurrence*

The aflatoxins are a family of highly substituted coumarins produced by *Aspergillus flavus* and *Aspergillus parasiticus* containing a fused dihydrofuran moiety (Structure 1). These fungi are ubiquitous but their ability to produce

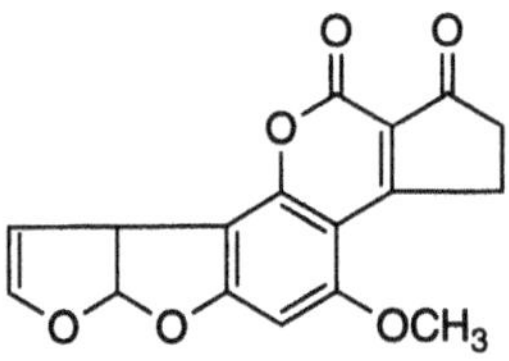

Aflatoxin B_1

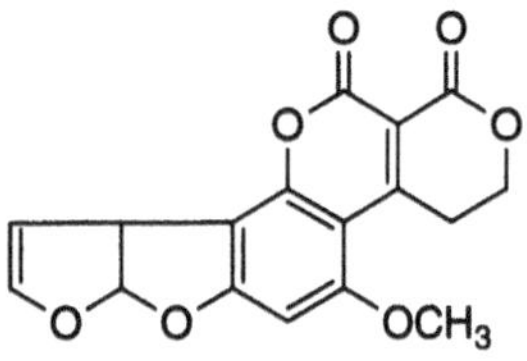

Aflatoxin G_1 (1)

aflatoxin seems to be limited by temperature, the nature of the substrate, and the genetic potential of the individual. Aflatoxin production is a much more serious problem in tropical and subtropical regions of the world, e.g., crops grown in the southeastern United States are more likely to contain aflatoxin than those grown in the northern regions of the country.[51] Similarly, in a recent study of primary liver cancer in Denmark, the authors were able to show that all of the aflatoxin in grains used in the preparation of animal feeds in Denmark came from commodities imported from overseas, i.e., aflatoxin was not detected in their domestic crops.[52] Although the Aspergilli are usually thought to be "storage fungi" with limited ability to attack healthy plants, *A. flavus* and *A. parasiticus* are sometimes transported into healthy plant tissue by boring insects.

2.1.3. Biological Activity and Toxicology

Aflatoxins are primarily metabolized by microsomal mixed-function oxidases in the liver,[46] leading to inhibition of DNA-dependent RNA polymerases, stimulation of DNA repair, increased permeability of mitochondria, and interruption of electron transport.[46] The primary organ affected by the aflatoxins is the liver, and considerable circumstantial evidence exists to suggest that aflatoxin is a primary liver carcinogen in humans.[46,53] Aflatoxin is a potent genotoxicant of the indirect type, i.e., it is not active without metabolic activation. Activation seems to involve the formation of an epoxide ring by the addition of a molecule of water across the difuran bond, forming derivatives that can react with guanidine residues in the DNA, leading to the formation of frameshift mutations and the transformation of liver cells.[54] Aflatoxin B_1 can be metabolized in the body to form several other derivatives, e.g., M_1, M_2, P_1, and Q_1. The aflatoxins are remarkably stable to heat and sterilization by autoclaving has little effect.

2.2. Ergotism

2.2.1. Description of the Disease

Knowledge of this disease has existed for approximately 5000 years.[55] Accurate descriptions of ergot epidemics characterized by necrosis of limbs and death were recorded in Germany in 857 and in Paris in 945, and innumerable epidemics occurred throughout Europe, but especially in France over the next 600 years.[55] During the great epidemics in Europe during the later Middle Ages, up to 40,000 people succumbed to ergotism and frequently experienced intense pain, the unbearable stench of rotting flesh, and necrosis and gangrene of extremities, which would turn black and rot before detachment.[55] Recent outbreaks have confirmed that approximately 100 g of ergot over a period of a few days is sufficient to cause death from gangrene.[55]

2.2.2. Toxigenic Fungi and Natural Occurrence

Ergotism is caused by compounds known as ergot alkaloids, which are produced by approximately 10 species of the genus *Claviceps*.[55] These fungi are pathogens of a wide range of grasses, of which rye and wheat are of the greatest concern in domestic animals and humans. *Claviceps purpurea* is the most common cause of ergotism. These fungi infect the developing fruit of the grass and produce a sclerotium, i.e., an enlarged and hardened fruit which is largely replaced by fungal tissue and which contains large quantities of the ergot alkaloids. It has been known for centuries that ergotism is more prevalent during unusually wet seasons, particularly if followed by warmer than usual summers. The ascospores of *C. purpurea* are discharged from the sclerotia in the spring after the latter have fallen to the ground during harvest. Conditions that prolong the flowering period favor the probability of infection. Early development of the infection leads to the production of large numbers of conidia in a sweet mucilaginous "honeydew," which attracts insects that spread the infection. Warm weather favors the maturation of sclerotia. In addition to *Claviceps* species, ergot alkaloids have been isolated from several species of *Aspergillus*, *Rhizopus*, and *Penicillium*, and from some species of higher plants (e.g., *Ipomoae* spp.). A single sclerotium may contain 10 different groups of substances, with a total of over 100 compounds of many diverse types.[55] The most important ergot alkaloids are classified either as derivatives of lysergic acid (simple lysergic acid amides or peptide derivatives of lysergic acid) or as clavine alkaloids, which differ from classical lysergic acid derivatives in that the carboxyl group of the lysergic acid has been reduced to a hydroxymethyl or methyl group. The ergot alkaloids can be detected by a variety of colorimetric tests and, more recently, thin layer chromatographic methods for the resolution and quantitation of ergot alkaloids have been developed.[55]

2.2.3. Biological Activity and Toxicology

The ergot alkaloids were instrumental in the development of the science of animal experimentation because the principles of experimental design did not exist prior to the discovery of this disease.[55] All smooth muscle, vascular and nonvascular, is contracted, leading to vasoconstriction. Among the properties of the ergot alkaloids, perhaps none is more significant that peripheral vasoconstriction. Ergot alkaloids have been used therapeutically in the treatment of migraine headaches and hemorrhage, and it is claimed that the ancient Chinese used ergot for obstetrical purposes 5000 years ago. The hallucinogenic effects of ergot derivatives are well known.[56]

2.3. Mycotoxic Nephropathy

2.3.1. Description of the Disease

For many years a disease involving kidney degeneration in pigs has been recognized in Denmark. This disease has also been called *mold nephrosis* and it occurred in up to 7% of the pigs slaughtered. This disease had been thought to be caused by mold contamination of the feed, and the disease was later induced in pigs and rats by feeding naturally moldy barley. The most toxic fungus isolated from this feed was a strain of *Penicillium viridicatum*, which subsequently was shown to produce both citrinin and ochratoxin A. Both of these mycotoxins were detected in Danish barley and oats fed to swine noted upon slaughter to suffer from mycotoxic nephropathy. Present evidence indicates a cause and effect relationship between the production of ochratoxin A and/or citrinin in feeds and moldy nephrosis in swine. Whether induced by nephrotoxic grains, ochratoxin-contaminated barley, or crystalline citrinin, lesions are observed only in the kidneys.[57] The kidneys are macroscopically enlarged, with a color change from the normal red-brown to grey. There is a diffusely distributed fibrosis in the cortex, which may also include small cysts.[58] Degeneration of the proximal tubules and formation of connective tissue in the interstitium can be observed microscopically, along with atrophy of epithelial cells and thickening of the basement membrane. There may also be sclerotic tufts and atrophy in the glomeruli.[58] Feeding experiments with naturally contaminated barley (ochratoxin A) led to the conclusion that the primary action of the toxin occurs in the renal tubular system, especially in the proximal tubules. Ochratoxin has been implicated in fatal endemic nephropathy in humans in Yugoslavia, Bulgaria, and Romania, although the cause of this disease has not been proven. Endemic nephropathy in the Balkans shows striking similarities with ochratoxin A-induced nephropathy in pigs.[59] The disease progresses slowly, with anemia, and without increased blood pressure and edema. The first sign of renal change is impairment of tubular function, followed by changes in glomerular function. In advanced cases the kidneys are contracted and histological tubular atrophy, glomerular hyalinization, and interstitial fibrosis are observed.

2.3.2. *Toxigenic Fungi and Natural Occurrence*

The ochratoxins are a small group of closely related compounds produced by several species of the genera *Aspergillus* and *Penicillium*.[59] As methods of analysis became available, ochratoxin was found in a variety of commodities, such as several species of cereal (corn, barley, wheat, oats, rye); in moldy coffee beans, beans, and peanuts; in hay[59]; and in tissues of bacon pigs and poultry in Denmark and in pork in Yugoslavia. Canadian workers noted much higher levels of ochratoxin in grains naturally heated due to rapid microbial growth than in other sources of grain. Commodities contaminated with ochratoxin are often found to be cocontaminated with citrinin and/or penicillic acid. These mycotoxins can act synergistically. Ochratoxin A (Structure 2) is a colorless crystalline compound composed of a dihydroisocoumarin moiety linked through its 7-carboxyl group to L-β-phenylalanine, and it contains an atom of chlorine attached to the isocoumarin ring.[57] It is

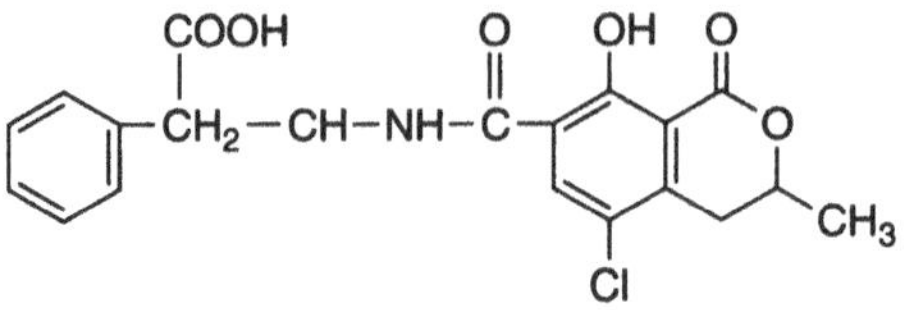

Ochratoxin A (2)

readily extracted from corn or from fungal cultures with mixtures of chloroform and methanol (1:1). Ochratoxin can be detected by thin layer chromatography by its strong fluorescence under UV light (340 nm). The color and intensity of fluorescence can be enhanced by treatment with ammonia vapors and other reagents, and these treatments can be used as confirmatory tests in the analysis of ochratoxin from natural substrates.

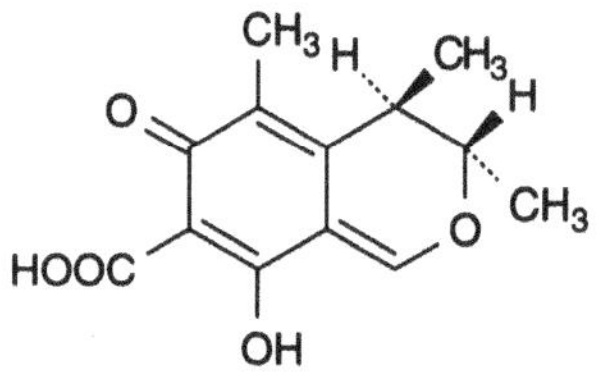

Citrinin (3)

Citrinin (Structure 3) is produced by at least two species of *Aspergillus* and by several species of *Penicillium*.[59] It was first isolated from *P. citrinum*. Citrinin is a yellow crystalline p-quinone compound that can be extracted from acidified grain with chloroform and detected on thin layer chromatography plates by its fluorescence under long-wave UV. Its strong tendency to streak on TLC plates can be diminished by impregnating the plates with oxalic acid before use, and quantitation can be accomplished by densitometry.

2.3.3. *Biological Activity and Toxicology*

The kidneys have been shown to be the target organ for ochratoxin and, although damage in other organs has also been demonstrated, it appears that this damage is secondary to the initial kidney damage. Renal damage has been demonstrated in several species of experimental animals, and spontaneous cases of nephropathy due to ochratoxin have been reported in pigs and poultry. Pharmacokinetic studies have shown that ochratoxin disappears exponentially from the tissues of bacon pigs, and it has been suggested that residue-free pigs can be provided, even if contaminated feed is used for pigs, provided that toxin-free feed is used for the final 4 weeks before slaughter.[59] This makes it possible to avoid major economic losses associated with unavoidable ochratoxin contamination.

2.4. Trichothecoses

2.4.1. *Alimentary Toxic Aleukia (ATA)*

2.4.1a. Description of the Disease. ATA is a disease that first came to the attention of the scientific community before and during World War II in the Soviet Union, although it has probably existed for centuries. At that time villagers in some parts of the Soviet Union were unable to harvest their crops before the onset of winter and, because of shortages of food, were sometimes forced to eat grain that had overwintered in the field. The grain was often contaminated by various fungi, including species of the genus *Fusarium* now known to produce toxins of the trichothecene family. ATA is a very serious disease, often leading to death, characterized by typical spots on the skin, necrotic angina, extreme leukopenia, and multiple hemorrhages, sepsis, and exhaustion of the bone marrow.[17] The course of the disease can be divided into four stages, which have been described in detail by Joffe.[17] The first stage occurs shortly after ingestion of toxic grains, when the patient experiences burning sensations in the mouth, tongue, throat, palate, esophagus and stomach. During the second stage the patient often feels well but begins to develop progressive leukopenia, granulopenia, lymphocytosis, anemia, and a decrease in platelet count. These changes lower resistance to bacterial infection and are often accompanied by disturbances in the central nervous system. The third stage may develop after 2–8 weeks but develops rapidly if consumption of toxic grains continues. Petechial hemorrhages develop on the skin of the trunk, in the axillary and inguinal areas, and other large areas of the body. Necrotic changes appear in the throat, leading to difficulty in swallowing, and may extend to the gums, buccal mucosa, larynx, and vocal cords, leading to bacterial infections of these necrotic areas. The fourth stage is a period of convalescence, which may last more than 2 months before the bone marrow returns to normal.[17] The primary target organs and tissues are in the immune system, and death is due to secondary infections resulting from the loss of immune function.[17]

2.4.1b. Toxigenic Fungi and Natural Occurrence. The principal etiologic fungi reported in ATA are *F. tricinctum* and *F. sporotrichioides*, although *F. tricinctum* is now regarded as a synonym of *F. sporotrichioides*.[60] This species occurs widely in nature, both as a saprobe and a parasite on grasses. The toxigenicity of *F. sporotrichioides* is greatly enhanced at low temperatures, such as occur during a Russian winter, thus explaining the severe toxicosis following overwintering in the field.

The trichothecenes are naturally occurring sesquiterpene metabolites that possess the tricyclic trichothecane skeleton with an olefinic group at carbon 9,10 and an epoxy group at carbon atoms 12,13 (Structure 4). Macrocyclic trichothecenes have a carbon chain between carbon atoms 4 and 15

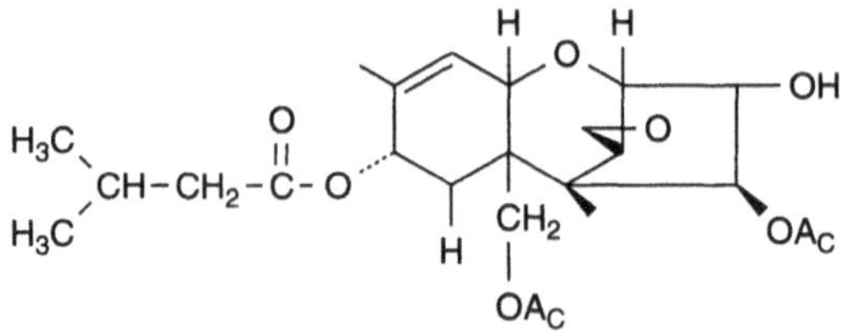

T-2 Toxin (4)

containing an ether or an ester linkage.[61] Structure–activity studies have demonstrated that the 12,13-epoxide ring, the double bond between carbons 8 and 9, and the presence of various free ester groups are essential for toxicity.[61]

2.4.1c. Biological Activity and Toxicology. T-2 toxin and many other trichothecenes have dermatitic properties causing severe local irritation, inflammation, and desquamation.[61] The trichothecenes are toxic to laboratory animals, causing such symptoms as diarrhea, hematuria, vomiting, loss of weight, anorexia, thirst, severe inflammation of the gastrointestinal tract, degeneration of nerve cells in the central nervous system, slitting and hemorrhaging of cardiac muscle, and severe lesions in the lymph nodes, testes, and thymus.[61] The trichothecenes are cytotoxic to a variety of mammalian cell types,[62,63] and they strongly inhibit protein and DNA synthesis in eukaryotic cells[62,64] by binding to eukaryotic ribosomes.[65,66]

2.4.2. *Stachybotryotoxicosis*

2.4.2a. Description of the Disease. Stachybotryotoxicosis is a severe toxic syndrome of large domestic animals first reported in the Ukraine in 1931.[67] The typical form of the disease occurs in a series of stages[68] similar to ATA. The disease is characterized by leukopenia, severe thrombocytopenia, hemorrhage, and arrhythmic heartbeat, and often leads to death. Russian investigators also reported that the disease occurs frequently in human populations where the disease is endemic in the horse. The disease was associated with contaminated straw and dust aerosols heavily laden with spores in cottonseed-oil processing plants, grain elevators, textile mills, and other grain-processing operations, including breweries.[68]

2.4.2b. Toxigenic Fungi and Natural Occurrence. Stachybotrys chartarum (= *S. atra*) is a cellulytic saprobe capable of producing several toxins of the trichothecene family (satratoxin G and H, and verrucarins B and J) as well as other mycotoxins (trichoverrins A and B). This fungus grows well on such materials as straw, hay, and ceiling tiles, and can be recognized by the black, sooty appearance of the substrate after extensive growth and sporulation. The toxins are found in the conidia.[12]

2.4.2c. Biological Activity and Toxicology. Stachybotryotoxicosis has been produced experimentally in laboratory animals, including dogs, rabbits, guinea pigs, mice, and chicks. Since the macrocyclic trichothecenes as a group are highly toxic and produce biological effects in experimental animals similar to those observed in stachybotryotoxicosis, they are regarded to be the chemical agents responsible for stachybotryotoxicosis.[61,69] Persons handling material heavily contaminated with this fungus describe symptoms of cough, rhinitis, burning sensations in the mouth and nasal passages, and cutaneous irritation at the point of contact with the toxins, especially in areas of heavy perspiration, e.g., the armpits and scrotum. Moist dermatitis was usually succeeded by a phase characterized by dried, crusted layers of serous exudate, followed by painful, severe pharyngitis and a burning sensation in the nose, which may also result in bloody nasal exudates. Symptoms appeared within 2–3 days in farm workers who handled contaminated straw. The major symptoms were strong itching and reddening of the skin laryngitis, pharyngitis, and conjunctivitis. It is likely that the toxins are directly responsible for these effects, since similar symptoms develop when solutions of pure trichothecenes are applied directly to the skin.

2.5. Yellowed Rice Toxicoses

Yellowed rice is natural yellowing of rice due to contamination by certain species of *Penicillium*.

2.5.1. Cardiac Beri-beri

Cardiac beri-beri is a disease formerly prevalent in the Orient for at least the past three centuries.[70] Originally thought to be a form of beri-beri, i.e., a vitamin-deficiency disease, current evidence indicates that it is caused by citreoviridin, a yellow crystalline compound produced principally by *Penicillium citreo-viride* Biourge.[70]

2.5.1a. Description of the Disease. The acute form of the disease usually progressed very rapidly with vomiting, convulsions, ascending paralysis, hypothermia, and respiratory arrest. Death usually ensued with excruciating pain within 3 days of onset.[70]

2.5.1b. Toxigenic Fungi and Natural Occurrence. The incidence of cardiac beri-beri typically increased in early summer and late autumn, and appeared in Japan, Korea, Taiwan, Indonesia, India, Hawaii, South America, and Africa. Miyaki[70] demonstrated that growth of *P. citreo-viride* can develop at 14.6% moisture, i.e., below the minimum moisture requirement for many competing fungi. Once established in the rice, maximum toxicity and maximum citreoviridin concentrations developed between 12 and 20°C and at high levels of moisture.[70] The disease has been seen rarely in Japan since

1910 and its disappearance correlates with the increase in rice inspection by the Japanese government. Wicklow *et al.*[71] reported the occurrence of citreoviridin in 6 of 8 samples of corn kernels in south Georgia, ranging in concentration from 19 to 2790 to 759 μg/kg. *Eupenicillium ochrosalmoneum* was consistently isolated from toxic kernels, which could be recognized by their bright lemon-yellow color. The concentration of citreoviridin in these "pick-out" kernels ranged from 53,800 to 759,900 μg/kg. Aflatoxin was also present in all of the bulk samples.

Citreoviridin (Structure 5) is composed of an alpha-pyrone chromophore, a conjugated polyene moiety, and a hydrofuran ring.[70] It emits a brilliant yellow fluorescence under ultraviolet light (366 nm).

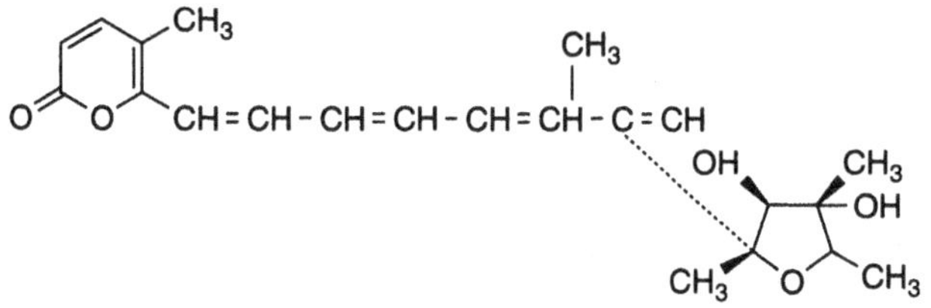

Citreoviridin (5)

2.5.1c. Biological Activity and Toxicology. Citreoviridin is a neurotoxin that causes progressive paralysis of the hind legs of experimental animals due to spinal and medullary depression.[70] It causes selective inhibition of motor neurons or internuncial neurones along the spinal cord. The cause of death is an attack of the respiratory center.[70,72] Citreoviridin inhibits mitochondrial F1-ATPases, ATP hydrolysis, and ATP synthesis.[73]

2.5.2. Penicillium islandicum Hepatotoxins

As a consequence of investigations in Japan on the "rice-fungus-toxin" hypothesis to explain cardiac beri-beri,[70] several hepatotoxic mycotoxins were isolated from *Penicillium islandicum*. The most important of these are luteoskyrin and rugulosin (anthroquinones), and cyclochlorotine (Structure 6)

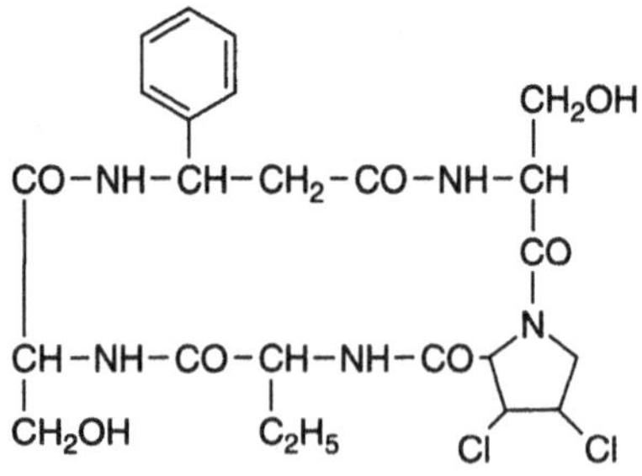

Cyclochlorotine (6)

and islanditoxin (chlorine-containing cyclic peptides).[74] Although there is limited evidence of human toxicosis due to these mycotoxins, some are potent carcinogens in experimental animals.[74] In spite of a low incidence of *P. islandicum* in rice in Japan, the chemical stability of the toxins and its occasional occurrence in human foodstuffs suggest a possible causative role in human hepatic cancers, which are prevalent in that country.[74,75]

2.6. Important Mycotoxins Not Associated with Episodes of Mycotoxicosis in Humans (Gliotoxin, Sporidesmin, Patulin)

There are many mycotoxins that have not yet been associated with episodes of human mycotoxicoses but that either are highly toxic, relatively common, or cause major economic losses in domestic animals. Three that have been shown to have immunomodulating effects are discussed below.

2.6.1. Gliotoxin

Gliotoxin is a member of the epipolythiodioxopiperazine class of fungal metabolites, which is a large group of compounds having a bridged polysulfide piperazine ring in common[39] (Structure 7) and possessing potent

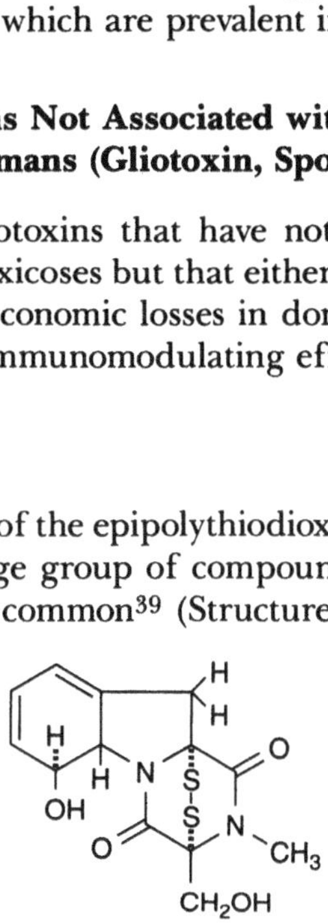

Gliotoxin (7)

immunomodulating activity.[39] Gliotoxin is produced by at least six genera of fungi, including six species of *Aspergillus* and *Penicillium*.[76] Structure–activity studies of the epipolythiodioxopiperazines suggest that the bridged disulfide moiety is the most biologically active portion of the molecule for their immunoregulatory activity *in vitro*,[41] and gliotoxin and other epipolythiodioxopiperazines can be reduced with dithiothreitol, glutathione, and mercaptoethanol, abrogating their biological activities.[39]

Gliotoxin has important biological properties, including inhibition of bacterial and fungal growth, antiviral activity, inhibition of phagocytosis by rodent macrophages,[77] mitogenic stimulation of lymphocytes, and induction of alloreactive and MHC-restricted cytotoxic T cells. The dithio derivative of gliotoxin and some related epipolythiodioxopiperazines cause single- and double-stranded breaks in plasmid and eukaryotic DNA. The antiproliferative action of gliotoxin may result from DNA damage caused by reactive oxygen species or other radicals generated through redox cycling.[78] Gliotoxin was found to cause double-stranded DNA breakage in spleen cells, which was specific to cell types and occurred in all cell types in which gliotoxin inhibited proliferation. The response was dose and time dependent.[79]

Production of gliotoxin appears to be common among cultures of *Aspergillus fumigatus* isolated from animal tissues. This finding suggests that gliotoxin could play a role in the etiology of disease caused by this species and other fungi that produce it. Waring *et al.*,[39] described an experiment in which mice immunocompromised by sublethal irradiation were given spores of *A. fumigatus* intraperitoneally. The fungus was detected in the peritoneal cavity, kidney, and spleen, and gliotoxin was isolated from peritoneal lavages from these animals. Similarly, *Thermoascus crustaceus* was isolated from monocytes of AIDS patients and was found to produce gliotoxin and related metabolites in culture.[39] Although conclusive evidence for the role of gliotoxin in the etiology of aspergillosis is lacking, it was recently shown to be produced in naturally infected tissue at concentrations higher than concentrations known to produce morphological changes in cells.[80]

2.6.2. *Sporidesmin*

Sporidesmins are produced by *Pithomyces chartarum* (Berk. and Curt.) M. B. Ellis, formerly known as *Sporidesmium bakeri* Sydow.[81] *P. chartarum* is most frequently associated with plant debris, particularly grasses, but has been found on leaves and stems of plants, floral parts, roots, seed, foil, air, plywood, paper, and human foodstuffs.[81] It has been a major problem in New Zealand and Australia, where it causes facial eczema in sheep and cattle, leading to significant economic losses.[81] The sporidesmins belong to the epipolythiodioxopiperazine class of mycotoxins and thus share many of the features and properties of chemical structure and biological activity with gliotoxin.[39]

The principal changes in the early stages of sporidesmin intoxication occur in the parenchymal cells of the liver, where vacuolation and the presence of large quantities of lipid in cytoplasm surrounding the vacules have been observed. Edematous inflammatory changes are reported around the bile ducts, the cystic ducts are severely affected with degenerate and necrotic epithelium, and liver function is severely impaired.[81]

2.6.3. *Patulin*

Patulin is a polyketide lactone (Structure 8) produced principally by species of *Penicillium* and *Aspergillus*.[82] It has been reported to be common in moldy corn silage and in naturally rotted apples.[36] The initial interest in patulin was for its antimicrobial properties, but subsequent work has shown it to be toxic in experimental animals, carcinogenic in rats, mutagenic in yeast, and teratogenic in chick embryos.[83] Patulin was also shown to be the etiologic

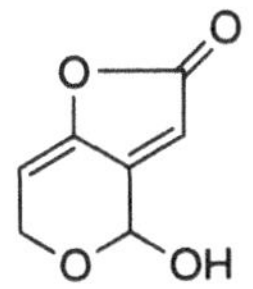

Patulin (8)

agent in an accidental epidemic of feed poisoning in Japan, resulting in the mass death of 118 dairy cows in 1952.[84] Patulin induces single- and double-strand breaks in HeLa DNA and causes a high percentage of polyploid cells in human leukocyte cultures. More recent studies have demonstrated that patulin inhibits transcription in RNA synthesis in a cell-free system from rat liver; inhibits translation of protein synthesis in rabbit reticulocyte lysates; inhibits synthesis of rRNA, tRNA, and probably mRNA in *Saccharomyces cerevisiae*; and inhibits protein synthesis and phagocytosis in alveolar macrophages.[36] Patulin also causes alterations in plasma membrane permeability in cultured renal cells.[85]

REFERENCES

1. Schieffer, H. B., 1988, *Mycotoxicoses: A Summary of Pertinent Facts*, Toxicology Research Centre, University of Saskatchewan.
2. Hamilton, P. B., 1978, Fallacies in our understanding of mycotoxins, *J. Food Protection* **41:** 404–408.
3. Yamazaki, M., 1978, Chemistry of mycotoxins, in: *Toxicology, Biochemistry, and Pathology of Mycotoxins* (K. Uraguchi and M. Yamazaki, eds.), Halsted Press, New York, pp. 65–106.
4. Frisvad, J. C., and Filterborg, O., 1989, Terverticillate penicillia: Chemotaxonomy and mycotoxin production, *Mycologia* **81:**837–861.
5. Betina, V., 1985, Thin-layer chromatography of mycotoxins, *J. Chromatogr.* **334:**2121–2276.
6. Coker, R. D., Jewers, K., Tomlins, K. I., and Blunden, G., 1988, Evaluation of instrumentation used for high performance thin-layer chromatography of aflatoxins, *Chromatographia* **25:**875–880.
7. Jaenchen, D. E., and Issaq, H. J., 1988, Modern thin-layer chromatography: Advances and perspectives, *J. Liquid Chromatogr.* **11:**1941–1965.
8. Frisvad, J. C., and Thrane, U., 1987, Standardized high-performance liquid chromatography of 182 mycotoxins and other fungal metabolites based on alkylphenone retention indices and UV-VIS spectra (diode array detection), *J. Chromatogr.* **404:**195–214.
9. Pestka, J. J., 1988, Enhanced surveillance of foodborne mycotoxins by immunochemical assay, *J. Assoc. Off. Anal. Chem.* **71:**1075–1081.
10. Ueno, Y., and Ueno, I., 1978, Toxicology and Biochemistry of Mycotoxins, in: *Toxicology, Biochemistry, and Pathology of Mycotoxins* (K. Uraguchi and M. Yamazaki, eds.), Halsted Press, New York, pp. 107–188.
11. Creasia, D. A., Thurman, J. D., Wannemacher, R. W., Jr., and Bunner, D. L., 1990, Acute inhalation toxicity of T-2 mycotoxinin the rat and guinea pig, *Fundam. Appl. Toxicol.* **14:**54–59.
12. Sorenson, W. G., Frazer, D. G., Jarvis, B. B., Simpson, J., and Robinson, V. A., 1987, Trichothecene mycotoxins in aerosolized conidia of *Stachybotrys atra*, *Appl. Environ. Microbiol.* **53:** 1370–1375.

13. Wicklow, D. T., and Shotwell, O. L., 1983, Intrafungal distribution of aflatoxin among conidia and sclerotia of *Aspergillus flavus* and *Aspergillus parasiticus*, *Can. J. Microbiol.* **29:**1–5.
14. Croft, W. A., Jarvis, B. J., and Yatawara, C. S., 1986, Airborne outbreak of trichothecene toxicosis, *Atmos. Environ.* **20:**549–552.
15. Tobin, R. S., Baranowski, E., Gilman, A. P., Kuiper-Goodman, T., Miller, J. D., and Giddings, M., 1987, Significance of fungi in indoor air: Report of a working group, *Can. J. Public Health* **78:**s1–s32.
16. Abarca, M. L., Bragulat, M. R., Bruguera, M. T., and Cabanes, F. J., 1988, Comparison of some screening methods for aflatoxigenic moulds, *Mycopathologia* **104:**75–79.
17. Joffe, A. Z., 1974, Toxicity of *Fusarium poae* and *F. sporotrichioides* and its relation to alimentary toxic aleukia, in: *Mycotoxins* (I. F. H. Purchase, ed.), Elsevier, Amsterdam, pp. 229–262.
18. Lafarge-Frayssinet, C., Cecloitre, F., Mousset, S., Martin, M., and Frayssinet, C., 1981, Induction of DNA-single-strand breaks by T-2 toxin, a trichothecene metabolite of *Fusarium*: Effect on lymphoid organs and liver, *Mutat. Res.* **88:**115–123.
19. Fromentin, H., Salazar-Mejicanos, S., and Mariat, F., 1981, Experimental cryptococcosis in mice treated with diacetoxyscirpenol, a mycotoxin of *Fusarium*, *Sabouraudia* **19:**311–313.
20. Tryphonas, H., Iverson, F., So, Y., Nera, E. A., McGuire, P. F., O'Grady, L., Clayson, D. B., and Scott, P. M., 1986, Effects of deoxynivalenol (vomitoxin) on the humoral and cellular immunity of mice, *Toxicol. Lett.* **30:**137–150.
21. Corrier, D. E., and Ziprin, R. L., 1987, Immunotoxic effects of T-2 mycotoxin on cell-mediated resistance to *Listeria monocytogenes* infection, *Vet. Immunol. Immunopathol.* **14:** 11–21.
22. Pestka, J. J., Tai, J.-H., Witt, M. F., Dixon, D. E., and Forsell, J. H., 1987, Suppression of immune response in the $B6C3F_1$ mouse after dietary exposure to the *Fusarium* mycotoxins deoxynivalenol (vomitoxin) and zearalenone. *Fundam. Chem. Toxic.* **25:**297–304.
23. Rosenstein, Y., Kretschmer, R. R., and Lafarge-Frayssinet, C., 1981, Effect of *Fusarium* toxins, T-2 toxin and diacetoxyscirpenol on murine T-independent immune responses, *Immunology* **44:**555–560.
24. Tryphonas, H., O'Grady, L., Arnold, D. L., McGuire, P. F., Karpinski, K., and Vesonder, R. F., 1984, Effect of deoxynivalenol (vomitoxin) on the humoral immunity of mice, *Toxicol. Lett.* **23:**17–24.
25. Friend, S. C. E., Babiuk, L. A., and Schieffer, H. B., 1983, The effects of dietary T-2 toxin on the immunological function and herpes simplex reactivation in Swiss mice, *Toxic. Appl. Pharmacol.* **69;**234–244.
26. Lafarge-Frayssinet, C., Lespinats, G., Lafont, P., Loisillier, F., Mousset, S., Rosenstein, Y., and Frayssinet, C., 1979, Immunosuppressive effects of *Fusarium* extracts and trichothecenes: Blastogenic response of murine splenic and thymic cells to mitogens, *Proc. Soc. Exp. Biol. Med.* **160:**302–311.
27. Gerberick, G. F., Sorenson, W. G., and Lewis, D. M., 1984, The effects of T-2 toxin on alveolar macrophage function *in vitro*, *Environ. Res.* **33:**246–260.
28. Tomar, R. S., Blakly, B. R., and De Coteau, W. E., 1988, *In vitro* effects of T-2 toxin on the mitogen responsiveness and antibody-producing ability of human lymphocytes, *Toxicol. Lett.* **40:**109–117.
29. Pestka, J. J., and Forsell, J. H., 1988, Inhibition of human lymphocyte transformation by the macrocyclic trichothecenes roridin A and verrucarin A, *Toxicol. Lett.* **41:**215–222.
30. Miller, K., and Atkinson, H. A. C., 1987, The *in vitro* effects of trichothecenes on the immune system, *Arch. Toxicol.* **Suppl. 11:**321–324.
31. Tai, J. H., and Pestka, J. J., 1988, Synergistic interaction between the trichothecene T-2 toxin and *Salmonella typhimurium* lipopolysaccharide in C3H/HeN and C3H/HeJ mice, *Toxicol. Lett.* **44:**191–200.
32. Luster, M. I., Germolec, D. R., Burleson, G. R., Jameson, C. W., Ackermann, M. F., Lamm, K.

R., and Hayes, H. T., 1987, Selective immunosuppression in mice of natural killer cell activity by ochratoxin A, *Cancer Res.* **47**:2259–2263.
33. Lea, T., Steiem, K., and Stormer, F. C., 1989, Mechanism of ochratoxin A-induced immunosuppression, *Mycopatholia* **107**:153–159.
34. Creppy, E., Stormer, F., Roschenthaler, R., and Dirheimer, G., 1983, Effects of two metabolites of ochratoxin A, (4R)-4-hydroochratoxin A and ochratoxin a, on immune response in mice, *Infect. Immunity* **39**:1015–1018.
35. Sorenson, W. G., Simpson, J., and Castranova, V., 1985, Toxicity of the mycotoxin patulin for rat alveolar macrophages, *Environ. Res.* **38**:407–416.
36. Sorenson, W. G., Gerberick, G. F., Lewis, D. M., and Castranova, V., 1986, Toxicity of mycotoxins for the rat pulmonary macrophage *in vitro*, *Environ. Health Persp.* **66**:45–53.
37. Bourdiol, D., Escoula, L., and Salvayre, R., 1990, Effect of patulin on microbicidal activity of mouse peritoneal macrophages, *Food Chem. Toxicol.* **28**:29–33.
38. Escoula, L., Thomsen, M., Bourdiol, D., Pipy, R., Peuriere, S., and Roubinet, F., 1988, Patulin immunotoxicology: Effect on phagocyte activation and the cellular and humoral immune system of mice and rabbits, *Int. J. Immunopharmacology* **10**:983–9⁹9.
39. Waring, P., Eichner, R. D., and Mullbachner, A., 1988, The chemistry and biology of the immunomodulating agent gliotoxin and related epipolythiodioxopiperazines, *Med. Res. Rev.* **8**:499–524.
40. Mullbacher, A., Moreland, A. F., Waring, P., Sjaarda, A., and Eichner, R. D., 1988, Prevention of graft-versus-host disease by treatment of bone marrow with gliotoxin in fully allogeneic chimeras and their cytotoxic T cell repertoire, *Transplantation* **46**:120–125.
41. Mullbacher, A., Waring, P., Tiwari-Palni, U., and Eichner, R. D., 1986, Structural relationship of epipolythiodioxopiperazines and their immunomodulating activity, *Mol. Immunol.* **23**: 231–235.
42. Reddy, R. V., and Sharma, R. P., 1989, Effects of aflatoxin B_1 on murine lymphocytic functions, *Toxicology* **4**:31–44.
43. Kadian, S. K., Monga, D. P., and Moel, M. C., 1988, Effect of aflatoxin B_1 on the delayed type hypersensitivity and phagocytic activity of reticuloendothelial system in chickens, *Mycopathologia* **104**:33–36.
44. Park, D. L., Lee, L. S., Price, R. L., and Pohland, A. E., 1988, Review of the decontamination of aflatoxins by ammoniation: Current status and regulation, *J. Assoc. Off. Anal. Chem.* **71**: 685–703.
45. Butler, W. H., 1974, Aflatoxin, in: *Mycotoxins* (I. F. H. Purchase, ed.), Elsevier, Amsterdam, pp. 1–28.
46. Dvorackova, I., 1990, *Aflatoxins and Human Health*, CRC Press, Boca Raton, FL.
47. Dvorackova, I., Aflatoxin inhalation and alveolar cell carcinoma, *Br. Med. J.* **1**:691.
48. Deger, G. E., 1976, Aflatoxin—human colon carcinogenesis? *Ann. Intern. Med.* **85**:204.
49. Hayes, R. B., Van Nieuwenhuize, J. P., Raatgever, J. W., and Kate, F. J. W., 1984, Aflatoxin exposures in the industrial setting: An epidemiological study of mortality, *Food Chem. Toxicol.* **22**:39–43.
50. Reye, R. D. K., Morgan, G., and Baral, J., 1963, Encephalopathy and fatty degeneration of the viscera: A disease entity in childhood, *Lancet* **2**:749.
51. Wood, G. E., 1989, Aflatoxins in domestic and imported foods and feeds, *J. Assoc. Off. Anal. Chem.* **72**:543–548.
52. Olsen, J. H., Dragsted, L., and Autrup, H., 1988, Cancer risk and occupational exposure to aflatoxins in Denmark, *Br. J. Cancer* **58**:392–396.
53. Yeh, F. S., Yu, M. C., Mo, C. C., Luo, S., Tong, M. J., and Henderson, B. E., 1989, Hepatitis B virus, aflatoxins, and hepatocellular carcinoma in southern Guangxi, China, *Cancer Res.* **49**: 2506–2509.
54. Skilleter, D. N., Mattocks, A. R., and Neal, G. E., 1988, Selectivity of different phases of the cell cycle to selected hepatotoxins in cultured liver-derived (B19L) cells, *Xenobiotica* **18**:699–705.

55. Van Rensburg, S. J., and Altenkirk, B., 1974, *Claviceps purpurea*—ergotism, in: *Mycotoxins* (I. F. H. Purchase, ed.), Elsevier, Amsterdam.
56. Hoffer, A., and H. Osmond, 1967, *The Hallucinagens*, Academic Press, New York.
57. Harwig, J., 1974, Ochratoxin A and related metabolites, in: *Mycotoxins* (I. F. H. Purchase, ed.), Elsevier, New York, pp. 345–367.
58. Krogh, P., 1974, Mycotoxic nephropathy, in: *Mycotoxins* (I. F. H. Purchase, ed.), Elsevier, Amsterdam, pp. 419–428.
59. Krogh, P., 1977, Ochratoxins, in: *Mycotoxins in Human and Animal Health* (J. V. Rodricks, C. W. Hesseltine, and M. A. Mehlman, eds.) Pathotox, Park Forest South, IL, pp. 489–498.
60. Nelson, P. E., Toussoun, T. A., and Marasas, W. F. O., 1983, *Fusarium Species. An Illustrated Manual for Identification*, The Pennsylvania State University Press, University Park, PA.
61. Bamburg, J. R., and Strong, F. M., 1972, 12,13-Epoxytrichothecenes, in: *Microbial Toxins*, Vol. 7 (A. Ciegler, S. Kadis, and S. J. Ajl, eds.), Academic Press, New York, pp. 207–292.
62. Gerberick, G. F., Sorenson, W. G., and Lewis, D. M., 1984, The effects of T-2 toxin on alveolar macrophage function *in vitro*, *Environ. Res.* **33**:246–260.
63. Vidala, D., and Mavet, S., 1989, In vitro and in vivo toxicity of T-2 toxin, a *Fusarium* mycotoxin to mouse peritoneal macrophages, *Infect. Immun.* **57**:2260–2264.
64. Ueno, Y., 1977, Trichothecenes: Overview address, in: *Mycotoxins in Human and Animal Health* (J. V. Rodricks, C. W. Hesseltine, and M. A. Mehlman, eds.), Pathotox, Park Forest South, IL, pp. 189–207.
65. McLaughlin, C. S., Vaughan, M. H., Campbell, I. M., Wei, C. M., Stafford, M. E., and Hansen, B. S., 1977, Inhibition of protein synthesis by trichothecenes, in: *Mycotoxins in Human and Animal Health* (J. V. Rodericks, C. W. Hesseltine, and M. A. Mehlman, eds.), Pathotox, Park Forest South, IL, pp 263–273.
66. Middlebrooke, J. L., and Leatherman, D. L., 1989, Binding of T-2 toxin to eukaryotic cell ribosomes, *Biochem. Pharmacol.* **38**:3103–3110.
67. Eppley, R. M., and Bailey, W. J., 1973, 12,13-Epoxy-delta 9-trichothecenes as the probably mycotoxins responsible for stachybotryotoxicosis, *Science* **181**:758–760.
68. Forgacs, J., 1971, Stachybotryotoxicosis, in: *Microbial Toxins*, Vol. 8 (A. Ciegler, S. Kadis, and S. J. Ajl, eds.), Academic Press, New York, pp. 95–128.
69. Rodricks, J. V., and Eppley, R. M., 1974, *Stachybotrys* and stachybotryotoxicosis, in: *Mycotoxins* (I. F. H. Purchase, ed.), Elsevier, Amsterdam, pp. 181–197.
70. Ueno, Y., 1974, Citreoviridin from *Penicillium citreo-viride* Biourge, in: *Mycotoxins* (I. F. H. Purchase, ed.), Elsevier, Amsterdam, pp. 283–302.
71. Wicklow, D. T., Stubblefield, R. D., Horn, B. W., and Shotwell, O. L., 1988, Citreoviridin levels in *Eupenicillium ochrosalmoneum*-infested maize kernels at harvest, *Appl. Environ. Microbiol.* **54**:1096–1098.
72. Nishie, K., Cole, R. J., and Dorner, J. W., 1988, Toxicity of citreoviridin, *Res. Commun. Chem. Pathol. Pharmacol.* **59**:31–52.
73. Sayood, S. F., Suh, H., Wilcox, C. S., and Schuster, S. M., 1989, Effect of citreoviridin and isocitreoviridin on beef heart mitochondrial ATPase, *Arch. Biochem. Biophys.* **270**;714–721.
74. Enomoto, M., and Ueno, I., 1974, *Penicillium islandicum* (toxic yellowed rice)-luteoskyrin-islanditoxin-cyclochlorotine, in: *Mycotoxins* (I. F. H. Purchase, ed.), Elsevier, Amsterdam, pp. 303–326.
75. Enomoto, M., 1985, How does Japan differentiate hyperplasia from neoplasia? *Toxicol. Pathol.* **13**:110–118.
76. Richard, J. L., Lyon, R. L., Fichtner, R. E., and Ross, P. F., 1989, Use of thin layer chromatography for detection and high performance liquid chromatography for quantitating gliotoxin from rice cultures of *Aspergillus fumigatus* Fresenius, *Mycopathologia* **107**: 145–151.
77. Eichner, R. D., Al Salami, M., Wood, P. R., and Mullbacher, A., 1986, The effect of gliotoxin upon macrophage function, *Int. J. Immunopharmacol.* **8**:789–797.

78. Eichner, R. D., Waring, P., Geue, A. M., Braithwaite, A. W., and Mullbacher, A., 1988, Gliotoxin causes oxidative damage to plasmid and cellular DNA, *J. Biol. Chem.* **263**:3772–3777.
79. Braithwaite, A. W., Eichner, R. D., Waring, P., and Mullbacher, A., 1987, The immunomodulating agent gliotoxin causes genomic DNA fragmentation, *Mol. Immunol.* **24**:47–55.
80. Bauer, J., Gareis, M., Bott, A., and Gedek, B., 1989, Isolation of a mycotoxin (gliotoxin) from a bovine udder infected with *Aspergillus fumigatus*, *J. Med. Vet. Mycol.* **27**:45–50.
81. Atherton, L. G., Brewer, D., and Taylor, A., 1974, *Pithomyces chartarum*, a fungal parameter in the aetiology of some disease of domestic animals, in: *Mycotoxins* (I. F. H. Purchase, ed.), Elsevier, New York, pp. 29–68.
82. Ciegler, A., Detroy, R. W., and Lillehoj, E. B., 1971, Patulin, penicillic acid, and other carcinogenic lactones, in: *Microbial Toxins*, Vol. 4 (A. Ciegler, S. Kadis, and S. J. Ajl, eds.), Academic Press, New York, pp. 409–434.
83. Ciegler, A., Beckwith, A. C., and Jackson, L. K., 1976, Teratogenicity of patulin and patulin adducts formed with cysteine, *Appl. Environ. Microbiol.* **31**:664–667.
84. Kurata, H., 1978, Current scope of mycotoxin research from the viewpoint of food mycology, in: *Toxicology, Biochemistry, and Pathology of Mycotoxins* (K. Uraguchi and M. Yamazaki, eds.), John Wiley, New York, p. 26.
85. Riley, R. T., Hinton, D. M., Showker, J. L., Rigsby, W. and Norred, W. P., 1990, Chronology of patulin-induced alterations in membrane function of cultured renal cells, *Toxicol. Appl. Pharmacol.* **102**:128–141.

22

Mushroom Poisoning

R. SCOTT PORE

1. INTRODUCTION

Mushroom poisoning is a subcategory of poisoning in general, and food poisoning in particular. The poisoned patient is initially evaluated and treated by a physician as an emergency case. Integral to choosing an appropriate treatment is an assessment by the physician of the type and potential severity of the poisoning. There is no simple toxin–antidote therapy for mushroom poisoning, rather, the toxicology is complex and a crucial bit of needed information is the correct identification of the mushroom. A mycologist can make this identification and can contribute other useful information as well.

1.1. Definition

Mushroom poisoning is also known by the term *mycetismus*, namely, ingestion of the sporocarp of the large fleshy poisonous Basidiomycetes or Ascomycetes under the mistaken impression that they are edible and nonpoisonous. Poisonous mushrooms are commonly called *toadstools* (suggested etymology: Gr., Todesstuhl = death's stool), but there are no distinguishing characteristics by which to separate toadstools from nonpoisonous mushrooms.

R. SCOTT PORE • Department of Microbiology and Immunology, West Virginia University School of Medicine, Morgantown, West Virginia 26506-9177.

Fungal Infections and Immune Responses, edited by Juneann W. Murphy *et al.* Plenum Press, New York, 1993.

1.2. Etiology

Between 1% and 2% of all species of mushrooms are estimated to contain toxins (in the United States, there are > 50 poisonous species). There are no proven reasons for the presence of toxins, but an analogy to the presence of toxins in some plants can be made. Relative to plant toxins, mushroom toxins are notorious among lay persons for their potency, but this is not supported by fact. Specific toxins are associated with specific species of fungi, while closely related species may be completely free of toxins, or even have a chemically different toxin. Some poisonous mushrooms contain two or more types of toxin. In some poorly documented instances, a single species of poisonous mushroom may have a quantitatively variable amount of toxin based upon ecological factors such as host–mycorrhizal association.

1.3. Incidence

Mushroom poisoning is considered a rare event; nevertheless, the American Association of Poison Control Centers (AAPCC) reported 7836 cases of mushroom poisoning in 1988.[1] Of these cases, which drew figures from an estimated 63% of the U.S. population, 924 patients experienced minor symptoms, 194 patients were moderately poisoned, and 7 had major or serious outcomes. The serious cases are often of amanitin-type poisoning. It is estimated that there are proportionally more mushroom poisonings in Europe and Asia, where mushrooms comprise a staple part of the diet. There are no statistics on mushroom poisoning fatalities worldwide, but estimates of >100/yr have been made. A cluster of four fatalities occurred in California in 1988.[2]

1.4. Epidemiology

The largest category of mushroom poisoning cases is not poisonings at all, but false alarms. Usually these are toddlers, age 0.5–3 years, who put everything in their mouth, hence, when turned loose to play in the lawn, they "graze" on the mushrooms. The distraught parents rush the child to the emergency room and the physician, and subsequently the mycologist, is confronted with a potential poisoning as if it were an actual poisoning, which it sometimes is. The identification of the suspected poisonous mushroom is no less difficult to make in a false alarm as in an actual poisoning.

The remaining categories are actual poisonings. Adult mushroomers, those who gather wild mushrooms for their meals, may inadvertently include poisonous species. Would-be experts are sometimes victims. Another category is the experimentalist, primarily individuals seeking hallucinogenic experiences and indiscriminately eating wild mushrooms. Mushroom poisoning is episodic. In temperate North America and Europe, mushroom poisoning is more common in the spring, summer, or fall, but in subtropical California and Asia, winter poisonings may be common. Home-canned poisonous mushrooms have no such seasonal restrictions. Additional epidemiological factors are geographic, latitude, elevation, rainfall, vegetation, and rural or urban settings. The cultural heritage of the victim is another point to be considered in the epidemiology of mushroom poisoning.

2. MUSHROOM IDENTIFICATION

2.1. Role of the Mycologist

The identification of mushrooms for the purpose of differentiating edible from toxic species is best left to the mushroom expert. There are a few professional mycologists who identify mushrooms, and there are a few amateur mycologists who, because of extensive practical experience, are also good at it. Practical experience obtained by frequent field identification of mushrooms is the current best means to develop the necessary skills.

Due to their affiliations with large medical centers, board certification, and unique training, medical mycologists could play an important role in the diagnosis and management of mushroom poisoning cases. Yet few have proficiency with the problems of mushroom poisoning. The training and experience in microscopic and chromatographic techniques gained with mycoses can be applied to the identification of mushrooms and toxins. The medical mycologist also may appreciate the seemingly arcane facts about mushroom morphology, taxonomy, physiology, genetics, and ecology, and can integrate this information with the symptomatologic findings familiar to the physician. As with the mycoses, mushroom poisonings are also replete with missing data and idiosyncratic findings.

2.2. Identification

The best way to learn to identify mushrooms is the way most "experts" do it. The first step is the purchase of a paperback field guide commonly available in most book stores. *The Peterson Field Guide*,[3] *The Audubon Society Field Guide*,[4] or *Mushrooms of North America*[5] are all excellent for this purpose. Although most people rely on the color pictures for identification, the medical mycologist will appreciate the information on size and shape of the basidiospores and other microscopic features, not to mention techniques for the chemical detection of starch or toxins. A field trip should be planned so that two or three locally available poisonous and nonpoisonous mushrooms can be collected for identification. For introductory purposes a simplified field key to some poisonous and nonpoisonous genera may be useful (Table I). Structural features useful for identifying mushrooms are illustrated in Fig. 1.

TABLE I
Key to Some Common Genera Containing Poisonous Species[a]

Sporocarp bearing lamellae, Hymenomycete	A
Sporocarp lacking lamellae	B
A. Lamellae and spores white (or brightly colored)	1
1 stipe off center, on wood	*Pleurotus*
1 stipe central, soil or base of tree	2
2 milky latex from cuts, stipe stout	*Lactarius*[a]
2 no milk, red pileus, stipe stout and brittle	*Russula*[a]
2 lamellae free from stipe	3
2 attached or decurrent lamellae	4
3 volva present, may have annulus	*Amanita*[a]
3 no volva, annulus, stipe delicate, pileus separable	*Lepiota*[a]
3 no annulus, stipe stout, pileus not separable	*Tricholoma*[a]
4 in bunches at base of tree, annulus	*Armillaria*
4 no annulus, stipe stout	*Clytocybe*[a]
AA. Lamellae and spores brown to black when mature	1
1 small brown pileus, fragile, hollow stipe, on wood	*Galerina*[a]
1 cobweb weil when young, rusty colored spores, soil	*Cortinarius*[a]
1 yellow-brown spores, fibrous pileus	*Inocybe*[a]
1 spores and lamellae turning darker with maturity	2
2 lamellae white → pink → cocoa-brown, annulus	*Agaricus*
2 lamellae white → inky black autolysis	*Coprinus*[a]
B. Soft fleshy fruiting structures	1
1 smooth cap, pores on underside, stipe central, soil	*Boletus*[a]
1 cap wrinkled	2
1 no cap or conspicuous stipe	3
2 funnel-shaped orange pileus with ribs on underside	*Cantharellus*
2 apothecium wrinkled, brown	*Gyromitra*[a]
3 puffball with opening on top, on rotten wood	*Scleroderma*[a]
3 bright red apothecium	*Peziza*
BB. Woody, pores, stipe lateral, on trees	1
1 bright yellow-orange, soft edge when young	*Polyporus*

This is an introductory key and is incomplete. Do not use this key to determine edibility.
[a]Genera containing poisonous species.

Terminology follows the dictionary by Ainsworth,[6] and additional useful illustrations of macroscopic and microscopic features are presented in the back of the dictionary. Once identified, the species should be confirmed by consulting with experienced local mushroomers. Better yet is affiliation with the local North American Mycological Association (NAMA) chapter and participation in forays attended by experts, or perhaps even attendance at a regional or national convention.

A scientific basis for mushroom identification requires a second step. Advanced species information about basidial and cystidial details, and generic monographs may be beyond the scope of the general medical mycologist, but some books specifically on poisonous species may present laboratory diagnostic methods familiar to the medical mycologist.[7,8]

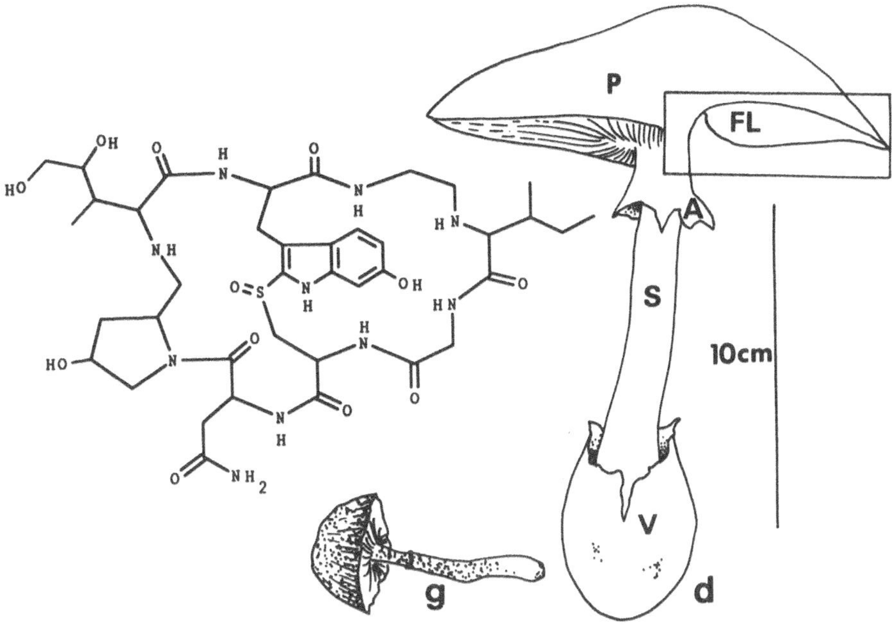

FIGURE 1. α-amanitin. d. *Amanita phalloides*, P = pileus, FL = free lamellae, A = annulus, S = stipe, V = volva. g. *Galerina autumnalis*.

3. DIAGNOSIS AND TREATMENT

The clinical classifications of mushroom poisoning adopted in several recent authoritative emergency medicine tests,[9–11] follows Mitchel.[12] It is based upon eight different pathophysiological types of poisoning that are caused by a similar number of toxins. The remaining format of this chapter follows such a mushroom poisoning classification scheme (Table II). This sort of classification is useful to the physician as a basis for emergency treatment. Two reasons for maintaining this priority are that, first, a mushroom may not be available for identification and, second, there is usually no one available to identify the mushroom.

Nevertheless, a mycologist should subsequently be consulted with 24 hours in all cases. If a presumed specimen of the ingested mushroom is available, further therapeutic considerations can be predicated upon an accurate identification, especially in the case of potentially lethal species. On the other hand, even in the absence of a specimen, there are many clues that a mycologist could use to contribute to the case evaluation; for example, eye witness descriptions of mushrooms, sites of picking, seasons of year, quantities consumed, and symptoms are all critical to proper assessment. Regarding this last point, most physicians are unfamiliar with the above-mentioned

TABLE II
Categories of Mushroom Poisoning

Type	Toxin
Rapid onset of symptoms (0.25–3 hr)	
Parasympathomimetic	Muscarine
Antabuselike	Coprine
Anticholinergiclike	Muscimol
Hallucinogenic	Psilocybin
Gastic only	Unknown gastrotoxins
Delayed onset of symptoms (6–24 hr)	
Hepatocellular necrosis	Amanitin
Hepatocellular necrosis	Gyromitrin
Renal necrosis	Orellanine

mushroom poisoning classification scheme because of the infrequency of cases and they may need to be reminded of the salient features. There is so much misinformation about mushroom poisoning that a mycologist is necessary to separate fact from fiction (Table III). One source of information available to both the mycologist and physician is Poisindex (Micromedex, Inc., 600 Grant St., Denver, CO 80204). Located in libraries and emergency rooms of many large medical centers and universities, Poisindex is a microfiche system of the most recent mushroom poisoning information. Color photographs of poison mushrooms may facilitate identification.

Amanitin is the archetypical and most notorious mushroom poison. Because it causes a late onset of poisoning symptoms, which may preclude early emergency emesis, it is important to differentiate it from the common early-onset poisonings. Early-onset poisonings often lack a specific antidote, so a general scheme for emergency treatment is useful for all types of mushroom poisonings. Following is a section on general measures to institute for all suspected mushroom poisonings.

3.1. Abbreviated Guideline for Emergency Care

Mushroom poisoning causes nonspecific gastrointestinal and neurologic symptoms. In cases with a history of mushroom ingestion, the time of onset of symptoms is important for prognosis and treatment.

3.1.1. Two Hours or Less Since Ingestion of Mushrooms

Induce emesis with ipecac (15–30 cc) and/or enhance elimination of the toxins with a cathartic such as sorbitol-charcoal. Early symptoms of poison

TABLE III
Facts About Mushrooms Poisoning

1. Frequent in toddlers, adolescents taking drugs, mushroomers, and adults of recent Asian or European extraction.
2. Cooking rarely destroys toxins; silver spoons do not turn black in the presence of poisonous mushrooms; garlic and other nostrums are useless.
3. Improperly canned mushrooms, like beans and other low acidity vegetables, may contain botulinum toxin, which is not a mushroom toxin.
4. Mushrooms poisoning symptoms are not specific. In addition to the common gastrointestinal symptoms, consider cholinergic and anticholinergic activity, hallucinations, ataxia, altered mental status, coma, hepatic or renal failure, and seizures.
5. Muscarine poisoning characterized by severe cholinergic symptoms is only one type of poisoning—*and the only one* calling for atropine. Otherwise, atropine may potentiate poisoning.
6. Symptoms can occur as late as 24–36 hours after ingestion (or even several days for orellanine).
7. Certain mushrooms sensitize to alcohol, and disulfiramlike symptoms appear when alcohol is consumed, even 5 days after eating mushrooms.
8. Fear of mushroom poisoning may produce poisoning symptoms.
9. In order to obtain entire mushrooms for ID, one must dig them up so the subterranean parts are preserved.
10. Gastric contents may provide evidence necessary for a provisional identification of poisonous mushrooms. This also will require the assistance of mycologist.
11. There is no efficacious antidote for α-amanitin. Experimental drugs such as penicillin may be tried.
12. Poisonous *Amanita* spp. have white gills, an annulus, and veil, and frequently a bulbous base surrounded by a sac or volva.
13. Some mushrooms have two or more toxins, e.g., hallucinogens, anticholinergics, and gastrotoxins are all found in *Amanita muscaria.*
14. Even the so called non deadly mushrooms may cause death in small children or adults if large amounts are ingested, the patient has cardiac or pulmonary disease, or has taken other drugs.
15. Worldwide, hundreds of people die and thousands are poisoned by mushrooms each year.
16. *Amanita phalloides*, the Death Cap, is not common in the United States, but the *A. virosa* Destroying Angel group that are abundant frequently encountered mushrooms on lawns and in the forest from spring through fall, also contain α-amanitin.
17. The edible *Agaricus campestris* has pink or chocolate-brown gills, but some poisonous mushrooms, such as *Galerina* spp., also have brown gills.
18. Poisonous mushrooms have been reported to taste very good. Taste bears no relevance to toxicity.
19. There is no test to differentiate poisonous from non poisonous mushrooms ("toadstools" from mushrooms).
20. A single mushroom, which could easily contain 7 mg of α-amanitin, is enough to kill an adult, but children under 10 may succumb to lesser amounts.
21. The United States is experiencing an epidemic of illegal experimental mushroom use. The poisonous, brown-gilled LBM (little brown mushrooms) *Galerina* or *Conocybe* have been confused with the hallucinogenic *Psilocybe.*
22. Poisonous mushrooms are not toxic to touch; in other words, there is no analogy to poison ivy.
23. Some mushrooms taste bitter, peppery, or cause a burning sensation in the mouth or upon contact with the eye. These sensations that commence immediately or within minutes are otherwise harmless and bear no relation to poisoning.
24. There are more than 50 species of poisonous mushrooms in the United States.

ing include abdominal cramping, vomiting, diarrhea, neuromuscular and neurological disturbances, or the patient may have lost consciousness. Use neurologic drugs or sedate as needed, but use atropine only for muscarine symptoms or severe hyperperistalsis. Most mushroom poisoning does not require atropine. Contact a mycologist. Maintain supportive care and expect recovery in 4–48 hours.

3.1.2. Over Three Hours (Most Frequently 6–12 Hours) Have Passed Since Mushrooms Were Ingested

If symptoms are of poisoning as listed above, then suspect a deadly species of mushroom such as the Death Cap, *Amanita phalloides*. The toxin is α-amanitin, a heat-stable, cyclic octapeptide for which there is no specific antidote. The poison is not excreted and despite a 1- to 4-day remission of the poisoning symptoms, the α-amanitin-caused hepatocellular necrosis progresses, and in 50% of the cases hepatic encephalopathy and death follow. Less than one mushroom may be fatal for a child. Therefore, obtain a specimen of the mushroom and contact a mycologist for positive identification. There is also a test (Meixner test, next section) for α-amanitin that can be done with a piece of the mushroom.

The initial stage of late-onset (or severe early-onset) mushroom poisoning can be treated with emetics, cathartics, gastric lavage, and activated charcoal (oral or by gastric tube) to eliminate poison. Vigorous prompt hydration and forced diuresis is of value, as some toxin may be cleared in the urine. Monitor volume status and urine output. Obtain baseline chemistries, blood counts, urinalysis, and liver function tests.

In anticipation of the second phase or renewed onset (24 hours to 5 days after ingestion), continue to monitor liver, renal, and central nervous systems and manage by standard methods. Rising bilirubin, prothrombin time, AST, and ALT values are early indications of grave complications. In view of the severity of the complications, experimental therapies should be tried. Some success has been recorded for penicillin G Na, thioctic acid, silibinin, cimetidine, corticosteroids, or charcoal hemoperfusion. Because these treatments are only experimental, toxicologists at poison control centers should be consulted.

Amanita spp. can be recognized as having white gills and a bulbus stipe arising from a volva. *Galerina* spp., which also contain α-amanitin, are small brown mushrooms with brown gills. *Gyromitra* spp., which contain gyromitrin, have no gills. Approximately 50 other poisonous mushrooms have a combination of usually nonlethal poisons, producing a variety of cardiovascular, neuromuscular, and gastrointestinal symptoms (onset less than 2 hours). Irrespective of the early course of the poisoning, the physician should refer to a poison control center and consult appropriate references in order to identify the mushroom involved and to institute appropriate treatment. The contents of this guideline have been excerpted and simplified to facilitate early recognition and treatment. There is no substitute for a detailed follow-up.

3.2. Meixner Test[13]

Several species of *Amanita* and *Galerina* contain α-amanitin, and the following simple chemical test may be used to detect it.

1. Choose a piece of cheap pulp newspaper that has a high lignin content. Lignin is essential for the test.
2. Squash a small piece of mushroom onto a piece of newspaper so that some juice soaks into the paper, peel off the mushroom and discard, mark the spot with a pencil, and then air-dry the paper for several minutes; do not overheat.
3. Add a drop of concentrated HCL to the spot where the juice was and in 1–10 minutes observe a blue color positive reaction for the presence of amanitin. No color or yellow is negative. Store known positive and negative mushroom specimens in the freezer for use as controls.
4. This test is about 90% accurate, with some false positives or negatives. A positive reaction suggests α-amanitin is in the mushroom, while a negative reaction from an otherwise suitable mushroom should be interpreted cautiously. Mushroom identification should supersede the Meixner test if doubt exits. This test does not detect other toxins, so it should not be relied upon to determine edibility.

3.3. Reporting Mushroom Poisonings

In North America there are several methods for case reporting. The American Association of Poison Control Centers (AAPCC) includes mushroom poisoning incidence and fatalities in its data base. Results are published every year.[1] Regional poison control centers should be contacted in all proven or suspected poisoning cases. The North American Mycological Association (NAMA) maintains a registry of mushroom poisoning cases. This requires filling out the NAMA Mushroom Poisoning Report Form. Data include patient data and mushroom identification data. Although fewer cases are included than for AAPCC, the data are more complete and useful for subsequent epidemiological analysis. For NAMA report forms, contact John H. Trestrail, III, R.Ph. (Chr. NAMA Toxicology Committee), Blodgett Regional Poison Center, 1840 Wealthy St. S.E., Grand Rapids, MI 49506. The Rocky Mountain Poison and Drug Center, 645 Bannock St., Denver, CO 80204, (800) 525-9083, is another information center that also maintains a registry of cases.

4. AMANITIN

Most fatal cases of mushroom poisoning have been attributed to *Amanita* spp. containing a family of cyclopeptide poisons, the most notorious of which is α-amanitin. All medical mycologists should learn to identify the *A. phalloides/virosa* mushrooms, even if they are otherwise uninterested in mushrooms.

4.1. Chemistry

The amanitins are a group of 15 cyclopeptides containing either seven or eight amino acids. All may be toxic, but α-amanitin is considered the primary toxin, and it has attracted the most attention by investigators. α-Amanitin ($C_{39}H_{54}N_{10}O_{13}S$) is a cysteine-containing bicyclic octapeptide that is heat stable (not destroyed by cooking) and is not digested appreciably in the human gut. It is adsorbed from the gut and manifests itself as an extremely potent hepatocellular toxin. α-Amanitin specifically inhibits eukaryotic nuclear RNA polymerase II, blocking DNA transcription, resulting in irreversible liver necrosis. The toxin is poorly detoxified or excreted by the human and participates in the enterohepatic recirculation as well as renal reabsorption. A single mushroom may contain 7 mg of α-amanitin (0.1 mg/kg is a potentially lethal dose). Amanitin is amenable to TLC,[14] RIA,[15] and HPLC[16] analysis from mushrooms or human tissue. A simple emergency room test for the toxin is the Meixner test[13,14,17,18] described earlier in this chapter. One of the cyclopeptides, phalloidin ($C_{35}H_{48}N_8O_{11}S$), is a heptapeptide toxin that has a potent affinity for cellular actin, but its capacity for human toxicity by the oral route is unconfirmed.

4.2. Mycology

Amanita phalloides	Death Cap
A. virosa	Destroying Angel
A. verna, A. biosporigera, A. ocreata, A. tenuifolia	
Galernia autumnalis	Autumn Skullcap
G. venerata	Deadly Galerina
G. marginata, G. sulciceps	
Lepiota josserandii and others	Deadly Lepiota
Conocybe filaris	Deadly Conocybe

A. phalloides (Fig. 1) which is thought to cause 90% of all mushroom poisoning fatalities, along with the *Amanita virosa* complex (*A. verna, A. bisporigera, A. ocreata*, and *A. tenuifolia*) and *Galerina* spp., contain significant quantities of α-amanitin. *Lepiota* spp. and *Conocybe* spp. may contain α-amanitin, but documentation and poisonings are few. The amanitin-containing species are worldwide in distribution, primarily in temperate regions. *Ama-*

nita spp. are found in soil in association with trees in urban areas and forests, while *Galerina* spp. grow on dead trees. All species fruit in the summer and early fall but, depending on geographic location, spring and winter fruiting is possible. *A. phalloides* was formerly thought to be a European species, but recent identification from the United States, not to mention poisonings, has led to a reassessment of its distribution.[19] It may have been introduced into the United States, or it may have been misclassified due to varietal differences—a common trait in all mushrooms.

The deadly *Amanita* spp. have white lamellae, white basidiospores, an annulus, saccate volva around a bulbous base, and a universal veil in the immature stage. The pileus of *A. phalloides* is pale olive-green, but those of the *A. virosa* complex are white. The unopened basidiocarp stage of *Amanita* spp. may be confused with the same stage of the common edible field mushroom, *Agaricus campestris*, but *A. campestris* lacks a volva and the lamellae turn pink and then brown with age. *A. phalloides* Fr.: pileus 6–16 cm, convex, smooth, viscid, greenish, lamellae white, free; stipe 7–14 cm, basal bulb large, white, annulus large, volva saccate; spores 6–10 μm spheroidal, smooth, white, amyloid. *A. virosa* (Fr.) Bert.: pileus 4–10 cm, convex, smooth, white (stains yellow with KOH); stipe 5–15 cm, slender, fibrose, annulus, saccate volva; spores 8–10 μm, spheroidal, smooth, white, amyloid.

Galerina spp. are small, brown mushrooms that usually grow on fallen trees but also may grow in soil. All structures are pale to dark brown or a variation of brown, such as orange-brown. *G. autumnalis* (Pk.) A. H. S. Sing.: pileus 3–6 cm, convex, brown, viscid; stipe 2–7 cm; spores 8–11 × 5–6 μm, ellipsoidal, rough, brown. The lamellae are attached or decurrent. There is an inapparent fibrillose annulus.

Lepiota spp. resemble *Amanita* spp. in many respects, but they differ by lacking a volva. They also may have tan scales on the white pileus. *Conocybe* spp. are small, delicate mushrooms growing in soil, with a conical cap and brown spores. The presence of α-amanitin in both *Lepiota* spp. and *Conocybe* spp. remains in question and harmless species are common.

4.3. Poisoning

Unlike most types of poisoning, the onset of α-amanitin symptoms characteristically occur 10–15 hours after ingestion of the mushrooms. The time of onset may vary from 6 to 24 hours, but any time after 3 hours is considered delayed onset and must be considered a potentially lethal α-amanitin case. Symptoms are initially gastrointestinal: mild or severe abdominal pain, nausea, emesis, and watery or bloody diarrhea with a duration of 12–24 hours. Then there is a seeming 12- to 72-hour remission, which belies the severity of the poisoning. Patients may be mistakenly discharged during this interval. Some patients may not seek treatment during the initial attack of poisoning symptoms, but do so for the more severe renewed onset, after the seeming remission. By this time, emesis, lavage, and catharsis are less effective or useless. During the initial 48 hours or longer, the α-amanitin is causing irreversible hepatocellular necrosis, complicated by enterohepatic recirculation and resorption of α-amanitin from the damaged renal tubules, which is finally manifest by a second onset of severe gastrointestinal symptoms plus jaundice, convulsions, cyanosis, muscular twitching, and hypoglycemia. The symptoms are dose related and may progress to hepatorenal failure, encephalopathy, hypovolemia, coma, and death, or slow recovery over a period of weeks or months. Kidney failure is common, but whether kidney toxicity is direct or secondary to hepatic dysfunction is unresolved. Cardiovascular and CNS effects are usually ascribed to hepatic dysfunction, but damage to these organs/systems must be included in diagnostic and treatment plans. Case reports include adults and children requiring liver transplants,[20,21] adult fatalities,[2] adults who survived amanitin poisoning,[22,23] and a series of 46 cases.[24] Amanitin poisoning was recently reviewed by Piqueras.[25]

Initial diagnosis is based upon a history of mushroom ingestion and delayed onset of gastrointestinal symptoms. Treatment follows the general principle of reducing absorption, facilitating excretion, and maintaining fluid and electrolyte balance. These methods include use of emetics, cathartics, gastric lavage, gastric charcoal lavage, and diuretics. At this point special efforts should be taken to establish the identification of the mushrooms ingested. A person should be designated to return to the original site and collect more mushrooms. If mushrooms are available, they should be refrigerated and sent to a mycologist or mushroom expert. Gastric lavage material, and blood and urine samples, should be saved for laboratory analysis for α-amanitin[16] or spores,[14] as appropriate. Upon positive (*Amanita* or *Galerina*) mushroom identification or renewed symptoms in 2–5 days, liver, renal, and blood values should be assessed.[23] Amanitin causes increased AST, ALT, BUN, amylase, creatine kinase, creatinine, and prothrombin time. Hypoglycemia and electrolyte imbalance are common. Tests should be repeated for 3 consecutive days.

If α-amanitin poisoning is confirmed, heroic efforts must be made because of the high mortality rate and the lack of an antidote. Within 24 hours of ingestion but no later than 48 hours, renewed efforts to prevent absorption and to enhance excretion may be tried. Such treatments have included forced diuresis, renal dialysis, charcoal hemoperfusion,[23] hemodialysis, blood replacement, plasma exchange, and biliary fistula. Experimental drugs should be tried. Two theories for high-dose penicillin G Na 250 mg/kg/day IV are that it either binds and inactivates α-amanitin[24] or lowers absorption of intestinal GABA.[27,28] High-dose corticosteroid, e.g., dexamethasone, may ameliorate liver toxicity. Thioctic acid (α-lipoic acid) is the most commonly used therapy,[23] but its efficacy is questionable.[26,29] A specific monoclonal antibody against α-amanitin has perversely enhanced toxicity.[30]

Silibinin is used experimentally.[21,26] Orthotopic liver transplantation is the only remedy for irreversible liver failure.[20,21] Klein *et al.*[20] review the treatment options for amanitin poisoning in detail.

5. COPRINE

Coprine itself is not toxic, it is only the incidental ingestion of it with ethanol that makes it a mushroom poison.

5.1. Chemistry

Coprine toxicity may result from altered neurotransmitter concentrations, as Ellenhorn[9] suggests, but the rationale for a disulfiramlike toxicity is also possible. Tetraethylthiuram disulfide (TETD), the disulfiram called *antabuse*, causes the incomplete metabolism of ethanol, resulting in acetaldehyde toxicity, but TETD has not been identified from mushrooms. Ellenhorn speculates that N^5-(1-hydroxycyclopropyl) glutamine found in the incriminated mushroom may be coprine, but it does not block acetaldehyde dehydrogenase, it interferes with dopamine-β-hydroxylase in catecholamine synthesis. In any regard, coprine does have a clinical disulfiramlike effect on the autonomic nervous system in conjunction with the ingestion of ethanol.

5.2. Mycology

Coprinus atramentarius	Inky Cap
Clitocybe clavipes	Clubfoot Mushroom

Coprinus atramentarius (Fig. 2) is caespitose in proximity to trees, often in grassy lawns. Upon opening, both the gray lamellae and the pileus autodeliquesce into an inky black slime. *C. atramentarius* Fr.: pileus 3–7 cm convex, flattened sides, striate, tan or gray; stipe 5–10 cm; spores ellipsoidal, 4–6 × 7–10 μm, smooth, pore, black.

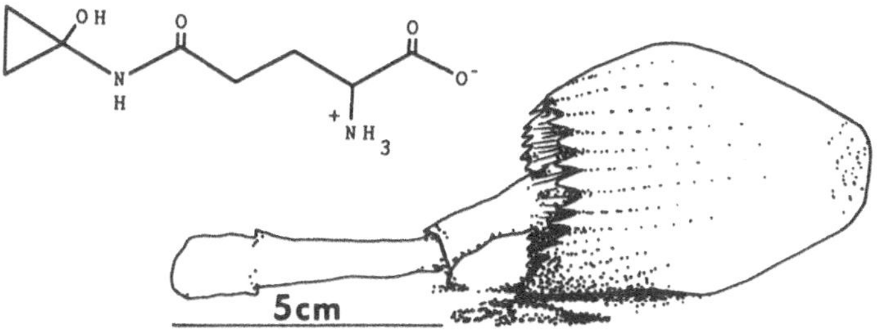

FIGURE 2. Coprine. *Coprinus atramentarius.*

5.3. Poisoning

A 15–30 minute time of onset of symptoms after ingestion of ethanol may occur for up to 5 days after ingestion of toxin mushrooms. Alternatively, an onset of 2 hours after ingestion of mushrooms in the presence of ethanol in the patient's system also will provoke the symptoms. The patient undergoes tachycardia, orthostatic hypotension, paresthesias, headache, flushing of the skin of head and neck, metallic taste, and vomiting. Spontaneous recovery usually occurs within 2–4 hours, but the patient should be treated for hypotension (fluids and noradrenaline) and arrhythmias (propranolol) in the case of underlying cardiovascular disease.[9] Case reports[31] always include a history of alcohol consumption. *C. atramentarius* alone is nontoxic and a commonly eaten mushroom, but the symptomatology is a function of the concentrations of both coprine and ethanol in the patient's system.

6. GASTROTOXINS

Gastrotoxins account for most of the mushroom poisonings among wild mushrooms hunters, because of look-alike edible and poisonous species. In some cases there is individual intolerance to otherwise edible species. Thus, any untoward reaction may result in a poisoning report, so that almost every mushroom, at one time or another, has been incriminated. There is better documentation for some species being poisonous than others (e.g., *Entoloma lividum* and *Tricholoma pardinum*), but few, if any, toxins have been characterized.

6.1. Chemistry

This is a diverse collection of mushroom poisons that have not been characterized (Fig. 3).

6.2. Mycology

Agaricus hondensis, A. placomyces
Amanita brunnescens
Boletus satanas, B. luridus, B. sensibilis, B. eastwoodiae
Chlorophyllum molybdites
Entoloma lividum
Gomphus floccosus
Hebeloma crustuliniforme, H. mesophaeum
Lactarius glaucescens, L. torminosus, L. deceptivus, L. naucin
Marasmius ureus
Naematoloma fasciculare
Omphalotus olearius (illudens)

Paxillus involutus
Pholiota aurea, P. squarrosa
Ramaria formosa, R. gelatinosa
Russula emetica, R. densifolia
Scleroderma aurantium
Tricholoma pessundatum, T. irium, T. paradinum
Tricholomopsis platyphylla
Verpa boehemica
Partial list only

Tricholoma pardinum Quel.: pileus 4–14 cm, convex to irregular flat, gray, squamose, splitting margin, lamellae off-white, adnexed; stipe 4–14 cm; spores 5–6 × 7–9 μm, ellipsoidal, smooth, white, nonamyloid. *T. pardinum* is a terrestrial forest species associated with conifers.

Entoloma lividum Fr.: pileus 8–14 cm, convex, gray, lamellae pink, adnexed; stipe 4–12 cm; spores 6–10 μm spheroidal, walls may collapse, pink. It is a terrestrial forest species.

Scleroderma aurantium Pers.: basidiocarp spheroidal, 3–9 cm dia.; peridium brown, warty, irregular pore on top; gleba gray to purple-black; basidiospores becoming black with age, 8–12 μm, spheroidal, verrucose. Grows in forest litter and may be confused with the edible puffball.

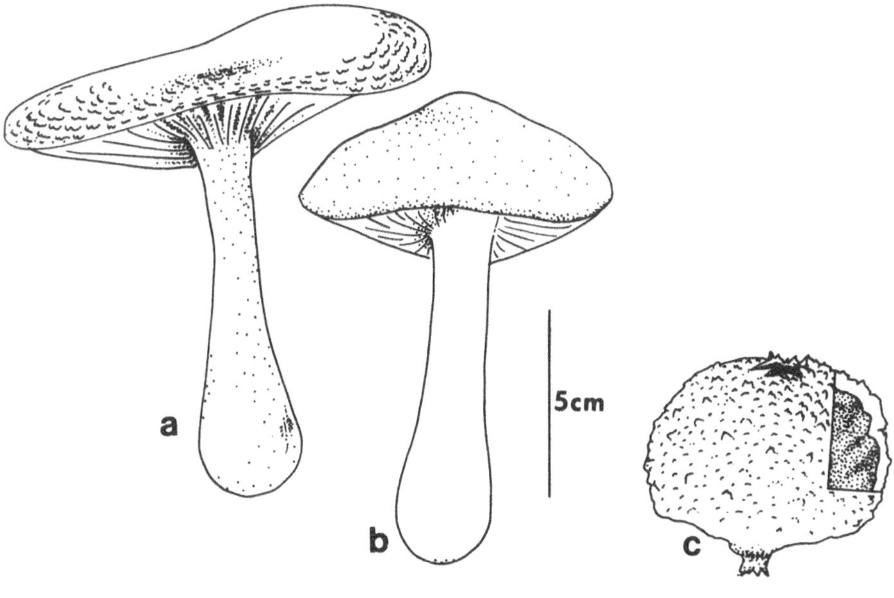

FIGURE 3. a. *Tricholoma paradinum*, b. *Entoloma lividum*, c. *Scleroderma aurantium.*

6.3. Poisoning

There are no absolutely specific symptoms in this catch-all category because of the potential diversity of the poisons involved. The gastrotoxins category usually has an onset 0.5–2 hours after ingestion, but symptoms may persist from hours to several days. The symptoms are mild or severe nausea, emesis, diarrhea, and abdominal pain. Treatment consists of an emetic (if needed) and supportive management, such as fluid and electrolyte replacement. Recovery is usually complete, but the patient should be contacted later to preclude late-onset symptoms due to mixed mushroom poisoning, i.e., simultaneous ingestion of late-onset amanitin-containing mushrooms. Case reports of *O. olearius*[29,32] and *E. lividum*[22] exemplify the heterogeneous gastrotoxins category.

7. GYROMITRIN

Not all people who eat *Gyromitra esculenta* (Fig. 4) are poisoned and the mushrooms may be detoxified by cooking. This mushroom may be confused with the edible morel as their seasonability and ecology overlap. In certain geographic locales, *G. esculenta* is the leading cause of mushroom poisoning.

7.1. Chemistry

Gyromitrin, N-methyl-N-formylhydrazine (MFH), found in certain *Gyromitra* spp., is hydrolyzed in the GI tract to monomethylhydrazine (MMH), also known as methylhydrazine CH_6N_2 (mol. wt. 46.07), a very strong reducing agent that is cytotoxic. It is heat labile, but sometimes enough MFH remains to cause poisoning because of insufficient cooking.

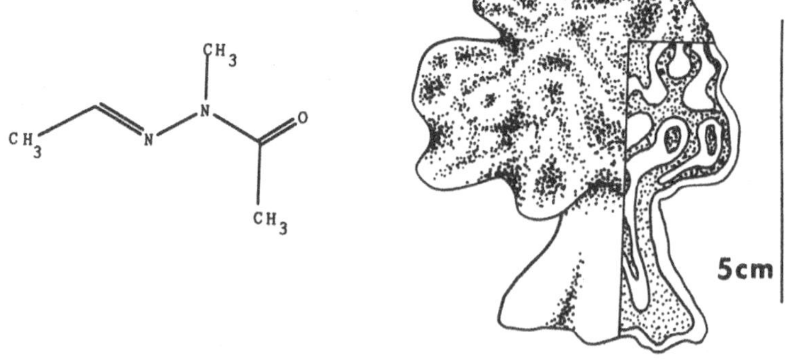

FIGURE 4. Gyromitrin. *Gyromitra esculenta.*

7.2. Mycology

Gyromitra esculenta False Morel
G. brunnea, G. californica, G. ambigua,
G. infula, G. underwoodii, G. lacunosa, G. gigas

Gyromitra spp. (Discomycetes) usually appear in the spring in temperate climates in association with conifers, such as the Great Lakes region of the United States and Northern Europe. *Gyromitra* spp. have a convoluted ascocarp, superficially resembling the edible *Morchella* spp. with which it may be confused. It lacks lamellae and the ascocarp is brown in color. *G. esculenta* Fr.: apothecium 2–10 cm, convoluted, brown; stalk 3–10 cm in length by 2–4 cm in width, hollow; ascospores 8–12 × 18–22 μm, ellipsoidal, smooth, hyaline. There is regional morphological variability or strain variability, which may be related to the quantitative level of gyromitrin.

7.3. Poisoning

The onset of gyromitrin symptoms is 6–10 hours after ingestion of mushrooms. It is intermediate between most toxins (0.5–2 hours) and the late-onset amanitin (10–15 hours), and therefore unique. The symptoms include nausea, emesis, diarrhea (watery or bloody), abdominal pain and cramps. The symptoms may persist for several days and convulsions, coma, and death due to hepatorenal failure may occur after 5–7 days in severe poisoning. Chromatographic detection of gyromitrin is possible[33] but not routinely available. Measures to decrease absorption, enhance excretion, and maintain supportive recovery are the most important initial features of patient care, as they are for all types of mushroom poisoning. Liver function methemoglobin and hemoglobin values are valuable for prognosis. There is no antidote, but megadose pyridoxine HCl 25 mg/kg IV adjusted to the patient is empirically administered.[21] Diazepam is used for convulsions and high-dose corticosteroids may be of value. A case report is available for study.[34]

8. MUSCARINE

Muscarine was the first mushroom toxin to be isolated and its mode of action described (Fig. 5).

8.1. Chemistry

Muscarine, tetrahydro-4-hydroxy-N,N,N,5-tetramethyl-2-furanmethanaminium (mol. wt. 174.26), is a heat-stable alkaloid with acetylcholinelike parasympathomimetic activity. It is a cholinergic drug that is competitive

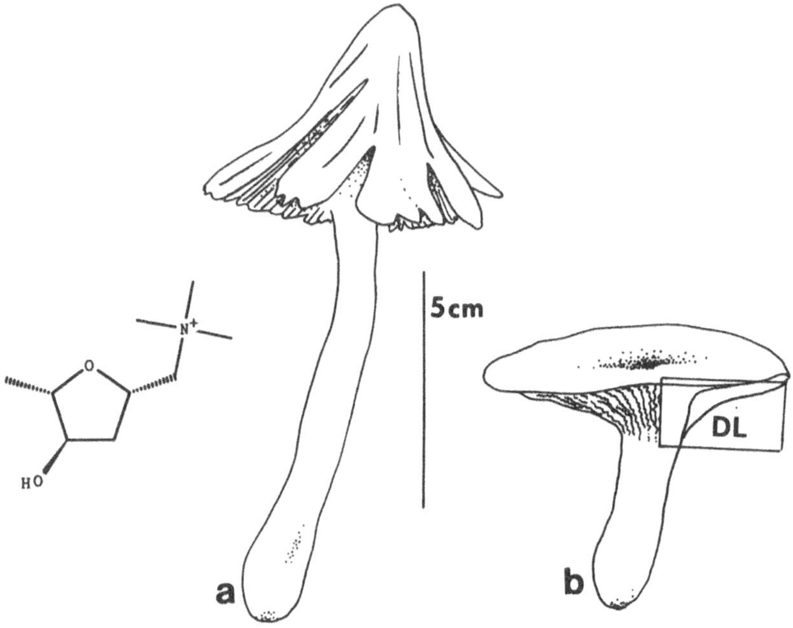

FIGURE 5. Muscarine. a. *Inocybe fastigiata*, b. *Clitocybe dealbata*, DL = decurrent lamellae.

with acetylcholine for muscarinic receptors[35] and because it is not hydrolyzed by cholinesterases, the toxic effects on the parasympathetic innervation of smooth muscles and glands are prolonged. Muscarinic receptors have also been found in the CNS.[36] Analogs of muscarine also may be present.[34]

8.2. Mycology

Inocybe fastigata	Inocybe
I. patouillardi, I. interaria, I. lanuginose, I. geophylla, I. pudica, I. lilicina, I. lacera, I. napipes	
Clitocybe dealbata	Sweating
C. dilatata, C. truncicola, C. rivulosa, C. cerussata, C. morbifera	Mushroom
Amanita muscaria, A. pantherina (see muscimol)	
Boletus luridus?	
Mycena pura?	
Paneolus separatus?	
Russula emetica?	

Nearly all *Inocybe* spp. containing muscarinic toxins have a flat to convex scaly or fibrous brown to lilac or white pileus with attached brown lamellae and brown spores. They are terrestrial or grow on rotten wood in a forest setting. The present classification of diverse species into one genus precludes an easy means for easy identification. *I. fastigiata* (Fr.) Quel.: pileus 2–7 cm, conical, umbonate, brown, fibrillose, striate, and developing a split margin with age; stipe 4–8 cm; spores 4–7 × 6–10 μm, ellipsoidal, smooth yellow-brown.

Clitocybe dealbata and other poisonous species are terrestrial in open or wooded areas in the summer and fall. There are several edible species as well, but they may be slightly poisonous if eaten raw. *C. dealbata* (Fr.) Kum.: pileus 2–5 cm irregular convex, flat, or concave, pallid gray to buff-white, lamellae, decurrent, off-white; stipe 2–4 cm; spores 3–4 × 4–6 μm, ellipsoidal, smooth, white, nonamyloid.

8.3. Poisoning

Muscarine poisoning is typified by early, 0.5–2 hours, after ingestion, onset of profuse sweating, salivation and lacrimation associated with the cholinergic syndrome.[37] There is much variability, but nausea, vomiting, diarrhea, abdominal cramps, tachycardia, miosis, and blurred vision are characteristic. For *C. dealbata* poisoning, sweating may be the only symptom. High doses lead to bradycardia, hypotension, convulsions, and shock. Coma and death cannot be discounted, and underlying cardiovascular complications may make the patient prone to serious consequences.

Atropine sulfate IV is the specific anticholinergic antidote for muscarine poisoning.[37] Dosage is adjusted to the patient's symptoms.[9] Be alert for atropine augmentation and potential overdose, because mushrooms may contain several toxins or a mixture of different poisons mushrooms may have ingested. Blood volume repletion and symptomatic care will lead to amelioration of symptoms and progressive recovery over 5–24 hours. Do a 24-hour follow-up on the patient because of the possibility of concurrent amanitin poisoning. Gas chromatographic detection of muscarine is possible,[35,38] but not generally available.

9. MUSCIMOL

The fly agaric is eaten by many, but cooking and food preparation are not well understood; most mycologists consider this mushroom poisonous. The same mushroom also has a reputation as a hallucinogen but its toxic manifestations dominate.

9.1. Chemistry

Muscimol, ibotenic acid, and its analogs muscazone and tricholomic acid have potent anticholinergiclike activity on the CNS. Specifically they are GABA-receptor agonists. They are heat stable in boiling water and still active in urine. Muscimol, 5(aminomethyl)-3(2H)-isooxazolone ($C_4H_6N_2O_2$), and its carboxylated metabolite ibotenic acid, α-amino-2,3-dihydro-3-oxo-5-isoxazoleacetic acid, possess potent CNS neuroexcitatory and depressant activity. Muscazone found simultaneously in the same mushrooms is slightly different chemically.

9.2. Mycology

Amanita muscaria	Fly Agaric
A. pantherina	Panther
A. cothurnata	Booted Amanita
A. gemmata, A. cokeri, A. porphyria	

Reference to the *A. muscaria* (Fig. 6) or *A. muscaria/pantherina* groups evokes thought of a type of mushroom poisoning much different from that of the *A. phalloides* type. *A. muscaria* contains potent neurotoxins. Further, caution should be exercised not to confuse muscimol with muscarine (cholinergic) poisoning. Although both toxins are found in *A. muscaria*, muscimol is quantitatively dominant and the symptoms are partly anticholinergic. *A. muscaria* and related species, like all *Amanita* spp., have white lamellae and spores, an annulus, and a bulbous base. In contrast to *A. phalloides*, the volva of *A. muscaria/pantherina* is not saccate but is adherent to the bulb and inconspicuous. The pileus is a distinctive feature but it may be completely white in the case of *A. cothurnata*, yellow-green in *A. gemmata*, and brown in *A. pantherina*, to a bright yellow, orange, or red in *A. muscaria*. The pileus is further distinguished by an adornment of white or off-white warts,

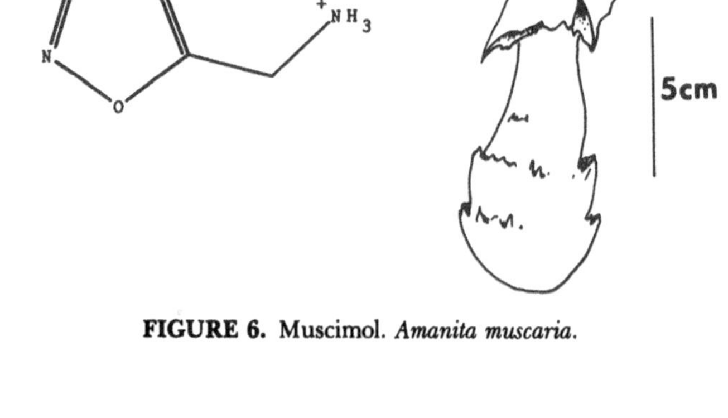

FIGURE 6. Muscimol. *Amanita muscaria.*

remnants of the universal veil. *A. muscaria* (L.) Hook.; pileus 5–20 cm convex, yellow, orange or red with white warts, lamellae white; stipe 5–15 cm, rough bulbous base, annulus; spores 6–8 × 10–12 μm, spheroidal or ellipsoidal, white, smooth, nonamyloid.

9.3. Poisoning

The onset of symptoms occurs 0.5–2 hours after ingestion. There is a great deal of individual variability based on the quantitative and qualitative mixture of muscimol-related toxins, also with respect to the presence of lesser amounts of muscarine. In general, the patient passes through stages of dizziness, staggering, ethanol-like intoxication, lack of coordination with muscular spasms, and hyperkinetic activity, but also periods of enhanced strength. Visual perception is distorted and hallucinations occur. The toxin generally has anticonvulsant activity, but convulsive activity may occur in children.[34] The high cholinergic tone of children may make them more sensitive to mushroom poisoning. Stupor and comalike deep sleep may precede or follow the above symptoms. Recovery takes place in 6–24 hours. Antidotes to counter anticholinergic symptoms are rarely used. In this regard, the value of physostigmine has not been proved. Do not give atropine, it may exacerbate symptoms. Indications for atropine use are found in the muscarine section. Treatment consists of general measures to reduce absorption and supportive maintenance of vital signs.[9] The patient may need to be restrained to protect himself and others. Muscimol is readily excreted and remains active in the urine. Extraction and detection by chemical and GLC methods have been developed[35] but they are not done clinically.

9.4. WVUH Emergency Medicine Case Report

M.P. is a 61-year-old man of Italian heritage in good health. At approximately noon on 7/16/85, the patient picked two large mushrooms in his yard and prepared "white fried mushrooms in wine." One hour after ingestion he developed nausea (M,I) and vomited (M,I), which progressed to diaphoresis (M), flushing, and delirium (I) over the next 4–6 hours. He was taken to a local hospital, where he was given ipecac and transferred to WVUH. En route the patient developed "twitching" (I), an apneic episode (M), and a decreased level of consciousness (I). Past medical history was remarkable for prostatic hypertrophy, allergy to penicillin, and possible alcoholism.

Physical examination revealed normal vital signs. Pupils were small and reactive (M). Lungs had scattered rhonchi. Cardiac exam revealed a systolic ejection murmur and S4. Bowel sounds were hypoactive. There was muscular twitching and his Glasgow scale was 10.

Laboratory data revealed ABG pH 7.35; PCO_2 42; pO_2 150; HCO_3 23 on 6L NC. ALT, AST, electrolytes, coagulation studies, CBC, and CXR were normal; KUB showed an ileus pattern. The Meixner hydrocholoric acid test was negative. EKG showed sinus rhythm with APCs. Mg was 0.9. Urinalysis revealed +1 protein and a trace of blood after placement of the Foley.

The patient was combative (I) during nasotracheal intubation and NG tube placement. The patient was treated with activated charcoal and magnesium citrate. He was vigorously hydrated and treated with pyridoxine and steroids. He received conventional supportive care.

The hospital course consisted of brief hypotension (M) and intense diaphoresis (M) with auditory and visual hallucinations (I) for the first 24 hours. Hallucinations were intermittent for 24–48 hours after he returned to his baseline mental status at 72 hours. He was discharged and promised never to eat mushrooms again.

Mushrooms obtained from the original site were a completely white *Amanitas* spp. with adherent pieces of the veil stuck to the pileus. They were identified as *A. cothurnata*, related to *A. muscaria*, and causing the muscimol type of poisoning. Note: (I) in case report = ibotenic acid or muscimol symptomatology, but (M) = muscarinic symptoms are concurrent.

Ron deAndrade, M.D., Robert Burrell, Ph.D., and R. Scott Pore, Ph.D.

10. ORELLANINE

10.1. Chemistry

Orellanine (orelline) is tetra-hydroxy bipyridine,N,N-dioxide.[39,40] It may be chemically related to the pesticide diquat, it is heat stable, light activated, causes nephritis, and is detectable by two-dimensional TLC.[39] A chemical test for orellanine has also been proposed.[41] Cortinarins, which are cyclic peptides, also may be present.[40]

10.2. Mycology

Cortinarius orellanus	Cortinarius
C. gentilus, C. rainierensis, C. speciosissimus, C. splendens, C. venenosus	

Cotrinarius spp. are found in forest soil and are difficult to differentiate from other small brown mushrooms. *C. orellanus* Fr.: pileus 2–8 cm, convex brown; lamellae widely separate, brown; stipe 2–9 cm; spores 6–7 × 8–11 μm, ellipsoidal, rough, brown. The distinctive feature is that the partial veil leaves cobweblike fibrose striae over the lamellae and stipe.[6]

10.3. Poisoning

Many instances of poisoning in Europe have been recorded.[40] Orellanine may cause delayed onset (> 3 days, average 8 days or longer), gastritis, and rarely subsequent acute renal failure. Treatment is symptomatic and supportive. Hemodialysis and hemoperfusion may be helpful. The cumulative effects of daily ingestion may be fatal.[9,40,41,42]

11. PSILOCYBIN

Psilocybin is a hallucinogenic toxin from *Psilocybe* spp., which in some ethnic cultures is ingested for its intoxicating effects (Fig. 7).

11.1. Chemistry

Psilocybin, 3-[2-(dimethylamino) ethyl]-1H-indol-4-ol dihydrogen phosphate ester (mol. wt. 284.27), is an analog of dimethyl tryptamine and serotonin. A less active metabolite is psilocin. It is a psychomimetic drug and controlled substance in the United States.

11.2. Mycology

Psilocybe cubensis — Magic Mushroom
P. mexicana, P. baeocystis, P. semilanceata, P. stuntzii,
P. cyanescens, P. pelliculosa, P. silvatica
Panaeolus subbaleatus, P. campanulatus,
P. foenisecii
Gymnopilus spectabilis
Conocybe cyanopus

Psilocybe spp. are found in temperate subtropical and tropical climates. Not all species contain psilocybin. *P. cubensis* and several other species will

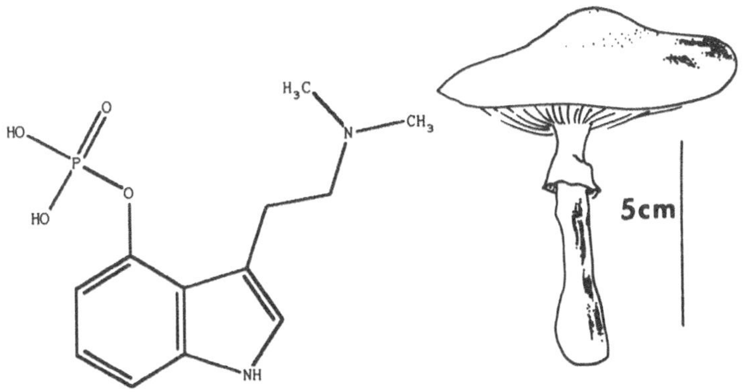

FIGURE 7. Psilocybin. *Psilocybe cubensis.*

stain blue when bruised, but this characteristic does not always correlate with the presence of psilocybin.[35] Another source of *Psilocybe* spp. are from home-grown culture and the drug culture trade. Sometimes the only available material for identification is dried mushroom parts. *P. cubensis* (Earl.) Sing.: pileus 2–8 cm, convex to flat, smooth, tan, lamellae gray, annulus; stipe 4–12 cm; spores 7–11 × 10–16 μm, ellipsoidal, smooth, apical pore, purple-brown. Fungal tissue stains blue when bruised. Grows in association with dung in cattle pastures and some forest settings.

11.3. Poisoning

Patients may have experimented with drugs. Generally *Psilocybe* spp. have a reputation of safety, but untoward reactions, overdoses, and even a psilocybin-related death have been reported. The onset of symptoms occurs in 0.5–1 hours, and the sequence of clinical effects has been timed.[34] The patient may have a pleasant or apprehensive mood, unmotivated laughter or hilarity, compulsive movement and hyperkinesis, and/or muscle weakness and drowsiness. Mydriasis and blurred vision are common signs. Repetitious visual hallucinations and vividly colored patterns with eyes closed, whether awake or asleep, are characteristic features of psilocybin intoxication. There may be LSD-like effects and psychotic sequelae, but recovery from the initial stages occurs in 2–12 hours. Benzodiazepine is rarely necessary for sedation. Chromatographic techniques have been used for psilocybin detection.[44]

12. SUMMARY

Mushroom poisoning is an infrequent event, but the role of the mycologist is essential for proper diagnosis and treatment. The mycologist provides specialized knowledge about poisonous mushrooms and performs the oftentimes difficult identification of the mushroom. The best way for the mycologist to develop the necessary skills is to use authoritative references on mushroom poisoning,[9–11,14,34,42,45] good field guides to mushroom identification,[3–5] and several years of practical experience. By and large, the toxins themselves have been chemically determined, their mode of action elucidated, and chromatographic assays developed, if only at the research level. Likewise, the symptomatology of mushroom poisoning, although not completely understood, is at least known and appreciated. Finally, although toxin antidotes are few, empiric treatment regimens are satisfactory for all but the deadly toxins. An antidote for amanitin is a grail for research scientists.

Many poisonous mushrooms can be identified macroscopically, but one area of research seems remarkably underdeveloped, namely, practical microscopic and chemical methods for detecting and identifying poisonous mushrooms. The medical mycologist has few laboratory tests available to simplify poisonous mushroom identification, and spore recognition, the basis for most fungal identifications, has not received the attention it deserves. Basically, spore size, appearance, color, and amyloid reaction are the existing parameters for spore recognition. Ammirati *et al.*[6] provide a useful point of departure in this regard. They describe the spore-bearing structures and paraphysoids. As medical mycologists apply themselves to these important problems, poisonous mushroom identification will develop into a laboratory science.

REFERENCES

1. Litovitz, T. L., Schmitz, B. F., and Holm, K. C., 1989, 1988 annual report of the American Association of Poison Control Centers national data collection system, *Am. J. Emerg. Med.* **7:** 495–545.
2. McClain, J. L., Hause, D. W., and Clark, M. A., 1989, *Amanita phalloides* mushroom poisoning: A cluster of four fatalities, *J. Forensic Sci.* **34:**83–87.
3. McKnight, K. H., and McKnight, V. B., 1987, *A Field Guide to Mushrooms: North America*, Houghton Mifflin, Boston.
4. Lincoff, G. H., 1981, *The Audubon Society Field Guide to North American Mushrooms*, Alfred A. Knopf, New York.
5. Miller, O. K., 1977, *Mushrooms of North America*, E. P. Dutton, New York.
6. Ainsworth, G. C., 1963, *Ainsworth and Bisby's Dictionary of Fungi*, 5th ed., Commonwealth Mycological Institute, Kew.
7. Ammirati, J. F., Traquair, J. A., and Horgen, P. A., 1985, *Poisonous Mushrooms of the Northern United States and Canada*, University of Minnesota Press, Minneapolis.
8. Gillman, L., 1978, Identification of common poisonous mushrooms, in: *Mushroom Poisoning: Diagnosis and Treatment* (B. H. Rumack and E. Salzman, eds.), CRC Press, Boca Raton, FL, pp. 27–65.
9. Ellenhorn, M. J., and Barceloux, D. G., 1987, *Medical Toxicology: Diagnosis and Treatment of Human Poisoning*, Elsevier, New York.
10. Arena, J. M., and Drew, R. H., 1986, *Poisoning: Toxicology, Symptoms, Treatment*, 5th ed., Charles C. Thomas, Springfield.
11. Brent, J., Kulig, K., and Rumack, B. H., 1990, Mushrooms, in: *Clinical Management of Poisoning and Drug Overdose* (L. M. Haddad and J. F. Winchester, eds.), W. B. Saunders, Philadelphia, pp. 581–590.
12. Mitchel, D. H., 1980, *Amanita* mushrooms poisoning, in: *Annual Review of Medicine*, Vol. 31 (W. P. Creger, C. H. Coggins, and E. W. Hancock, eds.), Annual Reviews Inc., Palo Alto, CA, pp. 51–57.
13. Meixner, A., 1979, Amatoxin-Nachweis in Pilzen, *Z. Mykol.* **45:**137–139.
14. Lampe, K. F., 1987, Differential diagnosis of poisoning by North American mushrooms, with particular emphasis on *Amanita phalloides*-like intoxication, *Ann Emergen. Med.* **16:**956–962.
15. Andres, R. I., Frei, W., Gautschi, K., and Vonderschmitt, J., 1986, Radioimmunoassay for amatoxins by use of a rapid, ^{125}I-tracer-based system, *Clin. Chem.* **32:**1751–1755.
16. Rieck, W., and Platt, D., 1988, High-performance liquid chromatographic method for the determination of alpha-amanitin and phalloidin in human plasma using the column-switching technique and its application in suspected cases of poisoning by the green species of Amanita mushroom (*Amanita phalloides*), *J. Chromatog.* **425:**121–134.
17. Beutter, J. A., and Vergeer, P. P., 1980, Amatoxins in American mushrooms: Evaluation of the Meixner test, *Mycologia* **72:**1142–1149.

18. Vergeer, P. P., 1986, The Meixner test evaluated, *McIlvainea* **7**:61–66.
19. Tanghe, L. J., 1983, Spread of *Amanita phalloides* in North America, *McIlvainea* **6**:4–8.
20. Bivins, H. G., Knopp, R., Lammers, R., McMicken, D. B., and Wolowodiuk, O., 1985, Mushroom ingestion, *Ann. Emergen. Med.* **14**:1099–1104.
21. Klein, A. S., Hart, J., Brems, J. J., Goldstein, L., Lewin, K., and Busuttil, R. W., 1989, *Amanita* poisoning: Treatment and the role of liver transplantation, *Am. J. Med.* **86**:187–193.
22. Harahan, J. P., and Gordon, M. A., 1984, Mushroom poisoning: Case reports and a review of therapy. *JAMA* **251**:1057–1061.
23. Piering, W. F., and Bratanow, N., 1990, Role of the clinical laboratory in guiding treatment of *Amanita virosa* mushroom poisoning: Report of two cases, *Clin. Chem.* **36**:571–574.
24. Sanz, P., Reig, R., Piqueras, J., Marti, G., and Corbella, J., 1989, Fatal mushroom poisoning in Barcelona, 1986–1988, *Mycopathology* **108**:207–209.
25. Piqueras, J., 1989, Hepatotoxic mushroom poisoning: Diagnosis and management, *Mycopathology* **105**:99–110.
26. Floersheim, G. L., 1987, Treatment of human amatoxin mushroom poisoning; myths and advances in therapy, *Med. Toxicol.* **2**:1–9.
27. Floersheim, G. L., 1983, Toxins and intoxications from the toadstool *Amanita phalloides*, *Trends Pharm. Sci.* **3**:263–266.
28. Jones, E. A., Schafer, D. F., Ferenci, P., and Pappas, S. C., 1984, The GABA hypothesis of the pathogenesis of hepatic encephalopathy: Current status, *Yale J. Biol. Med.* **57**:301–316.
29. Kulberg, A. G., Goldfrank, L. R., and Bresnitz, E. A., 1986, Mushrooms: Toxic and hallucinogenic, in: *Toxicologic Emergencies*, 3rd ed. (L. R. Goldfrank, N. E. Flomenbaum, N. A. Lewin, R. S. Weisman, M. A. Howland, and A. G. Kolberg, eds.), Appleton Century Crofts, Norwalk, CT, pp. 545–559.
30. Faulstich, H., Kirchner, K., and Derenzini, M., 1988, Strongly enhanced toxicity of the mushroom toxin alpha-amanitin by an amatoxin-specific Fab or monoclonal antibody, *Toxicon* **26**:491–499.
31. Hatfield, G. M., and Schaumberg, J. P., 1978, The disulfiram-like effects of *Coprinus atramentarius* and related mushrooms, in: *Mushroom Poisoning: Diagnosis and Treatment* (B. H. Rumack and E. Salzman, eds.), CRC Press, Boca Raton, FL, pp. 181–186.
32. French, A. L., and Garrettson, L. K., 1988, Poisoning with the North American Jack O'Lantern mushroom, *Omphalotus illudens*, *J. Toxicol. Clin. Toxicol.* **26**:81–88.
33. Giusti, G. V., and Carnevale, A., 1974, A case of fatal poisoning by *Gyromitra esculenta*, *Arch. Toxicol.* **33**:49–54.
34. Lampe, K. F., 1978, Pharmacology and therapy of mushroom intoxicants, in: *Mushroom Poisoning: Diagnosis and Treatment* (B. H. Rumack and E. Salzman, eds.), CRC Press, Boca Raton, FL, pp. 125–169.
35. Chilton, W. S., 1979, Chemistry and mode of action of mushroom toxins, in: *Mushroom Poisoning: Diagnosis and Treatment* (B. H. Rumack and E. Salzman, eds.), CRC Press, Boca Raton, FL, pp. 87–124.
36. Whittaker, V. P., 1990, The contribution of drugs and toxins to understanding of cholinergic functions, *Trends Pharm. Sci.* **11**:8–13.
37. Stallard, D., and Edes, T. E., 1989, Muscarinic poisoning from medications and mushrooms. A puzzling symptom complex, *Postgrad. Med.* **85**:341–345.
38. Eugster, H., and Schleusener, E., 1969, Stereomere Muscarine Kommen in der Natur vor Gas-Chromatographische Trennung der Norbasen, *Helv. Chin. Acta* **52**:708–715.
39. Rapior, S., Delpech, N., Andary, C., and Huchard, G., 1989, Intoxication by *Cortinarius orellanus*: Detection and assay of orellanine in biological fluids and renal biopsies, *Mycopathology* **108**:155–161.
40. Michelot, D., and Tebbett, I., 1990, Poisoning by members of the genus *Cortinarius*—a review, *Mycol. Res.* **94**:289–298.

41. Lampe, K. F., and Ammirati, J. F., 1990, Human poisoning by mushrooms in the genus *Cortinarius*, *McIlvainea* **9**:12–25.
42. Lincoff, G., and Mitchel, D. H., 1977, *Toxic and Hallucinogenic Mushroom Poisoning*, Van Nostrand Reinhold, New York.
43. Ammirati, J. F., 1983, *Cortinarius* I: Generic Description and comments on the subgenera, *McIlvainea* **6**:44–51.
44. Lee, R. E., 1985, A technique for the rapid isolation and identification of psilocin from psilocin/psilocybin-containing mushrooms, *J. Forensic Sci.* **30**:931–941.
45. Ramack, B. H., and Salzman, E., 1978, *Mushrooms Poisoning: Diagnosis and Treatment*, CRC Press, Boca Raton, FL.

23

Modulators of Immune Responses to Fungi

JULIE Y. DJEU

1. INTRODUCTION

Fungi are ubiquitous in the environment and survive as saprophytes in the soil, with a few, such as *Candida albicans*, finding a niche in the normal flora of the gut, skin, and oral cavity of humans. Despite constant exposure to these organisms, only a very limited number of them cause disease. Most are opportunists that can only find a foothold in individuals that are immunocompromised. The strong resistance to fungal infections is mediated by natural barriers and innate immune mechanisms, which can completely eliminate the entry of fungi or contain the invading organism at the early stages so that no apparent disease is detected. The phagocytes, particularly neutrophils, form the first line of defense against most pathogens, including fungi.[1–3] Monocytes and macrophages also play an important role in phagocytosis of fungi and have an added key role in processing of antigens for presentation to T cells and activation of B cells to generate specific immunity.[4–6] These specific T and B cells cannot serve as the end effector cells but participate in the destruction of fungi by the production of cytokines and antibodies, respectively. Antibodies facilitate ingestion of the opsonized microbes by phagocytes, while the cytokines act to simulate their microbicidal pathways. Elimination of the ingested fungi is via both oxidative and nonoxi-

---

JULIE Y. DJEU • Department of Medical Microbiology and Immunology, College of Medicine, University of South Florida, Tampa, Florida 33612.

Fungal Infections and Immune Responses, edited by Juneann W. Murphy *et al.* Plenum Press, New York, 1993.

dative mechanisms.[2–4,7–9] Highly reactive oxygen radicals are released either into the phagosome containing the ingested yeasts or externally at the site of contact with larger fungal hyphae that cannot be phagocytized. In addition, degranulation of lytic enzymes can take place, again into the phagosome or externally into the environment, to digest the pathogens.

Neutrophils and monocytes are ultimately responsible for the elimination of fungi but they are under the control of cytokines that regulate their differentiation, function, and migration. Cytokines provide a means of triggering an immune cascade, each with a unique set of properties to induce maturation, to activate function, or to recruit phagocytes to the site of infection. A variety of such cytokines are produced not only by T cells, but also by monocytes and natural killer (NK) cells with large granular lymphocytic (LGL) morphology. Most recently, neutrophils have also been found to produce certain cytokines involved in autocrine activation of function. Besides a defect in phagocyte function, a defect in the cells that produce neutrophil/monocyte activating and recruiting factors can lead to an inadequate defense system. Another effector cell that has the capacity to kill or inhibit fungal growth is the NK cell, although its range of fungal targets is quite restricted compared to phagocytes, which can destroy most pathogens. These NK cells are also under the influence of certain cytokines. Individuals with an immunocompromised immune system, either due to HIV infection or cytoreductive therapy associated with cancer or organ transplantation, can therefore fall prey to opportunistic fungal infections.

This chapter will mainly deal with the cytokines that are responsible for phagocyte activation and recruitment. In addition, the recent data on the direct antifungal role of NK cells and their interaction with phagocytes to control fungal growth will be discussed in detail. The literature of cytokine activation of phagocytes for functions other than fungicidal and fungistatic effects is vast and cannot be covered fully here. Emphasis will therefore be placed on information dealing mainly with antifungal immunity.

2. NEUTROPHIL ACTIVATING FACTORS

Their abundance in the circulation and their quick mobility to sites of infection have been the hallmark of neutrophil defense against microbes. Several cytokines have been identified that are involved in the various stages of neutrophil maturation and activation. Starting from their generation in the bone marrow to their release into the peripheral blood and migration to sites of infection, neutrophils are under the control of cytokines. Most studies concerned with neutrophil activation and recruitment have focused on their response to nonfungal agents, but increasing number of reports are now appearing related to antifungal immunity because of increased clinical interest associated with the higher frequency of opportunistic fungal infections in AIDS patients and cancer or transplant patients undergoing chemotherapy. The important cytokines studied so far in relationship to fungal immunity will be surveyed, and their mechanisms of action will be discussed.

2.1. Colony Stimulating Factors

Hematopoietic bone marrow stem cells that are the precursors of neutrophils are highly sensitive to differentiation signals, such as granulocyte-macrophage colony stimulating factors (GM-CSF), G-CSF, and interleukin 3 (IL3).[10–12] In fact, patients with neutropenia caused by chemotherapy or disease respond well to recombinant GM-CSF and G-CSF therapy, and routinely manifest a rapid increase in circulating neutrophil levels.[13–18] This type of immunotherapy has been associated with acceptable toxicity *in vivo* and has markedly reduced the time of hospital stay for the patient. There is some suggestive evidence that this therapy may also alleviate the host's ability to combat opportunistic infections. The colony stimulating factors (CSF) are, indeed, effective in the activation of function, besides eliciting the differentiation of neutrophils from the bone marrow. *In vitro* results confirm that CSF can activate mature circulating neutrophils, measured by the release of superoxide anions, increased binding to the strong chemotactic peptide, formyl–methionine–leucine–phenylalanine (FMLP), or increased antifungal activity against *Candida albicans*.[19–23]

A number of sources are available for the production of GM-CSF and G-CSF. Neutrophils are short lived in the circulation and survive less than 2–3 days, therefore, requiring constant replacement. Normal hematopoiesis to replenish neutrophils occurs daily and is probably aided by factors produced by bone marrow stromal cells;[24,25] however, much of the information regarding CSF comes from studies of activated mononuclear cells.[26,27] T cells and NK cells responding to either mitogens, phorbol esters, or calcium ionophores, can produce large amounts of cytokines, including GM-CSF, which can amplify the immune cascade by calling in phagocytes that respond with high antimicrobial activity. Monocytes, endothelial cells, and fibroblasts have also been shown to release GM-CSF and G-CSF.[27–32] These latter non-lymphocytic cells are triggered by interleukin 1 to produce CSF. With regards to *C. albicans*, only NK cells, but not T cells or monocytes from normal individuals, appear to react rapidly to exposure to this opportunistic fungus by releasing GM-CSF.[23] There appears to be a distinct NK/neutrophil axis for the control of *Candida* in humans. Studies have shown that human LGL, when cultured for 24–48 hours with *C. albicans*, can release cytokines into the supernatants that have potent neutrophil activating properties against the same pathogen.[33] Neutralization with specific antibodies indicated that the neutrophil activating factor can be abolished by anti-GM-CSF but not by an antibody against a related cytokine, IL3. The presence of GM-CSF was also detected in LGL supernatants by its biological activity on a CSF-dependent cell line. In addition, recombinant GM-CSF, at the doses detected in LGL supernatants, actively stimulated neutrophil function against *C. albicans*. These results suggest that opportunistic fungi, such as *C. albicans*, may be normally under the early control of NK cells and neutrophils. This pathway may be sufficient in effective containment of the organisms, but ultimately specific T-cell release of cytokines and B-cell production of specific antibodies are also involved, especially if the organisms are invasive and remain in the host system long enough for specific immune cells to recognize them.

2.2. Tumor Necrosis Factor

Tumor necrosis factor (TNF) was originally discovered as a serum factor from mice injected with mycobacteria that caused necrosis of tumors *in vivo* and the destruction of certain tumor cells *in vitro*.[34] It has subsequently been shown to have numerous properties, such as growth promotion of fibroblasts, cachexia, and activation of eosinophils.[35–37] Most importantly, it is now known to be a potent neutrophil activator.[38–41] TNF not only activates neutrophil killing of antibody-coated tumor cells, but also enhances phagocytosis and microbicidal activity. Both oxidative and nonoxidative anti-microbial mechanisms can be induced by TNF, especially if neutrophils are brought in contact with serum-opsonized fungi.

Classically, TNF has been considered a product of monocytes responding to antigenic stimulation;[42] however, TNF appears to be also released by NK cells and mast cells.[43–45] NK cells, responding to *C. albicans*, can quickly release TNF within 8 hours of exposure *in vitro*. Most recently, neutrophils have also been found to release TNF in response to *C. albicans* and bacterial lipopolysaccharide (LPS).[46] This discovery of neutrophil-associated TNF release, coupled with other observations of IL1, interferon alpha (IFNa), and G-CSF release, provides a new concept that neutrophils are not quiescent end effector cells, waiting for signals from other immune cells.[46–49] They are, in fact, active participants in the immune cascade and can provide key cytokines, either to self-regulate function or to influence other components of the immune system via the cytokines they produce.

2.3. Interferon Gamma

Interferon gamma (IFNg) has, in addition to its antiviral effects, a number of key properties for modulation of both immune and nonimmune cells.[50] An important property is the ability to induce Class I and Class II expression on antigen-presenting cells and tumor cells, rendering them more accessible to interaction with specific T cells. With regards to antifungal immunity, the most studied aspect of IFNg is its ability to activate neutrophils. Neutrophils, exposed to IFNg, are highly phagocytic and microbicidal.[51–55] In addition, IFNg can synergize with TNF to achieve even higher antimicrobial or antitumor activity in neutrophils.[39] The source of IFNg is, so far, limited to T cells and NK cells.[50,54] In the case of T cells, monocyte processing of antigen for presentation to T cells is required for T-cell production of IFNg; however, NK cells can apparently respond directly to antigen stimulation or to IL2 by the production of IFNg. The accumulating data, therefore, suggest that cytokines are critical in all aspects of immune amplification and effector cell function, and in order to ensure the availability of specific cytokines, alternate systems of production exist so that if one source is nonfunctional, another can be in place for proper activation of the immune system.

3. NEUTROPHIL CHEMOTACTIC FACTORS

In addition to neutrophil differentiation and activation factors, cytokines exist that are responsible for influencing the migration of neutrophils. A family of neutrophil chemoattractants has been identified, the most studied of which is interleukin 8 (IL8).[55,56] Initially purified from monocytes stimulated with LPS, IL8 is now found to be also released by T cells and NK cells. With NK cells, IL8 release can be induced directly by *C. albicans*.[57] Aside from its chemoattractant nature, IL8 can also activate mature neutrophils to inhibit *Candida* growth *in vitro*.[58] This activity is mediated specifically by IL8 induction of neutrophil degranulation of lytic enzymes. Production of IL8 locally at the site of infection can, therefore, quickly mobilize and activate neutrophils for destruction of the invading organisms.

It should be noted that other cytokines that activate neutrophil function, e.g., GM-CSF and TNF, can also act as a chemotactic factor for the same cells.[59,60]

4. MONOCYTE ACTIVATING FACTORS

Monocytes/macrophages present an alternate phagocytic system to neutrophils.[4,5] They also have similar oxidative and lytic-enzyme machineries to destroy microorganisms; however, several key differences exist that distinguish monocytes from neutrophils. Monocytes are long-lived circulating cells and exist in markedly smaller numbers than neutrophils in the blood. Tissue macrophages, however, abound that may make up the difference. These monocytes/macrophages ingest microbes and kill them, but another important aspect of this ingestion is the processing of small antigenic components of the microbes to be presented in association with Class I or Class II major histocompatibility complex molecules for the activation of antigen-specific $CD8^+$ or $CD4^+$ cells.[6] In addition to the initial destruction of microbes, monocytes play a key role in initiating the immunological cascade towards specific T- and B-cell immunity. Another feature of monocytes absent in neutrophils is the ability of some pathogens to establish a parasitic, intracellular existence within them. *Histoplasma capsulatum* is a typical fungal pathogen that can multiply intracellularly in monocytes/macrophages.[61,62]

Monocytes, like neutrophils, are under the control of cytokines. A number of overlapping cytokines activate monocytes and neutrophils for microbial activity and for differentiation from the bone marrow stem cells; however, some differences exist that will be discussed in relation to antifungal activity.

4.1. Colony Stimulating Factors

Signalling for differentiation of hemopoietic stem cells towards monocytes can be achieved by several cytokines produced by the bone marrow stromal cells.[24,25] The multipotential IL3 and GM-CSF are both involved, as is M-CSF.[10–12] These cytokines, as mentioned earlier, can be induced during antigen or mitogenic stimulation of T cells and NK cells.[26,27] Other cells, such as monocytes, B cells, fibroblasts, and endothelial cells, are also capable of releasing M-CSF.[63] All three cytokines have the additional ability to enhance the antifungal effects of mature human monocytes *in vitro*. The *in vivo* administration of these CSF may therefore cause recruitment and activation of not only neutrophils but also monocytes. In monocytes, GM-CSF has been shown to activate anti-candidal activity by induction of the oxidative pathway by several methods: 1) measurement of superoxide anion generation, 2) by inhibition of function with superoxide dismutase and catalase, and 3) inhibition of function inhibitors of myeloperoxidase and scavengers of oxygen radicals.[64] GM-CSF and IL3 also have the added ability to maintain anti-candidal function in cultured monocytes *in vitro*.[65] Fresh monocytes have high anti-candidal activity, which is progressively lost over time as the monocytes are cultured up to 5 days. This decrease in antifungal activity parallels the decreasing loss of tumoricidal activity in monocytes with *in vitro* culture. If the monocytes are cultured in the presence of GM-CSF or IL3, however, their function is maintained at a steady state without any apparent downregulation of function. How this is accomplished is not known and its mechanism awaits further investigation.

4.2. Interferon Gamma

Interferon gamma (IFNg) is one of the most potent stimulators of monocyte function. Its ability to activate macrophage killing of a number of fungal pathogens, e.g., *Blastomyces dermatitidis*, *Paracocidioides brasiliensis*, and *Candida albicans*, has been well documented.[65–67] It is a particularly important cytokine in activating monocytes/macrophages that harbor intracellular fungi, such as *Histoplasma capsulatum*.[68] These intracellular fungal pathogens are unable to survive and establish infection within monocytes/macrophages that are exposed to IFNg; however, in some cases IFNg can be detrimental to monocytes. In one study, monocytes plated on fibronectin-coated surfaces have been reported to be able to kill *Crytococcus neoformans*, but this function is defective in monocytes treated with IFNg.[69] With certain bacteria, such as *Mycobacterium avium intracellulare*, IFNg has also been reported to enhance the survival of the intracellular pathogens within monocytes, rather than to help eliminate them.[70] IFNg is a product of T cells and NK cells;[50,54] thus there is reciprocal regulation of T cells and monocytes, with monocytes producing cytokines such as IL1 that affect T-cell function and T/NK cells providing cytokines that control monocyte function.

5. NATURAL KILLER CELLS

Natural killer (NK) cells, first described to have the capacity to lyse certain tumor cells without prior immunization with the relevant antigens, have recently been investigated for their role in fungal immunity.[54] They have been extensively studied for their role in immunosurveillance against tumors, in graft vs. host rejection, and in early defense against viral and bacterial infections. These cells readily respond within hours to 3 days *in vivo* and *in vitro* to tumor cells, allogeneic implants, viruses, and bacteria. The functional response of NK cells is measured by lysis of tumor cells, allogeneic target cells, virus-infected fibroblasts, and bacteria-infected monocytes. The NK cells, therefore, possess the machinery to lyse a wide range of target cells. It is not surprisingly, then, that NK cells have recently been found to directly inhibit fungal growth *in vitro*. NK cells have been shown to be effective against *Crytococcus neoformans* and *Paracoccidioides brasilinesis*.[71,72] Binding of NK cells to the fungi appears to be a necessary initiating step in the process of fungal elimination. The actual destruction may be mediated by the release of degradative enzymes, similar to those found in phagocytes. Although NK cells are not phagocytic in nature, they possess an abundance of azurophilic granules containing a variety of hydrolases and proteases. Contact of NK cells with susceptible fungi may then trigger degranulation of these enzymes into the external microenvironment. In support of this, granules isolated from rat NK cells have been reported to interfere with *C. neoformans* growth *in vitro*. One of the enzymes purified from these granules, cytolysin, which lyses red blood cells and certain tumor cells within minutes, has also been observed to inhibit crytococcal growth.[73] There is also some suggestive evidence that NK cells may be effective against *Histoplasma capsulatum* and this might be facilitated via antibody-dependent cell cytotoxicity.[74]

NK cells, however, are not universally microbicidal against all fungi. In fact, *C. albicans* is totally resistant to inhibition by NK cells.[33] Some reports on NK killing of *C. albicans* exist but these should be interpreted with caution because contamination with monocytes/macrophages in the NK-cell preparations can lead to the observation of antifungal activity. In human NK cells that have been completely removed of monocytes, direct antifungal activity has not been obtained. Even in the presence of IL2 or IFN, which can both activate tumoricidal function, NK cells do not acquire the ability to inhibit candidal growth.

Despite the inability to directly inhibit *Candida* growth, NK cells has been found to indirectly influence host resistance against this pathogen. NK cells can be stimulated by *C. albicans* to produce neutrophil activating factors.[33] A wide variety of cytokines are induced from LGL that have been associated with NK function during exposure to *C. albicans*. TNF, IFNg, GM-CSF, and IL8 are all released and are responsible for activating and mobilizing neutrophils.[23,44,57] Thus, NK cells may provide an added immune circuit whereby NK cells can interact with neutrophils via specific cytokines to control fungal invasion and growth. Whether other microorganisms are eliminated via this circuit needs to be further investigated.

6. SUMMARY

Much is known about the regulation of immune function by cytokines. Cytokines that are produced in each step of the immune cascade are being defined and cloned, and their mechanisms of action are being analyzed. Their function in antifungal immunity has only begun to be studied, and this interest has recently gained ground due to the increasing fungal infections seen in immunocompromised hosts. Of the cytokines that have been studied with various fungal pathogens, it is becoming obvious that they work in the same manner as described for other microbes. The colony stimulating factors are potent activators of both neutrophil and monocyte functions, as is IFNg. TNF has also been shown to be a strong functional stimulator of neutrophils against fungi. Other cytokines, e.g., IL1, IL2, IL4, and IL6, that have been shown to activate phagocyte function against tumor cells or other microbes need to be evaluated against fungi. With the availability of recombinant technology to mass-produce these cytokines, future studies will provide interesting and important information on the mechanisms used by the host to combat fungal infections.

ACKNOWLEDGMENTS. The research activities related to *Candida albicans* immunology by the author was supported by NIH grant AI-24699. Because this is intended to be a brief overview of the interaction of cytokines and fungal immunity, apologies are extended to those investigators whose works have been inadvertently left out.

REFERENCES

1. Klebanoff, S. J., and Clark, R. A., 1975, *The Neutrophil: Function and Clinical Disorders*, North Holland, Amsterdam.
2. Clark, R. A., 1990, The human neutrophil respiratory burst oxidase, *J. Inf. Dis.* **116:**1140–1147.
3. Lehrer, R. I., and Ganz, T., 1990, Antimicrobial polypeptides of human neutrophils, *Blood* **76:**2169–2181.
4. Lehrer, R. I., 1970, The fungicidal activity of human monocytes: A myeloperoxidase-linked mechanism, *Clin. Res.* **18:**408–413.
5. Paul, W. E., 1984, *Fundamental Immunology*, Raven Press, New York.
6. Unanue, E. R., 1984, Antigen-presenting function of the macrophage, *Ann. Rev. Immunol.* **2:** 395–428.
7. Diamond, R. D., Clark, R. A., and Haudenshild, C. C., 1980, Damage to *Candida albicans* hyphae and pseudohyphae by the myeloperoxidase system and oxidative products of neutrophil metabolism *in vitro*, *J. Clin. Invest.* **66:**908–917.
8. Diamond, R. D., and Clark, R. A., 1982, Damage to *Aspergillus fumigatus* and *Rhizopus oryzae* by oxidative and non-oxidative microbicidal products of human neutrophils *in vitro*, *Infect. Immun.* **38:**487–495.
9. Spitznagel, J. K., 1986, Nonoxidative antimicrobial reactions of leukocytes, *Contemp. Top. Immunobiol.* **14:**283–343.
10. Morstyn, G., and Burgess, A. W., 1988, Hemopoietic growth factors: A review, *Cancer Res.* **48:** 5624–5637.
11. Weisbart, R. H., Gasson, J. C., and Golde, D. W., 1989, Colony stimulating factors and host defense. *Ann. Intern. Med.* **110:**297–303.
12. Groopman, J. E., Molina, J. M., and Scadden, D. T., 1989, Hematopoietic growth factors, biology and clinical applications, *N. Engl. J. Med.* **321:**1449–1459.
13. Antman, K. S., Griffin, J. D., Elias, A., Socinski, M. A., Ryan, L., Cannistra, S. A., Oette, D., Whitley, M., Frei, E., and Schnipper, L. E., 1988, Effect of recombinant human granulocyte-macrophage colony stimulating factor on chemotherapy-induced myelosuppression, *N. Engl. J. Med.* **319:**593–598.
14. Brandt, S. J., Peters, W. P., Atwater, S. K., Kritzberg K., Borowitz, M. J., Jones, R. B., Shpall, E. J., Bast, R. C., Gilbert, C. J., and Oette, D. H., 1988, Effect of recombinant human granulocyte-macrophage colony stimulating factor on hematopoietic reconstitution after high-dose chemotherapy and autologous bone marrow transplantation, *N. Engl. J. Med.* **318:** 869–876.
15. Morstyn, G., Campbell, L., Souza, L. M., Alton, N. K., Keech, J., Green, M., Sheridan, W., Metcalf, D., and Fox, R., 1988, Effect of granulocyte colony stimulating factor on neutropenia induced by cytotoxic chemotherapy, *Lancet* **2:**667–672.
16. Neidhart, J., Mangolik, A., Kohler, W., Stidley, C., Saiki, J., Duncan, P., Souza, L., and Downing, M., 1989, Granulocyte colony stimulating factor stimulates recovery of granulocytes in patients receiving dose-intensive chemotherapy without bone marrow transplantation, *J. Clin. Oncol.* **7:**1685–1692.
17. Vadhan-Raj, S., Keating, M., LeMaistre, A., Hittelman, W. N., McCredie, K., Trujillo, J. M., Broxmeyer, H. E., Henney, C., and Gutterman, J. U., 1987, Effects of recombinant human granulocyte-macrophage colony stimulating factor in patients with myelodysplastic syndromes, *N. Engl. J. Med.* **317:**1545–1552.
18. Antin, J. H., Smith, B. R., Holmes, W., and Rosenthal, D. S., 1988, Phase I/II study of recombinant human granulocyte-macrophage colony stimulating factor in aplastic anemia and myelodysplastic syndromes, *Blood* **72:**705–713.
19. Weisbart, R. H., Golde, D. W., Clark, S. C., Wong, G. G., and Gasson, J. C., 1985, Human granulocyte-macrophage colony stimulating factor is a neutrophil activator, *Nature* **314:** 361–363.

20. Burgess, A. W., Begley, C. G., Johnson, G. R., Lopez, A. F., Williamson, D. J., Mermod, J. J., Simpson, R. J., Schnitz, A., and DeLamarter, J. R., 1987, Purification and properties of bacterially-synthesized human granulocyte-macrophage colony stimulating factor, *Blood* **69:** 43–51.
21. Atkinson, Y. H., Lopez, A. F., Marasco, W. A., Lucas, C. M., Wong, G. G., Burns, G. F., and Vadas, M. A., 1988, Recombinant human granulocyte macrophage colony stimulating factor (rhGMCSF) regulates fmet-leu-phe-receptors on human neutrophils, *Immunology* **64:**519–521.
22. Nathan, C. F., 1989, Respiratory burst in adherent human neutrophils: Triggering by colony stimulating factors CSF-GM and CSF-G, *Blood* **73:**301–306.
23. Blanchard, D. K., Michelini-Norris, M. B., and Djeu, J. Y., 1991, Production of GM-CSF by large granular lymphocytes stimulated with *Candida albicans*: Role in activation of human neutrophil function, *Blood* **77:**2259–2265.
24. Broudy, V. C., Zuckerman, K. S., Jetmalani, S., Fitchen, J. H., and Babgy, G. C., 1986, Monocytes stimulate fibroblastoid bone marrow stromal cells to produce multi-lineage hematopoietic growth factors, *Blood* **68:** 530–534.
25. Fibbe, W. E., Van Damme, J., Billiau, E., Goselink, H. M., Voogt, P. J., Van Eeden, G., Ralph, P., Altrock, B. W., and Falkerberg, J. H., 1988, Interleukin 1 induces human marrow stromal cells in long term culture to produce granulocyte colony stimulating activity and macrophage colony stimulating activity, *Blood* **71:**430–435.
26. Yunis, A. A., Wu, M. C., and Miller, A. M., 1983, Colony stimulating factor, in: *Growth and Maturation Factors* (G. Guroff, ed.), John Wiley & Sons, New York, pp. 209–248.
27. Cuturi, M. C., Anegon, I., Sherman, F., Loudon, R., Clark, S. C., Perussia, B., and Trinchieri, G., 1989, Production of hematopoietic colony stimulating factors by human natural killer cells, *J. Exp. Med.* **169:**569–583.
28. Siefff, C. A., Tsai, S., and Faller, D. V., 1989, Interleukin 1 induces cultured endothelial cell production of granulocyte-macrophage colony stimulating factor, J. Clin. Invest. **79:**48–51.
29. Munker, R., Gasson, J., Ogawa, M., and Koeffler, H. P., 1986, Recombinant human tumor necrosis factor induces production of granulocyte-macrophage colony stimulating factor, *Nature* **323:**79–82.
30. Zucali, J. R., Dinarello, C. A., Oblon, D. J., Gross, M. A., Anderson, L., and Weiner, R. S., 1986, Interleukin 1 stimulates fibroblasts to produce granulocyte macrophage colony stimulating activity and prostanglandin E2, *J. Clin. Invest.* **77:**1857–1863.
31. Zsebo, K. M., Yuschenkoff, V. N., Schiffin, S., Chang, D., McCall, E., Dinarello, C., Brown, M. A., and Altrock, B., 1988, Vascular endothelial cells and granulopoiesis: Interleukin 1 stimulates release of G-CSF and GM-CSF, *Blood* **71:**99–103.
32. Piacibello, W. L., Lu, L., Wachter, M., Rubin, B., and Broxmeyer, H. E., 1985, Release of granulocyte-macrophage colony stimulating factors from major histocompatibility complex Class II antigen-positive monocytes is enhanced by gamma interferon, *Blood* **66:**1343–1351.
33. Djeu, J. Y., and Blanchard, D. K., 1987, Regulation of human polymorphonuclear neutrophil (PMN) activity against *Candida albicans* by large granular lymphocytes via release of a PMN activating factor, *J. Immunol.* **139:**2761–2769.
34. Carswell, E. A., Old, L. J., Kassel, R. L., Green, S., Fiore, N., and Williamson, B., 1975, An endotoxin-induced serum factor that causes necrosis of tumors, *Proc. Natl. Acad. Sci. USA* **72:** 3666–3670.
35. Vilcek, J., Palombella, J. V., Henriksen-Destefano, D., Swenson, C., Feinman, R., Hirai, M., and Tsujimoto, M., 1986, Fibroblast growth-enhancing activity of tumor necrosis factor and its relationship to other polypeptide growth factors, *J. Exp. Med.* **163:**632–643.
36. Beutler, B. D., Greenwald, D., Hulmes, J. C., Chang, M., Pan, C. E., Mathison, J., Ulevitch, R., and Cerami, A., 1985, Identity of tumor necrosis factor and the macrophage-secreted factor, cachetin, *Nature* **316:**552–554.
37. Silberstein, D. S., and David, J. R., 1986, Tumor necrosis factor enhances eosinophil toxicity to *Schistosoma mansoni* larvae, *Proc. Natl. Acad. Sci. USA* **82:**1055–1059.

38. Shalaby, M. R., Aggarwal, B. B., Rinderknecht, E., Sverdersky, L. P., Finkle, B. S., and Palladino, M. A., 1985, Activation of human polymorphonuclear neutrophil function by interferon-g and tumor necrosis factor, *J. Immunol.* **135:**2069–2073.
39. Djeu, J. Y., Blanchard, D. K., Halkias, D., and Friedman, H., 1986, Growth inhibition of *Candida albicans* by human polymorphonuclear neutrophils: Activation by interferon-g and tumor necrosis factor, *J. Immunol.* **137:**2980–2984.
40. Ferrante, A., 1989, Tumor necrosis factor alpha potentiates neutrophil antimicrobial activity: Increased fungicidal activity against *Torulopsis glabrata* and *Candida albicans* and associated increase in oxygen radical production and lysosomal enzyme release, *Infect. Immun.* **57:** 2115–2122.
41. Klebanof, S. J., Vadas, M. A., Harlan, J. M., Sparks, L. H., Gamble, J. R., Agosti, J. M., and Waltersdorph, A. M., 1986, Stimulation of neutrophils by tumor necrosis factor, *J. Immunol.* **136:**4220–4225.
42. Ruff, M. R., and Gifford, G. E., Tumor necrosis factor, in: *Lymphokines II* (E. Pick and M. Landy, eds.), Academic Press, San Diego, pp. 235–248.
43. Peters, P. M., Ortaldo, J. R., Shalaby, M. R., Sverdersky, L. P., Nedwin, G. E., Bringman, T. S., Halk, P. G., Aggarwal, B. B., Herberman, R. B., Goedell, D. V., and Palladino, M. R., 1986, Natural killer-sensitive targets stimulate production of TNFa but not TNFb by highly purified human peripheral blood large granular lymphocytes, *J. Immunol.* **137:**2592–2598.
44. Djeu, J. Y., Blanchard, D. K., Richards, A. L., and Friedman, H., 1988, Tumor necrosis factor induction by *Candida albicans* from human natural killer cells and monocytes, *J. Immunol.* **141:**4047–4052.
45. Richards, A. L., Okuno, T., Takagaki, Y., and Djeu, J. Y., 1988, Natural cytotoxic cell-specific cytotoxic factor (NCCF) produced by IL3-dependent basophilic/mast cells: Relationship to tumor necrosis factor, *J. Immunol.* **141:**3061–3066.
46. Djeu, J. Y., Serbousek, D., and Blanchard, D. K., 1990, Release of tumor necrosis factor by human polymorphonuclear leukocytes, *Blood* **76:**1405–1408.
47. Lindemann, A., Riedel, D., Oster, W., Meuer, S. C., Blohm, D., Mertelsman, R. H., and Herrmann, F., 1988, Granulocyte-macrophage colony stimulating factor induces interleukin 1 production by human polymorphonuclear neutrophils, *J. Immunol.* **140:**837–839.
48. Lindemann, A., Riedel, D., Oster, W., Ziegler-Heitbrock, H. W. L., Mertelsmann, R., and Herrmann, F., 1989, Granulocyte-macrophage colony stimulating factor induces cytokine secretion by human polymorphonuclear leukocytes, *J. Clin. Invest.* **83:**1308–1312.
49. Shirafuji, N., Matsuda, S., Ogura, H., Tani, K., Kodo, H., Ozawa, K., Nagata, S., Asaon, S., and Takaku, F., 1990, Granulocyte colony stimulating factor stimulates human mature neutrophilic granulocytes to produce interferon alpha, *Blood* **75:**17–19.
50. Trinchieri, G., and Perussia, B., 1985, Immune interferon: A pleiotropic lymphokine with multiple effects, *Immunol. Today* **6:**131–136.
51. Morrison, C. J., Brummer, E., Isenberg, R. A., and Stevens, D. A., 1987, Activation of murine polymorphonuclear neutrophils for fungicidal activity by recombinant interferon gamma, *J. Leuk. Biol.* **41:**434–440.
52. Perussia, B., Dobayashi, M., Rossi, M. E., Anegon, I., and Trinchieri, 1987, Immune interferon enhances functional properties of human granulocytes: Role of Fc receptors and effect of lymphotoxin, tumor necrosis factor, and granulocyte-macrophage colony stimulating factor, *J. Immunol.* **138:**765–774.
53. Morrison, C. J., Brummer, E., and Stevens, D. A., 1989, *In vivo* activation of peripheral blood polymorphonuclear neutrophils by gamma interferon results in enhanced fungal killing, *Infect. Immun.* **57:**2953–2958.
54. Trinchieri, G., 1989, Biology of natural killer cells, *Adv. Immunol.* **47:**187–376.
55. Matsushima, K., Morishita, K., Yoshimura, T., Lavu, S., Kobayashi, Y., Lew, W., Appella, E., Kung, S. F., Leonard, E. J., and Oppenheim, J. J., 1988, Molecular cloning of cDNA for a human monocyte-derived neutrophil chemotactic factor (MDNCF) and the induction of

MDNCF mRNA by interleukin 1 and tumor necrosis factor, *J. Exp. Med.* **167**:1883–1893.
56. Baggiolini, M., Walz, A., and Kunkel, S. L., 1989, NAP/IL8, a novel cytokine that activates neutrophils, *J. Clin. Invest.* **84**:1045–1049.
57. Blanchard, D. K., Pearson, C. A., Serbousek, D., Matsushima, K., Oppenheim, J. J., and Djeu, J. Y., 1991, Large granular lymphocytes produce interleukin 8 in response to opportunistic pathogens. Submitted for publication.
58. Djeu, J. Y., Matsushima, K., Oppenheim, J. J., Shiotsuki, K., and Blanchard, D. K., 1990, Functional activation of human neutrophils by recombinant monocyte-derived neutrophil chemotactic factor/IL8, *J. Immunol.* **144**:2205–2210.
59. Wang, J. M., Colella, S., Allavena, P., and Mantovani, A., 1987, Chemotactic activity of human recombinant granulocyte macrophage colony stimulating factor, *Immunology* **60**:439–444.
60. Ji-Ming, W., Bersani, L., and Mantovani, A., 1987, Tumor necrosis factor is chemotactic for monocytes and polymorphonuclear leukocytes, *J. Immunol.* **138**:1469–1474.
61. Howard, D. H., 1965, Intracellular growth of *Histoplasma capsulatum*, *J. Bacteriol.* **89**: 518–523.
62. Newman, S., Bucher, C., and Gootee, L., 1990, Intracellular growth of *Histoplasma capsulatum* yeasts in human monocytes, cultured macrophages and alveolar macrophages, *Abstr. Annu. Meet. ASM* **F98**:425.
63. Das, S. K., and Stanley, E. R., 1982, Structure-function studies of a colony stimulating factor (CSF-1), *J. Biol. Chem.* **257**:13679–13684.
64. Smith, P. D., Lamerson, C. L., Banks, S. M., Saini, S. S., Wahl, L. M., Calderone, R. A., and Wahl, S. M., 1990, Granulocyte macrophage colony stimulating factor augments human monocyte fungicidal activity for *Candida albicans*, *J. Infect. Dis.* **161**:999–1005.
65. Wang, M., Friedman, H., and Djeu, J. Y., 1989, Enhancement of human monocyte function against *Candida albicans* by the colony stimulating factors (CSF): IL3, granulocyte-macrophage-CSF, and macrophage-CSF, *J. Immunol.* **143**:671–677.
66. Brummer, E., and Stevens, D. A., 1989, Candidacidal mechanisms of peritoneal macrophages activated with lymphokines or gamma interferon, *J. Med. Microbiol.* **28**:173–181.
67. Brummer, E., Hanson, L. H., and Stevens, D. A., 1988, Gamma interferon activation of macrophages for killing of *Paracoccidioides brasiliensis* and evidence for non-oxidative mechanisms, *Int. J. Immunopharm.* **10**:945–952.
68. Wu-Hsieh, B., and Howard, D. H., 1987, Inhibition of the intracellular growth of *Histoplasma capsulatum* by recombinant murine gamma interferon, *Infect. Immun.* **55**:1014–1016.
69. Levitz, S. M., and Farrell, T. P., 1990, Growth inhibition of *Cryptococcus neoformans* by cultured human monocytes: Role of the capsule, opsonins, the culture surface, and cytokines, *Infect. Immun.* **58**:1201–1209.
70. Douvas, G. S., Looker, D. L., Vatter, A. E., and Crowle, A. J., 1985, Gamma interferon activates human macrophages to become tumoricidal and leismanicidal, but enhances replication of macrophage-associated mycobacteria, *Infect. Immun.* **50**:1–8.
71. Murphy, J. W., and McDaniel, D. O., 1982, *In vitro* reactivity of natural killer (NK) cells against *Cryptococcus neoformans*, *J. Immunol.* **128**:1577–1583.
72. Jimenez, B. E., and Murphy, J. W., 1984, *In vitro* effects of natural killer cells against *Paracoccidiodes brasiliensis* yeast phase, *Infect. Immun.* **46**:552–558.
73. Hidore, M. R., Nabavi, N., Reynolds, C. W., Henkart, P. A., and Murphy, J. W., 1990, Cytoplasmic components of natural killer cells limit the growth of *Cryptococcus neoformans*, *J. Leuk. Biol.* **48**:15–26.
74. Yamada, T., Khardori, N., and Tewari, R. P., 1982, Natural and antibody-dependent cellular cytotoxicity (ADCC) of macrophages and lymphocytes from normal and immune mice for yeast cells of *Histoplasma capsulatum* (HC), *Abstr. Annu. Meet. ASM*. **F9**, 328.

24

Nocardial Infections

BLAINE L. BEAMAN

1. INTRODUCTION

In 1882, Robert Koch isolated the etiology of tuberculosis, which he named *Bacterium tuberculosis*. In 1896 this organism was renamed *Mycobacterium tuberculosis* by Lehmann and Neumann. Following Koch's work on tuberculosis, Nocard isolated a strictly aerobic, branching filamentous organism from cases of bovine farcy in 1888. One year later, the genus *Nocardia* was created by Trevisan and *Nocardia farcinica* became the type species. In 1890, Eppinger isolated a similar organism from a fatal systemic disease in a human. This organism was initially named *Cladothrix asteroides*; however, in 1896 it was renamed *Nocardia asteroides*. Because of the confusion and uncertainty about Nocard's original isolates of *N. farcinica*, *N. asteroides* is currently considered the type species for the genus.[1]

Thus, by the late 1890s the stage was set for recognition of a group of filamentous microorganisms that resembled both fungi and bacteria. These organisms were collectively called Actinomycetes, which literally translates to "ray fungi." Over the ensuing 100 years, the original four genera (*Actinomyces*, *Streptomyces*, *Nocardia*, and *Mycobacterium*) that constituted the Actinomycetes have been expanded and modified and new genera have been described, so that by 1989 more than 40 genera of actinomycetes were described in detail by *Bergey's Manual of Systematic Bacteriology*,[1] and new Actinomycetes are being added to this list with increasing frequency.

Even though the Actinomycetes that cause disease are no more closely

BLAINE L. BEAMAN • Department of Medical Microbiology and Immunology, University of California School of Medicine, Davis, California 95616.

Fungal Infections and Immune Responses, edited by Juneann W. Murphy *et al.* Plenum Press, New York, 1993.

related to fungi than other Gram-positive bacteria, historically these organisms have been studied by medical mycologists. Because of this tradition, it is appropriate to include discussions of the pathogenic Actinomycetes in context with immune responses to fungal infections.

2. DISEASES CAUSED BY *NOCARDIA*

There are no specific clinical manifestations that are pathognomonic for nocardial infections; however, they can be broadly grouped according to the route of infection and the subsequent clinical and pathological response of the host. These broad categories include the following: pulmonary and systemic nocardiosis, localized-cutaneous and subcutaneous nocardiosis, lymphocutaneous infections (sporotrichoid disease), and actinomycetomas.[2–8] In dairy animals, an additional category is recognized in the form of a progressive and destructive mastitis.[9]

2.1. Pulmonary and Systemic Nocardiosis

The following conditions determine the progression of disease during exposure of the host to *Nocardia*: 1) The specific metabolic status of the microorganism (i.e., log-phase cells are more virulent then stationary-phase cells);[10] 2) the relative virulence of the microorganism for the host (this aspect is complex, not fully understood and depends upon many factors, such as strain differences);[11] 3) the relative number of organisms introduced into the host;[11,12] 4) the route of exposure of the organism to the host;[12] and 5) the susceptibility of the host at the time of exposure (i.e., compromised vs. noncompromised, good nutrition vs. poor nutrition, stressed vs. nonstressed, immune vs. nonimmune, etc.).[10–15]

Humans and a wide variety of animals are susceptible to pulmonary infections as a result of inhalation of nocardial cells from the environment.[6,16] Presumably, the nocardiae become airborne on dust particles and are inhaled. If the size of the particle is sufficiently small, it can reach either the lower bronchiolus or the alveoli. As the nocardial cell becomes adherent to the surface, it may initiate growth, which in turn attracts phagocytic cells, such as either alveolar macrophages or polymorphonuclear neutrophils (PMNs). Nocardiae are readily phagocytized by these phagocytes, but they are not easily destroyed by them.[17–19] Thus, the bacterial cell may initiate growth, especially within the macrophage. As this process progresses, an inflammatory response ensues with an infiltration of macrophages, lymphocytes, and PMNs. Over a period of several days, this continues until a sufficient host response results in phagocytic cells that are "activated" to inhibit or kill the nocardiae. If the host response is not sufficient to stop the nocardial growth (due to a variety of reasons), then a progressive infection develops that may manifest itself as an expanding pneumonia, or it may become circumscribed into a pulmonary abscess. The clinical and pathological picture in pulmonary nocardiosis is variable and not specific. The infection in the lungs may remain localized, with expansion by direct extension; however, in a significant number of cases (approximately 20–25%) the nocardiae gain access to the blood and become disseminated to other regions of the body. The brain represents the major target for both primary and secondary lesions during hematogenous dissemination, but many other regions, such as the skin, kidneys, eyes, and heart, may become involved.[6,8,20–23]

Disease caused by *Nocardia* may vary from a self-limited, benign or subclinical process that may go undiagnosed to an acute, fulminating, and aggressively destructive infection. Furthermore, nocardiosis may present as either a chronic abscess, a chronic fibronodular infection, or more rarely, a granuloma. Nocardiosis often mimics other diseases, such as tumors and various types of cancers, pyogenic infections, actinomycosis, tuberculosis, and a variety of fungal infections. Therefore, nocardial infections are easily misdiagnosed.[4,8,20–28]

2.2. Localized Cutaneous and Subcutaneous Nocardiosis

Traumatic inoculation of *Nocardia* into the skin usually results in a cutaneous or subcutaneous infection that may be superficial and self-limited, or it may develop into a persistent lesion that gradually enlarges, forming pustules, localized abscesses, or pyoderma, similar to staphylococcal infections. Spread is usually by direct extension; however, dissemination can occur by way of either the bloodstream or the lymphatic system.[29–33]

2.3. Lymphocutaneous Infections (Sporotrichoid Disease)

Lymphocutaneous nocardiosis may follow an infection that results from inoculation by an abrasion to the skin or a puncture wound (i.e., a thorn prick). At the site of inoculation, the initial lesion usually develops into a purulent, crusting ulcer (pyoderma). Nodules that are frequently red, hot, and tender develop along the lymphatics. There is regional adenopathy and the clinical picture appears to be identical to the disease sporotrichosis caused by the fungus, *Sporothrix schenckii*. Therefore, this form of nocardiosis is frequently called Sporotrichoid disease.[34–40]

2.4. Actinomycetoma

A relatively painless nodule may develop at the site of a localized injury, such as a thorn or splinter prick. Following a period from weeks to months, this nodular mass may discharge through a draining tract. The lesion may extend into the muscle and bone with a concomitant cellular infiltration that is composed of polymorphonuclear neutrophils, lymphocytes, and mononuclear phagocytes and circumscribed by fibroblasts. This expanding purulogranulomatous lesion results in swelling and enlargement of the surrounding area, with the emergence of multiple secondary nodules that exude a purulent serous and serosanguineous fluid. This discharge contains characteristic granules that are composed of inflammatory cells surrounding macroscopic colonies of the infecting organism. The size, shape, and color of the granule is different for each organism; therefore, a tentative etiologic diagnosis may be made on the basis of the appearance of the granule.

Mycetomas may be caused by a variety of microorganisms, which include several fungi (Eumycetoma) as well as most of the pathogenic Actinomycetes (Actinomycetoma). This disease may develop on any region of the body that has been inoculated; however, because most cases are the result of a thorn or splinter prick, the feet and hands are the most common areas affected. Mycetomas are generally progressive and remain active for the life of the individual, so that in some instances amputation may be required. They do not respond easily to chemotherapy and can result in significant debilitation of the patient.[6,41–50]

3. ETIOLOGY OF NOCARDIAL INFECTIONS

The genus *Nocardia* is composed of Gram-positive variably acid-fast, aerobic Actinomycetes, which form branching filamentous cells that fragment into irregular rods and coccoid units. This genus is characterized by having a cell wall containing meso-diaminopimelic acid, arabinose, galactose and mycolic acids, with a total chain length that varies between C_{44} and C_{60}. The cells also contain catalase, 10-methyl stearic (tuberculostearic) acid, and tetrahydrogenated menaquinone with nine isoprene units. The percent G+C content of the DNA ranges from 64 to 72%.[1]

The type species, *N. asteroides*, along with 8–10 additional species, define the genus. There is taxomic uncertainty of several organisms within this group. Some taxonomists include *N. farcinica* is a valid species, whereas others believe that because of the historical controversy regarding its true identity, this species name should not be used, especially since *N. farcinica*, as currently defined, shares many key properties with *N. asteroides* and appears to differ from Nocard's original description. Within this review, *N. farcinica* will be considered as a variant of *N. asteroides* recognizing that, at best, *N. asteroides* is a heterogenous species that may contain as many as a dozen distinct subgroups.[1,51–54]

Although other strains of *Nocardia* are occasionally isolated from clinical material, *N. asteroides*, *N. brasiliensis*, and *N. otitidiscaviarum* (*N. caviae*) are the three species that cause most human and animal infections (except for *N. seriolae* infections in fish and *N. crassostreae* infections in oysters).[6,16,53,55] In humans, about 80% of the cases of pulmonary and systemic nocardiosis are caused by *N. asteroides* and about 15% are caused by *N. brasiliensis*. *N. otitidiscaviarum* is recognized in about 4% of the cases of systemic nocardiosis, and other *Nocardia* species represent about 1% of those identified.[55] This pattern of infection is different for cutaneous and subcutaneous infections in that *N. brasiliensis* is recognized more frequently as the etiology than *N. asteroides*. *N. brasiliensis* is almost always the etiology of sportotrichoid disease; however, *N. caviae* has been reported to cause this syndrome.[29–40] Mycetomas resulting from infection with *Nocardia* are usually caused by *N. brasiliensis*, although *N. asteroides* and *N. otitidiscaviarum* also cause mycetomas.[41–50] Similar patterns of infection occur in naturally acquired nocardiosis in other mammalian species.[16] However, *N. asteroides* is the most common cause of bovine mastitis, with some cases caused by *N. otitidiscaviarum*. To our knowledge, *N. brasiliensis* has never been recognized to cause mastitis in dairy cattle.

4. EPIDEMIOLOGY OF NOCARDIAL INFECTIONS

Pulmonary and systemic nocardiosis, cutaneous and subcutaneous infections, Sporotrichoid disease, and nocardial mycetomas have been recognized worldwide; however, infections caused by *N. asteroides* appear to be more prevalent in the temperate regions of the world, while those caused by *N. brasiliensis* predominate in the warmer, tropical, and subtropical areas. Reports of infection by *N. otitidiscaviarum* appear randomly throughout the world literature, and it appears to have no geographic preference. Although nocardial infections have been recognized in Alaska and Siberia, there are no reports of infection in the far northern and southern polar regions of the world.[2–8,20–50]

Although these three pathogens have been isolated from soil samples in almost every country in the world, the incidence of infection in humans remains unknown. In 1976, Beaman *et al.*[55] attempted to determine the frequency of recognition of nocardial infections in the United States. Data were obtained that indicated between 500 and 1000 cases were recognized each year. Currently, this is believed to be an underestimate and that it is thought that significantly more infections are now being recognized (Beaman, unpublished data).

Many factors contribute to the apparent increase in nocardial infections during the past 10 years. Probably the most significant of these is greater awareness, combined with the use of more aggressive diagnostic procedures; however, there is also an actual increase in the number of patients infected. For example, *N. asteroides* infections in AIDS patients are being recognized with increasing frequency.[56–64] Our experience, combined with reports from the literature, suggest that 1–2% of AIDS patients are recognized to acquire disseminated nocardiosis. In addition, it is important to point out that nocardial infections occur in both the "normal" patient as well as those with a large variety of immunocompromising conditions.[65–73] Since more than half of the cases reported in the literature occur in compromised patients, nocardiosis is considered an opportunistic infection; however, it should be noted that the hospitalized patient who has a known debilitation, such as cancer or organ transplantation, is more rigorously monitored for infection than a previously healthy patient in a community hospital environment. Therefore, primary infection in otherwise healthy individuals would not be easily recognized. Nevertheless, several investigators have pointed out the importance of nocardial infections in both the normal and the compromised patient.[2–8,20–50]

There is general consensus that the incidence of nocardial infection in the United States and worldwide is increasing. This increase is due to an increasing population at risk. Organ transplantation has become routine, with an ever-increasing number of transplants each year. This is also true with cancer patients on long-term therapy, and there are large numbers of patients currently receiving a variety of immunosuppressive drugs to control various underlying diseases. All of these individuals are at increased risk for acquiring nocardial infection.[2–8,20–50,55–73]

Nocardiosis may also occur with mycobacterial infections and these two could easily be misdiagnosed. It has been estimated that 5% of all patients with tuberculosis also develop nocardiosis, and there are several reports of mixed nocardial and mycobacterial infections. The reasons for this combination of mixed infection are not known, but considering that mycobacterioses are still prominent worldwide, a significant number of nocardial infections probably occur in these patients.[16,55,68,73–75] There has been an increased recognition of *Nocardia* in IV drug users, which in turn is on the increase. Therefore, yet another population of individuals previously unrecognized are at risk for developing systemic nocardiosis.[76–78] Based on our own data (unpublished), as well as more than 1000 publications describing cases of nocardial infections, a conservative estimate would indicate that several thousand persons are infected with *Nocardia* each year in the United States alone, and perhaps several hundred thousand infections may occur worldwide.

Nocardial mycetomas occur in the United States; however, they are much more common in Mexico, Central and South America, Africa, and India. Nocardiosis is more frequently recognized in the United States, and if the nocardial infections remain localized within the lung, the response to chemotherapy is usually good, especially if diagnosis is made early. Disseminated and progressive nocardiosis often do not respond to therapy and result in a high rate of mortality. The rate of survival in patients with progressive brain infections is less than 50% (50–80% of these patients die).[2–8,20–50,55–85]

Generally, it has been thought that humans are infected by *Nocardia* from the soil and that there is no human-to-human transmission. However, a small epidemic occurred in a hospital in England. Seven out of nine patients that shared a single recovery room in a hospital developed pulmonary infections caused by a single strain of *N. asteroides*. The source of this outbreak appeared to be from one patient who had pulmonary nocardiosis prior to being placed in the recovery room.[86] There have been additional reports of small, point-source outbreaks of nocardial infections in humans. For example, the CDC reported that in a cancer immunotherapy clinic in Grand Bahama Island, there was an outbreak associated with subcutaneous injections of human serum proteins in which 7 of 16 (44%) patients developed cutaneous infections caused by *N. asteroides*.[5] Similar but more extensive outbreaks of nocardial infections have been reported for a variety of animals. Most notable of these are the epidemics of chronic nocardial mastitis in dairy cattle.[9]

5. HOST DEFENSES TO INFECTION BY NOCARDIA

Both innate resistance factors and specific acquired immunity are essential for host defense against nocardial infection.

5.1. Innate Host Resistance

The skin provides an important physical barrier to *Nocardia*. These organisms to not appear to be able to cause cutaneous or subcutaneous infections without either a traumatic breach of surface integrity of the skin or by dissemination to the skin from a lesion located elsewhere within the body. Furthermore, the nocardiae are unable to colonize healthy skin for extended periods of time.[2,3,30–50]

In general, mucus secreted along the inner surfaces of the body probably blocks nocardial adherence to the surface of epithelial cells. The nocardiae become trapped in the mucus and they are removed by the action of ciliated epithelial cells, as well as by sneezing and coughing. In addition, many body fluids, such as tears, saliva, urine, and nasal secretions, provide a washing action of body surfaces, as well as containing a variety of microbicidal substances.

Studies using germ-free mice indicate that the normal bacterial flora of the body play an important role in host resistance to infection by *Nocardia*.[15] The normal flora suppress growth at superficial sites by producing inhibitory substances, competition for nutrients, and blocking adherence sites. Furthermore, the normal flora serve to increase the level of nonspecific activation of phagocytic cells in the host. Therefore, germ-free animals are significantly more susceptible to lethal infections by nocardiae than are the conventionally grown littermate controls.[15]

Once nocardiae breach the physical barriers of the body, they encounter "professional" phagocytes, which consist of polymorphonuclear neutrophils (PMN) and the mononuclear phagocyte system. The interactions of *Nocardia* with human and animal phagocytes have been studied both *in vitro* and *in vivo*.[17–19,87–98]

The interactions of *N. asteroides* with macrophages have been studied extensively. All strains of *Nocardia* are readily phagocytized, even in the absence of serum components (i.e., antibody and complement). The intracellular fate of the nocardiae, however, depends upon the strain of organism (i.e., virulent vs. avirulent) and the specific metabolic condition (i.e., log phase vs. stationary phase) of the nocardial cells when they are phagocytized. Thus cells of virulent strains of *N. asteroides* are not killed following phagocytosis, but instead, they rapidly grow within these phagocytes, resulting in the destruction of the macrophage. When less virulent strains are phagocytized by normal macrophages, some of the nocardial cells are killed; however, the surviving intracellular bacteria ultimately grow out of the macrophage. In contrast, the strains of *N. asteroides* that are relatively avirulent are killed following phagocytosis. With some nocardial strains, the cell wall is damaged and these cell-wall-defective organisms can persist for long periods of time in macrophages. Therefore, pathogenic strains of *Nocardia* are facultative intracellular pathogens.[17–19,87–89]

Virulent cells of *N. asteroides* that grow within macrophages are relatively toxic for these phagocytes.[17,18,88] At the same time they effectively inhibit phagosome-lysosome fusion,[90–92] block or neutralize phagosomal acidification,[93] and modulate lysosomal enzyme content, probably by utilizing some of these enzymes, such as acid phosphatase, as a growth substrate.[94–98] Less virulent strains have an intermediate capacity to alter these phagocyte activities, whereas the relatively avirulent organisms do not inhibit phagosome–lysosome fusion, neutralize phagosomal acidification, nor readily grow within normal macrophages.[90–98]

Bone marrow promonocytes differentiate to monocytes in the blood, whereby they migrate to various tissues to mature into macrophages. Once they become localized at specific sites within the body, they differentiate further. For example, those that remain localized in the lung become alveolar macrophages, in the liver they are Kupffer cells, in the brain they are microglia, in the kidney glomerulus they are mesangial cells, in the spleen they are splenic macrophages, and in the peritoneal cavity they are peritoneal macrophages. Each of these populations of macrophages are relatively long lived, phagocytic cells that remain localized and differentiate into distinct subpopulations that have some functional and structural differences. Thus, alveolar macrophages are distinguishable from peritoneal macrophages by a variety of criteria.[99]

The simultaneous interactions of *N. asteroides* with four subpopulations of macrophages isolated from the same animal were studied using BALB/c mice.[95,96] Comparisons of alveolar, splenic, peritoneal macrophages with

Kupffer cells demonstrated that macrophages isolated from different anatomical sites differed functionally from each other with respect to their interactions with *N. asteroides*. Kupffer cells isolated from nonimmunized mice had a decreased capacity for phagosome–lysosome fusion, and both virulent and less virulent strains of *Nocardia* grew within these phagocytes. In contrast, peritoneal and splenic marcophages had a greater capacity to phagocytize nocardial cells with subsequent increased phagosome–lysosome fusion and increased nocardicidal activity. Virulent and less virulent strains of *N. asteroides* inhibited phagosome–lysosome fusion, reduced lysosomal acid phosphatase, and grew within alveolar macrophages. On the other hand, the relatively avirulent strains of *N. asteroides* were unable to inhibit phagosome–lysosome fusion, there was an increase in lysosomal acid phosphatase, and the growth of these organisms was inhibited.[95,96] Therefore, there was good *in vitro* correlation between the functional capacities of macrophages isolated from the liver, lungs, peritoneum, and spleen of normal mice as compared to either the clearance or growth of *N. asteroides* in these anatomical sites following either intravenous, intranasal, or intraperitoneal inoculation.[12–15,93–98]

Polymorphonuclear neutrophils (PMNs) are the dominant white cell in the bloodstream, and the interactions between human PMNs and *N. asteroides* have been studied *in vitro*.[19] It was shown that nocardial cells were readily phagocytosed by these phagocytes; however, the bacteria were not killed.[19] Even though virulent strains of *Nocardia* grew within PMNs, it was demonstrated that by continually adding fresh PMNs to the nocardial–PMN mixture, the growth of nocardiae was inhibited.[100] Therefore, it was concluded that PMNs retarded the growth of *Nocardia* long enough to allow sufficient time for macrophages to become activated to kill these organisms.[19,100]

A primary microbicidal mechanism used by PMNs involves oxidative metabolism. Upon contact with a microorganism, the phagocyte is stimulated to produce a burst of oxidative metabolism. The products of this metabolic activity are toxic oxygen-containing substances such as superoxide radical, hydroxyl ion, singlet oxygen, and hydrogen peroxide. In addition, PMNs contain myeloperoxidase, which catalyzes the reaction $Cl^- + H_2O_2 \rightarrow ClO^- + H_2O$, and the resulting oxidized halide is a potent microbicidal substance. In order for a microorganism to grow in PMNs it must be able to block, neutralize, or be resistant to these toxic substances.[101]

Virulent strains of *N. asteroides* induce an oxidative metabolic burst when added to human PMNs, and these organisms are resistant to the microbicidal effects of oxidative metabolites *in vitro*.[19,102] It was found that virulent *N. asteroides* synthesized and secreted a unique superoxide dismutase (SOD) that became cell-surface associated, whereas avirulent strains of *Nocardia* did not secrete SOD.[103] In addition to the secretion of SOD, these strains of nocardiae also contain high levels of intracytoplasmic catalase. It was shown that both of these enzymes were important in protecting nocardiae from oxidative killing *in vitro* by human PMNs.[104] Furthermore, by utilizing monoclonal antibodies that neutralize SOD activity, it was demonstrated that surface associated and secreted SOD was important for the survival and growth of *N. asteroides in vivo* in the murine host.[105] Therefore, the secretion of SOD, combined with high levels of intracytoplasmic catalase, represent an important mechanism whereby nocardiae resist oxidative microbicidal activities, and this contributes to the relative virulence of these organisms.[19,100–105]

5.2. Immune Responses to *Nocardia*

Both antibodies and delayed-type hypersensitivity (DTH) can be demonstrated in humans and animals following the development of disease by *Nocardia*.[9,106–112] Thus, humoral and cell-mediated responses are evoked by nocardial infection.[106–112] The roles of these responses in host resistance to nocardiae are not clear; however, since nocardia are relatively common environmental organisms, and because contact with these organisms probably occurs with most individuals, and yet clinically apparent disease does not have a high frequency, the immune response of the "normal" host against *Nocardia* must be effective.

5.2.1. The Humoral Response

The humoral immune response to nocardia has been studied. A variety of immunoglobulins, which include IgM, IgG, and IgA have been detected following experimental infection in animals and in humans with progressive nocardial infections.[106–112] Passive immunization studies utilizing sera from immunized animals injected into nonimmunized recipients, however, have failed to show significant protection *in vivo* when challenged with virulent nocardial cells (Beaman, unpublished data).

Furthermore, *in vitro* studies with sera from immunized animals failed to demonstrate significant opsonic activity, demonstrated no microbicidal activity in either the presence or absence of complement, and showed little or no enhancement of the microbicidal activities of PMNs, monocytes, or macrophages.[92,96,106,107] These results may reflect, in part, the fact that the titers of antibody following recovery from an infection or after standard immunization are relatively low, and they may not be directed against surface antigens relevant to the virulence or pathogenicity of the nocardial strain being studied.

There are studies that indicate antibodies may enhance progressive nocardial disease.[113] It was shown that chronic lesions induced by *Nocardia* contain large amounts of antigen–antibody complexes. It is suggested that these immune complexes exacerbate the severity and chronicity of the lesions and play a role in the extensive tissue damage that results from these lesions. Therefore, in these instances a strong humoral response would be detrimental to the host and may be important for the pathogenesis of the nocardial cells.[113]

The role of B cells in host resistance to nocardial infection was investigated by using congenitally B-cell-deficient CBN mice.[106] By immunizing these mice with virulent cells of *N. asteroides*, it was found that no detectable antibody was produced in the B-cell-deficient mice, whereas a good antibody titer was detected in the littermate controls that had normal B-cell functions. Growth and clearance data obtained from these mice following intravenous injection of nocardial cells demonstrated that the mice that had no antibody were more resistant to infection than the "normal" mice that had a good humoral response. This suggested that antibody might adversely affect host clearance mechanisms.[106,113] Therefore, functional B cells appeared not to be important in enhanced host resistance to *N. asteroides*.[106]

Studies utilizing purified antibodies specific for defined cell-surface antigens indicate that high titers of certain antibodies may be protective for the host. For example, it was demonstrated *in vitro* that polyclonal rabbit IgG that neutralized nocardial superoxide dismutase activity significantly enhanced the PMN killing of virulent cells of *N. asteroides*.[104] Furthermore, the passive transfer of murine monoclonal IgM antibody that neutralized nocardial SOD activity into mice significantly enhanced the ability of the mice to retard nocardial growth *in vivo*.[105] Therefore, it should be possible to use high titer antibody against specific nocardial components to enhance host resistance.

5.2.2. *Cell-Mediated Immunity*

Generally, humans who have a deficiency in cell mediated immunity are more susceptible to developing progressive nocardiosis than those who have an intact cell-mediated response; those who have an abnormal humoral response do not appear to be at increased risk.[2–8,20–50,55–86] These observations suggest that cell-mediated and not humoral immunity are most important in host defense against invasive nocardial infection. Research using a variety of animal models indicates that cell-mediated immunity is the mechanism whereby the host resists nocardial infection.[11–15,114–119]

It has been demonstrated both *in vitro* and *in vivo* that T lymphocytes are involved and important in the protection of the host against nocardial infection.[96,107,114,119,120] T cells appear to be involved in at least two distinct mechanisms of host resistance to nocardial infection. The first is their well-established role in immunologically specific interactions with subpopulations of other T cells and macrophages to augment macrophage microbicidal activity.[96,107,114,119] By using T-cell-deficient mice (nude mice), it was shown that the T cell is important for protection against chronic infection.[96,107,114,119] Furthermore, adoptive transfers of subpopulations of T cells from immunized animals into T-cell-deficient recipients demonstrated that T cells significantly enhanced host clearance of virulent strains of *N. asteroides* following IV inoculation.[107]

The function of cell-mediated responses to *Nocardia* has been studied *in vitro*.[18,92,95] Using *in vitro* maintained alveolar macrophages from normal and immunized rabbits, it was shown that "activated" macrophages alone were not sufficient to kill ingested cells of virulent strains of *N. asteroides*.[18,92,95] The most effective macrophage activity occurred when "activated" macrophages from immunized rabbits were incubated with primed T cells from lymph nodes of the same animal in the presence of immune serum and alveolar lining material. The physical presence of the T cells was essential, since their removal from the mixture eliminated the ability of the macrophages to kill virulent *Nocardia*.[92] Thus, the intact cell-mediated response, with all of its components, is necessary to adequately kill virulent cells of *N. asteroides*. Removal of any one of these components reduced the ability of macrophages to kill these bacteria.[92]

A second role for T lymphocytes in host resistance to *Nocardia* depends upon specific binding of the nocardial cell to the lymphocyte.[120] It was found that immunologically specific T lymphocytes could be isolated from the spleens of mice following immunization. These T cells were able to bind and kill *Nocardia in vitro*.[120] The direct T-cell-mediated killing of nocardia was specific for the organism used to immunize the animals, and normal T cells from nonimmunized mice could neither bind nor kill nocardial cells. The killing capacity of the lymphocyte from immunized animals was enhanced by *in vitro* priming of the lymphocytes with nocardial cell wall fragments 12–48 hours before challenge. Specific antibody had no effect on this killing; therefore, this lymphocyte–nocardial interaction was antibody independent. It was observed that the nocardial cell would tightly adhere to the surface membrane of the T lymphocyte, and at this point of contact the nocardial cell wall was damaged, resulting in bacterial lysis.[120] The role of this immunologically specific direct T-lymphocyte killing of *Nocardia asteroides* in human resistance to nocardial infections is not known.

5.2.3. Regulation of Immune Response

The regulation of the immune response of the host following exposure to *Nocardia* is not known. Generally, it is believed that macrophages phagocytize nocardial cells, and some undefined components of the nocardial cell serve to trigger a response. The macrophages then interact with T lymphocytes. Presumably, nocardial components processed by the macrophages interact with specific lymphocytes. The populations of T cells are expanded, and there is an elaboration of a variety of lymphokines and cytokines that regulate additional populations of T cells, B cells, macrophages and PMNs. The functional capacities of macrophages, and possibly PMNs, become modified, and these phagocytes are activated, with a resulting increase in microbicidal capacity (activation).

The specific nocardial components that are involved in this process are not known; however, it is suggested that the glycolipid, cord factor, may induce tumor necrosis factor (TNF),[121] while a cell-wall subunit structure may stimulate B cells directly (B cell mitrogen),[122] and additional cell-wall components may affect the regulation and enhancement of the host immune response, especially T-cell activity.[120–122] The relationships of the cytokines and lymphokines secreted by the host in regulating these responses are not known. However, studies with γ-interferon and TNF suggest that, contrary to other facultative intracellular pathogens, the growth of *Nocardia* is stimulated in macrophages activated with these factors. Thus, INF-γ and TNF do not activate either murine peritoneal and alveolar macrophages or human peripheral blood monocytes to kill virulent cells of *N. asteroides*.[123] These phagocytic cells activated by INF-γ or TNF, however, do kill *Coccidioides immitis*, *Listeria monocytogenes*, and *Toxoplasma gondii*.[123] Therefore, the mechanisms whereby macrophages are activated to kill virulent strains of *Nocardia* in the immune host require further study.

6. IMMUNODIAGNOSIS AND PROGNOSIS

There have been numerous studies on the serological relatedness of *Nocardia* species, and attempts to develop diagnostic procedures using serology and skin test reactions (DTH) have been reported.[108–113,124–139] (It is not possible to list all of the appropriate references in this review). The major problems with most of these studies are that the tests are either not specific enough, too specific, or lack sensitivity. As a consequence, highly accurate and specific serodiagnoses of nocardial infections have not been developed for routine use. More recently, there have been studies that have identified a major genus-specific extracellular antigen that has promise as a routine immunodiagnostic test.[112,124–127] A 55,000 molecular weight extracellular protein was partially purified by Sugar and colleagues.[112,124,125] This antigen appears to be specific for recognizing humans that have active disease caused by either *N. asteroides*, *N. brasiliensis*, or *N. otitidiscaviarum*.[125] Pier and colleagues in earlier studies found a similar extracellular antigen that was useful for diagnotic purposes in monitoring infected dairy cattle.[135–138] In our laboratory we found that Pier's diagnostic extracellular antigen appears to be the same as Sugar's 55-kDa extracellular diagnostic antigen. More recently, Boiron and Provost[126,127] reported the use of a 54-kDa immunodominant antigen for the diagnosis of nocardiosis. In all probability, this 54-kDa antigen is the same as the 55-dKa antigen used by Sugar and associates.[112,124–127] Based upon our own studies (Beaman et al., unpublished data), combined with the independent studies of Sugar and associates,[112,124,125] Boiron and associates,[126,127] and others,[128–139] it appears that several extracellular proteins secreted by *Nocardia* species (i.e., 31-, 43-, 55-, and 62-kDa proteins) are useful antigenic markers. These proteins permit serodiagnosis of patients who develop disease due to infection with the three pathogenic species of *Nocardia*.[112,124–139] Furthermore, the 55-kDa protein is the most promising diagnostic antigen that may be successfully used in the routine serodiagnosis of nocardial disease.[112,124–127]

Serodiagnosis of patients with nocardial infection permits an earlier and more rapid recognition of nocardial-induced disease. It provides a basis for earlier chemotherapy, which improves the prognosis of the disease. The following case underscores the usefulness of serodiagnosis. In 1988, a 68-year-old woman in Kansas presented with a chronically infected foot that resulted from an injury with a small rock caught between the shoe and the instep of the foot. The infection had gradually worsened over a period of several months, and all efforts to obtain a culture had failed. Bacteria were not observed in Gram-stained preparations of the purulent discharge. Serum from the patient was obtained and the reactivity of the serum to 55-kDa antigen was determined by ELISA. She was found to have a strong serological response to the 55-kDa antigen. Based upon this result, she was treated with trimethoprim–sulfamethoxazole. The infection in the foot rapidly cleared, and the patient has remained asymptomatic to this day. Even though *Nocardia* was never observed in or isolated from this patient's foot, based upon the serologic results and the rapid successful response to trimethoprim–sulfamethoxazole, it is almost certain that she had a nocardial infection that was recognized serologically and treated successfully (Beaman *et al.*, unpublished data).

7. THERAPY OF NOCARDIAL INFECTIONS

During the past 30 years, a variety of chemotherapeutic agents have been used to treat nocardial infection. (It is impossible to review all of these here.) These have often failed and, currently, nocardial infections are frequently difficult to treat. In general, the most successful therapy has traditionally relied upon the use of sulfonamides, often in combination with surgical intervention. More recently, several investigators have stressed that the combination of trimethoprim and sulfamethoxazole should be used as the drugs of choice for treating patients with nocardiosis.[140–143]

Unfortunately, all patients cannot tolerate sulfonamides or trimethoprim–sulfamethoxazole, and the infections often do not respond to these drugs (especially disseminated or well-established chronic infections).[144] Therefore, nocardial susceptibility to additional antibiotics have been studied both *in vitro* and *in vivo*.[145–148] Generally, the nocardiae are resistent to the penicillins and other β-lactam antibiotics because these organisms often produce β-lactamase or they are intrinsically resistant (probably due to decreased permeability of the cell wall).[149] It was demonstrated that some of the newer quinolones and β-lactam antibiotics were active against *Nocardia* by using *in vitro* assays. Thus, Ciprofloxacin, Ceftriaxone, Cefpirome, Cefotaxime, Ceftizoxime, Augmentin, and Timentin had good *in vitro* activity against either *N. asteroides* or *N. brasiliensis*.[145–148] Both *in vitro* and *in vivo* studies indicate that nocardiae are susceptible to amikacin and imipenem, and there are several reports of successful therapy of central nervous system infection and disseminated disease by using either amikacin[150–152] or imipenem.[153]

The mortality rate for nocardial infections is not known; however, based upon reviews of the literature, it appears that the overall rate of successful therapy has increased during the past 10 years. This increase is due to better and earlier diagnosis of disease, combined with more prompt administration of appropriate antibiotics, which now include not only the sulfonamides but also trimethoprim–sulfamethoxazole, amikacin, and imipenem.[141–153] Still, significant numbers of patients succumb to nocardial infections, even with early administration of these compounds, thus underscoring the need for more active antibiotics and better, more rapid diagnostic procedures.

8. CONCLUSIONS

There are more than 1000 publications in the world literature (most of these since 1960) describing several thousand human infections caused by the genus *Nocardia* (it is impractical to list them all in this review). Based upon the data presented in these publications, it can be concluded that nocardial infections are worldwide in occurrence; difficult to diagnose; underestimated in incidence; increasing in incidence; often unresponsive to therapy; carry a high rate of mortality; occur both as a primary pathogen in otherwise healthy people as well as behaving as an opportunistic pathogen in patients with a wide variety of predisposing conditions; involve multiple organs, which include the lung, brain, skin, muscle, bone, eyes, kidneys, heart, and virtually all other organ systems; localize in any of the above listed anatomical sites or frequently disseminate to numerous sites; affect males more frequently than females; affect middle-aged adults (40–60 years of age) more frequently than those <30 years or >70 years of age; and are relatively common in occurrence, which is contrary to the perception by the medical community that nocardial infections are rare.

ACKNOWLEDGMENTS. Much of the research reported in this review was supported by Public Health Service grant AI-20900 from the National Institute of Allergy and Infectious Diseases. I thank Theresa Andreozzi for typing this manuscript.

REFERENCES

1. Lechevalier, H. A., 1989, Nocardioform Actinomycetes, in: *Bergey's Manual of Systematic Bacteriology*, Vol. 4 (S. T. Williams, M. E. Sharpe, and J. G. Hold, eds.), Williams and Wilkins, Baltimore, pp. 2348–2404.

2. Hachisuka, H., Ichiki, M., Yoshida, N., Nakano, S., and Sasai, Y., 1989, Primary subcutaneous abscess caused by *Nocardia otitidiscaviarum*, *J. Am. Acad. Dermatol.* **21:**137–139.
3. Hay, R. J., 1983, Nocardial infection of the skin, *J. Hyg. Camb.* **91:** 385–391.
4. Mammana, R. B., Peterson, E. A., Fuller, J. K., Siroky, K., and Copeland, J. G., 1983, Pulmonary infections in cardiac transplant patients: Modes of diagnosis, complications and effectiveness of therapy, *Ann. Thor. Surg.* **36:**700–705.
5. Gregg, M. B. (ed.), 1984, Cutaneous nocardiosis in cancer patients receiving immunotherapy injections, *MMWR* **33:**471–472.
6. Schaal, K. P., and Beaman, B. L., 1983, Clinical significance of Actinomycetes, in: *Biology of the Actinomycetes* (M. Goodfellow, M., Mordarski, and S. T. Williams, eds.), Academic Press, London, pp. 389–424.
7. Schwartz, J. G., McGough, D. A., Thomer, R. E., Fetchick, R. J., Tio, F. O., and Rinaldi, M. G., 1988, Primary lymphocutaneous *Nocardia brasiliensis* infection: Three case reports and a review of the literature, *Diag. Microbiol. Infect. Dis.* **10:**113–120.
8. Wilson, J. P., Turner, H. K., Kirchner, K. A., and Chapman, W. W., 1989, Nocardial infections in renal transplant recipients, *Medicine* **68:**38–54.
9. Bushnell, R. B., Pier, A. C., Fichtner, R. E., Beaman, B. L., Boos, H. A., and Salman, M. D., 1979, Clinical and diagnostic aspects of herd problems with nocardial and mycobacterial mastitis, *Am. Assoc. Vet. Lab. Diag.* **22:**1–12.
10. Beaman, B. L., and Maslan, S., 1978, Virulence of *Nocardia asteroides* during its growth cycle, *Infect. Immun.* **20:**290–295.
11. Beaman, B. L., and Maslan, S., 1977, The effect of cyclophosphamide on experimental *Nocardia asteroides* infection in mice, *Infect. Immun.* **16:**995–1004.
12. Beaman, B. L., Maslan, S., Scates, S., and Rosen, J., 1980, Effect of route of inoculation on host resistence to *Nocardia*, *Infect. Immun.* **28:**185–189.
13. Beaman, B. L., Gershwin, M. E., and Maslan, S., 1978, Infectious agents in immunodeficient murine models: Pathogenicity of *Nocardia asteroides* in congenitally athymic (nude) and hereditarily asplenic (Dh/+) mice, *Infect. Immun* **20:**381–387.
14. Beaman, B. L., Goldstein, E., Gershwin, M. E., Maslan, S., and Lippert, W., 1978, Lung response of congenitally athymic (nude), heterozygous, and Swiss Webster mice to aerogenic and intranasal infection by *Nocardia asteroides*, *Infect. Immun.* **22:**867–877.
15. Beaman, B. L., Gershwin, M. E., Scates, S. M., and Ohsugi, Y., 1980, Immunobiology of germfree mice infected with *Nocardia asteroides*, *Infect. Immun.* **29:**733–743.
16. Beaman, B. L., and Sugar, A. M., 1983, Nocardia in naturally acquired and experimental infections in animals, *J. Hyg. Camb.* **91:**393–419.
17. Beaman, B. L., 1977, The *in vitro* response of rabbit alveolar macrophages to infection with *Nocardia asteroides*, *Infect. Immun.* **15:**925–937.
18. Beaman, B. L., 1979, Interaction of *Nocardia asteroides* at different phases of growth with *in vitro*-maintained macrophages obtained from the lungs of normal and immunized rabbits, *Infect. Immun.* **26:**355–361.
19. Filice, G. A., Beaman, B. L., Krick, J. A., and Remington, J. S., 1980, Effects of human neutrophils and monocytes on *Nocardia asteroides*: Failure of killing despite occurrence of the oxidative metabolic burst, *J. Infect. Dis.* **142:**432–438.
20. Frazier, A. R., Rosenow, E. C., and Roberts, G. D., 1975, Nocardiosis: A review of 25 cases occurring during 24 months, *Mayo Clin. Proc.* **50:**657–663.
21. Freese, J. W., Young, W. G., Sealy, W. C., and Conant, N. F., 1963, Pulmonary infection by *Nocardia asteroides*: Findings in eleven cases, *J. Thorac. Cardiovasc. Surg.* **46:**537–547.
22. Goodman, J. S., and Koenig, M. G., 1970, Nocardia infections in a general hospital, *Ann. N.Y. Acad. Sci.* **174:**552–567.
23. Neu, H. C., Silva, M., Hazen, E., and Rosenheim, S. H., 1967, Necrotizing nocardial pneumonitis, *Ann. Intern. Med.* **66:**274–284.
24. Barnicoat, M. J., Wierzbicki, A. S., and Morgan, P. M., 1989, Cerebral nocardiosis in immunosuppressed patients: Five cases, *Q. J. Med., New Series* **72:**689–698.

25. Fernandez, M., Paredes, L., Castillo, M., and Donoso, A., 1987, Nocardiosis: Aspectos clinicos y bacteriologicos experiencia en 16 casos, *Rev. Med. Chile* **115:**1053–1060.
26. Palmer, D. L., Harvey, R. L., and Wheeler, J. K., 1974, Diagnostic and therapeutic considerations in *Nocardia asteroides* infection, *Medicine* **53:**391–401.
27. Hodges, G. R., and Rosett, W., 1978, Recent experiences with nocardial infections, *Am. J. Med. Sci.* **276:**279–285.
28. Young, L. S., Armstrong, D., Blevins, A., and Lieberman, P., 1971, *Nocardia asteroides* infection complicating neoplastic disease, *Am. J. Med.* **50:**356–367.
29. Berd, D., 1973, *Nocardia brasiliensis* infection in the United States: A report of 9 cases and a review of the literature, *Am. J. Clin. Pathol.* **60:**254–258.
30. Cahen, R., Boibieux, A., Chomarat, M., Trolliet, P., Francois, B., and Peyramont, D., 1987, Nocardiose sous-cutanee primitive compliquant un syndrome nephrotique sous corticotherapie, *La Presse Medicale* **16:**1242–1243.
31. Haim, S., and Merzbach, D., 1979, Primary chancriform syndrome caused by *Nocardia asteroides*, *Mykosen* **22;**360–363.
32. Satterwhite, T. K., and Wallace, R. J., 1979, Primary cutaneous nocardiosis, *JAMA* **242:** 333–336.
33. Kalb, R. E., Kaplan, M. H., and Grossman, M. E., 1985, Cutaneous nocardiosis, *J. Am. Acad. Dermatol.* **13:**125–133.
34. Hessen, M. T., and Santoro, J., 1988, Lymphocutaneous nocardiosis in Pennsylvania, *Penn. Med.* **91:**54–58.
35. Landau, J., Feld, S., Frumkin, A., and Resnitzky, P., 1986, *Nocardia brasiliensis* skin infections, *Israel J. Med. Sci.* **22:**397–399.
36. Mitchell, G., Wells, G. M., and Goodman, J. S., 1975, Sporotrichoid *Nocardia brasiliensis* infection, *Amer. Rev. Resp. Dis.* **112:**721–723.
37. Moeller, C. A., and Burton, C. S., 1986, Primary lymphocutaneous *Nocardia brasiliensis* infection, *Arch. Dermatol.* **122:**1180–1190.
38. Moore, M., and Conrad, A. H., 1967, Sporotrichoid nocardiosis caused by *Nocardia brasiliensis*, *Arch. Dermatol.* **95:**390–393.
39. Wishnitzer, R., Berrebi, A., vonder Walde, J., Miskin, A., and Frumkin, A., 1984, *Nocardia brasiliensis* lymphocutaneous syndrome in a patient with lymphosarcoma, *Dermatologica* **169:**215–216.
40. Wlodaver, C. G., Tolomeo, T., and Benear, J. B., 1988, Primary cutaneous nocardiosis mimicking sporotrichosis, *Arch. Dermatol.* **124:**659–660.
41. Alteras, Il., and Feuerman, E. J., 1986, The second case of actinomycetoma due to *Nocardia caviae* in Israel, *Mycopathology* **93:**185–187.
42. Barnetson, R. S., and Milne, L. J. R., 1978, Mycetoma: A review, *Br. J. Dermatol.* **99:**227–231.
43. Carlisle, J. T., Greer, D. L., and Hyslop, N. E., 1988, Actinomycetoma of the hand caused by *Nocardia asteroides*, *J. Infect. Dis.* **158:**244–246.
44. Green, W. O., and Adams, T. E., 1964, Mycetoma in the United States: A review and report of seven additional cases, *Am. J. Clin. Pathol.* **42:**75–91.
45. Sanyal, M., Thammayya, A., and Basu, N., 1978, Actinomycetoma caused by organisms of the *Nocardia asteroides* complex and closely related strains, *Mykosen* **21:**109–121.
46. Saul, A., Bonifaz, A., Messina, M., and Andrade, R., 1987, Mycetoma due to *Nocardia caviae*, *Int. J. Dermatol.* **26:**174–177.
47. Serrano, J. A., Beaman, B., Mejia, M. A., Viloria, J. E., and Zamora, R., 1988, Histological and microbiological aspects of actinomycetoma cases in Venezuela, *Rev. Inst. Med. Trop. Sao Paulo* **30:**297–304.
48. Serrano, J. A., Beaman, B. L., Viloria, J. E., Mejia, M. A., and Zamora, R., 1986, Histological and ultrastructural studies of human actinomycetomas, in: *Biological, Biochemical and Biomedical Aspects of Actinomycetes* (G. Szabo, S. Biro, and M. Goodfellow, eds.) FEMS Symposium 34, Akademiai Kiado, Budapest, pp. 647–662.
49. Talivar, P., and Sehgal, S. C., 1979, Mycetomas in North India, *Sabouradia* **17:**287–291.

50. Tight, R. R., and Bartlett, M. S., 1981, Actinomycetoma in the United States, *Rev. Infect. Dis.* **3**:1139–1150.
51. Kurup, V. P., Piechura, J. E., Ting, E. Y., and Orlowski, J. A., 1983, Immunochemical characterization of *Nocardia asteroides* antigens: Support for a single species concept, *Can. J. Microbiol.* **29**:425–432.
52. Nocard, E., 1888, Note sur la maladie de boeufs de la Guadeloupe, connue sons le nom de farcin, *Ann. Inst. Pasteur*, Paris **2**:293–302.
53. Schaal, K. P., 1988, Actinomycetes as human pathogens, in: *Biology of Actinomycetes '88* (Y. Okami, T. Beppu, and H. Ogawara, eds.), Japan Scientific Societies Press, Tokyo, pp. 277–282.
54. Tsukamura, M., 1982, Rejection of the name *Nocardia farcinica* Trevisan 1889: Request for an opinion, *Int. J. Syst. Bacteriol.* **32**:235–236.
55. Beaman, B. L., Burnside, J., Edwards, B., and Causey, W., 1976, Nocardial infections in the United States, 1972–1974, *J. Infect. Dis.* **134**:286–289.
56. Adair, J. C., Beck, A. C., Apfelbaum, R., and Baringer, R., 1987, Nocardial cerebral abscess in the Acquired Immunodeficiency Syndrome, *Arch. Neurol.* **44**:548–550.
57. Gaspar, G., Alia, I., Esquinas, G., and Ballesteros, P., 1989, Primary cutaneous nocardiosis and human immunodeficiency virus infection, *Med. Clin.* **92**:598–599.
58. Holtz, H. A., Lavery, D. P., and Kapila, R., 1985, Actinomycetales infection in the Acquired Immunodeficiency Syndrome, *Ann. Intern. Med.* **102**:203–205.
59. Macher, A. M., 1988, The pathology of AIDS, *Pub. Health Rep.* **103**:246–254.
60. Macher, A. M., DeVinatea, M., Daly, M., and Stocker, J. T., 1986, AIDS: Case for diagnosis, *Milit. Med.* **151**:73–80.
61. Mallolos, J., Martinez, J. A., Gatell, J. M., and Soriano, E., 1988, Pulmonary nocardiosis as a presenting form of Acquired Immunodeficiency Syndrome, *Med. Clin.* **90**:91–92.
62. Pachar, J. V., Crooks, D. A., and Ermocilla, R., 1987, Disseminated mycobacteriosis and nocardiosis in a patient with Acquired Immunodeficiency Syndrome, *Rev. Med. Panama* **12**: 131–134.
63. Rodriquez, J. L., Barrio, J., and Pitchenik, A.E., 1986, Pulmonary nocardiosis in the Acquired Immunodeficiency Syndrome, *Chest* **90**:912–914.
64. Sharer, L. R., and Kapila, R., 1985, Neurological observations in Acquired Immunodeficiency Syndrome, *Acta Neuropathol.* **66**:188–198.
65. Abdi, E. A., Ding, J. C., and Cooper, I. A., 1987, Nocardial infection in splenectomized patients: Case reports and a review of the literature, *Postgrad. Med. J.* **63**:455–458.
66. Berkeley, P., and Bodey, G. P., 1989, Nocardial infection in patients with neoplastic disease, *Rev. Infect. Dis.* **11**:407–412.
67. Brechot, J. M., Capron, F., Prudent, J., and Rochemaure, J., 1987, Unexpected pulmonary nocardiosis in a non-immunocompromised patient, *Thorax* **42**:479–480.
68. Brown, R. B., Sands, M., and Ryczak, M., 1986, Community-acquired pneumonia caused by mixed aerobic bacteria, *Chest* **90**:810–814.
69. Camp, M., Mehta, J. B. and Whitson, M., 1987, Bronchiolitis obliterans and *Nocardia asteroides* infection of the lung, *Chest* **92**:1107–1108.
70. Curry, W. A., 1980, Human nocardiosis: A clinical review with selected case reports, *Arch. Intern. Med.* **14**:818–826.
71. Murray, J. F., Finegold, S. M., Froman, S., and Will, D. W., 1961, The changing spectrum of nocardiosis: A review and presentation of nine cases, *Am. Rev. Resp. Dis.* **83**:315–330.
72. Rosendale, D. E., Myers, C., Boyko, E. J., and Jafek, B., 1988, *Nocardia asteroides* cervical osteomyelitis in an immunocompetent host, *Otolarygol. Head Neck Surg.* **99**:334–337.
73. Simpson, G. L., Stinson, E. B., Egger, M. J., and Remington, J. S., 1981, Nocardial infection in the immunocompromised host: A detailed study of a defined population, *Rev. Infect. Dis.* **3**:492–507.
74. McQuown, A. L., 1955, Actinomycosis and nocardiosis, *Am. J. Clin. Pathol.* **25**:2–8.

75. Neimeister, R. P., Patterson, N. T., Cocklin, J. H., and Harada, G. F., 1971, The isolation of pathogenic fungi from sodium hydroxide-processed sputum specimens from patients suspected to have tuberculosis, *Am. J. Clin. Pathol.* **54**:201–203.
76. Hershewe, G. L., Davis, L. E., and Bicknell, J. M., 1988, Primary cerebellar brain abscess from nocardiosis in a heroin addict, *Neurology* **38**:1655–1656.
77. Siao, P., McCabe, P., and Yagnik, P., 1989, Nocardial spinal epidural abscess, *Neurology* **39**:996.
78. Vanderstigel, M., Leclercq, R., Brun-Buisson, C., Schaeffer, A., and Duval, J., 1986, Blood-borne pulmonary infection with *Nocardia asteroides* in a heroin addict, *J. Clin. Microbiol.* **23**: 175–176.
79. Cosnett, J. E., Moodley, M., Bill, P., and Bullock, R., 1988, Operculum syndrome a brain abscess in a left hander, *J. Neurol. Neurosurg. Psychiatry* **51**:307–308.
80. Dietlein, E., Firshing, R., and Peters, G., 1988, Therapy of brain abscess caused by *Nocardia farcinica*, *Med. Klin.* **83**:613–614.
81. Fried, J., Hinthorn, D., Ralston, J., Gerjarusak, P., and Liu, C., 1988, Cure of brain abscess caused by *Nocardia asteroides resistant to multiple antibiotics, South. Med. J.* **81**:412–413.
82. Hadley, M. N., Spetzler, R. F., Martin, N. A., and Johnson, P. C., 1988, Middle cerebral artery aneurysm due to *Nocardia asteroides*: Case report of aneurysm excision and extracranial-intracranial bypass, *Neurosurgery* **22**:923–928.
83. Hall, W. A., Martinez, A. J., Dummer, J. S., and Lansford, L. D., 1987, Nocardial brain abscess: Diagnostic and therapeutic use of stereotactic aspiration, *Surg. Neurol.* **28**:114–118.
84. Ludmerer, K. M., and Kissane, J. K. (eds.), 1989, Headache, mental status changes and death in a 36 year old woman with Lupus, *Am. J. Med.* **86**:94–102.
85. Schmid, A., Traupe, H., Todt, H. C., and Trittmacher, S., 1977, Intracerebral *Nocardia brasiliensis* infection, *Neurosurg. Rev.* **10**:315–319.
86. Stevens, D. A., Pier, A. C., Beaman, B. L., Morozumis, P. A., Lovett, I. S., and Houang, E. T., 1981, Laboratory evaluation of an outbreak of nocardiosis in immunocompromised hosts, *Am. J. Med.* **71**:928–934.
87. Bourgeois, L., and Beaman, B. L., 1974, Probable L-forms of *Nocardia asteroides* induced in cultured mouse peritoneal macrophages, *Infect. Immun.* **9**:576–590.
88. Beaman, B. L., and Smathers, M., 1976, Interaction of *Nocardia asteroides* with cultured rabbit alveolar macrophages, *Infect. Immun.* **13**:1126–1131.
89. Filice, G. A., Beaman, B. L., and Remington, J. S., 1980, Effects of activated macrophages on *Nocardia asteroides*, *Infect. Immun.* **27**:643–649.
90. Davis-Scibienski, C., and Beaman, B. L., 1980, Interaction of *Nocardia asteroides* with rabbit alveolar macrophages: Association of virulence, viability, ultrastructural damage, and phagosome-lysosome fusion, *Infect. Immun.* **28**:610–619.
91. Davis-Scibienski, C., and Beaman, B. L., 1980, Interaction of *Nocardia asteroides* with rabbit alveolar macrophages: Effect of growth phase and viability on phagosome-lysosome fusion, *Infect. Immun.* **29**:24–29.
92. Davis-Scibienski, C., and Beaman, B. L., 1980, Interaction of alveolar macrophages with *Nocardia asteroides*: Immunological enhancement of phagocytosis, phagosome-lysosome fusion, and microbicidal activity, *Infect. Immun.* **30**:578–587.
93. Black, C. M., Paliescheskey, B., Beaman, B. L., Donovan, R. M., and Goldstein, E., 1986, Acidification of phagosomes in murine macrophages: Blockage by *Nocardia asteroides*, *J. Infect. Dis.* **154**:952–958.
94. Black, C. M., Beaman, B. L., Donovan, R. M., and Goldstein, E., 1983, Effect of virulent and less virulent strains of *Nocardia asteroides* on acid phosphatase activity in alveolar and peritoneal macrophages maintained *in vitro*, *J. Infect. Dis.* **148**:117–124.
95. Black, C. M., Beaman, B. L., Donovan, R. M., and Goldstein, E., 1985, Intracellular acid phosphatase content and ability of different macrophage populations to kill *Nocardia asteroides*, *Infect. Immun.* **47**:375–383.

96. Beaman, B. L., and Black, C., 1985, Interaction of *Nocardia asteroides* in BALB/c mice: Modulation of macrophage function, enzyme activity, and the induction of immunologically specific T-cell bactericidal activity, *Current Topics Microbiol. Immunol.* **122:** 138–147.
97. Black, C., Paliescheskey, M., Beaman, B. L., Donovan, R. M., and Goldstein, E., 1986, Modulation of lysosomal protease-esterase and lysozyme in Kupffer cells and peritoneal macrophages infected with *Nocardia asteroides, Infect. Immun.* **54:**917–919.
98. Beaman, L., Paliescheskey, M., and Beaman, B. L., 1988, Acid phosphatase stimulation of the growth of *Nocardia asteroides* and its possible relationship to the modulation of lysosomal enzymes in macrophages, *Infect. Immun.* **56:**1652–1654.
99. Roitt, I., Brostoff, J., and Male, D., 1989, *Immunology*, Gower Medical Publishing, London; Harper and Row Publishers, New York, pp. 1.2–25.13.
100. Filice, G. A., 1985, Inhibition of *Nocardia asteroides* by neutrophils, *J. Infect. Dis.* **151:**47–56.
101. Beaman, L., and Beaman, B. L., 1984, The role of oxygen and its derivatives in microbial pathogenesis and host resistance, *Ann. Rev. Microbiol.* **38:**27–48.
102. Filice, G. A., 1983, Resistance of *Nocardia asteroides* to oxygen-dependent killing by neutrophils, *J. Infect. Dis.* **148:**861–867.
103. Beaman, B. L., Scates, S. M., Moring, S. E., Deem, R., and Mishra, H. P., 1983, Purification and properties of a unique superoxide dismutase from *Nocardia asteroides, J. Biol. Chem.* **258:**91–96.
104. Beaman, B. L., Black, C. M., Doughty, F., and Beaman, L., 1985, Role of superoxide dismutase and catalase as determinants of pathogenicity of *Nocardia asteroides*: Importance in resistance to microbicidal activities of human polymorphonuclear neutrophils, *Infect. Immun.* **47:**135–141.
105. Beaman, L., and Beaman, B. L., 1990, Monoclonal antibodies demonstrate that superoxide dismutase contributes to protection of *Nocardia asteroides* within the intact host, *Infect. Immun.* **58:**3122–3128.
106. Beaman, B. L., Gershwin, M. E., Ahmed, A., Scates, S. M., and Deem, R., 1982, Responses of CBA/N × DBA2/F mice to *Nocardia asteroides, Infect Immun.* **35:**111–116.
107. Deem, R. L., Beaman, B. L., and Gershwin, M. E., 1982, Adoptive transfer of immunity to *Nocardia asteroides* in nude mice, *Infect. Immun.* **38:**914–920.
108. Ortiz-Ortiz, L., and Bojalil, L. F., 1972, Delayed skin reactions to cytoplasmic extracts of *Nocardia* organisms as a means of diagnosis and epidemiological study of *Nocardia* infection, *Clin. Exp. Immunol.* **12:**225–229.
109. Bojalil, L. F., and Magnusson, M., 1963, Specificity of skin reactions of humans to *Nocardia* sensitins, *Amer. Rev. Resp. Dis.* **88:**409–411.
110. Shainhouse, J. Z., Pier, A. C., and Stevens, D. A., 1978, Complement fixation antibody test for human nocardiosis, *J. Clin. Microbiol.* **8:**516–519.
111. Blumer, S. O., and Kaufman, L., 1979, Microimmunodiffusion test for nocardiosis, *J. Clin. Microbiol.* **10:**308–312.
112. Sugar, A. M., Schoolnik, G. K., and Stevens, G., 1985, Antibody response in human nocardiosis: Identification of two immunodominant culture-filtrate antigens derived from *Nocardia asteroides, J. Infect. Dis.* **151:**895–901.
113. Ortiz-Ortiz, J., Melendro, E. I., and Conde, C., 1984, Host-parasite relationship in infections due to *Nocardia brasiliensis*, in: *Biological, Biochemical and Biomedical Aspects of Actinomycetes* (L. Ortiz-Ortiz, L. F. Bojalil, and V. Yakoleff, eds.), Academic Press, San Diego, pp. 119–133.
114. Beaman, B. L., 1983, Actinomycete pathogenesis, in: *The Biology of the Actinomycetes* (M. Goodfellow, M., Mordarski, and S. T. Williams, eds.), Academic Press, London, pp. 457–479.
115. Filice, G. A., and Niewoehner, D. E., 1987, Contribution of neutrophils and cell-mediated immunity to control *Nocardia asteroides* in murine lungs, *J. Infect. Dis.* **156:**113–121.

116. Mishra, S. K., Sandhu, R. S., Randhawa, H. S., Damodaran, V. N., and Abraham, S., 1973, Effect of cortisone administration on experimental nocardiosis, *Infect. Immun.* **7**:123–129.
117. Pomeroy, C., and Filice, 1988, Effect of intravenous silica on the course of *Nocardia asteroides* pneumonia, *Infect. Immun.* **56**:2507–2511.
118. Falb, P. I., Jaffe, R., and Altman, G., 1976, *Nocardia asteroides* and *Nocardia brasiliensis* infections in mice, *Infect. Immun.* **13**:1490–1496.
119. Folb, P. I., Timme, A., and Horowitz, A., 1977, *Nocardia* infections in congenitally athymic (nude) mice and other inbred mouse strains, *Infect. Immun.* **18**:459–466.
120. Deem, R. L., Doughty, F. A., and Beaman, B. L., 1983, Immunologically specific direct T-lymphocyte-mediated killing of *Nocardia asteroides*, *J. Immunol.* **130**:2401–2406.
121. Silva, C. L., Tincani, I, Brandao Filho, S. L., and Faccioli, L. H., 1988, Mouse cachexia induced by trehalose dimycolate from *Nocardia asteroides*, *J. Gen. Microbiol.* **134**:1629–1633.
122. Oritz-Ortiz, L., Parks, D. E., Lopez, J. S., and Weigle, W. O., 1979, B-lymphocyte activation with an extract of *Nocardia brasiliensis*, *Infect. Immun.* **25**:627–634.
123. Beaman, L., and Beaman, B. L., 1989, Effects of Interferon γ on mononuclear phagocyte interactions with *Nocardia asteroides*, Abst. Ann. Meeting Amer. Soc. Microbiol. E-36.
124. Angeles, A. M., and Sugar, A. M., 1987, Rapid diagnosis of nocardiosis with an enzyme immunoassay, *J. Infect. Dis.* **155**:292–296.
125. Angeles, A. M., and Sugar, A. M., 1987, Identification of a common immunodominant protein in culture filtrates of three *Nocardia* species and use in etiologic diagnosis of mycetoma, *J. Clin. Microbiol.* **25**:2278–2280.
126. Boiron, P., and Provost, F., 1988, Enzyme immunoassay on whole *Nocardia asteroides* cells for human nocardiosis, *Serodiagn. Immunother. Infect. Dis.* **2**:445–452.
127. Boiron, P., and Provost, F., 1990, Use of partially purified 54-kilodalton antigen for diagnosis of nocardiosis by Western blot (immunoblot) assay, *J. Clin. Microbiol.* **28**:328–331.
128. El-Zaatari, F. A., Reiss, E., Yakrus, M. A., Bragg, S. L., and Kaufman, L., 1986, Monoclonal antibodies against isoelectrically focused *Nocardia asteroides* proteins characterized by the enzyme-linked immunoelectrotransfer blot method, *Diagn. Immunol.* **4**:97–106.
129. Estrada-Parra, S., 1973, Immunochemical characteristics and differentiation of *Nocardia*, *Rev. Lat. Am. Microbiol.* **15**:87–93.
130. Humphreys, D. W., Crowder, J. G., and White, A., 1975, Serological reactions to *Nocardia* antigens, *Am. J. Med. Sci.* **269**:323–326.
131. Jimenez, T., Diaz, A. M., and Zlotnick, H., 1990, Monoclonal antibodies to *Nocardia asteroides* and *Nocardia brasiliensis* antigens, *J. Clin. Microbiol.* **28**:87–91.
132. Kurup, V. P., and Scribner, G. H., 1981, Antigenic relationship among *Nocardia asteroides* immunotypes, *Microbios* **31**:25–30.
133. Kwapinski, J. B. G., Kwapinski, E. H., Dowler, J., and Horsman, G., 1973, The phyloantigenic position of nocardiae revealed by examination of cytoplasmic antigens, *Can. J Microbiol.* **19**:955–964.
134. Ortiz-Ortiz, L., Contreras, M. F., and Bojalil, L. F., 1972, Cytoplasmic antigens from *Nocardia* eliciting a specific delayed hypersensitivity, *Infect. Immun.* **5**:879–882.
135. Pier, A. C., and Keller, R. F., 1965, Extracellular antigens of *Nocardia asteroides*: I. Production and immunologic characterization, *Am. Rev. Resp. Dis. 91*:391–399.
136. Pier, A. C., Thurston, J. R., and Larsen, A. B., 1968, A diagnostic antigen for nocardiosis: Comparative tests in cattle with nocardiosis and mycobacteriosis, *Am. J. Vet. Res.* **29**: 397–403.
137. Pier, A. C., and Fichtner, R. E., 1971, Serologic typing of *Nocardia asteroides* by immunodiffusion, *Am. Rev. Resp. Dis.* **103**:698–707.
138. Pier, A. C., Fichtner, R. E., 1981, Distribution of serotypes of *Nocardia asteroides* from animal, human, and environmental sources, *J. Clin. Microbiol.* **13**:548–553.
139. Ridell, M., 1981, Immunodiffusion studies of some *Nocardia* strains, *J. Gen. Microbiol.* **123**:69–74.

140. Adams, H. G., Beeler, B. A., Wann, L. S., Chin, C. K., and Brooks, G. F., 1984, Synergistic action of trimethoprim and sulfamethoxazole for *Nocardia asteroides*: Efficacious therapy in five patients, *Am. J. Med. Sci.* **287**:8–12.
141. Brochot, J. M., Capron, F., Prudent, J., and Rochemaure, J., 1987, Unexpected pulmonary nocardiosis in a non-immunocompromised patient, *Thorax* **42**:479–480.
142. Smego, R. A., Moeller, M. B., and Gallis, H. A., 1983, Trimethoprim–sulfamethoxazole therapy for *Nocardia* infections, *Arch. Intern. Med.* **143**:711–718.
143. Filice, G. A., 1987, Treatment of nocardiosis, *Lancet* **(8544)**:1261–1262.
144. Stamm, A. M., McFall, D. W., and Dismukes, W. E., 1983, Failure of sulfonamides and trimethoprim in the treatment of nocardiosis, *Arch. Intern. Med.* **143**:383–385.
145. Berkey, P., Moore, D., and Ralston, K., 1988, *In vitro* susceptibilities of *Nocardia* species of newer antimicrobial agents, *Antimicrob. Agents Chemother.* **32**:1078–1079.
146. Gombert, M. E., Aulicino, T. M., duBouchet, L., Silverman, G. E., and Sheinbaum, W. M., 1986, Therapy of experimental cerebral nocardiosis with imipenem, amikacin, trimethoprim-sulfamethoxazole, and minocycline, *Antimicrob. Agents Chemother.* **30**:270–273.
147. Gombert, M. E., Aulicino, T. M., duBouchet, L., and Berkowitz, L. R., 1987, Susceptibility of *Nocardia asteroides* to new quinolones and β-lactams, *Antimicrob. Agents Chemother.* **31**:2013–2014.
148. Wallace, R. J., Steele, L. C., Sumter, G., and Smith, J. M., 1988, Antimicrobial susceptibility patterns of *Nocardia asteroides*, *Antimicrob. Agents Chemother.* **32**:1776–1779.
149. Wallace, R. J., Nash, D. R., Johnson, W. K., Steele, L., and Steingrube, V. A., 1978, β-lactam resistance in *Nocardia brasiliensis* is mediated by β-lactamase and reversed in the presence of clavulanic acid, *J. Infect. Dis.* **156**:959–966.
150. Cockerill, F. R., Edson, R. S., Roberts, G. D., and Waldorf, J. C., 1984, Trimethoprim/sulfamethoxazole-resistant *Nocardia asteroides* causing multiple hepatic abscesses: Successful treatment with ampicillin, amikacin, and limited computed tomography-guided needle aspiration, *Am. J. Med.* **77**:558–560.
151. Godlstein, F. W., Hautefort, B., and Acar, J. F., 1987, Amikacin-containing regimens for treatment of nocardiosis in immunocompromised patients, *Eur. J. Clin. Microbiol.* **6**: 198–200.
152. Meier, B., Metzger, U., Muller, F., Siegenthaler, W., and Luthy, R., 1986, Successful treatment of a pancreatic *Nocardia asteroides* abscess with amikacin and surgical drainage, *Antimicrob. Agents Chemother.* **29**:150–151.
153. Krone, A., Schaal, K. P., Brawanski, A., and Schuknecht, B., 1989, Nocardial cerebral abscess cured with imipenem, amikacin and enucleation, *Neurosurg. Rev.* **12**:333–340.

Index